Microbe

SECOND EDITION

Microbe

SECOND EDITION

MICHELE SWANSON
Department of Microbiology &
Immunology
University of Michigan

GEMMA REGUERA
Department of Microbiology and
Molecular Genetics
Michigan State University

MOSELIO SCHAECHTER
Department of Biology,
San Diego State University
Division of Biological Sciences,
University of California at San Diego

FREDERICK NEIDHARDT
Department of Microbiology &
Immunology
University of Michigan

With Rachel Horak, Ph.D.
Curriculum Guidelines Editor

ASM
PRESS

Washington, DC

Library of Congress Cataloging-in-Publication Data

Names: Swanson, Michele, author. | Reguera, Gemma, author. | Schaechter, Moselio, author. | Neidhardt, Frederick C. (Frederick Carl), 1931- author.
Title: Microbe / Michele Swanson, Gemma Reguera, Moselio Schaechter, Frederick Neidhardt.
Description: 2nd edition. | Washington, DC : ASM Press, [2016] | 2nd edition of: Microbe / Moselio Schaechter, John L. Ingraham, Frederick C. Neidhardt. Washington, D.C. : ASM Press, ?2006. | Includes index.
Identifiers: LCCN 2016018071 (print) | LCCN 2016018432 (ebook) | ISBN 9781555819125 (print) | ISBN 9781555819132 (eISBN) | ISBN 9781555819132 (e-book)
Subjects: LCSH: Microbiology.
Classification: LCC QR41.2 .S93 2016 (print) | LCC QR41.2 (ebook) | DDC 579—dc23
LC record available at https://lccn.loc.gov/2016018071

10   9   8   7   6   5   4   3   2   1

All Rights Reserved
Printed in the United States of America

Address editorial correspondence to ASM Press, 1752 N St., N.W., Washington, DC 20036-2904, USA

Send orders to ASM Press, P.O. Box 605, Herndon, VA 20172, USA
Phone: 800-546-2416; 703-661-1593
Fax: 703-661-1501
E-mail: books@asmusa.org
Online: http://estore.asm.org

Cover image: Shutterstock.com
Cover and interior design: Debra Naylor, Naylor Design Inc.

This book is dedicated to

the many colleagues whose dedicated service to ASM benefits us all
- MSS

Kaz and Dario, for making it all worthwhile
- GR

Edith, with love
- ES

the memory of Boris Magasanik
- FN

Contents

Preface

This is a book with an attitude, a compelling one we hope. We assert that the microbial world does not just comprise a large taxonomic group of living things, but that microbiology influences the attributes of all living things on our planet, and many of the non-living too. Microbes make up about half the mass of everything alive, and they make and unmake our landscape and our atmosphere. They matter for our very existence. Thus, as much as anything, microbiology is a frame of mind.

As with the previous edition, our goal is to emphasize concepts, not facts. Throughout the book, we integrate key concepts, learning outcomes, and fundamental statements from the *ASM Recommended Curriculum Guidelines for Undergraduate Microbiology Education*. Using case studies to open each chapter, we tell stories about the way microbes are put together, what they must do to grow and survive, and how they interact with all living things. The new edition no longer includes a separate chapter about evolution but, rather, infuses each of the chapters with descriptive tales of microbes' evolutionary journey. In addition to updating a lot of the material, we also expanded the microbial diversity, ecology, and pathogenesis content by adding several new chapters and including extensive visual aids throughout. Importantly, we stimulate scientific thinking and encourage students to apply microbiology knowledge and skills throughout the text and with the Supplemental Activities and Resources provided with each chapter.

Our approach is far from encyclopedic, although this is a fairly thick book because there is so much to tell about microbes. Amidst the exciting advances in the microbial sciences, we have kept to our guns, using a conceptual framework based on the dogged belief that all living things, microbes included, can only be understood in the light of their evolution and ecology. For those of you who will enter the health professions, keep in mind that the interaction between host and pathogen is entirely an ecological one. For those who pursue other interests, the same worldview holds. All microbial interactions play out in the environment. In the words of the Spanish philosopher Ortega y Gasset, "I am I and my circumstances."

Indeed, the microbial world is the only world we humans inhabit. Please enjoy it.

Acknowledgments

We are indebted to ASM Press for their generous support since this book was but a mere idea. Their staff and the freelancers who worked on this book provided the skills, expertise, and patience needed to turn our manuscripts and figures into a book. Christine Charlip (Director) led a terrific team consisting of Larry Klein, Ellie Tupper, graphic artist Patrick Lane, and our wonderful production editor, John Hoey. Thank you all for your many insights and total dedication to the project.

An early meeting at Friday Harbor, Washington, materialized the vision for this book. We are thankful to Dr. Beth Thraxler for hosting the retreat and for her valuable contributions when this edition was in its early stages, to Carrie Harwood, Chair of ASM Press for her oversight and encouragement, and to our faculty guest Dr. Amy Siegesmund, who generously shared her ideas and expertise to ensure that the book would provide a quality educational tool to microbiology instructors and students alike. The thoughtful faculty who participated in our focus group at 2014 ASMCUE sharpened our focus. From these interactions, a fruitful collaboration was born with the ASM Press Director Christine Charlip, ASM Education Board Director Amy Chang, Board Chair Sue Merkel, and postdoctoral fellow Dr. Rachel Horak. We are especially grateful to Dr. Horak for expertly integrating the *Curriculum Guidelines* into *Microbe*, writing clear learning outcomes for each chapter section, mapping these learning outcomes to the Fundamental Statements, and overseeing development of Supplemental Activities to enhance the students' learning experience. Working closely with the authors, Dr. Horak brought the *Curriculum Guidelines* to life by demonstrating how to engage the readers as scientists in the making.

Our first retreat was followed by two more in Academy Village (Tucson, Arizona), gatherings that served as catalysts to bring the team project to successful completion. Our gratitude goes to our hosts at the Academy Village, Betty, Erika, and Bernd, for their generosity and hospitality. Their good humor and friendship made the long working hours easier to bear.

Special appreciation is extended to the many individuals who provided material for the Supplemental Activities: Dr. Rachel Horak provided leadership to the authors and to graduate students Mike Manzella and Becky Steidl at Michigan State University and a team of postdocs at the University of Michigan comprised of Drs. Zack Abbott, Melissa Smaldino, Philip Smaldino, Laura Mike, Melody Zeng, Kalyani Pyaram, and Marc Sze. We are also grateful to Dr. Jamie Henzy for her virology expertise and to Merry Youle for compiling the glossary and generously editing the chapters along the way, just because.

Last, but not least, we are grateful to the many colleagues who answered our queries, provided images, and so generously devoted their time to review critically each of the chapters: Drs. Mercedes Berlanga, Yves Brun, Lynn Enquist, Jeff Gralnick, Ricardo Guerrero, Neal Hammer, Daniel Haeusser, Jamie Henzy, Ian Hogue, John Howieson, Gary Huffnagle, Lee Kroos, Jake McKinlay, Stanley Maloy, Helene Marquis, Nik Money, Vincent Racaniello, Larry Reitzer, Tony Richardson, Irwin Rubenstein, Erik Skaar, Ashley Shade, Linc Sonenshein, Valley Stewart, Amy Vollmer, Daniel Wall, Christoph Weigel, Kevin Young, and Merry Youle. To all of you, many thanks.

About the Authors

Michele S. Swanson was born and raised in the Midwest, majored in Biology at Yale, and discovered biomedical research as a laboratory technician at Rockefeller University. She fell in love with microbial genetics while earning an M.S. at Columbia and Ph.D. at Harvard Medical School and became captivated by microbial pathogenesis as a postdoc at Tufts University School of Medicine. At the University of Michigan Medical School, her research group investigates how metabolic and environmental cues govern the life cycle of the intracellular pathogen *Legionella pneumophila*. Her email address is mswanson@umich.edu.

Gemma Reguera is a native of Spain, where she majored in Biological Sciences from the University of Oviedo. As an M.S. student at the University of Massachusetts in Amherst, she discovered her passion for environmental microbiology. And there she stayed, getting a Ph.D. in Microbiology before moving to Harvard Medical School as a postdoc to study the ecology of infectious diseases. At Michigan State University she studies microbial energy conversions and physiologies of environmental relevance, particularly those that can be harnessed for applications in bioenergy and bioremediation. Her email address is reguera@msu.edu.

Moselio Schaechter was born in Italy, lived in Ecuador as a teen, and obtained his Ph.D. at the University of Pennsylvania. He spends most of his academic life at Tufts University School of Medicine and moved to San Diego in 1995. His research interests involve aspects of bacterial physiology, including growth rate regulation, membrane biology, and chromosome transactions. His email address is elios179@gmail.com.

Frederick C. Neidhardt was born in Philadelphia, majored in Biology at Kenyon College in Ohio, and received his Ph.D. at Harvard University. He held academic posts at Harvard, Purdue University, and the University of Michigan. His research focuses on catabolite repression, growth rate regulation, aminoacyl-tRNA synthetases, and heat shock and other global networks. His email address is fcneid@umich.edu.

PART I

Fundamentals of Microbial Life

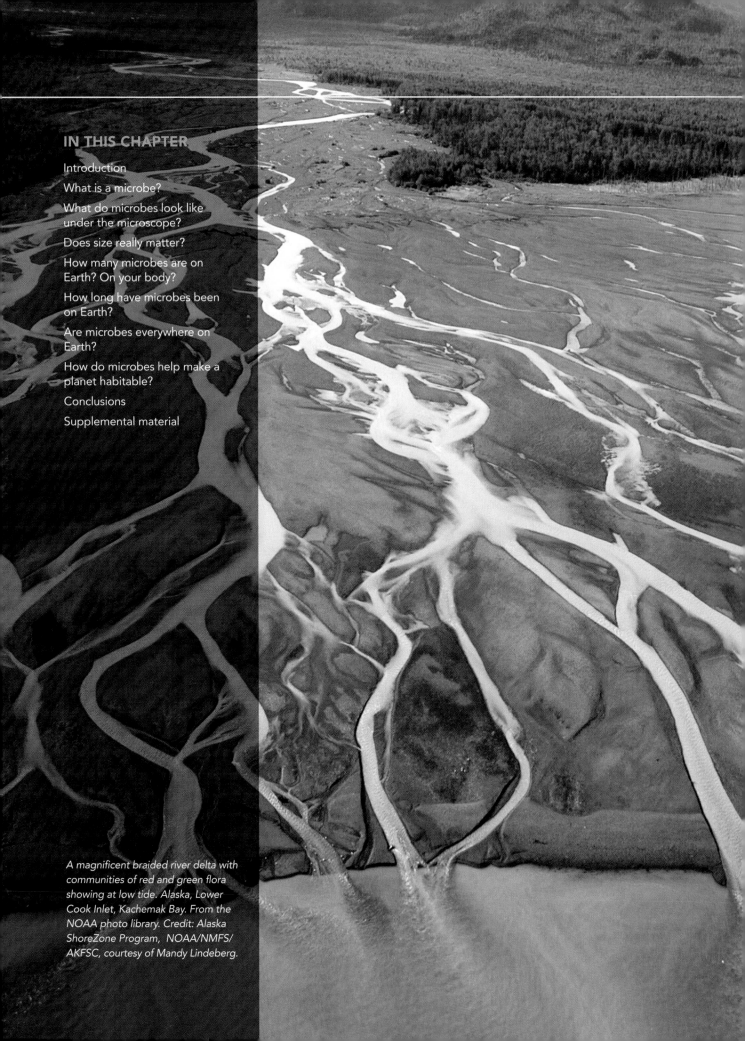

IN THIS CHAPTER

A magnificent braided river delta with communities of red and green flora showing at low tide. Alaska, Lower Cook Inlet, Kachemak Bay. From the NOAA photo library. Credit: Alaska ShoreZone Program, NOAA/NMFS/ AKFSC, courtesy of Mandy Lindeberg.

CHAPTER ONE

A Microbial Planet

KEY CONCEPTS

This chapter covers the following topics in the ASM Fundamental Statements.

EVOLUTION

1. Cells, organelles (e.g., mitochondria and chloroplasts), and all major metabolic pathways evolved from early prokaryotic cells.
2. Mutations and horizontal gene transfer, with the immense variety of micro-environments, have selected for a huge diversity of microorganisms.

CELL STRUCTURE AND FUNCTION

3. The structure and function of microorganisms have been revealed by the use of microscopy (including bright-field, phase-contrast, fluorescent, and electron).
4. While microscopic eukaryotes (for example, fungi, protozoa, and algae) carry out some of the same processes as bacteria, many of the cellular properties are fundamentally different.

METABOLIC PATHWAYS

5. The survival and growth of any microorganism in a given environment depend on its metabolic characteristics.

MICROBIAL SYSTEMS

6. Microorganisms are ubiquitous and live in diverse and dynamic ecosystems.
7. Microorganisms can interact with both human and nonhuman hosts in beneficial, neutral, or detrimental ways.

IMPACT OF MICROORGANISMS

8. Microbes are essential for life as we know it and the processes that support life (e.g., in biogeochemical cycles and plant and/or animal microflora).

INTRODUCTION

Life on Earth is much more abundant than what we can see with our own eyes. The organisms that are visible to us such as other humans, animals, and plants are just a small minority, the tip of a large, submerged iceberg. Most living organisms are in fact not visible to the unaided eye. Most likely, you use the broad term "microbes" to refer to those small organisms, though, as you will learn in this chapter, there are always exceptions and some microbes are large enough to be seen by the naked eye.

Why are microbes so abundant, you may ask? Well, for one, their relatively small size and simple design allow them to reproduce fast and adapt quickly to changes in their environment. On the other hand, they have lived on this planet the longest, giving them much time to evolve new properties and diversify. They were the first to colonize the planet, almost 4 billion (yes, "billion" with a "b," or 4,000 million!) years ago, at a time when conditions on Earth were inhospitable to most of life. Not only were microbes here first, they managed to colonize every corner of Earth and, while they were at it, adapted to the profound changes that the planet underwent throughout its long history. Microbes evolved in untold ways to interact with living and nonliving components of their environment. The impact of these interactions is truly staggering (Table 1.1).

You may be familiar with microbes that cause disease, but these are only a small fraction of all the microbial diversity existing today. Microbes are in fact our greatest allies: they sustain the life of all other organisms and help us advance technology and cope with the profound effects that our own activities have on the planet. They have become integral members of our lives and functioning. This is made evident by the interactions between humans and microbes (Table 1.2). In this chapter, we will introduce you to the microbes' most notable characteristics to understand the magnitude of their contribution to transforming Earth into the living planet of today and shaping our future world. Be prepared for a visit to a world that is stunning in its diversity, immense in size, and pervasive in its importance to all of us. You will not see the world in the same way again.

TABLE 1.1 Characteristics of microbes

Microbes:
- Are the source of all life forms
- Are more phylogenetically diverse than plants and animals
- Are enormously abundant
- Grow in virtually every place on Earth where there is liquid water
- Carry out transformations of matter essential for life
- Transform the geosphere
- Affect the climate
- Participate in countless symbiotic relationships with animals, plants, and other microbes
- Cause disease
- Influence the behavior of animals and plants

TABLE 1.2 Human use of microbes

- Carrying out chemical activities of major industrial importance
- Engineering for production of useful proteins, e.g., vaccines
- Enhancing food production and preservation
- Providing vital public health measures, such as sewage disposal
- Bioremediation of polluted sites
- Malevolent intent (biological warfare, bioterrorism)
- Source of heat-resistant enzymes

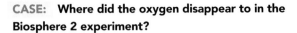

CASE: Where did the oxygen disappear to in the Biosphere 2 experiment?

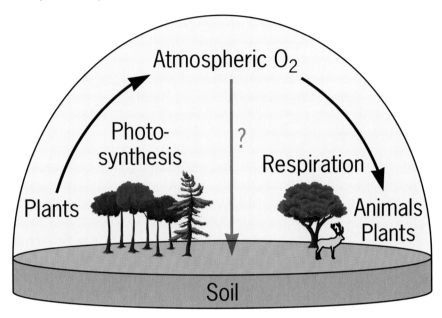

Microbes have a global impact on life on Earth. It is easy to appreciate their significance in examining what went wrong in the Biosphere 2 project. In the early 1990's, scientists sought to build a closed ecosystem in Arizona, where plants and animals could live in a self-sustained manner within an airtight, glass-and-steel enclosure. They included enough plants to produce plenty of oxygen during the day via photosynthesis so as to support the nighttime consumption by both plants and animals in the enclosure. To their dismay, the atmospheric oxygen disappeared at rates much faster than predicted and could not be maintained stably. So low was the oxygen in the enclosure's atmosphere that the crew members could not work inside even for short periods of time, and many of the plants and animals died.

- Can you form a hypothesis as to why oxygen levels in the sealed enclosure could not be maintained in Biosphere 2? (Hint: Scientists did not consider microbial activities in the soils!)
- How could you test this hypothesis?
- Can you suggest a potential solution to maintain oxygen concentrations in Biosphere 2?

WHAT IS A MICROBE?

LEARNING OUTCOMES:
- Explain why it is difficult to define what a microbe is.
- List exceptions to the definition that "microbes are so small that they are only visible under the microscope."

Traditionally, the term "microbe" has been used to describe free-living organisms so small (usually less than about 100 micrometers [μm]) that they are only visible under the microscope (hence, they are *microscopic*).

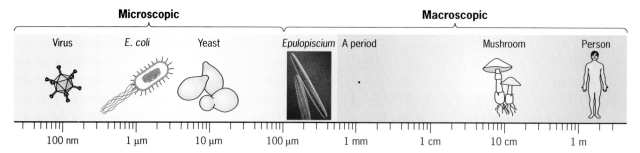

FIGURE 1.1 **Size range of microbes.** Microbes have a wide range of sizes, but most maintain a relatively small size.

However, as often happens with microbes, there are exceptions (many!) to the general rule and the size range for known microbes is quite broad (Fig. 1.1). Sure enough, most microbes are too small to be seen without a microscope. For example, *Escherichia coli*, the best-known organism on Earth, has a width of 0.5 µm and can reach 2 µm in length. Many other microbes fall within these ranges. On the lowest end of the size range are marine bacteria so minuscule (around 0.2 µm in diameter) that they are smaller in size than the largest viruses. But then there are microbes so large that you can see them with the naked eye. *Epulopiscium fishelsoni*, for example, is a giant bacterium whose cells can grow as big as the period at the end of this sentence. Under the microscope, an *E. coli* cell looks like a speck of dust next to a cell of *E. fishelsoni* (Fig. 1.2). In fact, you could fit several thousands of *E. coli* cells inside an "epulo" cell. And what about the "free-living" in our definition? Well, that is also not always the case. Many microbes are indeed free-living, but some exist in association with hosts. Our favorite giant bacterium *E. fishelsoni* lives in the gut of surgeonfish (a group of *Finding Nemo* fame) and never ventures outside its host. Some microbes take host dependency to the extreme and live inside the host cells themselves. This is the case of **obligate intracellular parasites**, such as the bacterium that causes leprosy (*Mycobacterium leprae*). So, while the concept of microbes as free-living organisms applies to many of them, remember that there are exceptions. Why is that? Because microbes have lived in this planet for so long that they have learned all the tricks to colonize new niches, whether outside or inside hosts.

Finally, what about the viruses? They can be as small as the smallest bacteria (Fig. 1.1). But are they microbes? Not really, because they lack one key quality of living cells: the *capacity to maintain their bodily integrity throughout their life cycle*. Viruses reproduce and mutate, as do cellular organisms, but they come apart during their reproductive cycle. The major constit-

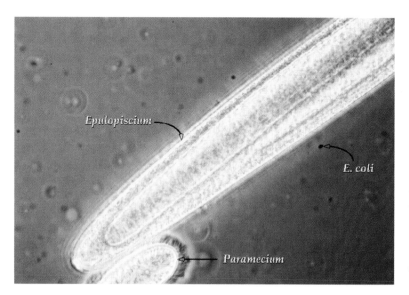

FIGURE 1.2 **How big can a microbe be?** Really big! *E. fishelsoni*, one of the largest prokaryotes known, greatly exceeds an average bacterium (*E. coli*), and even a large unicellular eukaryote (*Paramecium*), in size. (Courtesy of E. Angert.)

uents of viruses are nucleic acids and proteins, just like cellular organisms, but they are made separately and then assembled inside their host cells. Without the host cell, viruses could not produce progeny viral particles. Viruses, for example, do not have the capacity to make proteins (they have no ribosomes, the protein factories of all cells) and rely on the host to supply most, if not all, of the building blocks needed for biosynthesis of their macromolecules. Outside a host cell, a viral particle is inert and incapable of growth and reproduction. Yet despite not being microbes, they interact with microbial cells and are critical to their evolution. Viral biology is discussed in chapter 17.

WHAT DO MICROBES LOOK LIKE UNDER THE MICROSCOPE?

LEARNING OUTCOMES:
• Compare and contrast the general structure of *Bacteria*, *Archaea*, and eukaryotic microbes.

A misconception about microbes is to assume that they are simple, both on the outside (Fig. 1.3) and the inside (Fig. 1.4). Having a long past, microbes have diversified to have many shapes: spheres (cocci), rods (bacilli), ovals, bowling pins, spirals, comma shapes, and even corkscrews. You can appreciate the diversity of microbial shapes by examining a drop of pond water with a microscope (Fig. 1.5). Why so many shapes? Shape matters for many reasons: it affects the efficiency of nutrient uptake, motility, attachment to surfaces, and even reproduction. As you examine them with a microscope you will also realize that not all microbes are unicellular. Some form pairs, chains, or filaments or aggregate in some other

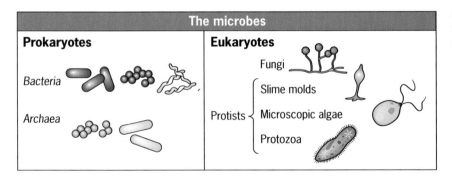

FIGURE 1.3 **Microbes.** Microbes, both prokaryotes and eukaryotes, come in many different shapes.

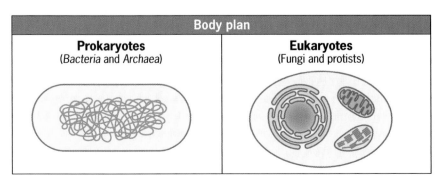

FIGURE 1.4 **Traditional classification of microbes as prokaryotes or eukaryotes.** The classification depends on whether the cells lack (prokaryotes) or have (eukaryotes) a true nucleus. The eukaryotes also have membrane-bound organelles of bacterial ancestry: mitochondria and plastids.

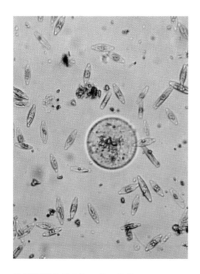

FIGURE 1.5 Microbes' diverse shapes and forms. A drop of pond water scum examined at ×400 magnification with a phase-contrast microscope reveals many different types of microbial cells.

unique fashions. In fact, in their natural environment, many of them colonize surfaces to form sophisticated three-dimensional communities, or **biofilms**, that come close to rivaling the complexity of our own tissues. In chapter 21, we will learn about microbe-microbe interactions and about multicellularity and the consequences of these social behaviors.

The internal structure of microbial cells revealed under the microscope can be diverse as well. The most notable difference is the absence of a nucleus in some (prokaryotes) and its presence in others (eukaryotes) (Fig. 1.4). The prokaryotes comprise two separate but very large and diverse groups, the *Bacteria* and the *Archaea*. The two prokaryotic groups may appear similar in size and morphology under the microscope, but, as we will learn later in this chapter, each has followed a distinct evolutionary path and, as a result, each has evolved unique defining features. Yet the body plan of the two is similar: in addition to lacking nuclei, they do not have membrane-bound organelles such as mitochondria or plastids.

Sizes and morphologies are even more variable among the eukaryotic microbes, which include algae, fungi, and a diverse group of organisms collectively called the protists. Some of these groups are especially hard to delineate, in part because they have both small and large relatives. For example, yeasts are certainly microbes, but mushrooms are not, yet both are fungi. The same goes for algae—some are microscopic, others are the giant seaweeds (kelp). Still, they all follow the general cell body plan of having a true nucleus and most also have membrane-bound organelles.

DOES SIZE REALLY MATTER?

LEARNING OUTCOMES:
- Describe some benefits and detriments to a microbe having a relatively small size.
- Explain why surface area-to-volume ratio is an important constraint in bacterial growth.

Throughout billions of years of evolution, microbes have diversified in many ways but most have maintained a relatively small size. There is a key reason for this: being small allows them to maintain a high surface-to-volume ratio to maximize chemical exchange with the surrounding environment (Fig. 1.6). It is easier to picture how small microbes really are in terms of volume (Fig. 1.7). The volume of a representative bacterium such as *E. coli* is about 1 µm³, or roughly 1/1,000 that of a human cell. Eukaryotic microbes tend to be larger. For example, the baker's yeast, a eukaryotic microbe that is used to make bread, beer, and wine, has an average volume of around 100 µm³, or 1/10 that of a human cell. What do all these numbers mean? This means that you would need about 10¹⁷ bacteria or 10¹⁵ yeast cells to occupy the volume of an adult human. Or, if you are fond of such comparisons, if an *E. coli* cell were the size of a human being, a yeast cell would be that of an elephant. The epulo, the giant of all microbes, would be the Godzilla of the microbial world!

As cells grow larger, the volume increases more rapidly relative to the surface area and the rates of chemical exchange (nutrients diffusing in

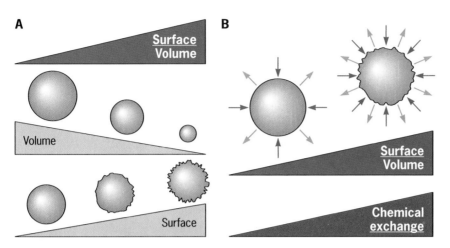

A   Surface / Volume

Volume

Surface

B   Surface / Volume

Chemical exchange

FIGURE 1.6 **The advantage of being small.** Being small allows microbes to keep a high surface area-to-volume ratio so as to maximize chemical exchange and growth. However, some microbes can compensate for their large volume by increasing their surface area (e.g., by folding the outer membrane).

and waste products diffusing out) decrease proportionally. To keep their metabolism functioning at the highest possible rate in order to grow and reproduce quickly, cells must be able to exchange chemicals efficiently with their surroundings. Thus, rapidly growing microbes process their nutrients at rates 10 to 1,000 times higher per gram than do mammalian cells. Hence, even relatively small increases in size can make a big difference in chemical diffusion and, potentially, in cell growth (Box 1.1).

Rapid chemical turnover has significant consequences, some of which are easy to grasp. One can get sick with a bacterial infection very quickly because the infective cells grow and reach great numbers before our immune system can eliminate the threat. Fruit, vegetables, and meats can spoil very

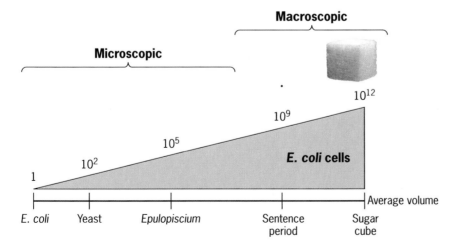

Macroscopic

Microscopic

10^{12}

10^9

10^5

E. coli cells

10^2

1

Average volume

E. coli   Yeast   *Epulopiscium*   Sentence period   Sugar cube

FIGURE 1.7 **Numbers, large and small, matter.** One billion of an average-sized bacterium such as *E. coli* occupy a volume comparable to the period at the end of this sentence. A sugar cube could accommodate a trillion of them! Larger microbes can also house bacteria inside and start amazing cooperative behaviors. Just look at how many *E. coli* cells fit in the volume of a yeast cell or a medium-sized epulo cell: hundreds to thousands. Yet, in some cases, exposure to only a few cells of bacteria can make us sick!

BOX 1.1

Size Matters!

Let us do this activity to understand how much cell size matters. Imagine two spherical cells, one with a radius of 1 μm and the other of 2 μm. If you were to examine the two cells under a microscope, they would still be tiny, right? Now let us calculate the surface-to-volume ratio with formulas shown in the figure. Note that the surface area increases as the *square* of its radius, but the volume increases as the *cube* of the radius.

- Did you come up with the surface/volume ratio of the two cells?
- Which one do you predict will grow faster?

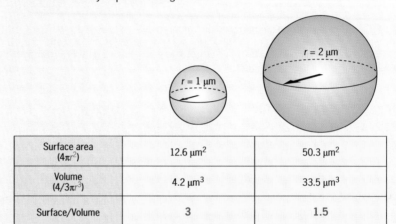

Surface area ($4\pi r^2$)	12.6 μm^2	50.3 μm^2
Volume ($4/3\pi r^3$)	4.2 μm^3	33.5 μm^3
Surface/Volume	3	1.5

fast outside the fridge because of the rapid growth and activities of microbes. And how long does it take for yeasts to grow and make bread rise? Hours!

A high surface-to-volume ratio is particularly important in an environment where competition for nutrients is high. Some of the smallest known microbes live in marine and freshwater environments. The smallest of all, designated "*Candidatus* Actinomarinidae minuta," has an average cell volume of about 0.013 μm^3 (about 60 times smaller than *E. coli*!). Although this minute bacterium has not been cultured, its genome has been sequenced directly from cells in the environment. With this information, scientists made probes specific to one of its ribosomal RNAs (16S rRNA) and used them to tag the cells and assess their abundance in seawater samples. Guess what, the ultra-small bacterium is widely distributed in oceans around the world and is very (very!) abundant (~4% of all the bacterioplankton). It pays off to be that small!

So how come some microbes are large? Do they compromise their growth efficiency for size? Some of them find ways to extend their surface area to compensate for the large volume, so they can continue to grow quickly. Let us look again at *E. fishelsoni*, the gargantuan bacterium of the surgeonfish's gut. Although scientists have not figured out how to grow epulo cells in the laboratory, they can remove them from the fish and then study them under the microscope. As the epulo cells grow larger, their membranes wrinkle, forming folds and pockets. This effectively increases the surface area of the cell to maximize nutrient uptake and growth when the fish is actively feeding during the day. Once the fish stops feeding, the large epulo cells reproduce and start the cycle of growth all over again, ready to start feeding the next day.

Note, however, that being small also has its disadvantages. Small microbes are an easy target for larger microbes and animals (think about whales feeding on plankton in the oceans!). Some microbes avoid getting eaten by forming larger aggregates, sometimes with other microbes. Such masses of cells are often visible to the naked eye and to a predator, not that easy to digest. Smart!

HOW MANY MICROBES ARE ON EARTH? ON YOUR BODY?

LEARNING OUTCOMES:
- Consider why the immensity of the microbial world was unknown until the very end of the early 20th century.

Small as individual microbes are, their total collective mass on Earth is staggering. Microbes are nearly ubiquitous, and they are found practically anywhere there is free liquid water. In the oceans alone, microbes account for more than 90% of the biomass. There are nearly 1 million bacterial cells in 1 ml (approximately 20 drops) of seawater, a total of about 10^{30} cells in all the world's oceans. And these numbers are just for bacteria. Archaea are also abundant there. Plus, eukaryotic microbes account for almost half of the microbial biomass in the ocean's surface waters. Microbes are even more abundant in soils, sediments, and regions below the two (the subsurface). Collectively, the microbial biomass on the planet is comparable to that of all other living things, and according to some estimates, it may be even greater. Pause to contemplate the importance of this stunning fact: because prokaryotes contain proportionally more nitrogen and phosphorus than plants, they are the largest reservoir of these elements in the biological world. In chapter 19, we will learn more about the contribution of microbes to the cycling of these and other elements. And here is another surprising fact: the immensity of the microbial world was not known until the very end of the 20th century. We, the authors, spent most of our early careers in blissful ignorance of the global impact of microbes.

Numbers, large and small, matter. A trillion of your average-sized bacterium weigh scarcely a gram and occupy a volume approximately that of a sugar cube (Fig. 1.7). Yet some diseases, such as bacterial dysentery, are acquired by swallowing just a few bacterial cells. For these germs, the number of cells contained in the sugar cube is sufficient to infect not only all humans but also all other susceptible vertebrates on Earth, with plenty to spare. How much fecal contamination would it take to pollute a large body of water and make even a few swallows of its water infectious? Not much, probably less than the bacteria you could fit in the sugar cube. And our bodies contain more bacteria than human cells, though, collectively, the bacteria would occupy a very small volume. The lower intestines alone house a kilogram or two of microbes, so many that they account for almost one half of the dry weight of our feces. Such crowding is not unusual in nature. As we mentioned earlier, microbes preferentially live in association with surfaces, where they build biofilms. All surfaces are colonized, from a rock to your intestine. The space between teeth and gums (the gingival crevice that dental hygienists probe for pockets), for example, contains wall-to-wall bacteria. They may be small, but they are there in great numbers.

HOW LONG HAVE MICROBES BEEN ON EARTH?

LEARNING OUTCOMES:
- State three characteristics of LUCA.
- Explain how the three-domain tree is different from the "ring of life" model.
- Describe the impact cyanobacteria had on early Earth.

The staggering abundance of microbes is the direct result of their long evolutionary path (Fig. 1.8). The two prokaryotic groups, *Bacteria* and *Archaea*, branched early on from a common ancestor, and from them all other life forms originated. Just how this last universal common ancestor, or LUCA, originated is anybody's guess (Box 1.2). Proof of our shared ancestry is the fact that all extant organisms, including humans, contain footprints of a unique ancestor in their DNA, pretty much the same way that our DNA contains information from our parents and grandparents and all the generations that preceded them. The most useful DNA sequences to reconstruct the evolution of life on Earth are those encoding processes shared by all extant life forms, such as ribosome-mediated protein synthesis. The now iconic three-domain tree of Carl Woese compared, for example, the relatedness of the small rRNA genes and compared the number of muta-

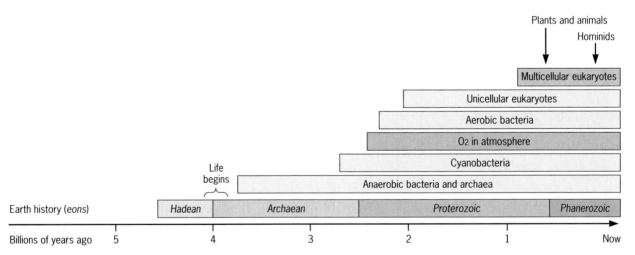

FIGURE 1.8 **Evolution of life forms on Earth.** *Bacteria* and *Archaea* branched early on after life began some 4 billion years ago. Microbial life was anaerobic until the emergence of cyanobacteria, which carried out a type of photosynthesis (oxygenic) that produced oxygen (O_2) as a by-product. This critical event oxygenated the planet and the atmosphere and provided the planetary conditions for the evolution of aerobic bacteria and then eukaryotic microbes. Microbial life dominated the planet for about the first 3 billion years, until the first multicellular eukaryotes emerged. The current eon, the Phanerozoic, is the shortest of all (542 million years), yet includes the eras when relatively simple animals and plants arose: the Paleozoic era (542 million to 248 million years ago); the age of dinosaurs, birds, and fish (Mesozoic era; 248 million to 65 million years ago); and the era of mammals (Cenozoic era; 65 million years ago until now).

BOX 1.2

Who Is "LUCA"?

Well, we really don't know, but that has not prevented scientists from speculating about its cellular features and origin. There seems to be agreement that LUCA had the cellular features that are universal to extant organisms:

- DNA as genetic material
- Proteins and RNAs to catalyze essential processes needed to grow and reproduce
- A lipid membrane to enclose all its components

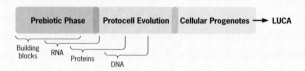

Everyone in the business also agrees that before LUCA was formed there was a **prebiotic phase** in the hot, early Earth when life's essential building blocks and molecules were generated in spontaneous chemical reactions. A popular theory suggests that RNAs came first, and they were able to self-replicate and catalyze chemical reactions. At some point the RNAs were engulfed in lipid micelles to make *primitive cellular entities, or protocells.* The protocells also evolved: the RNAs evolved the ability to make proteins from amino acids, and the primitive metabolisms were born. Proteins also catalyzed the synthesis of the first DNA molecules using RNA templates. At this time in life's history there may have been large populations of cellular progenotes containing DNA, RNA, and proteins. They were promiscuous and exchanged a lot of their genetic material via horizontal gene transfer to diversify rapidly. From this population only one cell, LUCA, reproduced so successfully that it outcompeted the rest. And it is from this common ancestor that all living forms, including us, originated. Thank you, evolution!

tions that accumulated in these genes to reconstruct the evolutionary history, or **phylogeny**, of various microbial and nonmicrobial organisms (Fig. 1.9). The tree showed three distinct evolutionary groups or domains (*Bacteria*, *Archaea*, and *Eukarya*) as if they were branches of a tree rooted in a common ancestor. Looking at the tree, it is easy to appreciate that the branch point of *Bacteria* and *Archaea* descent is ancient; furthermore, the *Eukarya* domain evolved from an ancestral archaeon before the archaeal lineage diversified. From phylogenetic analyses it is also easy to see that the term "prokaryote" does not have any evolutionary meaning. Hence, the term should be limited to describing a cell body type (mainly, the lack of a true nucleus).

Since Woese's pioneering phylogenetic studies, many other genes have been used to reconstruct the genealogy of all life forms based on the accumulation of mutations in the target genes. All supported the three-domain view, with the early branching of *Bacteria* and *Archaea* and the branching of *Eukarya* from the *Archaea* lineage. To some scientists, this helped explain why some archaeal processes such as DNA replication and transcription (to make RNAs) are more similar to the eukaryotic ones than to those of the *Bacteria*. However, improved and more accurate phylogenetic methods have been developed that challenged the validity of the three-domain scheme. Furthermore, in an era when entire genomes can be sequenced inexpensively and rapidly, even from organisms that are not available in pure culture, researchers have turned to analyzing whole genomes, rather than genealogy-specific genes such as rRNA. These analyses not only rely on mutational changes of genes to study the relatedness of organisms but also consider another major driver of variation: horizontal gene transfer (i.e., the lateral exchange of genetic material among

BOX 1.3

A Three-Domain Tree or a Ring of Life?

Horizontal gene transfer (i.e., the lateral transfer of one or more genes from one organism to another) has been an important source of genetic variation among organisms. In fact, rRNAs can be transferred this way as well. Hence, mutational changes in rRNAs, as used to measure relatedness in the iconic three-domain tree, provide only a partial glimpse of the evolutionary past of all organisms. Genome-scale analyses bypass this limitation and have revealed the three different evolutionary features of eukaryotes.

- Some eukaryotic genes are related to cyanobacteria, consistent with the symbiotic origin of eukaryotic photosynthetic organelles (such as chloroplasts) from a cyanobacterial ancestor.
- Another set of eukaryotic genes is closer to the bacterial group *Alphaproteobacteria*, consistent with the symbiotic origin of the eukaryotic mitochondrion from an ancestral alpha-type bacterium.
- One other set of genes places the eukaryotes within the *Archaea* domain, consistent with the acquisition of mitochondria and/or chloroplasts by an ancestral archaeal host cell.

The "ring of life" tree reconciles the genomic data by placing the *Bacteria* and *Archaea* as the two primary and ancient lineages evolved from a common ancestor. The eukaryotes are no longer a separate domain, but a hybrid group evolved from the *Bacteria* and *Archaea* groups after their diversification had already begun.

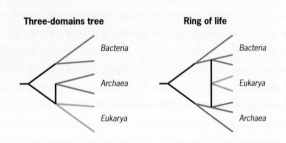

organisms). Such phylogenetic studies no longer support the three-domain phylogeny. Rather, genome-scale data support a "ring of life" model in which the *Eukarya* are no longer a primary lineage of life but a chimeric group of symbiotic origin from the *Bacteria* and *Archaea* lineages (Box 1.3).

Regardless of the phylogenetic approach, all trees reveal the enormous phylogenetic diversity of the microbial lineages, particularly the *Bacteria* and *Archaea* groups. This should not surprise us: prokaryotes had the planet to themselves for nearly 2 billion years and diversified their metabolisms to adapt to the myriad planetary changes they encountered along the way (Fig. 1.8). In the process, they transformed Earth into the planet that we know today. First to evolve were anaerobic metabolisms, which thrived in the oxygen-free early Earth. But once cyanobacteria arose almost 3 billion years ago, they carried out a new form of photosynthesis that released oxygen as a by-product. The ancestral cyanobacteria produced so much oxygen that they oxygenated the planet surface and its atmosphere. Let us pause here to discuss the impacts of releasing so much oxygen: the greenhouse gas methane was oxidized and global temperatures plummeted. It got so cold that it triggered the first glaciation: for the first time the planet was covered with snow! The cold temperatures killed many of the early microbes and selected for those adapted to the cold. Oxygen also killed many of the early anaerobic microbes, but the new conditions exerted selective pressure for the evolution of oxygen-tolerant microbes and aerobic bacteria. The latter grew abundantly while respiring the abundant O_2. Not long after this, the first unicellular eukaryotes emerged when an archaeal cell "ingested" an

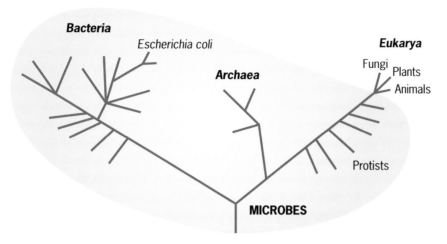

FIGURE 1.9 **Simplified representation of the classical three-domain tree.** The *Bacteria* and *Archaea* (prokaryotes) domains branched first, and the *Eukarya* (eukaryotes) branched out from the *Archaea*. Each domain is subdivided into a number of phyla (small branches in the tree). Note that there are microbes in the three domains and that they contribute with far more phyla than the macroscopic organisms. Animals, plants, and fungi are actually perched on a small branch of this "tree of life."

aerobic bacterium, which became the energy factory of eukaryotic cells: the mitochondrion.

By the time the first multicellular eukaryotes appeared some 750 million years ago (Fig. 1.8), microbes had been the planet's only "guests" for more than 3 billion years! Why is that? Microbes are the only forms of life that are self-sufficient; that is, only they can exist indefinitely without any other living things. They can also mutate, exchange genetic material, and reproduce rapidly, allowing them to diversify their metabolisms and adapt to the changing environment. Along the way, they transformed Earth and left numerous footprints of their past activities. It may be hard to believe, but some chemicals made by these ancient microbes have survived in the rocks to this day. Those are some very useful biomarkers that tell scientists what microbes lived then and what they were doing! In some cases, the microbes themselves have been preserved in mineralized form (microfossils) in the rocks. Together, the molecular (DNA footprints) and the geological (mineral footprints) records have allowed scientists to reconstruct chapters of the history of microbes on Earth with outstanding precision. Amazing!

ARE MICROBES EVERYWHERE ON EARTH?

LEARNING OUTCOMES:
- Give examples of how metabolism dictates where an organism can grow.
- Justify the following statement: "No niche on Earth that supports life is without microbes."

After all we have learned about microbes, it should not surprise us that no niche on Earth that supports life is without them. This includes environments considered extraordinarily harsh. Credibility is strained here:

TABLE 1.3 **The known extremes of life**

Some microbes *survive*:
 Very high temperatures (up to 130°C)
 5 megarads of gamma radiation (ca. 10,000 times what would kill a human)
 Very high pressures (ca. 8,000 atmospheres, or 117,000 pounds per square inch)

Some microbes *grow at*:
 Extremes of pH (0 to 11.4)
 Extreme temperatures (−15 to 121°C)
 High hydrostatic pressures (ca. 1,300 atmospheres, or 18,500 pounds per square inch)
 High osmotic pressures (5.2 M NaCl)

microbes not only survive but actually grow under extreme conditions, such as temperatures above that of boiling water, very acidic pH, pressures of thousands of atmospheres, and the salinity of concentrated brine (Table 1.3). The current world champion among **thermophiles** (as organisms that grow only at high temperatures are called) is strain 121, an archaeon that grows at (yes, you guessed it) 121°C. And this is not all this champion can do: it survives for several hours at 130°C. Such temperatures are only attainable at high pressures, such as those found in deep-sea hydrothermal vents. To grow such an organism in the laboratory requires a pressure cooker or an autoclave. Now, keep in mind that we routinely use autoclaves to sterilize microbiological media in the laboratory, yet not only do these conditions not kill this heat-loving microbe, but they support its growth. Equally amazing are microbes that grow and survive under extremes of pH or salinity that would make any other cell die almost immediately. Clearly, we have an anthropocentric view of life: we think that these organisms are bizarre, and we call them **extremophiles**. However, from the perspective of these organisms, a temperature of 37°C, a pH of 7, and low salinity are not simply uncomfortable but are actually incompatible with life.

Microbes have also diversified their metabolisms to use a narrow or a wide range of nutrients. The "picky eaters" can only utilize a narrow spectrum of carbon and energy sources, restricting themselves to environments where those particular nutrients are available. Many of the bacteria associated with the human body fall into this category, possibly because they have adapted to a rich and relatively constant environment. Others utilize a wide range of substrates and can colonize many environments. Certain species of bacteria called *Pseudomonas* can metabolize hundreds of organic compounds. Not surprisingly, they are abundant in soil, plant roots, bodies of water, and animals. And then there are microbes adapted to environments with few or no organic **nutrients**. They may be limited to a minimal diet, but they live in environments without competitors. It is a trade-off. Many of the microbes on this planet are photosynthetic; in fact, they invented photosynthesis some 2 billion years ago (Fig. 1.8). All they need is light to energize their metabolisms and atmospheric CO_2 to build their biomass, just like plants. Not surprisingly, these microbes (**phototrophs**) can colonize nutrient-poor environments, such as the surface waters of some lakes and oceans, as long as light is abundant. Other microbes have evolved ways to "extract" the energy not from light but from chemical reactions involving some intractable-sounding but energy-rich compounds. Imagine a diet of hydrogen and iron or sulfides? Well, that is the "normal" diet of many microbes. As they do not rely on the energy of the sun to grow, you can find these microbes in dark places such as deep-sea hydrothermal vents in the ocean and in cracks and fissures in rocks. So large are these communities that they are thought to outweigh all the oxygen-using organisms above ground. The extraordinary metabolic and ecological repertoire of such microbes has led to speculation that analogous life forms may even exist or have existed in other places in our solar system, such as Mars and one of Jupiter's moons, Europa.

The lesson to learn from all these examples is that microbes have diversified their metabolisms to occupy every planetary niche and to use any available resources. And the diversification continues to this day. New compounds are released into the environment through our industrial activities, yet even the most contaminated sites harbor numerous microbes. It has been said that for nearly any naturally occurring organic compound there is at least one kind of microbe that can break it down, as long as water is available. This is particularly useful for the purpose of cleaning up environmental pollutants with microbes, a technique known as **bioremediation**. It has been used successfully in the cleanup of oil spills and has continued to find other important applications to the present day. With this immense microbial metabolic versatility and the widespread presence of microbes, it would simply be too difficult for life to originate anew: the chemicals needed to make a cell would not accumulate long enough for a new cell to be assembled from scratch. The microbes in the environment would simply eat them up.

HOW DO MICROBES HELP MAKE A PLANET HABITABLE?

LEARNING OUTCOMES:
- State two ways in which microbes directly impact your life.
- Describe three locations where microbes live that are surprising to you.

Taking into account the metabolic versatility of microbes, it should not require a leap of faith to realize that they play a major role in global processes. How can they do this? Microbes link their metabolisms to carry out processes that they could never achieve individually, further expanding their metabolic potential. At a planetary scale, microbes carry out life-sustaining processes in the biosphere, including the recycling of essential elements such as carbon, oxygen, and nitrogen. Let us discuss oxygen one more time: photosynthetic bacteria oxygenated our planet, and, together with plants, they ensure that oxygen is constantly replenished to maintain stable levels in the modern atmosphere. Without it, all plants and animals and many microbes would not be able to respire and would die. Could cyanobacteria have helped scientists design a self-sustainable ecosystem in Biosphere 2 (case study, above)?

And what about carbon? Were it not for the breaking down of organic material by microbes, carbon would accumulate in dead plant matter and there would not be enough carbon dioxide for plant growth. To cycle carbon and other elements, microbes also team up with nonmicrobial hosts.

For example, mammals do not make enzymes that break down the cellulose in plant material. Instead, cattle, sheep, deer, and other ruminants harbor cellulose-digesting microbes in a large fermentation chamber called the **rumen**. The animal provides the microbes with a suitable habitat for growth; the microbes reciprocate by breaking down the cellulose into volatile fatty acids, which are used by the animal as a major energy source. Such interactions are part of the grand recycling system of all organic (carbon-containing) material on Earth. No animal or plant ever decomposes through the activities of a single microbial species. Food chains and interactive populations are the norm. Microbes also provide nitrogen for all living beings, thanks to their ability to fix it from the atmosphere. Besides nitrates used for fertilizers, which are man-made, prokaryotic microbes are the only significant source of utilizable nitrogen, essential for all of life. We estimated that if all bacteria went on strike and this microbial activity ceased abruptly, plants would run out of usable nitrogen in about one week. Sometimes, not enough attention is paid to such microbial activities. The case study highlights this fact. (By the way, have you figured out what microbial activities consumed O_2 in the enclosure's atmosphere so rapidly?)

For almost 4 billion years microbes have diversified their metabolisms to colonize every niche on Earth and carry out geological processes long thought to be abiological. Many large caverns owe their existence to microbes that oxidize hydrogen sulfide to produce sulfuric acid. The acid solubilizes the calcium carbonate rocks into calcium sulfate, which is dissolved away by water streams. This process is relatively slow, but microbes being so old and tough, these microbial miners have continued their activities steadily, transforming the underground landscape in impressive ways. Microbes also make rocks. Limestone accounts for about 10% of all rocks on Earth and is made largely by microbial members of the marine plankton, algae called coccolithophores. Each year, these organisms make 1.5 million tons of calcite, the main constituent of limestone. Similarly, diatoms, a type of protist, account for the huge deposits of silica-containing rocks. Microbes are also involved in the weathering of rocks, the process whereby rocks wear and break apart. The surfaces of rocks were commonly thought to harbor a few microbes, but microbes are present in large amounts in the crevices of rocks going down several kilometers from the surface. Their biomass is thought to exceed that found on the surface of Earth.

Even cloud formation can have microbial origins. Marine algae and bacteria combine to produce a volatile compound, dimethyl sulfide, to the tune of some 50 million tons annually. Dimethyl sulfide turns into sulfate in the atmosphere, and this acts as a nucleating agent for water vapor to become water droplets. The result is clouds. Note the feedback loop: the greater the cloud cover, the less light there is for photosynthesis, and the less production of cloud-forming sulfur compounds. In time, this leads to fewer clouds, and the cycle begins again. Such global feedbacks are not uncommon and should be kept in mind when designing self-sustained ecosystems such as the one described in the case study. But be aware of the impact that human activities can have on these processes. We release, for example, compounds in sewage that marine bacteria convert into dimethyl sulfide. Are we slowly changing the natural balance of cloud formation? Do you think microbes will find a way to counteract the effect? And what about the effect of rising global temperatures on microbial processes? The melting of the ice caps continues to stimulate the activities of many microbes that had previously been barely active or dormant. The activation of these microbial communities releases greenhouse gases such as methane and carbon dioxide to the atmosphere, raising global temperatures further. Microbes do indeed shape this planet. But let us not forget: it is up to us to capitalize on their activities to ensure the habitability of Earth for generations to come.

CASE STUDY REVISITED: Where did the oxygen disappear to in the Biosphere 2 experiment?

Did you come up with a hypothesis about microbial activity as a sink for oxygen in Biosphere 2?

The scientists hypothesized that soil microbes were responsible for the disappearance of atmospheric oxygen in Biosphere 2. The scientists tested this hypothesis by measuring the organic content of the soils in the enclosure. They soon realized that the soils in the enclosure had 2 to 5 times more compost and peat than regular soils. In other words, the soil was so rich in organic matter that the soil microbes had a feast consuming it while respiring the atmospheric oxygen in the process. So fast was oxygen consumption by the microbes that the plants could not keep up with them, and in 16 months the atmospheric oxygen levels dropped to half and the flora and fauna began to die. As a last resort, the organizers started pumping in oxygen from the outside. The exhibit was never self-sustained like our planet. On a small scale, this illustrates the impact of microbes on our environment and their essential role in making Earth habitable.

What could have been done differently? Perhaps the scientists could have reduced the compost in the soils to limit the food supply to the soil microbes. Alternatively, they could have included large ponds for photosynthetic microbes to grow. Their important activity contributes as much oxygen to Earth's atmosphere as the plants on land. This should have satisfied the demands of the soil microbes and maintained stable levels of oxygen in the enclosure's atmosphere. After all, this is exactly what these microbes did when they emerged on Earth almost 3 billion years ago: they oxygenated the planet and gave us a breathable atmosphere!

CONCLUSIONS

Microbes are small, ubiquitous, abundant, adaptable, and necessary for the continued existence of other life forms. They were the first forms of life on our planet and are the ancestors of all organisms. They play a vital role in shaping this planet. Thus, their study is essential for all who seek to understand life on Earth. In the next chapters, we will learn more about their defining features in anticipation of discussing their physiology, genetics, ecology, and evolution.

Representative ASM Fundamental Statements and Learning Outcomes for the Chapter

EVOLUTION

1. Cells, organelles (e.g., mitochondria and chloroplasts), and all major metabolic pathways evolved from early prokaryotic cells.

   a. State three characteristics of LUCA.
   b. Explain how the three-domain tree is different from the "ring of life" model.

2. Mutations and horizontal gene transfer, with the immense variety of microenvironments, have selected for a huge diversity of microorganisms.

   a. Describe some benefits and detriments of a microbe's relatively small size.

CELL STRUCTURE AND FUNCTION

3. The structure and function of microorganisms have been revealed by the use of microscopy (including bright-field, phase-contrast, fluorescent, and electron).

   a. Explain why it is difficult to define what a microbe is.
   b. List exceptions to the definition that "microbes are free-living organisms so small that they are only visible under the microscope."
   c. List three ways that cell shape influences cell metabolism.

4. While microscopic eukaryotes (for example, fungi, protozoa, and algae) carry out some of the same processes as bacteria, many of the cellular properties are fundamentally different.

   a. Compare and contrast the general structure of *Bacteria*, *Archaea*, and eukaryotic microbes.

METABOLIC PATHWAYS

5. The survival and growth of any microorganism in a given environment depend on its metabolic characteristics.

   a. Give examples of how cell metabolism dictates where a cell can grow.
   b. Explain why surface area-to-volume ratio is an important constraint in bacterial growth.

MICROBIAL SYSTEMS

6. Microorganisms are ubiquitous and live in diverse and dynamic ecosystems.

   a. Justify the following statement: "No niche on Earth that supports life is without microbes."
   b. Describe three locations where microbes live that are surprising to you.
   c. Predict why the immensity of the microbial world was unknown until the very end of the early 20th century.

7. Microorganisms can interact with both human and nonhuman hosts in beneficial, neutral, or detrimental ways.

   a. State two ways in which microbes directly impact your life.

IMPACT OF MICROORGANISMS

8. Microbes are essential for life as we know it and the processes that support life (e.g., in biogeochemical cycles and plant and/or animal microflora).

   a. Describe the impact cyanobacteria had on early Earth.

SUPPLEMENTAL MATERIAL

References

- **Lyons TW, Reinhard CT, Planavsky NJ.** 2014. The rise of oxygen in Earth's early ocean and atmosphere. *Nature* **506:**307–315. doi:10.1038/nature13068.
- **McInerney JO, O'Connell MJ, Pisani D.** 2014. The hybrid nature of the Eukaryota and a consilient view of life on Earth. *Nat Rev Microbiol* **12:**449–455. doi:10.1038/nrmicro3271.

- **Whitman WB, Coleman DC, Wiebe WJ.** 1998. Prokaryotes: the unseen majority. *Proc Natl Acad Sci U S A* **95:**6578–6583. http://dx.doi.org/10.1073/pnas.95.12.6578.

Supplemental Activities

CHECK YOUR UNDERSTANDING

1. Why are viruses not considered microbes? (One or more answers possible)

 a. They are smaller than 1 μm.
 b. They do not remain intact during replication.
 c. They require host cells to grow.
 d. They lack nucleic acids to store genetic information.
 e. They must possess lipid membranes.

2. Microbes are _____. Check the best definition, presenting arguments to support your answer.

 a. relatively small
 b. microscopic
 c. free-living
 d. unicellular
 e. prokaryotes

3. Bacteria range in volume over 1-million-fold. Discuss some of the consequences of being much larger or much smaller than the average *E. coli* cell.

DIG DEEPER

1. Phylogenetic trees are used to represent how related different species are to each other by comparing genealogy-specific genes such as the one encoding the small ribosomal subunit RNA (16S in prokaryotes and 18S in eukaryotes). Learn how to interpret a phylogenetic tree by first looking at the simplified version of the tree of life shown in Fig. 1.9. You'll recognize the two major features of a phylogenetic tree: the nodes and the branches. Identify a node and a branch in the tree.

 a. What does each of these features represent?
 b. Count the number of nodes between *E. coli* and the animals. Count the number of nodes between *E. coli* and the most terminal *Archaea* branch.
 c. What do these numbers tell us?
 d. How does this information constrain use of the term "prokaryote"?

2. Bacteria such as *Bacillus subtilis* are rod shaped; that is, they are like a cylinder capped with half spheres. As the bacillus cell grows, the cylinder section expands, keeping its diameter constant, as shown in the figure:

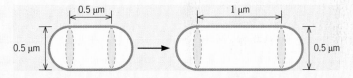

 a. Calculate the volume and surface area of a bacillus cell with an average diameter of 0.5 μm when the cylinder portion of the cell grows from 0.5 to 1 μm.
 b. Calculate the surface-to-volume ratio for each cell.
 c. Can you make predictions about what cell will grow faster? Why?
 d. What are the advantages and disadvantages of being small or large?

3. The image below is an alignment of the amino acid sequences corresponding to an RNA-modifying protein found in the budding yeast (*Saccharomyces cerevisiae*), humans (*Homo sapiens*), plants (*Arabidopsis thaliana*), and bacteria (*E. coli*). The shaded areas represent amino acids highly conserved in the protein sequences. These four proteins are considered homologous, meaning their amino acid sequences are generally conserved across species, consistent with them performing similar functions (tRNA chemical modification) in each species.

ATG/GTP binding

```
                 10        20        30        40        50        60        70
          ....|....|....|....|....|....|....|....|....|....|....|....|....|....|
S. cerevisiae  ----------MLKGPLKG-----CLNMSKKVVIVIAGTTGVGKSQLSIQLAQKFNGEVINSDSMQVYKDIP  55
H. sapiens     MASVAAARAVPVGSGLRG-----LQRTLPLVVILGATGTGKSTLALQLGQRLGGEVISADSMQVYEGLD   64
A. thaliana    -----MMMLNPSNGGIEGE----KMKKKAKVVVIMGPTGSGKSKLAVDLASHFPVEIINADAMQIYSGLD  61
E. coli        -----------MSDISKAS-------LPKAIFLMGPTASGKTALAIELRKILPVELISVDSALIYKGMD   51

                 80        90       100       110       120       130       140
          ....|....|....|....|....|....|....|....|....|....|....|....|....|....|
S. cerevisiae  IITNKHPLQEREGIPHHVMNHVE-WSEEYYSHRFETECMNAIEDIHRRGKIPIVVGGTHYYLQTLFNKRV 124
H. sapiens     IITNKVSAQEQRICRHHMISFVDPLVTNYTVVDRFNRATALIEDIFARDKIPIVVGGTNYYIESLLWKVL 134
A. thaliana    VLTNKVTVDEQKGVPHHLLGTVS-SDMEFTARDFRDFTVPLIEEIVSRNHIPVLVGGTHYYIQAVVSKFL 130
E. coli        IGTAKPNAEELLAAPHRLLDIRD-PSQAYSAADFRRDALAEMADITAAGRIPLLVGGTMLYFKALLEGLS 120

                150       160       170       180       190       200       210
          ....|....|....|....|....|....|....|....|....|....|....|....|....|....|
S. cerevisiae  DTKSSER-----------KLTRKQLDILESTDPDVIYNTLVKCDPDIATKYHPNDYRRVQRMLEIYYKT  182
H. sapiens     VNTKPQEMG---------TEKVIDRKVELEKEDGLVLHKRLSQVDPEMAAKLHPHDKRKVARSLQVFEET 195
A. thaliana    LDDAAEDTEECCADVASVVDQDMVVESVFGRDDLSHGYELLKELDPVAANRIHPNNHRKINQYLSLHASR 200
E. coli        PLPSADP-----------EVRARIEQQAAEQGWESLHRQLQEVDPVAAARIHPNDPQRLSRALEVFFIS  178
```

Amyloid domain   DMAPP binding

```
                220       230       240       250       260       270       280
          ....|....|....|....|....|....|....|....|....|....|....|....|....|....|
S. cerevisiae  GKKPSETFNEQKIT---------LKFDT-LFLWLYSKPERLFQRLDDRVDDMLERGALQEIKQLYEYYSQ 242
H. sapiens     GISHSEFLHRQHTEEGGGPLGGPLKFSNPCILWLHADQAVLDERLDKRVDDMLAAGLLEELRDFHRRYNQ 265
A. thaliana    GVLPSKLYQGKTAENWG--CINASRFDY-CLICMDAETAVLDRYVEQRVDAMVDAGLLDEVYDIYKPG-- 265
E. coli        GKTLTELTQTSGDA---------LPYQVHQFAIAPASRELLHQRIEQRFHQMLASGFEAEVRALFARG-- 237

                290       300       310       320       330       340       350
          ....|....|....|....|....|....|....|....|....|....|....|....|....|....|
S. cerevisiae  NK--FTPEQCENGVWQVIGFKEFLPWLTGK----------------------------TDDNT       275
H. sapiens     KNVSENSQDYQHGIFQSIGFKEFHEYLITEG---------------------------KCTLETSN     304
A. thaliana    -------ADYTRGLRQSIGVREFEDFLKIH----LSETCAGHLTSLSNDDKVMKENLRKILNFPKDDKLY 324
E. coli        ------DLHTDLPSIRCVGYRQMWSYLEGE---------------------------           261

                360       370       380       390       400       410       420
          ....|....|....|....|....|....|....|....|....|....|....|....|....|....|
S. cerevisiae  VKLEDCIERMKTRTRQYAKRQVKWIKK------MLIPDIKGDIYLLDATDLSQWDRNASQRAIAISNDFI 339
H. sapiens     QLLKKGIEALKQVTKRYARKQNRWVKN---RFLSRPGPIVPPVYGLEVSDVSKWEESVLEPALEIVQSFI 371
A. thaliana    IMLEEAIDRVKLNTRRLLRRQKRRVSR-LETVFGWNIHYIDATEYILSKSEESWNAQVVKPASEIIRCFL 393
E. coli        ISYDEMVYRGVCATRQLAKRQITWLRG-----------WEGVHWLDSEKPEQARDEVLQVVGAIAG--- 316
```

Zn-finger domain

```
                430       440       450       460       470       480       490
          ....|....|....|....|....|....|....|....|....|....|....|....|....|....|
S. cerevisiae  SNRPIKQERAPKALEELLSKGETTMKKLDDWTHYTCNVCRNADGKNVVAIGEKYWKIHLGSRRHKSNLKR 409
H. sapiens     QG------HKPTATPIKMPYN----EAENKRSYHLCDLC------DRIIIGDREWAAHIKSKSHLNQLKK 425
A. thaliana    ET------ETESGRDPTSGKS----IERDLWTQYVCEAC---GNKILR-GRHEWEHHKQGRTHRKRTTR 448
E. coli        ------------------------------------------------------------------- 316

                500       510       520       530
          ....|....|....|....|....|....|....|....|
S. cerevisiae  NTRQADFEKWKINKKETVE---------------------- 428
H. sapiens     RRRLDSDAVNTIESQSVSPDHNKEPKEKGSPGQNDQELKCSV 467
A. thaliana    HKNSQTYKNREVQEAE-VN---------------------- 466
E. coli        ---------------------------------------- 316
```

Source: http://www.sciencedirect.com/science/article/pii/S0378111914011032)

a. For each species, calculate a "conservation score" by giving 1 point for every amino acid that is conserved (shaded).

b. Based on their conservation scores, which proteins are most closely related evolutionarily?

c. From what species is the protein sequence least conserved? Where in the protein is it most different from the other three?

d. What would you predict the chronological order of evolution to be for these proteins? Draw a phylogenetic tree to answer this question.

4. **Team work:** This chapter describes only some of the many ways microbes impact the world. Working in groups, research one of the following ways that microbes affect life on this planet. Develop a 5-minute presentation with visual aids to teach your classmates about the interactions between the microbes and this world. How do the microbes impact their environment? Does this interaction benefit their environment? The microbes?

- Microbes in termites
- Sewage treatment plants
- Nitrogen cycling in the soil
- Microbes in our gastrointestinal tract
- Find your own interesting example

Supplemental Resources

VISUAL

- How can we see microscopic microbes? "The Virtual Microscope" (http://micro.magnet
.fsu.edu/primer/virtual/magnifying/index.html).

- Video illustrating what a protocell may have looked like and how it could have grown
and divided: http://exploringorigins.org/protocells.html.

- It is difficult to believe that essential life molecules such as RNAs could have been
formed spontaneously billions of years ago, right? This video illustrates how RNAs could
have been formed on clay minerals: http://exploringorigins.org/.nucleicacids.html

WRITTEN

- Moselio Schaechter tells us more about the largest microbes in the *Small Things Considered* blog post "The Microbe That Could Be Seen" http://schaechter.asmblog.org/
schaechter/2009/02/by-elio----the-microbe-is-so-very-small--you-cannot-make-him-
out-at-all--but-many-sanguine-people-hope--to-see-him-through-a.html.

- Gemma Reguera discusses what rocks "say" about the emergence of microbial life in
the *Small Things Considered* blog post "The Oldest Gem Tells Its Tale" (http://schaechter
.asmblog.org/schaechter/2014/04/the-oldest-gem-tells-its-tale.html).

- You know about dinosaur fossils. What do microfossils look like? http://evolution.berkeley
.edu/evosite/evo101/IIE2aOldestfossils.shtml.

- Graduate student Yana, also known as Psi Wavefunction, clarifies what different phylogenetic terms mean in the *Small Things Considered* blog post "Of Terms in Biology: Monophyletic, Paraphyletic . . ." (http://schaechter.asmblog.org/schaechter/2010/10/of-terms
-in-biology-monophyletic-paraphyletic.html).

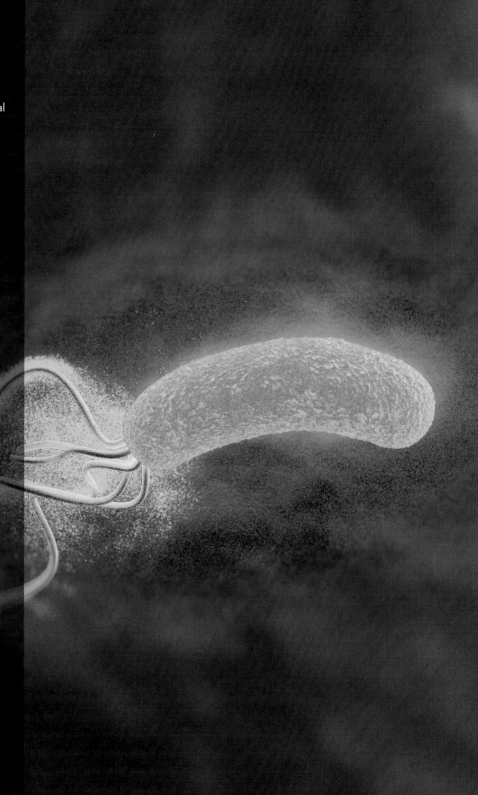

IN THIS CHAPTER

Helicobacter pylori *cell with polar flagella*

CHAPTER TWO

Prokaryotic Cell Exterior: *Envelopes and Appendages*

KEY CONCEPTS

This chapter covers the following topics in the ASM Fundamental Statements.

EVOLUTION

1. Cells, organelles (e.g., mitochondria and chloroplasts), and all major metabolic pathways evolved from early prokaryotic cells.
2. Mutations and horizontal gene transfer, with the immense variety of micro-environments, have selected for a huge diversity of microorganisms.

CELL STRUCTURE AND FUNCTION

3. The structure and function of microorganisms have been revealed by the use of microscopy (including bright-field, phase-contrast, fluorescent, and electron).
4. Bacteria have unique cell structures that can be targets for antibiotics, immunity, and phase infection.
5. *Bacteria* and *Archaea* have specialized structures (e.g., flagella, endospores, and pili) that often confer critical capabilities.

INFORMATION FLOW AND GENETICS

6. Although the Central Dogma is universal in all cells, the processes of replication, transcription, and translation differ in *Bacteria*, *Archaea*, and eukaryotes.

MICROBIAL SYSTEMS

7. Most bacteria in nature live in biofilm communities.

INTRODUCTION

This chapter and the next deal with structures of prokaryotic cells, focusing first on the external structure (this chapter) and then on the interior structure of the cell (chapter 3). The structural features of selected eukaryotic cells are explored in chapters 15 (fungi) and 16 (protists). Why focus on the cell exterior first? Despite their apparent simplicity, prokaryotes have evolved complex outer layers both to preserve the integrity of the macromolecules housed in their interior and to achieve efficiency for the metabolic reactions that take place in the cytoplasm. These external layers also equip the cell to interface with its environment, modifying its chemical properties and enabling new functions and colonization of new environments.

We will first focus on the prokaryotic cell membrane, its evolutionary significance, and its functions. Next we will discuss the additional layers that make the envelopes of prokaryotic cells, paying special attention to their structural and biochemical diversity and the specific properties they confer on the cell. Lastly, we will describe the coats and appendages that prokaryotic cells assemble only when needed, mainly capsules and appendages such as flagella and pili.

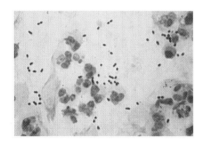

ASM MicrobeLibrary.org © Pfizer Inc.

CASE: **A simple Gram stain and a microscope to diagnose meningitis**

JH, a college sophomore, woke up with a fever of 104°F, headache, and a stiff neck. The staff at the student infirmary suspected that he might have bacterial meningitis and sent him in great haste to the nearest emergency room. There, the health care workers performed a lumbar puncture to collect a sample of his spinal fluid. The specimen was stained to differentiate Gram-positive (purple) from Gram-negative (red) bacteria under a microscope. As shown in the micrograph, the spinal fluids contained purple (Gram-positive) cocci, either singly or in pairs, and larger cells stained pale red but with brighter red large nuclei (human white blood cells). This image was all the doctor needed to confirm the diagnosis of bacterial meningitis and prescribe a high dose of the antibiotics vancomycin and cefotaxime. Once on antibiotics, the patient recovered quickly.

- Why did examination of the Gram stain suffice to suggest a diagnosis?
- Why was the antibiotic cocktail that the doctor prescribed effective?

HOW COMPLEX ARE PROKARYOTIC CELLS?

LEARNING OUTCOMES:
- Identify three ways in which prokaryotes and eukaryotes are structurally similar.
- Speculate on whether having a nucleoid and allowing ribosomes to directly attach to nascent mRNA transcripts, such as the case in prokaryotes, may be beneficial to survival in some cases.
- Compare and contrast the cell envelopes of Gram-negative bacteria and Gram-positive bacteria.

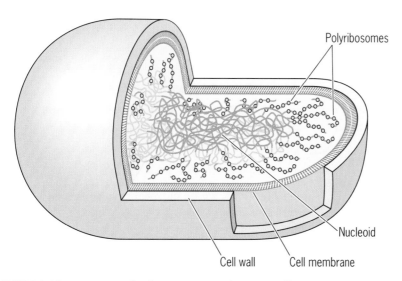

FIGURE 2.1 Ultrastructure of a characteristic prokaryotic cell. Inside the cell membrane, the DNA is packed, often in the center, to form a dense mass called the nucleoid. The DNA is unpacked locally and used as a template to make RNAs. Ribosomes attach to single mRNA molecules to make a polyribosome, thus coupling RNA transcription and protein synthesis. (Note that polyribosomes tend to occupy the areas around the nucleoid.) Outside the cell membrane is a cell wall and, sometimes, an additional layer called the capsule (not shown).

Microbes span the largest divergence in *cell type* within the living world. Take, for example, the coccoid bacterial cells of the case study and the yeast cells that make our beer and bread. Yeast cells have nuclei, just like the white blood cells shown from the case study. As we learned in chapter 1, the morphological attributes of cells—mainly, the absence or presence of a true nucleus—allow us to group them as prokaryotes and eukaryotes, respectively. (Remember that this division is unrelated to their phylogeny, with the prokaryotes, *Bacteria* and *Archaea*, constituting two of the three life domains, and the *Eukarya* the third.) Compared to the eukaryotes, prokaryotic cells are generally smaller and "look" simpler, due to the lack of membrane-bound organelles such as nuclei, mitochondria, and chloroplasts. But looks can be deceiving. Small though they are, prokaryotic cells, whether bacterial or archaeal, are not just simple, nondescript, tiny balls or sausages whose intracellular constituents rattle around like marbles in a jar. Far from it. Prokaryotic cells are surprisingly organized and structured (Fig. 2.1). Nothing is left to chance. Their chromosomes are neatly packed as a **nucleoid** in specific sites of the cell, often in a central position. Because the nucleoid is *not* surrounded by a nuclear membrane as in eukaryotes, *ribosomes can attach directly to nascent messenger RNA (mRNA) transcripts*, forming **polyribosomes**. Cellular organelles, if any, are also targeted to specific positions. To make sure all the internal cellular components stay put in their proper location, prokaryotic cells polymerize proteins to make a skeletal framework analogous to the eukaryotic cytoskeleton. And, by being small, the prokaryotes increase their surface-to-volume ratio and maximize the diffusion of nutrients and waste products in and out of the cell (chapter 1).

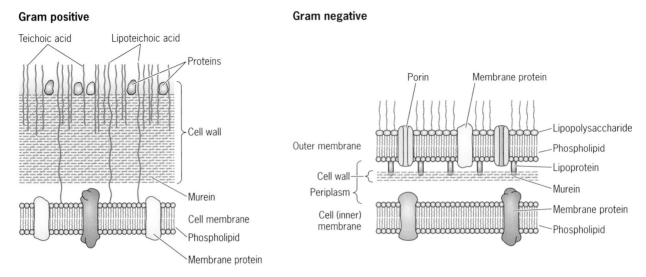

Gram positive

Teichoic acid Lipoteichoic acid

Proteins

Cell wall

Murein

Cell membrane

Phospholipid

Membrane protein

Gram negative

Porin Membrane protein

Outer membrane

Lipopolysaccharide

Phospholipid

Lipoprotein

Cell wall

Periplasm

Murein

Cell (inner) membrane

Membrane protein

Phospholipid

FIGURE 2.2 Envelopes of Gram-positive and Gram-negative bacteria. Gram-positive bacteria have a thick cell wall made up of many layers of murein. Several constituents (teichoic acids, lipoteichoic acids, and protein) protrude from this layer. Gram-negative bacteria have a thin cell wall and an outer membrane. The outer leaflet of the outer membrane is made up of lipopolysaccharide (LPS). The space between the two membranes is called the periplasm.

Also complex is the prokaryotic **cell envelope**. Most bacteria, for example, can be grouped as **Gram positive** or **Gram negative** based on the structure of their cell envelope: the cell membrane of Gram-positive cells is surrounded by a thick **cell wall**, whereas Gram-negative cells have a thin cell wall surrounded by an **outer membrane** (Fig. 2.2). Each component of the cell envelope serves a purpose. The cell wall, for example, provides structural support but also anchors molecules that change the chemistry of the cell's outer surface to influence transport and other critical cell functions. The outer membrane of Gram-negative bacteria positions proteins and other molecules that participate in the chemical interactions between the cell and the external milieu. Although archaea lack the typical bacterial cell wall, they do have other structures that surround the cell membrane and fulfill analogous structural and functional roles. Prokaryotic cell envelopes can also anchor extracellular appendages such as **flagella** for swimming or **pili** for adhesion and for motility on surfaces. In this chapter, we will introduce you to the complex "simplicity" of prokaryotic cells, emphasizing some of the major features of their external layers. So important are these structures that make up the prokaryotic envelope that we will continue to highlight examples of the sophisticated anatomy of the prokaryotic cell throughout the book.

HOW DO WE VISUALIZE THE STRUCTURAL DETAILS OF MICROBIAL CELLS?

LEARNING OUTCOMES:
- Explain why microbial cells must be stained for us to see them.
- List advantages and disadvantages of light microscopy, fluorescence microscopy, and electron microscopy.

As we learned in chapter 1, microbes are generally too small to be seen with the naked eye. The importance that microbiologists have placed on looking at microbes has waxed and waned over the years. At its beginning, microbiology was strictly an observational science, and microscopy led the way. In the 17th century, Antonie van Leeuwenhoek—possibly the first microbiologist—improved the available microscopes using carefully crafted lenses to magnify small objects and observe their structural make-up. Leeuwenhoek's curiosity was endless. He examined the gunk between his teeth and saw microbes with various morphologies. He also found parasitic *Giardia* cells in his stool during a bout of intestinal illness. Findings like this depended entirely on microscopic observation: when it came to the microbial world, all one could do was look.

Light Microscopy and Stains

Throughout the "golden age of microbiology" in the late 19th century, the light microscope remained one of the most valuable tools in microbiology. By then, microbiologists had overcome the inherent difficulty of observing the colorless prokaryotic cells: without color, they lack sufficient contrast and can't be seen readily under the microscope. Hence, dyes were used to stain the cells and make them clearly visible. Combinations of stains opened the way to major achievements. Robert Koch, the notable 19th century German microbiologist, discovered the cause of tuberculosis by using a differential stain that enabled him to see tubercle bacilli as small, dark blue cells within brown-stained host tissue. The stain procedure has changed over time: modern procedures stain these bacteria and tissues red and blue, respectively; but the differential approach remains the same. As a medical student, Christian Gram developed a differential staining procedure that still bears his name and that allowed him to sort bacteria into two groups based on the structure of their cell envelope: Gram-negative bacteria stain red, whereas Gram-positive bacteria stain purple (Fig. 2.3). The Gram stain is still used in laboratories all over the world. In the case study, we learned about its use to diagnose meningitis in the patient's spinal fluid specimen. Bacterial cells are rarely found in spinal fluid but become abundant during bacterial meningitis. Coccus-shaped bacteria are often the culprits in meningitis infections. The usual suspects are either pneumococcus (*Streptococcus pneumoniae*, which is Gram positive) or one of two Gram-negative bacteria, meningococcus (*Neisseria meningitidis*) or *Haemophilus influenzae* bacilli. Hence, knowing the result of the Gram stain (positive or negative) and the morphology of the infecting bacteria (cocci or bacilli) enables a physician to make a rapid presumptive determination of the causative agent and select the most effective antibiotic treatment. Now that you have the same information the doctor has, can you determine which bacterial pathogen likely caused this patient's meningitis?

Fluorescent stains can also be used with light microscopy. The technique relies on the fluorescence emitted from chemicals called **fluorochromes** when exposed to incident light of a shorter wavelength. A fluorescence microscope is equipped with a light source to excite the fluorochrome and a detector to observe the emission from the excited molecule. Fluorescent stains are more discriminating than traditional stains and can be used to visualize specific types of molecules. To stain specific

A

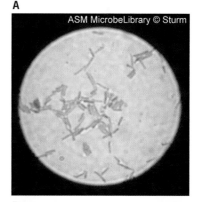

B

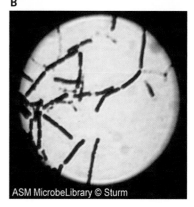

FIGURE 2.3 **Gram-stained bacterial cells visualized with a light microscope.** Gram-negative bacteria such as *E. coli* stain red, whereas Gram-positive bacteria such as *Bacillus subtilis* stain purple. (ASM Microbe Library © Tasha Sturm)

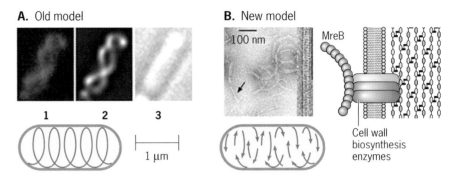

A. Old model

B. New model

FIGURE 2.4 Tracking the cytoskeletal protein MreB microscopically. (A) Fluorescence micrographs of *E. coli* cells expressing the MreB protein fused to GFP supported a model in which the protein forms helical filaments along the cell (images 1 and 2 are, respectively, the fluorescence micrographs before and after deconvolution, a computer processing method that sharpens the MreB signal; image 3 shows the outline of the cell). (B) Cryo-ET micrographs revealed short, discontinuous MreB filaments (red arrows), which are mostly anchored to the cell wall biosynthetic enzymes in the cytoplasmic side of the inner membrane (IM). (Panel A from Shu YL, Rothfield L, *PNAS* **100:**7865–7870, 2003, with permission; panel B from Celler K et al., *J Bacteriol* **195:**1627–1636, 2013, with permission.)

cellular components, scientists often attach the fluorochrome to a molecule that reacts specifically with the desired cell target. For example, specific antibodies or DNA probes can be modified to carry a fluorochrome. Such tagged components "light up" when excited by the right wavelength from the fluorescence microscope to reveal the location of the target component in the cell or tissues. The same approach can be used to localize a fluorescent protein. Gene fusions of the target and fluorescent proteins can be engineered and then used to express the chimeric gene in the cell. This technique allows scientists to track proteins in living cells, and, when coupled with computer-based methods, the fluorescence signal can be enhanced to provide amazing images of the protein's localization inside the cell (Fig. 2.4). The first fluorescent protein used for this type of study was derived from a jellyfish and is known as the **green fluorescent protein** (GFP). Several modified forms of this protein, such as yellow fluorescent protein and cyan fluorescent protein (a greenish blue color), have also been engineered. By using fluorescent proteins of different colors, multiple proteins can be tracked simultaneously in the same cell, live or fixed. We will see examples of this technique later in the chapter.

Even with traditional and fluorescent stains, the small size of most prokaryotic cells tests the limits of the standard light microscope. The resolving power (the minimum distance that permits two points to be perceived as being separate) of a light microscope is about 0.2 μm. To give you a sense of scale, this distance is approximately one-third the width of a typical prokaryotic cell. However, almost miraculous recent technological advances have pushed back the limits of resolution of the optical microscope, permitting us to distinguish very small nanostructures even in living cells. We use the general term **super-resolution microscopy** to refer to these novel techniques. In 2014, the Nobel Prize in chemistry was awarded for the development of super-resolved fluorescence microscopy. The light microscope lives on!

Beyond Light Microscopy: Electron and Scanning Probe Microscopes

The fine structural details of microbial cells are often studied by electron microscopy. Electron microscopes use accelerated electrons instead of light photons to visualize objects much smaller than those resolved optically. We have included several electron micrographs of cells and their components throughout this book to help you appreciate just how powerful this technique is. As you can imagine, shooting electrons at a biological sample can distort its structure. Because of this, samples are typically treated with strong chemicals (fixatives) that cross-link the biological molecules to preserve the cell's structural features during examination. Unfortunately, such pretreatment can introduce artifacts as well. Scientists have figured out ways to avoid this. A useful approach is to freeze the cells at temperatures so low that ice crystals cannot form in the sample. The frozen cells can then be imaged directly with a **cryo-electron microscope** (or cryo-EM). Some cryo-EM techniques can also generate three-dimensional images of the sample. A technique called **cryo-electron tomography** (or cryo-ET) tilts the sample during imaging to generate a series of images that, when superimposed, yields a three-dimensional reconstruction. This is analogous to how your organs are imaged during a computed tomography scan. As you can see, microscopists have many tricks up their sleeve to visualize cells and their components without artifacts and despite their small size.

But this is not all. Whereas light and electron microscopes use lenses (optical and electron, respectively) to image the sample, some very powerful microscopes called scanning probe microscopes use a simple metal tip. The tip is a small, pointed probe that scans the sample and reveals the fine details of its topography (height) or other local properties such as friction, magnetism, or conductivity. The most commonly used scanning probe microscopes are the scanning tunneling microscope, which probes the sample's conductivity, and the atomic force microscope, which measures the force needed to keep the tip in proximity to the sample. A laser beam tracks the tip's subtle movements as it scans the sample, and any deflections are recorded in a computer and used to generate an image. Here the limit of resolution is set by the size of the probe (tip) and not by the properties of lenses. The probe can be sharpened to a point as small as a single atom and held very close to the surface of the specimen, achieving atomic-level resolution. Yes, you heard right. Single atoms can be imaged with these microscopes! At the fast pace that technology is advancing, how much more will we be able to resolve in 10 years?

WHY DO ALL CELLS HAVE A MEMBRANE?

LEARNING OUTCOMES:
- Explain why all cells must have a membrane.
- Speculate on why membrane proteins are more abundant in *Bacteria* and *Archaea* than in mammalian cells.
- Compare and contrast lipid membranes in *Bacteria* with those in *Archaea*.
- Summarize why hopanoids are important biomarkers for *Bacteria*.
- Describe three ways in which membranes have adapted to confer survival in some harsh environments.

As we learned in chapter 1, life on Earth was preceded by a **prebiotic phase** during which primitive cellular entities, or protocells, were formed. From these protocells, the LUCA (remember, the last universal common ancestor of chapter 1) of all organisms evolved. A critical event in the formation of the first protocells must have involved the engulfment of self-replicating and catalytic biological molecules, most likely RNAs, by lipid micelles. The micelles provided a selective barrier for chemical diffusion and allowed the primitive biological molecules to replicate and catalyze reactions in a protected, favorable environment. So important was this event in the evolution of life that the presence of a lipid membrane, along with having DNA as genetic material and proteins and RNAs for catalytic and regulatory functions, is one of the universal cellular features of all extant organisms. No membrane, no cell.

As cells evolved, the chemical composition of the cell membrane changed to adapt its permeability and integrity to the cell's specific needs and the particular environment inhabited. The cell membrane of modern bacteria, for example, is a **phospholipid** bilayer. Phospholipid molecules contain a hydrophilic glycerophosphate "head" and a hydrophobic "tail" composed of two fatty acid chains linked to the head's glycerol molecule via *ester* linkages (Fig. 2.5). Because each glycerol is linked to two fatty acid chains via ester bonds, we call the phospholipids of the bacterial cell membrane *glycerol diesters*. The bipolar nature of phospholipids, with hydrophilic heads and hydrophobic chains, favors the spontaneous formation of a lipid *bilayer* in an aqueous solution: the hydrophobic fatty acid chains tend to aggregate away from the water, which they do best by being sandwiched between top and bottom layers of hydrophilic heads (Fig. 2.5). The fatty acid chains make the bacterial lipid bilayer quite flexible. To increase its rigidity and presumably to reduce its permeability even more, some bacteria also have sterol-like molecules called hopanoids in their membranes (Box 2.1). But there is more to the membrane: the lipid bilayer also serves as an anchor for some 200 different kinds of proteins, each of which contains hydrophobic domains to insert themselves into the bilayer. Collectively, proteins account for approximately 70% of the mass of the membrane and confer on the cell membrane myriad additional functions. This is a much higher proportion of membrane proteins than that found in mammalian cell membranes. Why do you think that is?

Like bacterial cell membranes, those of *Archaea* also contain many proteins, but the lipids are different. In the *Archaea*, the glycerophosphate head is attached to **isoprenoids**, rather than fatty acids, using *ether* (rather than *ester*) bonds (hence, they are *glycerol diethers*) (Fig. 2.5). Sometimes the lipid head is missing the phosphate groups; hence, the lipid is no longer a phospholipid. In some archaea, the isoprenoid chains are about as long as the fatty acid tails of bacterial phospholipids and spontaneously form a bilayer very similar in structure to that of bacterial membranes. Other archaea have lipids with isoprenoid chains that are twice as long and are capped by glycerol ether heads *on both ends* (hence, the lipid molecules are said to be *diglycerol tetraethers*). Because of their bipolar nature, these lipids spontaneously form monolayers rather than the customary bilayers (Fig. 2.5). The isoprenoid bilayers, and particularly the monolayers, are more stable than phospholipid membranes at low pH and at high temperatures

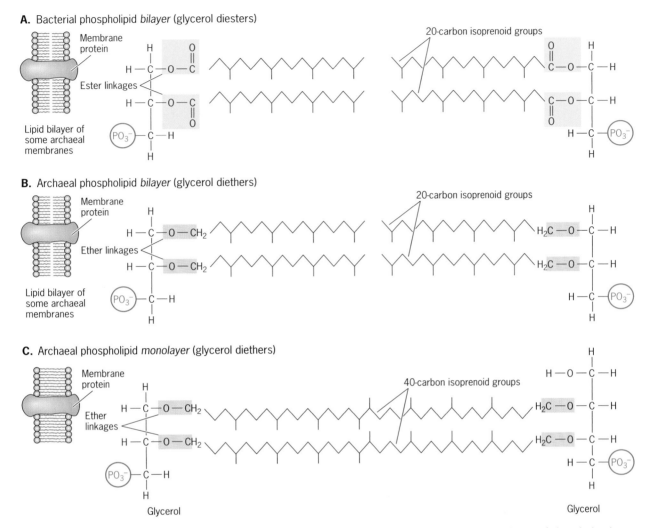

A. Bacterial phospholipid *bilayer* (glycerol diesters)

B. Archaeal phospholipid *bilayer* (glycerol diethers)

C. Archaeal phospholipid *monolayer* (glycerol diethers)

FIGURE 2.5 Types of bacterial and archaeal phospholipid membranes. (A) Bacterial membranes are bilayers of phospholipids, each containing a hydrophilic glycerophosphate head and a hydrophobic portion with two 20-carbon-long fatty acid chains linked to the glycerol via ester linkages. (B and C) Archaeal membranes are either phospholipid bilayers (B) or monolayers (C) of isoprenoids linked to glycerol by ether bonds. Note that the isoprenoids of archaeal monolayers (C) are twice as long as those in the bilayers (B) and have two glycerol heads at the ends, keeping the thickness and polarity of the membrane similar despite the different structure and composition.

and salt concentrations, making them the membrane choice for heat-loving archaea. This interesting fact reveals how the prokaryotic cell membrane has diversified chemically and structurally to adapt to different environments.

Whether bacterial or archaeal, prokaryotic cell membranes are busy places. They carry out functions that in eukaryotic cells are distributed among the plasma membrane and the membranes of intracellular organelles (Table 2.1). As in the original protocells, the lipid membrane functions as an osmotic and solute barrier to control the selective diffusion of chemicals in and out of the cell. But prokaryotic cell membranes also contain a high number of transport systems that concentrate specific substrates within the cell to a level 10^5 times higher than that in the surrounding medium (discussed in detail in chapter 7). In addition, the cell membrane

BOX 2.1

Bacterial Hopanoids: Lipids That Last "Forever"

Some bacteria make lipids of an unusual kind called hopanoids. They resemble the sterols that eukaryotes use in their cell membranes, such as cholesterol, but add an extra cyclic ring. And just like eukaryotic sterols, they add rigidity and stability to the cell membrane (see the figure). Not all bacteria make them, but they play a vital role for the ones that do. Guess how we know? If you treat hopanoid-containing bacteria with inhibitors of hopanoid synthesis, you inhibit their growth. You can find hopanoids in trace amounts in some plants. However, they have not been detected in any *Archaea*, and for this reason they have been used as biomarkers for *Bacteria*. As an example, the figure illustrates the chemical structure of 2-methylhopanoids, which serve as biomarkers for cyanobacteria.

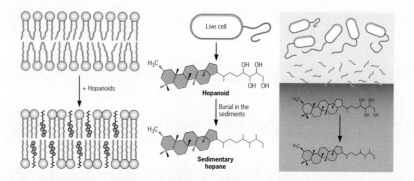

Hopanoids are complex and diverse molecules decked out with a large number of different side chains that are characteristic of the species that makes them. Why such a variety? Not much is known about that yet. What is puzzling is that within closely related species, such as some **cyanobacteria**, some have them and others don't. One interesting use for hopanoids is seen in a group of bacteria called the actinomycetes, which makes filaments that stick out into the air. This mode of life in the dry air calls for special membrane properties that may be related to their hopanoid content. Could the extra membrane rigidity conferred by hopanoids prevent desiccation?

If you're still not convinced that hopanoids are intriguing and need to be studied further, listen to this incredible fact: hopanoids are amazingly stable, even in extreme pH and at high temperatures, and have been preserved in rocks of great antiquity. Hopanoids were among the few kinds of organic molecules that survived the geological events that formed rocks and petroleum even as long ago as 1.8 billion years. In the process, the hopanoids were stripped of some of their side chains, and thus now look simpler (see the formation of sedimentary hopanes in the figure). Being so old and stable, hopanoids have accumulated in spectacular amounts, perhaps as much as 10^{12} tons, an enormous mass equal to that of the organic compounds of all living organisms. And since they are uniquely made by some bacteria, hopanoids serve as bacterial biomarkers (that is, indicators of the presence of extinct bacteria) in rocks formed nearly 2 billion years ago. What a treasure they are!

TABLE 2.1 **Some functions of prokaryotic and eukaryotic cell membranes and organelles**

Function	Structure	
	Prokaryotes	**Eukaryotes**[a]
Osmotic barrier	Cytoplasmic membrane	Cytoplasmic membrane
Transport of solutes	Cytoplasmic membrane	Mainly cytoplasmic membrane
Respiratory electron transport	Cytoplasmic membrane	Mitochondrial membrane
Protein synthesis	Polyribosomes in cytoplasm	Polyribosomes on endoplasmic reticulum
Synthesis of lipids	Cytoplasmic membrane	Smooth endoplasmic reticulum, Golgi apparatus
Synthesis of wall polymers	Cytoplasmic membrane	Golgi apparatus in plants
Protein secretion	Cytoplasmic membrane	Endoplasmic reticulum and secretory vesicles
Photosynthesis	Various membrane types, some continuous with, others independent of, the cytoplasmic membrane	Chloroplast membranes

[a]Main sites only.

anchors the electron transport chain that generates energy for cell growth and functioning. We'll say once more: without cell membranes, life as we know it would not exist.

WHY DO MICROBES NEED A CELL ENVELOPE?

LEARNING OUTCOMES:
- List three roles of the bacterial murein cell wall.
- Predict mechanisms that could allow Gram-positive bacteria to become resistant to penicillins.
- Speculate on why Gram-negative bacteria have porins to transport small (<700 daltons) hydrophilic molecules while larger molecules must enter the cell using a transporter.
- Compare and contrast the bacterial and archaeal cell walls.
- Contrast the effectiveness of secreted enzymes against target antibiotics for Gram-negative and Gram-positive cells, taking into account the structure of their cell envelopes.
- Discuss how acid-fast bacteria can withstand membrane-damaging chemicals.
- Explain why a bacterial or an archaeal cell would invest energy into making an S-layer when it already has a cell envelope.
- Justify the following statement: "The differences in composition of the envelopes of the *Archaea* and *Bacteria* can reflect their ecological distribution."
- Outline the limitations of using a Gram stain and light microscopy to differentiate *Bacteria* and *Archaea*.

Critical as membranes are to the cell's survival, microbes have evolved mechanisms to protect membrane integrity. Their lipid nature makes the prokaryotic cell membrane vulnerable to destabilizing chemicals, such as detergents and other amphipathic compounds (those containing both

TABLE 2.2 **The Gram stain**

1. Stain with crystal violet (purple); wash.
2. Mordant (bind) the dye with potassium iodide; wash.
3. Flush with alcohol; wash. Gram-negative bacteria are decolorized; Gram-positive bacteria remain purple.
4. Counterstain with safranin (red); wash. Gram-negative bacteria become red; Gram-positive bacteria remain purple.

polar and nonpolar domains), which are naturally present in many environments. The membranes must also be mechanically reinforced to withstand the propensity of its concentrated intracellular environment to take in water (osmotic pressure). Most prokaryotes maintain integrity by surrounding their membrane with a tough, corset-like jacket. Several variations of this sac have evolved to suit the lifestyle of particular microbes. You may already be familiar with the **murein** cell wall that most bacteria use to protect their cell membrane. Murein is a type of **peptidoglycan**, a complex polymer of sugars with side chains of amino acids. In Gram-positive bacteria, *the cell wall is quite thick*, consisting of several layers of murein. By contrast, Gram-negative cells have a *thin cell wall* that is *surrounded by an outer membrane* (Fig. 2.2). Such differences result in distinct properties that help differentially stain the cells using the Gram stain procedure (Table 2.2). Gram-positive bacteria stain purple because their thick cell wall retains a complex of a *purple* dye and iodine after a brief alcohol wash, whereas Gram-negative bacteria do not retain the complex, become translucent, and can then be counterstained with a *red* dye (Fig. 2.3). But the strategy is the same in both cell types: the membrane is protected by a polymeric armor. So tough and rigid is the murein cell wall in both Gram-positive and Gram-negative bacteria that it can be isolated as an intact sac (called a **sacculus**) that retains the original size and shape of the bacterium (Fig. 2.6). Not surprisingly, the cell wall sacculus also confers on bacteria their distinctive shape: rod (or bacillus; plural, bacilli), sphere (coccus; plural, cocci), spring-like or helical (spirillum; plural, spirilla), or a number of other shapes, including triangles and squares.

Shape matters for many reasons, most significantly because it influences nutrient uptake, motility, attachment to surfaces, and even reproduction. Helping the cell wall sacculus keep the cell's shape as it grows is an internal structural framework provided by the bacterial cytoskeleton. The cytoskeletal protein MreB, which we discussed earlier (Fig. 2.4), confers on most bacterial rods their distinctive elongated shape. How this protein establishes cell shape has been the subject of long, heated debate. Recall that fluorescence microscopy of MreB tagged to GFP first suggested that it polymerizes as a long, continuous helical structure (Fig. 2.4). However, cryo-electron tomography (cryo-ET) revealed instead short MreB filaments, most of them attached to the cytoplasmic cell membrane (Fig. 2.4). Furthermore, the MreB filaments were found to be anchored to the membrane-associated cell wall biosynthetic enzymes, which guide the movement of the underlying MreB filaments (Fig. 2.4). So important is the rod-shaping function of MreB that mutants lacking this protein become spherical. Interestingly, MreB is but one of many proteins that form

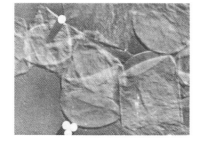

FIGURE 2.6 **Peptidoglycan (or murein) sacculi.** The electron micrograph shows flattened murein sacculi from *E. coli*. In three dimensions, the structure would have the same dimensions as the cell from which it is derived. The white spheres are latex beads 0.25 mm in diameter, included to show the scale. (Courtesy of M. R. J. Salton.)

BOX 2.2

Does the Eukaryotic Cytoskeleton Have Bacterial Ancestry?

Bacterial cytoskeletal proteins

Name	Known function	Eukaryotic counterpart
FtsZ	Division septum formation	Tubulin
MreB	Cell shape in rods	F-actin
ParM	Plasmid segregation	F-actin
Crescentin	Cell curvature	Intermediate filaments

The peptidoglycan cell wall is not the only thing that determines the shape of bacterial cells. Many rod-shaped bacteria contain an actin-like protein called MreB that is critical to produce the rod shape. In eukaryotes, actin forms a complex intracellular network of thin filaments that control the shape of the cell and participate in cellular processes such as migration, division, and endocytosis. The eukaryotic cytoskeleton also includes microtubules made of a protein called **tubulin** and intermediate filaments made of several different proteins. Interestingly, prokaryotes have several homologs of eukaryotic cytoskeletal proteins (see table). The characteristic curvature of curved and spiral cells is due, for example, to a cytoskeleton made of crescentin, a homolog of the proteins that make the eukaryotic intermediate filaments. The cell division protein **FtsZ**, on the other hand, is a homolog of tubulin. ParM is, like MreB, an actin-like protein but its role is to provide intracellular tracks for the segregation of plasmids during cell division. The presence of homologs of the eukaryotic cytoskeletal proteins in bacteria suggests that the eukaryotic cytoskeleton may have a bacterial ancestry. Bacterial and eukaryotic actin molecules differ greatly in amino acid sequence (their identity is only about 15%) but have nearly the same three-dimensional structures, suggesting that *the protein architecture* is what is worth conserving.

the bacterial skeletal framework, and many of these proteins have eukaryotic homologs, suggesting that eukaryotic cytoskeletal proteins originated from bacterial ancestors (Box 2.2).

To appreciate how important it is for cells to protect the cell membrane with a tough armor like the peptidoglycan cell wall, consider this simple experiment. Treat the Gram-positive bacterium *Bacillus megaterium* with the enzyme **lysozyme** to hydrolyze the cell wall layer; the rod-shaped bacterium will now become spherical (Fig. 2.7). Such spherical cells lacking a cell wall are called **protoplasts** when generated from Gram-positive bacteria or **spheroplasts** if they originate from Gram-negative organisms. But back to our experiment: reduce the osmolarity (salt content) of the aqueous solution just a tiny bit and the protoplasts of *B. megaterium* will get filled with so much water that they eventually burst. After bursting, all that is left of the cells are remnants of the cell membrane, like cell ghosts (Fig. 2.7). What does this experiment tell you about the role of the cell wall of most bacteria? Now, think about where the peptidoglycan-destroying enzyme lysozyme could be found in nature. (*Hint:* it is present in many mammalian tissues and secretions, including tears. What role does lysozyme play at such mammalian sites?)

Many bacteria follow either the Gram-positive or Gram-negative architectural style, but others rely instead on specialized outer layers (e.g., the **acid-fast** envelope) or lack a cell wall altogether (as in **mycoplasmas**). Furthermore, *Archaea* have their own unique cell envelopes, consistent with their distinct evolutionary path, something that we will discuss below. But let us not forget that these strategies, as different as they may seem, serve the same purpose: to protect the cell membrane and maximize the growth and survival of the microbe in a particular environment.

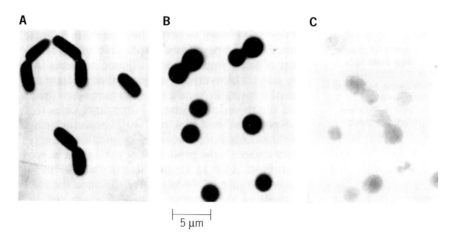

FIGURE 2.7 **Protoplasts of a member of the genus *Bacillus* (*B. megaterium*).** (A) Intact cells. (B) Protoplasts formed by the action of lysozyme in the presence of 0.5 M sucrose. (C) Ghosts (i.e., empty membranes) formed after osmotic lysis of protoplasts in a hypotonic medium. (Courtesy of C. Weibull. Reprinted from Neidhardt FC et al., *Physiology of the Bacterial Cell*, Sinauer Associates, Inc., Sunderland, MA, 1990.)

The Gram-Positive Bacterial Solution: A Thick Peptidoglycan Cell Wall

As we learned above, Gram-positive bacteria have evolved a thick peptidoglycan (or murein) cell wall and no outer membrane (Fig. 2.2). The peptidoglycan of the bacterial cell wall is made of linear chains of repeating disaccharide subunits composed of *N*-acetylglucosamine alternating with an *N*-acetylmuramic acid that carries a side chain of D-amino acids; the sugars are linked by β-1,4 bonds to form linear chains stacked parallel to each other (Fig. 2.8). The overall structure of the cell wall peptidoglycan is similar in all bacteria, but the presence of the amino acid lysine (rather than the diaminopimelic acid, or DAP, of Gram-negative cell walls) is the distinctive chemical feature of all Gram-positive cell walls (Fig. 2.9). These are also the amino acids (lysine in Gram-positive and DAP in Gram-negative bacteria) that cross-link the peptide chain with a D-alanine residue from a neighboring peptide, effectively bonding the layers to each other (Fig. 2.9). Adding many layers of peptidoglycan enriches for polar molecules such as phosphates, sugars, and charged amino acids. The resultant cell wall forms a thick, polar barrier that limits the passage of hydrophobic compounds, including some noxious chemicals such as the bile salts found in our intestines.

Gram-positive cell walls contain an additional component, **teichoic acids**, which are chains of glycerol phosphate or ribitol phosphate linked by phosphodiester bonds and bound covalently to the peptidoglycan cell wall (Fig. 2.10). The teichoic acid chains can be very long, so long that they reach the underlying cell membrane, where they are anchored (Fig. 2.2). Teichoic acids give rigidity to the cell wall but also change the chemistry of its external face, promoting adherence of bacteria to specific surfaces. This is how organisms such as streptococci attach to tissues of the host, from teeth to heart valves. Hence, teichoic acids contribute to pathogenesis.

But the Gram-positive cell wall can also be an Achilles' heel for these organisms. Antibiotics that inhibit peptidoglycan synthesis (such as the entire class of β-lactams that includes penicillin) are effective against Gram-

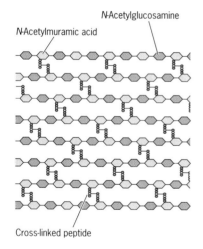

N-Acetylglucosamine

N-Acetylmuramic acid

Cross-linked peptide

FIGURE 2.8 **Structure of the peptidoglycan cell wall.** As the name indicates, the peptidoglycan cell wall consists of peptide-modified glycan chains. The glycan chains are made of alternating units of *N*-acetylglucosamine and *N*-acetylmuramic acid to which a short peptide is connected. Some of these peptides are cross-linked to the peptides of another glycan chain, making a three-dimensional network.

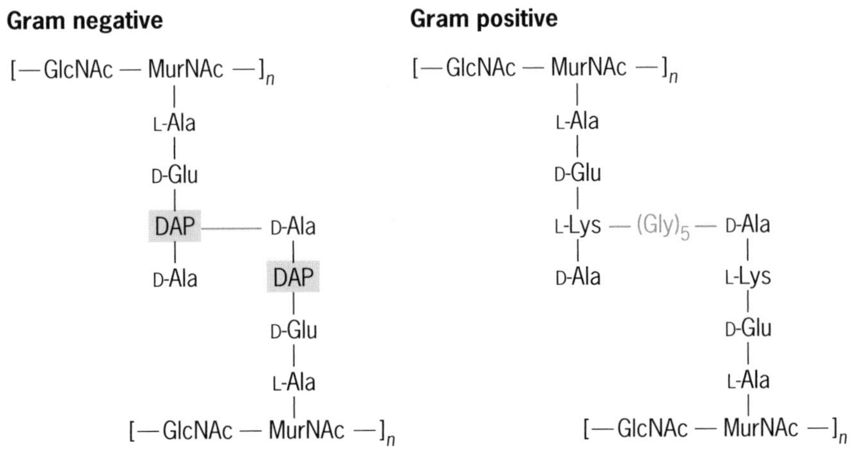

Gram negative
[—GlcNAc — MurNAc —]n
L-Ala
D-Glu
DAP
D-Ala
D-Ala
DAP
D-Glu
L-Ala
[—GlcNAc — MurNAc —]n
Gram positive
[—GlcNAc — MurNAc —]n
L-Ala
D-Glu
L-Lys — (Gly)5 — D-Ala
D-Ala
L-Lys
D-Glu
L-Ala
[—GlcNAc — MurNAc —]n

FIGURE 2.9 **Peptidoglycan composition.** The bacterial peptidoglycan has several unusual components, including D-amino acids and, in Gram-negative bacteria, DAP. Typically, peptide cross-links are between DAP and D-alanine in Gram-negative bacteria and between an L-lysine and a D-alanine residue, in this example via a glycine pentapeptide (but other connections are possible), in Gram-positive bacteria.

positive pathogens such as staph (*Staphylococcus*) and strep (*Streptococcus*). How do they interfere with cell wall synthesis? Consider a growing cell: it needs to remodel its cell wall to accommodate the growing cell volume while keeping its shape. This feat is achieved by cooperation between enzymes anchored on the membrane that make new peptidoglycan units and another set of secreted enzymes that first cut the existing peptide cross-links in the peptidoglycan cell wall and then fill the spaces with the newly synthesized units. Penicillin binds and inactivates the enzyme that bonds the peptides of the peptidoglycan layers to each other. Without an active enzyme to cross-link the peptidoglycan layers together, the cell wall armor weakens so much that the internal osmotic pressure eventually causes the cell to burst. Can you deduce what mechanisms render Gram-positive bacteria resistant to penicillins?

The Gram-Negative Bacterial Solution: A Thin Peptidoglycan Cell Wall and an Outer Membrane

Gram-negative bacteria have evolved a radically different mechanism to protect their cell membranes. Their peptidoglycan cell wall is thin and surrounded by an outer membrane (Fig. 2.2). Thin though it may be, the cell wall of Gram-negative bacteria wraps the cell in a rigid armor that

Glycerol teichoic acid
Murein
Ribitol teichoic acid
Murein

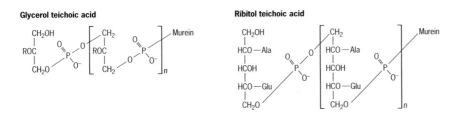

FIGURE 2.10 **Teichoic acid structure.** The repeating units of two kinds of teichoic acids, glycerol and ribitol, are shown. The lengths of chains are variable among species.

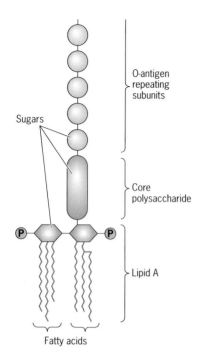

FIGURE 2.11 LPS structure. LPS consists of three parts: lipid A, a phosphorylated disaccharide to which fatty acids are attached; a core polysaccharide region; and the O antigen, made up of repeating sugars. This region is highly variable among strains and is the main reason for antigenic variety among Gram-negative bacteria.

determines the shape of the cell. (Remember the cell wall sacculus in Fig. 2.6?) But here is where Gram-negative cell envelopes are unique: they surround the peptidoglycan cell wall with an *outer membrane that is chemically distinct from all other biological membranes* and is especially *resistant to harmful chemicals.*

The Gram-negative bacterial envelope has an *inner leaflet* composed of the customary phospholipids, whereas its *outer leaflet* contains a distinctive component called **lipopolysaccharide (LPS)** (Fig. 2.11). As the name indicates, LPS is a lipid modified with sugar: it consists of one glycolipid (**lipid A**) and one or two sugar portions (**LPS core** and **O antigen**). Here are the key features of each LPS component.

1. In lipid A, the short-chain fatty acid chains of the bacterial phospholipids are attached to a disaccharide. Being a glycolipid comes in handy: the hydrophobic fatty acid chains are embedded in the outer leaflet of the outer membrane, like the roots of a plant in the ground, whereas the hydrophilic disaccharides stick out of the membrane to anchor the sugars of the core region. This unusual glycolipid also makes LPS easy for eukaryotes to recognize as foreign; we'll learn in chapter 22 that it induces fever and shock in vertebrates. For this reason, LPS is also known as **endotoxin** (never mind the "endo-," which is a historical misnomer).
2. Attached to the lipid A is a short series of sugars that form the LPS core. The composition of the core is relatively constant among Gram-negative bacteria and includes two somewhat unusual sugars, keto-deoxyoctanoic acid and heptose.
3. The O antigen is variable and specific to the organism. It is usually a long carbohydrate chain (up to 40 sugar residues in length) bound to the underlying core of the LPS. The hydrophilic carbohydrate chains of the O antigen swathe the bacterial surface and exclude hydrophobic compounds.

To appreciate the importance of the O-antigen chains of the LPS, consider this simple genetic experiment: inactivating different steps in their biosynthesis results in mutants that make shorter, less, or no O antigen. The sensitivity of these mutant cells to bile salts and antibiotics skyrockets! As the name "antigen" implies, the O antigen is highly immunogenic; it elicits a strong antibody response when introduced into a vertebrate animal. O antigens are also very diverse, differing among species and sometimes among strains within the same species. For example, the unique O antigen (O157) carried by the *Escherichia coli* O157:H7 strain makes the cells particularly infectious. Not surprisingly, this is the very *E. coli* strain associated with severe infections after ingestion of contaminated hamburger meat and the one that gives this bacterium its bad reputation. Because of the O antigens' variability, antibodies against one type will not protect an individual against infection from strains containing another kind. (Chapter 26 will vividly illustrate how O-antigen variability contributes to epidemics of cholera.) In addition, some O antigens are toxic, accounting for some of the virulence (disease-causing properties) of certain Gram-negative bacteria. But not all Gram-negative bacteria have O antigen. Some bacteria have what is known as **rough LPS**, a type of LPS that lacks the O antigen and is, as a result, more hydrophobic than the **smooth LPS**

A

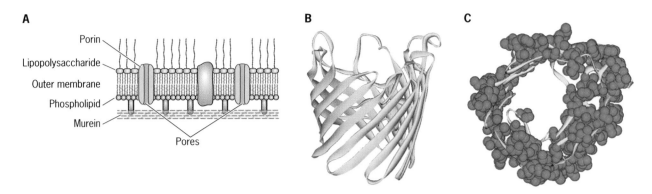

Porin

Lipopolysaccharide

Outer membrane

Phospholipid

Murein

Pores

B

C

FIGURE 2.12 Porins. (A) The outer membrane of a Gram-negative bacterium shows pores made by porin molecules. (B) The three-dimensional structure of a porin (side view). Porins form a 16-strand β-barrel inserted in the outer membrane. (C) Top view of a porin showing the hydrophilic residues (in purple) that make a water-filled channel.

varieties, which carry the O antigen portion. Can you think of reasons why a rough LPS may be advantageous to some bacteria? (*Hint:* think about how rough or smooth LPS can affect adhesion and transport.)

While having an outer membrane that excludes deleterious hydrophobic compounds is clearly advantageous to the Gram-negative organisms, it also reduces permeability and nutrient uptake, which would slow growth. To bypass this limitation, Gram-negative bacteria decorate their outer membrane with protein channels called **porins** (Fig. 2.12). As the name suggests, porins make pore-like structures that permit the diffusion of many hydrophilic compounds, such as sugars, amino acids, and certain ions. Looking at their structure will reinforce their function in transport: porins are formed by triads of proteins that leave an inner channel just large enough to permit the entry of compounds of up to 600 to 700 daltons, about the size of a trisaccharide (Fig. 2.12). To transport larger hydrophilic compounds such as vitamins, larger polysaccharides, and chelating molecules importing essential metals (e.g., iron), the outer membrane is equipped with additional proteins that translocate these compounds. Equipped with porins and dedicated transport systems, the outer membrane allows the passage of small hydrophilic nutrients; excludes hydrophobic compounds, large or small; and allows the specific entry of larger hydrophilic molecules it may need.

The outer membrane is anchored to the underlying peptidoglycan cell wall via lipoproteins. Whether these two are also anchored to the cell membrane is not well known. Regardless, the architecture of the Gram-negative cell envelope leaves an aqueous space between the inner and outer membrane, called the **periplasm**, where many essential processes take place (Fig. 2.2). The periplasmic compartment contains the peptidoglycan cell wall and a gel-like solution of cell wall precursors and proteins that assist in nutrition. Also residing in this space are degradative enzymes, such as phosphatases, nucleases, and proteases that break down molecules to sizes small enough to be transported across the inner membrane. In addition, the periplasm contains proteins with high binding affinity for specific sugars and amino acids that equip the cell to soak up nutrients from the medium, much like a sponge. The periplasm also contains enzymes called β-lactamases, which protect the cell by inactivating antibiotics such as penicillins and cephalosporins. Now think again about what Gram-positive bacteria are missing: they do not have a periplasmic compartment, so they secrete all these enzymes into the environment. How effective would these enzymes be against their target antibiotics once secreted outside the cell? Compare this to the Gram-negative strategy, which concentrates these valuable enzymes in the periplasmic space.

The Acid-Fast Solution of Tubercle Bacilli and Their Ilk

Some bacteria, notably the tubercle bacillus (*Mycobacterium tuberculosis*) and its relatives, have developed yet another way to protect their cell membranes from environmental challenges, the so-called **acid-fast** cell envelope (Fig. 2.13). The cell walls of these organisms contain large amounts of **waxes** (yes, you heard right!) called **mycolic acids**. These waxy molecules are made of fatty acids, as are the phospholipids of the cell membrane, but these fatty acids are longer (60 to 90 carbons), branched, and complex. The mycolic acids orient themselves as a highly ordered lipid bilayer membrane to keep their hydrophobic tails in the middle and away from water. Proteins are embedded in this layer, where they form water-filled pores through which nutrients and certain drugs can pass slowly. The mycolic acid bilayer is thick and covalently attached to the murein cell wall via layers of complex sugars.

With such a robust hydrophobic cover over their murein cell wall, the acid-fast bacteria are impervious to many harsh chemicals, including disinfectants and strong acids. You can test this with the acid-fast stain (Table 2.3): any bacteria, including those with the waxy cell wall, can be stained red with a dye such as hot fuchsin by brief heating or transitory treatment with detergents. The dye can then be removed from most bacteria by dilute hydrochloric acid treatment, but not from mycobacteria, whose waxy cell walls make them resistant to the acid. Hence their name "acid fast"; they retain the red stain and remain red after the preparation is counterstained with a blue dye. You can use this procedure to easily differentiate cells with

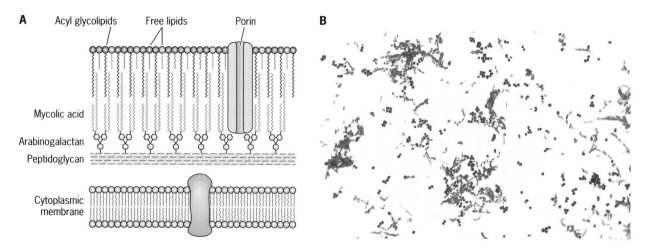

FIGURE 2.13 (A) Structure of the acid-fast cell envelope, with the outer mycolic acid bilayer and its attachment to the murein cell wall via an underlying layer of complex sugars. (B) Light micrograph showing the differential staining of acid-fast bacilli of *Mycobacterium smegmatis* (red) compared with the Gram-positive cocci of *Staphylococcus aureus* (blue). The waxy mycolic bilayer of acid-fast cells prevents the release of the hot fuchsin red dye after acid treatment. (MicrobeWorld © Tasha Sturm.)

TABLE 2.3 **The acid-fast stain**

1. Stain with hot fuchsin (red); wash.
2. Decolorize with acid alcohol; wash. Only acid-fast bacteria remain red.
3. Counterstain with methylene blue; wash. All other material becomes blue.

a waxy envelope (red) from other bacteria (blue) (Fig. 2.13). The waxy envelope of acid-fast bacteria also makes them resistant to common soap-based disinfectants. For this reason, surfaces and materials contaminated with tubercle bacilli require something stronger than just soap and water for disinfection, typically certain quaternary detergents. In clinics, you will see such detergents labeled as mycobactericidal. The mycobacteria resist not only strong chemicals but also white blood cells. All this protection comes at a cost: the waxy armor reduces the rate of nutrient uptake. For example, the permeability for hydrophilic molecules is 100- to 1,000-fold lower than for the Gram-negative *E. coli*. Perhaps for this reason, these organisms grow very slowly. Some acid-fast bacteria, such as the human tubercle bacillus, divide only once every 24 hours. Yet do not underestimate these slow growers: millions of people are diagnosed with tuberculosis every year worldwide, with immunocompromised patients being particularly susceptible. Remember Aesop's fable about the tortoise and the hare? Well, think about acid-fast bacteria as the slow tortoise that ends up winning the race. Chapter 24 will discuss how the waxy mycobacterial envelope contributes to tuberculosis.

Bacteria without Cell Walls: The Mycoplasmas

After our singing the praises of the peptidoglycan and acid-fast cell walls, you may be surprised to learn that some bacteria called mycoplasmas do not have a cell wall of any type. Some of these wall-less bacteria are human pathogens and difficult to treat in clinical settings. Why? Because common antibiotics that target cell wall biosynthesis are ineffective against these organisms. Thus, mycoplasmas are resistant to penicillins and their relatives the cephalosporins. So how do these wall-less bacteria protect their cell membrane? Some, like *Mycoplasma pneumoniae*, the agent of "walking pneumonia," contain sterols in their membranes, the same strategy used by eukaryotic cells to make their cell membranes more rigid and preserve their integrity. Although delicate in culture, mycoplasmas are remarkably persistent in the human body.

In addition to being insensitive to some commonly used antibiotics, mycoplasmas are very small and spread fast. They can attach to the surface of many types of host cells, including cells of the immune system, and some can even get inside host cells. And so mycoplasma bacteria grow either outside or inside the host cells, sometimes evading the host immune system for long periods of time. Interestingly, phylogenetic analysis places the mycoplasmas close to the Gram-positive bacteria, from which they may well have been derived. Clearly, losing the peptidoglycan cell wall in the course of evolution was advantageous to them. To date, mycoplasmas appear to be the only bacteria that lack peptidoglycan. Many archaea are also wall-less. But, unlike the mycoplasmas, wall-less archaea are protected by a protein S-layer (see below).

Crystalline Surface Layers of *Bacteria* and *Archaea*: S-Layers

Many bacteria and almost all archaea surround their cells with an external layer of protection called the **crystalline surface layer**, or **S-layer**. In the *Bacteria*, the S-layer is anchored to the murein cell wall of Gram-positive bacteria or to the outer membrane of Gram-negative bacteria. In the *Archaea*, S-layers can be attached to the cell walls, if present, or directly to the cell membrane, in which case they serve as the only wall component. The widespread occurrence of the S-layer structure among prokaryotes attests to its ability to protect the cells under a wide range of environmental conditions. In most cases, the S-layer is a lattice of a single protein or glycoprotein that self-assembles via noncovalent interactions to form intricate, highly ordered arrays (Fig. 2.14). Sometimes carbohydrates are also attached. Scientists have been able to isolate S-layers and dismantle them into the individual subunits. The isolated molecules can then recrystallize on their own (i.e., they *self-assemble*) to make sheets with similar or identical lattice patterns to those originally present on the cells.

The type of protein subunit determines the S-layer's thickness (5 to 25 nm thick), pore size (generally 2 to 8 nm), chemical properties (charge and hydrophobicity), and lattice symmetry (oblique, square, or hexagonal). Because of its rigidity, the S-layer provides structural support and enables the cell to withstand the internal turgor pressure and maintain its shape. Also, as can be expected for proteins that are exposed to the environment, S-layers are quite resistant to proteolytic enzymes and protein-denaturing agents, providing a protective shield for the cell within. In some organisms (e.g., the intestinal pathogen *Campylobacter*), S-layers help prevent phagocytosis. In addition, the defined size and shape of the pores allow S-layers to function as selective permeability barriers that control the flux of chemicals in and out of the cells. Some microbes also use their S-layer to anchor extracellular enzymes that would otherwise be diluted out and lost in the external environment. S-layers can also protect the cell from viral infections, or, paradoxically, they can promote infection by functioning as receptors for viruses. Some bacteria even use their S-layer to adhere to surfaces. Clearly, S-layers play many important roles, which may be the

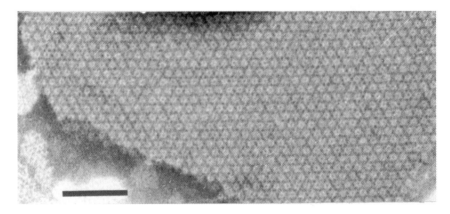

FIGURE 2.14 The S-layer of some archaea and bacteria. A fragment of an S-layer shows the regular arrangement of the subunits as seen under an electron microscope. Bar, 100 nm. (Courtesy of R. G. E. Murray.)

reason why so many prokaryotes invest a lot of energy in making an S-layer even when they already have a cell envelope.

More on the Archaeal Way

The cell envelopes of the *Archaea* differ significantly from those of the *Bacteria*, which points to the evolutionary distance between the two groups of prokaryotes. Not only is the archaeal cell membrane chemically different from that of *Bacteria*, but the *Archaea* do not have murein cell walls. Instead, most archaea protect their cell membrane with just an S-layer (Fig. 2.14), often made up of glycoproteins. Although archaeal S-layers are more sensitive to detergents than are bacterial S-layers, the archaeal isoprenoid membranes are more stable than bacterial phospholipid membranes. Hence, the combination of a sturdy cell membrane and a para-crystalline surface layer suffices for these organisms. In fact, some archaea, like the bacterial mycoplasmas, are wall-less, yet thrive in acidic, high-temperature environments. Species of *Thermoplasma*, for example, have no cell wall, but their cell membranes contain the atypical tetraether lipids that increase their stability under acidic conditions and high temperatures. These wall-less organisms are certainly no weaklings: one strain was isolated from a slow-burning coal mine in Indiana; another, from acidic hot springs in Indonesia. Notably, in these wall-less *Archaea* the S-layer is attached to the cell membrane via hydrophobic protrusions, thus forming a gap (up to 70 nm wide) between the membrane and the S-layer. This unexpected compartment is functionally similar to the periplasmic space of Gram-negative bacteria. Do you remember why a periplasmic space is advantageous to the cells? (*Quiz spoiler:* remember how many proteins can be secreted to this external compartment to process nutrients, for example.)

Some *Archaea* do have cell walls made up of polysaccharides or of a type of peptidoglycan called **pseudomurein** (or pseudopeptidoglycan). The latter differs from the bacterial peptidoglycan by having L- (rather than D-) amino acids and being made of disaccharide units of *N*-acetylglucosamine and *N*-acetyltalosaminuronic acid (rather than *N*-acetylmuramic acid) bonded by β-1,3 (rather than β-1,4) linkages. With all these chemical differences, do you think lysozyme would hydrolyze the archaeal pseudomurein cell wall? (*Hint:* lysozyme hydrolyzes β-1,4 glycosidic bonds.) Pseudomurein has only been found in a few methanogenic archaea. It plays a similar role as the bacterial peptidoglycan, but the two cell walls appear to have evolved independently in the bacterial and archaeal groups. This is what is known as **convergent evolution**: two pathways that arose independently yet led to a similar structure and function. But beware: archaeal cells (those with a pseudopeptidoglycan cell wall) stain Gram positive! What does this tell you about the limitations of using a Gram stain and light microscopy to differentiate *Bacteria* and *Archaea*?

HOW DO PROKARYOTES MODIFY THEIR CELL ENVELOPE?

LEARNING OUTCOMES:
- Provide evidence for two cases of convergent evolution with regard to bacterial and archaeal cell exteriors.
- Discuss how a slime layer allows cells to form biofilms.
- Describe three main features of the structure and function of bacterial flagella.
- Infer a reason to explain why animal cells have receptors that specifically recognize the flagellar filament of many bacterial pathogens.
- Compare and contrast the structure and function of bacterial flagella in Gram-positive and Gram-negative bacteria.
- Propose an experiment to determine if the flagellar rod is physically retained on the cell surface.
- Compare and contrast the composition and assembly of flagella and pili (or fimbriae) in Gram-negative bacteria.

Much as we add layers of clothes or accessories to withstand the cold or rain, prokaryotes have evolved solutions to modify their cell envelope to suit their needs. Depending on the environmental conditions, some bacteria and archaea can make additional exterior layers such as **capsules** and **slime layers**. Others assemble (and subsequently dismantle) appendages such as **flagella** and **pili** for specific functions. Hence, these components are *not always present* but their production is essential for survival under certain circumstances. Next we discuss what they are and why they matter.

Capsules and Slime Layers

Many prokaryotes, bacteria and archaea alike, surround themselves with a coat of slime. When the slime is tenacious and remains attached to the cells, it is called a capsule (Fig. 2.15). If it is loose, it is known as a slime

FIGURE 2.15 Bacterial capsule. The capsule is the fuzzy material surrounding the cell envelopes in this electron micrograph thin section. Note that its thickness is about one-fourth of the cell's diameter. Some bacteria have considerably thicker capsules. (Courtesy of T. J. Beveridge.)

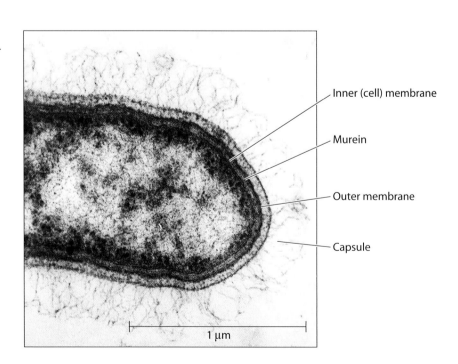

Inner (cell) membrane

Murein

Outer membrane

Capsule

1 μm

layer, but this is not a sharp distinction. In most cases, this layer is made up of a high-molecular-weight polysaccharide, either heteropolymeric (i.e., containing more than one kind of sugar) or homopolymeric (one type of sugar only). In some bacteria, the layer is, oddly, not a polysaccharide but a polymer of amino acids, e.g., glutamate in the anthrax bacillus (*Bacillus anthracis*). As we mentioned earlier, these additional coats are only needed under certain conditions, so the cells only synthesize the material in response to particular environmental cues. Capsules and slime layers allow microbes to retain water and nutrients and can protect the cells from desiccation. They also provide cells with a protective layer that limits the diffusion of harmful chemicals. Some microbes, for example, make a capsule in response to antibiotics or bactericides and to maximize survival when living at high temperatures. The "Bag City" hydrothermal vent in the Pacific Ocean is famous for its accumulation of copious amounts of exopolysaccharide on the sediment surface, material that is produced by some of the resident thermophilic bacteria as a protective mechanism.

A slime layer also helps microbes adhere to surfaces and build multicellular communities known as **biofilms**. *Streptococcus mutans*, for example, attaches to teeth and initiates tooth decay (caries). To better adhere to the enamel of teeth and form plaque, it feeds on the available sucrose (a disaccharide of glucose and fructose) and uses the sucrose-derived glucose to make a capsule of a sticky polymer called dextran. For pathogens like *S. mutans* the capsule is also a major line of defense. This is how many microbes avoid being "eaten up" by white blood cells or other microbes. We will learn more about cannibalistic microbes in chapter 16 when we discuss the protists, a group that includes many members that feed on other microbes by ingesting them via phagocytosis, like white blood cells do (chapter 22). Many pathogens that must travel through the blood to reach their target organs are indeed encapsulated. Examples are the agents of bacterial meningitis, such as the meningococci (*N. meningitidis* and *H. influenzae*) or the pneumococcus (*S. pneumoniae*) of the case study we described above. No wonder so many microbes resort to slime when conditions call for additional defense!

Flagella

Locomotion, or the ability to move from one place to another, is a key trait of many microbes, yet one that requires the assembly of specific appendages on the cell envelope when needed. Many bacteria and archaea are propelled through liquids and across wet surfaces by the action of extracellular filaments called flagella (singular, flagellum). Depending on the species, a single cell may have one or many flagella, either at the end of the cell (**polar**) or at random points around the periphery (**peritrichous**, or "hairy all over") (Fig. 2.16). This distinction is useful in taxonomy and diagnostic microbiology.

Archaeal and bacterial flagella differ in their assembly and structure. The archaeal flagellum (also known as the **archaellum**) is similar to bacterial appendages known as pili, which we discuss in more detail below. However, unlike bacterial pili, the archaellum does not move the cell by cycles of extension and retraction. Rather, it propels the cell by rotating, just like bacterial flagella. This again is a case of convergent evolution:

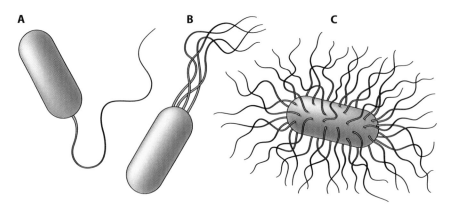

FIGURE 2.16 Arrangements of flagella in some types of bacteria. (A) Single polar flagellum. (B) Multiple polar flagella. (C) Flagella arising from all around the cell (peritrichous).

archaeal and bacterial flagella could not be structurally more different, yet they function the same way to promote their rotation. This contrasts with eukaryotic flagella, which move in a two-dimensional, whip-like motion.

As the bacterial flagellum is the best studied of the prokaryotic flagella, we will focus this final section on its structure. A flagellum is composed of three parts (Fig. 2.17): a long helical filament, which may extend outside the cell 5 to 10 μm (several times the length of the cell); a connecting hook; and a basal body equipped with a rotor to turn the flagellum. To fuel this rotor, bacteria use ATP, the universal energy currency of all cells. Because ATP is generated inside the cell, bacteria anchor their flagellar

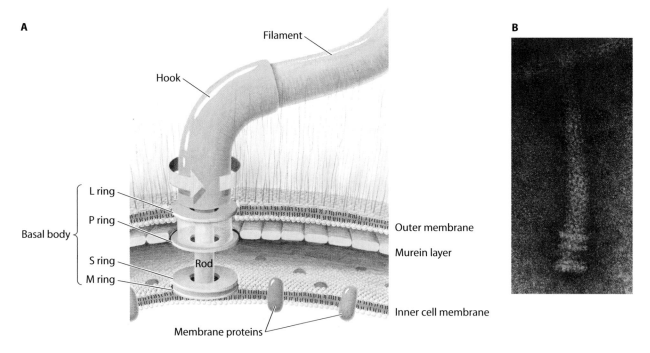

FIGURE 2.17 Structure of flagella. (A) Insertion of a flagellum into the double membrane of Gram-negative bacteria. Flagella of Gram-positive bacteria lack the outer two rings. (B) Electron micrograph of the portion of a flagellum of *E. coli* closest to the cell.

basal body in the cell membrane in such a way that part of the basal body is exposed to the cytoplasm, accessible to the ATP fuel (Fig. 2.17). Here are the key features of each flagellar part.

1. The bacterial flagellar filament is a helical, hollow structure often composed of a single protein, **flagellin**. Mechanically speaking, the flagellin filament is strikingly rigid, about 100 times stiffer than the F-actin of muscle. How do the flagellins assemble? Flagellin molecules *aggregate spontaneously* to form the structure characteristic of the flagellar filament. As we learned for the S-layer, flagellar filaments can also be isolated and dissociated into their subunits, which in the right ionic environment can aggregate to form filaments indistinguishable from the original one. Since they make up the most exposed part of the flagellum, it may not surprise you to hear that flagellins, like the O antigen of the Gram-negative LPS, are highly antigenic. In addition, flagellins are recognized by specific receptors on animal cells, which then elicit specific immune defenses (chapter 22). Why do you think animal cells have receptors that specifically recognize the flagellar filament of many bacterial pathogens?

2. The hook is the "joint" between the filament and the basal body. Like the flagellar filament, the hook is made up of a single type of protein, which polymerizes as a short, curved hook structure, slightly larger in diameter than the filament, to attach it to the cell envelope while still permitting its rotation.

3. The basal body is a complex structure composed of some 15 proteins that aggregate to form a rod with attached rings. The rings act as bushings, or stators, that anchor the structure in the various layers of the cell envelope while allowing the rod (the rotor) to rotate. As might be expected, Gram-positive and Gram-negative bacteria have different numbers of rings: the Gram-positive bacteria have only two rings (M and S), one embedded in their cell membrane and the other associated with the teichoic acid component of their wall, respectively; the Gram-negative bacteria have two additional rings (P and L) to anchor the basal body to the cell wall and the outer membrane (Fig. 2.17). Despite many years devoted to the study of the flagellum structure, it is not yet known how the rod is physically anchored on the cell surface. Can you propose a way?

Assembling a flagellum is challenging. To do so, cells assemble each component in a stepwise fashion, from the bottom (the basal body first, then the hook) to the top (the filament). The three components are hollow, allowing the flagellin subunits to travel from the cytoplasm, where they are synthesized, all the way to the tip of the nascent flagellar filament, where they assemble. In chapter 9, we will learn in detail how this complex apparatus assembles and works to propel the cell.

Intact flagellar motor structures have been isolated from various bacterial species and examined by transmission electron microscopy. Most of them follow the *E. coli* body plan (Fig. 2.17), but structural variations are common that allow the cell to adapt its motility apparatus to its particular needs. As we learned in chapter 1, the diversity of microbes is so large that

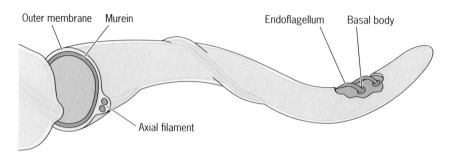

FIGURE 2.18 **Arrangement of the axial flagellum within the periplasm of a spirochete.**

you can almost always find exceptions to general rules. The same applies to flagella. Helical bacteria such as spirochetes, a group that includes the agents of syphilis and Lyme disease, take flagellar variations to the extreme. In these bacteria, the flagella do not protrude from the cell. Rather, they are contained within the periplasm, attached near the cell poles (Fig. 2.18). When the filaments rotate, the cell turns in a corkscrew-like manner. This internal rotation allows spirochetes to move in media so viscous that motility of externally flagellated bacteria would be impeded. This is a fine example of how microbes have evolved structures for specific functions that are also adapted to their environment. Without locomotion in highly viscous media, spirochetes would not be able to colonize host tissues.

Pili

Many prokaryotic cells also decorate their cell envelope with filamentous appendages known as pili (also called **fimbriae**) (Fig. 2.19). Pili (singular, pilus) comprise a diverse group of filamentous structures that protrude from the cell to contribute a wide range of physiological functions, such as attachment to host cells and other surfaces, transfer of proteins and nucleic acids to other cells, and motility. Some pili are also engaged in bacterial sex, allowing cell contact and transfer of DNA between donor and recipient. Not all pili do all these things—most are specialized and restricted in their functions. Pili are thinner and often shorter than flagella (although some are longer) and distributed, sometimes in large numbers, over the surfaces of some bacteria (Fig. 2.19). In some species, they occur in tufts at one or both poles of the cell, and in others, on one side.

Virtually all Gram-negative bacteria possess pili, but relatively few Gram-positive bacteria do (why is anybody's guess). In the human body, pili allow bacteria to adhere to mucosal surfaces (e.g., in the gastrointestinal and urinary tracts). Sticking is often mediated by specific **adhesins**, proteins *located at the tips or the sides of the pili*. Pili also have other roles in disease. Like the capsules, they inhibit the phagocytic ability of white blood cells. They are also antigenic and elicit an immune response in the host. The proteins of pili, **pilins**, are highly changeable, permitting some bacteria to put on a succession of disguises that enable them to outflank the immune system. For example, gonococci, the cause of gonorrhea, have a large number of genes that code for variants of pilin. Each version of pilin is antigenically distinct and elicits the formation of different antibodies by the host (just like the flu virus, which changes antigenically every year).

A   **B**

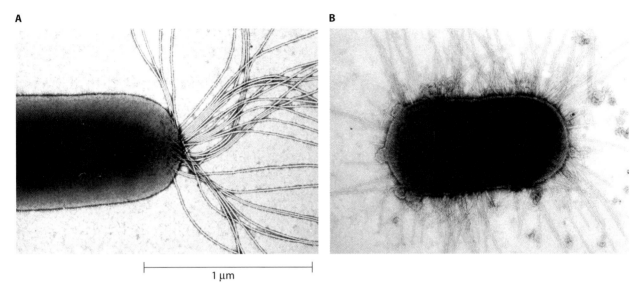

1 µm

FIGURE 2.19 **Flagella and pili seen under the electron microscope.** (A) Flagella inserted at the pole of a bacterium. In other organisms, flagella are inserted all over the cell or a single flagellum may be inserted at a cell's pole. (B) Pili surrounding a Gram-negative bacterium.

In the presence of antibodies to one type of pilin, there is rapid selection for strains that have switched to the synthesis of another antigenic type of pilin. Thus, in this quick-change scenario, they keep one step ahead of the host immune response. It is easy to see why vaccines against gonococci containing pilins have failed so far.

Motility via pili is completely different from flagellar motion. Thus, bacterial pili do not rotate. Their different mode of motility reflects their different ways of assembly. Unlike the flagella, they are formed by adding each pilus subunit, a pilin, to the base of the pilus filament rather than at the growing tip of the filament. And the process can be reversible, with pilins undergoing cycles of polymerization and depolymerization that cause the pilus filament to extend and retract. If, as a pilus grows and extends, its tip touches a surface some distance from the cell, the adhesins at the tip, if present, can strongly adhere to the surface. The pilus can then depolymerize at its base, effectively shortening the filament and dragging the cell forward. In this respect, pili resemble grappling hooks (like those used by ninjas). The result is that, during repeated rounds of this process, the bacterium moves in the direction of the adhering tip. This kind of surface motility is called **twitching motility** and is quite widespread among piliated bacteria. However, as we mentioned earlier, structures similar to bacterial pili function as flagella in *Archaea*. In these organisms pili rotate, just like the bacterial flagella, although their structure and assembly are pili-like. Observation of these overlapping structures and functions suggests that the evolution of cellular appendages such as flagella and pili may have been modular, i.e., may have involved the acquisition of sets of proteins (modules) for specific functions (chapter 9).

CASE REVISITED   **A simple Gram stain and a microscope to diagnose meningitis**

As you probably already know, symptoms such as high fever, headache, and a stiff neck are often linked to meningitis, a disease caused when viruses or bacteria infect the fluids around the brain and the spinal cord, triggering inflammation of the protective membranes that surround them. Though both viruses and bacteria can cause meningitis, the most severe forms are those involving bacteria. Bacterial meningitis is life-threatening and must be treated promptly with effective antibiotics. As most bacteria causing meningitis are either Gram positive or Gram negative, health care workers use the Gram stain to test for the presence of bacteria in spinal fluids. If bacteria are present (remember, they are smaller than the large, nucleated white cells present in the sample), the result of the Gram stain can confirm a diagnosis of bacterial meningitis and identify the likely causative agent as *N. meningitidis*, *S. pneumoniae*, or *H. influenzae*. In the case study, micrographs of the patient's spinal fluids confirmed the presence of Gram-positive (purple) cocci, either singly or in pairs. This pattern is consistent with the cell envelope and morphology of *S. pneumoniae* (the other two bacteria that commonly cause the disease are Gram negative). Hence, the doctor immediately prescribed a cocktail of antibiotics recommended to treat this type of bacterial meningitis. The diagnosis was correct, and the patient recovered quickly.

CONCLUSIONS

As small as most bacteria and archaea are, they are endowed with a sophisticated set of cell envelopes that play a major role in their survival in the environment. The wall, capsule, and outer membrane of bacterial cells contain molecules and residues—odd sugars, D-amino acids, and unique lipids—not found elsewhere in the biological world. Bacterial flagella and pili have no direct counterparts in eukaryotes. This makes "microbial dermatology" a specialized area of study. It is clear from what we have learned in this chapter that prokaryotes evolved their structural complexity to acquire new functions, allowing them to diversify metabolically, increase their fitness, and colonize new niches.

Representative ASM Fundamental Statements and Learning Outcomes for the Chapter

EVOLUTION

1. Cells, organelles (e.g., mitochondria and chloroplasts), and all major metabolic pathways evolved from early prokaryotic cells.

   a. Explain why all cells must have a membrane.

2. Mutations and horizontal gene transfer, with the immense variety of microenvironments, have selected for a huge diversity of microorganisms.

   a. Provide evidence for two cases of convergent evolution with regard to *Bacteria* and *Archaea* cell exteriors.

CELL STRUCTURE AND FUNCTION

3. The structure and function of microorganisms have been revealed by the use of microscopy (including bright-field, phase-contrast, fluorescent, and electron).

   a. Identify three ways in which prokaryotes and eukaryotes are structurally different.
   b. Explain why microbial cells must be stained in order to see them.
   c. List advantages and disadvantages of light microscopy, fluorescence microscopy, and electron microscopy.
   d. Compare and contrast the cell envelopes of Gram-negative bacteria and Gram-positive bacteria.
   e. Speculate on why membrane proteins are more abundant in *Bacteria* and *Archaea* than in mammalian cells.
   f. List three roles of the bacterial murein cell wall.
   g. Speculate on why Gram-negative bacteria have porins to transport small (<700 daltons) hydrophilic molecules while larger molecules must enter the cell using a transporter.
   h. Compare and contrast the bacterial and archaeal cell walls.
   i. Justify the following statement: "The differences in composition of the envelopes of *Archaea* and *Bacteria* can reflect their ecological distribution."
   j. Outline the limitations of using a Gram stain and light microscopy to differentiate *Bacteria* and *Archaea*.

4. Bacteria have unique cell structures that can be targets for antibiotics, immunity, and phase infection.

   a. Predict mechanisms that could allow Gram-positive bacteria to become resistant to penicillins.
   b. Contrast the effectiveness of secreted enzymes against target antibiotics for Gram-negative and Gram-positive cells, taking into account the structure of their cell envelopes.
   c. Infer a reason to explain why animal cells have receptors that specifically recognize the flagellar filament of many bacterial pathogens.

5. *Bacteria* and *Archaea* have specialized structures (e.g., flagella, endospores, and pili) that often confer critical capabilities.

   a. Compare and contrast lipid membranes in *Bacteria* with those in *Archaea*.
   b. Summarize why hopanoids are important biomarkers for *Bacteria*.
   c. Describe three ways in which membranes have adapted to confer survival in some harsh environments.
   d. Discuss how acid-fast bacteria can withstand membrane-damaging chemicals.
   e. Explain why a bacterial or an archaeal cell would invest energy into making an S-layer when it already has a cell envelope.
   f. Describe three main features of the structure and function of bacterial flagella.
   g. Compare and contrast the structure and function of bacterial flagella in Gram-positive and Gram-negative bacteria.
   h. Propose an experiment to determine if the flagellar rod is physically retained on the cell surface.
   i. Compare and contrast the composition and assembly of flagella and pili (or fimbriae) in Gram-negative bacteria.

INFORMATION FLOW AND GENETICS

6. Although the Central Dogma is universal in all cells, the processes of replication, transcription, and translation differ in *Bacteria*, *Archaea*, and eukaryotes.

   a. Speculate on whether having a nucleoid and allowing ribosomes to directly attach to nascent mRNA transcripts, such as the case in prokaryotes, may be beneficial to survival in some cases.

MICROBIAL SYSTEMS

7. Most bacteria in nature live in biofilm communities.

   a. Discuss how a slime layer allows cells to form biofilms.

SUPPLEMENTAL MATERIAL

References

- **Fagan RP, Fairweather NF.** 2014. Biogenesis and functions of bacterial S-layers. *Nat Rev Microbiol* **12:**211–222. doi:10.1038/nrmicro3213.
- **Ingerson-Mahar M, Gitai Z.** 2012. A growing family: the expanding universe of the bacterial cytoskeleton. *FEMS Microbiol Rev* **36:**256–266. doi:10.1111/j.1574-6976.2011.00316.x.
- **Minamino T, Imada K.** 2015. The bacterial flagellar motor and its structural diversity. *Trends Microbiol* **23:**267–274. doi:10.1016/j.tim.2014.12.011.

Supplemental Activities

CHECK YOUR UNDERSTANDING

1. **You have isolated a new organism that is a small coccus and stains purple in a Gram stain. Based on this, you can conclusively identify the cells as _____.**
   a. bacteria
   b. Gram-positive bacteria
   c. Gram-negative bacteria
   d. acid-fast bacteria
   e. none of the above

2. **In which condition are the isoprenoid bilayers of archaeal cell membranes *not* more stable than phospholipid membranes?**
   a. High temperatures
   b. High salt concentrations
   c. Neutral pH
   d. Low pH

3. **Cell membranes contain many kinds of proteins (about 70% of the mass of the membrane). Why?**

DIG DEEPER

1. The localization of MreB inside the cells is critical to understand its role in shaping prokaryotic rod cells. This was a thorny issue for many years. Fluorescence microscopy of MreB-GFP in *E. coli* suggested, for example, that this protein polymerizes and forms helical filaments inside the cell; however, cryo-ET revealed short, discontinuous MreB filaments attached to the cell wall (Fig. 2.4). Read the paper "Multidimensional view of the bacterial cytoskeleton" (K. Celler et al., *J Bacteriol*, **195:**1627–1636, 2013; doi:10.1128/JB.02194-12) and answer the following questions.

   a. Why are the helical MreB-GFP filaments now considered to be artifacts?

   b. What is the advantage of using cryo-ET over fluorescence microscopy?

   c. Why is MreB associated with the cell wall biosynthetic enzymes anchored to the membrane?

2. Find a peer-reviewed research article that analyzes a microbe using a chimeric GFP fusion protein as a tool.

    a. How was the fusion protein generated?

    b. What was the biological question addressed using the GFP fusion protein?

    c. What was the primary scientific finding made possible by the GFP fusion protein?

3. The *Planctomycetes* were long believed to lack a peptidoglycan cell wall, just like the mycoplasmas, and to surround their cell membrane simply with an S-layer. However, the sequenced genomes of several *Planctomycetes* available in pure culture were discovered to contain genes encoding enzymes for peptidoglycan cell wall synthesis.

    a. A team of scientists digested cell pellets of the model representative *Gemmata obscuriglobus* with an enzyme cocktail that cleaves the sugars (*N*-acetylmuramic acid [MurNAc] and *N*-acetylglucosamine [GlcNAc]) of the peptidoglycan cell wall, if present. The soluble fraction resulting from the enzymatic digestion was then treated with a kinase enzyme, MurK, that specifically attaches a radioactive phosphate group to the carbon 6 of any of the two cell wall sugars. The samples were then spotted immediately (0 h) or after 1 h of incubation with the kinase and radioactive phosphate on a resin plate to separate the sugars in the sample by thin-layer chromatography (TLC). The sugars were then visualized as dark spots based on the intensity of the radioactive phosphate label, as shown in the figure.

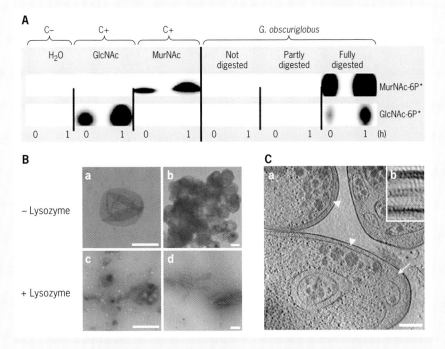

(Modified from Jeske O et al., *Nat Commun* **6:**7116, 2015, with permission.)

    i. As a negative control (C–) they repeated the labeling reaction with just water. Why?

    ii. They also included positive controls (C+) whereby aqueous solutions of GlcNAc and MurNAc were also labeled with the radioactive phosphate and separated by TLC. Why did they need these controls?

    iii. The *G. obscuriglobus* sample had strong signals for GlcNAc-6P and MurNAc-6P but only after full digestion of the cell pellet. Why?

    b. The researchers also boiled the cell pellets in the presence of detergent and isolated cell sacculi, as shown in negative-stained preparations visualized by transmission electron microscopy (TEM) (panels a and b in the figure). However, the integrity of the sacculi was lost after incubation with lysozyme (panels c and d). Why did the researchers treat the sacculi with lysozyme?

c. One of the amino acids extracted from the cell pellets and identified by mass spectrometry was DAP. Is this evidence for a peptidoglycan cell wall?

d. Cryo-ET micrographs of another planctomycetal species, *Planctomyces limnophilus*, revealed a cell wall layer contained between inner and outer membranes (see the figure: arrowheads in a; magnified in b). Also apparent in this tomographic slice are invaginations of the inner membrane (arrow) that are typical of *Planctomycetes* (scale bar, 200 nm). Is this enough evidence to conclude that this is a Gram-negative cell envelope?

4. **Team work:** Cell shape matters for many reasons. Read the article "The selective value of bacterial shape" (K. D. Young, *Microbiol Mol Biol Rev*, **70:**660–703, 2006; doi:10.1128/MMBR.00001-06) to understand the evolutionary pressures that may have selected for different cell shapes. Work in groups to answer the following questions, and present your conclusions to the class.

a. The author proposes that shape is a significant element in the physiological adaptations of microbes to their environment. What arguments support this notion?

b. Propose an experiment to demonstrate that cell shape influences nutrient diffusion.

c. Phylogenetic evidence suggests that the earliest cells were rods or filaments. What selective advantage could they have had over spherical cells in an early Earth?

Supplemental Resources

VISUAL
- Flagellar assembly animation: https://www.youtube.com/watch?v=GnNCaBXL7LY.
- Gram stain animation: http://www.microbelibrary.org/library/gram-stain/3018-the-gram-stain-an-animated-approach.

SPOKEN
- iBiology lecture "Electron Microscopy" by Eva Nogales (47 min): http://www.ibiology.org/ibioseminars/techniques/eva-nogales-part-1.html.
- iBiology lecture "The Spatial Organization of Bacterial Cells" by Christine Jacobs-Wagner (27 min): http://www.ibiology.org/ibioseminars/christine-jacobs-wagner-part-1.html.

WRITTEN
- Moselio Schaechter discusses the classical microbiology technique in the *Small Things Considered* blog post "The Gram Stain: Its Persistence and Its Quirks" (http://schaechter.asmblog.org/schaechter/2013/02/the-gram-stain-its-persistence-and-its-quirks.html).
- Moselio Schaechter discusses billions-of-years-old bacterial lipids in the *Small Things Considered* blog post "Bacterial Hopanoids—The Lipids That Last Forever" (http://schaechter.asmblog.org/schaechter/2014/09/bacterial-hopanoids-the-lipids-that-last-forever.html).
- Moselio Schaechter discusses structural variations in the basic plan of the bacterial flagellar motor in the *Small Things Considered* blog post "Self-Assembly for Me" (http://schaechter.asmblog.org/schaechter/2014/08/self-assembly-for-me.html).
- Sonja-Verena Albers and Ken F. Jarrell discuss why the archaeal flagellum may be more similar to a bacterial pilus than a flagellum in the *Small Things Considered* blog post "The Archaellum—The Motility Structure of Archaea" (http://schaechter.asmblog.org/schaechter/2015/02/the-archaellum-the-motility-structure-of-archaea.html).
- Moselio Schaechter highlights the first visual proof that Gram-negative bacteria have two membranes in the *Small Things Considered* blog post "Pictures Considered #34: The First Look at the Two Membranes of Gram-Negatives" (http://schaechter.asmblog.org/schaechter/2016/02/pictures-considered-34-the-first-look-at-the-two-membranes-of-gram-negatives.html).

- Moselio Schaechter highlights the structure of the basal body and hook in the *Small Things Considered* blog post "Pictures Considered #19: The Basal End of Bacterial Flagella" (http://schaechter.asmblog.org/schaechter/2014/08/pictures-considered-19-the-basal-end-of-bacterial-flagella.html).
- Learn more about how scanning probe microscopes work: http://education.mrsec.wisc.edu/background/STM/

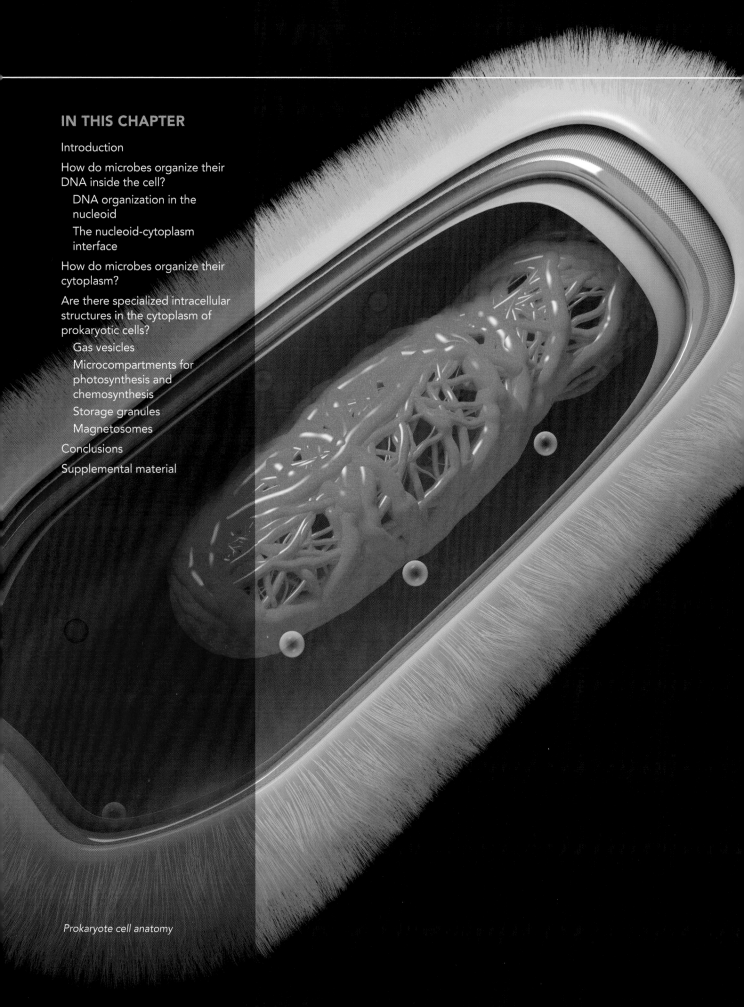

IN THIS CHAPTER

Prokaryote cell anatomy

CHAPTER THREE

Prokaryotic Cell Interior

KEY CONCEPTS

This chapter covers the following topics in the ASM Fundamental Statements.

CELL STRUCTURE AND FUNCTION

1. The structure and function of microorganisms have been revealed by the use of microscopy (including bright-field, phase-contrast, fluorescent, and electron).
2. *Bacteria* and *Archaea* have specialized structures (e.g., flagella, endospores, and pili) that often confer critical capabilities.
3. While microscopic eukaryotes (for example, fungi, protozoa, and algae) carry out some of the same processes as bacteria, many of their cellular properties are fundamentally different.

INFORMATION FLOW AND GENETICS

4. Genetic variations can impact microbial functions (e.g., biofilm formation, pathogenicity, and drug resistance).
5. Although the Central Dogma is universal in all cells, the processes of replication, transcription, and translation differ in *Bacteria*, *Archaea*, and eukaryotes.

INTRODUCTION

The cell envelope and surface structures of prokaryotes were the focus of chapter 2. We described the most notable features of their architecture and biochemical composition that are critical for their function. What about the cell interior? Is it also different from that of eukaryotic cells? Not surprisingly, the answer is yes. In this chapter, we examine the two major components inside the prokaryotic cell envelope:

the **nucleoid** and the cytoplasm. Why just two compartments? Mitochondria, nucleus, Golgi apparatus, endoplasmic reticulum, plastids, mitotic apparatus—all the cellular elements that are almost universal among eukaryotic cells—are absent from *Bacteria* and *Archaea*.

Is there so little of interest within the cell envelope of prokaryotes? Guess again! Though lacking the apparent complexity of the eukaryotic cell interior, prokaryotic cells have evolved a clever interior design with some unique and fascinating structural features that confer unsuspected functions (Table 3.1). Think of a eukaryotic cell as a large cruise ship with many activities carried out in distinct compartments (engine room, galley, dining rooms, casino, etc.). By contrast, a prokaryotic cell brings to mind a compact enclosure with little room and equipment to spare—more like a spaceship. But look how far a spaceship can take us!

Let us explore the remarkable prokaryotic cell interior and its hidden treasures. We focus first on the unusual nature of its nucleoid and cytoplasm and then briefly review some of the less prominent but critical interior structures of certain prokaryotes. Though focusing on the interior of bacterial cells, which we know a good deal about, keep in mind that much of what you will learn here also applies to the *Archaea*.

CASE 1: What is all the viscous "stuff" that comes out of *Escherichia coli*?

A researcher treated cells of *E. coli* with the murein-hydrolyzing enzyme lysozyme and a chelator of divalent cations (EDTA) in the presence of 1 M sucrose. Under these conditions, the cells were converted into round **spheroplasts** (membrane-bound cells devoid of cell wall; chapter 2). Treatment of the spheroplasts with a mild detergent lysed the cells, producing a very viscous lysate. After fixing and staining a sample of the viscous

TABLE 3.1 What's inside the cell membrane of prokaryotes?

Structure or component	Functions
Nucleoid	Repository of genome
	Transcription
Cytosol	
Polyribosomes	Protein synthesis
Enzymes	Metabolism
Regulatory proteins	Regulation of gene expression
Metabolites, precursors, energy compounds, salts	Participate in metabolism
Cytoskeleton	Chromosome segregation and cell division
Vesicles (in some only)	
Gas vesicles	Cell buoyancy
Photosynthetic vesicles	Photosynthesis
Chemosynthetic vesicles	Chemosynthesis
Carboxysomes	Enhance CO_2 fixation in heterotrophs?
Enterosomes	Metabolism of propanediol, others
Storage granules (in some only)	
Acidocalcisomes	Store energy-rich compounds:
Others	Polyphosphates, PHAs
	Glycogen-like compounds, sulfur compounds
Other structures	
Magnetosomes	Involved in directional orientation with respect to magnetic field

liquid, the researcher examined the preparation with an electron microscope. What he saw was striking: many fibrous and star-like structures, like the one in the photograph.

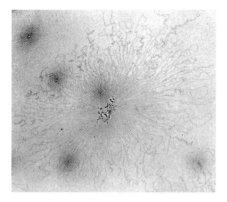

(From Wang X et al., *Nat Rev Genet*
14:191–203, 2013, reproduced from
© Designergenes Posters Ltd.)

- Generate a hypothesis about the composition of the material in the viscous liquid.
- How could you collect evidence and test your hypothesis?

CASE 2:   What is stored inside prokaryotic cells?

Ms. G, a microbiology student, isolated a soil bacterium belonging to the genus *Pseudomonas*. When examined with a phase-contrast microscope, the cells appeared as dark rods, some of which were dividing (as in the photograph's left panel). But when she stained the cells with the fluorescent dye Nile blue A and examined them with a fluorescence microscope, a large number of intracellular yellow granules were apparent. Eager to learn what was stained in the cells, she researched Nile blue A and found that it binds poly-β-hydroxybutyrate (PHB), a carbon-containing polymer.

- How can cells accumulate such large granules in the small cell interior?
- Form a hypothesis about the biological function of intracellular PHB granules.
- Design an experiment to test your hypothesis.

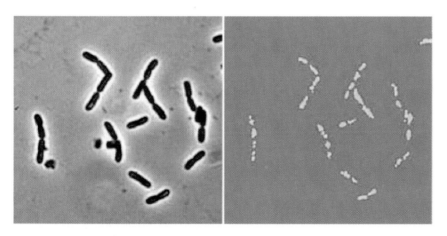

(Modified from MicrobeLibrary © William Ghiorse.)

HOW DO MICROBES ORGANIZE THEIR DNA INSIDE THE CELL?

Compared to eukaryotes, prokaryotes' major distinction is the absence of a true nucleus (chapter 1). Rather than confining their DNA within a membrane-bounded compartment, prokaryotes compact their DNA within the cytoplasm in a structure known as the **nucleoid**. The lack of a nuclear membrane offers prokaryotes a unique opportunity to *couple the two essential processes of* **transcription** *and* **translation**. Thus, as soon as a prokaryotic gene begins to be transcribed into a messenger RNA (mRNA) molecule, the ribosomal subunits latch onto the nascent transcript and translate the information encoded in its sequence to make a protein. These two dynamic processes occur simultaneously with remarkable precision and efficiency (chapter 8).

Under an electron microscope, the prokaryotic nucleoid looks like an irregular blob of DNA that is separated from the cytoplasm, despite lacking its own membrane (Fig. 3.1). Are there exceptions to this general rule? None so far, although the *Planctomycetes* did fool us for some time. Electron micrographs of *Planctomycetes* nucleoids easily trick the eye, giving the false impression of a double-membrane nuclear boundary (Fig. 3.2). In fact, an invagination of the cell membrane partially envelops the nucleoid, masquerading as a "nuclear envelope."

In addition to lacking true nuclei, prokaryotes differ from eukaryotes in the way they replicate their DNA during cell division. Most prokaryotes have a single circular chromosome that does not segregate via a eukaryotic-type mitotic apparatus. As yet, how nucleoids partition into the two daughter cells produced by cell division is not entirely understood (chapter 8). Strange, isn't it, that such a fundamental process is understood for eukaryotes (mitosis) but so far has escaped elucidation in the otherwise well-

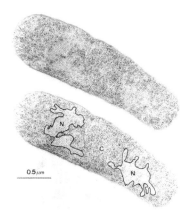

0.5 μm

FIGURE 3.1 Electron micrograph of a thin section through an *E. coli* cell. The clear irregular area corresponding to the nucleoid (N) has been outlined in the lower picture. The granular appearance of the rest of the cell, the cytoplasm (C), reflects its high content of ribosomes. (Courtesy of E. Kellenberger.)

Nucleoid   Nuclear envelope

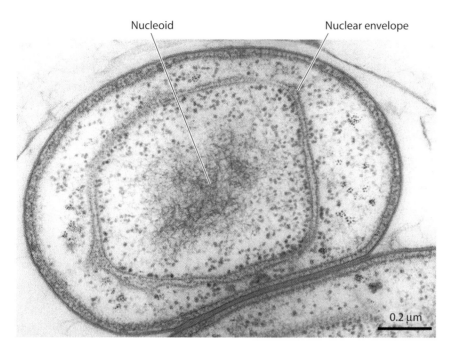

0.2 µm

FIGURE 3.2 **The prokaryotic nuclear envelope that is no such thing.** Electron micrograph of a thin section of the planctomycete *Gemmata obscuriglobus* shows the nucleoid surrounded by what appears to be a nuclear envelope consisting of two membranes. The "nuclear envelope" turned out to be an invagination of the inner membrane that partially surrounds the nucleoid. (Careful with what electron micrographs show! The harsh chemical treatments that the cells receive to prepare for electron microscopy can change the structure of the cell and its components and produce artifacts. Lesson learned!) Note that there are ribosomes (small, dark granules) within and outside this membranous structure. (Reprinted from Lindsay MR et al., *Arch Microbiol* **175**:413–429, 2001, with permission.)

studied bacteria? We do know that, in the absence of a nuclear membrane, prokaryotes evolved an entirely different mechanism to keep the DNA packed tightly inside the cell while retaining its function. To learn how this is accomplished, read on.

DNA Organization in the Nucleoid

The DNA "spaghetti" in the figure for Case 1 illustrates a key fact: prokaryotes pack a lot of DNA in their nucleoid. The sizes of genomes for *free-living* microbes within the *Bacteria* and *Archaea* vary considerably (Fig. 3.3). (For comparison, eukaryotic chromosomes range from 2.9 million to over 4 billion base pairs.) The largest is found in the *Bacteria* and contains almost 15 million base pairs. The smallest corresponds to an archaeon aptly named *Nanoarcheum*, whose genome contains some 490,000 base pairs. The genomes of some **endosymbiotic** bacteria are even more "reduced." The smallest known genome, with about 139,000 base pairs, belongs to a bacterium that lives within specialized cells of sap-feeding insects. (We will learn about such symbiotic interactions in chapter 21.) In general, the size of a prokaryote's genome reflects the number of genes it needs to prosper in its particular range of habitats (chapter 1). For example, *Pseudomonas* species can metabolize a wide range of nutrients and therefore live in many different habitats; their genomes must be large to accommodate all the necessary metabolic genes.

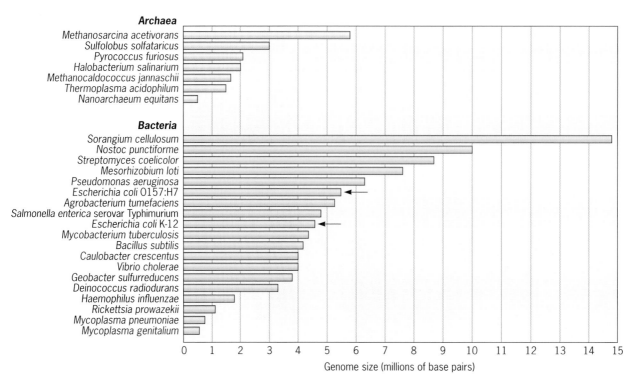

FIGURE 3.3 Genome sizes of *Archaea* and *Bacteria*. The overall range of genome sizes for free-living organisms is about 30-fold. Note that the genomes of two strains of *E. coli* (O157:H7 and K-12; arrows) differ by about 20%, or 1 million base pairs.

Nucleoids may be visible in electron micrographs of bacterial cells (Fig. 3.1), but several important features of the chromosome are not. Many, perhaps most, *Bacteria* and *Archaea* have a single chromosome, but there are well-known exceptions. The cholera bacillus (*Vibrio cholerae*; chapter 26) has two different chromosomes; some rhizobia (nitrogen-fixing bacteria associated with the nodules of legumes) have three; and yet others have four. No matter the number, in general, prokaryote chromosomes are circular molecules. It makes sense to circularize the chromosome, because prokaryotic cells contain <u>exo</u>nucleases that could otherwise chew up the ends of linear DNA molecules. Yet there are also exceptions to this rule in disparate bacterial species, such as the agent of Lyme disease (*Borrelia burgdorferi*) and various antibiotic-producing soil bacteria in the genus *Streptomyces*. In these microbes, as in eukaryotes, the ends of the linear chromosome are protected either by double-stranded hairpin structures or by specific proteins covalently bound to each end of the chromosome. Why are not all bacterial chromosomes circular? Your guess is as good as ours.

Just like eukaryotic cells (and DNA viruses!), prokaryotes must solve a demanding topological problem: how to organize their long, thin DNA into what is essentially a ball. The universal feature of DNA is that it consists of two complementary strands wrapped around each other as a helix (thus the name "double helix"; Fig. 3.4). Its helical conformation helps pack the DNA a bit, but not enough to fit within the small cell compartment. If stretched out, the DNA molecule of *E. coli* would be about 1,000

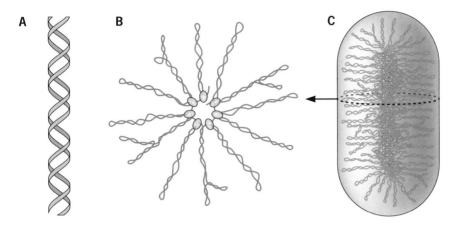

FIGURE 3.4 The structure of the prokaryotic chromosome. Although the structure of the prokaryotic nucleoid is still mysterious, many scientists believe that the DNA double helix (A) in the typically circular chromosomes is folded by DNA-binding proteins into supercoiled loops that protrude from a denser central core (B). This results in the packing of the chromosome like a bottlebrush (C).

times the length of the cell. Imagine: if the DNA double helix were as thick as spaghetti, one *E. coli* chromosome would fill 200 dinner plates! To pack such a long DNA molecule in a small space, it must be very tightly organized.

The physical state of the DNA molecule within the nucleoid is somewhat mysterious. We do know that when bacteria are broken open, the nucleoid "explodes" and the lysate becomes highly viscous, almost like a gel. Consider the experiment described in Case 1. Even when diluted 100-fold, so much DNA is released that it makes the lysate viscous! The "explosion" of DNA that is observed after lysing the cell is in part due to electrostatic repulsion of the highly negatively charged strands of DNA. However, add to that lysate high enough concentrations (e.g., 0.1 M) of positively charged ions (cations), such as magnesium, and it would no longer be viscous. Why? (*Quiz spoiler:* the nucleoids will remain condensed!) This cation dependence is a clue to the state of DNA inside the cell: *cations bind the DNA* to neutralize its intrinsic negative charge. Once the charge is neutralized, dedicated DNA-binding proteins *fold the DNA into loops* and organize them around a central core, like bristles of a bottlebrush (Fig. 3.4). Hence, in living cells, the ionic environment inside and around the nucleoids, working together with DNA-binding proteins, induces a special state of condensation of the DNA. For reasons that are not yet entirely known, the nucleoid excludes the cell's numerous ribosomes from its interior. Indeed, electron micrographs of prokaryotic cells (Fig. 3.1) reveal a fibrous-appearing nucleoid (the tightly folded DNA strands) and a granular cytoplasm (due to the many ribosomes).

The Case 1 micrograph also suggests a key feature of the nucleoid: most of the DNA loops are coiled (like the annoying twists of a coiled telephone cord; remember those?). Actually, they are called **supercoils** to emphasize that these coils are twists of an already twisted DNA molecule (the double helix). It is easy to demonstrate how supercoils form in a circular,

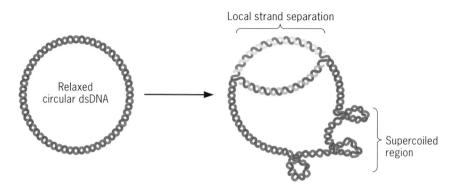

FIGURE 3.5 Supercoiling of circular DNA molecules. Separating the two strands of DNA in a relaxed circular DNA molecule tightens the double helix in the unopened region and induces the formation of supercoils, which simultaneously compact the molecule.

double-stranded DNA (dsDNA) molecule (Fig. 3.5). If you separate the two strands of DNA in one region (as happens when a cell's DNA is replicated or transcribed into RNA), the rest of the double helix tightens. To alleviate the torsion created by these extra twists of the double helix, right-handed supercoils form. The molecule is now said to be *negatively* supercoiled. Looking at Fig. 3.5 you may have noticed another interesting fact: *supercoiling helps compact the DNA in the nucleoid!* Linear chromosomes are also supercoiled, but somehow the ends must be constrained, or the coils would unwind. (Notice how much more we have to learn about prokaryotic linear chromosomes?)

Supercoiling has another noteworthy consequence: it stores energy in the chromosome, which is essentially spring-loaded. In particular, supercoiling lowers the energy required to separate the two complementary strands of DNA, a process needed for both DNA replication and transcription into RNA. So important is DNA supercoiling that prokaryotic cells devote many proteins to keeping it under control. To maintain the optimal level of supercoiling, two enzymes counterbalance each other's activity (Fig. 3.6). Both are **topoisomerases**, enzymes that change the topology of the DNA. **DNA gyrase** uses energy from ATP hydrolysis to *cut the double helix* and *introduce negative (right-handed) supercoils*. If left uncontrolled, DNA gyrase would introduce supercoils to the point that the strands of the DNA double helix would separate and break. But this does not happen because the second enzyme, **topoisomerase I**, counteracts gyrase *by making single-strand breaks* and *relaxing the negative supercoils*. And it does so without any ATP energy!

By counteracting each other's activity, these two topoisomerases ensure that the dsDNA in the nucleoid is, overall, slightly negative supercoiled: just enough to facilitate strand separation during replication and transcription yet not enough to trigger a catastrophic separation of the two strands. How is this delicate balance achieved? Gyrase and topoisomerase I are themselves regulated by the level of DNA supercoiling: too much negative supercoiling induces the synthesis of topoisomerase I; too little (or positive) supercoiling induces the making of DNA gyrase. This elegant fine-tuning mechanism highlights just how important DNA supercoiling

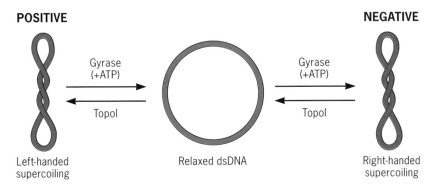

POSITIVE

NEGATIVE

Gyrase
(+ATP)

Topol

Gyrase
(+ATP)

Topol

Left-handed
supercoiling

Relaxed dsDNA

Right-handed
supercoiling

FIGURE 3.6 **Control of supercoiling by gyrase and topoisomerase I.** Gyrase uses energy from the hydrolysis of ATP to cut the double helix and introduce a right-handed (negative) supercoil. Topoisomerase, on the other hand, cuts only one of the strands to remove a negative supercoil and does not require ATP.

is to the cell. Another strategy to ensure that supercoiling is held in check is to divide the prokaryotic chromosome into individually supercoiled domains. By managing the level of supercoiling of each region independently, the cell can better control the separation of the DNA strands when it is time to replicate and/or transcribe that segment of the chromosome. Thus, far from a bowl of spaghetti, the nucleoid is organized along its length to permit local regulation of its activity. The implications of nucleoid structure and supercoiling for DNA replication, segregation, and transcription are discussed further in chapter 8.

The Nucleoid-Cytoplasm Interface

Clearly, microbes are equipped to keep their DNA molecule compacted in the cell yet allow it to function. For example, how does the transcription machinery access DNA that is stored in a tightly packed nucleoid? Every time a gene is transcribed, the nucleoid is "unpacked" in that region and the two strands of DNA are separated so one can serve as a template for transcription. Now picture many genes scattered on the chromosome being simultaneously transcribed. How does such a long DNA molecule (often 1,000 times longer than the cell) avoid becoming hopelessly tangled? One solution is that DNA is transcribed into RNA specifically *at the nucleoid-cytoplasm interface* (Fig. 3.7). This process can be visualized in live cells by genetically engineering a fusion between the gene encoding the **green fluorescent protein** (GFP; chapter 2) and the gene you choose to track. For example, the target protein could be RNA polymerase (RNAP), the enzyme that transcribes DNA into RNA. You can now visualize where the RNAP molecules localize in the cell using a fluorescence microscope. To increase your resolution, you can also stain the DNA with a fluorescent dye that emits a different-color light (e.g., red). In this experiment, RNAP-GFP fluorescence is concentrated around the nucleoid (Fig. 3.7), consistent with the model that transcription occurs at the nucleoid-cytoplasm interface. You will also notice that the distribution of RNAP-GFP fluorescence around the nucleoid is not uniform; some areas are brighter than others. These foci are thought to represent transcription factories,

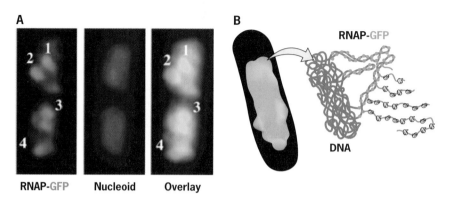

FIGURE 3.7 Localization of transcription in the periphery of the nucleoid in *E. coli*.
(A) An overlay image of the fluorescence emitted from RNAP-GFP (green) and the stained nucleoid (red) suggests that transcriptional activity preferentially occurs on the nucleoid periphery (orange color). However, areas with more intense transcriptional activity (yellow spots in overlay) are also detected. (B) Model consistent with the presence of transcriptional foci in the periphery of the nucleoid. By transcribing the exposed DNA strands of the nucleoid, RNAP makes the RNA transcripts (green) accessible to the ribosomes (brown beads) and promotes coupled transcription and translation. (From Jin DJ, Cabrera JE, *J Struct Biol* **156**:284–291, 2007, with permission.)

i.e., sites where RNAP and other factors are concentrated to maximize transcription efficiency and facilitate the translation into proteins (Fig. 3.7B). Neat, isn't it?

What are the advantages of concentrating the transcriptional activity at the nucleoid-cytoplasm interface? First, this location maximizes exposure to the transcriptional machinery and to the ribonucleotide triphosphate pool needed to make the transcripts (chapter 8). Moreover, at the nucleoid-cytoplasm interface the cytoplasmic ribosomes have ready access to the nascent mRNA transcripts even before they have been fully transcribed. Prokaryote protein synthesis begins when ribosomes recognize and attach to specific sequences in the nascent mRNA molecules (chapter 8). As a result, *transcription and translation occur concurrently* in prokaryotes, reducing the time required to make proteins from the transcripts. In eukaryotes, translation is delayed while the mRNA transcripts travel from the nucleus to the cytoplasm.

One more question about the nucleoid: how are the genes in its interior transcribed and translated? Realize that the nucleoid is not a static tangle of DNA. On the contrary, it is a dynamic skein. Different portions of the DNA thread are perpetually wriggling, snake-like, from the interior to the nucleoid surface and back, enabling each gene to be transcribed. In fact, the transcriptional and translational activities in the nucleoid periphery actually "pull" DNA from the interior. When it is time for the cell to divide, this dynamic property of the nucleoid also helps prokaryotes replicate their chromosome. Clearly, the nucleoid is much more than a "blob" of DNA.

HOW DO MICROBES ORGANIZE THEIR CYTOPLASM?

LEARNING OUTCOMES:
- Compare and contrast the structure of cytoplasm in *Bacteria* and *Archaea* with that of eukaryotes.
- State three benefits of having cell structures crowded in bacterial cytoplasm.
- Distinguish between the circular chromosomes of *Bacteria*, *Archaea*, and plasmids.
- Speculate as to why GFP travels 4 times more slowly in bacterial cytoplasm than in the interior of a mitochondrion (as measured by FRAP).
- Identify a potential drawback of crowding high concentrations of soluble molecules in bacterial and archaeal cells.

For many years (even after the advent of the electron microscope), the bacterial cytoplasm was glibly assumed to be a rather plain bag of soluble enzymes, ribosomes, and other molecular machines. All the intriguing structures of prokaryotes were thought to be located on the outside of the cell. That view has changed significantly, thanks to the development of sophisticated electron and fluorescence microscopy techniques. So what is hiding in the cytoplasm of a prokaryotic cell? Let's imagine we were so small (about 200 nm high) that we could travel inside a prokaryotic cell 10 times our size (a 2 µm long bacterial rod). The nucleoid would look like an irregular ball, roughly twice the size of the traveler. The cytoplasm would feel viscous and full of uniformly sized rocks, each about the size of the traveler's head. The viscosity is due not to the DNA, which we have seen is tightly confined, but to the *high concentration of soluble molecules*, large and small. The "rocks" are the numerous ribosomes (20,000 to 200,000 in a single *E. coli* cell, depending on how fast it is growing). In addition, the traveler would encounter long cables made up of cytoskeletal proteins that span great distances to carry out processes such as cell division and chromosome separation.

How crowded is the prokaryotic cytoplasm? Typically, the concentration of macromolecules in the bacterial cytoplasm is about 30% (weight per volume), which is 3 times that of a chicken's egg white. Moreover, these macromolecules bind water and ions, which increase their effective molecular volume. In other words, the cytoplasm is really crowded (Fig. 3.8A). Because of such molecular crowding, chemical reactions within a bacterial cell likely take place in an environment totally different from the conditions we provide in test tubes in the lab. A warning is called for: extrapolations from *in vitro* studies to the living cell are risky! Neither enzyme kinetics nor rates of macromolecular movement can be inferred from test tube measurements. If only we could provide our miniature traveler the right equipment to collect direct measurements . . . wishful thinking!

Amazingly, the crowded cytoplasm of prokaryotic cells actually facilitates myriad highly organized processes. Crowding, for example, aids protein folding. Folding reactions that would take a long time to occur in a dilute aqueous solution are stimulated by the electrostatic interactions with the many neighboring molecules. The dense cytoplasmic environment also promotes interactions between nascent proteins and chaperones,

A

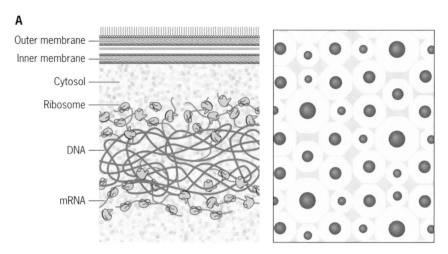

Outer membrane
Inner membrane
Cytosol
Ribosome
DNA
mRNA

FIGURE 3.8 The crowded cytoplasm. (A) Rendering of the crowded prokaryotic cytoplasm (with ribosomes, DNA, and RNA) confined within the cell envelope. (B) Steric repulsion between molecules (brown) creates void aqueous spaces (white) for the circulation of small molecules and solutes.

the proteins that assist other proteins to acquire their correct conformation (chapter 8). Crowding also helps confine the DNA in the nucleoid and *creates the interface* that promotes the interactions between the DNA, RNAP, and ribosomes that couple transcription with translation. The first ribosome attaches to an mRNA molecule soon after its synthesis begins. As transcription continues and the mRNA elongates, more and more ribosomes attach, forming what are known as **polyribosomes**, or polysomes. In fact, in the cytoplasm of growing prokaryotes, few ribosomes float freely: most are attached to mRNA and are engaged in protein synthesis.

Also compacted in the crowded cytoplasm are **plasmids**, extrachromosomal DNA genetic elements carried by many prokaryotes. These genetic entities usually encode dispensable functions: a prokaryotic cell that loses a plasmid may remain viable, but one that loses a chromosome does not. Some plasmids are very large and are stably inherited; the cells must package these and ensure they are replicated and segregated during cell division as well. As with the nucleoid, condensing the plasmids is easier in the crowded environment of the cytoplasm because of its physical constraints and because it facilitates interactions with DNA-binding proteins.

The synthesis of other macromolecules is also more efficient in the cytoplasm because the **building blocks** and the proteins that polymerize them find each other more easily when they are concentrated. Fibrous structures also assemble more readily in a crowded medium. This is significant for the assembly of cell division proteins, such as the **FtsZ** protein that forms the division septum, or proteins that make up the cell's cytoskeleton (chapter 4). But the opposite effect is also possible. Electrostatic interactions can cause two molecules with the same charge to repel each other.

Furthermore, the water that surrounds cytoplasmic molecules creates a "buffer zone," or a space around them. This phenomenon, called *steric repulsion*, has an unsuspected effect: it establishes a dense network of void spaces throughout the cytoplasm that permits the circulation of solutes and small molecules (Fig. 3.8B). We can get a sense of how fast such small molecules circulate using a technique called **fluorescence recovery after photobleaching (FRAP)**. After staining cells fluorescently, a small zone is zapped with a jolt of intense light, which abolishes its fluorescence (the zone is said to have been "photobleached"). Now the experimentalist measures how fast small fluorescent molecules travel into the bleached zone, refilling it with fluorescence. By this method, it is possible to calculate the diffusion rates of the molecules in the cytoplasm of live cells. FRAP experiments have revealed that fluorescent proteins such as GFP typically diffuse about 10 times more slowly in the cytoplasm of *E. coli* than in water and about 4 times more slowly than in the interior of mitochondria. In other words, a small protein travels the length of a bacterium such as *E. coli* (say, a 2 μm distance) in about 200 milliseconds. Such time scales are reasonably fast for many biological responses. For example, that rate is just about high enough to account for how fast genes are turned on by regulatory molecules made elsewhere in the cell. Nevertheless, such diffusion rates may be too slow to supply reactants for some chemical reactions to proceed at their maximum rates. Therefore, it is likely that the diffusion rate of certain molecules limits some intracellular transactions, which in turn determines how fast a cell can grow. Some large protein complexes, such as those involved in **chemotaxis** (chapter 12), do not diffuse through the cytoplasm; instead, they concentrate their activities in compartmentalized regions of the cytoplasm. Hence, prokaryotic cells have achieved a workable compromise between how much they can accumulate inside the cell and what functions they can carry. Their cytoplasm may be packed to the seams, but the cells operate with a high level of organization and function.

ARE THERE SPECIALIZED INTRACELLULAR STRUCTURES IN THE CYTOPLASM OF PROKARYOTIC CELLS?

LEARNING OUTCOMES:
- Identify an argument for why specialized bacterial and archaeal intracellular structures are not considered to be organelles.
- Give an example of how structure supports unique function for an intracellular structure.
- Speculate on the composition of gases in a typical gas vesicle of a photosynthetic bacterium.
- Discuss why it is beneficial for autotrophs to have specialized and separate microcompartments for energy and carbon metabolism in the cytoplasm.
- Explain how the strategy of amplifying a membrane structure benefits microbes.
- Compare and contrast carboxysomes and enterosomes.
- State a benefit and potential drawback for a bacterium that hosts inclusion bodies.
- State a benefit and potential drawback for a bacterium to host magnetosomes.

Although crowded, the cytoplasm of some prokaryotes contains a number of specialized internal structures that house particular branches of the cell's physiology. It may not surprise you to learn that the composition and function of these intracellular structures are diverse. Some internal structures (**inclusion bodies**) store carbon and energy resources, which can be re-mobilized should nutrients become scarce. Some specialized domains reside within bilayered lipid-protein membranes, while others are demarcated by protein shells. Some scientists refer to these internal structures of pro-karyotes as organelles. But are they? Not quite. One feature that distin-guishes them from the mitochondria and chloroplasts of eukaryotic cells is that *they do not contain DNA and probably never did.* In Case 2, for example, Ms. G isolated a soil bacterium filled with inclusions of PHB, a carbon-containing polymer. If she had stained the inclusions with a DNA-specific dye, her results would have been negative. These inclusions harbor no DNA, just PHB. In fact, the bacterium invests a great deal of energy to ac-cumulate that carbon polymer in its already crowded cytoplasm. The poly-mer must play an important function, don't you think? (Have you guessed why storing a carbon polymer would be important for this bacterium?) The PHB inclusions described in Case 2 are just one of many examples of intra-cellular structures contained in the prokaryotic cytoplasm (Table 3.1). Typ-ically, each structure is characteristic of a particular group of prokaryotes. Why? As we'll see in the next section, these structures contribute a spe-cific function that equips the cell to succeed in its distinct environment.

Gas Vesicles

Aquatic photosynthetic bacteria face a problem: buoyancy. The greatest light intensity is, of course, at the surface of the water. Though they are small, light-requiring bacteria would eventually sink into darker waters did they not have a way to achieve sufficient buoyancy. Their solution is **gas vesicles**, structures filled with a gas similar to that dissolved in their cytoplasm (Fig. 3.9). Gas vesicles are a common trait of aquatic photosyn-thetic microbes belonging to a large group known as the blue-green bacte-ria, or **cyanobacteria**. Various members use light of different wavelengths for photosynthesis. Since the spectrum of the sunlight is altered as the light penetrates to greater depths, different cyanobacteria favor particular depths in the water column. Gas vesicles are also generated by some halo-philic (salt loving) archaea; for example, the gas vesicles shown in Fig. 3.9 belong to a halophilic archaeon. Rather than use photosynthesis, these organisms contain light-sensitive pigments that harvest light energy to pump protons across the membrane; the energy in the transmembrane proton gradient is then harnessed to make ATP to support growth and other energy-consuming cellular activities. The gas vesicles increase buoyancy so the cells can remain close to the surface of their brine pool habitat, maximiz-ing their exposure to light to generate energy when needed.

Whether bacterial or archaeal, prokaryotic gas vesicles have a similar structure and composition. A lipid membrane would be a hindrance since it would not permit gas to pass in and out of the vesicle. Instead, they are surrounded by a protein shell, which is freely permeable to gases and also keeps water at bay. Any gas dissolved in the cytoplasm will diffuse inside the vesicle until its concentration inside the vesicle reaches equilibrium. Hence, *gas vesicles do not store gas*; rather, they prevent water in the cyto-plasm from filtering in. Whether filled with gas or in a vacuum, the pro-tein shell maintains the vesicle's structure. Only when cells are subjected to a sudden increase in hydrostatic pressure do the shells irreversibly col-

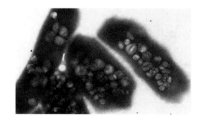

FIGURE 3.9 Cells of a *Halobacte-rium* archaeon containing large numbers of gas vesicles. (Courtesy of S. S. DasSarma.)

lapse. To maintain vesicle stability, microbes that live in deep lakes have evolved narrower gas vesicles (like footballs rather than soccer balls). The oval vesicles better withstand the hydrostatic pressure—a beautiful example of how microbial structures adapt to their environment for optimal function!

Microcompartments for Photosynthesis and Chemosynthesis
Carbon is needed in great amounts to build precursor metabolites and, from them, building blocks and essential macromolecules. These biosynthetic reactions also need a hefty supply of chemical energy. Hence, many prokaryotes have evolved specialized microcompartments for harvesting energy and assimilating carbon. Here we will focus on two notable examples: the specialized structures used by the **phototrophic** cyanobacteria to increase the efficiency of photosynthesis (**thylakoids**) and those used by **autotrophs** to fix carbon dioxide (**carboxysomes**). Note that many phototrophs are also autotrophs. Hence, thylakoids and carboxysomes often coexist inside the photosynthetic prokaryotes.

Thylakoids. Photosynthetic bacteria maximize light harvesting by enlarging the surface of the membranes that house the photosynthetic machinery. In some cases, the plasma membrane is folded to accommodate more of the photosynthetic machinery and harvest more light energy. Cyanobacteria take this strategy of membrane amplification to new heights: they have evolved complex membranous structures called thylakoids, which contain all the pigments, proteins, and cofactors needed to capture light energy and convert it into chemical energy. Thylakoids are stacks of membranous sacs, interconnected to form a common internal space, or lumen. The stacks are underneath the plasma membrane and often organized as concentric shells in the periphery of the cytoplasm, much like the layers of an onion (Fig. 3.10A). Computational three-dimensional models of thylakoids reveal a labyrinthine compartment comprising concentric and intercommunicated shells with perforations that allow the cytoplasm to flow through them (Fig. 3.10B). This arrangement greatly increases the surface area available for the photosynthetic reactions without compromising molecular exchange with the cytoplasm.

The phototrophic bacteria are not the only microbes to adopt the strategy of amplifying membrane structures. **Chemolithotrophs**, which obtain their energy from chemical reactions involving inorganic compounds such as reduced iron, sulfide, or hydrogen (chapter 6), also amplify the membrane surfaces where these processes take place.

Carboxysomes. Autotrophic bacteria fix carbon dioxide (CO_2) and use it as a source of carbon to make their biochemical building blocks. Carbon dioxide fixation requires specialized enzymes that autotrophic microbes often confine into dedicated structures called carboxysomes (Fig. 3.11A). These structures are polyhedral protein shells (again, no lipid membrane) that contain the key enzyme in carbon fixation, ribulose bisphosphate carboxylase, or **RuBisCo**. The structured environment inside the carboxysomes may enhance the catalytic power of this enzyme, but how this works is not known. Most phototrophs, such as the cyanobacteria that we

A

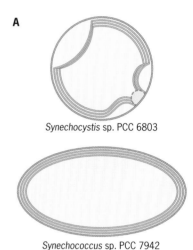

Synechocystis sp. PCC 6803

Synechococcus sp. PCC 7942

B

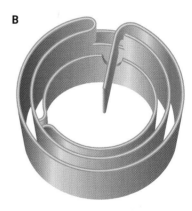

FIGURE 3.10 **The cyanobacterial thylakoid.** (A) Thylakoids are double-membranous structures that function as specialized sites for photosynthesis in cyanobacteria. Their architecture changes with the species. Shown are thylakoids from species of *Synechocystis* and *Synechococcus*. (B) Model of a thylakoid showing its concentric architecture and perforations, which are filled by the cytosol.

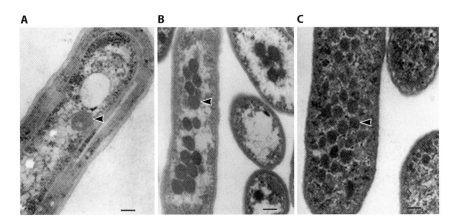

FIGURE 3.11 Various bacterial carboxysomes. (A) Carboxysomes in a photosynthetic cyanobacterium, *Synechococcus*. The concentric membrane layers underneath the plasma membrane are the thylakoids, where light harvesting takes place. (B) Carboxysomes in a sulfide-reducing autotroph, *Halothiobacillus*. (C) Enterosomes in a heterotroph, *Salmonella*. The arrowheads point to typical structures. Bars, 100 nm. (From Cannon GC et al., *Appl Environ Microbiol* **67**:5351–5361, 2001, with permission.)

described earlier, are also autotrophs. These organisms confine their light energy metabolism to the thylakoids and their carbon fixation to the carboxysomes. Fixing carbon dioxide is energy-intensive and requires a generous supply of ATP and reducing power (in the form of reduced molecules such as NADH and NADPH) to make precursor metabolites and, from them, the building blocks such as amino acids, nucleotides, sugars, and fatty acids. Chemolithotrophs, which harness energy from reactions involving inorganic chemicals, also fix carbon dioxide to assimilate carbon and, therefore, have carboxysomes in their cytoplasm (Fig. 3.11B). Whether phototrophs or chemolithotrophs, these organisms have specialized and separate cytoplasmic microcompartments for energy harvesting and carbon fixation to maximize the efficiency of both processes. Pretty smart!

Surprisingly, structures similar to carboxysomes are also found in some **heterotrophic** bacteria, which use organic carbon sources as both a carbon and energy source. Enterobacteria such as *Salmonella* and *E. coli*, for example, do not fix carbon dioxide, yet they have carboxysome-like structures called **enterosomes** (so named because they were first found in enteric bacteria). Under the electron microscope, enterosomes appear to be similar to carboxysomes (Fig. 3.11C). Unlike carboxysomes, they do not contain RuBisCo, but they do contain proteins that are homologous to some carboxysome components. These enzymes metabolize specific compounds, such as propanediol and ethanolamine. Indeed, these structures assemble only when the organisms are grown on these substrates. What role do such substrates play in these bacteria, and why are complex enzymatic arrays needed to metabolize them? It is thought that enterosomes equip bacteria to cope with a toxic intermediate in propanediol metabolism, propanaldehyde. These structures must confer a fitness advantage, since their synthesis requires considerable energy. We do know that propanediol is a metabolite of fucose, a sugar found on the intestinal wall of mammals that can be degraded and consumed by intestinal bacteria, including *Salmonella*, which requires propanediol metabolism for virulence.

Storage Granules

Some bacteria have refractile inclusion bodies in their cytoplasm (Fig. 3.12) that store sulfur, calcium, organic polymers such as glycogen, and phosphate in several chemical forms, including polymeric pyrophosphates. Studying inclusion bodies with a light microscope does not provide any information about their composition. Sometimes their presence is only revealed after staining with a dye. In Case 2, for example, Ms. G did not see any remarkable feature inside the cells of a soil *Pseudomonas* isolate by phase-contrast microscopy. However, after staining for carbon-containing polymers with the fluorescent dye Nile blue A, she observed numerous inclusion bodies in the cells' cytoplasm. Nile blue A specifically stains inclusion bodies of PHB, a type of polyhydroxyalkanoate (PHA). PHAs are a group of carbon polymers that many bacteria make and accumulate in inclusion bodies when carbon is abundant and then use as a carbon and energy source when external sources of carbon are low and growth-limiting (Fig. 3.12). PHAs also have industrial uses. Some of these compounds (PHB being one of them) make stiff and brittle plastics; others make rubber-like substances. Because they are inherently biodegradable, they are regarded as an attractive source of nonpolluting plastics.

Magnetosomes

Microbes accumulate compounds in their cytoplasm for many purposes, not only for use as nutrients. Some motile aquatic bacteria, for example, make **magnetosomes**. As their name implies, these membrane-bounded, iron-containing crystals function as tiny magnets. These structures enable bacteria to orient themselves by responding to the magnetic field of the Earth. Each cell contains 15 to 20 crystals of a magnetic iron mineral, typically magnetite (Fe_3O_4) or greigite (Fe_3S_4). The crystals are chained together in a row, making a long, thin structure that acts like the needle of a compass. A graduate student discovered these fascinating structures after noticing that, once the bacteria were placed on a microscope slide, he could use a magnet to control their movement!

What holds the chain of magnetic crystals together? The particles do not float freely in the cytoplasm. Rather, they are encased in an invagination of the cytoplasmic membrane that gives them their characteristic shape (Fig. 3.13). The result is an elongated vesicle, the magnetosome, which is attached by a pedicle (Latin, "small foot") to the cell membrane. Ferric ions are transported into this vesicle, where they are reduced to ferrous ions and **biomineralized** (that is, converted into a mineral with a specific chemical composition) to form the components of the magnetic crystals. Often, the chains are aligned along the major axis of the cell to give it a favorable magnetic orientation. Aligning along Earth's magnetic field with sufficient force provides the bacteria with a sense of direction. Being motile, they will swim toward a favored portion of their environment, usually where the oxygen concentration is quite low, something these organisms prefer.

Magnetic bacteria and their magnetosomes have some industrial uses for making gadgets that depend on magnetism. They have been used to make magnetic antibodies and to immobilize enzymes. DNA-coated magnetosomes are also shot into cells that are otherwise difficult to genetically engineer. Yet another use is for enriching for host cells that contain

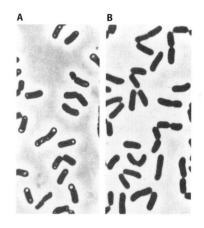

FIGURE 3.12 **Storage granules in a member of the genus *Bacillus* (*B. megaterium*; phase-contrast microscopy).** (A) Cells growing at a high concentration of glucose and acetate. The light areas are granules of PHA. (B) Cell from the same culture after having been incubated for 24 hours in the absence of a carbon source. (Courtesy of J. F. Wilkinson. Reprinted from Neidhardt FC et al., *Physiology of the Bacterial Cell*, Sinauer Associates, Inc., Sunderland, MA, 1990.)

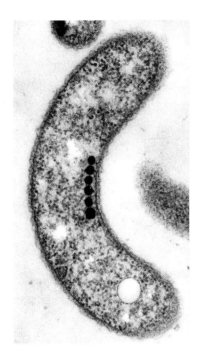

FIGURE 3.13 *Magnetospirillum magnetotacticum* containing magnetosomes (dark bodies inside cell). (Courtesy of T. J. Beveridge.)

magnetic bacteria, usually engineered to carry a certain protein. Such cells can be retrieved by simply placing a magnet on the side of the test tube that holds the cell suspension. Practical obstacles exist, such as the need to grow these bacteria in large quantities, but they have potential for diverse applications.

CASE 1 REVISITED   What is all the viscous "stuff" that comes out of *Escherichia coli*?

Did you form the hypothesis that the viscous "stuff" that was released after mild detergent treatment of *E. coli* spheroplasts was the cell's DNA? Remember the spaghetti analogy? If the *E. coli* DNA double helix were as thick as spaghetti, you would have enough pasta to fill 200 plates! To keep so much DNA inside the small cell volume and still have some room for the cytoplasm and its components, prokaryotes condense the DNA and pack it into a structure called the nucleoid. They face a problem, though: the negative charge of DNA would create repulsion forces so strong that the packing of the DNA would be impossible. Hence, the first step is to neutralize the DNA charge with cations. Then DNA-binding proteins start looping the DNA to pack it like a bottlebrush. When the researcher lysed the cells with detergent, he also denatured some of the proteins that were keeping the nucleoid condensed. Furthermore, the cations got diluted in the solution of the test tube. Without the cations, repulsion forces of the negatively charged DNA caused it to "explode," releasing more DNA and unpacking the nucleoid even more. In the end, what remained of the nucleoid was a star-like structure formed by long DNA loops and some proteins holding it together in a central core.

How did you decide to collect evidence to test your hypothesis? One way might be to apply a DNA-binding fluorescent stain (such as DAPI [4′,6-diamidino-2-phenylindole], which binds to AT-rich DNA regions) to both intact and detergent-treated cells, and examine the differences in fluorescence between the two treatments using microscopy.

CASE 2 REVISITED   What is stored inside prokaryotic cells?

Did you form the hypothesis that the PHB granules that the student stained with Nile blue A were carbon reserves that the *Pseudomonas* cells stockpiled for tougher times? Based on how much area of the cytoplasmic space was occupied by the PHB inclusions, you can make some educated guesses about the environment inhabited by this bacterium. Storage of carbon as PHB indicates that their carbon sources likely fluctuate a lot, with times of abundance alternating with periods of scarcity. That these cells pack so much PHB inside their already crowded cytoplasm suggests that these carbon reserves must be a matter of life and death for them.

How did you decide to collect evidence to test your hypothesis and design your experiment? One way to test the hypothesis that PHB granules are carbon reserves is to culture the cells with carbon (C-replete) and without carbon (C-deplete). You could periodically sample the cultures and use microscopy to quantify and compare the number of PHB inclusions formed by cells subjected to the two treatments. Which treatment would have more inclusions? (Hint: see Fig. 3.12 for an example.)

CONCLUSIONS

Prokaryotic cells generally lack a nuclear membrane, and their genomic DNA forms an irregular mass called the nucleoid. Although prokaryotic cells lack the organelles that characterize eukaryotic cells, their interiors are surprisingly complex, densely packed with metabolic machinery, and quite viscous. In the next chapter, we examine some characteristics of microbial growth and reproduction, in preparation to learn how such deceptively simple cells are put together.

Representative ASM Fundamental Statements and Learning Outcomes for the Chapter

CELL STRUCTURE AND FUNCTION

1. The structure and function of microorganisms have been revealed by the use of microscopy (including bright-field, phase-contrast, fluorescent, and electron).

   a. Identify a consequence of *Bacteria* and *Archaea* having their DNA in the cytoplasm rather than in a separate membrane-bounded compartment.
   b. Describe how DNA is packed and folded tightly inside nucleoids.
   c. State a topological problem of having to confine a lot of DNA inside a small bacterial or archaeal cell.
   d. State three benefits of having cell structures crowded in bacterial cytoplasm.
   e. Identify a potential drawback of crowding high concentrations of soluble molecules in *Bacteria* and *Archaea*.
   f. Speculate on the composition of gases in a typical gas vesicle of a photosynthetic bacterium.
   g. Explain how the strategy of amplifying a membrane structure benefits microbes.

2. *Bacteria* and *Archaea* have specialized structures (e.g., flagella, endospores, and pili) that often confer critical capabilities.

   a. Identify an argument for why bacterial and archaeal specialized intracellular structures are not considered to be organelles.
   b. Give an example of how structure supports unique function for an intracellular structure.
   c. Discuss why it is beneficial for autotrophs to have specialized and separate microcompartments for energy and carbon metabolism in the cytoplasm.
   d. Compare and contrast carboxysomes and enterosomes.
   e. State a benefit and potential drawback for a bacterium hosting inclusion bodies.
   f. State a benefit and potential drawback for a bacterium hosting magnetosomes.

3. While microscopic eukaryotes (for example, fungi, protozoa, and algae) carry out some of the same processes as bacteria, many of their cellular properties are fundamentally different.

   a. Compare and contrast the nucleoid of *Bacteria* and *Archaea* with the nucleus of a eukaryote.
   b. Compare and contrast the structure of cytoplasm in *Bacteria* and *Archaea* with that of eukaryotes.
   c. Speculate as to why GFP travels 4 times more slowly in cytoplasm than in the interior of a mitochondrion (as measured by FRAP).

INFORMATION FLOW AND GENETICS

4. Genetic variations can impact microbial functions (e.g., biofilm formation, pathogenicity, and drug resistance).

   a. Predict why endosymbiotic bacteria generally have reduced genomes that are smaller than the smallest genome of a free-living bacterium.
   b. Distinguish between the circular chromosomes of *Bacteria*, *Archaea*, and plasmids.

5. Although the Central Dogma is universal in all cells, the processes of replication, transcription, and translation differ in *Bacteria*, *Archaea*, and eukaryotes.

   a. Discuss the differences in general chromosomal structure between *Bacteria* and *Archaea* on the one hand and eukaryotes on the other.
   b. Contrast the site of transcription in *Bacteria* and *Archaea* with that of eukaryotes.
   c. Compare and contrast the processes of transcription and translation in *Bacteria* and *Archaea* with that of eukaryotes.
   d. Speculate on an advantage of having bacterial replication proceed in both directions from the origin of replication.
   e. Identify the benefits of supercoiled chromosomes in *Bacteria* and *Archaea*.
   f. Describe how DNA gyrase and topoisomerase act together to regulate chromosome supercoiling.

SUPPLEMENTAL MATERIAL

References

- **Jin DJ, Cabrera JE.** 2006. Coupling the distribution of RNA polymerase to global gene regulation and the dynamic structure of the bacterial nucleoid in *Escherichia coli*. *J Struct Biol* **156**:284–291. doi:10.1016/j.jsb.2006.07.005.
- **Kerfeld CA, Erbilgin O.** 2015. Bacterial microcompartments and the modular construction of microbial metabolism. *Trends Microbiol* **23**:22–34. doi:10.1016/j.tim.2014.10.003.
- **Spitzer J.** 2011. From water and ions to crowded biomacromolecules: *in vivo* structuring of a prokaryotic cell. *Microbiol Mol Biol Rev* **75**:491–506. doi:10.1128/MMBR.00010-11.

Supplemental Activities

CHECK YOUR UNDERSTANDING

1. **How is the prokaryotic nucleoid different from the eukaryotic nucleus?**
   a. It is condensed.
   b. It is supercoiled.
   c. It is circular.
   d. It lacks a membrane boundary.

2. **Which of the following is *not* true regarding DNA supercoiling in the prokaryotic chromosome?**
   a. It helps pack the DNA in the small space of the nucleoid.
   b. Too much supercoiling induces the synthesis of DNA gyrase.
   c. It lowers the energy to separate the two DNA strands.
   d. Supercoiling is possible because the prokaryotic chromosome is circular.

3. **What keeps the DNA in a prokaryotic cell nucleoid in an organized and condensed form?**
   a. Nuclear envelope
   b. Cations
   c. Ribosomes
   d. DNA-binding proteins

4. **Which of these is *not* found in any prokaryotic cell?**
   a. Plasmids that are stably inherited
   b. Storage granules containing DNA
   c. Circular DNA with regulated supercoiling
   d. Coexisting thylakoids and carboxysomes
   e. Gas vesicles and carboxysomes surrounded by protein shells

DIG DEEPER

1. Perform an NCBI genome or other search to determine the genome size, GC content, and topology (circular or linear) of the chromosome of the following bacterial species: *Streptomyces coelicolor*, *Chlamydia trachomatis*, and *Staphylococcus aureus*.

    a. Can you make predictions about the type of environmental niches each species inhabits based on its genome size?

    b. Can you make predictions about the type of environmental niches each species inhabits based on its GC content?

    c. Can you make predictions about the genome topology (circular or linear) of any of these microbes?

2. According to the structure of the DNA double helix elucidated by Watson and Crick, the two strands of DNA are twisted around each other every 10.5 base pairs.

    a. Calculate the number of twists in a relaxed circular dsDNA plasmid that is 5,250 base pairs long.

    b. After gyrase action, the plasmid acquires two negative supercoils. What is the twist of the double helix in this molecule?

    c. Would it be easier to separate the two DNA strands in the relaxed or the negative-supercoiled plasmid?

    d. Transcription of one of the genes *in the relaxed plasmid* induces positive supercoils ahead of the RNAP and negative supercoils behind the enzyme. What topoisomerases do you expect to act on each type of supercoil?

3. What specialized intracellular components would increase the fitness of prokaryotes dwelling in these particular microenvironments? Work alone or in groups and justify your answers.

    a. Aerophilic magnetic bacteria in aquatic sediments
    b. Nonphotosynthetic ocean-dwelling archaea utilizing light as an energy source
    c. Gut-inhabiting enterobacteria
    d. Photosynthetic cyanobacteria in damp soil
    e. Chemotrophs in subsurface compartments of stratified lakes

4. **Team work:** In groups, investigate one of the cellular processes below. Summarize for your classmates your findings in a 5-minute presentation with schematics.

    • DNA packing in the nucleoid
    • Coupling of transcription and translation
    • Chromosome replication
    • Buoyancy control by gas vesicles
    • Light energy harvest by thylakoids
    • Carboxysome or enterosome contribution to energy metabolism
    • Direction provided by magnetosomes

Supplemental Resources

VISUAL

• A simple demonstration of DNA supercoiling (21 sec): https://www.youtube.com/watch?v=qj_eYWgQ90k.
• Topoisomerase 1 and 2 (1:45 min): https://www.youtube.com/watch?v=EYGrElVyHnU.
• A simulation of the crowded bacterial cytoplasm (30 sec): https://www.youtube.com/watch?v=2fobDHHl11c.
• Dancing magnetotactic bacteria (1:50 min): https://www.youtube.com/watch?v=3uUL4ooM6KI.

SPOKEN

• iBiology video: Margaret Gardel explores the physical properties of the cytoplasm (11:59): https://www.youtube.com/watch?v=kl_LBcIKQvQ&ebc=ANyPxKptLnnUlVOf_8Xm1Lr1QQ3UWoINAkoyIGQ_1RA_kqyAXy4M3o_3hEzsPJHkB1hATWn11lgHByDfOTWyYbcMX81rh82i2Q.
• iBiology lecture "Dynamics of the Bacterial Chromosome" by Lucy Shapiro (35 min): http://www.ibiology.org/ibioseminars/microbiology/lucy-shapiro-part-1.html.

WRITTEN

- Conrad Woldringh discusses whether or not the nucleoid-cytoplasm interface exists in the *Small Things Considered* blog post "Where Art Thou, O Nucleoid?" (http://schaechter.asmblog.org/schaechter/2008/10/where-art-thou.html).
- Merry Youle tells us some interesting facts about linear prokaryotic chromosomes in the *Small Things Considered* blog post "Some Like It Linear" (http://schaechter.asmblog.org/schaechter/2008/04/some-like-it-li.html).
- Merry Youle discusses organelle-like structures in bacteria in the *Small Things Considered* blog post "Three Roads to Cellular Compartments" (http://schaechter.asmblog.org/schaechter/2015/04/three-roads-to-cellular-compartments.html).
- Daniel Haeusser discusses how bacteria sense Earth's magnetic field to their advantage in the *Small Things Considered* blog post "The Attraction of Magnetotactic Bacteria" (http://schaechter.asmblog.org/schaechter/2015/03/the-attraction-of-magnetotactic-bacteria.html).
- An article on FRAP: "In Vivo Biochemistry in Bacterial Cells Using FRAP: Insight into the Translation Cycle" (http://www.ncbi.nlm.nih.gov/pmc/articles/PMC3491719/).
- Diffusion coefficients of several molecules through the cytoplasm: http://book.bionumbers.org/what-are-the-time-scales-for-diffusion-in-cells/.

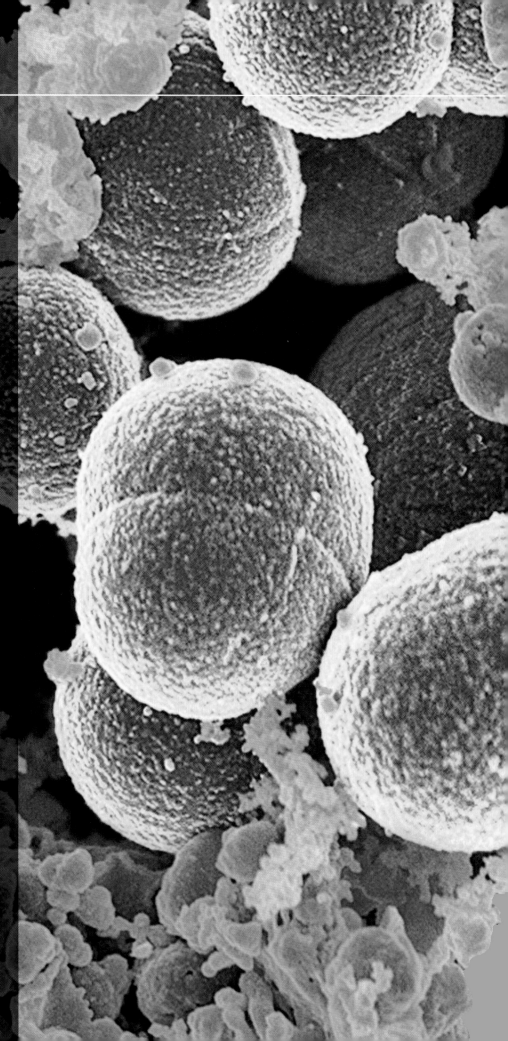

IN THIS CHAPTER

Colorized scanning electron micrograph of Staphylococcus aureus (golden) evading human white blood cells (blue). (Frank DeLeo, NIAID; CDC–Public Health Image Library)

CHAPTER FOUR

Microbial Cell Growth and Division

KEY CONCEPTS

This chapter covers the following topics in the ASM Fundamental Statements.

EVOLUTION
1. Cells, organelles, and all major metabolic pathways evolved from early prokaryotic cells.

CELL STRUCTURE AND FUCNTION
2. The structure and function of microorganisms have been revealed by the use of microscopy (including bright-field, phase-contrast, fluorescent, and electron).

METABOLIC PATHWAYS
3. The survival and growth of any microorganism in a given environment depend on its metabolic characteristics.
4. The growth of microorganisms can be controlled by physical, chemical, mechanical, or biological means.

IMPACT OF MICROORGANISMS
5. Microorganisms provide essential models that give us fundamental knowledge about life processes.

INTRODUCTION

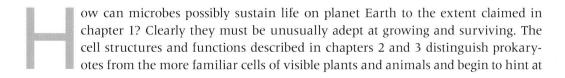

How can microbes possibly sustain life on planet Earth to the extent claimed in chapter 1? Clearly they must be unusually adept at growing and surviving. The cell structures and functions described in chapters 2 and 3 distinguish prokaryotes from the more familiar cells of visible plants and animals and begin to hint at how microbial dominance has been brought about. In this chapter we will gain a fuller understanding of the basis of microbial success: **growth** and reproduction. We will first examine growth—what it is, how it is measured, and how it contributes to microbial domination of the biology of Earth. Later in the chapter, we shall introduce the remarkable ability of microbes to grow under harsh conditions to such an extent that Earth's **biosphere** is virtually defined as the presence of viable microbes. Lastly, we will learn how cells multiply, some by dividing the growing cell in the middle and others with creative approaches that reflect adaptations of each microbe to their unique environment.

CASE: A simple experiment to understand microbial growth

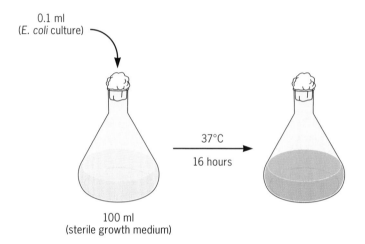

In lab you fill a flask with 100 ml of a simple liquid medium that supports the growth of the common bacterium *Escherichia coli*. The ingredients of this "minimal medium" are listed in Table 4.1, but the key components are a mixture of inorganic salts, which provide the elements essential for cell growth and function and also help maintain the pH of the medium. Then you add to the medium a sugar, glucose, which *E. coli* uses as its source of carbon and energy. After sterilizing the flask and its contents to kill any contaminating microbes, you aseptically add a very small inoculum: 0.01 ml from a previous culture of *E. coli*. (For the purposes of this exercise, assume that this dilute inoculum contains a single cell.) You place the culture on a shaking device in an incubator at 37°C and go meet a friend for dinner, as it is now 5:00 PM. When you return the next morning, some 16 hours later, you find that what started out as a clear, transparent liquid has become cloudy—not milky white—but definitely cloudy.

• Why is the morning culture cloudy?
• What happened to the single cell you inoculated in the flask during 16 hours of incubation?
• Do you expect the rate of growth to be constant throughout the incubation period? Design an experiment to test your hypothesis.

TABLE 4.1 **A minimal medium known as MOPS medium for the growth of E. coli**[a]

Constituent	Concentration (mM)
Potassium phosphate (as K_2HPO_4)	1.32
Ammonium chloride	9.52
Magnesium chloride	0.523
Potassium sulfate	0.276
Ferrous sulfate	0.01
Calcium chloride	5×10^{-4}
Sodium chloride	50
MOPS (a buffer)	40
Molybdenum, boron, cobalt, copper, manganese, zinc	Trace amount
Carbon source (e.g., 0.2% glucose)	

[a]MOPS, morpholinepropanesulfonic acid. To make a **rich medium** that supports high growth rates, the following supplements are added: all 20 amino acids, purines and pyrimidines, and several vitamins. Note that minimal media are totally defined and hence reproducible. In addition, they permit the use of isotopes (^{14}C, ^{35}S, and ^{32}P) and organic precursors for efficient labeling of cell components.

WHAT IS MICROBIAL GROWTH?

LEARNING OUTCOMES:
- Learn what "growth" in a microbe means compared to "growth" in any other organism, for example, a human.

When cattle ranchers or pediatricians speak of growth, the conversation is usually how many pounds or ounces their subjects gained in a week. Farmers talk of their crops' growth of inches or feet. But in the domain of microbes, growth means something quite different. For microbiologists, growth usually refers to the *increase in numbers of a population of cells*, as illustrated in the case study. Obviously, increase in numbers requires the increase in mass of each individual microbial cell: a newborn cell synthesizes its constituents (proteins, nucleic acids, carbohydrates, and lipids) until it reaches a point—usually a doubling of its mass—at which it divides into two cells, which can each then repeat the process (Fig. 4.1). But seeing is believing: Fig. 4.2. Amazing! This *increase in new cells* is what microbiologists typically measure and call growth.

As we learned in chapter 1, being small helps microbes grow fast, increasing in numbers rapidly when conditions are favorable. In the lab, the entire process of synthesizing a single *E. coli* cell takes only about 20 minutes at 37°C in rich media. And *E. coli* is not even the speed champion. A bacterium isolated from salt marshes, *Vibrio natriegens*, can double in less than 10 minutes. In less than 1 day, this single bacterium could multiply to a volume greater than that of planet Earth! What prevents this speed demon from taking over the world? As you can imagine,

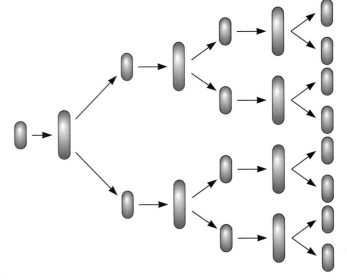

FIGURE 4.1 **The cycle of growth and cell division.** Cells grow until they reach a critical mass, then they divide into two daughter cells, and the cycle starts again, generating four cells from two, and so on.

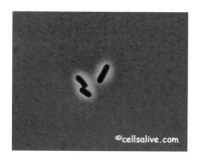

FIGURE 4.2 Bacterial growth.
Growth of *Escherichia coli* by binary fission is shown at 1,760 times normal speed in this video from cellsalive .com. (From http://www.cellsalive .com/ecoli.htm.)

the bacterial population quickly runs out of food. Even if food was plentiful, the microbe's **metabolism** would cause growth-inhibiting toxic products to accumulate. Fast growth may be required for success in some environments but, surprisingly, not in others. Indeed, some microbes grow exceedingly slowly even under optimal conditions. *Mycobacterium tuberculosis*, the bacillus that causes tuberculosis (chapter 24), divides only about once a day inside the human host, and sometimes it stops growing for long periods of time, despite the availability of nutrients. This strategy allows the pathogen to go undetected by the immune system and persist within the host for decades. It may take extraordinary patience for scientists to culture the tubercle bacillus, but its slow growth is a part of the microbe's survival strategy.

HOW DO WE MEASURE MICROBIAL GROWTH?

LEARNING OUTCOMES:
- Learn the usefulness of measuring microbial growth with turbidity, viable cell counts, and total particle counts.
- Describe what the cells are doing during each of the four phases of growth.
- Calculate the number of generations, generation time, and specific growth rate for a given culture.

How can we study how microbes grow and divide when they are so small and reproduce so fast? Consider the case study experiment. If you were to return to the lab after a few hours of incubation, you would see a hint of cloudiness in the liquid. This turbidity indicates that the single cell you inoculated in the flask has started the cycle of growth and cell division (Fig. 4.1), and many cells are now growing in the culture. How can you measure the number of cells now present in the culture? Below we describe a few of the methods most commonly used in the microbiology lab.

Measuring Growth in the Lab
Measuring the turbidity of a culture. **Turbidity** is the most widely used measure of microbial growth as it can be determined nearly instantaneously using a spectrophotometer. In addition to convenience, this measurement is quite accurate. You measure cell density by passing a ray of light through the culture. We typically measure its turbidity as the **absorbance** or **optical density** at a particular wavelength of light. But the measurement does not depend on the absorption of light by bacteria—they are virtually transparent. Instead, *the bacterial cells scatter part of the incident light*, reducing the amount of light that is detected. The more cells (and/or the greater their volume) the more scattering and the less amount of light that passes through the culture. Hence, indirectly, turbidity is a measure of the bacterial biomass (weight/culture volume) present in the culture. There are instances when such a measurement is not practical, e.g., when the medium itself is cloudy. However, if attention is paid to the chemical composition of the medium and the temperature, this measure-

ment of growth rate can be reproducible and accurate, easily to within an error of 1 to 2%.

Measuring the number of viable cells. The turbidity of the culture measures the total cell biomass but does not discriminate between cells that are alive and those that are dead. To determine the number of living cells in a population, microbiologists can grow the cells on the surface of a thin layer of agar-solidified medium in lidded plastic dishes (called petri dishes, after their inventor, Julius Richard Petri). Once deposited on the surface of a suitable solid medium, a single, living cell will grow and reproduce, producing a colony visible to the naked eye. Of course a dead cell will not grow into a colony. To carry out such a **viable count**, one simply dilutes the sample sufficiently and spreads a measured volume of the dilution on the surface of the agar to ensure that single cells are separated from each other. After incubating the petri dish for an appropriate time, the experimenter counts the number of colonies, each of which developed from an individual cell. The technique takes patience and also has potential for errors that in some cases are difficult to avoid. For example, if the bacteria naturally clump together, the number of colonies reveals the number of clumps and underestimates the number of individual cells.

Measuring cell numbers with a microscope. Cell growth in a culture can also be measured with a microscope by counting the total number of cell particles at different times. A simple wet mount is prepared by depositing a drop of liquid containing microbes on a glass slide and then examining the sample with a microscope. If you want higher contrast, you can use a phase-contrast microscope to see cell particles as different shades of gray (Fig. 4.3). In particular, a counting chamber device is used to count the cells with a microscope (Fig. 4.3). The chamber consists of a modified glass slide with a central depression of known depth, the bottom of which is ruled into squares of known area. The depression is filled with the cell suspension, covered with a coverslip, and allowed to stand until the cells have settled to the bottom, so they will all be in the same plane of focus. The number of cells per unit area (here, proportional to the volume) reveals the total number of cells in the suspension. An analogous device used for counting blood and other animal cells is called a hemocytometer.

Counting cells with a flow cytometer. Specialized instruments are also available that detect specific physical or chemical properties of the cells and provide a total particle count. One of the most versatile technologies to do this is **flow cytometry**, as it can simultaneously count and analyze many physical and chemical characteristics of individual cells (or any other particles) (Fig. 4.4). As individual cells pass through the laser beam, they interrupt the light path; lateral (side) and forward scattered light is then collected to gather information about each cell particle, including its size, shape, surface topography, and even its internal complexity (e.g., presence or absence of nucleus, internal granules, etc.). Prior to flow cytometry, cells can be differentially stained with fluorescent dyes that bind to specific components of the cell, or with fluorescently labeled antibodies or DNA probes, to collect quantitative fluorescence data about the genetic and/or

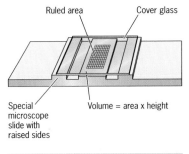

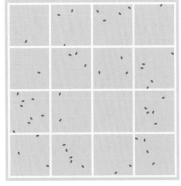

FIGURE 4.3 Chamber for counting cells under the microscope (also known as a hemocytometer). The ruled portion defines an area, and the space between the raised cover glass and the slide defines the height (for bacteria, usually 0.02 mm). Counting the particles over a certain area gives an estimate of their number per unit volume.

FIGURE 4.4 **Flow cytometry.** As a dilute suspension of particles, e.g., microbial cells, is passed through the laser beam of a flow cytometer, light scattering from individual cells is analyzed. The amount and direction of the scattered light reveal information about each cell particle. In this example, two subpopulations are shown that were differentially stained with red and green fluorescent dyes, quantified (plot at right), and separated (sorter) for further analyses.

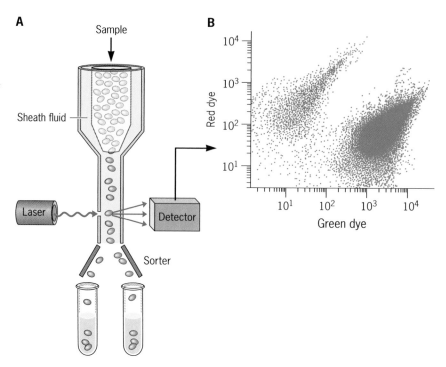

biochemical composition of the cells. Additionally, flow cytometers can be equipped with a sorter: as individual cells pass through the laser beam and exit the flow chamber, they are separated into distinct cell subpopulations that are now ready for further analysis (Fig. 4.4).

The Phases of Microbial Growth

Congratulations! You have now recorded the turbidity of your culture as a function of time, and you are ready to plot the data (Fig. 4.5). What do you see in the graph? It is obvious that the culture grew over a period of many hours but stopped growing by around 12 hours. The culture is now said to be in the **stationary phase** of growth (Fig. 4.5). What made growth stop? Simple: the cells in the culture either used up some essential nutrients or secreted noxious waste products that accumulated to an inhibitory concentration. In any case, growth could no longer be sustained, and the total number of cells in the culture reached a plateau. Under most natural conditions, microbes do in fact alternate between periods of growth and nongrowth; that is, they are constantly entering into and coming out of exponential growth. However, the non-growth state is likely more prevalent than exponential growth. So important is the physiology of the stationary phase cell for survival that we will discuss it in detail in chapter 12. Furthermore, the non-growth state can lead in some microbes to the differentiation of resistant and dormant structures (chapter 13). Indeed, some form of dormancy is a successful strategy to survive fluctuating and often suboptimal environmental conditions.

If you now transfer the stationary phase culture to fresh medium, growth will not start right away (Fig. 4.5). It will be delayed for a time period that depends on the organism, the particular medium, and the length of time the culture was in stationary phase. The initial still period is called

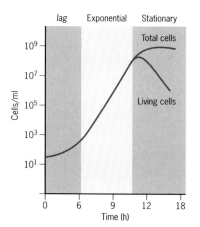

FIGURE 4.5 **A typical bacterial growth curve.** Measurements were carried out for the total number of cells and for the number of viable cells. Note that the two are the same during the initial adaptation phase (*lag*) and exponential growth, indicating that viability was nearly 100%. However, as growth slowed down during entry in stationary phase, some cells died. Some lyse as their growth slows down. Some may still be intact, however, and thus contribute to the total number.

the **lag phase**; it reflects the time the cells need to adapt to the new conditions and resume growth. In spore-forming bacteria, such as *Bacillus subtilis*, as well as others that form dormant cell types (chapter 13), the lag phase has special attributes. Because spores are made during the stationary phase, subsequent growth of the culture in fresh medium can resume only after the lengthy process of **spore germination**. *E. coli* also undergoes physiological and structural rearrangements during the stationary phase (chapter 12), although the morphological changes are less dramatic than in **sporulation** (chapter 13). Hence, *E. coli* cells also will undergo a lag phase while they prepare to grow again.

Once growth resumes at a steady rate, the cells grow and divide exponentially (as in the case study). During this **exponential phase** (Fig. 4.5), the cell growth rate is maximal. The exponential phase growth rate is a reproducible characteristic of the particular organism, the specific medium, and the temperature. A culture in the exponential phase grows at a constant rate as long as the concentration of substrate being consumed does not fall below a utilizable level, often in the micromolar range. When the substrate is depleted, the growth rate decreases to zero. The relationship between growth rate and substrate concentration approximates first-order kinetics (Fig. 4.6). Thus, it resembles the relationship between the velocity of an enzyme-catalyzed reaction and the concentration of the substrate for that reaction, as described by the Michaelis-Menten equation, a familiar biochemistry concept.

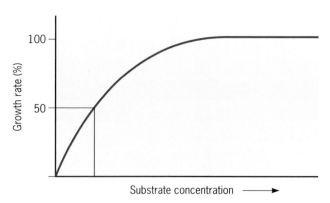

FIGURE 4.6 **Change in growth rate as a function of concentration of an essential limiting substrate.** The growth rate is constant over a wide range of substrate concentrations but is lower when the substrate is scarce. For measurements to be carried out in this region, the bacterial cultures must be highly dilute, or else the cells would readily consume the small amount of substrate and thus alter its concentration over the time of the measurement. For a perspective, the concentration of glucose that results in the half-maximal growth rate (marked by the horizontal and vertical lines) in *E. coli* is 10^{-6} M.

Law of Growth by Binary Fission

Have you heard the saying "When cells divide, they multiply?" It's true! At each division, one cell becomes two, these become four, etc. (Fig. 4.1, 4.2). Thus, the initial cell that was inoculated in the culture in the case study grew into 2, 4, 8, 16, . . . , 2^n. This series describes the increase in cell numbers, where n is the number of generations (1, 2, 3, 4, etc.). But what happens when a culture is started with more than one cell? The number of cells, N, after n generations beginning with N_0 cells, can be calculated from the following equations:

$$N = 2^n N_0 \quad (1)$$

and

$$\log N = n \times \log_{10} 2 + \log N_0 \quad (2)$$

If the numbers of cells at the beginning and at any particular time thereafter are known, the number of intervening generations can be calculated from the equation:

$$n = \frac{\log N - \log N_0}{0.301} \quad (3)$$

Let us put this knowledge in practice. Back in the lab, we inoculate our 100-ml culture with 0.01 ml containing not one but 100,000 bacteria. Therefore, the cell concentration in the inoculum was 1,000 (or 10^3) bacteria/ml. How many generations are needed to reach a cell concentration of 1 billion (or 10^9) cells/ml? From equation 3, we know that the number of generations n is equal to $(\log 10^9 - \log 10^3)/0.301$; thus, n is equal to 6/0.301, or about *20 generations*.

Any rigorous scientist knows to record in their lab notebook the time when the experiment was started and when the final cell counting was done. The time interval can then be used to calculate how much time (t) it takes for the cells to double in number (one **generation time** [g]) by using the equation:

$$g = t/n \quad (4)$$

For a time interval of 16 hours, as in the case study, the generation time g would be 16/20, which is 0.8 hour or *48 minutes*. This experiment was performed with glucose in a minimal medium, which slows down the growth of *E. coli*. If we had used rich medium instead, the cells would have grown faster, doubling evey 20 minutes, and the billion figure would have been reached in just 6 hours 40 minutes!

Even without mathematics, it should be obvious that, in a growing culture, the more cells there are, the more cells are made in a particular interval of time. Thus, exponential growth, also known as **logarithmic growth** (*now you know why!*), mimics an autocatalytic, or first-order, chemical reaction. The fundamental law of growth (Box 4.1, equation 1) is the same principle as compound interest, which Benjamin Franklin is said to have called the "eighth wonder of the world."

WHY IS EXPONENTIAL GROWTH "BALANCED"?

LEARNING OUTCOMES:
- Explain what balanced growth is.
- Compare and contrast growth during exponential versus stationary phase.
- Explain in general terms what a chemostat is.
- Compare and contrast batch culture and chemostat for the following properties over time: growth rate, nutrient conditions.

Why is the study of exponentially growing cells so prevalent in laboratory research? As long as a culture is in the exponential phase, *all cell constituents increase by the same proportion over any given interval of time*. In the time it takes to double the number of cells in the culture, the amount of DNA, number of ribosomes, number of individual enzyme molecules, and so on also double. This state is called **balanced growth**. Although this condition does not usually continue for long in the environment, it can be reproducibly maintained over a considerable time in the laboratory. And experimental reproducibility is key to doing science! So microbiologists pay careful attention to medium composition and physical parameters such

BOX 4.1

How is growth rate calculated?

Typically we think of microbial growth as cells doubling. But to compare different growth conditions, it is useful to calculate the *rate* of growth. Here we explain how, and give you a chance to apply what you learned in a calculus course (perhaps for the first time).

In mathematical terms:

$$dN/dt = kN \qquad (1)$$

where dN is the change in the number (N) of cells per unit volume over time (t), and k is a proportionality constant called the **specific growth rate**. The units of k are reciprocal time, usually expressed as hours^{-1}. To help us calculate the growth rate, can we simplify this equation (1)? Yes, indeed. Just follow the steps from equation 1 to equation 4 below. As any mathematical aficionados will know, this is how we solve the equation for k, the specific growth rate:

1. Integrate equation 1 to yield: $\qquad N/N_0 = e^{kt} \qquad (2)$

2. Take the logarithms of both sides: $\quad \ln N/N_0 = \ln N - \ln N_0 = \ln(e)kt \qquad (3)$

3. Finally, convert to log base 10: $\qquad k = 2.303 (\log N - \log N_0)/t \qquad (4)$

The specific growth rate (k) of the culture describes how fast a particular bacterium grows in a particular environment. For example, if a culture contains 10^3 cells at time zero and 10^8 cells 6 hours later, the specific growth rate is $(8-3)2.303/6$, or $k = 1.92$ hours^{-1}. In this hypothetical culture, that's how fast one thousand cells become one hundred million!

From equation 4, we can also determine the relationship between growth rate (k) and generation time (g). Because g is the time for N_0 to double, i.e., for one cell to become two, we see that $k = 2.303(0.301 - 0)/g$, or the equation: $k = 0.693/g \qquad (5)$

---

as temperature, osmotic pressure, and pH. By simply diluting the culture with the same medium before it enters stationary phase, scientists can keep their cells in the exponential (logarithmic or log) phase of growth (Fig. 4.7).

During balanced growth, the *mean* cell size also remains constant. Of course when microbes grow, individual cells do increase in size until eventually they divide (Figs. 4.1, 4.2). Therefore, we emphasize "mean," the statistical average. Balanced growth refers to the *average* behavior of cells in a population (e.g., cell size), not to growth of individual cells as they go through their cell cycle.

Other terms related to balanced growth, such as "unchanging environment" and "steady state," must be taken as approximations only. For example, some cell properties change early during the growth of a culture, long before the increase in mass slows down, and some of these changes are imperceptible. Therefore, *balanced growth conditions can be approximated only at low cell densities*. This fact can be inconvenient; for example, to measure a biochemical parameter of cells during balanced growth may require a large culture volume. Alas, many experimenters fail to appreciate this subtle but important point.

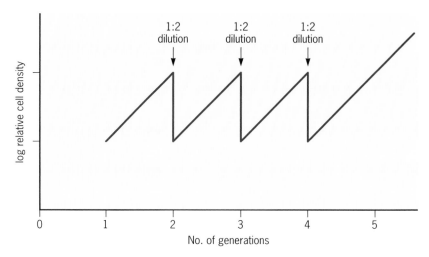

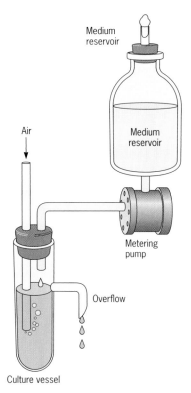

FIGURE 4.7 **Maintaining balanced growth.** A culture will grow exponentially, i.e., be in balanced growth, as long as it does not exceed a certain cell density and is transferred periodically. In this example, a 2-fold dilution was carried out after each generation. Although tedious, this procedure can keep a culture in a steady state, in principle forever.

FIGURE 4.8 **Continuous-culture device, or chemostat.** Fresh medium from a reservoir is fed at a constant rate into a culture vessel by using a metering pump. The volume of the culture is maintained constant by the removal of excess through an overflow at the same rate as new medium is added. Not shown is a device for mixing the culture vessel contents. The size of the culture vessel may vary from a few milliliters to thousands of gallons.

How Is Balanced Growth Maintained in the Lab?

In the lab, a culture can be maintained in balanced growth by diluting it at set intervals with fresh medium (Fig. 4.7). As long as someone is willing to do this careful, tedious work, the cells will remain in the exponential phase of growth indefinitely. Fortunately, with automation, cultures can be maintained in a state of balanced growth over a range of growth rates, using a continuous-culture device, or **chemostat** (Fig. 4.8). These specialized culture vessels are kept at the desired temperature and with a gas atmosphere suitable for the organism under investigation. For example, aeration promotes growth of aerobes, but oxygen-free gases are used to grow anaerobes. The volume in the culture vessel is kept constant by a metering pump or a valve, which adds fresh medium at the same rate as culture medium is removed through an overflow device. To ensure that the fresh medium rapidly equilibrates with the contents of the culture vessel it is mixed vigorously.

For a chemostat to function properly, the cell density should not exceed that which allows *balanced growth* in a batch culture. This condition is achieved by maintaining a limiting concentration of some essential nutrient, such as glucose, ammonia, phosphate, or a required amino acid. By doing so, the chemostat becomes a *self-correcting system.* Follow this logic, and you will understand balanced growth. If you were to increase the rate of addition of fresh medium, the rate of outflow would increase, and cells would be lost at a greater rate than they are formed. The density of cells in the vessel will decrease, so the limiting nutrient will be used at a lower rate. As the concentration of that nutrient in the vessel increases, the growth rate will increase until it matches the rate of cell loss through the outflow. On the other hand, if you add fresh medium at a new, lower rate, the opposite cascade of events would ensue: the growth rate of the culture would decrease until it matches the slower rate of cell loss from the vessel. In either scenario, the culture would self-correct to maintain a constant density. Beautiful! Thus, there are two critical properties of chemostats:

- The rate of addition of fresh medium (per volume) determines the growth rate in the culture vessel (up to the maximum possible for that organism in that particular medium).
- The density of cells in the culture vessel is constant and is determined by the concentration of a limiting nutrient.

How are chemostats used in microbiology laboratories? In addition to studies of bacterial mutagenesis and evolution, the device also has applications in large-scale industrial fermentations. In fact, no other laboratory culturing technique is this reproducible.

HOW IS THE PHYSIOLOGY OF A CELL AFFECTED BY ITS GROWTH RATE?

LEARNING OUTCOMES:
- Describe how the growth rate of a particular microbe can influence its macromolecular composition.
- Explain why each of the following parameters would or would not change the macromolecular composition of a growing cell: nutritional properties of the medium, temperature, pH.

All organisms grow faster and function best when provided an adequate diet. (Note that "adequate" does not mean plentiful: some microbes actually require low nutrient conditions to grow.) Thus, as the nutritional environment changes, the cell must undergo profound changes to fine-tune its growth rate. As the cells grow faster, they become bigger and cell mass, protein, DNA, and RNA (mainly in ribosomes) increase (Fig. 4.9). The reasons are fairly obvious. For example, if a microbe is to grow faster, it needs more protein-synthesizing capacity. Unused machinery is a waste of energy, so to grow at the maximum rate in a particular medium, a cell must maintain the optimal amount of the protein-synthesizing components. Since extra macromolecules take up space, fast-growing cells are larger.

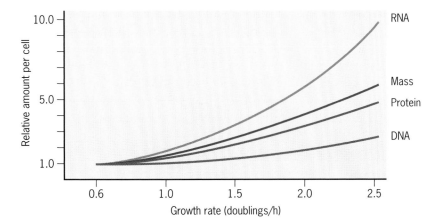

FIGURE 4.9 Compositions of _E. coli_ cultures growing at different rates. Individual cultures were grown at the same temperature in a variety of media, each supporting a different growth rate. The relative amounts of various constituents are plotted as values per cell.

See for yourself: under the microscope (Fig. 4.10A), fast- and slow-growing cells look so different that it is hard to believe they are the same species!

But nutrient availability does not affect each macromolecule in the same way (Fig. 4.9). RNA increases much faster than mass, protein increases somewhat more slowly, and DNA increases even more slowly. Furthermore, the increases are not linear. This is easily appreciated when the ribosome content is plotted as a percentage of the total mass, rather than on a per-cell basis (Fig. 4.10B). When growth is slowest (less than 0.5 doubling per hour), the proportion of the cell mass composed of ribosomes does not increase linearly with growth rate. Likely, cells retain a minimum number of ribosomes even when conditions do not favor growth, possibly to be able to resume growth rapidly once conditions become favorable. Yet once a threshold growth rate is reached, the relative concentration of ribosomes increases linearly with the growth rate. Note that *each ribosome* (and its accompanying components of the protein-synthesis apparatus) synthesizes proteins at the same rate in fast- and slow-growing cells. Hence, to grow faster, cells need to make more ribosomes. This property also happens to provide an efficient form of regulation (chapter 11).

HOW DO MICROBES GROW IN EXTREME ENVIRONMENTS?

LEARNING OUTCOMES:
- Describe how very high and very low temperatures, pH, and salt concentration inhibit growth.
- Describe adaptations of thermophiles, psychrophiles, barophiles, acidophiles, and alkaliphiles that permit them to exist in their optimal growth conditions.
- Explain two strategies that are used in human food preparation to minimize microbial growth during food storage.

Microbes can grow wherever there is liquid water, even in environments with extremes of temperature, pH, osmolarity, pressure, radiation, etc. Such microbes, called **extremophiles**, not only tolerate but also grow (*thrive!*) under such extreme conditions, sometimes several at once (Table 4.2). The field of astrobiology—the study of the possibility of life outside planet Earth—focuses heavily on the properties of Earth's extremophiles. How individual microbes adjust to changes in their environment (referred to as **adaptation**) is the subject of chapter 12. Here, we present an overview of how certain microbes have evolved to live *permanently* in extreme environments.

Temperature

Why does temperature affect microbial growth? Easy: microbes need water to grow. On Earth, the range of temperatures for liquid water extends well beyond the freezing and boiling points of pure water at sea level. Water is kept liquid below the freezing point by dissolved solutes and above the boiling point by high pressure. Consequently, some microbes can grow in high salt environments at −15°C and probably lower. This is why household freez-

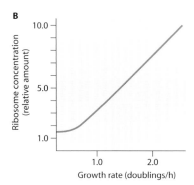

A

B

FIGURE 4.10 **Change in cell size and ribosome content with growth rate.** (A) Electron micrograph of a mixture of *E. coli* cells grown in different media. The large cells grew at 2.3 doublings per hour, and the smaller ones grew at 0.8 doubling per hour. Bar, 1 μm. (Courtesy of C. L. Woldringh.) (B) Relationship between the ribosome concentration (ribosome content per total mass) and the growth rate.

TABLE 4.2 **Temperature responses of microbes**

Class	Typically grows at:
Psychrophile	5°C
Facultative	Maximum at 20°C or above
Obligate	Maximum 20°C
Mesophile	37°C
Thermophile	50 to 70°C
Eurythermophile	Broad range of temperatures; can grow at 37°C
Stenothermophile	Narrow range of temperatures; cannot grow at 37°C
Hyperthermophile	70°C

ers are set at temperatures below −15°C. (But if you have had hams spoil after some weeks or months in the freezer, you will know that even such low temperatures do not ensure that salty food remains palatable and safe.) At the other end of the temperature scale are microbes that can grow at well above the temperature of boiling water, which exists under high pressure in the depths of the ocean or in fissures of deep rocks.

Microbes that grow at extreme temperatures are definitely customized. Broadly speaking, bacteria and fungi dominate the lowest temperatures and archaea dominate the highest, but every microbe specializes in one particular range of temperature. Few species of prokaryotes, if any, can grow over a range of more than 40°C, and many are restricted to considerably narrower ranges.

Using the *optimum temperature* that supports their growth, we can classify microbes as **psychrophile** (optimum is low), **mesophile** (moderate), **thermophile** (high, 60 to 80°C), and **hyperthermophile** (very high, >80°C) (Table 4.2). The champion psychrophile *Psychromonas ingrahamii* grows at −12°C, whereas the hyperthermophilic archaeon strain 121 grows at 121°C! Despite the wide range of temperatures that support microbial growth, some generalizations are possible. Over a certain range ("normal" in Fig. 4.11), temperature influences growth rate the same way it affects the rate of a chemical reaction. Indeed, growth rate obeys the *Arrhenius equation*, which describes the logarithm of a chemical reaction rate as a linear function of the reciprocal of absolute temperature. Thus, the plot passes through a maximum (the optimum growth temperature) and declines progressively through the normal range (Fig. 4.11). Outside this range are the high- and low-temperature ends of the plot, which indicate where growth becomes increasingly slower than is predicted by the Arrhenius equation. These ends of the plot are the maximum and minimum temperatures that support growth, respectively. Although the temperature range varies with the organism, the general shape for this plot is conserved in all microbes.

At the maximum growth temperature, chemical reactions proceed more slowly and eventually stop because essential macromolecules, such as proteins, denature and thus cease to function. On the other hand, as the temperature decreases, protein activity slows down but does not stop. Hydrophobic interactions in proteins do weaken at low temperatures because the properties of solvating water change and interfere with these interactions. Weakened hydrophobic interactions cause conformational changes in proteins, some of which affect activity and thus the speed of the reactions that they catalyze. Conformational changes can

A

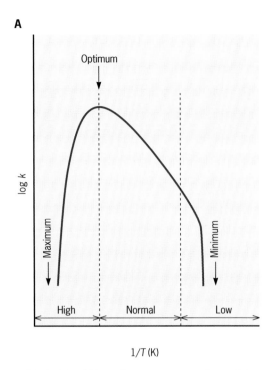

B

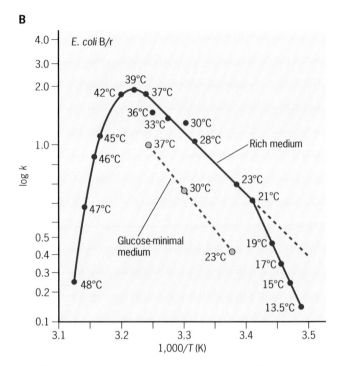

FIGURE 4.11 **Effect of temperature on bacterial growth.** (A) General form of an Arrhenius plot. Temperature (T) as the reciprocal of absolute temperature (Kelvin [K]) is plotted against the logarithm of the growth rate as k. (B) Specific values of the growth rate of *E. coli* B/r, a radiation-resistant derivative of strain B. Note how it grows more slowly in minimal medium than it does in a rich medium, but the slope of the Arrhenius plot in the normal range is nearly the same for both media. The minimum temperature of growth for *E. coli* is 7.8°C.

also interfere with the assembly of proteins into complex structures, such as ribosomes. In fact, much evidence indicates that initiation of protein synthesis by ribosomes is the vital cold-sensitive step for *E. coli*. At the limits of their growth range, microbes can adapt by synthesizing special proteins and adjusting the fatty acid composition of their membranes, a topic for chapter 12,

Growth at high temperatures. The growth rate declines and eventually stops at the limits of a microbe's growth range when vital cellular components, such as proteins, nucleic acids, and lipids, become thermally inactivated. In mesophiles, for example, proteins denature at temperatures much lower than those that support the growth of (hyper)thermophiles. On the other hand, enzymes isolated from (hyper)thermophiles are active at temperatures close to the optimal growth temperature, but are often inactive at room temperature. When cloned in mesophilic hosts, these thermophilic enzymes generally retain their thermal properties, suggesting that heat resistance, at least in proteins, is genetically encoded. How? Protein thermostability is influenced by hydrogen bonding, hydrophobic internal packing, and salt bridges and thus depends on the protein's tertiary structure. Accordingly, proteins of (hyper)thermophiles typically have a larger proportion of charged amino acids, such as glutamic acid and arginine, on their molecular surfaces because these residues promote salt bridges. In addition, (hyper)thermophiles contain a great amount of **heat shock proteins** and **chaperones**, which are proteins that have the noble role of protecting other proteins from heat denaturation or facilitating their folding and refolding, respectively. When grown at 108°C, for example, the archaeon *Pyrodictium occultum* produces a specific chaperone that accounts for up to 80% of the dry weight of the cell. Hence, thermal resistance is based partly on inherent properties of the proteins and partly on their protection by chaperones. We shall see in chapter 12 how all prokaryotes, not just (hyper)thermophiles, produce heat shock proteins and chaperones to cope with shifts to higher temperature.

As we learned in chapter 1, the ability of (hyper)thermophiles to produce thermostable proteins has been immensely useful in industry. Not only do these proteins function at the high temperatures often required in industrial processes, they have a longer shelf life and a higher activity (thus, they can be used in smaller amounts). For example, thermophilic proteases and starch degrading enzymes called amylases are often added to laundry detergent formulations to break down "dirt" from your clothes, even in hot water cycles. In the laboratory, heat-stable DNA polymerases permit the successive cycles of DNA synthesis that are the basis for product amplification in the polymerase chain reaction (PCR) technology (chapter 10).

What about other macromolecules? How do nucleic acids, for example, maintain their integrity at high temperatures? In solution, DNA melts (its two strands separate) at temperatures well below 80°C, so some mechanism must keep this from happening in (hyper)thermophiles. These extremophiles have both thermoprotective DNA-binding proteins and relatively high magnesium concentrations that neutralize the phosphate groups in the DNA molecule and stabilize its structure. In addition, (hyper)thermophiles have a unique enzyme, a **reverse gyrase**, which is the only topoisomerase known to introduce **positive supercoils** into DNA (chapter 3). Recall that most other microbes maintain their DNA in a negatively supercoiled state to facilitate the separation of the two strands during replication and transcription. Positive supercoils actually tighten the double helix, making it more difficult to separate the strands. Positive supercoiling increases the thermostability of DNA at high temperatures that would otherwise melt DNA within mesophilic organisms. Thus, as for proteins, increased thermostability in nucleic acids is due to a number of fairly subtle alterations to standard practices.

As for cell membranes, it is notable that *Archaea*, the domain that contains many thermophiles and all of the known hyperthermophiles (Fig. 4.12), do not have **phospholipid** bilayers. Instead, their cell membranes have isoprenoids linked to glycerol by highly stable ether bonds, rather than ester bonds as in ordinary lipids (chapter 2). In some cases, pairs of these molecules align their hydrophobic tails to form a lipid bilayer; in others, single double-headed molecules span the membrane as monolayers (chapter 2). Since ethers are more stable than esters and monolayers are more stable than bilayers, these unique molecular properties suggest that archaeal membranes are well adapted to high temperatures. But why do nonthermophilic *Archaea* also have this unusual architecture? A reasonable guess is that the *Archaea* first emerged in a high-temperature environment and later evolved into forms that grow at lower temperatures. Indeed, one theory is that life itself originated at the high temperatures characteristic of Earth's early periods. Although this

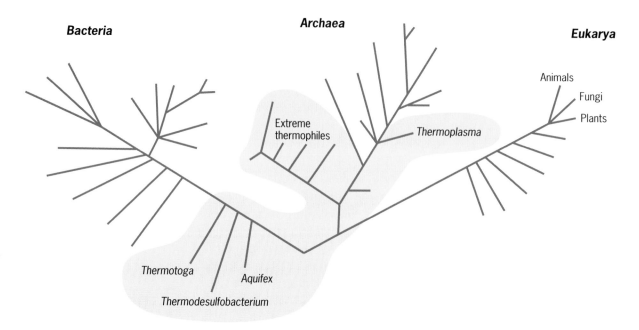

FIGURE 4.12 Distribution of (hyper)thermophiles in the tree of life. *Bacteria* and *Archaea* that can grow at temperatures above 60°C are found in relatively few groups (shaded) that tend to be clustered in the older branches of the phylogenetic tree.

notion is still the subject of debate, the most ancient lineages in both the *Archaea* and *Bacteria* are in fact (hyper)thermophilic (Fig. 4.12).

Lethal effects. Temperatures just a few degrees higher than those that stop growth may eventually kill the microbe (Fig. 4.13). This is why prolonged exposure to high temperatures is often used for sterilization. A curious observation about microbial death after exposure to high temperature, or any other lethal treatment, is that not all cells die at once. Instead, they die in a stepwise fashion: the few most sensitive first, then those with average sensitivity, and finally, the most resistant ones. This sequential death also occurs at a constant rate: if 90% of the population dies in the first 5 minutes, 90% of the remaining population will die in the next 5 minutes, and so on. Put mathematically, the logarithm of the number of survivors is a linear function of time of exposure (Fig. 4.13). The curve describes **single-hit kinetics**—the response one would expect if paintballs were fired randomly at a constant rate into a crowd and if each hit were individually visible. Unfortunately, we don't yet know what "targets" are hit when microbes are exposed to the lethal agent (e.g., heat) that causes them to die by this unique kinetic progression.

Fans of statistics will be interested to learn that plots such as the one in Fig. 4.13 can help determine how prolonged a treatment must be to eliminate a particular microbial population. To design an effective sterilization treatment, we need values for three criticial parameters: (i) the time necessary to decrease the viable population by 1 log unit, i.e., to kill 90% of the survivors, a value known as the **decimal reduction time**, or **D value** for that particular microbe at that treatment temperature; (ii) the size of the population; and (iii) the assurance we require that all cells will be killed.

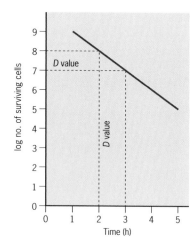

FIGURE 4.13 Lethal effect of temperature on microbes. The number of surviving cells in the population declines 90% (1 log unit) for each D value. In this case, the D value is 60 minutes.

At the other end of the spectrum, low temperature itself is not lethal to microbes; however, sudden shifts of growing cultures to low temperature can kill. The phenomenon is called **cold shock**. For example, if a culture of *E. coli* growing rapidly at 37°C is suddenly shifted to 5°C, 90% of the population will die. Freezing also kills microbes, but that's not a consequence of low temperature; instead, as the suspending medium freezes, the cells experience a transient exposure to high osmotic strength that is lethal.

Hydrostatic Pressure

Another physical extreme that some microbes have evolved to tolerate is hydrostatic pressure. High hydrostatic pressures, such as those encountered at great depths in the ocean, can crush and kill plants and animals, but not microbes. Most microbes are protected from its lethal effects because their outer cell barrier, including the cell membrane, is freely permeable to water. Thus, the pressure within a microbial cell rapidly balances the external pressure. Hydrostatic pressure does affect microbes at the molecular level, since increases in molecular volume are resisted while decreases in volume are favored. Nevertheless, many ordinary microbes withstand extraordinary pressure. As examples, *E. coli* thrives at pressures up to 300 atmospheres (about 4,400 pounds per square inch), but some other microbes can withstand about four times that pressure (Table 13 of chapter 1). Some prokaryotes, called **piezophiles** (or **barophiles**), grow faster at increased pressure than they do at 1 atmosphere. In fact, some piezophiles are obligate; they can grow only when exposed to pressures higher than 1 atmosphere. Yeasts are a curious exception. The growth of baker's yeast, *Saccharomyces cerevisiae*, stops at 8 atmospheres, the pressure that a sturdy champagne bottle can withstand.

As you can imagine, high hydrostatic pressure generally compacts—or reduces the volume—of the cell and its macromolecules. Lipid membranes are particularly sensitive to pressure, progressively losing their fluidity and reducing the activity of membrane-associated proteins, which control the permeability of the membrane to solutes. Not surprisingly, the membranes of piezophiles have a relatively high concentration of unsaturated fatty acids, which increase the fluidity of membranes. Increases in hydrostatic pressure also destabilize non-covalent chemical bonds in proteins, progressively denaturing them. Non-covalent bonds contribute to the formation of protein complexes, including those key to the assembly of the small and large ribosomal subunits during protein synthesis (chapter 8). Thus, ribosomes and other protein complexes dissociate under pressure and lose their function. To overcome this limitation, piezophilic microbes produce proteins with increased pressure resistance as well as protein chaperones that maintain the architecture of other proteins, despite the pressure. Piezophiles also accumulate in their cytoplasm low-molecular-weight osmolytes, which help stabilize macromolecules under pressure. In stark contrast to proteins, DNA is stabilized under pressure. As a result, strand separation during DNA replication and transcription (chapter 8) is more difficult. To compensate for these effects, the DNA of piezophiles is more negatively supercoiled than in microbes adapted to low pressure. Recall that negative supercoils are right-handed twists of the double helix that reduce the number of twists between the two strands of DNA and thus facilitate strand separation during replication and transcription (chapter 3). Clearly, major adaptations have equipped piezophiles for life under pressure.

Osmotic Pressure

For centuries, humans have preserved food by adding salt to it. In the 19th century, salt pork was the sailor's staple on long voyages; it was also the westward American pioneer's major sustenance in winter. High concentrations of salt completely stop the growth of many of the microbes that are likely to be present in food. But other microbes tolerate such environments to live in apparently hostile places such as salt marshes. Microbes can thrive at a range of salt concentrations that runs the gamut from distilled water (with nutrients being slowly supplied from the air) to more than 5.2 M sodium chloride, which is close to the salt's saturation point. Archaea, such as *Halobacterium*, grow at the highest concentrations of salt, but some bacteria, for example, *Halomonas*, are not far behind. (As an aside, here "*bacterium*" indicates that the *Halobacterium* genus was named before the *Archaea* were determined to be a separate domain of life.)

High osmotic pressure places two stresses on microbes: (i) it decreases the activity of water and hence its availability to the cells, effectively drying them out, and (ii) it diminishes the cell's turgor pressure by decreasing the difference in osmotic pressure between the inside and the outside. Most prokaryotes are dependent on a *high turgor pressure*, presumably *to expand their cell walls as they grow.* (An exception is wall-less species such as mycoplasmas; chapter 2.) When exposed to a high-osmotic external environment, microbes maintain turgor by increasing the solute concentration of their cytoplasm, either by pumping solutes in or by synthesizing more of them. Such compensatory solutes include potassium ion, glutamate, proline, choline, betaine (trimethylglycine), and the sugar trehalose. As the internal concentration of solutes rises, the microbe faces another problem, namely, inactivation of its proteins. However, certain solutes are less damaging than most others and may even be protective; these **compatible solutes** include proline, choline, and betaine. Trehalose is a bit magical: this sugar protects membranes from salt-induced and other forms of drying. Baker's yeast can survive desiccation thanks to high concentrations of trehalose. This strategy is not confined to microbes—a remarkable small soil-dwelling animal called the water bear, or tardigrade, contains high concentrations of trehalose and can withstand complete desiccation and then return to life when rehydrated!

pH

Most microbial species grow over a somewhat narrow range of pH, but collectively, they span a remarkably broad pH range. **Acidophiles** are abundant in the leachings of mines at pH 1.0, and **alkaliphiles** grow in soda lakes in deserts of the American West at pH 11.5. In general, fungi tend to prefer a slightly acid pH, and bacteria a slightly alkaline one. Archaea, on the other hand, have representatives at both extremes of pH. It is remarkable that prokaryotes can grow over a wider range of pH than their proteins can tolerate. How? They actually withstand the unfavorable pH by pumping protons in or out of their cells, thereby maintaining a nearly constant internal pH. For example, between pH 6.0 and 8.0, *E. coli* can grow at its full rate by maintaining an intracellular pH close to 7.8. Prokaryotes can also adapt to and survive exposure to extreme values of pH. A pH of 5.7 is lethal to an unadapted culture of *Salmonella enterica* serovar Typhimurium, but simply culturing the bacteria at pH 5.8 for one generation renders the population 100- to 1,000-fold more resistant to a subsequent exposure to pH 3.3.

HOW DOES A MICROBIAL CELL DIVIDE IN TWO?

LEARNING OUTCOMES:
- Describe the process of binary cell division.
- Compare and contrast the steps involved in initiating cell division in Gram-positive versus Gram-negative cells.
- Describe the role of the FtsZ protein in cell division.
- Explain how scientists visualized the formation and localization of the Z-ring with a microscope.
- Present evidence that supports the hypothesis that FtsZ and eukaryotic tubulins evolved from a common ancestor.
- Describe the role of Min proteins.
- Predict the phenotype of a Min mutant.
- Explain the nuclear occlusion model.

As long as the environment is favorable, most prokaryotes will reproduce continuously. In this somewhat monotonous life style, each cell grows and eventually divides into two equal progeny cells, which then repeat the process (Figs. 4.1, 4.2). Each generation closely resembles the previous one in composition and structure (Fig. 4.14). This process of **binary cell division** can be incredibly swift. How cells synthesize their multitude of components while they grow is the subject of the next five chapters (chapters 5 to 9). Here we focus on the moment in time when the cell has doubled its protoplasm and prepares to divide into two daughter cells. Although most *Bacteria* and virtually all *Archaea* grow by binary cell division, there are some very interesting exceptions, which we have saved for the end.

Morphological Considerations

Binary cell division involves invagination of the cell membrane, the cell wall, and, in the case of Gram-negative bacteria, the outer membrane as well. Only when these processes are completed can progeny cells separate (Fig. 4.14). Cell division, therefore, is a multistep process, and it differs in detail among species. Generally speaking, most Gram-negative bacteria, including the *E. coli* of the case study, divide by making a constriction of their envelope layers at midcell. The constriction deepens progressively inward until two hemispherical caps are formed and the daughter cells separate (Fig. 4.14). In some Gram-positive cocci and rods, cell division proceeds without apparent constriction of their girth. Here, a division septum is formed by the envelopes invaginating at 90° to the surface (Fig. 4.15). The poles of such cells are usually more blunt and less rounded than those of the rods that divide with constrictions, giving these cells a squared-off cylindrical shape rather than a sausage-like profile. Morphological

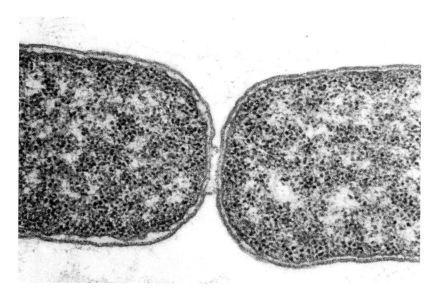

FIGURE 4.14 **Cell division in a Gram-negative bacterium.** This thin section shows two *E. coli* cells about to separate after the division of the mother cell. (Courtesy of T. J. Beveridge.)

differences aside, they all divide symmetrically into two halves, with each carrying the essential cellular components and becoming almost identical daughter cells.

One protein to start it all: FtsZ. When a microbial cell is growing, what induces its envelopes to begin building a partition? The breakthrough came in 1991 with a surprising finding: a protein called **FtsZ** (<u>f</u>ilamentous <u>t</u>emperature <u>s</u>ensitive protein <u>Z</u>) had the unexpected ability to make a constricting ring at the point where the cell would eventually partition (Box 4.2). Like the iris of the eye, this structure, called the **Z-ring**, closes gradually as the cell divides (Fig. 4.16).

Once the *ftsZ* gene was identified using one of the isolated *fts* mutants (Box 4.2), scientists wanted to catch the FtsZ protein in the act of cell division. The investigators generated antibodies that would specifically bind to FtsZ and labeled them with gold particles, which can be visualized as electron-dense particles under the electron microscope. After incubating the anti-FtsZ antibodies with ultrathin sections of bacteria and examining the samples with the electron microscope, they observed that the FtsZ protein preferentially localized at the septum, the site of division (Fig. 4.16A). However, only longer, therefore older cells, had the septum-associated FtsZ. Thus, they predicted that the Z-ring forms only in the late portion of the cell cycle. Subsequently, their deduction was confirmed by tagging FtsZ with fluorescent antibodies and tracking the formation of the FtsZ ring in growing cells with a fluorescence microscope (Fig. 4.16B).

The assembly of the Z-ring requires first the cytoplasmic polymerization of FtsZ into short filaments, or protofilaments, which then aggregate and associate with the cell membrane protein FtsA to form a contractile ring structure (Fig. 4.16C). This initial stage in cell division appears to be conserved in most prokaryotes. Nearly all *Bacteria* and many *Archaea* encode an *ftsZ* gene, as do chloroplasts and most mitochondria, eukaryotic

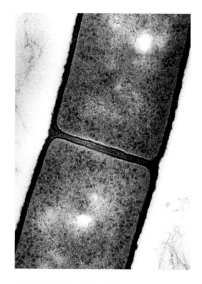

FIGURE 4.15 **Cell division in a Gram-positive bacterium.** This thin section of *Bacillus subtilis* shows no cell constriction during cell division, but a completed septum that is made before the progeny cells begin to detach. (Courtesy of T. J. Beveridge.)

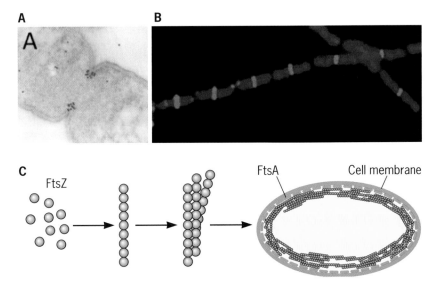

FIGURE 4.16 **Z-rings in growing bacterial cells.** (A) Transmission electron microscopy of a thin section of an *E. coli* cell showing the immunodetection of FtsZ at the septum. (From Erfei IB, Lutkenhaus J, *Nature* **354:**161–164, 1991, with permission.) (B) Anti-FtsZ antibodies (red) were also used to detect the Z-rings at the division site of cells of the Gram-positive *Bacillus subtilis* by fluorescence microscopy; the DNA of the cells was stained blue with a dye called DAPI (49,69-diamidino-2-phenylindole). (Courtesy of J. Pogliano.) (C) Model for cytoplasmic polymerization of FtsZ and association to the membrane protein FtsA to form the Z-ring.

organelles that originated from ancestral bacteria. This pattern suggests that FtsZ-mediated cell division was already present in prokaryotes before the endosymbiotic events that led to the emergence of the eukaryotic organelles. But the story doesn't end there: FtsZ is also a prokaryotic homolog of the eukaryotic cytoskeletal protein **tubulin** (chapter 2), and FtsA is homologous to another eukaryotic cytoskeletal protein, actin. Although their amino acid sequences have only about 20% identity, the three-dimensional structure of the FtsZ and tubulin proteins align very well. In addition, both

BOX 4.2

How FtsZ got its name

FtsZ owes its name to the way it was discovered. Researchers identified mutants of *E. coli* which *grew* normally at either 30°C or at 42°C, but could *divide* only at the lower temperature. At the higher temperature, the mutants, being unable to divide, grew as filaments rather than individual cells. Because of their filamentous and temperature-sensitive phenotype, the mutants were called "<u>f</u>ilamentous <u>t</u>emperature <u>s</u>ensitive" or *fts*. (Note that by convention, names of prokaryotic genes are italicized and those of proteins are capitalized but not italicized.) Interestingly, *ftsZ* is but one of the many genes that, when mutated, exhibit the same filament-forming phenotype. Indeed, there are over a dozen *fts* genes. What does this tell you? Simply, that many proteins are involved in binary cell division, FtsZ being one of them, and that all are essential because mutating any one of them inhibits the process.

cytoskeletal proteins hydrolyze GTP (guanosine triphosphate) to energize their polymerization. Hence, both proteins likely share an evolutionary ancestor. Is the Z-ring the functional equivalent of the contractile ring that mediates cell division in eukaryotic cells? It sure looks that way!

What happens after the FtsZ ring forms? Once the Z-ring is formed, FtsZ recruits other key cell division proteins. These proteins are not particularly conserved in prokaryotes but all play analogous functions, an example of convergent evolution. Some regulate the division process, others stabilize the protein assemblies at the division site, and yet others synthesize the peptidoglycan **murein** needed to make new cell wall at the septum. Among the division proteins involved in peptidoglycan synthesis is PBP3, a **penicillin-binding protein** that, as its name indicates, binds the antibiotic penicillin (and other ß-lactam antibiotics). We will learn more about this group of proteins and the synthesis of the peptidoglycan cell wall in chapter 9.

How does FtsZ find the cell's midline? Rod-shaped bacteria such as *E. coli* or *B. subtilis* tend to divide quite precisely in the middle (Fig. 4.16). What sort of ruler do they use to locate this position along the cell? There are several suggestive hypotheses, all of them still under study. However, exciting and surprising facts have been discovered. The story begins in 1969, when researchers found that certain mutants of *E. coli* divide abnormally near their ends, producing small or **minicells**, devoid of chromosomes (Fig. 4.17). They called these mutants "*min*," for minicell formers (and identified the *minC*, *minD*, and *minE* genes). An insightful deduction from this phenomenon was that a rod-shaped bacterium has three sites where division can potentially take place: one in the middle and one near each of its two poles. The last two may be residues or "scars" of previous septum sites that were shut off when the previous division was completed. The *min* mutants lack this shutting-off mechanism and so divide abnormally at a polar division site.

The Min proteins are like the good cops of the cell, patroling its interior from one pole to the other to prevent the polymerization of FtsZ in these "forbidden" regions (Fig. 4.18). MinD polymerizes at one pole of the cell and binds MinC, which depolymerizes the cytoplasmic FtsZ protofilaments. The movement of MinD is remarkably rapid: it takes about 20 sec-

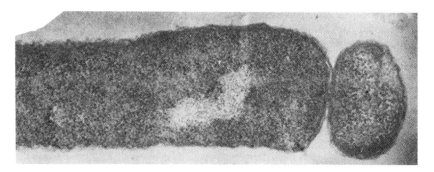

FIGURE 4.17 **Minicell production in *E. coli*.** A thin section shows an abnormal division at a pole, leading to minicell formation. (Courtesy of Y. Hirota. Reprinted from Neidhardt FC et al., *Physiology of the Bacterial Cell*, Sinauer Associates, Inc., Sunderland, MA, 1990.)

onds for the MinD aggregates to dissolve in one pole and polymerize in the opposite pole. The proteins do not move through the cytoplasm; instead, they slither along the cell membrane, making coiled structures that wind around the cell from pole to pole. MinE, on the other hand, forms a ring closer to these polar regions, approximately in the quarter position of the cell, and prevents MinCD from polymerizing further into the middle region. In this way, FtsZ is prevented from polymerazing at the poles. Not surprisingly, mutants with defective MinC or MinD proteins cannot block FtsZ from polymerizing and forming the Z-ring in these polar regions. As a result, the cell divides at the pole and a minicell is formed (Fig. 4.17).

Thanks to the Min proteins, the cell poles are protected from FtsZ. But what induces FtsZ to polymerize exactly in the middle of the cell? Surprisingly, the *state of chromosome replication and segregation* controls the formation of the Z-ring in the midcell. As the chromosome replicates in the middle of the cell, the newly replicated regions are segregated to the poles and the amount of DNA filling the center region is reduced progressively. This frees space in the middle region for FtsZ to polymerize and form the ring. Thus, this **nuclear occlusion** model proposes that the zone of the cell occupied by the nucleoid is a forbidden zone for cell division. Only when the newly replicated chromosomes are close to being fully formed and segregated to the poles does FtsZ find the space needed to form its ring, triggering cell division. This notion is alluring because, in the absence of such a mechanism, the cell division apparatus may guillotine the nucleoid and kill the cell. To avoid this lethal event, a protein called NO (for <u>n</u>uclear <u>o</u>cclusion, although "no!" has a nice ring to it) is thought to inhibit unwanted cell division in the region where the nucleoid is localized. The nuclear occlusion model is consistent with the prokaryotes' tightly coupling of cell division with chromosome replication and partioning. In fact, unlike eukaryotes, these fundamental processes are not sequential in most prokaryotes. They take place concurrently, particularly in fast-growing cells, and efficiently. In fact, prokaryotes very rarely generate anucleated cells, which would arise if segregation were unregulated during cell division. Regarding chromosome partioning, although we know a great deal about the eukaryotic process (mitosis) and the machinery that separates the segregated chromosomes (microtubular gliding), surprisingly little is known about the equivalent process in prokaryotes, which we discuss in chapter 8.

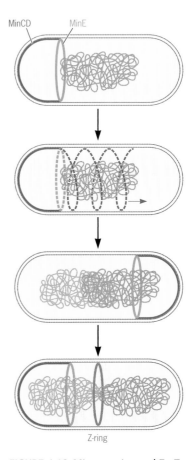

FIGURE 4.18 **Min proteins and FtsZ in action.** The polymerization of the Min proteins (MinCD and MinE) from pole to pole and polymerization in or close to the polar regions prevents the formation of the FtsZ ring in these areas. As soon as most of the chromosome has been replicated and segregated to the cell poles, enough space is freed in this mid region for the FtsZ ring to form.

IS BINARY CELL DIVISION THE ONLY WAY?

LEARNING OUTCOMES:
- List two examples of microbes that do not divide into two equal progeny cells.
- Speculate about why some microbes may have evolved alternative ways to produce progeny cells (examples, budding in yeast, vertical division in ectosymbionts).

Most "ordinary" microbes divide by binary fission into two equal progeny cells. The majority of these microbes use FtsZ to start it all, others use FtsZ-like proteins, yet all follow the standard binary fission mode of division in which the cell grows and divides in the middle. However, some microbes have different and more intricate division patterns. For example, the bacterium *Caulobacter* divides into two cells that are unequal with respect to size, shape, flagella, and a stalk-like structure (chapter 13). Other microbes divide not by fission but rather by budding (resembling yeasts in this regard; chapter 15). Then there are microbes in which the progeny cells do not separate from one another upon division but rather stick together, forming characteristic arrays such as pairs or chains of cocci, chains of rods, grapelike bunches, square sheets, or cubes, depending on the planes of division (Fig. 4.19). Even more intricate patterns are seen in prokaryotes that undergo complex developmental cycles, such as the myxobacteria, the cyanobacteria, and the actinomycetes (chapter 13). Some of these bacteria resemble fungi and algae in the morphological complexity and arrangements of their progeny cells.

Why do it differently? Consider the following scenario. Some marine nematodes, such as the worm *Laxus oneistus*, secrete a mucus around their body to attract and bind specific rod-shaped bacteria (Fig. 4.20). The worms live in environments where poisonous sulfide gas is produced, but the bacterial **ectosymbionts** metabolize the sulfide before it reaches the worm. Effective protection of their host requires the bacteria to form a thick layer around the nematode's body. To do so, the bacterial cells attach vertically to the mucus coating and in close proximity to each other (Fig. 4.20). Rather than growing in length, they grow wider, increasing their diameter (pretty much the way we humans gain weight around the waist). Once the bacteria reach their critical mass, they divide longitudinally to ensure that the descendants remain attached to the worm and the cell-cell proximity is maximized. They still use FtsZ to initiate cell division, but, in this case, the

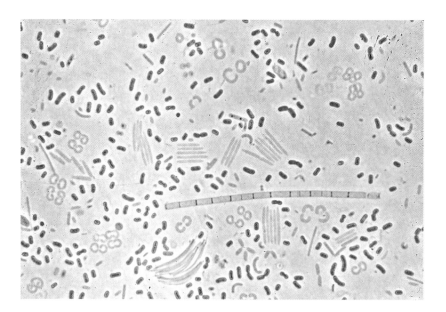

FIGURE 4.19 Cell shapes visualized with a microscope. The microbial community shown here was found in the sediment-water interface of Burke Lake, near East Lansing, Michigan. Note the rich variety of shapes and how some cells are in the process of dividing. The darker rod-shaped cells divide in the middle. A long multicellular filament with septa separating the individual cells is also revealed. (Courtesy of D. E. Caldwell. Reprinted with permission from *International Microbiology*.)

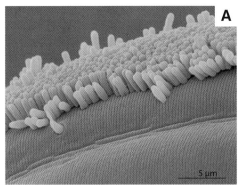
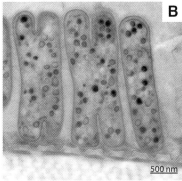

FIGURE 4.20 Bacterial ectosymbionts that divide longitudinally. (A) Bacterial ectosymbionts of the worm *Laxus oneistus* attach vertically to the marine roundworm's cuticle to form a thick bacterial coat. (B) The bacteria metabolize poisonous sulfide gas to protect the worm and store the sulfur as granules (white) within their cytoplasm. Shown are two cells caught in the act of dividing longitudinally. (From Bulgheresi S, *Symbiosis* **55:**127–135, 2011.)

FtsZ ring splits the cell in half along its length. Should FtsZ form in the middle, one of the daughter cells would be cut loose and lost! Hence, this mode of cell division ensures that the daughter cells remain attached to the worm, tightly packed against each other to preserve the protective bacterial armor. In turn, the bacteria get a free ride and easy access to food. Without the host, they cannot survive. *Worth the evolutionary twist, yes?*

CASE REVISITED: **A simple experiment to understand microbial growth**

Think back to the case study when the broth in the flask went from clear to cloudy in just 16 hours of incubation at 37°C with agitation. What does

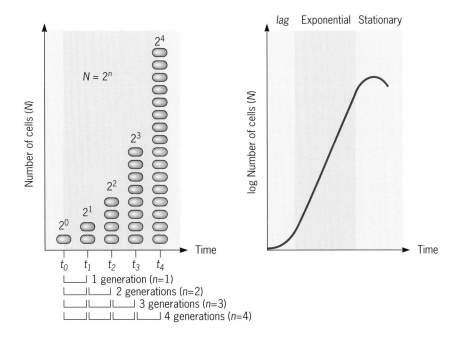

this mean? A single cell invisible to our eyes grew so fast that the broth became visibly turbid. In this chapter we learned that the initial cell adapted to the medium before starting to grow (*lag* phase), and then grew exponentially (exponential phase) until nutrients became scarce and waste products accumulated that slowed down and eventually stopped growth (stationary phase). The figure in this case study (left panel) illustrates these three phases and shows how quickly one cell grew and divided exponentially, following the equation $N = 2^n$. If you were to plot the cell number in a semi-logarithmic plot (figure, right panel), the exponential phase of growth would correlate linearly with the time of incubation.

CONCLUSIONS

Most microbes grow by binary fission, which results in exponential increases in the population up to the limit permitted by the environment. Methods to measure the total number of cells in a growing culture can be used to tally both those that are living and those that are not, or the total cell biomass. This information can be used to determine the rate of growth, a valuable parameter in microbial physiology. Eons of selection have led to the evolution of microbial species able to live under extreme conditions of temperature, pressure, pH, and salinity, to name just a few. In the next chapters we will describe the metabolic (chapters 7 to 9), genetic (chapter 10), and regulatory processes (chapter 11) that make growth possible.

Representative ASM Fundamental Statements and Learning Outcomes for the Chapter

EVOLUTION

1. Cells, organelles, and all major metabolic pathways were evolved by early prokaryotic cells.

   a. Present evidence that supports the hypothesis that FtsZ and eukaryotic tubulins evolved from a common ancestor.

CELL STRUCTURE AND FUNCTION

2. The structure and function of microorganisms have been revealed by the use of microscopy (including bright-field, phase-contrast, fluorescent, and electron).

   a. Explain how scientists visualized the formation and localization of the Z-ring with a microscope.

METABOLIC PATHWAYS

3. The survival and growth of any microorganism in a given environment depend on its metabolic characteristics.

   a. Learn what "growth" in a microbe means compared to "growth" in any other organism, for example, a human.

   b. Learn the usefulness of measuring microbial growth with turbidity, viable cell counts, and total particle counts.

   c. Describe what microbial cells are doing during each of the four phases of growth.

    d. Calculate the number of generations, generation time, and specific growth rate for a given culture.

    e. Explain what balanced growth is.

    f. Compare and contrast growth during exponential versus stationary phase.

    g. Describe how the growth rate of a particular microbe can influence its macromolecular composition.

    h. Explain why each of the following parameters would or would not change the macromolecular composition of a growing cell: nutritional properties of the medium, temperature, pH.

    i. Explain in general terms what a chemostat is.

    j. Compare and contrast batch culture and chemostat for the following properties over time: growth rate, nutrient conditions.

4. The growth of microorganisms can be controlled by physical, chemical, mechanical, or biological means.

    a. Describe how very high and very low temperatures, pH, and salt concentration inhibit growth.

    b. Describe adaptations of thermophiles, psychrophiles, barophiles, acidophiles, and alkaliphiles that permit them to exist in their optimal growth conditions.

    c. Explain two strategies that are used in human food preparation to minimize microbial growth during food storage.

IMPACT OF MICROORGANISMS

5. Microorganisms provide essential models that give us fundamental knowledge about life processes.

    a. Describe the process of binary cell division.

    b. List two examples of microbes that do not divide into two equal progeny cells.

    c. Speculate why some microbes may have evolved alternative ways to produce progeny cells (examples: budding in yeast, vertical division in ectosymbionts).

    d. Describe the role of the FtsZ protein in cell division.

    e. Describe the role of Min proteins.

    f. Predict the phenotype of a Min mutant.

    g. Explain the nuclear occlusion model.

SUPPLEMENTAL MATERIAL

References

- **Rothschild LJ, Mancinelli RL**. 2001. Life in extreme environments. *Nature* **409**:1092–1101. doi: 10.1038/35059215.
- **Rowlett VW, Margolin W**. 2015. The bacterial divisome: ready for its close-up. *Philos Trans R Soc Lond B Biol Sci* **5**:370(1679). doi: 10.1098/rstb.2015.0028.
- **Wang JD, Levin PA.** 2009. Metabolism, cell growth and the bacterial cell cycle. *Nat Rev Microbiol* **7**(11):822–827. doi: 10.1038/nrmicro2202.

Supplemental Activities

CHECK YOUR UNDERSTANDING

1. **An exponentially grown culture of *E. coli* contains of 2×10^6 cells and has a doubling time of 30 minutes. How many hours of incubation are needed for the culture to contain 2.56×10^8 cells?**

    a. 4

    b. 6

    c. 8

    d. 12

2. **A rod-shaped bacterium is preparing to undergo cell division. Select the correct statements.**

    a. The newly replicated nucleoids are occluding each pole.

    b. The newly replicated nucleoids are not occluding the mid-center.

   c. The FtsZ ring assembles at the mid-center.

   d. The Min proteins prevent the FtsZ ring from assembling at the mid-center.

3. Which of the following biochemical properties is NOT critical for adaptation of thermophiles to high temperature?

   a. Thermoprotective DNA binding proteins such as heat shock proteins.

   b. High intracellular magnesium concentrations.

   c. Highly stable ether bonds linking phospholipids and glycerol in the cell membrane.

   d. Positive supercoiling of DNA by reverse gyrase.

DIG DEEPER

1. To determine the number of viable cells in a stationary culture of *E. coli*, you serially dilute the culture (1:10, 1:100, 1:1,000, 1:10,000, and 1:100,000) and spread three 10 µl samples from each dilution on LB ("lysogeny broth") agar plates. After overnight incubation at 37°C, you quantify the number of colonies on each plate:

   • 1:100,000 dilution: 0, 0, and 0 colonies

   • 1:10,000 dilution: 48, 121, and 64 colonies

   • 1:1,000 dilution: 552, 1324, and 845

   • 1:100 dilution: colonies too crowded to count

   • 1:10 dilution: lawns of cells

   a. What plates would you use to determine the total viable counts in the culture?

   b. Why do you plate each dilution in triplicate?

   c. How many CFUs (colony-forming units)/ml were in your culture?

   d. Do you expect the total viable count to match the total number of cell particles estimated with a hemocytometer? Why?

2. In the case study, a student first grew a 100-ml culture and then took a 1-ml sample to inoculate 100 ml of fresh medium. Assume that you examined a sample of the initial turbid culture with a hemocytometer and counted 10,000,000 cells of *E. coli* per ml. Calculate:

   • the number of generations produced from 5 PM to 9 AM, assuming that the culture underwent a *lag* phase of only 1 hour and entered stationary phase at 8 AM

   • generation time

   • specific growth rate

3. Prokaryotic FtsZ and eukaryotic tubulin likely share a common ancestor.

   a. What evidence supports this claim?

   b. Using the Protein Data Bank (PDB) at www.rcsb.org, use the Needleman-Wunsch Method to compare the sequence homology of *Bacillus subtilis* FtsX protein (2VAM) and *Homo sapiens* tubulin (1Z5V).

     i. What are the similarity and identity scores of the two protein sequences?

     ii. Use the jFATCAT method to calculate the % identity and similarity for the structure of the two proteins. How similar are the protein structures?

     iii. Investigate the overlap of the two proteins in the Jmol viewer. How similar are the 3D structures?

     iv. Discuss the implications of your results in the context of the evolutionary pressures exerted on the amino acid sequence and structure of the two proteins.

4. Read the following open access article about strain 121, the champion hyperthermophile that grows at autoclave temperatures:

   Kashefi K, Lovley DR. 2003. Extending the upper temperature limit for life. *Science* **301**(5635): http://www.sciencemag.org/content/301/5635/934.full.pdf

   a. What isolation strategy and culture conditions allowed the recovery of strain 121 in pure culture?

b. What is balanced growth and why is it important? How did the investigators ensure that their cultures were in a period of balanced growth? How common is balanced growth for microbes in the environment?

c. Evaluate panel D of the figure and answer the following:

   i.   What is a "generation time"?

   ii. What is the range of temperatures that supports the growth of strain 121?

   iii. What is its optimal temperature for growth?

d. What are the highest temperature and exposure time tolerated by strain 121?

e. How many doublings do you expect this archaeon to have when grown for 24 hours at 108°C? And at 121°C?

f. Strain 121 can withstand sterilization in standard pressurized autoclaves. Should this be of concern? Would strain 121 be infectious to humans? What data in the article address this question?

g. The article states that "the factors that permit strain 121 to grow at such high temperatures are unknown." Based on what you read from the chapter, hypothesize what factors may contribute to its survival.

5. **Team work:** Working in groups, search the Web to learn more about the following types of extremophiles. Share your findings with classmates using visual aids in a 5-minute presentation.

   - Acidophile
   - Barophile (also called piezophile)
   - Halophile
   - Alkaliphile
   - Hyperthermophile
   - Osmophile
   - Psychrophile (was called cryophile)
   - Xerophile
   - Thermoacidophile

6. **Team work:** Food preservatives are used to inhibit the spoiling of food by microbial organisms. Working in groups, research how one of the following classes of food preservatives prevents bacterial/fungal growth. Share your findings with the class. You may find the following resource useful: http://www.eufic.org/article/en/food-safety-quality/food-additives/artid/preservatives-food-longer-safer/.

Substance/class	Foods
Sorbic acid and sorbate compounds	Cheese, wines, dried fruit, fruit sauces, toppings
Benzoic acid and benzoate	Pickled vegetables, low sugar jams and jellies, candied fruits, semipreserved fish products, sauces
Sulfur dioxide and sulfite compounds	Dried fruits, fruit preserves, potato products, wine
Nitrite and nitrate compounds	Sausage, bacon, ham, foie gras, cheese, pickled herring

Supplemental Resources

VISUAL

- A movie of growing bacteria (15 sec): https://www.youtube.com/watch?v=gEwzDydciWc
- Making and autoclaving agar media (7:40 min): https://www.youtube.com/watch?v=74Ndr9AVowM
- A gallery of bacterial colonies: http://www.microbiologyinpictures.com/bacterial%20colonies.html
- Counting viable cells in a soil sample (4:40 min): https://www.youtube.com/watch?v=nKFzlAuEyII

- Khan Academy explains the Arrhenius equation and what information it provides (9:23 min): https://www.khanacademy.org/science/chemistry/chem-kinetics/arrhenius -equation/v/arrhenius-equation

SPOKEN

- "It Came from Outer Space: Hyperthermophiles" video (2:36 min) from the Learning Channel discusses the possibility that life arose from microbes that tolerate extreme heat: http://www.microbeworld.org/component/jlibrary/?view=article&id=2434
- In "The Evolution of *Sulfolobus*," *Meet the Scientist* episode #8 (16 min), Merry Buckley interviews Rachel Whitaker, who analyzes *Sulfolobus islandicus*, an archaeon that thrives in geothermal springs: http://www.microbeworld.org/component/content/article?id=250.
- "Plasmid Pirates Piezophile Particles" *BacterioFiles* podcast #215 (11.2 min) discusses horizontal gene transmission of a deep-sea thermophile reported by Lossouarn et al. in "Ménage à trois: a selfish genetic element uses a virus to propagate within *Thermotogales*," *Environ Microbiol* **17**:3278–3288, 2015: http://www.microbeworld.org/component /content/article?id=1905.

WRITTEN

- Marvin Friedman discusses how bacteria cope when food runs out in the *Small Things Considered* blog post "Bacterial Antidepressants: Avoiding Stationary Phase Stress" (http://schaechter.asmblog.org/schaechter/2013/02/bacterial-antidepressants-avoiding -stationary-phase-stress.html).
- Marvin Friedman discusses strategies thermophiles use to thrive in extreme temperatures in the *Small Things Considered* blog post "Some Like it Hot" (http://schaechter .asmblog.org/schaechter/2011/01/some-like-it-hot.html).
- Veronic Rowlett discusses Min system control of *E. coli* cell division in the *Small Things Considered* blog post "The Min System: All the Places You'll Go!" (http://schaechter .asmblog.org/schaechter/2014/04/the-min-system-all-the-places-youll-go.html).
- Nanne Nanninga explains how the bacterial symbionts of some marine worms divide longitudinally in the *Small Things Considered* blog post "A Bacterium Learns Long Division" (http://schaechter.asmblog.org/schaechter/2012/11/a-bacterium-learns-long -division.html).
- Moselio Schaechter discusses how cocci find their "middle" during cell division in the *Small Things Considered* blog post "Pictures Considered #7. Cocci Divide at The Equator" (http://schaechter.asmblog.org/schaechter/2013/08/pictures-considered-7-cocci -divide-at-the-equator.html).
- Moselio Schaechter ponders how some bacteria grow as filaments in the *Small Things Considered* blog post "Why Do Bacteria Filament?" (http://schaechter.asmblog.org /schaechter/2008/02/why-do-bacteria.html).
- Gemma Reguera reviews binary cell division and describes how nanotechnology is revealing how it is coupled to DNA replication and chromosome segregation in the *Small Things Considered* blog post "Cell Division Through DNA Curtains" (http://schaechter .asmblog.org/schaechter/2012/05/cell-division-through-dna-curtains.html).

IN THIS CHAPTER

Molecular model of small metabolites

CHAPTER FIVE

Microbial Metabolism

KEY CONCEPTS

This chapter covers the following topics in the ASM Fundamental Statements.

EVOLUTION

1. Cell, organelles (e.g., mitochondria and chloroplasts), and all major metabolic pathways evolved from early prokaryotic cells.

METABOLIC PATHWAYS

2. *Bacteria* and *Archaea* exhibit extensive, and often unique, metabolic diversity (e.g., nitrogen fixation, methane production, anoxygenic photosynthesis).
3. The survival and growth of any microorganism in a given environment depends on its metabolic characteristics.

INFORMATION FLOW AND GENETICS

4. Genetic variations can affect microbial functions.

MICROBIAL SYSTEMS

5. Microorganisms are ubiquitous and live in diverse and dynamic ecosystems.

IMPACT OF MICROORGANISMS

6. Microbes are essential for life as we know it and the processes that support life (e.g., in biogeochemical cycles and plant and/or animal microbiota).
7. Microorganisms provide essential models that give us fundamental knowledge about life processes.

INTRODUCTION

In the last chapter we introduced the topic of microbial growth and learned how to study it in the laboratory. We also learned about the cycles of growth and cell division that allow microbes to multiply exponentially when conditions are favorable. How do microbes assimilate nutrients and harness energy to fuel their growth? And, in one miraculous way or another, every living creature seems to disobey the second law of thermodynamics: the rule that order degrades into disorder (entropy). Cells acquire inert matter from the environment and use it in highly organized intracellular reactions to produce life. *Wild!*

To resolve this apparent paradox, we need to understand the chemistry of cell growth, a complex array of life-generating processes collectively known as **growth metabolism**. In this chapter, we will present an overall framework that organizes the many details of metabolism. We will highlight the roles of the myriad individual biochemical processes that, together, generate living cells. We will also summarize how energy is obtained to drive the synthesis of living material and thus explain the thermodynamic paradox of creating order from disorder. This overview of metabolism will set the foundation for the next four chapters, which will explore the metabolic reactions that sustain energy generation (chapter 6) and the synthesis of cellular components (chapters 7 to 9) in living cells. Though our focus will be on prokaryotic metabolism, recall that the ancestor of all planetary life was a prokaryotic microbe. Thus, the metabolism of all cells follows universal "prokaryotic" principles.

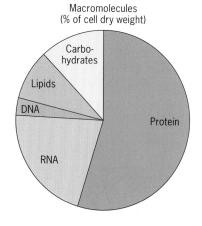

Macromolecules
(% of cell dry weight)

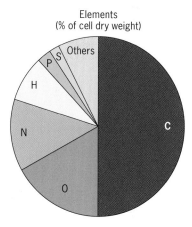

Elements
(% of cell dry weight)

CASE: How do we search for life in other planets?

In 2012, NASA's Mars Rover Curiosity landed on the surface of planet Mars with a mission: to determine if the red planet ever harbored microbial life. Since its landing, the rover has scooped many rocks and soil samples from places in the planet which may have contained water to support life. Curiosity analyzes the chemical composition of the samples, with state-of-the-art robotic equipment, seeking for evidence of life. What exactly is the Rover looking for? Suppose you were a NASA scientist. You know that all living cells on Earth have one thing in common: they all rely on essential carbon-containing macromolecules (proteins, carbohydrates, nucleic acids, and lipids) to grow and reproduce. In bacteria, proteins often account for more than half of a cell's dry weight, enriching the cell in carbon, as well as oxygen, nitrogen, hydrogen, and, to a lesser extent, sulfur (figure). Nucleic acids (DNA and RNA) account for almost a quarter of a typical cell's dry weight, enriching the cell in phosphorus plus the same elements that make up proteins.

Living cells may also be selective about their use of the **isotopes** for each element. For example, cells enrich carbon of the ^{12}C type over ^{13}C. By contrast, inanimate matter does not discriminate between the two isotopes. This unique and universal cellular chemistry distinguishes life from nonlife.

- What evidence would you look for to convince yourself that Mars had life?

- Would you search for the macromolecules common to all life?
- Would it be better to search for specific elements or isotopes of elements?

HOW IS LIFE MADE FROM INERT COMPONENTS?

LEARNING OUTCOMES:
- List four features that are unique to living cells, compared to inert chemicals.
- Explain how each of the overarching biochemical principles of life contributes to a cell's ability to reproduce.
- Distinguish between growth metabolism and nongrowth metabolism.
- Explain the connection between the ability to reproduce and entropy.
- Speculate about a different life form (on an imagined planet) that might not satisfy one or more of the four principles of chemical processes known to all Earth life.

Life sure is different from nonlife. Although nonliving objects may be chemically active on a small scale or geologically active on a grand scale, *they do not take command* of these events; by and large, they appear inert. Living things, on the other hand, breathe, move, respond to changes in their environment, modify their surroundings, and—*their hallmark and universal feature—reproduce themselves.* These attributes of living organisms are the product of highly organized chemical reactions collectively called metabolism. Through these reactions, the chemistry of the cells becomes markedly different from the inanimate matter that surrounds them. NASA's Curiosity Rover is looking for these unique chemical signatures of life on Mars (case study).

Living systems are implausible arrays of improbable molecules. Neither the individual macromolecules—proteins, carbohydrates, nucleic acids, and lipids—nor the complex cell structures made from them are likely to accumulate spontaneously, at least under present Earth conditions. (If any should form, the microbes already present would rapidly consume them!) Rather, *cells utilize information embedded in their existing structures to guide reactions* that synthesize and assemble these molecules into cells. This unique ability to control biochemical reactions allows cells to grow and to reproduce. So masterful is the chemistry of life that it has allowed microbes to diversify and colonize every niche in the planet for nearly 4 billion years!

We will first focus on the universal feature of all living systems, reproduction, which we studied in chapter 4. Reproduction is made possible by four overarching principles (Table 5.1):

  i. Enzyme catalysis: Cells use specific catalysts—**enzymes**—to accelerate otherwise slow reactions and support rapid cycles of cell growth and division.
 ii. Energy harvesting from redox reactions: Cells harvest energy from organic, inorganic, or photochemical oxidation-reduction (redox) reactions and concentrate it inside the cells in the form of high-energy adenosine triphosphate (ATP) and **reducing power** (reduced molecules such as nicotinamide adenine dinucleotide or NADH).

TABLE 5.1 **Chemical processes that form the basis of all cellular metabolism**

Chemical process	Function
Enzyme-mediated catalysis	Catalysts accelerate chemical reactions, in both the forward and reverse directions, by bringing reactant molecules together and by lowering the activation energy needed for the reaction. Most biocatalysts are proteins, but some, called ribozymes, are RNAs.
Reaction coupling	Coupling of a highly favorable reaction (e.g., the hydrolysis of ATP to AMP plus inorganic phosphate) to drive an energetically unfavorable reaction
Energy harvesting by redox reactions Organic substrates Inorganic substrates Photochemical reactions	Oxidation (the removal of electrons, alone or with protons, from organic or inorganic molecules) is used to harness energy in a metabolically usable form, such as a transmembrane proton gradient or ATP. Alternatively, the energy of sunlight can be used to raise electrons to a higher energy level, from which the electrons can generate a transmembrane proton gradient or ATP by returning to their original energy level.
Use of membranes to form gradients of charge and chemical concentration	Biological membranes make it possible to transduce energy, whether harvested by chemical or photochemical oxidation, into metabolically useful forms. Phospholipid membranes of *Bacteria* and isoprenoid membranes of *Archaea*, and the special proteins they contain, generate ionic gradients from the harvested energy, and these in turn are used to form ATP or to perform work directly.

iii. <u>Energetically coupled reactions</u>: Cells rely on a design strategy called **reaction coupling** that links energetically favorable to unfavorable reactions to drive essential chemical processes. The process transfers energy from favorable to unfavorable reactions in the high-energy bonds of molecules, such as ATP. In this manner, high-energy phosphate bonds are made in favorable reactions that, once hydrolyzed, release the energy needed to power unfavorable reactions.

iv. <u>Transduction of energy from transmembrane ion gradients into ATP</u>: Many cells use biological membranes to transduce energy harvested from oxidation reactions to generate ion gradients, which can be used directly to perform work (such as turning flagella) or to generate the universal energy currency, ATP. (Think of mitochondria, the powerhouse of eukaryotic cells.)

All together, enzyme catalysis, energy harvesting from chemical molecules or light, energetically coupled reactions, and energy transducing into ATP *distinguish life from nonlife*. These processes equip organisms to create order out of disorder. Metabolism is the sum of all the chemical processes that living systems use to achieve order. Many (though by no means all) of the several thousand reactions in a microbial cell directly contribute to the formation of new cells. This ensemble of reactions is what we call growth metabolism. We shall focus on the growth metabolism of microbes in this and the next three chapters. The nongrowth aspects of metabolism contribute to such vital cellular activities as maintenance of intracellular metabolite pools and turgor, repair of cellular structures, secretion, motility, and other responses to environmental stress. We call these activities **maintenance metabolism**.

This chapter passes through some general biology and biochemistry territory because *many details of metabolism are common to all organisms*. Not only do we all share a microbial ancestry, microbes are also easier to study. *Bacteria* and the eukaryotic laboratory powerhouses, the yeasts (chapter 15), are especially easy to grow, manipulate genetically, label with isotopes, and extract for biochemical analysis. Consequently, most of the pioneering studies of metabolic pathways were carried out with microbes. Not surprisingly, much of what we know of metabolism (particularly biosynthesis) in plants and animals has been learned through research on microbes. And so, let us begin . . .

HOW DO CELLS METABOLIZE SUBSTRATES TO GROW?

LEARNING OUTCOMES:
- List two reasons why *E. coli* is a model organism for growth metabolism.
- List the two most common elements in a typical bacterium.
- List the two most abundant macromolecules in a typical bacterium.
- Explain why all cells need each of the four basic building blocks, regardless of what that cell is using as an energy or carbon source.
- List four cell functions that require energy.
- Explain why obligate parasites often have relatively small genomes.
- Speculate why the vast majority of a cell's reducing power is used for protein synthesis.
- The broad processes of metabolism (fueling, biosynthesis, polymerization, assembly, and cell division) are interrelated by the fact that each requires starting materials that are the products of the preceding step. Identify the starting materials for each phase of metabolism.
- Speculate which processes of metabolism (fueling, biosynthesis, polymerization, assembly, and cell division) could be dispensable under certain circumstances (Hint: consider the effect of nutrition).
- Use the flowchart in Fig. 5.1 to explain how the metabolism of a cell would change if it was growing on different media (for example, a medium of simple salts and glucose compared to a rich medium with a mixture of peptides called tryptone, and yeast extract).

As we learned in chapter 1, microbes have diversified in many different ways throughout their 4-billion-year-long journey on planet Earth, yet almost all of them have maintained a relatively small size. In addition to promoting chemical exchange with their surroundings, the streamlined structural design of most microbial cells (chapters 2 and 3) is well-suited for speedy reproduction and rapid metabolism. Indeed, compared to plant or animal cells, many microbes have metabolic rates that are faster by at least an order of magnitude. Some microbes metabolize at such great rates that the cells double in just a few minutes! Such efficiency contributes to the rapid progression of some bacterial infections, like pneumonia caused by *Streptococcus pneumoniae* and necrotizing fasciitis due to *Streptococcus pyogenes*, the flesh-eating bacterium.

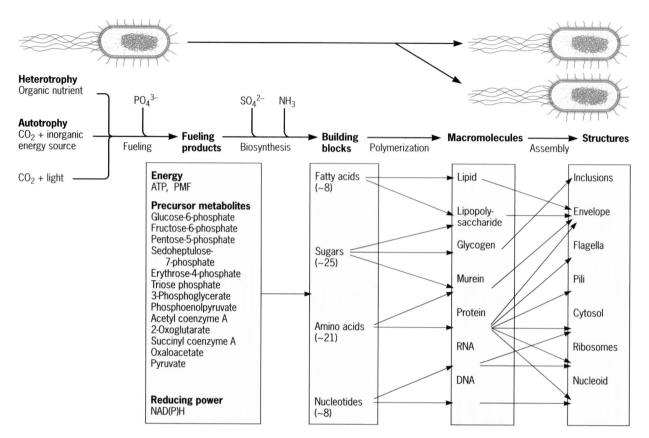

FIGURE 5.1 Framework of bacterial growth metabolism leading to the production of two cells from one. The diagram illustrates the biochemical flow that converts organic substrates (heterotrophy) or CO_2 (autotrophy) into the structures of a bacterial cell through the sequential processes of fueling, biosynthesis, polymerization, and assembly.

Reproduction is the triumph of a large number of chemical reactions. We can now estimate how many reactions contribute to reproduction, thanks to the biochemical knowledge refined over the past half-century, augmented by present-day computational analysis of the microbial genomes. Although the genomes of thousands of microbes are now being analyzed, the fundamental biochemical, genetic, and physiological knowledge gathered for *Escherichia coli* far exceeds that for any other organism. Hence, we shall rely on this bacterium as our guide here and elsewhere in the book. This *coli*-centric view is also justified by the fact that, in genome size and metabolic complexity, *E. coli* is rather average. Nearly half of the over 4,200 genes in the genome of some laboratory strains encode enzymes whose functions are known (Table 5.2). Many are engaged in converting glucose and inorganic salts into protoplasm, i.e., the cytoplasm and its components.

Growth Metabolism Made Simple

The list of known metabolic genes and the enzymes they encode is impressive . . . and overwhelming! As a practical framework, we offer a flowchart that begins with foods, proceeds through each step of growth metabolism, and ends with the cell dividing into two new cells (Fig. 5.1). Reaction flowcharts may be familiar from your chemistry classes, but this one is different. In organic chemistry, different starting materials typically

TABLE 5.2 **Gene products of *E. coli* associated with various metabolic processes**

Functional category	No. of genes
Metabolism of small molecules	
Degradation and energy metabolism	316
Central intermediary metabolism	78
Broad regulatory function	51
Biosynthesis	
Amino acids and polyamines	60
Purines, pyrimidines, nucleosides, and nucleotides	98
Fatty acids	26
Metabolism of macromolecules	
Synthesis and modification	406
Degradation	69
Cell envelopes	168
Cell processes	
Transport	253
Other; e.g., cell division, chemotaxis, mobility, osmotic adaptation, detoxification, and cell killing	118
Miscellaneous	107
Total	1,894

yield different products. But remember: microbes have diversified their metabolism to process specific nutrients available in their environment, yet all lead to a conserved internal "order" and chemistry. Just look at the flowchart (Fig. 5.1): you can begin with any one of several starting materials and still end up with essentially the same products. For example, take a **heterotroph**, a type of microbe that uses organic carbon sources. It matters not whether its diet is glucose, succinate, maltose, or any other usable organic carbon substrate; the end product is always the same cell, whatever its unique make-up.

Another fundamental truth is that microbial cells couple chemical reactions to drive the essential processes that culminate in growth and reproduction. Hence, production of a living cell depends not only on the chemistry of the metabolic reactions but also on the cooperative interplay of those reactions (Fig. 5.1). For simplicity, our flowchart omits the myriad feedback loops and other control devices for reactions that, operating in orderly unison, produce a living cell. The chart would never flow to its conclusion without these controlling forces. How these hundreds of reactions are *regulated* to ensure they function cooperatively in a grand synthesis— the everyday miracle of making a new cell—will be discussed in chapter 11.

Steps in Growth Metabolism, in Reverse

The framework in Fig. 5.1 organizes the myriad reactions of growth metabolism. Now, to bring cell metabolism to life, we shall begin at the end. Starting with the triumphant new cell located at the right side of the flowchart, we will proceed stepwise leftward, back to the food that sustains this organic synthesis. By proceeding backwards, each step will illuminate the tasks required of the preceding step.

Making two from one. Once a cell has successfully doubled in mass and size, having made every cellular compound and structure, it must divide into two living units. You learned about the process of cell division, usually by binary fission, in chapter 4. The key point here is that, because the daughters are each clones of their mother, all the maternal components must first be duplicated. That's where growth metabolism comes into play.

Assembling cell structures. To sustain its lineage, each new cell must have the same structures as its progenitor. Each daughter must inherit (i) a complex envelope, adorned in some cases with a flagellar apparatus and other appendages; (ii) a nucleoid; and (iii) a cytoplasm rich in enzymatic machinery and **polyribosomes** (multiple ribosomes bound to individual molecules of messenger RNA, mRNA). The reactions that assemble macromolecules into each of the cell's structures are discussed in chapters 8 and 9. In some cases (i.e., ribosomes or flagellar filaments), structures self-assemble as if by condensation. In others (i.e., outer membrane of Gram-negative bacteria), more elaborate processes have evolved to build the structure. Some structures are assembled by enzyme-catalyzed reactions; others are not. In addition, some cellular components must be translocated from their points of manufacture to their ultimate locations (a focus of chapter 9).

Making macromolecules. Using our knowledge of microbial structure (chapters 2 and 3), we can list the chemical composition of the major cellular structures: nucleic acids (DNA and RNA), proteins, carbohydrates, and lipids, plus hybrids of these, such as lipoproteins, lipopolysaccharides, and murein (Fig. 5.2). The approximate quantity of each of these

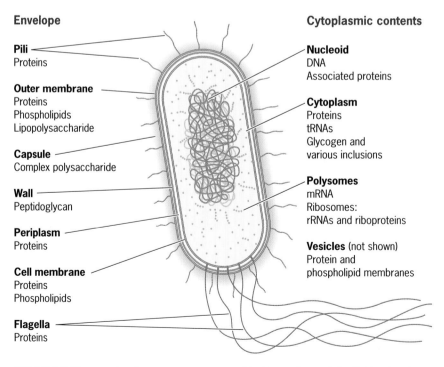

Envelope

Pili
Proteins

Outer membrane
Proteins
Phospholipids
Lipopolysaccharide

Capsule
Complex polysaccharide

Wall
Peptidoglycan

Periplasm
Proteins

Cell membrane
Proteins
Phospholipids

Flagella
Proteins

Cytoplasmic contents

Nucleoid
DNA
Associated proteins

Cytoplasm
Proteins
tRNAs
Glycogen and
various inclusions

Polysomes
mRNA
Ribosomes:
rRNAs and riboproteins

Vesicles (not shown)
Protein and
phospholipid membranes

FIGURE 5.2 Macromolecular components assembled into a bacterial cell. The distribution of macromolecules forming each of the major cellular structures or compartments is indicated, using a Gram-negative cell as an example.

components, in one strain of *E. coli* growing in a well-defined environment, are summarized in Table 5.3 and the case study pie charts. (We use *E. coli* as the example simply because more is known about it.) Realize that the quantities of these macromolecular components will vary under different growth conditions and also among different microbes. Indeed, the outlier *Archaea* lack a peptidoglycan (murein) cell wall and make related compounds. But all cells make their macromolecules from **building blocks**: proteins from amino acids, nucleic acids from nucleotides, lipids from fatty acids, polysaccharides from simple sugars, and a score of other related compounds (Fig. 5.1). (Note that although lipids are included with the macromolecules, and fatty acids with the monomeric building blocks, fatty acids are not simply polymerized to make lipids the way that amino acids and nucleotides form chains of proteins and nucleic acids, respectively.) Macromolecular synthesis requires large amounts of energy. To incorporate an amino acid into a growing polypeptide chain, the reaction consumes four high-energy phosphate bonds in the form of a molecule called guanosine triphosphate or GTP. The addition of a nucleotide to a growing nucleic acid chain expends two high-energy bonds in ATP per residue. In addition, energy is needed to proofread, correct errors, modify the completed macromolecule chains, fold proteins into their mature forms, and fold and coil the DNA.

Synthesizing building blocks. Since the building blocks for macromolecular synthesis are not available from the environment, microbes must rely on biosynthetic reactions to make them *de novo*. What is amazing is that all of their biosynthetic pathways begin with one or more of just *13 compounds*, the common **precursor metabolites** (Fig. 5.1). Typically, building blocks are larger and more complex than their precursor metabolites. They are also more reduced than the substrates used to make them. Indeed, cells consume large amounts of energy and reduced molecules, such as nicotinamide adenine dinucleotide phosphate, or NADPH, to make precursor metabolites and building blocks. As you can see, biosynthesis is expensive. Would you guess that microbes that inhabit nutrient-rich environments, including those that live in or on animal hosts, have lost much of their biosynthetic capacity during evolution? *Yes!* In a competitive world, why waste energy carrying genes that make unnecessary, costly enzymes?

Fueling reactions. Getting the precursor metabolites, energy, and reducing power the cell needs for biosynthesis is the work of the **fueling reactions** (Fig. 5.1). However, the energy demands of the cell go well beyond the manufacture of building blocks. Even a nongrowing cell is engaged in myriad other activities that require energy (Table 5.4). The quantities of energy needed for these activities vary with environmental conditions and from organism to organism, but they can be significant. For cells in a favorable environment, growth exerts, by far, the greatest demand for energy. The amount of energy and reducing power needed for growth (mostly in the form of ATP and NADH or NADPH, respectively) can be calculated once you know (i) the composition of a cell under specified growth conditions and (ii) details of its pathways of biosynthesis and

TABLE 5.3 Overall composition of an average *E. coli* cell

Substance	% of total dry wt.
Macromolecules	
Protein	55.0
RNA	20.4
23S RNA	10.6
16S RNA	5.5
5S RNA	0.4
Transfer RNA (4S)	2.9
Messenger RNA	0.8
Miscellaneous small RNAs	0.2
Phospholipid	9.1
Lipopolysaccharide	3.4
DNA	3.1
Peptidoglycan	2.5
Glycogen and other storage material	2.5
Total macromolecules	**96.1**
Small molecules	
Metabolites, building blocks, vitamins, etc.	2.9
Inorganic ions	1.0
Total small molecules	**3.9**

TABLE 5.4 Some cellular activities requiring energy

Cellular activity
Growth related
Entry of nutrients
Biosynthesis of building blocks
Polymerization of macromolecules
Modification and transport of macromolecules
Assembly of cell structures
Cell division
Growth independent
Motility
Secretion of proteins and other substances
Maintenance of metabolite pools
Maintenance of turgor pressure
Maintenance of cellular pH
Repair of cell structures
Sensing the surroundings
Communication among cells

TABLE 5.5 **Energy and reducing power used in making macromolecules**

Macromolecular component to be made from glucose	Energy cost (mol of ~P[a]/g of cells)	Reducing power cost (mol of NADPH/g of cells)
Protein		
Synthesis of amino acids	7,287	
Polymerization, mRNA, other steps	21,970	
Total	29,257	11,253
RNA		
Synthesis of ribonucleotides	6,540	
Polymerization, modification, other steps	256	
Total	6,796	427
Phospholipid		
Total	2,578	5,270
DNA		
Synthesis of deoxyribonucleotides	1,090	
Polymerization, methylation, coiling	137	
Total	1,227	200
Lipopolysaccharide		
Synthesis of nucleotide conjugates	470	
Total	470	564
Murein		
Synthesis of amino acids and sugars	248	
Cross-linking	138	
Total	386	193
Glycogen		
Synthesis of ADP-glucose	154	
Total	154	

[a]~P, high-energy phosphate bonds.

polymerization (Table 5.5). How do microbes obtain such a huge amount of chemical energy? The cells derive all their energy from chemical (organic or inorganic) or photochemical redox reactions. Animals and fungi use exclusively organic redox reactions, plants use photochemical redox reactions, and protists use both organic and photochemical redox reactions. Thanks to their long life on the planet, prokaryotic microbes have evolved myriad strategies that exploit all three of these reactions. Through their fueling reactions, living organisms seem to create order. They do not increase order—or decrease entropy—to make themselves: rather, they use ingenious strategies to *concentrate energy from their surroundings*. Thus, their increased order does not defy the second law of thermodynamics; instead, it reveals how chemical and physical energy can be gathered from the environment to achieve the organized concentration of energy we call life.

HOW IS GROWTH FUELED IN *BACTERIA* AND *ARCHAEA*?

LEARNING OUTCOMES:
- Compare and contrast each of the following, with respect to source of energy and carbon: chemoheterotroph, photoheterotroph, chemoautotroph, and photoautotroph.
- Identify an example species that is a (i) chemoheterotroph and (ii) photoautotroph.
- Describe how the habitats of chemoheterotrophs and photoautotrophs might differ.
- Describe three processes that reducing power fuels in all life, regardless of energy source.

Cell division, the assembly of cell structures, the polymerization of macromolecules, and the biosynthesis of building blocks are all fundamentally the same in all prokaryotes. Not so for fueling. How microbes gather energy from the organic and inorganic parts of our planet will be explored in chapter 6; here we will consider some key aspects of the fueling reactions in prokaryotes. *Bacteria* and *Archaea* display a variety and versatility not seen in the rest of the living world. Skeptical? Consider two points: (i) as their sole carbon source, prokaryotes as a group can use inorganic carbon (mainly carbon dioxide) or practically any organic compound on Earth (save for a few man-made plastics), and (ii) they can obtain energy and reducing power by oxidizing either inorganic or organic compounds or by harvesting energy from sunlight. Some of these fueling modes are unexpected and even seem weird if your view is limited to the metabolic activities of plants and animals. But it should not surprise us: microbes have lived in this planet for *billions* of years—plenty of time to learn all the metabolic tricks needed to colonize every niche on Earth.

So diverse is fueling in the prokaryotic world that people have devised terms to group prokaryotes based on their carbon and energy sources (Fig. 5.3). In this metabolic classification, the first label indicates the *nature of the carbon source*. Recall that microbes that obtain their carbon from organic compounds are called heterotrophs, and those using inorganic sources of carbon (such as carbon dioxide) **autotrophs**. (These terms also apply to plants, animals, fungi, and protists.) The second distinction is the *source of energy and reducing power*. If chemical sources are used, the microbes are said to be **chemotrophs**, and the prefix "chemo-" is added to heterotroph or autotroph. If light is used, the microbes are said to be **phototrophs**, and the prefix "photo-" is added.

Things get more complicated when one considers the diversity of chemical reactions that chemotrophs can use to harness energy for growth metabolism. Chemoautotrophs, for example, utilize inorganic molecules (including H_2, CO, NH_3,

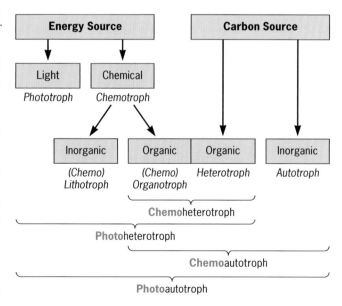

FIGURE 5.3 **Classification of microbes based on the energy and carbon sources they use to fuel their growth metabolism.**

TABLE 5.6 **Patterns of fueling reactions among microbes**

Class	Source of carbon	Source of energy
Heterotrophs[a]		
Chemoheterotrophs	Organic compound	Organic compound
Photoheterotrophs	Organic compound	Light
Autotrophs		
Chemoautotrophs[b]	CO_2	Inorganic compound
Photoautotrophs	CO_2	Light

[a]Organotrophs.
[b]Lithotrophs.

NO_2, H_2S, S, $S_2O_3^{2-}$, and Fe^{2+}) as their source of energy and reducing power. Because of the inorganic nature of the chemicals used as energy sources, these autotrophs are also called **lithotrophs** (stone eaters). In fact, it is quite common to employ the terms "lithotroph" and "autotroph" as synonyms. To make it more confusing, some microbiologists today are replacing "heterotroph" with "organotroph," because the organic chemical used for energy is also the carbon source. All these definitions are summarized in Table 5.6 and illustrated in Fig. 5.3. This new language may seem cumbersome, but do not worry. In chapter 6 we will review these terms as we describe some of the unique fueling reactions of each class of microbes.

Regardless of the type of metabolism, heterotrophic or autotrophic, fueling in prokaryotes begins with the *entry of the carbon and energy source*s and is followed by "processing" to make the precursor metabolites (Fig. 5.1). In many cases, entry requires that cells expend energy to concentrate certain nutrients that are dilute in the environment. Whatever the food source and type of metabolism, ultimately the food substrate is converted into the 13 precursor metabolites (unless the core precursor can be scavenged directly from the cell's environment). Some of the same enzymatic pathways that convert food substrates into the precursor metabolites also generate energy, thereby producing ATP. Ultimately, these conversions are achieved by the central pathways of metabolism (chapter 7).

Typically, cells convert energy into a useful form by the *oxidation of chemicals*, whether they are inorganic (e.g., molecular hydrogen) or organic (e.g., glucose, acetate). During oxidation, electrons are removed from the inorganic compounds or hydrogen atoms (protons) are pulled from the organic substrates. This is why *precursor metabolites are generally more reduced than the starting substrates* and why reducing power is often generated in the process. Microbes use some of the reducing power generated in the fueling pathways to synthesize building blocks. Prokaryotes, as well as all other living things, also use the reducing power generated in the oxidation reactions to *export protons* across the membrane and create an **electrochemical gradient** (AKA **proton motive force**), which they use to generate ATP. To establish this gradient, every time a proton is exported across the membrane, the matching electron is passed through dedicated electron carriers within the cell membrane and toward an electron acceptor, such as oxygen, nitrate, sulfate, and Fe^{3+} compounds, to name a few. We humans use oxygen, but many microbes living in oxygen-free environments "breathe" other chemical compounds. These and other modes of energy transduction, including harvesting energy from light, will be examined in chapter 6.

We close this introductory chapter on metabolism with another truth: metabolic prowess dictates the environment(s) microbes can inhabit. Organisms that rely on photochemistry to obtain energy live in the light; those equipped for oxidation of inorganic forms of sulfur, iron, or other elements live where those elements abound and so on. Environmental resources, fueling reactions, and microbial life style are inseparable.

CASE REVISITED: How do we search for life in other planets?

The Mars Rover Curiosity is part of NASA's Mars Science Laboratory, a robotic program that seeks to determine whether Mars ever had conditions suitable for habitability. Among the science objectives of the mission are three related to biology:

- The rover is analyzing the type of organic carbon compounds on Mars.
- It is also performing an inventory of the chemical building blocks of life by analyzing the elemental composition of samples of interest (C, H, N, O, P, S).
- It is also looking for biological markers; i.e., signatures of past microbial activities that still remain in the environment.

So, if you hypothesized that the Mars Rover Curiosity should look for specific chemical compounds or elements that are important to life, you were right! Since macromolecules degrade easily, they are poor targets for study. Building blocks, on the other hand, are more resilient and likely to endure the passage of time. For instance, the RNA macromolecule is fragile and degrades easily, but the building blocks that make it (ribonucleotides) would probably survive after the polymer degraded. For this reason, the rover is making an inventory of all of the chemical building blocks by measuring their unique elemental composition and ratios. Curiosity has an ambitious goal: to collect, grind, and analyze approximately 70 samples of soil and rock from the red planet. To do this, the rover is equipped with an instrument called SAM (Sample Analysis at Mars) which separates compounds by mass and measures their elemental composition and relative abundance. As you probably have already guessed, the SAM can measure the elements associated with life, such as C, H, O, and N. The instrument can also analyze the carbon and oxygen isotopes from carbon dioxide (CO_2) and methane (CH_4) gases trapped in the soil and rock samples. Because biological processes, unlike abiotic reactions, are selective about the isotopes they use for biosynthesis, a particular isotopic ratio of some elements can reveal the microbial origin of the gases.

CONCLUSIONS

This chapter laid out a conceptual framework to aid our understanding of how thousands of individual metabolic reactions allow a cell to grow and reproduce. But one must not come away with the notion that all microbes have identical metabolic activities and capabilities. Far from it: microbes are successful colonizers of Earth precisely because each group has evolved specialized metabolic programs to deal with specific environmental circumstances. The startling array of fueling reactions, for example, is possible because of the enormous variety of often quite specialized microbes. Yet, despite the staggering metabolic versatility of microbes, the principles that define microbial life share universal features that are also conserved with all other organisms, a testament of our shared ancestry to a now-extinct microbe.

The next four chapters will follow the route from food to a new living cell. We will not delve into the biochemical details of the individual reactions and pathways; many of them can be viewed on the EcoCyc website. Instead, we will emphasize the general features and special principles of each stage of metabolism: fueling, biosynthesis, polymerization, and assembly. Our focus will be on the unique prokaryotic way of harvesting and concentrating energy to create life. In the end, we hope that you, too, will admire the essential biochemical transformations that these dominating, but largely invisible, creatures contribute to the Earth's biosphere.

Representative ASM Fundamental Statements

EVOLUTION

1. Cell, organelles (e.g., mitochondria and chloroplasts), and all major metabolic pathways evolved from early prokaryotic cells.

   a. List four features that are unique to living cells, compared with inert chemicals.
   b. Explain how each of the overarching biochemical principles of life contributes to a cell's ability to reproduce.
   c. Explain the connection between the ability to reproduce and entropy.
   d. List the two most common elements in a typical bacterium.
   e. List the two most abundant macromolecules in a typical bacterium.
   f. List four cell functions that require energy.
   g. Describe three processes that reducing power fuels in all life, regardless of energy source.
   h. Explain why all cells need each of the four basic building blocks, regardless of what that cell is using as an energy or carbon source.

METABOLIC PATHWAYS

2. *Bacteria* and *Archaea* exhibit extensive, and often unique, metabolic diversity (e.g., nitrogen fixation, methane production, anoxygenic photosynthesis).

   a. Compare and contrast the following, with respect to source of reducing power and habitat: chemoheterotroph, photoheterotroph, chemoautotroph, photoautotroph.

3. The survival and growth of any microorganism in a given environment depends on its metabolic characteristics.

   a. Distinguish between growth metabolism and nongrowth metabolism.
   b. Speculate why the vast majority of a cell's reducing power is used for protein synthesis.
   c. The broad processes of metabolism (fueling, biosynthesis, polymerization, assembly, and cell division) are interrelated by the fact that each requires starting materials that are the products of the preceding step. Identify the starting products for each phase of metabolism.
   d. Speculate which processes of metabolism could be dispensable under certain circumstances. (Hint: consider the effect of nutrition.)

e. Use Fig. 5.1 to explain how the metabolism of a cell would change if it was growing on different media (for example, a medium of simple salts and glucose compared with a rich medium of tryptone and yeast extract).

INFORMATION FLOW AND GENETICS

4. Genetic variations can affectc microbial functions.

   a. Explain why obligate parasites often have relatively small genomes.

MICROBIAL SYSTEMS

5. Microorganisms are ubiquitous and live in diverse and dynamic ecosystems.

   a. Identify an example species that is a chemoheterotroph and one that is a photo-autotroph.

   b. Describe how the habitats of chemoheterotrophs and photoautotrophs might differ.

IMPACT OF MICROORGANISMS

6. Microbes are essential for life as we know it and the processes that support life (e.g., in biogeochemical cycles and plant and/or animal microbiota).

   a. Speculate about a different life form (on an imaginary planet) that might not satisfy one or more of the four principles of chemical processes known to all Earth life.

7. Microorganisms provide essential models that give us fundamental knowledge about life processes.

   a. List two reasons why *E. coli* is a model organism for growth metabolism.

SUPPLEMENTAL MATERIAL

References

- **Chen J, Gomez JA, Höffner K, Phalak P, Barton PI, Henson MA.** 2016. Spatiotemporal modeling of microbial metabolism. *BMC Syst Biol* **10:**21. doi: 10.1186/s12918-016-0259-2.
- **Des Marais DJ, Nuth JA III, Allamandola LJ, Boss AP, Farmer JD, Hoehler TM, Jakosky BM, Meadows VS, Pohorille A, Runnegar B, Spormann AM.** 2008. The NASA Astrobiology Roadmap. *Astrobiology* **8:**715–30. doi: 10.1089/ast.2008.0819.
- **Schönheit P, Buckel W, Martin WF.** 2015. On the origin of heterotrophy. *Trends Microbiol* **24:**12–25. doi: 10.1016/j.tim.2015.10.003.
- **Stephanopoulos G.** 2012. Synthetic biology and metabolic engineering. *ACS Synth Biol* **1:**514–25. doi: 10.1021/sb300094q.

Supplemental Activities

CHECK YOUR UNDERSTANDING

1. **Methanogens are a group of anaerobic *Archaea* that gain energy for growth by using hydrogen to reduce carbon dioxide and generate methane. How would you classify this organism?**
   a. Chemoheterotroph
   b. Chemoautotroph
   c. Photoheterotroph
   d. Photoautotroph

2. **Order the following macromolecules according to the energy a typical prokaryotic cell consumes for its biosynthesis, from highest to lowest:**
   a. DNA
   b. RNA
   c. Proteins
   d. Glycogen
   e. Phospholipid

3. **In the redox fueling reaction, Succinate + NAD ↔ Fumarate + NADPH, the succinate molecule is undergoing which process?**
   a. Reduction
   b. Catalysis
   c. Oxidation
   d. Protein synthesis
   e. Glycolysis

DIG DEEPER

1. Identify an example of one of the following processes and design visual aids to teach how the pathway captures energy that supports cellular reproduction.

   a. enzyme catalysis
   b. energetically coupled reactions
   c. harvesting energy from redox reactions
   d. energy transduction using membranes

2. Use Table 5.5 to estimate how much energy, in the form of ATP and NADPH, is required for a single *E. coli* cell to replicate. Express your answer in Joules and assume that: 1×10^{12} *E. coli* cells weigh 1 gram, the hydrolysis of ATP releases 32 J/mol, and the oxidation of NADPH releases 21 J/mol.

3. *Chlamydomonas reinhardtii* is a eukaryotic unicellular flagellate that can grow photoautotrophically in mineral medium or chemoheterotrophically in the same medium supplemented with acetate. These low nutritional requirements and versatile energy metabolism equip it to grow in diverse environments, such as soil, freshwater, seawater, and even in snow. In addition, this eukaryotic microbe is easy to grow in the laboratory due to its minimalistic diet, short generation times, and amenability to genetic manipulation.

   a. Design an experiment that compares the chemotrophic and phototrophic growth of the microbe using laboratory mineral medium. What culturing conditions will be needed for each?

   b. Identify any controls you will need.

   c. How many replicates would you need for each experimental group (treatment and controls)?

   d. Write a general experimental protocol that considers detailed culturing parameters (such as culture vessels and volume, inoculum size, incubation temperature, and light exposure) and method used to monitor growth in the cultures.

   e. Draw a hypothetical graph of your results illustrating growth in the photoautotrophic and chemoautotrophic cultures and the controls.

4. The prokaryotic world can be grouped into different classes according to their fueling strategy. Draw an organizational chart that first classifies microbes based on the carbon source and then on the energy source used to support their growth. Include all the names used to refer to each microbial group (note than some groups can be designated by different names).

5. Synthetic biology, a field that aims to create new biological molecules and possibly novel life forms, is an exciting new area of research that is beginning to contribute new knowledge about some of life's fundamental properties and processes. Read the following press release about a paper in *Nature* by Taylor et al. (Catalysts from synthetic genetic polymers, *Nature* **518**:427–430, published online December 1, 2014): http://www.independent.co.uk/news/science/major-synthetic-life-breakthrough-as-scientists-make-the-first-artificial-enzymes-9896333.html.

Summarize the press release and address the following questions:
- What did the study contribute to new knowledge of basic biological characteristics?
- How could this new knowledge possibly benefit humans?
- Identify possible implications for extraterrestrial life.
- Discuss possible ethical considerations of creating biomolecules, such as synthetic nucleic acids and enzymes, in a laboratory.

Supplemental Resources

VISUAL

- Khan Academy lecture: "Overview of Metabolism - Anabolism and Catabolism" (8:40 min): https://www.youtube.com/watch?v=ST1UWnenOo0.
- iBiology lecture by Jens Nielsen discusses how to engineer the metabolism of yeasts for the sustainable production of diverse products (23:15 min): http://www.ibiology.org/ibioeducation/metabolic-engineering-and-synthetic-biology-of-yeast.html.
- iBiology lecture by Kristala L. J. Prather on the use of synthetic biology and metabolic engineering to teach an old bacterium (*E. coli*) new tricks (26:11 min): https://www.youtube.com/watch?v=ndThuqVumAk.

WRITTEN

- Mars Explorations Rovers mission: http://mars.nasa.gov/mer/overview/.
- *National Center for Case Study Teaching in Science* "ELVIS Meltdown!" by Richard Steward, Ann Smith, and Patricia Shields teaches the fundamental concepts of culture, growth, and metabolism: http://sciencecases.lib.buffalo.edu/cs/collection/detail.asp?case_id=248&id=248.
- "Life in the Universe: An Assessment of U.S. and International Programs in Astrobiology" by the National Research Council (US) Committee on the Origins and Evolution of Life: http://www.nap.edu/catalog/10454/life-in-the-universe-an-assessment-of-us-and-international.
- Moselio Shaechter describes how precipitating light-sensitive mineral particles on a chemotrophic bacterium turned it into a phototroph in the *Small Things Considered* blog post "How to Make a Photosynthetic Bacterium in One Simple Step" (http://schaechter.asmblog.org/schaechter/2016/03/how-to-make-a-photosynthetic-bacterium-in-one-simple-step-.html).
- Rachel and Elie Diner ponder whether a mixotroph, that is, an organism that can grow phototrophically or chemotrophically, requires either or both trophic modes for survival in the *Small Things Considered* blog post "Dine In or Take Out?: The Dilemma of the Mixotroph" (http://schaechter.asmblog.org/schaechter/2015/08/dine-in-or-take-out-the-dilemma-of-the-mixotroph.html).

Molecular model of ATP

CHAPTER SIX

Bioenergetics of Fueling

KEY CONCEPTS

This chapter covers the following topics in the ASM Fundamental Statements.

EVOLUTION

1. Cells, organelles (e.g., mitochondria and chloroplasts), and all major metabolic pathways evolved from early prokaryotic cells.

CELL STRUCTURE AND FUNCTION

2. While microscopic eukaryotes (for example, fungi, protozoa, and algae) carry out some of the same processes as bacteria, many of their cellular properties are fundamentally different.

METABOLIC PATHWAYS

3. *Bacteria* and *Archaea* exhibit extensive, and often unique, metabolic diversity (e.g., nitrogen fixation, methane production, anoxygenic photosynthesis).
4. The survival and growth of any microorganism in a given environment depend on its metabolic characteristics.

MICROBIAL SYSTEMS

5. Microorganisms are ubiquitous and live in diverse and dynamic ecosystems.

IMPACT OF MICROORGANISMS

6. Microbes are essential for life as we know it and the processes that support life (e.g., in biogeochemical cycles and plant and/or animal microbiota).
7. Humans utilize and harness microorganisms and their products.

INTRODUCTION

Like all living beings, microbes need energy to sustain their activities, including locomotion, reproduction, transportation of nutrients, and synthesis of **building blocks** and macromolecules, just to name a few. They harness energy from light or chemical reactions primarily in the energy-rich phosphate bonds of molecules such as adenosine triphosphate (ATP), as well as in the form of reduced molecules that provide so-called **reducing power**. Without mechanisms to capture energy as ATP and reducing power, microbes would not have emerged and diversified, and the Earth would have been sterile and void of life. Why? Envision the power of ATP to concentrate the energy dispersed in the cell's surroundings and put it to use inside the cell to drive energetically unfavorable reactions and to make the necessary **precursor metabolites**, building blocks, and macromolecules. The quest for energy has exerted tremendous pressure on microbes throughout the almost 4 billion years of their existence. And the microbes have been successful; indeed, nothing beats them in this pursuit. They have evolved multiple mechanisms to use every available energy source in the planet. By doing so, microbes have colonized every corner of the Earth, exploiting the available resources and transforming the Earth into a habitable planet for other life forms. In this chapter, we will introduce you to the diverse strategies that microbes employ to gain their precious energy. *Buckle up!* We are about to explore the field of bioenergetics.

CASE: How do microbes harness energy with and without oxygen?

Ms. G, a microbiology student, isolated two bacteria (isolates #1 and #2), each from a different pond sediment. The two isolates grew well aerobically (i.e., with oxygen) at 30°C in a mineral medium containing glucose, as indicated by the cultures rapidly becoming turbid. When she replaced the air in the tube's headspace with dinitrogen gas to grow the isolates under anaerobic conditions she made the following observations:

1. Isolate #1 grew anaerobically in the mineral medium with glucose, and accumulated acetate, lactate, and ethanol.
2. Isolate #2 could only grow anaerobically if nitrate was added to the medium, in which case, it accumulated nitrite.

Questions emerging:

- What type of energy metabolism(s) allowed the two isolates to grow anaerobically?
- Are they chemotrophs or phototrophs? Why?
- Can you make predictions about what types of energy source are available in each of the pond sediments?

WHY DO MICROBES NEED ENERGY?

LEARNING OUTCOMES:
- Support this statement with evidence: "All major metabolic pathways were evolved by the early prokaryotic cells."
- Identify a reason why reducing power is equivalent to energy.
- Compare and contrast energy harvesting during fueling reactions for heterotrophs and autotrophs.

All organisms need energy to function. Consider us humans. We get energy from organic compounds, such as carbohydrates, proteins, and fats, and use it to carry out essential processes such as breathing, pumping blood, maintaining our optimal body temperature, as well as for moving, talking, and many more actions. We talk about calories in the food to estimate their energy content and exercise to "burn excess calories"; otherwise our bodies store energy as fat. In other words, we concentrate energy from our environment inside our cells, spend some in our metabolism, and store any excess in molecules (such as fat) for a rainy day. The capacity to harvest, use, and store energy is a universal feature of all organisms—microbes and non-microbes alike—inherited from our common prokaryotic ancestor (remember LUCA, the Last Universal Common Ancestor, from chapter 1?). Although the strategies to harness energy are extraordinarily diverse, all converge into one: energy is conserved intracellularly in the energy-rich phosphate bonds of molecules such as ATP (Box 6.1). Indeed, the universality of ATP as the energy currency of all

---

BOX 6.1

One Molecule To Rule Them All

What is it about ATP that all living cells, microbial or otherwise, use it as their primary molecule for energy transfer? It is small, but it contains a great amount of energy within the bonds of its three phosphates. Imagine each of the phosphate groups in ATP as a ball of negative charge and the chemical bonds between them as the force that pulls them close together despite their natural repulsion. It takes a lot of force to pull those three negative balls together and keep them bonded. That's how much energy there is in the phosphate bonds of ATP and related molecules. We can quantify the amount of energy needed to make those bonds via the Gibbs free-energy change (ΔG), which we express in Joules (abbreviated J and referring to the energy needed to produce mechanical work) or calories (heat energy). The hydrolysis of the first (most ex-

Energy in bonds

ternal) phosphate bond in ATP generates ADP (adenosine diphosphate) and releases ~31 kJoules of energy per mol of ATP. This is the same amount of energy, in calories, as that contained in two medium-sized strawberries! If the second bond is cleaved, AMP (adenosine monophosphate) and pyrophosphate (PP_i) are generated and ~46 kJ of energy are released (the equivalent of two large strawberries). You may have heard that ATP "stores" energy in the phosphate bonds. Not really, because the ATP pool inside the cell is recycled so fast that the energy is not really stored but is in continuous flux. Energy derived from energy-yielding reactions is "packed" in the phosphate bonds of ATP inside the cell and used to drive energetically unfavorable reactions and to make the cell's building blocks and macromolecules. Hence, ATP is the energy currency of all cells. Furthermore, ATP can also transfer its high-energy phosphate bonds to other molecules such as GTP (guanosine triphosphate) and CTP (cytosine triphosphate), which then carry the precious energy load to drive other unfavorable reactions. Indeed, ATP is the one molecule to rule them all!

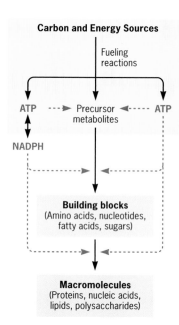

Carbon and Energy Sources

Fueling reactions

ATP ---▶ Precursor ◀---- ATP
metabolites

NADPH

Building blocks
(Amino acids, nucleotides, fatty acids, sugars)

Macromolecules
(Proteins, nucleic acids, lipids, polysaccharides)

FIGURE 6.1 **Steps in the biosynthesis of macromolecules.** The synthesis of essential macromolecules is a stepwise process initiated with fueling reactions that generate precursor metabolites, ATP, and reducing power (NAD(P)H). The products of the fueling stage are then used to make building blocks and, from them, macromolecules: proteins from amino acids, nucleic acids from nucleotides, carbohydrates from sugars, and lipids from fatty acids.

cells speaks volumes about its central role in energy metabolism and hints at an important evolutionary fact: it was already present in LUCA. From there, cells evolved many metabolic networks to harness energy, but all had the common goal of generating ATP.

Energy Harvesting during Fueling

Growth is probably the cellular activity that demands the highest energy investment. To grow, microbes must assimilate chemicals containing the elements (carbon, nitrogen, phosphate, and sulfur, among others) needed to make the cell's 13 precursor metabolites and, from them, the building blocks and essential macromolecules (proteins, nucleic acids, carbohydrates, and lipids). As we learned in chapter 5, these biosynthetic processes proceed in a stepwise fashion and are initiated with the **fueling reactions** that make the precursor metabolites from substrates taken up from the environment (Fig. 6.1). Energy-intensive reactions then combine the precursor metabolites in various ways to make the building blocks needed for the synthesis of the essential macromolecules: proteins from amino acids, nucleic acids from nucleotides, carbohydrates from sugars, and lipids from fatty acids.

Because energy demand for biosynthesis is high, microbes have evolved strategies to harness energy at the fueling stage (the term "fueling" already hints at this!). How? The fueling reactions often move electrons with protons to generate reducing power, a collective name given to the reduced molecules (such as NADH and NADPH) that carry the electron-proton pair (Fig. 6.1). In general, *fueling reactions* produce reducing power in the form of NADH but *biosynthetic pathways largely use reducing power in the form of* NADPH. Thus, cells must convert the NADH produced during fueling into the NADPH needed during biosynthesis; to do so, they use transhydrogenases. These enzymes ensure that there is always sufficient NAD^+ to receive electrons and protons from the fueling reactions and sufficient NADPH for biosynthetic reactions. (This is why the cell keeps its ratio of $NADPH/NADP^+$ higher than its $NADH/NAD^+$ ratio.) As we will learn below, many microbes can regenerate $NAD(P)^+$ from NAD(P)H in reactions catalyzed by dehydrogenase enzymes that ultimately lead to ATP generation. Given their interconvertibility, *ATP and reducing power are often used interchangeably to refer to energy in cell biology.* Collectively, ATP and reducing power are the driving force of life.

Heterotrophic versus Autotrophic Fueling

Energy harvesting is different in **heterotrophs** and **autotrophs**, reflecting the different substrates these microbes use for fueling (Fig. 6.2). Heterotrophs use organic carbon sources which can also be used to extract electrons and harness energy. For this reason, they intertwine their carbon and energy metabolism and simultaneously generate precursor metabolites, ATP, and reducing power (Fig. 6.2). Hence, heterotrophs are primarily **chemotrophs** (Fig. 6.3). Autotrophic fueling, on the other hand, relies on specific entry and feeder pathways to fix carbon dioxide and make the precursor metabolites in energy-consuming reactions. Thus, autotrophs have separate fueling reactions to harness energy from the energy source (chemical reactions or light) and drive carbon assimilation (Fig. 6.2). As a result, autotrophs can be chemotrophs or **phototrophs** (Fig. 6.3).

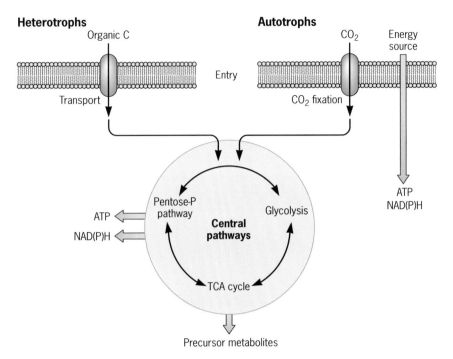

FIGURE 6.2 **Energy harvesting in heterotrophic and autotrophic fueling.** Heterotrophs and autotrophs have different entry pathways to assimilate the carbon source (organic substrates or CO_2, respectively), but both have feeder pathways to assimilate carbon in the central pathways and generate precursor metabolites. As organic carbon substrates are also energy sources, heterotrophs intertwine carbon and energy metabolism and generate ATP and reducing power (NAD(P)H) in the central pathways of fueling. Autotrophs, on the other hand, have separate pathways to harness energy from the energy source (chemical reactions or light) and generate ATP and NAD(P)H.

Despite their differences, all microbes—heterotrophs and autotrophs— use ATP and reducing power to make building blocks, and then macro- molecules, from the precursor metabolites. ATP and reducing power are also critical for a number of other essential cellular processes, such as cor- recting errors during DNA replication, repairing damaged DNA, post- translational modification and folding of proteins, transport, etc. Even cells undergoing starvation and stalled in growth, or those in a dormant state, need a minimum supply of energy to stay alive. Clearly, life would not have emerged without the evolution of mechanisms for energy con- servation and transduction. So diverse is energy metabolism and so impor- tant to the survival and evolution of microbes that it deserves a separate chapter in this book. Let us learn some of the fascinating ways microbes have mastered energy harvesting!

HOW IS ENERGY CONSERVED DURING FUELING?

LEARNING OUTCOMES:

- Compare and contrast ATP generation by transmembrane ion gradients and substrate level phosphorylation with respect to the mechanism of ATP generation and number of ATPs generated.

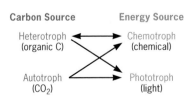

FIGURE 6.3 **Trophic designations based on carbon and energy metabolism.**

- Speculate a reason that life on Earth has sustained two distinct pathways to generate ATP instead of a single pathway.
- Compare and contrast the oxidation and dehydration pathways for fueling reactions.
- Decide whether you would use ΔG, $\Delta G°$, or $\Delta G°'$ to estimate the change in Gibbs free energy of a reaction *in vivo*.
- Predict the mechanism by which pH can alter the Gibbs free energy change.
- Give examples of bioenergetics processes and structures that all living organisms share.
- Explain why it is incorrect to state that energy is "stored" in the cell as ATP and reducing power.
- Discuss the process of fermentation. Summarize the mechanism by which fermentative microbes make valuable chemicals and products for humans.
- Describe the process of ATP generation by F_1F_o ATP synthase.
- List four ways to generate transmembrane ion gradients.

A. Oxidative phosphorylation

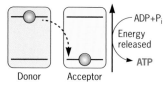

B. Photophosphorylation

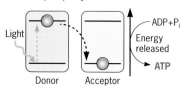

FIGURE 6.4 Harvesting energy from oxidation reactions in chemotrophs versus phototrophs. Chemotrophs and phototrophs alike harness the energy released in oxidation reactions to make ATP. A. In chemotrophs a chemical compound serves as an electron donor for the oxidative reaction (oxidative phosphorylation). B. In phototrophs a light-sensitive molecule is excited with light energy to push an electron to a higher enough energy level so it can serve as the electron donor in the reaction (photophosphorylation).

At the most fundamental level, all energy in living systems is derived from the movement of electrons down an energy gradient. Quoting the Nobel laureate Albert Szent-György: *"Life consists of electrons looking for a place to rest."* To understand the concept of the "restless" electron, remember that electrons can exist at *different energy levels* and that their spontaneous *movement from a higher to a lower energy level releases energy*. Thus, in chemotrophs, the electron moves from a high energy level in a chemical molecule (electron donor) to a lower energy level in another molecule, the electron acceptor (Fig. 6.4). Phototrophs, on the other hand, use light energy (i.e., photons) to excite photosynthetic pigments and move an electron to a higher energy orbital (Fig. 6.4). This "excited" state is unstable: the pigments rapidly dump the "restless" electron onto a series of molecules in a spontaneous, "downhill" energetic process. Whether using oxidation reactions involving chemical substrates (chemotrophs) or light (phototrophs), energy is harvested to phosphorylate ADP and generate ATP. Thus, the terms **oxidative phosphorylation** and **photophosphorylation**, respectively, are often used to designate each energy-harvesting strategy (Fig. 6.4).

Recall that microbes typically use molecules such as NAD^+ and $NADP^+$ to move electrons inside the cell. Because the electron is extracted together with a proton (i.e., they extract a hydrogen atom), the reaction is called a **dehydrogenation** (Fig. 6.5). As a result, the reducing power of the cell is generated: NADH, the primary reduced molecule generated in the fueling reactions, and NADPH, which transhydrogenases make from NADH (Fig. 6.1). Because the reduction of $NAD(P)^+$ to NAD(P)H is reversible, cells can use the reducing power to infuse electrons into organic molecules and thus drive fueling and biosynthetic reactions that would otherwise not occur spontaneously. Indeed, to make beer or wine, sugars released from the cereals or grapes (such as glucose) provide the carbon and energy source during fueling to make precursor metabolites, ATP and NADH; the NADH is then recycled in reactions that produce various compounds, including an alcohol (ethanol). We call this metabolism **fermentation**. (Could this be how isolate #1 from the case study converted glucose into ethanol and other organic acids such as lactate anaerobically?)

How Much Energy Is Needed To Make ATP and Reducing Power?

To understand the energetics of ATP and reducing power generation, first we need to review some basic concepts. We can calculate the amount of "usable" energy released or consumed in any reaction as the change in Gibbs free energy (or ΔG) (to honor the physical chemist Gibbs who worked out the thermodynamics of reactions). The ΔG unit is J/mol, where J is a Joule, a measure of mechanical work that can be performed with that energy, and mol refers to the molar concentration of reactants and products. The ΔG is positive if the reaction *consumes energy* (**endergonic**) and negative if it *releases energy* (**exergonic**). It is important to emphasize that the change in Gibbs free energy indicates the energy associated with a particular reaction, but it does not provide information about the rate of the reaction, which can be increased by enzymes that facilitate interactions between the reactants. Thus, ΔG is always the same, regardless of the speed of the reaction.

However, ΔG is affected by the concentration of substrates and products, and also by temperature, pH, and pressure. These parameters are not always easy to determine inside the cells; instead, scientists often calculate ΔG under controlled, standard conditions (25°C, standard atmospheric pressure [1 atm], and equimolar [1 M] concentrations of all the reactants and products). In some cases a pH of 7 is also used. To make sense of all the variables, scientists use the following designations:

> ΔG = change in Gibbs free energy for a particular reaction under defined conditions (here, the experimenter defines the conditions used, e.g., at 30°C, pH 6, and reactants and products at 2 M concentrations)
>
> ΔG^{o} = ΔG at 25°C, 1 atm, and 1 M concentrations of reactants and products (standard reaction conditions)
>
> $\Delta G^{o\prime}$ = ΔG^{o} at pH 7

Thus, in the reaction that phosphorylates ADP to make ATP:

$$ADP + P_i \rightarrow ATP + H_2O \qquad (\Delta G^{o\prime} = +31 \text{ kJ/mol})$$

The positive change in free energy associated with the reaction tells us that energy is needed to make ATP. The positive value also indicates that the reverse reaction (hydrolyzing ATP to make ADP) would occur spontaneously, as this is the direction for energy release (negative $\Delta G^{o\prime}$). However, in cells with an excess concentration of reactants (ADP and P_i), the reaction would eventually be pushed to the right, making ATP.

How accurately does $\Delta G^{o\prime}$ represent the change in Gibbs free energy of this reaction in live cells? (*Quiz spoiler*: reactant concentrations of 1 M are rarely reached inside microbes because chemicals are in continuous flux and rarely accumulate to these levels.) First, these physicochemical parameters fluctuate greatly in the environment. Furthermore, microbes have diversified their metabolism to operate at different ranges of temperature, pH, and pressure. Perhaps ΔG, if calculated under conditions that mimic those relevant to a particular microbe, is the most reliable estimate.

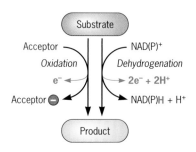

FIGURE 6.5 Oxidation and dehydrogenation reactions during fueling. The diagram shows the two types of reactions that the central pathways of fueling use to extract electrons from the substrate and make the precursor metabolites: oxidation, when an electron is extracted alone and passed to an electron acceptor, and dehydrogenation, when an electron is removed together with a proton and transferred to H-accepting molecules such as NAD or NADP, thus generating reducing power (NADH or NADPH).

Studies of this type suggest that for some microbes, ATP hydrolysis may generate even more energy. But every microbe is different, and each has unique mechanisms to manage its energy budget.

Let us now consider how much energy is needed to make reducing power (e.g., NADH) under standard conditions and a pH of 7 ($\Delta G^{o'}$):

$$NAD^+ + 2H^+ + 2\ e^- \rightarrow NADH + H^+ \qquad (\Delta G^{o'} = +219.25 \text{ kJ/mol})$$

The positive change in free energy that occurs in these reactions tells us that energy is also needed to make NADH (more so than to make ATP [$\Delta G^{o'} = 31$ kJ/mol]). Hence, only those exergonic reactions that release at least 219.25 kJ/mol can be used to make NADH. But then the cell can use the NADH to energize the phosphorylation of ADP and make ATP. In fact, it has sufficient energy to power this conversion nine times! Now you understand why reduced molecules such as NADH are called the cell's *reducing power (powerful indeed!)*.

Energy harvesting as ATP and NADH accomplishes an incredible feat: *energy from the environment has been concentrated inside the cell in these molecules*, and it can be used to infuse energy into just about any other reaction. In fact, the turnover of ATP and reducing power is so fast that energy is not really "stored" but is in continuous transfer. Any excess NAD(P)H not used in fueling reactions and for biosynthesis is recycled to maintain the NAD(P)$^+$ pool. As there is so much energy in NAD(P)H, some microbes have evolved strategies to use it to pump protons across the membrane and then use the transmembrane proton gradient to make ATP. (We will learn more about this neat trick in the next section.) ATP is mostly used for biosynthesis. Every nucleotide that is incorporated during DNA replication requires the energy generated from the hydrolysis of two of its bonds ($\Delta G^{o'} = -46$ kJ/mol). ATP is also used to make other high-energy molecules. Hydrolyze it to ADP, and you generate sufficient energy to make guanosine triphosphate (GTP) from the diphosphate form (GDP) and power protein synthesis. Ditto for the synthesis of cytosine triphosphate (CTP), which cells use to synthesize membrane **phospholipids**. And so energy is conserved as ATP and rapidly reused inside the cell. Remember the key point: ATP is the energy molecule that rules them all!

What Tricks Do Microbes Use To Make ATP?

All organisms, microbes and non-microbes alike, make ATP via one of two mechanisms: **substrate level phosphorylation** and **transmembrane ion gradients** (Fig. 6.6). These processes are highly conserved, reinforcing the microbial ancestry of all extant organisms and ATP as the universal currency of cellular energy. While sharing these general strategies, prokaryotes stand out for the remarkable creativity their two energy-harvesting processes acquired through evolution. Their diversity of energy-yielding reactions accounts for the astonishing ability of prokaryotes to exploit all of the Earth's life-sustaining habitats. Interestingly, some of these reactions, taken singly, yield only minuscule quantities of energy, far below that needed to make ATP. But collectively, they sustain life. Next we discuss how this is accomplished.

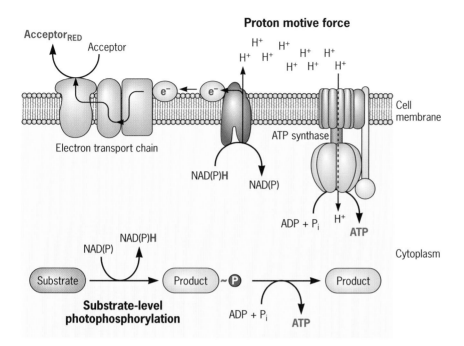

FIGURE 6.6 Modes of ATP generation via transmembrane ion gradients or substrate level phosphorylation. The diagram shows the two modes of ATP generation by substrate level phosphorylation or using transmembrane ion gradient (a proton motive force scheme is shown). In substrate level phosphorylation, reducing power can be generated in oxidation reactions of a phosphorylated substrate, which converts the low-energy phosphoryl bond into a high-energy phosphoryl bond that can be transferred to ADP to form ATP. The reducing power generated during fueling can be used to transport electrons down an energy gradient (electron transport chain) and to an electron acceptor, which gets reduced (RED). Protons are simultaneously pumped across the membrane to generate a proton gradient (proton motive force), which the ATP synthase uses to harness energy in the form of ATP.

Substrate level phosphorylation. The simplest and neatest way to generate ATP is substrate level phosphorylation. At first sight, it may seem a strange name for a process that generates ATP from ADP, but closer inspection reveals the rationale (Fig. 6.6). In this process, an organic substrate first becomes phosphorylated with inorganic phosphate (PO_4^{3-}) in a reaction that requires no energy input. The resulting phosphoryl group is not attached to the substrate by a high-energy bond yet; the crucial step comes next. The phosphorylated substrate is dehydrated or oxidized, and the energy released in the reaction is trapped in the phosphate bond. Thus, a low-energy phosphoryl bond is transformed to a *high-energy phosphoryl bond* that can be transferred to ADP to form ATP. To emphasize that the phosphorylation of ADP to make ATP uses a metabolic substrate, the reaction is termed substrate level phosphorylation.

A classic example of substrate level phosphorylation is found in energy harvesting in one of the most conserved fueling pathways, glycolysis (Fig. 6.7). In this process, 3-phosphoglyceraldehyde is phosphorylated by the enzyme phosphoglyceraldehyde dehydrogenase in a dehydrogenation reaction that generates NADH+H+. The enzyme phosphoglycerate kinase then hydrolyzes the high-energy phosphoryl bond

A

Phosphorylation/oxidation step

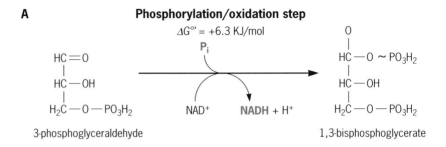

$\Delta G^{\circ\prime} = +6.3$ KJ/mol

3-phosphoglyceraldehyde   1,3-bisphosphoglycerate

B

ATP synthesis step

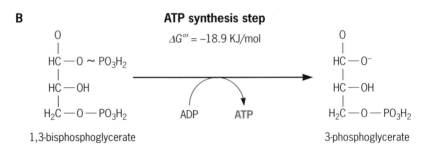

$\Delta G^{\circ\prime} = -18.9$ KJ/mol

1,3-bisphosphoglycerate   3-phosphoglycerate

FIGURE 6.7 **Substrate level phosphorylation during glycolysis.** The synthesis of 3-phosphoglycerate from 3-phosphoglyceraldehyde includes (i) a phosphorylation and dehydrogenation step in which a substrate with a high-energy phosphate bond is formed (1,3-biphosphoglycerate), and (ii) the phosphorylation of ADP to make ATP and the 3-phosphoglycerate product.

in 1,3-biphosphoglycerate and uses the energy released to phosphorylate ADP and make ATP. Another reaction in the glycolytic pathway converts the 3-phosphoglycerate to 2-phosphoglycerate, which is then dehydrated by the enzyme enolase to form phosphoenol pyruvate. The "phospho" in phosphoenol pyruvate is in fact a phosphoryl group attached to the substrate by a high-energy bond. Phosphoenol pyruvate can then perform substrate level phosphorylation on ADP and make another ATP. In all, glycolysis generates two ATPs for each molecule of glucose that enters the pathway and is converted into pyruvate, all thanks to substrate level phosphorylation!

Substrate level phosphorylation is also an integral event of the tricarboxylic acid (TCA) cycle, another fueling pathway (Fig. 6.2). Since many microbes rely on the glycolysis and TCA cycle fueling pathways, the substrate level phosphorylation mode of ATP generation is widespread in the microbial world. Eukaryotic cells also use substrate level phosphorylation: glycolysis generates ATP in the cytoplasm, while the TCA cycle generates ATP inside the mitochondria (another reminder of the microbial ancestry of eukaryotic cell organelles). For fermentative microbes, substrate level phosphorylation is the main, if not the only, way to generate ATP. During fermentation, glycolysis converts sugars such as glucose into pyruvate, thus generating reducing power (NADH) and ATP via substrate level phosphorylation. But here is the catch: fermentative microbes cannot regenerate NAD$^+$ by recycling excess reducing power while making a transmembrane proton gradient and generating ATP. So, fermentative microbes have evolved other metabolic reactions to "dump" their excess reducing power. Later in the chapter, we will learn more about these auxiliary reactions and the reduced organic products that they generate. Some of these fermentative products will be familiar: lactate (as in sauerkraut), acetate (as in vinegar), and ethanol (in alcoholic beverages or ethanol-containing transportation fuel). Have you noticed that these are the same chemicals that accumulated in the anaerobic cultures of isolate #1 described in the case study?

Transmembrane ion gradients. In most microbes, energy can be harvested from an ion gradient generated across the cell membrane. A common way of creating such a gradient is to extract electrons and protons from NAD(P)H at the membrane. The electrons are transferred sequentially to several membrane proteins (the **electron transport chain**), while the protons are exported across the inner membrane to generate the **proton motive force (PMF)** that provides energy to make ATP (Fig. 6.6). The donor molecule that supplies electrons to the electron transport chain can be a molecule other than NAD(P)H. And the proton gradient is sometimes established by consuming protons in the cytoplasm or releasing them outside the membrane, rather than pumping them. But the effect is always the same: a transmembrane proton gradient forms. An enzyme complex of the membrane, the **ATP synthase**, uses this gradient to energize the phosphorylation of ADP and make ATP (Fig. 6.6). Mitochondria and chloroplasts have an ATP synthase too, further testament to their microbial ancestry.

The ATP synthase. ATP synthase has another name, F_1F_0 synthase, derived from its two protein components (F_1 and F_0). To envision how this complex interconverts two forms of energy—electrochemical (ion gradient) and chemical (ATP)—imagine how a turbine rotates and generates electricity as water flows through a dam from a higher to a lower level. The F_1F_0 complex has the shape of an inverted mushroom embedded in the membrane through a hydrophobic "stem" (the F_0 portion of the enzyme) and exposed to the cytoplasm through its F_1 "head" (Fig. 6.8). A flexible shaft joins the two components. The F_0 component has an open cavity that takes in the protons (or other ions) and a lower portion that *rotates* in response to the spontaneous movement of protons down the concentration gradient across the membrane. The rotation activates the cytoplasmic F_1 component so it can bind and phosphorylate ADP to make ATP. *Voila!* Thus, the ATP synthase energizes ATP synthesis from the flow of protons. How many ions do you think must be transported to harness sufficient energy to generate one molecule of ATP? The best estimate is that three or four protons must pass through the ATP synthase to generate one molecule of ATP. Doesn't that strike you as efficient, considering how much energy is contained in the high-energy bond of each mole of ATP that forms? Remember, two medium-sized strawberries per mole (Box 6.1)!

Also remarkable is the fact that the F_1F_0 ATP synthase can work in the opposite direction, hydrolyzing ATP and driving the discharge of protons. This activity explains why the enzyme complex was first named **ATPase**. But what determines which direction the machine runs? As you may have guessed, like any reaction, the direction is influenced by the con-

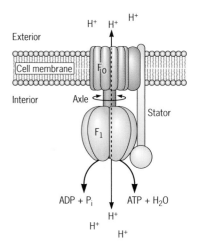

FIGURE 6.8 Membrane ATP or F_1F_0 synthase. This multi-subunit enzyme consists of two large complexes. One, F_0, forms a proton channel through the membrane; the other, F_1, is on the inner surface of the membrane. Passage of three or four protons from the exterior through F_0 of the energized membrane drives F_1 to form one ATP molecule from ADP. The reaction is poised to proceed in either direction, depending on the ratio of ATP to membrane potential: ATP hydrolysis can produce a proton gradient when the ratio is insufficient, and proton entry can produce ATP when the ATP concentration is too low for cellular functions (hence the alternative names for this complex).

centration of reactants and products. In this case, a low ATP concentration inside the cell and a steep ion gradient across the membrane promote the entry of ions and generation of ATP, whereas a high internal ATP concentration and a shallow gradient lead to the export of protons at the expense of ATP.

Variations on the general theme. Given the importance of transmembrane ion gradients, their generation is clearly central to life processes. In another impressive display of diversity, prokaryotes have evolved no fewer than four ways to generate such gradients: (i) **respiration**, (ii) photosynthesis, (iii) enzyme pumps, and (iv) scalar reactions. Respiration and photosynthesis couple electron transport to proton pumping to generate the transmembrane ion gradient. So important are these modes of energy conservation that we will devote separate sections to learn more about them.

Enzyme pumps and scalar reactions are somewhat rare, but they help microbes boost transmembrane ion gradients and make more ATP. Enzyme pumps are membrane proteins not associated with the electron transport chain that can pump protons or other ions across the membrane, creating a gradient that can be used for energy-requiring processes. Scalar reactions, on the other hand, have an intriguing name and an equally interesting role in energy metabolism: they consume or produce protons, rather than move them. Though these two reactions have been found somewhat infrequently, we can relate to them from daily processes. The microbes that acidify and enrich the complexity of red wines (malolactic bacteria) do so by taking malate into the cell and decarboxylating it to lactate; the lactate is then secreted in a reaction that consumes one proton in the cytoplasm. In this way, the reaction contributes to establishing a proton gradient, provides additional energy to the cell, and helps make a more delicious wine (*thank you, scalar reactions of malolactic bacteria!*).

Outliers are the *Bacteria* and *Archaea* living in alkaline or high-sodium environments (e.g., marine prokaryotes and bacteria that inhabit the rumen of cattle), which simply cannot form proton gradients. Can you figure out why? (*Quiz spoiler*: Any protons exported would react with the excess of hydroxyl ions and abolish a transmembrane proton gradient.) To bypass this barrier, these microbes establish transmembrane gradients with *other cations*, most often Na^+. In this case, the ion gradient is said to constitute a *sodium motive force*. As with the proton motive force, the sodium cations can be pumped into the cell by a modified version of the ATP synthase in an energy-yielding reaction that also phosphorylates ADP and makes ATP.

Energizing the membrane with an ion gradient provides benefits beyond making ATP. Some microbes use these gradients for (i) active transport across the cell membrane (called ion-coupled transport), (ii) maintaining the cell's turgor (water pressure), (iii) providing the proper cell's interior pH inside the cell, and (iv) powering the rotation of flagella for locomotion. Thus, the cell membrane is used in creative ways to transduce energy.

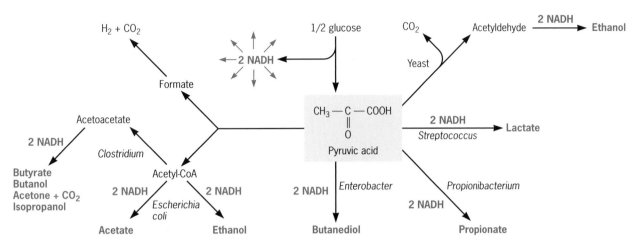

FIGURE 6.9 **Fermentation products from pyruvate in various microbes.** The NADH produced during glycolysis is used to reduce pyruvate and its derivatives, forming fermentation products distinctive for each organism.

HOW IS REDOX BALANCE ACHIEVED DURING FERMENTATION?

LEARNING OUTCOMES:
- Identify the main mode of energy conservation during fermentation.
- State why fermentative fueling needs auxiliary pathways to make fermentation end products to achieve redox balance.
- List five products of microbial fermentation that you have enjoyed.

Fermentative microbes can generate all the ATP they need via substrate level phosphorylation in reactions that also generate reducing power (Fig. 6.6). Although some of the reducing power is used in biosynthesis, most of it accumulates and must be recycled in some way to continue to supply NAD⁺ for substrate level phosphorylation reactions. Thus, fermentative microbes face a big problem: how to achieve redox balance. This feat is achieved by passing most of the hydrogen atoms from NADH first to *pyruvate*, the final product of the glycolytic pathway, and then to some compound derived from it. Thus, pyruvate works as the "hub" of a wide range of *auxiliary fueling pathways* that make fermentation products (Fig. 6.9). This is easy to appreciate in **homolactic fermentations**, whereby glucose is converted into pyruvate and then to lactate to consume the excess NADH and achieve redox balance (Fig. 6.10). In fact, this is how lactic acid bacteria make sauerkraut: they ferment sugars to lactate, which accumulates and acidifies the medium, preventing other bacteria from growing and spoiling the cabbage and giving the sauerkraut its distinctive "pickle" flavor and long shelf life.

Evolution has been enormously inventive in devising fermentation paths that equip prokaryotes to use a

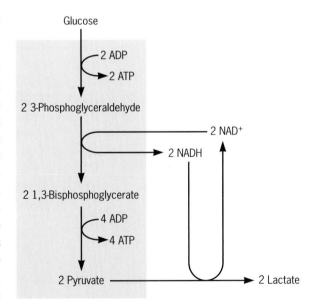

FIGURE 6.10 **Homolactic fermentation.** A shortened version of the glycolytic pathway combined with the reaction catalyzed by lactate dehydrogenase, an enzyme that makes lactate for pyruvate in a reaction that recycles the reducing power (NADH) generated during glycolysis. The reduction and reoxidation of the NAD molecule permits sustained ATP generation from glucose during fermentation.

TABLE 6.1 **Examples of microbial fermentations**

Substrate	Products	Organism(s)
Glucose and related sugars	Lactate, ethanol, acetate, butyrate, butanediol, etc.	*Lactobacillus* spp., *E. coli*, yeast, and many other organisms
Amino acids	Acetate (from glycine)	*Clostridium sporogenes*
H_2 and CO	Acetate	*Clostridium aceticum*
Oxalate	Formate+CO_2	*Oxalobacter formigenes*
Malonate	Acetate+CO_2	*Malonomonas rubra*
Lactate	Propionate+acetate	*Clostridium propionicum*
Pyruvate	Acetate	*Desulfotomaculum thermobenzoicum*

huge array of substrates to generate many kinds of reduced end products (Table 6.1). Note that the substrate used for fermentation is always an organic molecule (sugars, amino acids, compounds from the central pathways of metabolism, and the like), because it serves as both the carbon and energy source. Hence, the fermentative mode of fueling is *restricted to heterotrophs*. Furthermore, most of the substrate is used to harness energy as ATP and reducing power, because there are no alternative modes of energy generation other than substrate level phosphorylation reactions. Hence, only a small fraction of substrate is used to synthesize cell material; most is converted into the end products of fermentation.

Even if you did not know how microbes ferment substrates, surely you have enjoyed the products of these activities. Fermentations have been used by humans throughout history to produce beverages (the whole array of alcoholic products made from grain or fruit) and to process food for preservation and flavor (for example, pickled fruits and vegetables). Various bacteria and eukaryotic microbes such as yeast have been exploited for thousands of years for their superior fermentative abilities, and we still use them to make bread, beer, and wine. The yeast strains used to make beers and wines make their own unique combo of fermentation products, which gives alcoholic beverages their distinctive flavor and taste. Fermentative microbes are also of interest for many industrial scale fermentations that produce biofuels such as ethanol and butanol, as well as chemical intermediates for the chemical industry. Industrial microbiologists often rely on genetic engineering approaches to redirect the metabolic networks in such a way that all of the excess reducing power is rerouted toward the reactions that make the desired product. The cells grow happily because there is sufficient NAD^+ for the substrate level phosphorylation reactions and ATP continues to be generated. We, the humans, harvest the valuable chemicals and products. Everybody wins.

HOW ARE TRANSMEMBRANE ION GRADIENTS GENERATED DURING RESPIRATION?

LEARNING OUTCOMES:
- Describe the role of the electron transport chain in respiration.
- Identify the difference in location of respiration between eukaryotes and prokaryotes.
- Compare and contrast aerobic and anaerobic respiration with respect to electron acceptor diversity and energy yielded.

- Discuss how metabolic diversity has allowed microbes to colonize and harness energy from almost any resource on the planet.
- State the implications of the hierarchic placing of electron acceptors along the redox tower and the amount of energy harnessed from their respiration.
- Distinguish fermentation from anaerobic respiration.
- Speculate a reason why a microbe would ferment using substrate level phosphorylation, rather than respire glucose, when the energy yield for glucose respiration is so much higher.
- Review the major differences between chemoautotrophs and chemoheterotrophs during respiration.

During respiration, electrons, either alone or as hydrogen atoms, are passed from an electron donor to a terminal electron acceptor while simultaneously generating a transmembrane proton gradient to make ATP (Fig. 6.11). A number of membrane-bound electron carriers, collectively known as the electron transport chain or the **electron transfer system (ETS)**, hand off the electrons (or hydrogens) from one to another down the electrochemical gradient between the donor and the acceptor. In the ETS system, electron movement is favored by the oxidation-reduction (**redox) potentials** of the components of the chain. Flavoproteins, quinones, cytochromes, and other electron carriers are poised so that each member can spontaneously receive electrons from the previous one and pass them to the next, and so on. Team work at its best! Thus, the electrons (alone or

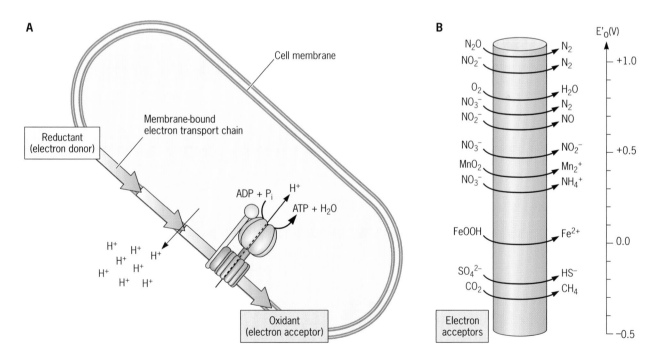

FIGURE 6.11 Electron transport and generation of a transmembrane ion gradient during respiration. Electrons removed from substrates (electron donors) travel down a redox gradient from carrier to carrier in a membrane, while protons (H^+) are simultaneously pumped across the membrane. This generates a transmembrane proton gradient, which the ATP synthase uses to phosphorylate ADP and make ATP. The magnitude of the redox gradient depends on the donor and acceptor driving the reaction. The redox tower at left shows, for example, the electron acceptors used for microbial oxidation. Their position in the tower matches the amount of energy yielded during its respiration: note that energetic electron acceptors such as O_2 are closer to the top, whereas the least energetic one, CO_2, is at the bottom.

TABLE 6.2 **Compounds that can replace O$_2$ as electron acceptor in respiration**

Organic compounds	Inorganic compounds
Fumarate	Nitrate (NO_3^-)
Dimethylsulfoxide (DMSO)	Nitrite (NO_2^-)
Trimethylamine *N*-oxide (TMAO)	Nitrous oxide (N_2O)
	Chlorate (ClO_3^-)
	Perchlorate (ClO_4^-)
	Manganic ion (Mn^{4+})
	Ferric ion (Fe^{3+})
	Gold (Au^{3+})
	Selenate (SeO_4^{2-})
	Arsenate (AsO_4^{3-})
	Sulfate (SO_4^{2-})
	Sulfur (S^0)

as hydrogen atoms) flow through the chain of carrier molecules to a terminal electron acceptor. When oxygen is the terminal electron acceptor, it is reduced to water and respiration is said to be **aerobic**. However, many microbes, unlike animals and plants, use alternative terminal electron acceptors, either inorganic or organic, in a process collectively known as **anaerobic respiration** (Table 6.2). We will discuss this important activity of the microbial world below.

ETS carriers transfer protons across the membrane in different ways. Some act as proton pumps, expelling the protons at the expense of some of the energy derived from the flow of electrons down the energy gradient. Others (the quinones) pick up protons from the cell interior and release them externally while passing on electrons to the next system member, a cytochrome. And recall that scalar reactions can also generate or boost the transmembrane ion gradient by consuming protons in the cytoplasm (e.g., NADH dehydrogenase) or producing them outside the cell membrane (e.g., formate dehydrogenase). Regardless of the mechanism, the outcome is the same: protons accumulate outside the cell. This proton shuffle creates a transmembrane proton gradient, which the ATP synthase can then use to translocate the protons back to the cytoplasm to energize the phosphorylation of ADP and make ATP (Fig. 6.11). If electron transport stops, proton pumping and ATP generation would stall, and the cell would stop growing and eventually die.

One final point: in eukaryotes, respiration occurs within the mitochondria but the general theme of harnessing energy is the same. And respiration takes place in the mitochondria membrane complex, just as the cell membrane is the site of prokaryotic respiration. This parallel should not be surprising, since mitochondria are highly evolved progeny of bacterial cells that had been engulfed by some ancestral archaeal host.

Aerobic and Anaerobic Respiration

The word respiration should immediately evoke oxygen, the electron acceptor used by all animals. However, oxygen only became available for respiration when the cyanobacteria emerged some 2.8 billion years ago and oxidized the Earth. Hence, the *most ancestral forms of respiration were anaerobic*. Many parts of our planet are still anaerobic. Therefore numerous microbes from these environments respire by using alternative terminal electron acceptors such as nitrate, nitrite, or sulfate (Table 6.2). The consequences of anaerobic respiration for life on this planet are striking. Respiration by nitrogen-containing electron acceptors, such as nitrate and nitrite, releases most of the dinitrogen gas (N_2) that accumulates in the planet's atmosphere. (Remember the two pond sediment bacteria isolated by the student in the case study? One isolate required nitrate to grow anaerobically and accumulated nitrite in the medium. Was the isolate respiring nitrate?) Have you ever smelled the unpleasant odor of rotting eggs (due to sulfide, H_2S) and of dead ocean fish (due to trimethylamine)? Blame it on microbes that use sulfur compounds and trimethylamine oxide, respectively, as terminal electron acceptors for their anaerobic respiration!

But each of the various terminal electron acceptors used for respiration does not harness the same amount of energy. The affinity of electron acceptors for the electrons (aka, their redox or reduction potential) can be used to align them hierarchically on a "redox tower," with the more electropositive potentials corresponding to those electron acceptors that have a greater affinity for the electron (Fig. 6.11). The greater the affinity for the electron, the greater the driving force that moves the electrons down the chain of ETS carriers, and the greater the energy harvested during respiration. For example, oxygen has a higher redox potential (it is more electropositive than sulfate). Hence, more energy can be harvested when microbes respire oxygen than sulfate. Not surprisingly, the competition for oxygen as the terminal electron acceptor for respiration is high. This is easily appreciated in soils and aquatic sediments, where oxygen is rapidly consumed within a thin top layer. Since microbes residing in lower layers are robbed of the opportunity to use oxygen, they respire using the "second best" available to them, often nitrate. Once nitrate is exhausted, they go for the "third best" (like sulfate) and so on, until all available electron acceptors are used. Many environments are stratified based on the electron acceptor used for respiration and, consequently, according to the amount of energy harnessed during respiration.

Regardless of which electron acceptor they use, microbes that respire gain a tremendous energy benefit: they generate more ATP than can be produced solely by substrate level phosphorylation (as in fermentation). One mole of glucose processed in fueling reactions by substrate level phosphorylation yields only *2 to 4 moles of ATP*, whereas respiration can yield up to *30-some moles* (depending on the species, growth conditions, and electron acceptor). Why would a microbe ever ferment rather than respire glucose? Metabolic cooperation! Fermentative microbes often feed their end products to respiratory microbes. If their fermentation end products were to accumulate, fermentation would slow down and eventually stop. However, their respiratory neighbors can use these end products as electron donors for respiration using a wide range of electron acceptors. More energy is harnessed as a team than individually. This type of metabolic network will be discussed in chapter 21.

Respiration and Autotrophy (Chemoautotrophs)

Recall that heterotrophs intertwine their carbon and energy metabolisms because they can harness energy from the substrate that serves as the

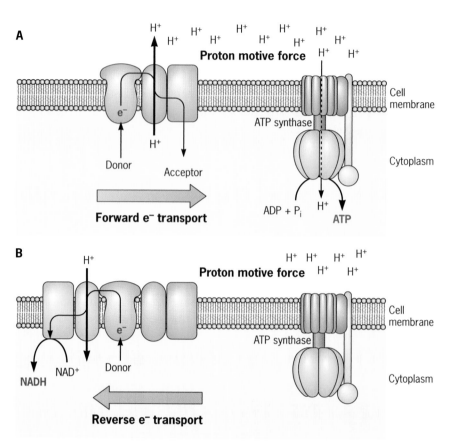

FIGURE 6.12 Respiration in chemoautotrophs. Chemoautotrophs can fix CO_2 to make precursor metabolites using energy derived from respiration. Their energy metabolism generates ATP in the standard way of respiration: by coupling the transport of electrons to proton pumping to generate a transmembrane proton gradient that drives the translocation of protons back into the cytoplasm and ATP generation by the ATPase/ATP synthase membrane enzyme complex. The electron transport chain can also revert the flow of electrons to make reducing power.

carbon source (Fig. 6.2). However, the pathways by which autotrophs fix carbon dioxide and make precursor metabolites *use vast amounts of energy or reducing power.* Some autotrophs rely on respiration to energize carbon dioxide fixation: they feed electrons to the electron transport chain *directly from inorganic chemicals* (thus, they are chemo<u>litho</u>trophs) and generate a proton motive force to make ATP (Fig. 6.12). When needed, chemoautotrophs use the energy in the proton motive force to pump the protons back inside the cell and reverse the flow of electrons in the ETS, thus generating reducing power (Fig. 6.12). *Isn't that a neat trick?*

Most chemoautotrophs carry out aerobic respiration (that is, they use oxygen as the terminal electron acceptor), but some also perform various types of anaerobic respiration (Table 6.3). For example, sulfur bacteria use sulfide (HS^-) or elemental sulfur (S^0) as electron donors for aerobic respiration, whereas nitrifying and iron-oxidizing bacteria oxidize nitrite and Fe^{2+}, respectively, using oxygen as the terminal electron acceptor. Hydrogen bacteria use hydrogen to respire oxygen, whereas methanogenic archaea use hydrogen to reduce carbon dioxide and make methane. Not all

TABLE 6.3 **Examples of inorganic compounds that can serve as electron donor and acceptor pairs for chemoautotrophs**

Group of chemoautotrophs	Electron donor	Electron acceptor
Hydrogen bacteria	H_2	O_2, SO_4^{2-}
Sulfur bacteria	HS^-, S^0	O_2
Carboxydotrophic bacteria	CO	O_2
Iron oxidizers (iron bacteria)	Fe^{2+}	O_2
Ammonia oxidizers	NH_4^+	O_2, NO_2^-
Nitrite oxidizers	NO_2^-	O_2
Methanogens	H_2	CO_2
Phosphite bacteria	PO_3^{3-}	SO_4^{2-}

the redox pairs release the same amount of energy, but all have a change in free energy ($\Delta G^{0\prime}$) greater than that needed to phosphorylate ADP into ATP. Certainly in the fueling area, chemoautotrophy is unsurpassed in expanding the niches prokaryotes can colonize.

HOW ARE TRANSMEMBRANE ION GRADIENTS GENERATED DURING PHOTOSYNTHESIS?

LEARNING OUTCOMES:
- Compare and contrast respiration and photosynthesis with respect to energy generation and generation of transmembrane ion gradients.
- Compare and contrast cyclic and noncyclic photosynthesis.
- Identify the role of electron acceptors and electron donors in energy harvesting via photosynthesis.
- Discriminate between anoxygenic and oxygenic photosynthesis.
- Identify a reason why anoxygenic phototrophs are well adapted to environments that are frequently subjected to fluctuations in energy and carbon availability.
- Describe the impact of oxygenic photosynthesis upon the evolution of the planet and life on Earth (including impacts on ancestral microbes and extant microbes).

Phototrophs harvest energy from light to generate a proton motive force and make ATP with an ATP synthase in a manner basically similar to respiration. A major difference is that phototrophs contain light-sensitive pigments, **chlorophylls,** that harness the energy in the light to push an electron to a higher energy level so it can flow back down through the electron transport chain to the terminal electron acceptor (Fig. 6.13). Hence, think of chlorophylls as antennas that capture light. (After all, light, radio waves, and TV waves are all part of the electromagnetic spectrum!) These pigments are often associated with proteins and cofactors in **reaction centers** that promote the transfer of electrons to the electron transport chain. In some phototrophs, light harvesting is maximized by peripheral chlorophylls and other light-sensitive pigments that capture light and funnel the energy to the chlorophylls positioned in the reaction center. Collectively, all of the components of the light harvesting apparatus are known as a **photosystem**.

Among prokaryotes, photosynthesis is restricted to the *Bacteria*; accordingly, their chlorophylls are often called **bacteriochlorophylls** (Table 6.4).

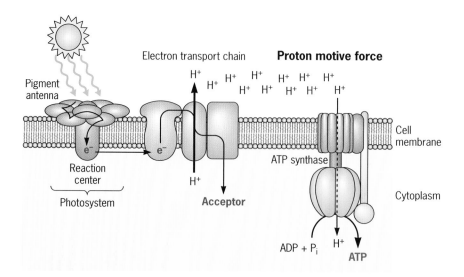

FIGURE 6.13 ATP generation during photosynthesis. The light-sensitive pigment antenna of the photosystem transfer energy from light to the chlorophylls of the reaction center to initiate electron transport through the ETS carriers to an electron acceptor, which can be their own chlorophyll (cyclic phosphorylation) or NADP+ (non-cyclic phosphorylation). Electron transport is coupled to proton translocation across the membrane to generate a proton motive force for the ATPase/ATP synthase to make ATP.

(Why is light harvesting not found in the *Archaea*?) Bacterial chlorophylls are diverse, varying in their sensitivity to different wavelengths of light. Since each wavelength contains a different amount of energy, the amount harvested from photosynthesis depends on the type of chlorophyll. Phototrophic communities in oceans and lakes are usually stratified in the water column based on the light wavelength that fuels their metabolism. Some phototrophs are so "picky" about wavelength that they cannot grow and will eventually die when exposed to light of other wavelengths. Another interesting fact about photosynthetic bacteria is that nearly all of them are autotrophs. Hence, similar to the respiratory systems of chemoautotrophs, photoautotrophs use their light-harvesting systems to capture energy as ATP and NADPH (Fig. 6.14). Below we discuss in more detail the light-harvesting strategies of bacteria, which will help us appreciate just how diverse phototrophic fueling is.

TABLE 6.4 Distinctive features of some prokaryotic phototrophs

Group	Electron donor	Chlorophyll type(s)	Example(s)
Anoxygenic			
Purple nonsulfur bacteria	Many substrates, including H_2, alcohols, organic acids, and Fe^{2+}	Bacteriochlorophylls *a* and *b*	*Rhodobacter* spp. and *Rhodopseudomonas* spp.
Purple sulfur bacteria	Reduced sulfur compounds (H_2 and also certain acids)	Bacteriochlorophylls *a* and *b*	*Cromatium spp.*
Green sulfur bacteria	Reduced sulfur compounds (H_2S and $S_2O_3^{2-}$)	Bacteriochlorophylls *c*, *d*, and *e*	*Chlorobium* spp.
Heliobacteria	Lactate, other organic acids	Bacteriochlorophyll *g*	*Heliobacillus* spp.
Oxygenic			
Cyanobacteria	H_2O	Chlorophyll *a* and phycobilins	*Synechococcus* spp., *Oscillatoria* spp., and *Nostoc* spp.

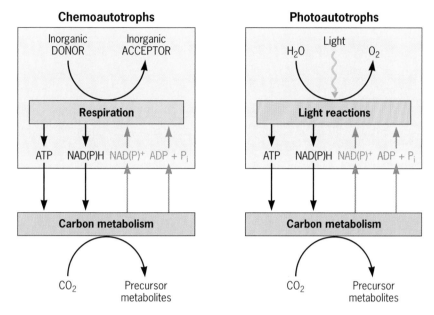

FIGURE 6.14 **Phototrophic fueling in autotrophs compared to chemoautotrophs.** Like chemoautotrophs, photoautotrophs also use carbon dioxide (CO_2) as carbon source to make precursor metabolites. However, the energy (ATP and reducing power) needed for CO_2 fixation is not harnessed in respiration but in light-dependent reactions (photosynthesis) that oxidize a donor such as water (H_2O). (Note that anoxigenic photosynthesis uses an inorganic donor other than water and, therefore, does not release oxygen as a byproduct.)

Cyclic versus Non-Cyclic Photosynthesis

During photosynthesis, light excites the chlorophyll to initiate transport of the "restless" electron down the electron transport chain to a terminal electron acceptor while, simultaneously, generating the proton motive force needed to make ATP. In cases where the electron acceptor is the same chlorophyll that was photoexcited, the process is called **cyclic photosynthesis** (Fig. 6.15). By accepting back the electron, the chlorophyll returns to the ground state, and it is once again ready to initiate a new round of photoactivation and electron transport. To make the reducing power for biosynthesis, these phototrophs can reverse the electron transport at the expense of the proton motive force, just as we learned for chemoautotrophs (Fig. 6.12).

In most phototrophs, however, the electrons do not return to the light-sensitive pigment but flow to $NADP^+$ to make reducing power, a process of **non-cyclic photosynthesis** (Fig. 6.15). The reducing power can then be used for biosynthesis, while any excess can be used to boost the transmembrane proton gradient and generate ATP. But this one-way street creates a problem: once the light-sensitive pigment in the reaction center transfers the electron to the first ETS carrier, it cannot start a new cycle of excitation. The chlorophyll needs an electron donor to replenish the missing electron, bringing the pigment back to its ground state. Non-cyclic phototrophs have diversified their metabolism to use a variety of electron donors (Table 6.4). Some use inorganic electron donors such as hydrogen, sulfur compounds, and iron (Fe^{2+}). Cyanobacteria have a unique strategy: they use water as an electron donor

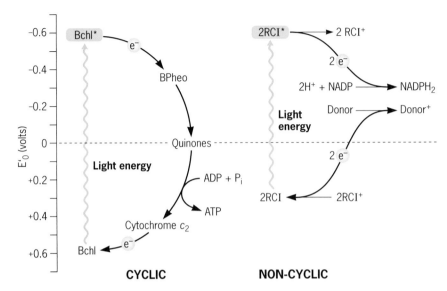

FIGURE 6.15 Anoxygenic photosynthesis. Anoxygenic phototrophs rely on cyclic or non-cyclic electron transport chains to move electrons down an energy gradient and generate ATP. The flow of electrons down the energy gradient (electrochemical potential, E'_0) is indicated by the arrows. In cyclic photosynthesis, light energy excites bacteriochlorophyll (Bchl) and raises it to an activated state (Bchl*), which is readily oxidized by passing two electrons through an electron transport chain (bacteriopheophytin, Bpheo; quinones, and cytochrome c2) that generates ATP and returning them to the bacteriochlorophyll. In non-cyclic photosynthesis, the electrons flow from the light-excited chlorophyll of the reaction center (RCI) through the electron transport chain to NADP, thus generating reducing power (NADPH$_2$). The oxidized reaction center (RCI$^+$) receives electrons from a donor to go back to its ground state (RCI) so it can start a new cycle of light excitation and electron transfer.

and generate oxygen as a byproduct. Because oxygen is released, these microbial phototrophs—like plants and algae—are said to carry out **oxygenic photosynthesis**. By contrast, microbes that recycle the electron back to the chlorophyll (non-cyclic) are said to carry out **anoxygenic photosynthesis**.

Anoxygenic versus Oxygenic Photosynthesis

Anoxygenic photosynthesis. While you may be most familiar with the oxygenic photosynthesis of plants and algae, remember that the first phototrophic bacteria to emerge on Earth were anoxygenic: they harvested energy from light without generating oxygen. Today, there are two major groups of anoxygenic phototrophs, **purple** and **green bacteria,** named for the distinctive color of their light-sensitive pigments (carotenoids and chlorophylls, respectively). Each group contains "sulfur" and "nonsulfur" representatives, terms that designate whether or not the light-sensitive pigments can use sulfur compounds as electron donors (Table 6.4). Most common are microbes that use sulfur-containing donors, such as elemental sulfur (S^0), to donate electrons to chlorophyll and bring it back to its ground state. In turn, this step generates hydrogen sulfide gas that diffuses out of the cell (and has the distinctive rotten egg smell). Anoxygenic phototrophs also include some heterotrophic representatives (i.e., those that use organic electron donors). Some, such as the purple nonsulfur bacteria, can shift from anoxygenic photosynthesis to respiring in the presence of oxygen. As you can imagine, these bacteria can live in many different environments, dark or light, with or without oxygen.

Oxygenic photosynthesis. Oxygenic phototrophic bacteria employ chlorophyll *a* as their light-sensitive pigment (Table 6.4), but they organize it in two separate photosystems that contain reaction centers I and II (RCI and RCII; Fig. 6.16). RCI follows the typical non-cyclic photosynthetic strategy: it gets photoexcited and transports electrons to NADP$^+$. However, the missing electron in RCI is replenished not by a chemical electron donor but a second photosystem (RCII). Once RCII donates electrons to RCI, RCII is left with the same problem: it needs an electron donor. RCII has the unique ability to extract electrons from water to bring its own chlorophyll *a* back to the ground state. This is a complex but clever system: water (H$_2$O) replenishes the electrons to RCII, and this photosystem, in turn, replenishes the electrons to RCI so it can continue to generate reducing power.

The impact that oxygenic photosynthesis had on the Earth and all of its life is immense. This microbial pathway oxygenated the planet and the atmosphere. Thanks to oxygenic photosynthesis, we are alive and breathe! But do not be fooled. The relationship of living organisms with oxygen is complex, since oxygen and its metabolic derivatives are extraordinarily toxic to cells (Box 6.2). Not surprisingly, oxygen killed many of the

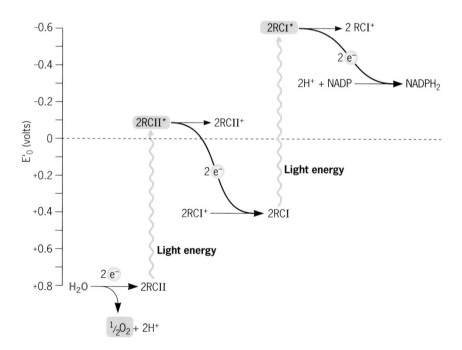

FIGURE 6.16 Oxygenic photosynthesis. In cyanobacteria (as in algae and plants), the chlorophylls of the reaction centers (RCII and RCI) undergo two sequential photoexcitations (Z-scheme). The photoexcitation of RCI promotes the transfer of electrons to NADP and makes reducing power (NADPH), whereas the photoexcitation of RCII replenishes the missing electrons in RCI. The oxidized RCII (RCII$^+$) returns to the ground state with electrons from water (H$_2$O), a reaction that generates oxygen (O$_2$).

BOX 6.2

Microbial Tricks To Keep Oxygen at Bay

The capacity of an organism to withstand the toxic effects of oxygen depends on two factors: the intrinsic sensitivity of its enzymes to oxygen and the ability of the cell to break down two highly toxic metabolic products of oxygen, hydrogen peroxide (H_2O_2) and the superoxide anion (O^{2-}). Due to variations in both of these factors, there is an enormous spectrum of bacterial responses to oxygen. At one extreme are the **strict anaerobes**, which are so exquisitely sensitive to oxygen that even a brief exposure to ambient air is lethal. At the other end of the spectrum are **strict aerobes** that are completely addicted to oxygen (including the cells of our body). Organisms that can tolerate oxygen generally have protective enzymes such as:

- **superoxide dismutase**, which converts superoxide into oxygen and hydrogen peroxide

- **catalase**, which converts hydrogen peroxide to water and oxygen (the fizz you see when you apply hydrogen peroxide to a wound are bubbles of oxygen).

Every organism that grows with oxygen has oxygen-detoxification enzymes (Table). By offering this protection, the enzymes equip the microbe to respire the oxygen. It is a fine balance. Then there are the organisms that cannot respire oxygen but can tolerate it, thanks to their detoxification enzymes. Among these "indifferent" microbes are opportunistic pathogens such as *Streptococcus pneumoniae* and *S. pyogenes*, which cause pneumonia and strep throat, respectively. They grow by fermentation, which they can carry out in the presence or absence of oxygen. *Streptococcus* species are catalase negative, but they do produce superoxide dismutase, which provides enough protection from the poisonous oxygen.

Electron acceptors used in respiration

Class	Synonym	Growth in air	Growth without oxygen	Possession of catalase and superoxide dismutase	Description	Example(s)
Aerobe	Strict aerobe	Yes	No	Yes	Requires oxygen; cannot ferment	*Mycobacterium tuberculosis, Pseudomonas aeruginosa, Bacillus subtilis*
Anaerobe	Strict anaerobe	No	Yes	No	Killed by oxygen; ferments in absence of O_2	*Clostridium botulinum, Bacteroides melaninogenicus*
Facultative		Yes	Yes	Yes	Respires with O_2; ferments or uses anaerobic respiration in absence of O_2	*Escherichia coli, Shigella dysenteriae, Staphylococcus aureus*
Indifferent	Aerotolerant anaerobe	Yes	Yes	Yes	Ferments in presence or absence of O_2	*Streptococcus pneumoniae, Streptococcus pyogenes*
Microaerophilic		Slight	Yes	Small amounts	Grows best at low O_2 concentration; can grow without O_2	*Campylobacter jejuni*

microbes existing at the time that cyanobacteria emerged. The survivors had evolved detoxification mechanisms to exclude oxygen from their cell interior or to keep its concentration very low (Box 6.2). Even today, microbes that use oxygen as a terminal electron acceptor for respiration must have strategies in place to keep toxic oxygen at bay. But oxygenic photosynthesis transformed the planet and set a new course in the evolution of life on Earth.

CASE REVISITED: How do microbes harness energy with and without oxygen?

Did you speculate that the aerobic cultures of both pond sediment isolates were respiring oxygen under aerobic conditions but, under anaerobic conditions, were fermenting (isolate #1) or respiring nitrate (isolate #2)? If so, you figured out the type of metabolism that each isolate relied on to grow with and without oxygen. In all cases, the microbes harnessed energy from chemical reactions to grow, so they are chemotrophs. When oxygen was available, both isolates respired (just like we humans do). Recall that during respiration, electrons are transported down the ETS carriers to the terminal electron acceptor (oxygen in this case) while simultaneously pumping protons across the membrane. This flux creates the transmembrane proton gradient that the ATP synthase uses to make ATP. Although pond water creates a barrier to the diffusion of atmospheric oxygen, some may reach the sediments. In the summer, for example, water evaporates more rapidly and the sediments may be exposed to substantial amounts of oxygen. Hence, it makes sense for sediment microbes to capitalize on the availability of this electron acceptor and grow via aerobic respiration.

What about anaerobic growth? Picture the ponds in the winter. The top water layer may freeze and deprive the sediments of any oxygen. Hence, the only microbes that will survive in the ponds throughout the year are those with the dual ability to grow aerobically and anaerobically (**facultative anaerobes**). Not surprisingly, Ms. G isolated two microbes from two different pond sediments, and both were facultative anaerobes. Both used glucose as the carbon and energy source for anaerobic growth. However, isolate #1 fermented the glucose into acids (acetate and lactate) and alcohols (ethanol), whereas isolate #2 needed to have an alternative electron acceptor (nitrate) for anaerobic respiration. Isolate #2 also produced nitrite, a reduced product of nitrate.

One can also predict that nitrate respiration would be prevalent in pond sediments that contain high concentrations of this electron acceptor. Fermentation makes more sense in anaerobic sediments poor in nitrate and other alternative electron acceptors and also in sediments where competition for these energy sources is high. Why fight for electron acceptors and

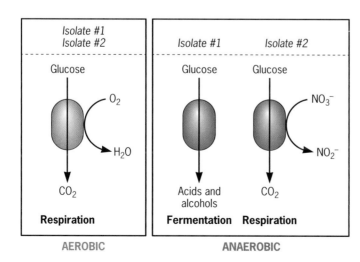

risk dying in the process? Fermentation is a great alternative. This pathway may not generate a lot of ATP, but it is enough to get by. Furthermore, fermentative microbes often live in consortia, i.e., teams of microbes that cooperate metabolically. In particular, fermentative microbes produce organic acids and alcohols, which "feed" the metabolisms of other microbes (coincidentally, the ones that carry out anaerobic respiration!). They are most efficient when they work as a team, and everybody wins!

CONCLUSIONS

In this chapter, we examined how ATP and reducing power are harvested during fueling to satisfy the energy demands for growth and other cellular activities. Here are some key points to remember:

1. An exotic variety of organic and inorganic substances serve one microbe or another as sources of energy. As we will learn in the next chapter, this diverse diet is handled by an enormous number of entry and feeder paths.
2. By integrating carbon and energy metabolism, heterotrophs can make the 13 precursor metabolites in the central pathways of fueling while harnessing energy as ATP and reducing power.
3. Unique adaptations to harvest energy from chemical and light sources provide autotrophs all of the ATP and reducing power they need to assimilate carbon from carbon dioxide.
4. The cell membrane works hard and in clever ways to transduce energy (e.g., transmembrane ion gradients).
5. Microbes use specific strategies to cope with the toxic effects of oxygen. Anaerobes have gained freedom from oxygen addiction.

Consequently, microbes play an essential role in recycling all living things and all their products. This last point deserves special attention. All of life's energy is obtained by oxidation of reduced compounds, be it familiar ones, such as glucose, or, in the case of microbes, an unexpected collection of virtually every organic compound on Earth, as well as most inorganic compounds. The list strains the imagination, as it includes hydrogen, ferrous iron, sulfide, ammonia, and many others. Microbes seem to "eat" anything that has energy: they are the ultimate scavengers for energy on the planet. Oxidation requires that something else become reduced, and here again, the microbial world continues to surprise us. Animals and plants use oxygen, but microbes use, among others, sulfate, nitrate, and ferric iron. Since we humans refer to the taking up of oxygen as breathing, microbes, by analogy, breathe rocks! Microbes alter our conventional view of energy metabolism, expanding it to include esoteric-sounding biochemical reactions. The biotechnological potential of microbial energy metabolisms, which are already used to generate industrially important chemicals, could also be harnessed to produce biofuels, clean up toxic compounds, and reduce greenhouse emissions (stay tuned for chapter 28). But their impact is far greater: the metabolic processes of the microbial world take place on a wide scale and are an essential component of the cycles of matter in nature, a topic that will be dealt with in chapter 20. Without their strange biochemistry, life on Earth could not be sustained.

Representative ASM Fundamental Statements and Learning Outcomes for the Chapter

EVOLUTION

1. Cells, organelles (e.g., mitochondria and chloroplasts), and all major metabolic pathways evolved from early prokaryotic cells.

   a. Support this statement with evidence: "all major metabolic pathways were evolved by early prokaryotic cells."
   b. Give examples of bioenergetics processes and structures that all living organisms share.
   c. Speculate a reason that life on Earth has sustained two distinct pathways to generate ATP instead of a single pathway.

CELL STRUCTURE AND FUNCTION

2. While microscopic eukaryotes (for example, fungi, protozoa, and algae) carry out some of the same processes as bacteria, many of their cellular properties are fundamentally different.

   a. Identify the difference in location of respiration between eukaryotes and prokaryotes.

METABOLIC PATHWAYS

3. *Bacteria* and *Archaea* exhibit extensive, and often unique, metabolic diversity (e.g., nitrogen fixation, methane production, anoxygenic photosynthesis).

   a. Compare and contrast energy harvesting during fueling reactions for heterotrophs and autotrophs.
   b. Discuss the process of fermentation.
   c. Compare and contrast aerobic and anaerobic respiration with respect to electron acceptor diversity and energy yielded during respiration.
   d. Review the major differences between chemoautotrophs and chemoheterotrophs during respiration.
   e. Distinguish fermentation from anaerobic respiration.
   f. Compare and contrast respiration and photosynthesis with respect to energy generation and transmembrane ion gradients.
   g. Discriminate between anoxygenic and oxygenic photosynthesis.
   h. Identify a reason why anoxygenic phototrophs are well adapted to environments that frequently are subjected to fluctuations in energy and carbon availability.

4. The survival and growth of any microorganism in a given environment depend on its metabolic characteristics.

   a. Identify a reason why reducing power is equivalent to energy.
   b. Compare and contrast ATP generation by transmembrane ion gradients and substrate level phosphorylation with respect to the mechanism of ATP generation and number of ATPs generated.
   c. Compare and contrast the oxidation and dehydration pathways for fueling reactions.
   d. Decide whether you would use ΔG, $\Delta G°$, or $\Delta G°'$ to estimate the change in Gibbs free energy of a reaction *in vivo*.
   e. Predict the mechanism by which pH can alter the Gibbs free energy change.
   f. Explain why it is incorrect to state that energy is "stored" in the cell as ATP and reducing power.
   g. Describe the process of ATP generation by F_1F_0 ATP synthase.
   h. List four ways to generate transmembrane ion gradients.
   i. Describe the role of the electron transfer system in respiration.
   j. Speculate a reason why a microbe would ferment using substrate level phosphorylation rather than respire glucose, when the energy yield for glucose respiration is so much higher.
   k. Compare and contrast cyclic and non-cyclic photosynthesis.
   l. Identify the role of electron acceptors and electron donors in energy harvesting via photosynthesis.
   m. Identify the main mode of energy conservation during fermentation.
   n. State why fermentative fueling needs auxiliary pathways to make fermentation end products to achieve redox balance.

MICROBIAL SYSTEMS

5. Microorganisms are ubiquitous and live in diverse and dynamic ecosystems.

   a. Discuss how metabolic diversity has allowed microbes to colonize and harness energy from almost any resource on the planet.

   b. State the implications of the hierarchic placing of electron acceptors along the redox tower and the amount of energy harnessed from their respiration.

IMPACT OF MICROORGANISMS

6. Microbes are essential for life as we know it and the processes that support life (e.g., in biogeochemical cycles and plant and/or animal microbiota).

   a. Describe the impact of oxygenic photosynthesis upon the evolution of the planet and life on Earth (including impacts on ancestral microbes and extant microbes).

7. Humans utilize and harness microorganisms and their products.

   a. Summarize the mechanism by which fermentative microbes make valuable chemicals and products for humans.

   b. List five products of microbial fermentation that you have enjoyed.

SUPPLEMENTAL MATERIAL

References

- **Burgin AJ, Yang WH, Hamilton SK, Silver WL.** 2011. Beyond carbon and nitrogen: how the microbial energy economy couples elemental cycles in diverse ecosystems. *Front Ecol Environ* **9**:44–52. doi:10.1890/090227.
- **Grüber G, Manimekalai MS, Mayer F, Müller V.** 2014. ATP synthases from archaea: the beauty of a molecular motor. *Biochim Biophys Acta* **1837**(6):940–952. doi: 10.1016/j.bbabio.2014.03.004.
- **Schoepp-Cothenet B, van Lis R, Atteia A, Baymann F, Capowiez L, Ducluzeau AL, Duval S, ten Brink F, Russell MJ, Nitschke W.** 2013. On the universal core of bioenergetics. *Biochim Biophys Acta* **1827**(2):79–93. doi: 10.1016/j.bbabio.2012.09.005.

Supplemental Activities

CHECK YOUR UNDERSTANDING

1. **The cellular reducing power of NADH and NADPH is used to generate**
   a. precursor metabolites
   b. macromolecules
   c. ATP
   d. all of the above

2. **Some bacteria use a cytoplasmic NADH dehydrogenase to catalyze the following reaction: NADH+H$^+$+acceptor →NAD$^+$+reduced acceptor. In addition to generating a proton membrane gradient and ATP, this chemical reaction generates a transmembrane ion gradient through**
   a. respiration
   b. photosynthesis
   c. enzyme pumps
   d. scalar reactions

3. **Glucose fermentation by some microbes results in the accumulation of acids such as acetate and lactate in the medium. The acid byproducts are necessary to . . .**
   a. increase ATP production by substrate level phosphorylation
   b. increase ATP production by respiration
   c. achieve redox balance
   d. reverse electron transport to generate reducing power
   e. disrupt energy transport

4. **Which of the following statements comparing and contrasting heterotrophs and autotrophs is NOT true?**
   a. Both heterotrophs and autotrophs synthesize precursor metabolites.
   b. Heterotrophs use organic carbon sources as energy sources whereas autotrophs use carbon dioxide as carbon sources only.
   c. Heterotrophs cannot be phototrophs.
   d. Heterotrophs and autotrophs can harness energy from light.

5. **Why is oxygen toxic?**

DIG DEEPER

1. Some microbes use enzymes called transhydrogenases to catalyze the reversible conversion of NADH into NADPH. Enterobacteria such as *E. coli* contain two transhydrogenases: one membrane bound (PntAB) and one cytoplasmic (UdhA) transhydrogenase (Figure).

   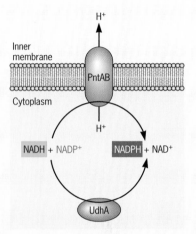

   a. The ratio of $NAD^+/NADH$ is higher than that of $NADP^+/NADPH$ when *E. coli* is grown aerobically with glucose. Why?
   b. Why do you think most microbes have these two forms of reducing power instead of just one?
   c. The *in vitro* reduction of NAD^+ to NADPH by cell extracts of *E. coli* from glucose-grown cultures was reduced in mutants where either UdhA or PntAB was inactivated and abolished in the double mutant, as shown in the graph.

      i. Based on the results presented in the graph, speculate about the likelihood that pathways not mediated by transhydrogenases operate in *E. coli* to maintain redox homeostasis between the NADH and NADPH pool.
      ii. The growth of the UdhA mutant with glucose was comparable to the wild type, but it was significantly reduced for the PntAB and UdhA-PntAB mutants. Why?
   d. Some cells also use $FAD/FADH_2$ as redox cofactor for the oxidation of alkenes/alkanes, leaving oxidation reactions of alcohols (-COOH) to aldehydes (-OH)/ketones(-C=O) to the $NAD(P)^+/NAD(P)H$ pairs. Suggest what parameters could control the redox cofactor specificity in these reactions.

2. Microbes can use the energy from transmembrane ion gradients for processes other than making ATP. Research how these gradients energize the processes listed below. Present your findings to the class using diagrams that illustrate the energy transduction mechanism.
   a. Active transport across the cell membrane (ion-coupled transport)
   b. Maintenance of the cell's turgor (water pressure)
   c. Intracellular pH control
   d. Flagellar rotation
   e. Reverse electron transport
   f. Efflux of antibiotics

3. Over the last century, it has become possible to make beers with ever-higher alcohol content. (http://www.cnbc.com/2015/08/07/craft-beers-get-heavy-on-the-alcohol.html)
   a. What limits the alcohol content of beer?
   b. How have brewers increased the alcohol content of beers?
   c. A new microbrewery opened in town, and the brewmaster contacts you, the local microbiologist, to help him develop a new yeast strain that will increase the alcohol content of his brews. Using your understanding of microbial bioenergetics, what strategy would you use?

4. **Team work:** Working in groups, research one of the following products of fermentation:
   - Sour cream
   - Yogurt
   - Korean soy sauce
   - Salami
   - Swiss cheese
   - Sauerkraut
   a. What are the starting ingredients and how does each contribute?
   b. What microbes contribute?

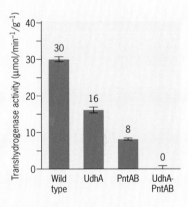

(From Sauer U et al., *J Biol Chem* **279:**6613–6619, 2003.)

c. What taste and texture characteristics result from fermentation?
d. What reactions and byproducts generate the characteristic taste?
e. How does fermentation increase the product's shelf life?

Supplemental Resources

VISUAL

- Oxidation and reduction review from a biological point of view by Khan Academy (13:28 min): https://www.youtube.com/watch?v=orl2m6IarJg.
- Video explaining how biofuels are made in industrial fermentations (3:03 min): https://www.youtube.com/watch?v=ctNrVfXzw9o.
- Video showing how to measure the respiratory activity of microbes in soil (5:07 min): https://www.youtube.com/watch?v=RjnCf0weSUQ.

SPOKEN

- *Microbes After Hours* session "The Microbiology of Beer," Charles Bamforth and Rebecca Newman teach the general public about harnessing microbial metabolism to make beer (1:13:41 hours): http://www.microbeworld.org/podcasts/microbeworld-video/archives/1488-mwv-episode-79-the-microbiology-of-beer.
- Everything you wanted to know about fermentation and more: The fermentation podcast: http://fermentationpodcast.com.

WRITTEN

- FAQ *If the Yeast Ain't Happy, Ain't Nobody Happy: The Microbiology of Beer*, a colloquium report from the American Academy of Microbiology to educate the general public about the microbial metabolic processes that produce beer: http://academy.asm.org/index.php/faq-series/435-beer.
- *Microbial Energy Conversion*, a 2006 colloquium report from the American Academy of Microbiology: http://academy.asm.org/index.php/browse-all-reports/176-microbial-energy-conversion.
- Gemma Reguera discusses how some microbes team up to respire various electron acceptors in the *Small Things Considered* blog post "Love at First Zap" (http://schaechter.asmblog.org/schaechter/2012/12/lov.html).
- Gemma Reguera reviews what is known about microbes that respire gold and the artistry of the process in the *Small Things Considered* blog post "The Art of Microbial Alchemy" (http://schaechter.asmblog.org/schaechter/2013/04/the-art-of-microbial-alchemy.html).
- Suzanne Winter reviews research about bacteria that respire metals, even toxic ones, in the *Small Things Considered* blog post "*Geobacter*: Microbial Superhero" (http://schaechter.asmblog.org/schaechter/2011/03/geobacter-microbial-superhero.html).
- National Center for Case Study Teaching in Science "Cow of the Future: Genetically Engineering a Microbe to Reduce Bovine Methane Emissions" by Richard Stewart, Daniel Stein, Kevin McIver, John Buchner, and Ann Smith incorporates principles of bacterial genetics, biotechnology, and methane oxidation as well as the impact of science on society (http://sciencecases.lib.buffalo.edu/cs/collection/detail.asp?case_id=738&id=738).

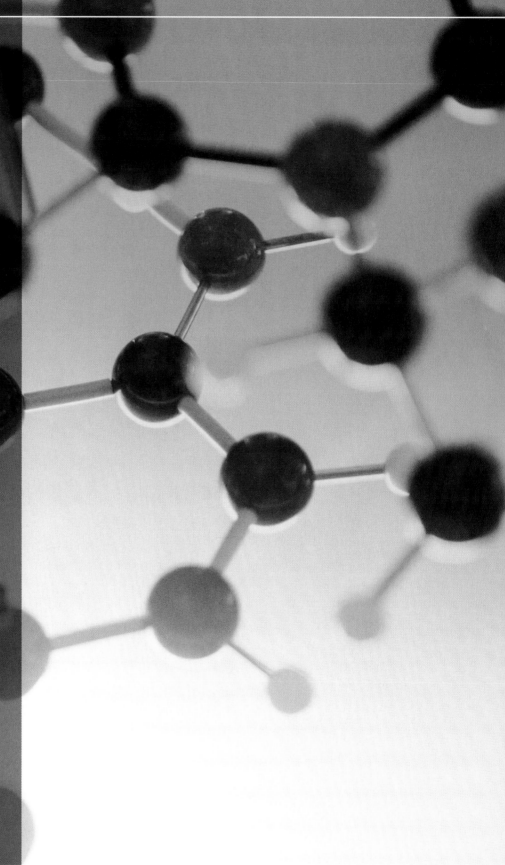

Molecular model of DNA structure

CHAPTER SEVEN

Synthesis of Building Blocks

KEY CONCEPTS

This chapter covers the following topics in the ASM Fundamental Statements.

EVOLUTION

1. Cells, organelles, and all major metabolic pathways evolved from early prokaryotic cells.

CELL STRUCTURE AND FUNCTION

2. *Bacteria* and *Archaea* have specialized structures that often confer critical capabilities.

METABOLIC PATHWAYS

3. *Bacteria* and *Archaea* exhibit extensive, and often unique, metabolic diversity.

4. The interactions of microorganisms among themselves and with their environment are determined by their metabolic abilities.

5. The survival and growth of any microorganism in a given environment depend on its metabolic characteristics.

IMPACT OF MICROORGANISMS

6. Microorganisms provide essential models that give us fundamental knowledge about life processes.

INTRODUCTION

In chapter 6, we learned how the bioenergetics of fueling generates energy (ATP and reducing power) for biosynthesis and other processes cells need to grow and reproduce. In this and the next chapters we will focus on how that energy is spent to assimilate substrates from the environment and build biomass. For convenience, we first consider the synthesis of **precursor metabolites** and **building blocks**, leaving macromolecules (nucleic acids, proteins, polysaccharides) and lipids to chapters 8 and 9. Consider the challenge: microbes assimilate a wide range of substrates, yet have evolved creative ways to convert them into the same 13 precursor metabolites needed for the synthesis of building blocks. Not surprisingly, the metabolic designs evolved by microbes are diverse and complex. It has taken billions of years for cells to evolve the thousands of linked enzyme-catalyzed reactions, arrayed in one grand interlocking pattern, that comprise cellular metabolism. Elucidation of this biochemical maze, now almost a completed task, was one of the triumphs of the 20th century. The pathways of bacteria proved closely applicable to plants and animals, underscoring the essential biochemical unity of all life on this planet.

CASE: **A bacterium that cannot make three essential amino acids**

−Aromatics   +Aromatics

(Courtesy Gemma Reguera.)

G, a microbiology student investigating the nutritional requirements of a new bacterial isolate, observed that it grew in glucose mineral medium but only when supplemented with three aromatic amino acids (phenylalanine, tyrosine, and tryptophan).

Questions arising:

- Why are those amino acids essential for growth of the bacterium?
- Based on your knowledge of how amino acids are synthesized, can you hypothesize what biosynthetic reactions may be missing or inactive in this bacterium?
- Describe an experiment to test your hypothesis.
- Can you speculate on the type of environment inhabited by this bacterium based on its nutritional requirements?

WHY DO CELLS NEED BUILDING BLOCKS?

LEARNING OUTCOMES:
- Identify two strategies microbes use to obtain building blocks, such as essential amino acids, from exogenous sources.
- Contrast biosynthetic capacity and nutrient requirements for a bacterium that normally inhabits nutrient-rich environments with that of a bacterium that normally lives in nutrient-poor environments.

TABLE 7.1 Major building blocks needed to produce a typical Gram-negative bacterium[a]

Protein amino acids	DNA nucleotides	Glycogen monomers
Alanine	dATP	Glucose
Arginine	dGTP	
Asparagine	dCTP	**Polyamine units**
Aspartate	dTTP	Ornithine
Cysteine		
Glutamate	**Lipid components**	**Coenzymes**
Glutamine		NAD
Glycine	Glycerol phosphate	NADP
Histidine	Serine	CoA
Isoleucine	Fatty acids (several)	CoQ
Leucine		Bactoprenoid
Lysine	**LPS components**	Tetrahydrofolate
Methionine		Cyanocobalamine
Phenylalanine	UDP-glucose	Pyridoxal phosphate
Proline	CDP-ethanolamine	Nicotinic acid
Selenomethionine	Hydroxymyristic acid	Other coenzymes
Serine	Fatty acid	
Threonine	CMP-KDO	
Tryptophan	NDP-heptose	
Tyrosine	TDP-glucosamine	
Valine		
	Murein monomers	
		Prosthetic groups
	UDP-*N*-acetylglucosamine	
	UDP-*N*-acetylmuramic acid	FMN
	Alanine	FAD
	Diaminopimelate	Biotin
	Glutamate	Cytochromes
		Lipoic acid
RNA nucleotides		Thiamine pyrophosphate
ATP		
GTP		
CTP		
UTP		

[a]GTP, guanosine triphosphate; UTP, uridine triphosphate; dTTP, deoxyribosylthymine triphosphate; UDP, uridine diphosphate; CDP, cytidine diphosphate; CMP, cytidine monophosphate; KDO, 2-keto-3-deoxyoctulosonic acid; NDP, nucleoside diphosphate; TDP, ribosylthymine diphosphate; FMN, flavin mononucleotide; FAD, flavin adenine dinucleotide; LPS, lipopolysaccharide.

As we learned in chapter 5, the essential macromolecules of the cells are synthesized from building blocks: nucleic acids from nucleotides, proteins from amino acids, carbohydrates from sugars, and lipids from fatty acids. Whether growing in the nutritionally sparse environment of a mountain stream or the rich environment of chicken broth, a microbial cell cannot make its macromolecules unless it has the same complete array of building blocks (Table 7.1). If the cell cannot make one of the building blocks, it must acquire it, preformed, from its environment. Most of the building blocks can be transported into cells, albeit at some energy cost. Indeed, this bit of energy expenditure makes the difference between life and death. Consider the bacterium that G isolated in the case study. It could only grow when three essential amino acids (the aromatic triplet) were provided in the medium. By investing energy to transport them into the cell, the microbe acquired the essential building blocks needed to make proteins. Without them, the bacterium could not grow. Instead, this bacterium would be restricted to environments where these amino acids are

TABLE 7.2 **Nutritional needs of two heterotrophic bacteria**

| | **Medium components** | | | |
| | | | **Organic** | |
Organism	Inorganic	Trace[a]	Carbon/energy source	Required nutrients[b]
Escherichia coli	K, NH_4^+, Mg^{2+}, Fe^{2+}, Cl, SO_4^{2-}, PO_4^{3-}	Mn^{2+}, Mo^{6+}, Cu^{2+}, Co^{2+}, Zn^{3+}, B^{3+}	Glucose	None
Streptococcus agalactiae[c]	Same as above	Same as above	Same as above	Alanine, arginine, aspartic acid, asparagine, cystine, cysteine, glutamic acid, glycine, glutamine, histidine, leucine, lysine, isoleucine, methionine, phenylalanine, proline, serine, tryptophan, tyrosine, valine, nicotinic acid, pantothenic acid, pyridoxal, thiamine, riboflavin, biotin, folic acid, adenine, guanine, xanthine, uracil

[a]These substances are usually present in sufficient amounts as contaminants of glassware, distilled water, and the major components of the medium.
[b]In many cases, only the natural enantiomorph (e.g., the L form of an amino acid) will suffice.
[c]Data from N. P. Willett and G. E. Morse, *J. Bacteriol.* **91:**2245–2250, 1966.

readily available. Alternatively, it could live in close association with another microbe, scavenging their excess (chapter 21).

Although it may seem advantageous to make your own building blocks to live independently, a number of microbes rely exclusively on an exogenous source of one or more essential building blocks. Some can be quite needy and forage for just about every building block. For example, the dairy bacterium *Leuconostoc citrovorum* requires 19 amino acids, two purines, one pyrimidine, and eight vitamins. (A diet more demanding than mammals' fare!) At the other end of the nutritional spectrum, many species are complete synthetic chemists: their extensive array of biosynthetic pathways can synthesize every one of the building blocks. This is quite an achievement, considering how many components are needed (Table 7.1). Between these two extremes exist a variety of intermediate states of nutritional competency (Table 7.2).

As you may have guessed, these two metabolic strategies—importing building blocks or making them—come with advantages and disadvantages. Biosynthetic capability and nutritional requirements are reciprocal: the greater the cell's biosynthetic capacity, the fewer its nutritional needs. Not surprisingly, microbes that reside in environments rich in organic matter, such as those that live in or on animal hosts, often lose unneeded pathways through natural selection. What benefit is carrying genes and making costly enzymes when the environment is brimming with a steady supply of the essential building blocks? On the other hand, in environments scarce in nutrients, microbes that make their own building blocks have the fitness advantage. These microbes may have a more complex metabolic network, but, with fewer growth requirements, they can colonize more environmental niches, including nutrient-poor ones. (Another of life's trade-offs.) Often, you can deduce the metabolic ability of an organism from the profile of its habitats.

WHAT IS NEEDED TO SYNTHESIZE BUILDING BLOCKS *DE NOVO*?

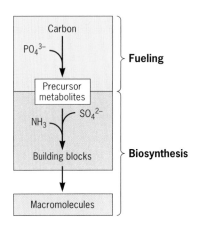

FIGURE 7.1 **Steps in the synthesis of macromolecules.** The synthesis of essential macromolecules is a stepwise process initiated with fueling reactions that generate precursor metabolites from carbon substrates and phosphate (PO_4^{3-}). The products of the fueling stage are used to make building blocks in biosynthetic reactions that also involve the assimilation of sulfur (SO_4^{2-}) and nitrogen (NH_3) sources. The building blocks are then used to make macromolecules: proteins from amino acids, nucleic acids from nucleotides, carbohydrates from sugars, and lipids from fatty acids.

LEARNING OUTCOMES:
* Describe three major requirements needed for *de novo* synthesis reactions.
* Summarize the purpose of each of the three stages of fueling reactions.
* Identify potential negative consequences for a microbe that can make all of its own building blocks.

If building blocks are not available in the environment, microbes must make them *de novo*. Easier said than done. Microbes do encounter and exploit a wide range of substrates in the environment, yet all must be converted into the same building blocks. To simplify the task, microbes make building blocks in two steps (Fig. 7.1). First, a variety of fueling reactions convert the substrates into the same 13 precursor metabolites (Table 7.3). Because making building blocks requires reducing power and energy in the form of ATP, cells also harness energy from chemical molecules or light at the fueling stage (chapter 6). As you can guess, making ATP and some of the precursor metabolites requires phosphorous; thus, cells assimilate this essential element during fueling as well; thus, the process is known as **assimilation** (Fig. 7.1). Having obtained these fundamental precursor metabolites and energy, cells can now use conserved biosynthetic reactions to make the building blocks for the cell's essential macromolecules. Because some building blocks (e.g., amino acids and nucleotides) contain nitrogen and sulfur groups, these elements are also assimilated in the biosynthesis stage (Fig. 7.1). Together, these reactions accomplish two essential feats: they generate ATP and make the monomeric units (building blocks) from which macromolecules and cell structures are formed. Next we consider individually the two major steps—fueling and biosynthesis.

Making Precursor Metabolites during Fueling

As we learned in the previous two chapters, the role of fueling reactions is to synthesize the 13 precursor metabolites and provide the energy and

TABLE 7.3 The 13 precursor metabolites generated in the central pathways of fueling that are used to synthesize building blocks

Precursor metabolites	Central pathway
Glucose-6 phosphate	Glycolysis
Fructose-6 phosphate	Glycolysis
Dihydroxyacetone-phosphate	Glycolysis
3-Phosphoglycerate	Glycolysis
Phosphoenolpyruvate	Glycolysis
Pyruvate	Glycolysis
Acetyl coenzyme A	Glycolysis-TCA connecting reaction
Oxaloacetate	TCA cycle
2-Oxoglutarate	TCA cycle
Succinyl coenzyme A	TCA cycle
Pentose-5-phosphate	Pentose phosphate cycle
Sedoheptulose-7-phosphate	Pentose phosphate cycle
Erythrose-4-phosphate	Pentose phosphate cycle

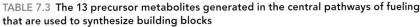

reducing power needed for biosynthesis. During fueling, living systems seem to create order from their disordered environment:

- Precursor metabolites: Microbial fueling is remarkably versatile. A myriad of enzymatically catalyzed reactions can convert many different substrates obtained from the environment into the 13 precursor metabolites (Table 7.3) the cell needs to synthesize building blocks.

- ATP: The energy demand for biosynthesis and other cellular activities is very high. Hence, chemical energy (mostly in the form of ATP) must also be obtained during fueling. Recall from chapter 6, there are two ways to make ATP during fueling: by **substrate-level phosphorylation** and by generating **transmembrane ion gradients** (usually proton gradients, which energize ADP phosphorylation by the **ATP synthase**). Some of this energy is used to assimilate essential nutrients and convert them into the 13 required precursors (Table 7.3). Note that ATP production is also the major pathway for phosphorous assimilation.

- Reducing power: Cell components are more reduced (i.e., carry more hydrogen atoms in their molecules) than the environmental substrates they are made from. Hence, microbes have evolved fueling reactions that generate reduced molecules such as NADH and NADPH. Recall that reduced molecules "pack" a lot of energy (chapter 6). In many cases, reduced molecules are oxidized to gain energy for ATP generation, either by substrate level phosphorylation or by a transmembrane proton gradient. NADPH also provides the reducing power needed for biosynthesis. Nothing is left to waste.

The fueling reactions that generate the 13 precursor metabolites and energy (ATP and reducing power) for biosynthesis can be conveniently divided in three stages (Fig. 7.2). First, the substrates available in the environment enter the cell either by passive diffusion through the cell envelope layers or by **active transport** using specialized transporters. Second, both **heterotrophs** and **autotrophs** use feeder pathways to generate the intermediate organic molecules needed for central metabolism and (third) to make the core precursor metabolites. However, the type of pathways used by these organisms differs, reflecting differences in the substrate(s) used as carbon source (organic carbon in heterotrophs and carbon dioxide in autotrophs). Versatility in the transport systems and feeder pathways enables a wide variety of substrates to be converted into a narrow range of intermediates that ultimately emerge as the 13 precursor metabolites common to all organisms.

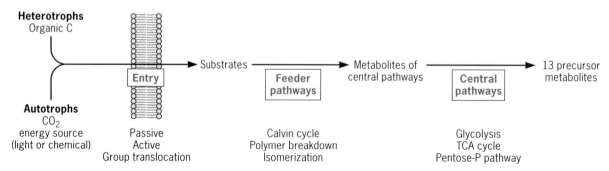

FIGURE 7.2 **Overview of the fueling stages leading to the synthesis of precursor metabolites. C, carbon; P, phosphate.**

Making Building Blocks from Precursor Metabolites

A microbial cell must supply to its biosynthetic machinery the same complete array of building blocks (Table 7.1). If it cannot acquire them from the environment, it must make them using the 13 precursor metabolites. Consider as an example the following central pathways of fueling that many microbes use for this purpose: **glycolysis**, **pentose-6-phosphate pathway**, and the tricarboxylic acids or **TCA cycle** (Fig. 7.3). Glycolysis generates several precursor metabolites needed to make sugars, amino acids, and fatty acids. It also feeds metabolites to the pentose-6-phosphate pathway, which synthesizes the precursor metabolites needed to make nucleotides and some amino acids. Glycolysis also generates pyruvate to make acetyl-coenzyme A (also known as acetyl-CoA), the precursor of fatty acids and some amino acids. Microbes being so metabolically diverse have evolved alternative pathways to these general ones, but all achieve the same goal: to supply the building-block-making biosynthetic machinery with the same precursor metabolites.

Note that some of the precursor metabolites and ATP also contain phosphate groups, which is why this element is essential. During fueling, phosphorus is incorporated into metabolites (Fig. 7.1). Nitrogen and sulfur are also a component of many cellular constituents; hence, these two elements must also be assimilated during the synthesis of the building blocks (Fig. 7.1), which we will consider later in the chapter.

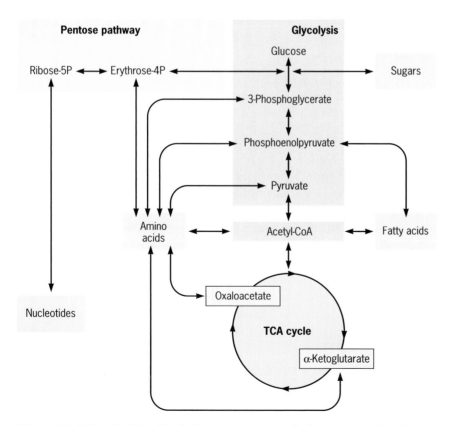

FIGURE 7.3 Making building blocks from precursor metabolites generated in the central pathways of fueling. Examples of precursor metabolites generated in some central pathways of fueling and the building blocks (nucleotides, amino acids, sugars, and fatty acids) used.

TABLE 7.4 **Families of amino acids and their precursor metabolites**[a]

Amino acid family	Precursor metabolite(s)
Serine family	
Serine	3-Phosphoglycerate
Glycine	3-Phosphoglycerate
Cysteine (and selenocysteine)	3-Phosphoglycerate
Aspartate family	
Aspartate	Oxaloacetate
Asparagine	Oxaloacetate
Threonine	Oxaloacetate
Methionine	Oxaloacetate
Isoleucine	Oxaloacetate, pyruvate
Lysine	Oxaloacetate
Glutamate family	
Glutamate	2-Oxoglutarate
Glutamine	2-Oxoglutarate
Arginine	2-Oxoglutarate
Proline	2-Oxoglutarate
Pyruvate family	
Alanine	Pyruvate
Valine	Pyruvate
Isoleucine	Pyruvate, oxaloacetate
Leucine	Pyruvate, acetyl-CoA
Aromatic family	
Tyrosine	Erythrose-4-PO_4, phosphoenolpyruvate
Phenylalanine	Erythrose-4-PO_4, phosphoenolpyruvate
Tryptophan	Erythrose-4-PO_4, phosphoenolpyruvate, pentose-5-PO_4
Histidine family	
Histidine	Pentose-5-PO_4

[a]Note that the amino acid selenocysteine can also be incorporated into polypeptide chains when its tRNA recognizes the termination codon UGA within a specific RNA sequence and structure.

Table 7.4 shows how amino acids are made from just a handful of precursor metabolites. Eight out of the 13 precursors are used in the cell's biosynthetic reactions to make the 21 amino acids. Each amino acid family often derives from only a single precursor metabolite. This apparently simple scheme involves many reactions but the final goal is achieved: to make amino acids from the same precursor metabolites regardless of which nutrients the cell is metabolizing. And this is just for the synthesis of amino acids. The cell will reroute precursors to other reactions that make nucleic acids, sugars, and fatty acids. It is easy to understand that microbes that can make their own building blocks will have greater metabolic complexity. This will require larger genomes to accommodate the genes needed to encode for the biosynthetic enzymes and the proteins that regulate the networks. More energy will be devoted to maintain the larger genome and to fuel the biosynthetic reactions. Yet there is a big pay-off: the greater a cell's biosynthetic capabilities, the fewer its nutritional requirements. As we saw in the case study, sometimes this makes the difference between life or death.

HOW ARE PRECURSOR METABOLITES MADE DURING FUELING?

LEARNING OUTCOMES:
- Describe three methods of molecular transport across the outer membrane of Gram-negative bacteria.
- Compare and contrast simple diffusion and facilitated diffusion.
- Compare and contrast ion-coupled transport, ABC transport, and group translocation.
- Describe the role of siderophores in microbes.
- Identify the purpose of feeder pathways for heterotrophs and autotrophs.
- Identify the role of RuBisCO in the autotrophic feeder pathway.
- Describe how glycolysis, the pentose phosphate pathway, and the TCA cycle are interconnected in metabolic pathways.
- List three auxiliary fueling pathways of central metabolism.
- Describe how adaptations in metabolic pathways allow some bacteria to grow and survive under varying environmental conditions.

Building on the foundation of how microbes make precursor metabolites during fueling (Fig. 7.2), let us examine in more detail the three processes—entry, feeder pathways, and central pathways—that equip microbes to bring in substrates from the environment and convert them into the 13 precursors.

Entry Mechanisms

The job of the cell membrane is to contain the cell and preserve the integrity of the cytoplasm where essential processes take place. Keeping cell components, especially low-molecular-weight metabolites, from diffusing out of the cell and into the environment is essential for life. At the same time, the membrane must also permit nutrients to get into the cell and waste products and other molecules to be excreted. In addition, remember that Gram-negative bacteria have two barriers, the outer membrane and the cell membrane. To meet these challenges, the cell membrane is equipped with transport and secretion proteins (chapters 2 and 9). Indeed, a considerable fraction of many microbial genomes is dedicated to transport of solutes.

Of course, microbes must harvest nutrients from their surroundings. And the rapid growth of microbes has a natural corollary: suitable food must enter at high rates. Yet many microbes grow rapidly even in highly dilute nutrient solutions. Thus, mechanisms to concentrate food molecules are crucial. Given the wide variety of molecules microbes ingest and the different architectures of their cell envelopes, it will come as no surprise that a number of different mechanisms to transport solutes have evolved. But each is a response to three challenges—*fast entry, selective admission*, and the *need to concentrate food molecules.*

Transport across the outer membrane of Gram-negative bacteria. The hydrophobic outer membrane of Gram-negative bacteria presents an impenetrable barrier for water and dissolved ions, were it not for porin-formed channels (chapter 2). By traversing the membrane, these pores are a conduit

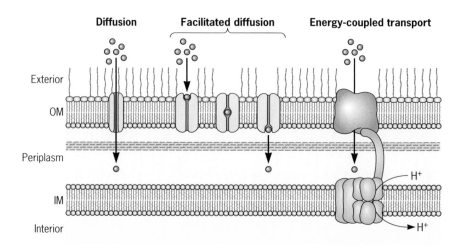

FIGURE 7.4 Solute transport across the Gram-negative outer membrane. Molecules can diffuse through porins of the outer membrane (OM) by simple or facilitated diffusion and always in the direction of the concentration gradient. Some porins can form complexes with proteins of the inner membrane that use the energy of the proton motive force to pump protons across the inner membrane (IM) to energize the porin-mediated transport across the outer membrane.

for water molecules and hydrophilic solutes to diffuse readily into the periplasm, as long as they are no larger than 600 to 700 daltons. Therefore, sugars, amino acids, and most ions can diffuse through the pores into the periplasm (Fig. 7.4). However, the hydrophilicity and size selectivity of the outer membrane porins means that hydrophobic or large (more than 600 to 700 daltons) hydrophilic molecules will be excluded. Instead, they require *dedicated transporters* on the outer membrane.

In some instances, porins transiently interact with specific molecules to assist their transport across the outer membrane (Fig. 7.4). Here, recognition between the porin and the molecule to be transported confers specificity to the transporter. However, the molecules still rely on diffusion to cross the outer membrane. Do you recognize drawbacks of this design? For solutes to diffuse through the outer membrane porins, they must be present at a greater concentration in the external medium than in the periplasm. To overcome this limitation, Gram-negative cells consume energy to transport specific molecules into the periplasm (Fig. 7.4). For example, the large vitamin B_{12} (cobalamin) molecule is transported across the outer membrane by **energy-coupled transport**. ATP is confined to the cytoplasm and cannot energize outer membrane transport. For this reason, the cobalamin porin forms a *complex with inner membrane proteins* that can use the proton motive force to pump protons from the periplasm back into the cytoplasm and thus energize the porin-mediated transport. The strategy is so effective that other porins use it to energize the transport of various large molecules, such as disaccharides and the organic iron chelates that enable the cell to scavenge this vital metal.

Transport across the cell membrane. Once past the Gram-negative outer membrane, solutes still have to get across the cell (inner) membrane. In the other bacteria and in archaea, the cell membrane is the only hydrophobic barrier to solute transport. Gram-positive bacteria do have a thick

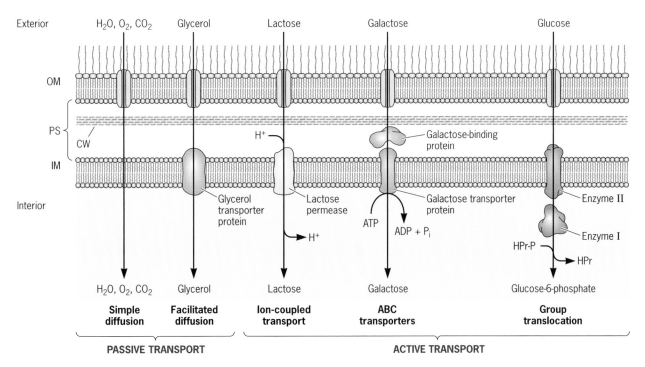

FIGURE 7.5 Solute transport across the cell membrane. The various modes of cell membrane transport are illustrated in *E. coli*, which contains a Gram-negative cell envelope with an outer membrane (OM), a periplasmic space (PS) with a thin peptidoglycan cell wall (CW), the cell membrane (inner membrane, IM), and a histidine-containing protein (HPr).

cell wall around the cell membrane, but its peptidoglycan lattice is readily permeable to hydrophilic compounds and does not significantly affect their diffusion. Few molecules can passively diffuse through the cell (inner) membrane (Fig. 7.5); most require *selective transport*. A variety of mechanisms have evolved to admit solutes selectively and rapidly into the cell, presumably because no single process would serve for the solutes in all environmental circumstances. There are three general types of transport (passive transport, active transport, and group translocation) in addition to specialized transporters that follow unique mechanisms adapted to the specific molecules they transport.

Passive transport. This type of transport relies on diffusion, uses no energy, and hence operates only when the solute is at higher concentration outside than inside the cell. **Simple diffusion** is sufficient for very few nutrients: dissolved oxygen, carbon dioxide, some ammonia, and water itself just about complete the list (Fig. 7.5). But this type of passive transport is neither fast nor selective. A more selective process is **facilitated diffusion**, which also requires no energy input and still depends on a concentration gradient. For example, *E. coli* and other enteric bacteria use selective protein channels to transport glycerol using this facilitated mechanism (Fig. 7.5). Facilitated diffusion is common in eukaryotic microbes (this is how yeast cells obtain sugar), but it is rare in prokaryotes.

Active transport. Active transport mediates the entry of virtually all other nutrients. Although the mechanisms vary widely (Fig. 7.5), all *use energy* to pump molecules into the cell at high rates, often against a concentration gradient. As a result, nutrients are concentrated inside the cell, sometimes more than a thousand fold! Each type of active transport mechanism uses a different source of energy:

- **Ion-coupled transport** is driven by the electrochemical gradient (proton or sodium motive force) established across the cell membrane either by electron transport or by hydrolysis of ATP by membrane-bound ATPases. In essence, this is a preload scheme: the *membrane is first energized* by exporting protons (primary event); then, the energy is *used for transport* (secondary event). Transport can occur in either direction (in or out) and translocate either one (uniport) or two solutes (symport if in the same direction and antiport if in opposite). Ion-coupled transport is particularly common among aerobic organisms, which can generate an ion motive force more easily than anaerobes can.

- TonB-dependent transporters also use the proton motive force to drive cargo across the inner membrane. To do so, the namesake outer membrane TonB collaborates with a protein complex that resides in the inner membrane and transduces energy to TonB. These transporters equip a wide variety of Gram-negative bacteria to take up iron and vitamin B_{12}, nickel, carbohydrates, and other cargo. *Pseudomonas aeruginosa* and *Caulobacter crescentus* each encode dozens of TonB-dependent transporters, and many species rely on them to transport siderophores, proteins that avidly chelate extracellular iron.

- **ABC (ATP binding cassette) transport** uses energy from the hydrolysis of ATP to pump solutes into the cell. Specific binding proteins, located in the periplasm of Gram-negative cells or attached to the outer surface or the cell membrane of Gram-positive cells, confer specificity by carrying their particular ligand to a protein complex on the membrane. Next, hydrolysis of ATP is triggered, and the energy is used to open the membrane pore and permit unidirectional flow of the substrate into the cell. This mode of transport is extremely common among bacteria. For example, dozens of specific binding proteins handle nearly half of the substrates transported by *E. coli* (including galactose, Fig. 7.5).

Group translocation. This transport mechanism is in a class by itself. A **phosphotransferase** system (PTS) transports a solute while *simultaneously phosphorylating it* (Fig. 7.5). This type of transport is found only in some bacteria and is responsible for their acquisition of a variety of sugars (Fig. 7.6). One of the common sources of the high-energy phosphate group is phosphoenolpyruvate (Fig. 7.6). By transferring such a molecule, energy is liberated to transport the sugar across the membrane. One ingenuity of group translocation is that energy is expended not for the transport process, but rather to form an *intracellular derivative* that is membrane impermeable and thus trapped within the cell. This phosphorylation step also makes transport of glucose by the PTS essentially free, energetically. Why? Since the first step of glycolysis is the phosphorylation of glucose to glucose-6-phosphate, the cell gets a two-for-one deal: a ~P would have to be consumed, even if glucose was transported with-

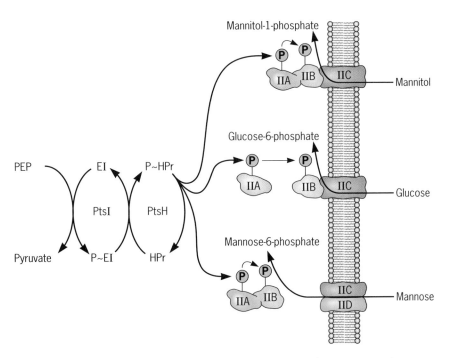

FIGURE 7.6 PTS transporters for three sugars in *E. coli.* EI and HPr are general proteins shared by all phosphotransferase (PTS) systems and serve to phosphorylate the sugar-specific transport components. PEP, phosphoenolpyruvate; IIA to IID, proteins specific for a given sugar.

out being phosphorylated. For this reason, the PTS strategy is particularly common among energetically challenged microbes, such as those living in the absence of oxygen. (Note that, because no concentration gradient is involved, group translocation is not active transport in the strict sense.)

Specialized transporters. Special transport processes supplement the three main mechanisms described above. Perhaps the most prominent is the one that equips many bacteria, including important pathogens, to steal iron (Fe^{3+}) from their mammalian hosts (Fig. 7.7). Iron is as essential for bacterial cells as it is for eukaryotic organisms, yet is sparingly soluble in its oxidized form (Fe^{3+}). To prevent this precious metal from precipitating, the internal compartments of animals *sequester it* in complexes with proteins such as transferrin and lactoferrin. Bacteria solve this dilemma by secreting powerful chelators of Fe^{3+} called **siderophores**. For example, *E. coli* secretes a siderophore called **enterochelin** to bind Fe^{3+} (Fig. 7.7). The extracellular enterochelin-Fe^{3+} complex is then recognized and actively transported into the cell by the cooperative actions of eight proteins that collectively span the outer membrane, periplasm, and inner membrane of the bacterium. It is an energy-demanding process driven by the proton motive force and ATP hydrolysis (Fig. 7.6). Once inside the cell, the Fe^{3+} is cleaved from the enterochelin by an esterase that simultaneously reduces it to the more soluble Fe^{2+} form. We'll learn more about a master pickpocket of iron, the opportunistic pathogen *Staphylococcus aureus,* in chapter 23.

Bear in mind that active transport of substances (certain end products and toxic materials) *out* of the cell is also extremely important, occupying a significant fraction of the prokaryotic cell's membrane machinery (chapter 9).

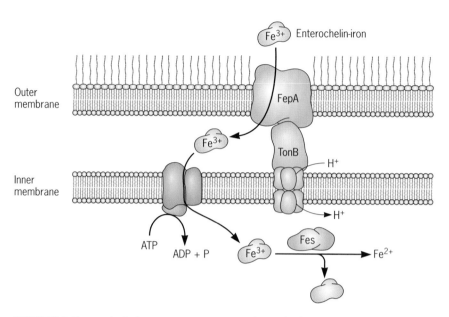

FIGURE 7.7 Enterochelin iron transport system of *E. coli*. The enterochelin molecule with its bound Fe passes through the FepA protein of the outer membrane in a process energized by the translocation of protons by TonB complex using the proton motive force. Next, the complex passes across the cell membrane through other Fep proteins, driven by ATP hydrolysis. An esterase (Fes) cleaves the enterochelin and reduces the Fe^{3+} to the soluble Fe^{2+} ion.

Hence, the cell devotes substantial amounts of resources to transporting molecules both in and out. However, the price is fair, since microbes exploit what is available in their environment to grow and reproduce.

Feeder Pathways

Once nutrients enter the cell, microbes must convert them into the substrates suitable for their central pathways of metabolism. Since heterotrophs use organic carbon for their metabolism, whereas autotrophs use carbon dioxide, their feeder pathways are different. We will examine each separately.

Heterotrophs. In heterotrophs, feeder pathways convert organic carbon compounds into intermediate molecules that can be introduced into the central pathways of metabolism. This is easier said than done. An enormous variety of substances can be growth substrates for heterotrophs. For example, *Salmonella* can utilize more than 80 organic compounds as its *sole source of carbon and energy*. The diet of organisms that inhabit the soil—such as *Pseudomonas* species—is even more expansive. Feeder pathways convert a vast variety of substances into the metabolites of the central pathways, from which the cell can make its 13 precursor metabolites. Some feeder pathways consist of a single reaction (the conversion of sucrose to two molecules of hexose, for example), others of many sequential enzyme-catalyzed steps.

Autotrophs. Whether plant or microbial, the defining characteristic of autotrophs is their ability to fix carbon dioxide (they convert the inorganic carbon molecule into organic compounds) to supply carbon for their pre-

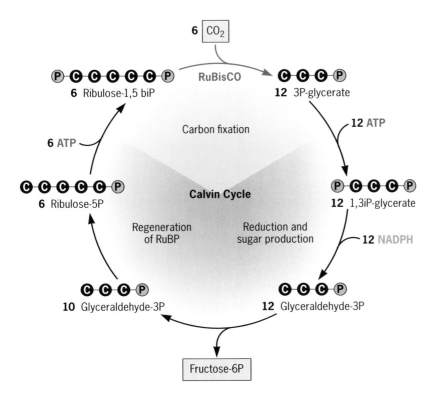

FIGURE 7.8 The Calvin cycle for carbon dioxide (CO_2) fixation. The cycle depicted shows the production of one hexose molecule (fructose-6-phosphate) from CO_2. In one round of the cycle six molecules of CO_2 are fixed, leading to the formation of one molecule of fructose-6-phosphate. See text for further explanation.

cursor metabolites. Fixing carbon dioxide rapidly enough to permit growth at a reasonable rate is difficult; it is also energetically expensive. Recall that autotrophic microbes have evolved separate energy metabolisms (photosynthetic and chemosynthetic) to generate ATP and reducing power to fuel carbon dioxide fixation (chapter 6). Here we will learn about the predominant feeder pathways that equip autotrophs to fix carbon dioxide.

The Calvin cycle. The overwhelming majority of autotrophs fix carbon dioxide via **ribulose bisphosphate carboxylase** (**RuBisCO**), a remarkable enzyme that catalyzes the addition of carbon dioxide to the five-carbon sugar-phosphate compound, ribulose bisphosphate. This reaction produces a six-carbon intermediate that splits spontaneously, yielding two molecules of 3-phosphoglycerate (Fig. 7.8). The RuBisCO is an integral component of the Calvin cycle, which, along with the pathways of central metabolism, is sufficient to synthesize all precursor metabolites from carbon dioxide. But this remarkable pathway is energetically costly, requiring considerable amounts of ATP and reducing power, as shown in the overall reaction:

$$6\ CO_2 + 12\ NADPH + 18\ ATP \Rightarrow C_6H_{12}O_6(PO_3H_2) +$$
$$12\ NADP^+ + 18\ ADP + 17\ P_i$$

RuBisCO, the functional heart of the Calvin cycle, is responsible for producing the overwhelming majority of organic material on the planet. It is

TABLE 7.5 **Some pathways of CO_2 fixation other than the Calvin cycle**[a]

Pathway	Key reaction	Distribution
Reductive TCA cycle	Fixes CO_2 by running the TCA cycle backwards; instead of releasing two molecules of CO_2, it fixes two.	*Chlorobium* and certain other photoautotrophic bacteria; some hydrogen-oxidizing bacteria
Hydroxypropionate pathway	Fixes CO_2 via two reactions in which acetyl-CoA and propionyl-CoA are CO_2 acceptors; in a cyclic series of reactions, acetyl-CoA is regenerated.	*Chloroflexus*, a photoautotrophic bacterium
Acetyl-CoA pathway	Fixes CO_2 in a fermentation in which H_2 is used to reduce CO_2 and ultimately form acetate	*Desulfobacterium autotrophicum*, certain clostridia, and other aceto-genic bacteria and archaea

[a]Note that archaea can fix carbon using the Calvin cycle in addition to the hydroxypropionate–hydroxybutyrate cycle and the dicarboxylate–hydroxybutyrate cycle.

also the most abundant enzyme in nature, partly because of its central role in autotrophic metabolism, but also because it is not a very efficient enzyme. Speculation about why a more proficient RuBisCO has not evolved is rife. Genetic engineers have even attempted, so far unsuccessfully, to give nature a boost. Possibly, RuBisCO is just faced with catalyzing a difficult chemical reaction.

Other carbon dioxide-fixing cycles. Although the Calvin cycle is the most widespread route for autotrophs to fix carbon dioxide, it is not unique. Some autotrophs enlist the reductive TCA cycle or the hydroxypropionate pathway (Table 7.5). Each of these routes is restricted to a relatively small group of prokaryotes. Some archaea use two variants of the hydroxypropionate pathway to fix carbon. Additionally, a sixth pathway exists in several unrelated anaerobic bacteria and archaea that make acetate from carbon dioxide and hydrogen (Table 7.5). All of these pathways for carbon dioxide fixation produce one or another metabolite of the central pathways or some compound easily converted to a central metabolite. The existence of six pathways for carbon dioxide fixation illustrates a fact often encountered in microbial metabolism: the development of multiple ways to achieve what seems like the same end. This seeming redundancy may reflect convergent evolution, or our incomplete understanding of the advantages of particular pathways in specific ecological niches.

Central Pathways

Whether heterotrophic or autotrophic, the key principle to keep in mind about feeder pathways is that organic compounds imported as food or made by carbon dioxide fixation must be converted, by however many sequential reactions it may take, to derivatives that can enter at least one of the pathways of central metabolism. The most common central pathways are glycolysis, pentose phosphate, and TCA cycle. These three processes are widely distributed among cellular organisms and together account for the synthesis of all 13 precursor metabolites. Figure 7.9 highlights the key products of these three central pathways and the precursor metabolites they generate. Although a number of species-specific auxiliary pathways exist, here we will summarize the core pathways that are common to many microbes.

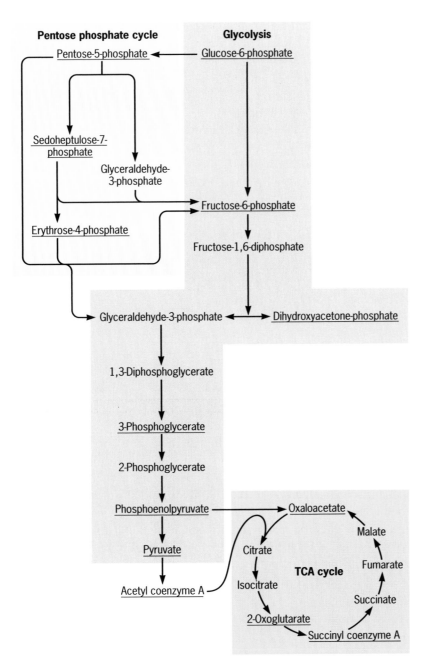

FIGURE 7.9 Three common central fueling pathways: glycolysis, the pentose phosphate cycle, and the TCA cycle. The 13 precursor metabolites (underlined) are generated in these three prominent central pathways of fueling. The chemical structures of the components of the pathways are shown in the Appendix figures at the end of this chapter.

Glycolysis. In the glycolytic pathway, glucose is phosphorylated and then split into two molecules of a triose phosphate (glyceraldehyde-3-phosphate), eventually yielding two molecules of pyruvate, a precursor metabolite (Fig. 7.9). Along the way, three other precursor metabolites, reducing power in the form of two molecules of NADH, and a net of two molecules of ATP are formed. (Note this process is also known as the Embden-Meyerhof-Parnas pathway, for its discoverers.)

Pentose phosphate pathway. Sometimes called the hexose mono-phosphate shunt, this pathway appears to be an alternate route for converting glucose-6-phosphate to glyceraldehyde-3-phosphate. It also supplies two additional precursor metabolites (Fig. 7.9) and, in the process, it generates two molecules of reducing power *in the form of NADPH*, the reduced molecule most commonly used for biosynthesis.

TCA cycle. The TCA cycle (tricarboxylic acid, also called the citric acid or Krebs cycle) is fed from pyruvate, the product of glycolysis, through the precursor metabolite acetyl-coenzyme A (Fig. 7.9). The TCA cycle can form three precursor metabolites and four units of reducing power while producing two molecules of carbon dioxide. Many strict anaerobes have only a portion of the cycle, functioning in the reverse of the normally depicted direction. By this means, the precursor metabolites needed for biosynthesis can be produced without the concomitant production of NADH—an advantage, since disposal of NADH is a challenge in the absence of oxygen. Common facultative organisms, such as *E. coli*, may possess the entire cycle but operate it anaerobically as two separate arms working in opposing directions, a maneuver discussed below.

Auxiliary pathways. Numerous auxiliary pathways of central metabolism occur throughout the microbial world. Three are particularly prevalent:

Entner-Doudoroff pathway. The Entner-Doudoroff pathway is simply an alternative link between an intermediate of the pentose phosphate pathway (6-phosphogluconate) and two compounds of glycolysis (glyceraldehyde-3-phosphate and pyruvate) (Fig. 7.10). But to operate this pathway, cells must also use glycolysis to produce ATP and to form precursor metabolites. Nevertheless, the pathway is widely distributed among diverse bacteria. For some species, it is the major route of sugar metabolism. Indeed, since a variety of sugar acids feed into a single intermediate (2-keto-3-deoxy-6-phosphogluconate), this pathway is a champion collector of metabolites from sugar acid feeder pathways.

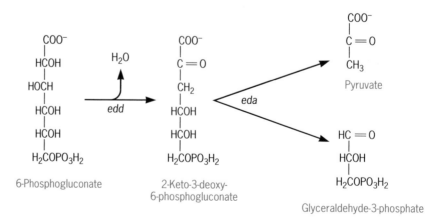

FIGURE 7.10 **The Entner-Doudoroff auxiliary pathway.**

Glyoxylate cycle. The glyoxylate cycle is unique to bacteria, protozoa, and plants and equips these organisms to grow on acetate (or fatty acids converted into acetate). This cycle is a modification of the TCA cycle that bypasses the cycle's two decarboxylation reactions (Fig. 7.11). In doing so, the cells retain the carbon, rather than burn it as carbon dioxide, and use it to make building blocks.

Fermentation pathways. Recall that many anaerobic heterotrophs generate all the ATP they need by substrate level phosphorylation, but only if they find some way to reoxidize NADH to NAD^+ (chapter 6). This feat, called **fermentation**, is achieved by passing most of the hydrogen atoms from NADH to pyruvate or some compound derived from pyruvate in short auxiliary fueling pathways of central metabolism. Note that fermentation uses soluble, intracellular electron carriers (NAD^+) to make reduced molecules such as organic acids and alcohol byproducts, which are secreted. By contrast, anaerobic respiration uses membrane-bound electron carriers (the **electron transport system**) to reduce an external electron acceptor (e.g., oxygen, nitrate, etc.) while simultaneously generating a transmembrane ion gradient for ATP generation.

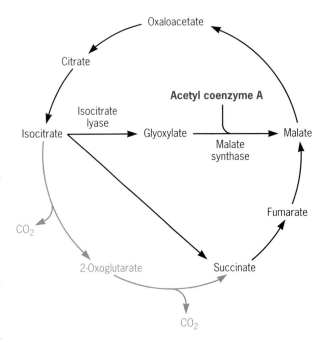

FIGURE 7.11 **Glyoxylate shunt.**

Diversity and flexibility of central metabolism. Microbes customize these fundamental pathways in astonishing ways. Bacteria have long been admired for their ability to grow without oxygen or to grow on two-carbon (or even on one-carbon) compounds and other non-sugars. Many other unique adaptations permit microbes to thrive in diverse circumstances. Let us consider, as an example, how facultative aerobic bacteria tune their central metabolism when growing aerobically or anaerobically. *E. coli* possess the entire TCA cycle but can also replenish oxaloacetate to run the cycle in reverse; by doing so, they make two other precursor metabolites, but do not generate energy (Fig. 7.12A). When growing anaerobically, the TCA cycle is operated as two separate arms working in opposing directions, one to make precursor metabolites and the other to generate energy (Fig. 7.12B).

How can cells acquire the fueling products if they are fed only with a fatty acid, or a pentose, or malate as carbon and energy source instead of glucose? For example, *E. coli* can grow aerobically by feeding malate into the TCA cycle and running the reactions of the central pathways *in reverse* to make the precursor metabolites (Fig. 7.12C). Hence, depending on environmental conditions, the central pathways operate in *either the forward or reverse direction.* After all, the precursor metabolites must be generated, no matter where a feeder pathway pours metabolites into a particular central pathway. In some cases, organisms can reverse the central pathways to make hexoses such as glucose from one-, two-, three-, or four-carbon

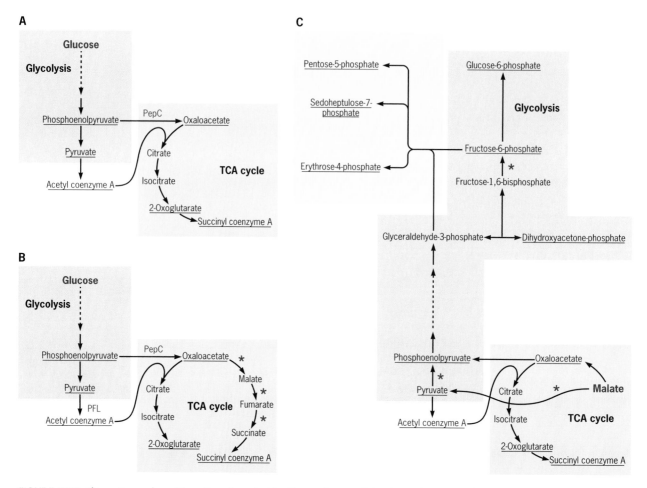

FIGURE 7.12 Alternate modes of function of central fueling pathways. A facultative heterotroph such as *E. coli* can make the 13 precursor metabolites (underlined) with alternative reactions of the central pathways. (A) During aerobic growth on glucose, PepC (phosphoenolpyruvate carboxylase) replenishes the oxaloacetate to drive a reduced TCA pathway that provides the three precursor metabolites without generating energy. (B) Anaerobic (fermentative) growth on glucose uses an open TCA path in which PepC replenishes oxaloacetate and a new enzyme, PFL (pyruvate-formate lyase), makes acetyl-CoA. The reactions marked with a star are catalyzed by enzymes that replace the aerobic enzymes. (C) Aerobic growth on malate reverses the direction of the reactions of the central pathways to make the precursor metabolites and also replaces the enzymes in some of the reactions (marked with a star). An anaerobic branch of the pentose phosphate cycle is also called into play.

substrates, a metabolic pathway called **gluconeogenesis**. Such flexibility in central pathway operations demands that, at some key steps, the same chemical reaction must be catalyzed by separate enzymes, subject to separate controlling factors, so as to achieve the direction of metabolite flux that is appropriate in a given circumstance (Fig. 7.12).

ARE PRECURSOR METABOLITES ALL THAT IS NEEDED TO MAKE BUILDING BLOCKS?

LEARNING OUTCOMES:
- Summarize how researchers have used auxotrophic mutants to elucidate biosynthetic pathways.
- Differentiate the role of NADH and NADPH in metabolism.

- Define allostery.
- Explain how biosynthetic pathways are regulated.
- Identify a reason why assimilation of phosphorus occurs at the fueling stage.
- Contrast the assimilation of nitrogen with that of phosphorus.
- Describe why ammonia is the preferred form of nitrogen for microbes.
- List ways in which microbes can obtain nitrogen from environmental sources.
- Identify a reason why nitrogen fixation is so energetically expensive.
- Describe how environmental conditions prescribe the pathway of nitrogen assimilation.
- Compare and contrast nitrogen and sulfur assimilation.
- Identify reasons why many bacterial pathogens and obligatory parasites have evolved mechanisms to import all preformed sulfur-containing building blocks.
- State how amino acids are grouped by families.
- Explain why the synthesis of DNA and RNA building blocks consumes much energy in a fast-growing cell.
- Describe how diverse fatty acids are synthesized to respond to the physical environment.
- State reasons why microbiology research has been essential to elucidating biochemical pathways of all life.
- Justify the following statement with biochemistry-related evidence: "Cells, organelles, and all major metabolic pathways evolved from early prokaryotic cells."

With few exceptions, all the biosynthetic reactions that generate building blocks are now known. But how were the pathways of biosynthesis deciphered? In the mid-20th century, scientists used bacteria having no nutritional requirements for building blocks (**prototrophs**) to isolate and characterize mutants with one or another nutritional requirement. Such mutants (called **auxotrophs**) provided the clues needed to unravel the complex networks of biosynthesis. Auxotrophic mutants, largely derived from *E. coli* and *Salmonella*, could not make one or more of the building blocks; it had to be provided in the medium (note the similarities with the natural isolate from the case study). However, the mutant defect was rescued in cultures that provided intermediates *past the blocked reaction*. Auxotrophic mutants also have a distinctive phenotype: they accumulate large quantities of the intermediate made *before the blocked reaction*. These intermediates, in turn, could feed mutants blocked at prior steps and so on. Hence, like a jigsaw puzzle, scientists used these mutants to piece together the biosynthetic pathways.

A second productive approach was to feed the mutant cells an **isotope** labeled metabolite. By monitoring the metabolism of the labeled substrate, scientists were able to measure enzyme activities in live cells and then deduce the enzyme missing in each mutant. One by one, the steps responsible for making each building block were identified. This achievement in microbiology carried a bonus for all of biology: the pathways discovered in bacterial studies have been largely copied by other bacteria, microbes, plants, and animals. Thus, the biosynthetic pathways that lead to the synthesis of building blocks are broadly conserved.

From 13 Precursor Metabolites to Nearly 100 Building Blocks

Until now, we have considered the major building blocks: nucleotides, amino acids, fatty acids, and sugars. But a cell also needs enzyme cofactors and a number of other building blocks essential to run its biosynthetic reactions. If we lay out on a very large sheet the structures (or even just the names) of all 75 to 100 building blocks (including the enzyme cofactors) and then add their enzymatically catalyzed reactions leading to their synthesis, the resulting metabolic map will be overwhelming. Hundreds of enzymatically catalyzed reactions will be diagrammed on the chart! A display of this size and complexity is more easily viewed on a dynamic computer screen. Biosynthetic metabolism is displayed in all its glory in a number of databases, including on the EcoCyc website.

On inspection, however, a simplifying organization does emerge from the maze of biosynthetic reactions (Fig. 7.13). Remember, nearly 100 building blocks are made from only 13 precursor metabolites—the intermediates of fueling pathways identified in chapter 5. Thus, from a mere baker's dozen of familiar starting compounds, parallel and branching pathways emerge to form the overall pattern of the biosynthetic chart. Complex as this picture is, it still misses two essential components of biosynthetic pathways: energy, in the form of reducing power and ATP, and regulatory control.

Reducing power and ATP. The building blocks—end products of each pathway—are generally more reduced than the precursor metabolites, and therefore the reactions that lead to their synthesis consume reducing power. Recall that *fueling reactions* produce reducing power primarily in the form of *NADH* (the reduced nicotinamide adenine dinucleotide), whereas *biosynthetic pathways* consume reducing power *largely in the form of NADPH* (the reduced coenzyme nicotinamide adenine dinucleotide phosphate) (chapter 6). Thus, cells must convert the NADH produced during fueling into the NADPH needed for biosynthesis. This critical task is the work of **NADP transhydrogenases**; these enzymes also ensure that the ratio of NADPH/NADP$^+$ inside the cell is kept higher than the ratio of NADH/NAD$^+$. This way NADPH is readily available for biosynthetic reactions and NAD$^+$ for dehydrogenation reactions in the fueling paths.

Of course biosynthesis also requires energy, in the form of ATP. ATP drives the otherwise thermodynamically unfavorable flow of material from precursor to building block in most of the pathways. This is why ATP also needs to be generated during fueling. And to make ATP, cells assimilate phosphorus at the fueling stage (Fig. 7.1). On the other hand, nitrogen and sulfur are assimilated during biosynthesis (Fig. 7.1).

Enzymatic regulation. One critical feature of biosynthetic pathways is not depicted in our metabolic map (Fig. 7.13). Enzymes that mediate key reactions in each pathway can assume two forms, *one catalytically active and one inactive.* How is such an allosteric enzyme switched on and off? Its activity is controlled by either binding or lack of binding of a particular molecule, called the **ligand**. Often these ligands are low molecular weight metabolites. They recognize and bind to a specific site on the protein, distinct from the enzyme's catalytically active site. When the ligand is bound,

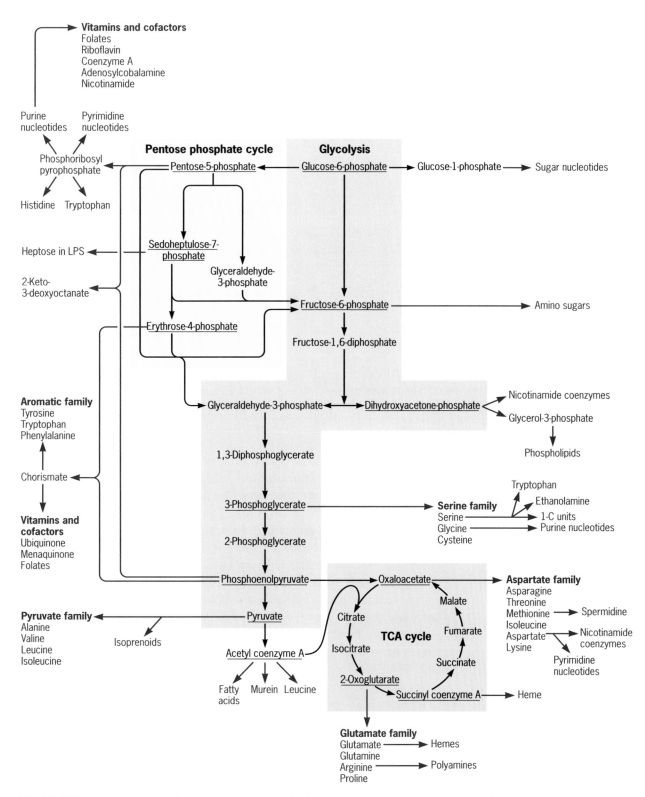

FIGURE 7.13 **Paths from central metabolism to biosynthetic end products.** The 13 precursor metabolites are underlined. The biosynthetic end products are shown in red.

A Generalized features of a pathway

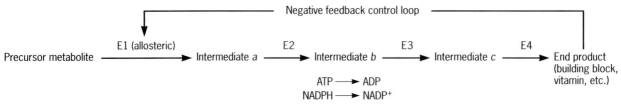

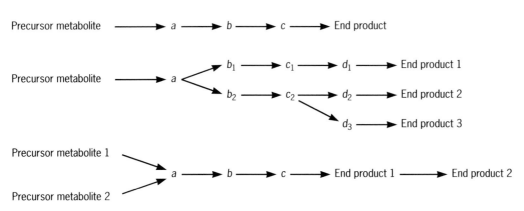

B Some pathway patterns

FIGURE 7.14 Characteristics of microbial biosynthetic pathways. (A) Generalized features of a biosynthetic pathway. E1 to E4, enzymes in the pathway. (B) Patterns of different pathways, illustrating multiple precursors and pathway branching.

the enzyme is forced into a form that is *inactive*. It also works the other way: a ligand *activates* some allosteric enzymes.

In many biosynthetic pathways, the allosteric ligand for the first enzyme in its biosynthetic pathway is actually its *end product*, a building block (Fig. 7.14). What an elegant design! Once a cell has produced sufficient quantities of, say, an amino acid, that same amino acid then inactivates the first enzyme in its biosynthetic pathway. By "biting the hand that feeds it," the end product tells the cell when to stop making more.

But such feedback control by end products does create an unexpected situation: these biosynthetic pathways are governed not by the supply of their starting materials, but by the rate their end products are consumed in polymerization reactions! Consider the curious implication for growth: cells *do not grow as fast as their building blocks can be made; instead, they make their building blocks as fast as they can grow!* We shall return to this topic and to allostery in chapter 11.

But one fact should already be clear: biosynthetic pathways are real biological entities, not drawings in a book to organize the biosynthesis network into bite-size pieces. The microbial cell, not the microbiologist, has defined the pathway as a metabolic unit. Scientists continue to discover compelling evidence that the enzymes of a pathway are bound to each other, physically associated within the cell, to facilitate their coordinated action. For example, close proximity can accelerate processes by minimizing the time it takes metabolites to diffuse from one catalyst to the next. Clearly, the cell is a highly ordered enzymatic machine, not some bag of enzymes.

Assimilation of Phosphorus

Phosphorus is a major component of nucleic acids, phospholipids, many co-enzymes, as well as phosphorylated sugars and proteins. Not surprisingly, all cells have a hearty appetite for this element. (Farmers know this all too well: phosphate is limiting in many environments and its availability often determines whether crops or weeds will grow.) Thus, evolutionary pressures have driven microbes to acquire effective strategies to assimilate this vital component, despite competition with other organisms.

Phosphorus usually enters the cell as inorganic phosphate, though organic phosphate compounds are another environmental source. Since most cells are impermeable to organic phosphates, many bacteria hydrolyze phosphate esters outside the cell, or, in the case of Gram-negatives, in the periplasm. When phosphate is abundant in the environment, the enzymes that hydrolyze organic phosphates are repressed. Next, phosphate enters the cell via either general or specific transport systems. Phosphorus assimilation occurs at the fueling stage (Fig. 7.1) in reactions of feeder and central pathways. The ATP produced by these reactions then serves as the major phosphate donor in the cell. The phosphorus assimilation reactions are numerous; prominent and familiar ones occur in the central pathways, and some were described in chapter 6. We will see how phosphate is sensed and its acquisition regulated in chapter 12.

Assimilation of Nitrogen

Nitrogen is a significant component of the cell, being the main element by weight after carbon (chapter 5). Nitrogen is found in amino acids, nucleotides, and other compounds, and comprises about 15% of the dry weight of most cells. Cells assimilate this critical element in three major steps (Fig. 7.15): (i) they first acquire environmental sources of nitrogen such as dinitrogen gas (N_2), the major component of Earth's atmosphere, nitrate and nitrite, and use them to make ammonia (NH_3) or ammonium (NH_4^+); (ii) the ammonium is then used to synthesize glutamate and

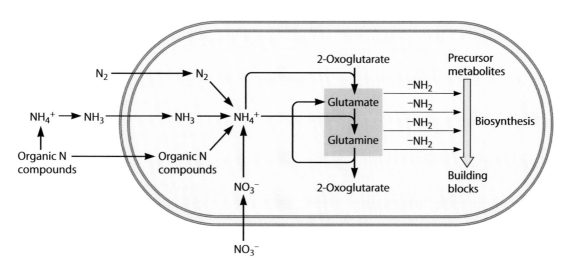

FIGURE 7.15 **Biosynthetic assimilation of nitrogen.**

glutamine, and (iii) these two amino acids then feed the nitrogen into the various pathways that generate building blocks. Let us consider the three major steps of nitrogen assimilation in more detail.

Making ammonia from environmental nitrogen sources. The Earth's nitrogen resides in various chemical forms and ecological locations. The ability to scavenge all possible sources of nitrogen in the environment has been a potent force in evolution. As a result, different microbes can acquire nitrogen from almost any molecule, organic or inorganic, or draw it directly from the great reservoir of atmospheric dinitrogen gas (N_2) to make the ammonium ion or its uncharged form, ammonia (NH_3) (Fig. 7.15). To a great extent, the interconversion of different forms of nitrogen and their transfer between physical locations is driven by microbial activity. The nitrogen cycle is described in more detail in chapter 20.

Inorganic and organic sources of nitrogen. In lab cultures, microbes can easily take up and use ammonium ion (NH_4^+) to make glutamate and glutamine. Dissolved gaseous ammonia (NH_3) diffuses into most microbial cells through transmembrane protein channels. But ammonia is relatively rare in natural environments (except where anhydrous ammonia has been used as a fertilizer). Thus, microbes often process nitrogenous compounds enzymatically to make ammonia. A variety of enzymes achieve this. Some microbes secrete, for example, deaminases into their environment or periplasm to release ammonia from the nitrogen sources. *Helicobacter pylori* and others secrete ureases to hydrolyze urea to ammonia and carbon dioxide, whereas *Klebsiella pneumoniae* produces a periplasmic asparaginase that hydrolyzes asparagine to ammonia and aspartate.

Each of these cells then takes up the resulting ammonia. Some microbes can also obtain ammonia from nitrate ion in a two-step pathway catalyzed by assimilatory nitrate reductase (nitrate to nitrite) and assimilatory nitrite reductase (nitrite to ammonia) (chapter 20).

N_2 fixation. Some microbes have the remarkable ability to make ammonia (or nitrate) from atmospheric dinitrogen gas (N_2), a process known as **nitrogen fixation**. Unique to prokaryotes, this skill is carried out by groups within the *Bacteria* and *Archaea*. Indeed, microbes are critical to the cycling of nitrogen on Earth (chapter 20). Biological nitrogen fixation (Fig. 7.16) is mediated

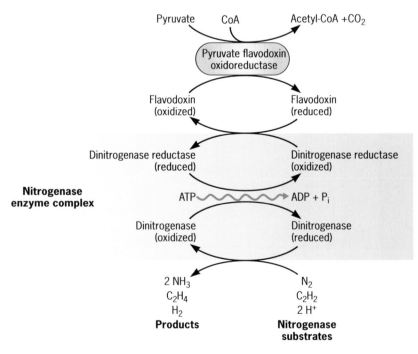

FIGURE 7.16 **Nitrogen fixation.**

by the **nitrogenase** enzyme complex. Dinitrogenase mediates the actual eight-electron reduction of dinitrogen gas, working in conjunction with an iron-molybdenum cofactor and using electrons supplied by nitrogenase reductase. The reaction generates two molecules of ammonia:

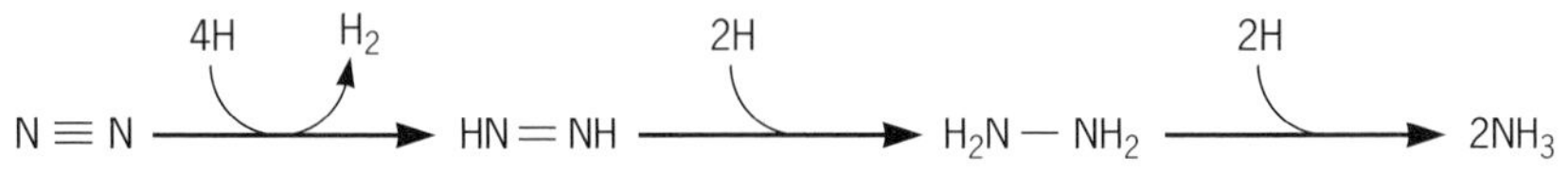

$$N \equiv N \xrightarrow[]{4H \quad H_2} HN = NH \xrightarrow[]{2H} H_2N - NH_2 \xrightarrow[]{2H} 2NH_3$$

Nitrogen fixation is quite energy intensive—it takes high activation energy to break the very strong triple bond that joins the two nitrogen atoms in N_2. Typically 16 molecules of ATP are needed to reduce just 1 molecule of dinitrogen to ammonia:

Overall reaction

$$8\ H^+ + 8\ e^- + N_2 \longrightarrow 2\ NH_3 + H_2$$

$$(16–24\ ATP \rightsquigarrow 16–24\ ADP + 16–24\ P_i)$$

The nitrogenase reductase complex has some noteworthy features. It is highly conserved, with a primary sequence that is 70% identical among the *Bacteria* and *Archaea*. Moreover, mixtures of dinitrogenase and dinitrogenase reductase proteins isolated from different species, or even from species in different domains, can nevertheless interact to reduce dinitrogen. Perhaps it evolved just once! The nitrogenase enzyme complex is also extremely sensitive to oxygen—it is among the most highly oxygen-labile proteins known. Consequently, nitrogen-fixing prokaryotes have evolved a variety of strategies to protect their enzyme complexes. For example, filamentous cyanobacteria produce specialized cells called **heterocysts**, which are nitrogen fixation factories engulfed in thick coats that minimize oxygen intrusions that could poison the nitrogenase enzymes (chapter 13).

Using ammonia to make glutamate and glutamine. Nitrogen assimilation has one goal: to make glutamate and glutamine from ammonia. Perhaps not surprisingly, these two synthetic pathways are interconnected. Some bacteria can synthesize glutamate in one of two ways, depending on the nutritional conditions—a neat example of how ecology prescribes physiology. Glutamate can be synthesized by incorporating ammonia into a precursor metabolite of the TCA cycle, 2-oxoglutarate, in a reaction catalyzed by glutamate dehydrogenase:

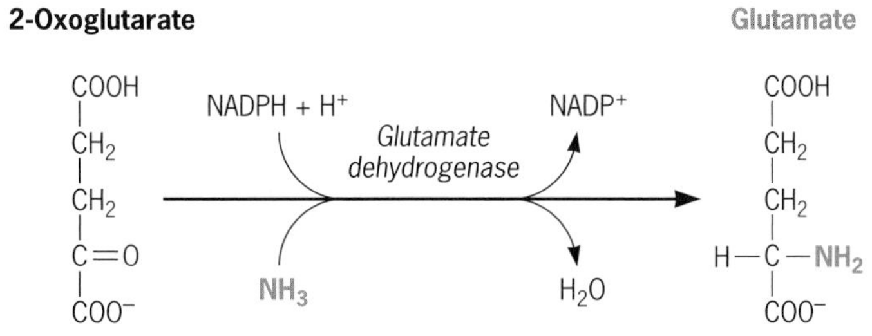

Some microbes instead make glutamate from glutamine and 2-oxoglutarate in a reaction catalyzed by glutamate-oxoglutarate amido transferase (GOGAT):

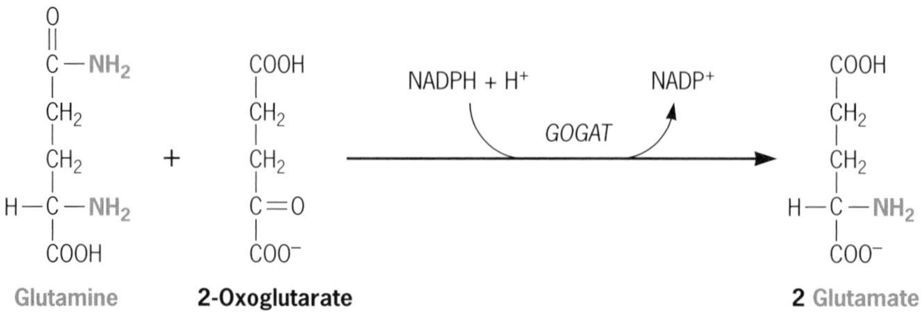

Next, glutamine can be synthesized from glutamate by incorporating a second molecule of ammonia in an ATP-driven reaction catalyzed by glutamine synthetase (GS):

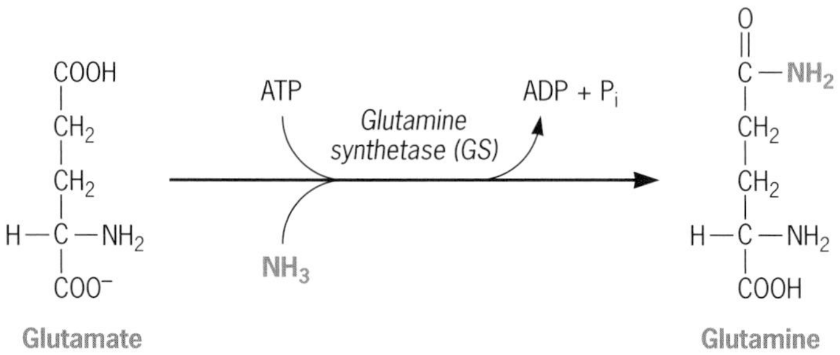

What advantage might microbes gain by having two routes for synthesizing glutamate? Note that making glutamine requires energy from ATP hydrolysis. Hence, making glutamate directly from a precursor metabolite and ammonia (reaction 1) requires less energy than beginning with glutamine (reaction 2). So there may be a trade-off between scavenging low concentrations of environmental ammonia and conserving ATP. Although the glutamate dehydrogenase route requires no expenditure of ATP, the reaction requires substantial concentrations of ammonia. The second route (GS plus GOGAT) has the opposite set of properties: it requires a greater expenditure of ATP (one molecule per glutamate), but it functions well at low concentrations of ammonia. If ammonia is abundant in the environment, the cell saves ATP by using the first route; if ammonia is scarce, it pays the energy price to use the second route.

Using glutamate and glutamine to make other building blocks. Glutamate is by far the more important source of nitrogen for biosynthesis: about 75% of the cell's nitrogen flows through it. Its role is to donate its *amino group* by **transamination** for the biosynthesis of other amino acids, such as phenylalanine. Without transamination, the cell would have to consume electrons to convert the alpha keto acid into an amino acid.

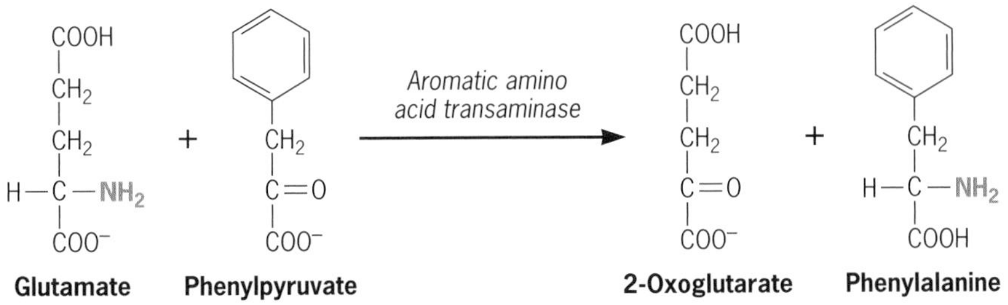

Glutamine, on the other hand, donates its *amido group* to various other biosynthetic reactions, including that of the nucleotide cytidine triphosphate (CTP):

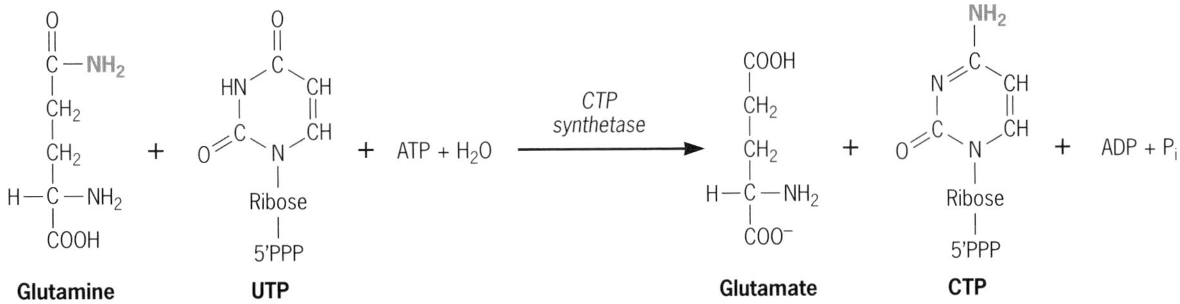

The cell now enlists its duo of nitrogen donors—glutamate and glutamine—to make any building block that needs nitrogenous groups. As a bonus, glutamate and glutamine together sequester three amino groups, protecting the cell from toxic free ammonia.

Assimilation of Sulfur

Sulfur is another essential element that all cells need for their building blocks. The amino acids methionine and cysteine, for example, contain sulfur; so do cofactors such as coenzyme A and many enzymes involved in energy metabolism. How do prokaryotes incorporate sulfur? Like nitrogen, once reduced, this element is assimilated during biosynthesis of building blocks from precursor metabolites (Fig. 7.17). Sulfur always enters as an inorganic form, sulfide (as H_2S or S^{2-}). Again, like nitrogen, sulfur assimilation starts with the entry of organic or inorganic sources of sulfur, their convergence into an ionic elemental species (in this case, S^{2-}, the ionic form of H_2S), and then its assimilation in reactions involving amino acids. In the case of sulfur metabolism, L-serine is first converted into an intermediate (*O*-acetyl-serine) in a reaction catalyzed by the serine acetyltransferase (SAT) enzyme. A second reaction, catalyzed by the enzyme *O*-acetylserine sulfhydrylase (OASS), incorporates the sulfide to make the sulfur-containing amino acid L-cysteine:

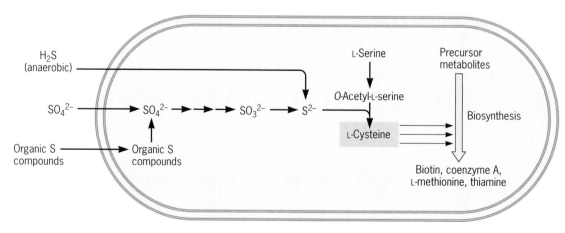

L-Cysteine then serves directly or indirectly as the source for most other sulfur-containing compounds in the cell, including L-methionine, biotin, thiamine, and coenzyme A (CoA).

Having enough sulfide available to feed this important reaction can be a challenge. Because sulfide is spontaneously oxidized by oxygen, it exists primarily in anaerobic environments. Strict anaerobes, which live in environments with low oxygen levels, face no problem: if sulfide is available, they can use it for biosynthesis. Aerobes, on the other hand, must reduce oxidized sulfur compounds to sulfide for assimilation (Fig. 7.17). The major form of sulfur in aerobic soil is organic (organic sulfates, amino acids, and other carbon-bonded sulfur compounds). Acquiring sulfur from organic sulfur compounds requires oxidizing them to sulfate (SO_4^{2-}) first. What happens next is energetically expensive, because sulfate must be converted into sulfite (SO_3^{2-}) and then sulfide in a series of four reactions requiring the consumption of three high-energy phosphate bonds, three NADPH, and a molecule of reduced **thioredoxin**. Without this expensive pathway, organisms are restricted to a diet of sulfide and therefore are chiefly anaerobes. Of course, there is an alternative: import all the preformed sulfur-containing building blocks. Many bacterial pathogens and obligatory parasites do just that!

FIGURE 7.17 **Biosynthetic assimilation of sulfur.**

**Using Energy, Reducing Power, and Precursor Metabolites
To Make Building Blocks**

Now that our microbial cells have secured supplies of carbon, phosphorous, nitrogen, and sulfur, let's see how they can exploit just about any substrate in their environment to make the building blocks for macromolecules. For clarity, we will simplify the biosynthetic pathways that equip microbial cells to convert just 13 precursor metabolites into almost 100 building blocks. Using Fig. 7.13 as a guide, we will describe some general features of the pathways leading to the main classes of building blocks. You can study these pathways in all their glorious detail at the EcoCyc website.

Amino acids. Microbes incorporate 21 amino acids into their proteins. These 21 amino acids can be grouped into six families, each defined by a common precursor metabolite, or a combination of precursors (Table 7.4). For example, all of the amino acids in the serine family are synthesized from a single precursor metabolite (3-phosphoglycerate). Similarly, 2-oxoglutarate is the sole precursor for all the amino acids in the glutamate family. The odd ball is histidine: because it is built from a carbon and nitrogen from the six-membered ring of a purine, ATP in this case is actually a precursor metabolite, rather than an energy source. Making aromatic amino acids is the most complex process, requiring two or three precursor metabolites. In the case study, G isolated a bacterium that could not make any of the aromatic amino acids. What precursor metabolite(s) do you think G's isolate failed to make? In nature, some microbes synthesize all the amino acids, but not all. Others make only some, and still others none at all. Will this affect the type of environment they can live in?

Nucleotides. The building blocks for nucleic acids (DNA and RNA) are manufactured in an activated state, ready for polymerization. In particular, the end products of the purine and pyrimidine biosynthetic pathways are ribonucleoside and deoxyribonucleoside triphosphates (Table 7.1). The purine nucleotides—both the ribose and deoxyribose forms—are made from the precursor metabolite ribose-5-phosphate in a multistep network catalyzed by 21 enzymes. The pyrimidine nucleotides are made from two precursor metabolites, ribose-5-phosphate and oxaloacetate, in a branching network catalyzed by 24 enzymes. Needless to say, these pathways consume much energy (as high-energy phosphate), reducing power (from both NADH and NADPH), and assimilated nitrogen. Since most fast-growing bacteria are rich in nucleic acids (up to 25% of the cell's total dry weight), the demand for these building blocks is high. To grow economically and efficiently, microbes must conserve the resources they need to synthesize RNA. To do so, they modulate the amounts of RNA made at different growth rates (chapter 4).

Sugars and sugar-like derivatives. The specialized capsules, walls, and outer membranes of many microbes depend on the synthesis of sugars and sugar-like derivatives. Since infectious microbes must evade recognition by their host's immune system (chapter 22), these selective pressures have contributed to the evolution of dozens of different sugar derivatives in various bacteria. Few, if any, of these components can be assimilated ready-made from the environment. Instead, almost all must be made from the precursor metabolites, in this case four-, five-, six-, and seven-carbon sugars (Fig. 7.13). Sugars also contribute to storage. Like humans, many prokaryotes store carbon reserves as glycogen. Some cells hold these polymer reserves in organelle-like structures called inclusion bodies (chapter 3).

Fatty acids and lipids. The biological surfaces that help define Earth's living organisms are membranes composed of phospholipid bilayers. The membrane structure of the *Archaea* is a variation on this theme (chapter 2). Recall that phospholipids are diglycerides that contain fatty acids in ester linkage with glycerol. A few other compounds also contain fatty acids, such as lipoprotein, the protein that anchors the murein wall to the cell membrane in Gram-negative bacteria. However, the amounts of fatty acids in these compounds are minor compared with those in the phospholipids of bacterial membranes and in lipid A of the outer membrane in Gram-negative bacteria. In spite of the vast diversity of fatty acids made by bacteria, all of them are synthesized by the same modular scheme, rather like making many different structures by stacking the same kind of bricks. The brick for making fatty acids is a two-carbon unit derived from acetyl-CoA. Units are added over and over until a fatty acid of the appropriate length is made.

Fatty acids differ in the *number of carbon atoms*, the *number and placement of double bonds*, and whether they are *branched*. In some cases, the diversity of phospholipids reflects specific responses to distinct physical environments (chapter 13). For example, the fluidity of lipids is increased when there are shorter chains and more double bonds and branching. Microbes have evolved mechanisms to modulate these characteristics during fatty acid synthesis to respond to fluctuating environmental temperature and pressure. Branched chain and unsaturated fatty acids are synthesized from a common pathway which then diverges at various points, depending on the species. The fatty acid end products of these pathways remain attached to CoA until they are incorporated into one of the several membrane phospholipids, lipid A, or lipoproteins.

Parting Thoughts about Biosynthesis

We close this chapter on biosynthesis by emphasizing two microbial themes. First, what we know of the ways all living creatures synthesize their building blocks has been learned largely by studying bacteria. The reasons are simple. Many bacteria can make all the building blocks, including some that humans cannot (those we call "essential" nutrients). In the middle of the 20th century, bacteria provided the best practical opportunity for a concerted genetic and biochemical exploration of biosynthesis. Indeed, the very concept of a biochemical pathway was born in the microbial sciences. Second, the discovery of biosynthetic pathways in microbes verified that the "unity of biochemistry" theme enunciated by biochemists early in the 20th century was indeed correct. All life on this planet has a common ancestry, and this underlying relatedness is reflected in similar, if not identical, biochemistry. Biosynthesis, then, reflects the essential relation of microbes to plants and animals. While the fueling metabolism of microbes is a testament to the potential for life forms to scavenge for food and thereby recycle organic material, their biosynthetic metabolism illuminates what we have lost in evolving into beings higher in the food chain.

CASE REVISITED: A bacterium that cannot make three essential amino acids

As we learned in this chapter, the three amino acids (phenylalanine, tyrosine, and tryptophan) in the aromatic family are essential building blocks to make proteins. A bacterium unable to make them *de novo* would die unless able to acquire them from the environment. That is why G could not grow her bacterial isolate unless she supplemented the growth medium with sufficient amounts of the three aromatic amino acids. Did you come up with a hypothesis to explain why the bacterium could not make the three aromatic amino acids? Two precursor metabolites (erythrose-4-phosphate and phosphoenolpyruvate) are needed for the synthesis of the three aromatic amino acids (tryptophan synthesis also requires pentose-5-phosphate) (Table 7.4). One can hypothesize that either the bacterium was unable to make enough quantities of the precursor metabolites or lacked the ability to make the aromatic amino acids from them. Supplementing the growth medium with these precursor metabolites, alone or in combinations, could help you test these hypothesis:

- Outcome 1: adding the precursor metabolites to the growth medium does not rescue the growth requirement for aromatics (i.e., aromatic amino acid supplementation is still needed to support growth). This experimental result would suggest that the bacterium cannot make the aromatics *de novo* because it lacks one or more of the biosynthetic reactions that make these amino acids *de novo* from the precursor metabolites. Alternatively, the pathways may be present, but they may just be too slow to support the high demands of building blocks needed to support cell growth. Or, the microbe may fail to transport the precursor metabolite.
- Outcome 2: adding the precursor metabolites rescues the growth requirement for aromatics. This result would suggest that the bacterium has the biosynthetic capacity to make these aromatics *de novo* but cannot make the precursor metabolites in sufficient quantities to satisfy the biosynthetic demands for these amino acids.

Either outcome would suggest that this bacterium lives in an environment where aromatic amino acids may be abundant. Over time, the bacterium reduced its aromatics biosynthetic capacity to capitalize on the availability of these building blocks in the environment.

CONCLUSIONS

In this chapter, we have examined how nutrients enter the cell and provide the substrates for fueling reactions that produce ATP, reducing power, and the 13 precursor metabolites needed for the synthesis of building blocks. The cell now has everything it needs to make the essential macromolecules: amino acids for proteins, nucleotides for nucleic acids, sugars for carbohydrates, and fatty acids for lipids, along with the considerable amounts of ATP needed for their synthesis and activity.

The following points bear repeated emphasis.

1. Using clever strategies for selective and fast transport of solutes across otherwise impermeable membranes, microbes exploit an exotic variety of organic and inorganic substrates as sources of carbon.

2. The cell membrane is used in cunning ways to transduce energy and to drive various forms of transport.

3. An enormous number of feeder paths process the incoming substrates into intermediates that are fed to the central pathways to generate the precursor metabolites. At this step, the metabolic versatility of microbes is on full display.

4. The central pathways are flexible. These pathways are reversible, depending on the substrates provided by the environment, and yet all lead to the generation of all the energy, reducing power, and precursor metabolites needed to make the building blocks.

5. Their complex biosynthetic machinery equips many microbes to be self-reliant: they make all of their essential building blocks *de novo*.

Representative ASM Fundamental Statements and Learning Outcomes for the Chapter

EVOLUTION

1. Cells, organelles, and all major metabolic pathways evolved from early prokaryotic cells.

    a. Justify the following statement with biochemistry-related evidence: "Cells, organelles, and all major metabolic pathways evolved from early prokaryotic cells."

CELL STRUCTURE AND EVOLUTION

2. *Bacteria* and *Archaea* have specialized structures that often confer critical capabilities.

    a Describe three methods of molecular transport across the outer membrane of Gram-negative bacteria.

    b. Compare and contrast simple diffusion and facilitated diffusion.

    c. Compare and contrast ion-coupled transport, ABC transport, and group translocation.

METABOLIC PATHWAYS

3. *Bacteria* and *Archaea* exhibit extensive, and often unique, metabolic diversity.

    a. Identify a reason why nitrogen fixation is so energetically expensive.

4. The interactions of microorganisms among themselves and with their environment are determined by their metabolic abilities.

    a. Identify two strategies microbes use to obtain building blocks, such as essential amino acids, from exogenous sources.

    b. Describe how adaptations in metabolic pathways allow some bacteria to grow and survive under varying environmental conditions.

    c. List ways in which microbes can obtain nitrogen from environmental sources.

    d. Describe how environmental conditions prescribe the pathway of nitrogen assimilation.

5. The survival and growth of any microorganism in a given environment depends on its metabolic characteristics.

a. Contrast biosynthetic capacity and nutrient requirements for a bacterium that normally inhabits nutrient-rich environments with that of a bacterium that normally lives in nutrient-poor environments.

b. Describe three major requirements needed for *de novo* synthesis reactions.

c. Summarize the purpose of each of the three stages of fueling reactions.

d. Identify potential negative consequences for a microbe that can make all of its own building blocks.

e. Describe the role of siderophores in microbes.

f. Identify the purpose of feeder pathways for heterotrophs and autotrophs.

g. Identify the role of RuBisCO in the autotrophic feeder pathway.

h. Describe how glycolysis, the pentose phosphate pathway, and the TCA cycle are interconnected in metabolic pathways.

i. List three auxiliary fueling pathways of central metabolism.

j. Differentiate the role of NADH and NADPH in metabolism.

k. Define allostery.

l. Explain how biosynthetic pathways are regulated.

m. Identify a reason why assimilation of phosphorus occurs at the fueling stage.

n. Contrast the assimilation of nitrogen with that of phosphorus.

o. Describe why ammonia is the preferred form of nitrogen for microbes.

p. Compare and contrast nitrogen and sulfur assimilation.

q. Identify reasons why many bacterial pathogens and obligatory parasites have evolved mechanisms to import all preformed sulfur-containing building blocks.

r. State how amino acids are grouped by families.

s. Explain why the synthesis of DNA and RNA building blocks consumes much energy in a fast-growing cell.

t. Describe how diverse fatty acids are synthesized to respond to the physical environment.

IMPACT OF MICROORGANISMS

6. Microorganisms provide essential models that give us fundamental knowledge about life processes.

a. Summarize how researchers have used auxotrophic mutants to elucidate biosynthetic pathways.

b. State reasons why microbiology research has been essential to elucidating biochemical pathways of all life.

SUPPLEMENTAL MATERIAL

References

- **Merchant SS, Helmann JD.** 2012. Elemental economy: microbial strategies for optimizing growth in the face of nutrient limitation. *Adv Microb Physiol* **60:**91–210.
- **Spaans SK, Weusthuis RA, van der Oost J, Kengen SW.** 2015. NADPH-generating systems in bacteria and archaea. *Front Microbiol* **6:**742. doi: 10.3389/fmicb.2015.00742.
- **Wilkens S.** 2015. Structure and mechanism of ABC transporters. *F1000Prime Rep* **7:**14. doi: 10.12703/P7-14.
- **Noinaj N, Guillier M, Barnard TJ, Buchanan SK.** 2010. TonB-dependent transporters: regulation, structure, and function. *Annu Rev Microbiol* **64:**43–60.
- **Frawkey ER, Fang FC.** 2014. The ins and outs of bacterial iron metabolism. *Mol Microbiol* **93:**609–616.
- **Berg IA, Kockelkorn D, Ramos-Vera WH, Say RF, Zarzycki J, Hügler M, Alber BE, Fuchs G.** 2010. Autotrophic carbon fixation in archaea. *Nat Rev Microbiol* **8:** 447–460.

Supplemental Activities

CHECK YOUR UNDERSTANDING

1. **Compared to making amino acids and sugars, the biosynthesis of nucleic acid building blocks is more costly because**

a. more building blocks are required.

b. the concentration of ATP in the nucleoid is lower than in the cytoplasm.

c. microbes can more readily scavenge peptides and sugars from the environment.

d. the building blocks for nucleic acid contain energy to drive polymerization.

e. nucleic acids are less stable than proteins and sugars.

2. **In the biosynthesis phase of growth metabolism, glutamate and cysteine are unique because they**

a. are the major source of nitrogen and sulfur, respectively, for the cells' macromolecules.

b. have dedicated active transporter for entry.

c. are subject to end product feedback control.

d. are generated by enzymes that are destabilized by oxygen.

3. **What selective pressure might have led to the evolution of two coenzyme carriers of hydrogen atoms (NAD$^+$ and NADP$^+$) rather than a single one?**

4. **In both biosynthesis and catabolism, large carriers ferry small "precious" cargo: CoA hangs on to acetate; NAD$^+$ and NADP$^+$ cling to electrons. How do these large carriers benefit the cell?**

5. **Cite a major way in which assimilation of phosphorus differs from that of sulfur or nitrogen.**

DIG DEEPER

1. Unlike the parental strain, growth of a mutant strain of the intracellular pathogen *Listeria monocytogenes* in laboratory medium required supplementation with 10 mM threonine or 2.5 mM of a peptide that contained threonine. However, the mutant cells replicated at nearly the same rate as the wild-type parent bacteria once they infected host cells.

a. Describe experiments to test the following hypotheses:

   i. This is an auxotrophic mutant that cannot transport threonine.

   ii. This is an auxotrophic mutant that cannot produce a critical precursor metabolite needed to synthesis threonine.

   iii. This is an auxotrophic mutant that cannot produce threonine from its precursor metabolite.

b. The mutant required only 2.5 mM of a threonine peptide but 10 mM of threonine to grow optimally. Why?

c. Based on the observation that the mutant replicated similarly to wild-type *L. monocytogenes* within host cells, what hypotheses can you form about this pathogen? This host?

2. Since 1923, clinical microbiologists have enriched for *Salmonella* serovars by culturing fecal samples in a broth supplemented with tetrathionate, a terminal electron acceptor. In 2014, Andreas Baumler and his colleagues performed an *in silico* analysis to compare the genomic content of 15 *Salmonella enterica* strains, some isolated from the gastrointestinal tract and others from body sites outside of the intestine. Read their report, "Comparative analysis of *Salmonella* genomes identifies a metabolic network for escalating growth in the inflamed gut" (Nuccio SP, Baumler AJ, *mBio* **5:**e00929-14, 2015, http://mbio.asm.org/content/5/2/e00929-14.full), to answer the following questions.

a. Why is tetrathionate useful for diagnostic tests of patients with diarrhea?

b. What major class of genes appears to be subjected to strong selective pressure in the gastrointestinal tract?

c. How does this class of genes increase the fitness of microbes in the gastrointestinal tract?

d. How could knowledge generated by this comparative genomic study inform the design of molecular tests to improve food safety? Patient care?

3. **Team work:** Working in groups, check the website EcoCyc (http://ecocyc.org) to find the pathways described below. Summarize your findings in a short 5-minute presentation to the class describing the pathway and its biological role.

a. The longest biosynthetic pathway leading to an amino acid.
b. The shortest biosynthetic pathway leading to an amino acid.
c. A pathway for the synthesis a nucleotide.
d. A pathway for the assimilation of nitrogen.
e. A pathway for the synthesis of a carbon storage compound.
f. A pathway that is regulated by an allosteric enzyme.

Supplemental Resources

VISUAL

- *The iBiology* lecture "Cooperation between bacteria and plants for protein nutrition" by Sharon Long discusses symbiosis between nitrogen-fixing bacteria and legumes (19 min): (http://www.ibiology.org/ibioseminars/plant-biology/sharon-long-part-1.html).
- Generation of *Enterobacter* sp. YSU auxotrophs using transposon mutagenesis (13:31 min): (http://www.jove.com/video/51934/generation-enterobacter-sp-ysu-auxotrophs-using -transposon).
- Six pathways of carbon fixation (10:21 min): (https://www.youtube.com/watch?v=EBG qduKvuLw).

WRITTEN

- National Center for Case Study Teaching in Science "ELVIS Meltdown!" by Richard Steward, Ann Smith, and Patricia Shields teaches the fundamental concepts of culture, growth, and metabolism: (http://sciencecases.lib.buffalo.edu/cs/collection/detail.asp?case _id=248&id=248).
- Amy Cheng Vollmer explains why understanding metabolism matters in the *Small Things Considered* blog post "Hello Again, Metabolism!" (http://schaechter.asmblog .org/schaechter/2010/07/hello-again-metabolism.html).
- Moselio Schaechter describes how two bacterial endosymbionts team up to make tryptophan for their insect host in the *Small Things Considered* blog post "A Bug in a Bug in a Bug" (http://schaechter.asmblog.org/schaechter/2011/09/a-bug-in-a-bug-in-a-bug.html).
- Moselio Schaechter discusses how some strains of *Salmonella enterica* grow on ethanolamine as carbon and energy source while respiring a sulfur compound in the human gut in the *Small Things Considered* blog post "*Salmonella*'s Exclusive Intestinal Restaurant" (http://schaechter.asmblog.org/schaechter/2012/03/salmonellas-exclusive-intestinal -restaurant.html).
- Marvin Friedman discusses how a pathogen steals food from its host in the *Small Things Considered* blog post "*Shigella* Steals Host Nutrients . . . Economically" (http://schaechter .asmblog.org/schaechter/2014/09/shigella-steals-host-nutrients-economically.html).

APPENDIX FIGURES

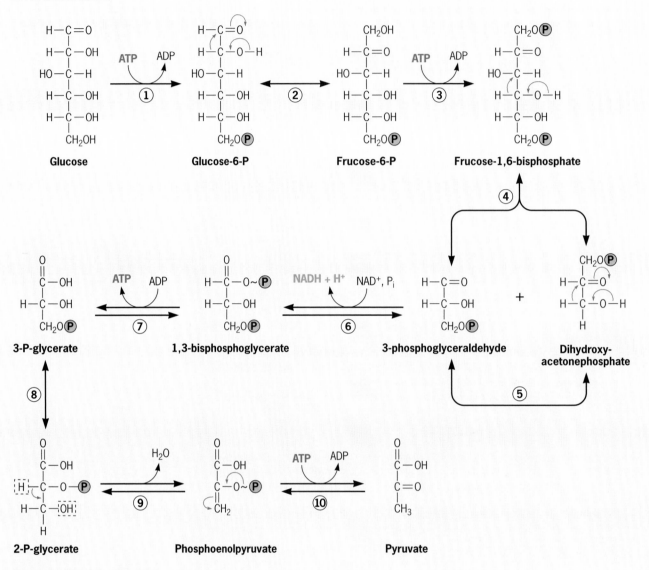

Appendix Figure 1. Glycolysis

Appendix Figure 2. Pentose phosphate cycle

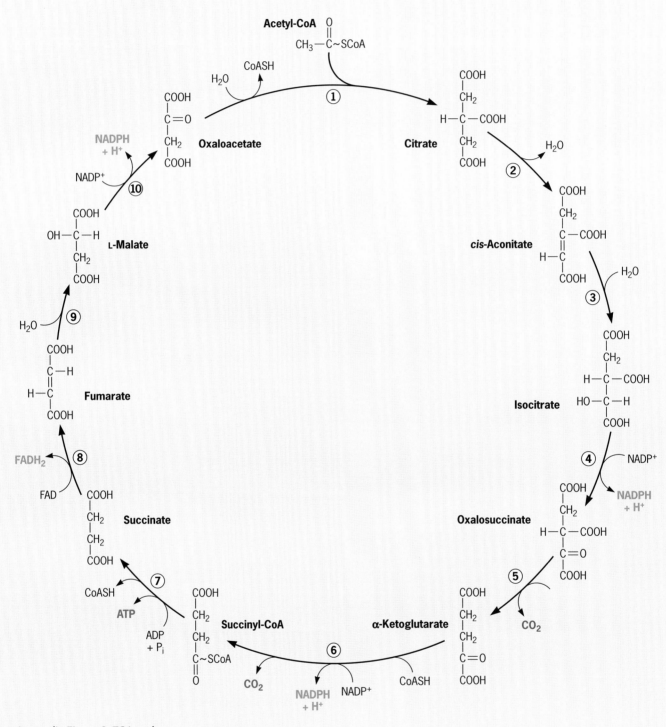

Appendix Figure 3. TCA cycle

Model of DNA replication fork

CHAPTER EIGHT

Building Macromolecules

KEY CONCEPTS

This chapter covers the following topics in the ASM Fundamental Statements.

EVOLUTION

1. Cells, organelles, and all major metabolic pathways evolved from early prokaryotic cells.

INFORMATION FLOW AND GENETICS

1. Although the Central Dogma is universal in all cells, the processes of replication, transcription, and translation differ in *Bacteria, Archaea* and eukaryotes.

IMPACT OF MICROORGANISMS

1. Microorganisms provide essential models that give us fundamental knowledge about life processes.

INTRODUCTION

Do you sense that the preceding two chapters about energy and metabolism have been leading to something important? If so, you are on target. We are now ready to see how the cell uses ATP and the products of the biosynthetic pathways (building blocks such as amino acids and nucleotides) to build its heart: the macromolecules DNA, RNA, and proteins. The steps of macromolecule synthesis may seem complex, but keep in mind that you are learning about the central processes that distinguish life from nonlife. Be prepared to encounter pathways that have been honed to near perfection by billions of years of evolution. They may not be simple, but they work wonders.

CASE: Antibiotics and Macromolecules

Mr. M, the 75-year-old grandfather of microbiology student MJ, came down with a severe case of pneumonia caused by the pneumococcus, *Streptococcus pneumoniae*. Mr. M was treated successfully with the antibiotic ciprofloxacin. MJ thought his grandfather should be happy with this outcome because pneumococcus is known to develop resistance to this and a number of other antimicrobial drugs. MJ also offered to explain to his grandfather how ciprofloxacin works but was told to wait until after his nap.

- What makes ciprofloxacin an effective antibiotic?
- How did the antibiotic work to clear the bacterial infection?
- Why was this drug not toxic to Mr. M?
- Is there a choice of other drugs for treating a pneumonia like Mr. M's?
- How does a pneumococcus develop resistance to this drug?

WHY DO CELLS HAVE NUCLEIC ACIDS AND PROTEINS?

LEARNING OUTCOMES:
- Explain why cells need both proteins and nucleic acids as opposed to needing ribozymes only.
- State a reason why DNA, not RNA, evolved to become the preferred macromolecule for inheritance.
- Explain how proteins and enzymes demonstrate high selectivity for substrates.

Most likely you are so accustomed to the universal presence of nucleic acids and proteins in living matter that the simple question of the reason for their existence never arises. But now is a good time to appreciate why all cells have these macromolecules, lest the details of their synthesis seem tedious rather than invoking awe. In chapter 1 we learned about the origin of the first cellular forms of life and the prebiotic phase that preceded it. Spontaneous chemical reactions generated the building blocks of life, such as ribonucleotides to make RNA, deoxyribonucleotides to make DNA, and amino acids to make proteins. The simplicity of RNAs and the ability of some to catalyze their own replication suggest to many that an "RNA world" preceded the formation of proteins and DNA. Some RNAs function as enzymes (we call them **ribozymes**) and may have built the first proteins from amino acids. The catalytic properties of some RNAs are evident in today's ribosomes. Strip the proteins from a ribosome and the overall structure does not substantially change (Fig. 8.1) because the two ribosomal subunits are made mostly of RNAs (ribosomal RNAs or rRNAs). Here is how they subdivide their work: The RNA in the **small ribosomal subunit** (30S in prokaryotes, 40S in eukaryotes) recognizes and binds a protein-encoding messenger RNA (mRNA) to initiate its translation; an RNA in the **large ribosomal subunit** (50S and 60S in prokaryotes and eukaryotes, respectively) bonds incoming amino acids to make a peptide. (By the way, the "S" stands for "Svedberg units," the sedimentation coefficient of these particles when spinning in an ultracentrifuge.)

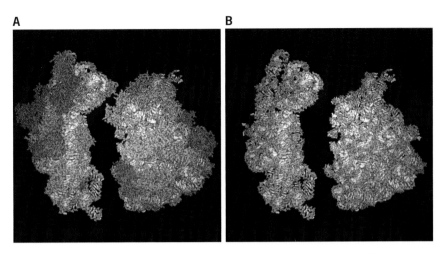

FIGURE 8.1 Structure of the prokaryotic ribosomal subunits with and without ribosomal proteins. A prokaryotic ribosome (70S) is formed when the small (30S) and large (50S) subunits come together to translate a mRNA. (A) Both subunits are made of ribosomal RNAs (blue and white) and proteins (green). (B) Strip the ribosomal subunits off their proteins and you can appreciate they are practically made of rRNAs. (Modified from Exploring Life's Origins, http://exploringorigins.org/ribozymes.html.)

Primitive ribosomes synthesized proteins from amino acids that were produced chemically, not by a living entity. The key flow of information from mRNA to amino acids was only possible once transfer RNAs (tRNAs) evolved. Now, specific triplets of ribonucleotides (**codons**) could be translated into an amino acid, and protein synthesis, as we know it, was born. This critical event in evolution equipped cells to make the same protein, with its defined sequence of amino acids, using instructions encoded in mRNA. Since proteins are *more efficient catalysts* of chemical reactions than RNAs and display *exquisite substrate specificity*, they likely replaced the ancestral ribozymes in the catalysis of chemical reactions. These catalytic proteins allowed primitive metabolisms to diversify and increase in complexity. Cellular life was about to emerge.

Once proteins took charge of biosynthetic reactions, the first DNA macromolecules could have been synthesized from RNA templates. Sound far-fetched? Not really: retroviruses (HIV being one notable member of the group) use RNA as their genetic material and rely on the catalytic activity of a protein called **reverse transcriptase** to make a DNA copy of the RNA template (see chapter 17). DNA is more chemically stable than RNA, which makes it the preferred macromolecule for inheritance. Thus, information flows from DNA to protein via mRNAs with precision and efficiency. This process is a remarkable feat, even more so in prokaryotes, where transcription of mRNA and its translation into protein take place *simultaneously*. And although the stability of DNA preserves the genetic information from generation to generation, changes in its nucleotide bases still occur, allowing for genetic variation. The fluctuating environment then selects the fittest cell of all. The door to evolution is never closed!

HOW DO CELLS MAKE COPIES OF THEIR CHROMOSOMES?

Information about the amino acid sequence of a protein must (i) be available continuously during cell growth and (ii) be transmitted faithfully to daughter cells. DNA fulfills these roles efficiently. It is the cellular library that stores the sequences of all the proteins, with each codon specifying a particular amino acid. This precious information is passed faithfully to daughter cells via the process of *DNA replication*. One mistake during replication could give a daughter cell a defective copy of a protein. The resulting mutant may be better off than its ancestor, or it may be worse—even unable to survive. Such is life.

What follows is an account of DNA replication in *Escherichia coli*, where the process has been most thoroughly studied. It is complex. We will leave the story about chromosome condensation and segregation for the end. Here we will focus on how cells overcome two major challenges: how to unwind the double helix and how to replicate each DNA strand simultaneously.

DNA Replication

LEARNING OUTCOMES:
- Define replisome.
- Identify the role of DNA polymerase III in DNA replication.
- Summarize the reactions during replication initiation.
- Identify the roles of DnaA in replication.
- State the mechanism by which the cell ensures that it only replicates DNA and divides when there is sufficient ATP.
- Integrate how DnaA initiates replication at the origin.
- Synthesize how bacteria prevent premature initiation of replication.
- Compare and contrast DNA strand elongation on the leading and lagging strand.
- Describe how the problem of positive supercoiling in DNA strands due to unwinding of the helix is solved.
- Explain the replication termination steps in bacteria that are unique as a result of having a circular chromosome.
- Describe an example of a replication process that was discovered and described in bacteria and is widespread among living organisms.

DNA replication relies on proteins to unwind the double helix and replicate the two strands simultaneously. Figure 8.2 illustrates how this is accomplished in *E. coli*, mainly:

1. DNA replication starts at the origin of replication (*oriC*), where a specific protein (**DnaA**) binds to and melts the two DNA strands.
2. Specific proteins are recruited at each of the two replication forks, thus forming two active replicative complexes, called **replisomes,** moving away from the origin bidirectionally.
   a. The replisome's **helicase** enzyme unwinds the two strands.
   b. The replisome's **DNA polymerase III** (Pol III) polymerizes nucleotides in the *5' to 3' direction*.

   c. Because the two template strands of DNA have opposite polarities, the replisome contains a flexible protein arm that loops one of them around to reverse its polarity and ensure that its replication can obey the 5'-3' law (Fig. 8.2). As a result, one strand is replicated continuously (**leading strand**) while the other (the **lagging strand**) is replicated discontinuously into short DNA stretches called **Okazaki fragments** (named after the discoverer).

3. Replication stops when the replisomes enter the terminus region (*terC*).

Keep in mind that this description does not apply universally because, even in this most fundamental of cellular processes, microbes exhibit unexpected diversity. (For example, not all prokaryotic chromosomes are circular.) Replication in the *Archaea* follows the general outline except that many of the proteins involved are more similar to those found in eukaryotes than to the bacterial proteins (chapter 1, Fig. 1.9). In addition, some archaea, like eukaryotes, start replication at more than one origin. Yet the program for DNA replication follows this common pattern (with interesting exceptions in some plasmids; see chapter 10). Below we describe the sequence of events (initiation, elongation, and termination) in more detail, using *E. coli* as a paradigm.

Initiation of replication. Replication is initiated when the initiator protein, DnaA, recognizes and binds *oriC* and melts it locally to separate the two strands and recruit the replisome proteins at each replication fork. Several questions arise immediately:

How does DnaA recognize, bind, and separate strands at the origin? DnaA proteins recognize and bind several sequences scattered throughout the origin (collectively known as DnaA boxes; see Fig. 8.3). However, only DnaA proteins carrying an ATP molecule (DnaA-ATP) can melt the origin because DnaA-ATP molecules hydrolyze the ATP to energize their polymerization and form a helical filament wrapped around the DNA. This filament exerts torsion on the DNA, melting the origin's AT-rich region to form a "bubble" (Fig. 8.3). To prevent the separated strands of DNA from reannealing, tetramers of **single-stranded binding protein** (SSB) coat each of the single strands.

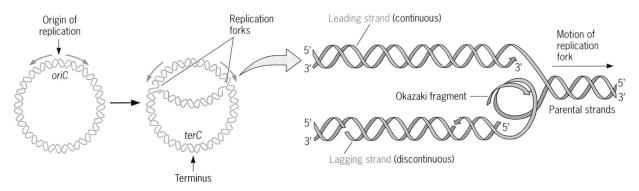

FIGURE 8.2 **Overview of DNA replication in *E. coli*.** See the text for a description.

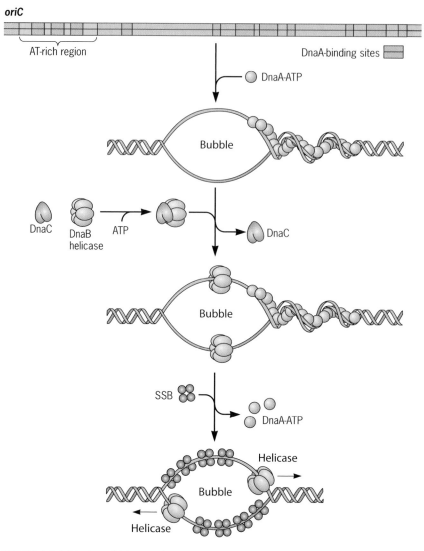

FIGURE 8.3 Initiation of DNA replication in *E. coli.* DnaA-ATP binds to DnaA boxes (blue) scattered through *oriC* and hydrolyzes ATP to energize its polymerization and the melting of the AT-rich region. With help from the DnaC chaperone, the helicase DnaB is loaded onto each of the single strands of DNA in the melted region (bubble), and they start moving in opposite orientation to unwind the DNA in the 5'-3' direction. Single strandedness is maintained by the binding of SSB proteins. DnaA can now dissociate and a functional replisome can be assembled onto each of the helicases.

How are the replisomes assembled at the replication forks? Opening up the DNA duplex at the origin's AT-rich region allows the first protein of the replisome, its helicase (or **DnaB**), to be recruited to each of the single strands (Fig. 8.3). The chaperone protein, cleverly called DnaC, interacts with DnaA to load the helicase at the origin and not anywhere else. The two helicases now move in opposite directions, away from the origin in the *5' to 3' direction*. The DnaB helicase also has the noble role of serving as a scaffold for the other proteins of the replisome: an RNA polymerase called the **primase** is followed by the DNA polymerase enzyme (Pol III) (Fig. 8.4). With a functional replisome assembled at each of the replication forks, DNA replication is ready to start.

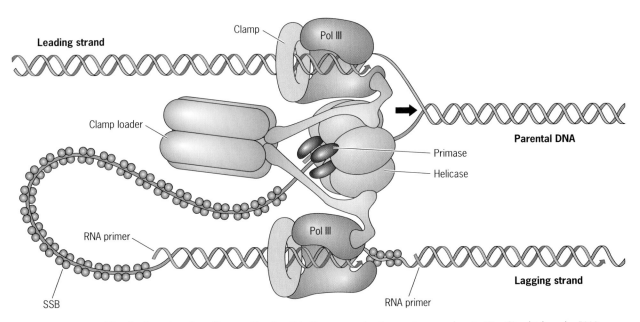

FIGURE 8.4 Assembly of a functional replisome. The DnaB helicase recruits three primase subunits (DnaG), which make RNA primers during replication. The components of the DNA polymerase III enzyme complex are assembled next to loop the lagging strand and replicate the two template strands simultaneously in the 5'-3' direction.

How does the cell know it is time to initiate a new round of replication? Replication must be well regulated, lest false starts mess up the proper structural organization of the chromosome. Timing is of the essence. Obviously, the cell needs to have *enough ATP.* (A starving cell will not replicate its DNA. Too risky!) Furthermore, a cell needs to make two identical copies of the chromosome, one for each daughter cell, prior to cell division. Thus, replication must be *timed to cell growth and division.* DnaA ensures that the timing is right. How? DnaA accumulates as the cell grows, until there is enough of it to bind all of the DnaA boxes in the replicative origin (Fig. 8.2). As DnaA at the origin is only functional when bound to ATP, DNA replication will only start when the DnaA boxes are bound by DnaA-ATP. Thus, the cell will only replicate its DNA when it has grown sufficiently to reach a threshold volume of DnaA and it has sufficient energy (ATP) to ensure that enough DnaA-ATP is bound to *oriC* to polymerize and force the DNA strands apart. Together, these mechanisms help "set the clock" for the initiation of replication.

How does the cell prevent premature initiation? Preventing premature initiation also matters because too many rounds of initiation would make the chromosome a tangled mess. One piece of this puzzle is a protein called SeqA, which transiently sequesters the origin. This so-called **origin sequestration** is regulated by methylation. Scattered throughout the *oriC* sequence are multiple GATC sequences that a **methylase** (DNA adenosine methyltransferase or Dam methylase) recognizes and methylates in the "A" position. As replication begins at the origin, the GATC in the newly replicated strands are not methylated yet. The origins are, therefore, hemimethylated. The SeqA protein happens to have a high affinity for the hemimethylated GATC sequences so it binds them tightly, effectively sequestering the *oriC* DNA and preventing new initiations.

DNA strand elongation. The elongation of the DNA chain during replication is a complicated business, requiring the cooperative involvement of the many proteins that form the replisome plus some additional ones.

The replisome in action. Try to picture the replisome in action (Fig. 8.4), with its helicase rotating to unwind the two strands and the two DNA polymerase III enzymes replicating each of the DNA template strands. Each parental strand is guided to the Pol III enzymes by a sliding bracelet (or clamp) and replicated at a rapid rate. Because DNA polymerases, such as Pol III, can extend a polynucleotide chain but cannot initiate one, an RNA polymerase within the replisome (the primase) makes an RNA primer segment on which Pol III can add nucleotides. (You add primers to a PCR reaction for the same reason! [chapter 10]). Each step of elongation involves the base-pairing of free deoxynucleoside triphosphate molecules with complementary nucleotides in the template strand of DNA ahead of Pol III and formation of a phosphodiester bond between the deoxynucleoside and the 3'-OH of the nascent strand, as follows:

$$DNA_n + dNTP \rightarrow DNA_{n+1} + PP_i$$

The replisome faces one additional challenge. Because DNA synthesis can only take place in the 5'-3' direction, only one of the template strands (the leading strand) is in the "right" direction for nucleotide polymerization. The polarity of the other so-called lagging strand is reversed by looping (Fig. 8.3). A flexible arm of the replisome, the clamp loader, loads a clamp while looping the lagging strand, so it can be replicated simultaneously to the leading strand. Periodically, the lagging strand loop dissociates and has to be reloaded and primed with a new RNA segment. Thus, elongation of the lagging strand is distinct: it requires repeated RNA priming, and it is synthesized discontinuously as short 1,000-bp 5'-3' segments, known as Okazaki fragments (Fig. 8.3).

And where are the two replisomes located? Are they motoring their way around the chromosome away from each other? Or are these magical complexes fixed in place while the duplex DNA feeds through them? Scientists believe that the two replisomes remain close to each other in the mid-cell, while the newly replicated DNA is forced to each pole. How they are confined to the mid position is not known. But we do know that this is the place where the chromosome is unpacked prior to replication. We will revisit this topic later when we discuss chromosome segregation.

Processing the lagging strand. Initially, the lagging strand is a patchwork of numerous 1,000-bp-long Okazaki fragments, each with an RNA primer and separated by gaps. But, as the replicating forks move, the RNA primers in the hybrid RNA-DNA stretches are removed by two enzymes: ribonuclease H (RNase H) and a versatile DNA polymerase called **DNA poly-**

merase I or Pol I. Pol I can also extend the Okazaki fragments to fill the gaps left after removing the RNA primers. Finally, the extended DNA fragments are ligated by an enzyme called **DNA ligase** (Fig. 8.5).

Removing positive supercoils ahead of the replisome. In many prokaryotes, particularly bacteria, the chromosome is condensed by "knotting" the double helix against the direction of the right-handed helix, creating **negative supercoils** (chapter 3). This twisting of the double helix reduces the tension on the two strands, facilitating DNA melting. However, as the replisome advances and the helicase unwinds the double helix, the DNA ahead of the fork rotates in the opposite orientation and forms **positive supercoils** (Fig. 8.6). Left alone, these twists would block the replisome's progress on the chromosome (and also expression of genes). To the rescue comes **DNA gyrase!** It cleaves the double helix at the supercoil crossover, passes the unbroken one through this break, and then reseals the cut to remove the "knotty" supercoil, allowing the replisome to continue its journey. In fact, DNA gyrase is so critical to DNA replication that bacterial infections can be treated by antibiotics that inhibit this enzyme. The drug ciprofloxacin, used to treat Mr. M in our case study, inhibits gyrase. Given the importance of this enzyme during DNA replication and its presence in virtually all bacteria, it is not surprising that ciprofloxacin is classified as a **broad-spectrum antibiotic.** (A joke: What do you give the person who has everything? Answer: A broad-spectrum antibiotic!)

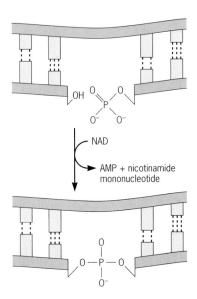

FIGURE 8.5 **Ligation of nicked DNA**. To make a continuous lagging-strand molecule, Okazaki fragments are ligated by DNA ligase, using NAD in the process.

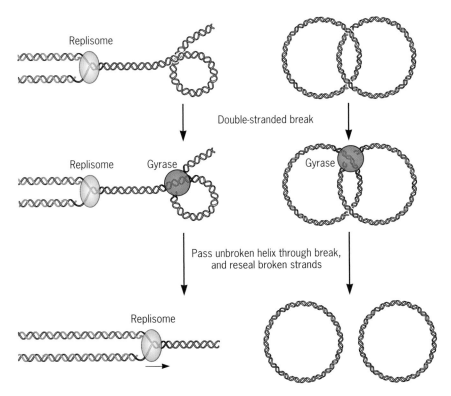

FIGURE 8.6 **The two roles of gyrase: removal of positive supercoils and decatenation of sister chromosomes.** See the text for details.

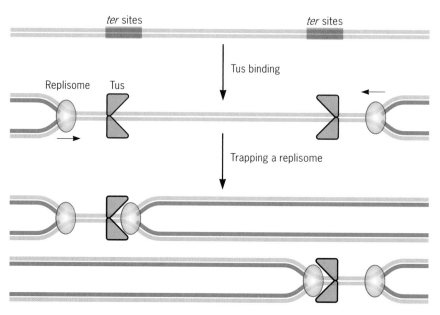

FIGURE 8.7 Replication termination. The binding of Tus proteins to the *ter* sites creates traps for the advancing replisomes so they stop in the termination region.

Termination of replication. DNA replication uses a well-regulated termination process to prevent over-replication and replisome collisions (Fig. 8.7). As the replisomes near completion of their task, the two complexes converge halfway around the chromosome (180° from *oriC*). Several specific *ter* sequences are arranged in opposite polarity in this region. One blocks the replisome moving counterclockwise, the other, clockwise. Binding to *ter* sites is the protein Tus (terminator utilization substance), which halts the replisome's progress by inhibiting its helicase.

Because most prokaryotic chromosomes are circular, one more maneuver is needed to complete the replication process. At the close of termination, the completed daughter chromosomes are topologically locked, like two links in a chain. This **concatenate** structure is the inevitable result of unwinding a helical double-stranded closed circle (Fig. 8.6). **Decatenation** is carried out by enzymes called **topoisomerases II**, such as topoisomerase IV and gyrase, which make a double-strand break, permit the uncut double helix to pass through the gap, and then reseal the broken one. Until it's resolved into separate circular chromosomes, the concatenate is susceptible to **homologous recombination;** i.e., crossover events between homologous sequences of the two chromosomes (see chapter 10). Since a single-crossover event generates a dimer of two chromosomes—a lethal event if not reversed—special **recombinases** (enzymes that cut and splice DNA) come into play here.

Repair of Errors in Replication

LEARNING OUTCOMES:
- Compare and contrast mismatch repair by DNA polymerases and the proteins of the mismatch repair system.
- Describe how methylation guides the repair of mismatches in *E. coli*.

Complex as DNA replication is, it occurs with amazing speed and fidelity. *E. coli* replicates its chromosome at a rate approaching 1,000 nucleotides per second, making errors only once in every 10 billion base pairs. Incredible! Can you imagine typing 8 million pages of text without a single error? The whole chromosome, a DNA molecule of two entwined coiled strands that are a thousand times the length of the cell, is copied in about 40 minutes at 37°C, all while transcription and coupled protein synthesis continue.

Proofreading by DNA polymerases. Spontaneous changes in DNA (mutations) are due mainly to **mismatches**, that is, matches other than *A with T* and *G with C*. If the wrong base is inserted through a mismatch, pairing will not occur with its complement on the other DNA strand. Spontaneous mutations in *E. coli* arise approximately only once for every 10^{10} bases synthesized. Faithful copying of DNA relies on several factors. Base-pairing of dNTPs to the parental strands separated during DNA synthesis is itself very precise. Accuracy is augmented by a special property of Pol III called the **3'–5' exonuclease activity**. DNA polymerases cannot progress unless there is a properly matched nucleotide pair behind them. If not, they pause and move backwards, use their exonuclease activity to remove the mismatched base, and then proceed. Is this proofreading activity in all probability the reason that Pol III requires a 3'-OH end on a DNA or an RNA primer? (*Quiz spoiler:* Since polymerases cannot start from scratch, synthesis will not proceed unless a properly matched base pair behind them provides the correct end of the molecule.)

The mismatch repair system. All cells, not just bacteria, have mechanisms for detecting and repairing mismatches that may have escaped proofreading. Collectively, this set of proteins is known as the **mismatch repair system**. The *E. coli* system, called **methyl-directed mismatch-repair**, contains proteins that (i) *recognize* the mismatch, (ii) *excise* the misincorporated base and its surrounding DNA, and (iii) *fill the gap* with newly synthesized DNA (Fig. 8.8). Strains that are defective in any of the proteins have a high mutation rate and are, therefore, called mutator strains. Three mutator proteins (Mut), called MutS, MutL, and MutH, take care of much of the repair. Here's how.

First MutS (S stands for *specificity*) scans the DNA until it recognizes the mismatch (Fig. 8.8). It then binds the DNA in this region and recruits the MutL and MutH proteins. The Mut complex that assembles at the mismatch site then moves along the DNA until it reaches a hemimethylated GATC. There (and only there), MutL stimulates MutH to cut the nonmethylated GATC site. How does methylation dictate which strand to cut? (*Hint:* Think about SeqA, the protein that sequesters the origin of replication.) The Mut complex also distinguishes the new from the old (template) strand by the methylation state of the GATC sequence. Once the Mut complex cuts the newly replicated strand at the GATC site, an exonuclease chews the cut strand back, effectively removing the mismatched base. Finally, Pol III replaces the missing segment, and the repaired strand is made whole by DNA ligase. *Voila!*

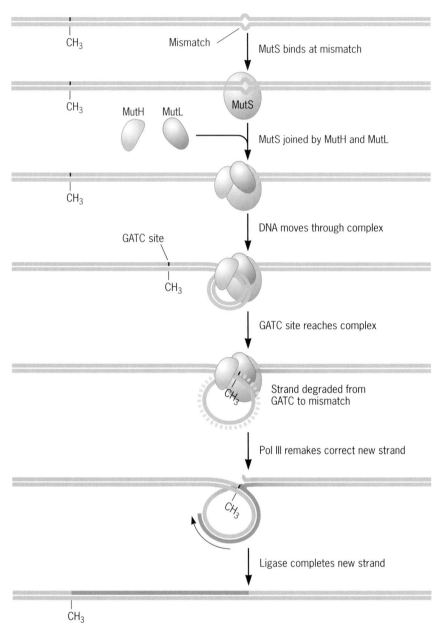

FIGURE 8.8 **Mismatch repair.** A mismatch is recognized by the protein MutS. MutH and MutL then bind to the MutS-DNA complex. DNA moves through this complex until it reaches a GATC site (the site for methylation by the Dam methylase). There the as-yet-unmethylated new strand with the mismatch is degraded, and the gap is filled by Pol III and DNA ligase.

Protecting the DNA

LEARNING OUTCOMES:
- Identify the role of endonucleases in microbial cells and in recombinant DNA technology.

During DNA synthesis, many bacterial and other microbial cells mark their DNA to identify it as their own (akin to branding cattle). To what purpose? Microbial cells are frequently invaded by "foreign" DNA intro-

TABLE 8.1 **Actions of certain restriction endonucleases**

Class	Enzyme	Producing microorganism	Recognized DNA sequence[a]
Six base pairs recognized; complementary single-stranded ends produced	EcoRI	*Escherichia coli* (R)[b]	$\begin{pmatrix} G\ A\ A\ T\ T\ C \\ C\ T\ T\ A\ A\ G \end{pmatrix}$
	HindIII	*Haemophilus influenzae*	$\begin{pmatrix} A\ A\ G\ C\ T\ T \\ T\ T\ C\ G\ A\ A \end{pmatrix}$
Six base pairs recognized; blunt ends produced	HpaI	*Haemophilus parainfluenzae*	$\begin{pmatrix} G\ T\ T\ A\ A\ C \\ C\ A\ A\ T\ T\ G \end{pmatrix}$
	HindII	*Haemophilus influenzae* Rd	$\begin{matrix} C\ A \\ \begin{pmatrix} G\ T(T)\ (G)A\ C \\ C\ A(A)\ (T)T\ G \end{pmatrix} \\ G\ C \end{matrix}$
Four base pairs recognized; complementary single-stranded ends produced	HhaI	*Haemophilus haemolyticus*	$\begin{pmatrix} G\ C\ G\ C \\ C\ G\ C\ G \end{pmatrix}$
	MboI	*Moraxella bovis*	$\begin{pmatrix} G\ A\ T\ C \\ C\ T\ A\ G \end{pmatrix}$
Four base pairs recognized; blunt ends produced	HaeIII	*Haemophilus aegypticus*	$\begin{pmatrix} G\ G\ C\ C \\ C\ C\ G\ G \end{pmatrix}$

[a]Arrowhead indicates site of single-strand cleavage. Upper sequence of bases is written in the 5'→3' direction.
[b]Encoded by genes that are plasmid borne.

duced as plasmids or by viruses. Such invasion is seldom beneficial to the cell (we'll learn about viral invasion in chapter 17). Bacteria have evolved powerful DNA-hydrolyzing endonucleases that destroy unwanted invading DNA. Many hundreds of these **restriction endonucleases** have been discovered among bacteria (thanks to their ability to *restrict* the growth of bacterial viruses). Each recognizes a specific sequence of bases (recognition site) where it produces double-strand breaks, triggering the complete degradation of the DNA by nucleases that attack these exposed ends (Table 8.1). Cells protect their own chromosomal DNA by methylating an adenine or a cytosine residue in the recognition site. This *"branding" by methylation* marks the DNA as self and protects it from its own restriction endonucleases.

In all likelihood, you associate restriction enzymes with recombinant DNA technology, where their many uses can make us forget their important biological role. Every laboratory engaged in genomic analysis and engineering relies on a tool box of purified restriction enzymes that cleave DNA specifically and reproducibly at specified sequences. DNA cloning, which gave birth to biotechnology, depends heavily on restriction enzymes.

Chromosome Condensation and Segregation

LEARNING OUTCOMES:
- List proteins and factors that help structure the nucleoid and play roles in chromosomal condensation and replication.

The chromosome of prokaryotic cells is extraordinarily tightly packed—as is the DNA of all cells. In eukaryotic cells, chromosomal DNA is tightly wound around a complex of histone protein molecules to form nucleosomes and then further compacted by coiling. In chapter 3 we learned that in prokaryotes the story is somewhat different, though perhaps analogous. Instead of nucleosomes, the prokaryotic chromosomal DNA possesses physical domains of supercoiled, compacted loops, somewhere between 30 and 200 total. The loops are sufficiently isolated from one another that single-strand breaks in any one of them will not affect supercoiling in the neighboring loops. This modular arrangement is convenient when it is time to replicate the DNA: the chromosome is decondensed locally to allow the replisome to work its magic. Then the new DNA strands are repackaged (Fig. 8.9). As the newly replicated regions are condensed, a force is generated that pulls more of the newly synthesized DNA away from the mid-center. Thus, condensation promotes chromosome segregation.

How the nucleoid is condensed is largely unknown, but some factors have been characterized. The Muk proteins, for example, resemble tweezers with sticky ends, which bind DNA regions located far apart on the chromosome (Fig. 8.9). They then hydrolyze ATP and close their extended arms, bringing distant regions together and condensing the DNA. These proteins have been found in almost all organisms and are quite conserved, suggesting that they originated early in evolution and play a key role. Collectively, they are known as SMC (for structural maintenance of chromosome) proteins.

Let us not forget that the whole point of replicating the genome is to prepare for the birth of two daughter cells! Segregation of the duplicated chromosomes is imperative, and sequestration of the origin to the membrane by SeqA may contribute. The assumption here is that SeqA bound to *oriC* associates with the cell membrane. As the cell membrane and wall

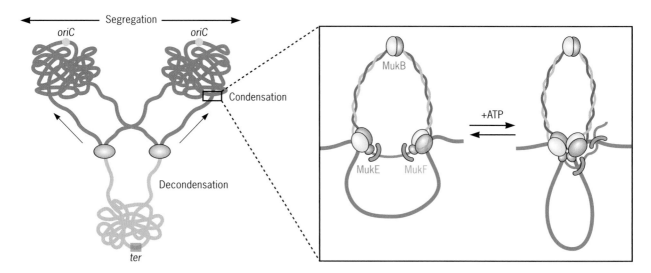

FIGURE 8.9 Chromosome segregation and the condensation "pull." DNA is decondensed and replicated in the mid-cell, where the replisomes (in orange) are located, and the newly replicated DNA (in red) is pulled towards the poles of the cell as it is condensed again. Condensation involves SMC proteins such as MukBEF, which bind distant DNA regions and bring them close together using energy from the hydrolysis of ATP.

grow by adding new material to the mid regions, the SeqA foci and their sequestered origins are pushed away from each other. By segregating the chromosomes towards the poles, space is freed in the mid-center to allow the **FtsZ** ring to form in this location and recruit the proteins needed to make the division septum (chapter 4). The cell is now ready to divide.

HOW IS DNA TRANSCRIBED INTO RNA?

LEARNING OUTCOMES:
- State the consequence for transcription by having operons.
- Identify the role of sigma (σ) factors in transcription.
- Summarize the process of RNAP binding to the promoter during transcription.
- Identify the condition under which abortive transcription takes place.
- State how transcription can be regulated after initiation.
- Name two mechanisms that signal to RNAP to terminate transcription.
- List benefits and costs of the short half-life of mRNA.
- Explain how ribosomes are assembled.
- Identify the role of enzymes in the processing steps of tRNA that make an active product.
- Explain reasons why there is a bias in many species to having genes on the leading strand as opposed to the lagging strand.
- Name factors that influence genome organization.

Transcription means synthesis of an RNA molecule, called a **transcript**, whose base sequence is complementary to one strand of the DNA. Eukaryotic cells have several RNA-polymerizing enzymes; prokaryotes have only one, known as **DNA-dependent RNA polymerase** (abbreviated as **RNAP**). This enzyme synthesizes all the cellular species of RNA, the *stable* ones (rRNA, tRNA, and small regulatory RNAs) and the *unstable* mRNAs. The latter also have a distinctive prokaryotic feature: almost all messenger transcripts are polycistronic—they encode several peptides because it is here that most protein-encoding genes of related function usually cluster as single transcriptional units called **operons**. As we will see in chapter 11, polycistronic mRNAs also help the cell coordinate the expression of genes whose products work together.

As with all macromolecular syntheses, transcription consists of three distinct phases: initiation, elongation, and termination. RNAP is involved in all three:

1- Initiation involves the association of RNAP to a subunit called the **sigma (σ) factor**, so it can bind and partially melt a *specific region* on the DNA called the **promoter**.

2- Elongation of the transcript occurs when the RNAP releases the σ subunit and moves past the promoter. Along the way, the RNAP, all by itself (unlike the story for DNA replication), unwinds the DNA double helix and links ribonucleoside triphosphates in the *5'-3' direction* (*again!*) in an order dictated by the template strand of DNA.

3- Transcription ends at a specific sequence called the **terminator**, with or without the assistance of other proteins.

With its many functions during transcription, it may not surprise you to learn that RNAP molecules are rather complicated (though with fewer components than the DNA replisome). Bacterial RNAPs share the same core subunit structure ($\alpha_2\beta\beta'\omega$) but associate with one of several σ factors to recognize different promoters and transcribe specific sets of genes. Transcription in the *Archaea* is also performed by a single RNAP, but one quite different from the bacterial one. Archaeal RNAP has as many as 14 subunits and more closely resembles that found in eukaryotes. One consequence is that antibiotics (such as rifampin) that inhibit the bacterial enzyme do not affect the archaeal polymerase. Much is known about transcription in *E. coli*, the model bacterium we will consider here.

Initiation of Transcription

The core RNAP (without a σ factor) has relatively low affinity for DNA and binds indiscriminately to any region of the chomosome. Think of the core RNAP as a closed clamp. But once it binds the σ factor, the RNAP clamp opens and interacts with the promoter region (Fig. 8.10). Once the σ factor is released, the clamp shuts again, holding RNAP on the DNA to clear the promoter and move along. These steps, promoter binding and clearance, control the strength of a promoter, that is, the amount of transcripts that RNAP can produce from that sequence. Thus, the cell zeros in on both steps to regulate transcription initiation and ensure that only the right genes are transcribed at the right rate.

Binding to the Promoter. The α and σ subunits of RNAP recognize and bind specific DNA sequences in the promoter called the −35 and −10 hexamers (the minus signs indicate distance upstream from the start of the transcript) (Fig. 8.10). No universal sequence exists for promoters and their −35 and −10 hexamers. There are, however, consensus sequences, consisting of the most statistically probable base, to be found at each nucleotide residue. Figure 8.11 shows, as an example, the consensus sequence recognized by the major σ subunit of RNAP, σ^{70} (named after its molecular weight, 70 kDa). This sequence differs from the consensus sequences favored by other σ subunits. Indeed, the nucleotide sequence of the promoter determines the specificity and efficiency of binding by RNAP.

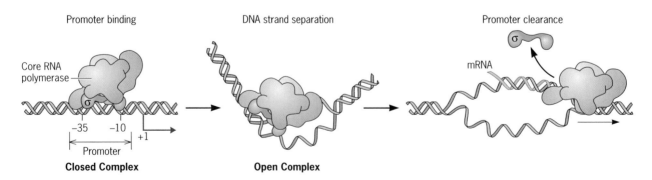

FIGURE 8.10 Initiation of transcription. The binding of the σ factor to the RNAP core opens up its structure and allows it to bind the −10 and −35 promoter regions forming a DNA-protein closed complex (step 1). Binding bends the DNA and separates the two strands forming the so-called open complex (step 2). The transition from a closed to an open complex allows RNAP to synthesize short RNA transcripts, but to continue elongation the σ factor must be released (step 3).

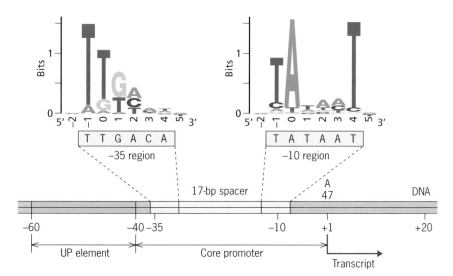

FIGURE 8.11 Structure of the consensus promoter recognized by σ⁷⁰ in *E. coli*. The core promoter includes the −35 (TTGACA) and −10 (TATAAT) hexamers separated by a 17-bp spacer. An upstream UP element (from −40 to −60) is also shown. The negative numbers assigned to these positions of bases in the DNA are numbered relative to the first base transcribed into mRNA (+1), which is an A in 47% of promoters. The plots (top) are sequence logos and show the relative frequency of nucleotide bases −35 and −10 hexamer sequences in known σ⁷⁰ promoters. The height of the stack of letters at each position is proportional to the sequence conservation (measured in bits), whereas the height of each letter within the stack reflects its relative frequency at that position. (Note: all bases indicated are those in the non-template strand of DNA.)

Escaping from the Promoter. Guided by the σ subunit, RNAP reversibly binds the promoter to form a *closed complex.* Then, in a relatively slow isomerization step, the two DNA strands separate and an open complex forms (Fig. 8.10). This transcription bubble of locally melted DNA duplex is about 12 bp in length. The RNAP then advances, enlarging the bubble to its mature length of 18 bp and making short (up to 10-bp-long) transcripts. However, yet obscure interactions of RNAP in the open complex prevent it from clearing the promoter right away; as a result, transcription is aborted prematurely and restarted. Such cycles of **abortive transcription** are widespread among promoters *in vivo* and delay transcription initiation. However, once RNAP can make a longer transcript, it is able to escape from the promoter, release the σ factor, and elongate the transcript.

To prevent delays in promoter escape by RNAP, some promoters include a −40 and −60 upstream sequence called an UP element (Fig. 8.11). This element interacts with the RNAP, bends the DNA locally, and weakens the interactions that prevent transcript elongation. Other tricks help RNAP escape the promoter. In chapter 11, we will learn about transcriptional activators that help RNAP clear the promoter. However that happens, loss of σ factor clamps the RNAP to the DNA and moves it down the template strand. The σ factor is now freed to associate with another RNAP core to guide it to a promoter.

RNA Chain Elongation

Once RNAP clears the promoter, it releases the σ factor and moves along the template DNA strand, simultaneously elongating the transcript in the

5'-3' direction and *unwinding* the DNA double helix (Fig. 8.12). Picture the bubble moving with RNAP, with a nascent RNA transcript trailing behind, still paired to the template DNA strand. Each step of elongation involves ribonucleoside triphosphate base-pairing with the template strand ahead of RNAP and formation of a phosphodiester bond between the ribonucleoside and the 3'-OH of the nascent chain, as follows:

$$RNA_n + NTP \rightarrow RNA_{n+1} + PP_i$$

Certain antibiotics, e.g., rifampin, bind to RNAP and prevent it from making new phosphodiester bonds. Because this step is universal in bacterial transcription, these drugs have a broad spectrum of targets. In fact, rifampin or one of its cousins could have been used to treat Mr. M's pneumonia (case study).

RNA chain elongation is not a constant, boring process. It has its own share of drama. When *E. coli* is cultured at 37°C, each second about 45

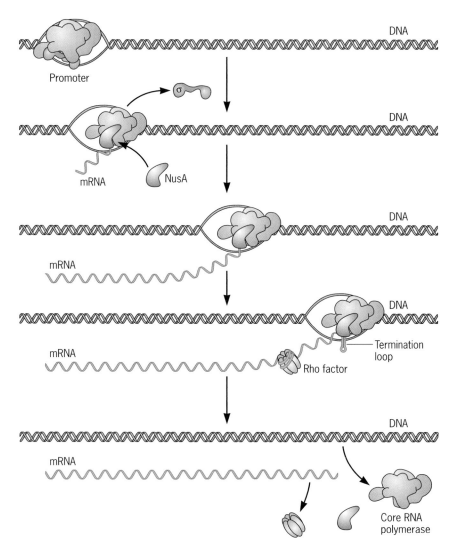

FIGURE 8.12 Transcription elongation and Rho-dependent termination. See the text for details.

nucleotides are added to mRNA chains and close to double that on rRNA, tRNA, and other **stable RNA** species. Other proteins contribute to RNA chain elongation. For example, NusA associates transiently with RNAP (Fig. 8.12) to assist when **hairpin loops** (GC-rich regions of dyad symmetry) form in the growing RNA strand. These **pausing sites** increase the chance that transcription will be interrupted prematurely. Orderly termination is our next topic.

Termination of Transcription

Usually, transcription ends when the RNAP reaches the end of the gene or operon and encounters a specific sequence in the template DNA that signals termination (Fig. 8.12). The terminator signal is often a *hairpin in the RNA*, that is, a double-stranded stem and a single-stranded looped head. RNA terminator hairpins form when RNAP transcribes a GC-rich region containing an inverted repeat followed by a string of A residues. The RNAP stalls at the hairpin and, as the successive AU pairs between transcript and template are weak, the transcript breaks free.

In some cases, termination of mRNAs is assisted by proteins such as Rho, which binds specific sequences (called *rut*) in the RNA and moves towards its 3' end like a helicase (Fig. 8.12). During Rho-dependent termination, RNAP pauses at a hairpin and Rho catches up with it, helping release the mRNA from its DNA template. But here is the catch: because transcription and translation are coupled in prokaryotes, ribosomes often block Rho's access to its recognition sequence on the mRNA. Thus, Rho-dependent termination of transcription is tuned to translation. This is one of the advantages of *coupled transcription and translation in prokaryotes:* there is cross-talk between the two processes so they can be regulated coordinately. In chapter 11, we will learn about one other example of coordinated transcription and translation, a process called **attenuation**, whereby the timing of transcription and translation controls whether a truncated or a full RNA transcript is made.

The Fate of Transcripts

RNAs that encode information to make a protein (mRNAs) are the target of ribosomes, even while chain elongation proceeds. In fact, many ribosomes can latch on to a single mRNA simultaneously, as in a factory assembly line. This amplifies the impact of mRNA, as each transcript directs the synthesis of many protein molecules. But their work is brief: All cells have evolved mRNAs that are relatively short-lived, **unstable RNAs**. By contrast, the stable RNAs include the RNA components of the protein-making machinery (rRNAs and tRNAs) and small regulatory RNAs (chapter 11). Next we focus on the fates of mRNA transcripts and then consider the rRNA and tRNA molecules of protein synthesis.

Life and death of mRNA. A hallmark of prokaryotic organisms is their rapid turnover of mRNA. For example, in *E. coli* the average mRNA half-life (time for 50% of the molecules to be destroyed) is about 1 minute at 37°C. Why so short-lived? By continually erasing its instructions for making proteins, the bacterial cell is poised to respond quickly to both its physiological status and its environment. Although hydrolyzing mRNA so liberally may seem wasteful, realize it is more economical than making useless proteins.

Assembly and modification of rRNAs and tRNAs. RNAP also transcribes the nearly 50 genes that encode tRNAs and each of the rRNAs. And the expression of this small subset of genes is carefully modulated to match the potential of each environment to promote growth. As an example, consider the synthesis of rRNA transcripts, which account for 82 to 90% of all of the RNA transcripts in the cell. One way cells synchronize their transcription is by clustering the rRNA genes in an operon (the *rrn* operon). *E. coli* has seven *rrn* operons scattered in its genome, each transcribed by multiple highly active promoters. Furthermore, in addition to the three rRNA molecules (5S, 16S, and 23S), these operons also encode select tRNAs (Fig. 8.13).

Transcription of an *rrn* operon produces a giant polycistronic RNA molecule that must be processed into usable RNAs (Fig. 8.13). As transcription proceeds, a group of ribonucleases cut and trim the growing product into individual rRNA and tRNA precursors. The rRNAs then bind the various ribosomal proteins sequentially in steps that are, to large extent, spontaneous to make each of the two subunits of the bacterial ribosome, 30S and 50S.

As with rRNAs, tRNAs also need processing to produce an active product (Fig. 8.13). First, individual tRNA precursors are cleaved from the large polycistronic transcripts, and then RNases cleave any extra nucleotides at each end. If not already in place, terminal CCA ends are added (to which amino acids become attached), and the tRNA intermediates un-

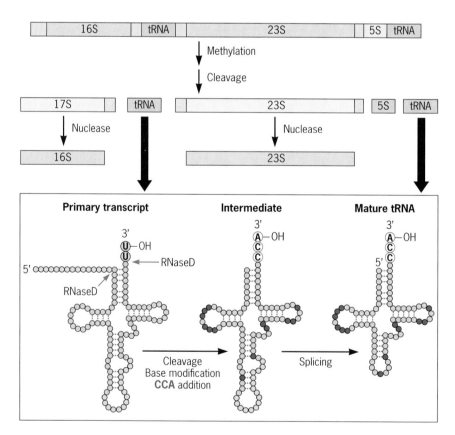

FIGURE 8.13 Processing of rRNA and tRNA transcripts. See text for description.

dergo extensive chemical modifications. (*E. coli* enlists more than 45 different enzymes to generate at least 30 different modified nucleosides for its tRNAs; other microbes employ more than 80 processing factors!) Modifications include methylating bases and incorporating moieties practically unheard of in nucleic acids, such as inosine and pseudouridine (Fig. 8.14). These modified ribonucleosides have subtle functions in mature tRNAs. Although many can be mutated without destroying the cell's ability to grow in the laboratory, there is good reason to believe that they contribute to the efficiency of tRNA's role in translation in some growth conditions. Indeed, they must be important because the cell invests more genetic information encoding the modifying enzymes than encoding the RNA itself. Clearly there is more biology to discover!

Polymerase Collisions

Picture the machines at work in a growing bacterial cell. Many RNAP enzymes simultaneously transcribe numerous genes scattered throughout the genome. Several RNAPs recognize each promoter and sequentially transcribe nascent mRNA transcripts. As one ribosome moves down an mRNA, a new one assembles and initiates translation. Not only is translation coupled to transcription, many ribosomes simultaneously process each mRNA as a **polyribosome** (Fig. 8.15). Now add to the mix a replication fork with its dual replisomes moving along the chromosome, replicating both strands of DNA at 1,000 nucleotides per second. But its track is not clear: Far from it! Ahead of the replisomes are RNAP molecules—perhaps thousands of them—transcribing genes on one or another strand. Some of these RNAP molecules are running away from the replisomes, while others are moving toward them. Collisions are inevitable: RNAP polymerizes only 45 nucleotides per second, some 20 times slower than the replisome's Pol III. Thus, even an RNAP traveling in the same direction as a replisome will quickly be overtaken.

Are such collisions catastrophic? Same-direction collisions are usually not; at most, they slow replication. However, head-on collisions halt replication, if only briefly, and abort transcription. Because same-direction collisions are fewer and have milder consequences, there is an advantage in having genes oriented in the same direction as replication. Analysis of

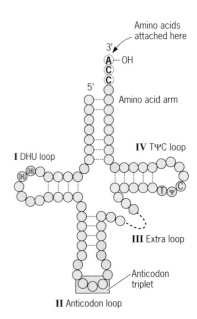

FIGURE 8.14 **Generalized structure of tRNA.** The molecule contains four loops and an amino acid arm with a terminal 3' CCA end for amino acid attachment. The anticodon loop contains the base triplet encoding the specific amino acid that is attached to the 3'-OH end. Loop IV (or TΨC loop) contains a pseudouridine (Ψ) flanked by a thymidine (T) and a cytosine (C) and loop I (DHU loop) contains dihydroxyuridine (DHU).

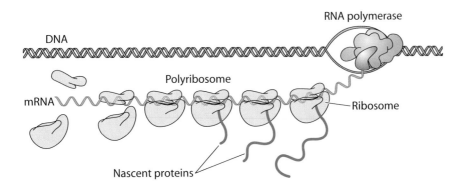

FIGURE 8.15 **Coupling of transcription and translation.** Translation (protein synthesis) takes place as the mRNA molecule is being made (transcription). This simultaneous synthesis can take place only in prokaryotes, where there is no nuclear membrane separating the two processes.

bacterial genomes confirms there is a strand bias in the locations of various genes. In many species, it appears there is a strong tendency toward having genes on the strand that orients them in the direction of replication (the leading, not lagging, strand). In some cases, this bias chiefly favors genes that are either highly expressed or are essential. Although, generally speaking, the topology of replication and transcription is linear, as you can see it is not all that simple.

WHAT IS NEEDED TO MAKE A FUNCTIONAL PROTEIN?

LEARNING OUTCOMES:
- Identify a consequence upon translation of *Bacteria* and *Archaea* not having a nucleus.
- Compare and contrast prokaryotic ribosomes and eukaryotic ribosomes.
- Summarize the steps of loading and preparing the tRNA molecule.
- Describe the role of the Shine-Dalgarno sequence in protein synthesis initiation.
- Outline the major reactions for each of three steps in polypeptide chain elongation: aminoacyl-tRNA binding, peptide bond formation, and translocation.
- Explain how polypeptide chain termination is triggered.
- Give examples of covalent modification in nascent polypeptide chains.
- Identify the role of chaperone proteins in the protein folding pathways.

Rapidly growing prokaryotic cells are bustling protein-making factories. Over half the cell's dry weight is proteins, and, depending on growth rate, another 30 to 60% of the cell is the machinery for making them. The entire **protein-synthesizing system** (PSS) consists of ribosomes; tRNAs; enzymes that charge amino acids onto each tRNA; proteins for initiation, elongation, and termination; enzymes that modify the completed product; and proteins that aid the folding and translocation of newly made proteins. On a weight-for-weight basis, few eukaryotic cells can match the rate of protein synthesis of fast-growing bacteria. Coupled transcription and translation is a strictly prokaryotic property (Fig. 8.15). Another is polycistronic mRNA, a mechanism to increase the efficiency and speed of protein synthesis. Prokaryotes also have a streamlined PSS. For example, eukaryotic cells use 10 protein initiation factors; bacteria need only 3, as we will discuss below. Not surprisingly, the molecular structure and mode of action of the bacterial PSS are different from their eukaryotic counterparts. These differences are exploited to treat bacterial infections, as components of the bacterial PSS, especially the ribosome itself, are targets of many antibiotics used clinically and in agriculture. One could have helped treat Mr. M's pneumonia (case study).

What about protein synthesis in the *Archaea*? The story is interesting, because it is a blend of bacterial and eukaryotic features, along with some unique archaeal components. Archaeal ribosomes resemble those of bacteria in size, but not in detailed structure. Yet, despite targeting different sites and sequences in the two ribosomes, some antibacterial antibiotics are also effective against archaeal ribosomes. We shall note other differences in translation along the way.

Translation of mRNA Transcripts by Ribosomes

Protein synthesis involves translation—a term, like transcription, also used in language. Whereas transcription in linguistics refers to rewriting information in the same language (e.g., copying a voice recording into a written document), translation involves expressing information in a different language (Swahili to Japanese, for example). In biology, transcription rewrites the language of nucleic acids from DNA to RNA; translation converts the information encoded in the nucleotide sequence of mRNA into a sequence of amino acids (the language of proteins) in a peptide chain. The biological translator is not one individual but a team integrated by a large set of proteins called **aminoacyl-tRNA synthetases**.

In most organisms, there are 20 aminoacyl-tRNA synthetases, each in charge of matching a particular amino acid to the **anticodon** of a cognate tRNA molecule (Fig. 8.14). By attaching the proper amino acid listed to each of the nearly 50 different tRNA molecules, the aminoacyl-tRNA synthetases match amino acids with specific codons in mRNA, following a defined genetic code (Fig. 8.16). This is what translation is about. The number of tRNA molecules varies between microbes. In a few species, there are fewer tRNA molecular species than there are codons, in which case the cell depends on **wobble** (the ambiguous reading of the third nucleotide) to utilize other codons. In these situations, a promiscuous tRNA anticodon can recognize more than one codon in the mRNA.

	Second letter				
	U	**C**	**A**	**G**	
U	UUU Phe	UCU Ser	UAU Tyr	UGU Cys	**U**
	UUC Phe	UCC Ser	UAC Tyr	UGC Cys	**C**
	UUA Leu	UCA Ser	UAA Stop	UGA Stop	**A**
	UUG Leu	UCG Ser	UAG Stop	UGG Trp	**G**
C	CUU Leu	CCU Pro	CAU His	CGU Arg	**U**
	CUC Leu	CCC Pro	CAC His	CGC Arg	**C**
	CUA Leu	CCA Pro	CAA Gln	CGA Arg	**A**
	CUG Leu	CCG Pro	CAG Gln	CGG Arg	**G**
A	AUU Ile	ACU Thr	AAU Asn	AGU Ser	**U**
	AUC Ile	ACC Thr	AAC Asn	AGC Ser	**C**
	AUA Ile	ACA Thr	AAA Lys	AGA Arg	**A**
	AUG Met	ACG Thr	AAG Lys	AGG Arg	**G**
G	GUU Val	GCU Ala	GAU Asp	GGU Gly	**U**
	GUC Val	GCC Ala	GAC Asp	GGC Gly	**C**
	GUA Val	GCA Ala	GAA Glu	GGA Gly	**A**
	GUG Val	GCG Ala	GAG Glu	GGG Gly	**G**

First letter — Third letter

FIGURE 8.16 The genetic code. The possible triplet codons of mRNA are listed with the amino acids they encode. Nonsense codons, called ocher (UAA), amber (UGA), and opal (UGA), which cause termination of translation, are shaded and labeled Stop. The typical starting codon in prokaryotes codes for methionine (Met) and is shaded in green.

Matching an amino acid to a particular tRNA is an amazing feat that aminoacyl-tRNA synthetases have perfected (Fig. 8.17). First, the enzyme binds its cognate amino acid and reacts it with ATP to form an enzyme-bound molecule of aminoacyl-AMP. Second, a transfer reaction releases the AMP and charges the tRNA with the amino acid specified by the anti-codon, generating an aminoacyl-tRNA. The second reaction is reversible, which means that the synthetase can get rid of an incorrect amino acid that is mistakenly attached to a tRNA. Such proofreading is critical, since an incorrect amino acid on the tRNA becomes an error in the protein, which could be deleterious.

Now that we have all the components needed for protein synthesis, we are ready to see how they each contribute to the process. As in replication and transcription, there are initiation, elongation, and termination phases.

1. Initiation of protein synthesis. Initiation starts when the two ribosomal subunits come together onto a nascent mRNA and the first aminoacyl-tRNA is loaded, forming the **initiation complex** (Fig. 8.17) Several questions arise:

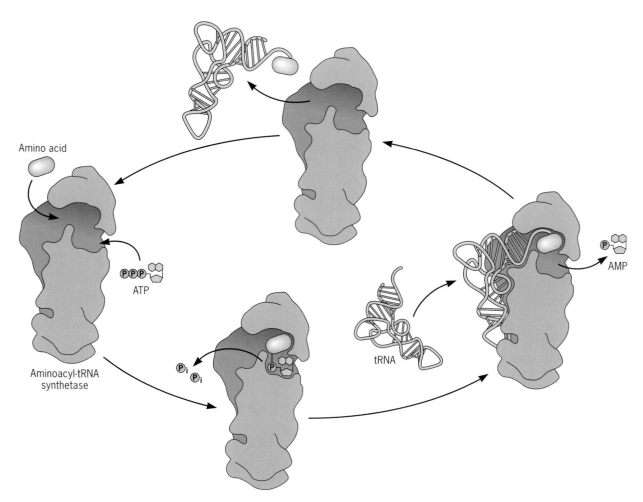

Amino acid

ATP

Aminoacyl-tRNA
synthetase

P_i P_i

tRNA

AMP

FIGURE 8.17 Amino acid activation and attachment to cognate tRNA by a dedicated aminoacyl-tRNA synthetase. See text for details.

How are the two ribosomal subunits prevented from reassociating in the cytoplasm? The small subunit (prokaryotic 30S) is bound to two initiation factors (IF1 and IF3) that prevent it from associating with the large 50S subunit prematurely.

How are mRNAs recognized? The same initiation factors that prevent the small ribosomal subunit from associating with the large one also promote pairing the 3′ end of its 16S rRNA with a site in the mRNA called the **Shine-Dalgarno sequence** (Fig. 8.18). This 4- to 6-base recognition sequence is positioned just upstream of the start codon. Stable RNAs lack this sequence, so they are not translated. What an efficient way to prevent these structural RNAs from making unusable proteins!

How is the 50S ribosomal subunit assembled? Docking of the large subunit onto the small subunit-mRNA complex occurs once a special initiator tRNA pairs with the mRNA's first codon. In bacterial mRNAs, this codon is usually AUG, which encodes for methionine (Fig. 8.16): the initiator tRNA is usually a tRNA charged with a derivative of that amino acid, formyl-methionine (fMet). Loading fMet-tRNA onto the start codon is not spontaneous but requires another initiation factor (IF2) bound to GTP (IF2-GTP) (Fig. 8.18). This GTP is hydrolyzed to eject the initiation factors and promote docking of the large ribosomal subunit. Now the mature 70S initiation complex is ready to go!

How conserved is the initiation of translation in Bacteria *and* Archaea*?* Very conserved! Among the *Archaea*, start signals are the same as their bacterial counterparts, but the initial amino acid is methionine, not fMet. Other than that, the formation of the initiation complex proceeds similarly. Also, as in *Bacteria*, soon after a ribosome has cleared the start site, another initiation complex forms. Repeated initiations generate a polyribosome consisting of a still-growing mRNA chain to which 70S ribosomes are continually being added at the translation start site (Fig. 8.15).

2. Peptide chain elongation. Once the initiation complex has formed, an amino acid is added to a nascent peptide, which grows by an elongation cycle in which the mRNA moves one codon at a time through the 70S ribosome (Fig. 8.19). The number of rounds of the elongation cycle equals the number of amino acids in the protein, and each round requires elongation factors and the expenditure of high-energy bonds (supplied by hydrolysis of GTP). To visualize the cycle, think of an mRNA attached to a 70S ribosome, trailed by a partially completed peptide chain (Fig. 8.19). The ribosome has three tRNA binding sites, designated E, P, and A. Incoming aminoacyl-tRNAs enter the ribosome at its **A site**. The **P site** is occupied by the tRNA carrying the growing peptide. The **E site**, or exit site, is occupied by an uncharged tRNA. At all three sites, the anticodons of the tRNA molecules are paired with their codons of the mRNA. Here is how the peptide, now in the P site, elongates:

(i) Entry in A site: An elongation factor bound to GTP (EF-Tu-GTP) loads the A site with an aminoacylated tRNA with the appropriate anticodon,

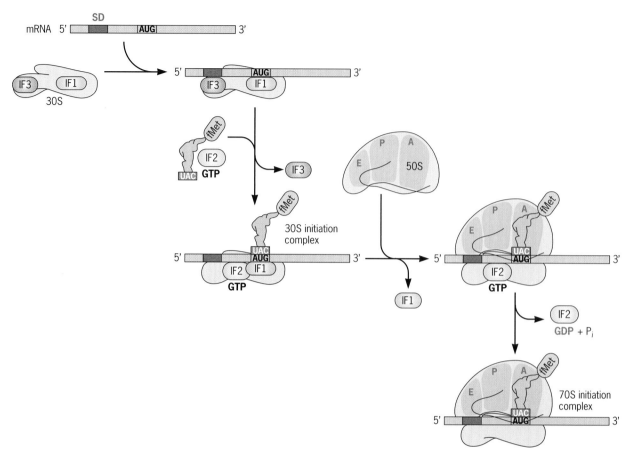

FIGURE 8.18 Initiation of bacterial protein synthesis. See the text for details. SD, Shine-Dalgarno sequence.

using energy provided by GTP hydrolysis. EF-Tu-GDP is now released and recharged with GTP by another elongation factor (EF-Ts), a process known as the "EF-Tu cycle."

(ii) Peptide bond formation: Once the aminoacylated tRNA is in the A site, the amino group of the attached amino acid is close enough to the terminal acid group of the growing peptide (in the P site), and they can be linked with a **peptide bond**. This reaction is catalyzed not by proteins but rather by a segment of the 23S RNA of the 50S ribosomal subunit, a fine example of a ribozyme in action. The growing peptide (now one amino acid longer) is now attached to the tRNA at the A site and an uncharged tRNA is in the P site.

(iii) Translocation: A new elongation factor bound to GTP (EF-G-GTP) binds the A site and transfers the peptide-bearing tRNA to the P site, dragging the attached mRNA along and translocating it by one codon. This process not only empties the A site but also pushes the uncharged tRNA from the P site to the E site. Translocation requires the hydrolysis of one of the phosphates in the GTP bound to EF-G, but the EF-G cycle regenerates the GTP bound form, priming it for a new round of translocation.

For all its complexity, the cycle is speedy, adding approximately 15 amino acid residues to the nascent chain per second (in *E. coli* growing at 37°C).

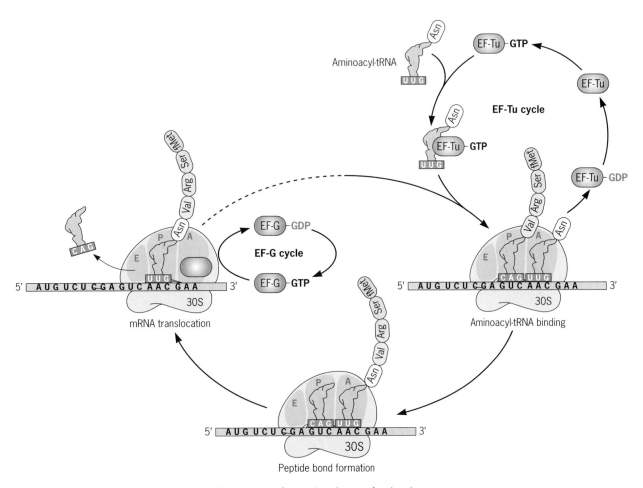

FIGURE 8.19 **Elongation cycle in bacterial protein synthesis.** See the text for details.

There is beauty in this number: The syntheses of RNA and proteins are synchronized. Recall that mRNA is synthesized at the rate of about 45 nucleotides per second, and a codon consists of three nucleotides. Thus, each mRNA molecule is coated with a parade of ribosomes busily translating the growing mRNA, and the ribosome leading the parade keeps pace with RNAP at the transcription bubble—an elegant process (Fig. 8.12). An advantage of this synchrony is that, as long as translation is proceeding and ribosomes are bound to the mRNA, mRNA is protected from nucleases.

Elongation of nascent peptides proceeds similarly in the *Archaea*, though once again, the components of the system differ structurally from their bacterial counterparts. For example, the archaeal EF-G is more like the eukaryotic than the bacterial factor.

3. Termination of translation. Translation does not terminate spontaneously but is triggered when the ribosome encounters one of three termination codons (UAA, UAG, or UGA; also called **nonsense codons**; see Fig. 8.16). The stop codons do not have tRNAs with complementary anticodon and hence they convey no sense. (One fascinating exception, evolved only in bacteria, is the use of UGA for the tRNA for a rare amino acid, selenocysteine, i.e., cysteine with a selenium atom in place of sulfur. Apparently, the context in which UGA appears, that is, the kinds of codons surrounding it, determines whether it will function as a stop codon or as a sense codon recognizing selenocysteinyl tRNA.)

Once a stop codon reaches the A site, it remains empty for a while. This gives the ribosome time to release factors that enter the A site and induce hydrolysis of the peptidyl-tRNA and the release of the completed peptide. The 70S ribosome then dissociates into its two subunits. Two initiation factors (IF-1 and IF-3) bind to the small subunit to prevent its reassociation, and a new cycle of translation can begin.

Making a Protein Functional

To be precise, translation of an mRNA produces a peptide—a specific linear array of amino acids in peptide linkage. Converting a peptide chain into a working protein (whether it has a catalytic, regulatory, or structural role) requires additional steps. Covalent modifications cleave residues from either end of the peptide chain and may add chemical groups to the primary peptide structure. Before or after such modifications, the chain must fold into its mature, active, three-dimensional shape. In some cases, the protein must associate with others to form a functional multiprotein complex. Finally, the protein must be translocated to its proper location in, on, or outside the cell. Molding a peptide into a protein does not wait until translation terminates; some folding and modification occur as the peptide chain grows.

Covalent modification. Cleavage of amino acid residues from the N termini is fairly common in proteins. In some cases, only the initial fMet residue is cleaved; in other cases, sizeable **leader peptides** such as the **signal peptides** that direct protein translocation are removed (chapter 9). Disulfide (S-S) bonds may also form within the peptide. In *E. coli*, and probably in most other bacteria, S-S bonds do not form within the cytoplasm because it is highly reduced. When Gram-negative bacteria secrete proteins, these bonds can form in the periplasm, which is less reduced. (S-S bonds also form within the cytoplasm of cyanobacteria, which is more oxidizing.)

Many other covalent modifications are possible (Table 8.2). It is quite common for cells to modify proteins with phosphate (phosphorylation), methyl groups (methylation), acyl groups (acylation), nucleotides, fatty acids, and sugars. In general, alterations fall into two functional classes. "Assembly modifications" simply add chemical groups the protein needs to function (examples are lipoproteins and glycoproteins). On the other hand, "modulation modifications" regulate the activity of the mature protein. These flexible adjustments alter protein structure and activity, and most are reversible. More will be said about them in chapter 12.

Protein folding. To this point, we have implied that, embedded in its primary amino acid sequence, is information to direct folding of the peptide into a mature, biologically active protein. This notion is too simple. Most peptides can fold into any of several final three-dimensional forms, only one of which may be functional. The paths from linear peptide chains to active folded proteins have not yet been thoroughly mapped. However, much has been learned from studying how bacteria respond to heat shock (see chapter 12). This stress unfolds many proteins but also activates pro-

TABLE 8.2 **Examples of bacterial protein modifications**

Modification type	Protein[a]
Chiefly assembly reactions	
Signal peptide cleavage	Secreted proteins
Formation of S-S bonds	Many proteins
Addition of lipid moiety	Murein lipoprotein
Addition of sugar moiety	Membrane glycoproteins
Attachment of prosthetic group	Many enzymes
Chiefly modulation reactions	
Phosphorylation	Ribosomal protein S6; isocitric dehydrogenase; many proteins with regulatory functions
Methylation	Chemotactic signal transducers
Acetylation	Ribosomal protein L7
Adenylylation	Glutamine synthetase

[a]Examples are drawn from *Escherichia coli.*

cesses that refold the damaged proteins to restore their function. Here we summarize some general aspects of protein folding in bacteria.

Two principles govern protein folding (Fig. 8.20). First, left to its own devices, a peptide will likely misfold. Second, misfolding must be corrected quickly, lest the misfolded protein gum up vital functions. Should misfolded peptides accumulate, whether through synthesis errors or environmental damage, the misfits are speedily destroyed by proteases and their amino acids reused (Fig. 8.20). Clearly, peptides should not be left alone for long. Bacteria enlist **chaperones**, factors that stick with peptides as they are maturing, to modulate the speed and path of their folding. Chaperones are so critical to a healthy life that all cells, prokaryotic and eukaryotic alike, encode several families of these molecules, and they have been strongly conserved in all living organisms.

In *E. coli*, the first chaperone encountered by a peptide is the **ribosome-associated trigger factor**. This multidomain protein interacts with proline residues, which introduce bends in protein chains. Trigger factor recognizes the prolines in nascent protein chains and rotates their peptide bonds.

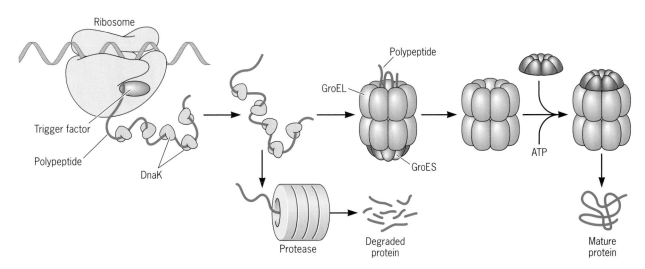

FIGURE 8.20 **Protein-folding pathways.** See the text for details.

Other chaperone pathways assist proteins after they have been synthesized. Once off the ribosome, a peptide still needing assistance is bound by the chaperone DnaK, which cooperates with other chaperones to repair small hydrophobic regions that have misfolded. Still other maturing peptides need the support of a folding machine. These large molecular chambers house young peptides as they fold, protected from proteases. In particular, GroEL forms a barrel where the peptide is inserted, and GroES is its lid that controls entry of the peptide. Once the protein is inside, the GroEL chamber releases the GroES lid and binds ATP. A conformational change then energizes binding of a new GroES lid at the opposite end of the barrel. Finally, ATP hydrolysis drives the release of the mature protein to the cytoplasm, and the GroEL/ES complex is ready for a new client.

Protein translocation. Many newly synthesized and folded proteins do not stay put in the cytoplasm, but are targeted to the cell envelope or outside the cell. In fact, the bacterial cell membrane contains over 300 proteins (chapter 2), and the Gram-negative cell's periplasm and outer membrane house another 100 or more. In addition, bacteria secrete proteins (enzymes, toxins) into their environments; others inject proteins directly into eukaryotic host cells. In the next chapter, we will explore how these feats are accomplished.

CASE REVISITED: Antibiotics and macromolecules

The patient, Mr. M, was treated with ciprofloxacin, an antibiotic known to be effective against his "bug," the pneumococcus. Like all antimicrobial

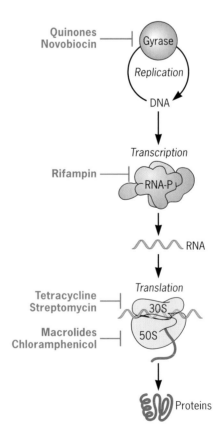

drugs, this one is selectively toxic to the bacterium and innocuous to the patient because the drug inactivates an essential protein of the bacterial DNA synthesis machinery, which differs significantly from that of the host cells. More specifically, ciprofloxacin inactivates DNA gyrase, a typically prokaryotic enzyme that removes extra positive supercoils that form ahead of the replisome and impede its advance. By contrast, Mr. M's own DNA replication machinery, typical of eukaryotes, does not use a gyrase (instead nucleosomes remove extra positive supercoils). Thus the drug was not toxic to Mr. M.

The attending physician had a wide choice of antibiotics to choose from. As shown in the figure, ciprofloxacin, a type of quinolone, is not the only antibiotic that inhibits the bacterial DNA gyrase. Other antibiotics that inhibit RNA and protein synthesis (see figure) could have cleared his infection. Here again, the machinery targeted in the bacteria differs from that of the host. For example, various components of the protein synthesizing system of bacteria (ribosomes, the various protein factors involved) are quite different from the mammalian ones, making them sensitive to certain drugs that do not affect the latter.

How does a pneumococcus develop resistance to any of these antibiotics? The quick answer is that spontaneous mutations to drug resistance are selected for in the presence of the drug. Resistance to quinolones such as ciprofloxacin can arise when mutations occur in the gene encoding DNA gyrase that prevent antibiotic binding but not its key contribution to replication. Mutations in membrane proteins that prevent antibiotic passage into the cell or in pumps that promote drug extrusion can also confer resistance to the antibiotic. Even if these specific mutations are rare, the sheer number of bacteria subjected to the drug's selective pressure is often great enough to permit mutants to arise and take over the population.

CONCLUSIONS

We have reviewed, in the past chapters, the metabolic processes that equip a newly formed bacterial cell to double its components. Here, we focused on how the cell builds the essential macromolecules DNA, RNA, and proteins. Are these the only macromolecules needed to build a cell? They are not. The cell also needs lipids for membranes and carbohydrates for cell walls, capsules, and exopolysaccharides. As we learned in chapter 2, appendages such as flagella and pili may also increase the cell's fitness. Biosynthesis of macromolecules such as DNA, RNA, and proteins is at the core of the complex biosynthetic machinery that is needed to build a functional cell envelope. The intricate events that bring about this feat are the next chapter in our story.

Representative ASM Fundamental Statements and Learning Outcomes for the Chapter

EVOLUTION

1. Cells, organelles, and all major metabolic pathways evolved from early prokaryotic cells.

   a. Explain why cells need both proteins and nucleic acids, as opposed to having ribozymes.
   b. State a reason why DNA, not RNA, evolved to become the preferred macromolecule for inheritance.

INFORMATION FLOW AND GENETICS

2. Although the Central Dogma is universal in all cells, the processes of replication, transcription, and translation differ in *Bacteria, Archaea*, and eukaryotes.

   a. Explain how proteins and enzymes demonstrate high selectivity for substrates.
   b. Define replisome.
   c. Identify the role of DNA polymerase III in DNA replication.
   d. Summarize the reactions during replication initiation.
   e. Identify the roles of DnaA in replication.
   f. State the mechanism by which the cell ensures that it only replicates DNA and divides when there is sufficient ATP.
   g. Summarize how DnaA initiates replication at the origin.
   h. State how *Bacteria* prevent the initiation of replication from happening prematurely.
   i. Compare and contrast DNA strand elongation on the leading and lagging strand.
   j. Describe how the problem of positive supercoiling in DNA strands due to unwinding of the helix is solved.
   k. Describe the replication termination steps in *Bacteria* that are unique as a result of having a circular chromosome.
   l. Compare and contrast mismatch repair by DNA polymerases and the proteins of the mismatch repair system.
   m. Describe how methylation guides the repair of mismatches in *E. coli*.
   n. Identify the role of endonucleases in microbial cells and in recombinant DNA technology.
   o. List proteins and factors that help structure the nucleoid and play roles in chromosomal condensation and replication.
   p. State the consequence for transcription by having operons.
   q. Identify the role of sigma (σ) factors in transcription.
   r. Summarize the process of RNAP binding to the promoter during transcription.
   s. Identify the condition under which abortive transcription takes place.
   t. State how transcription can be regulated after initiation.
   u. Name two mechanisms that signal to RNAP to terminate transcription.
   v. List benefits and costs of the short half-life of mRNA.
   w. Explain how ribosomes are assembled.
   x. Identify the role of enzymes in the processing steps of tRNA that make an active product.
   y. Explain reasons why there is a bias in many species to having genes on the leading strand as opposed to the lagging strand.
   z. Name factors that influence genome organization.
   aa. Identify a consequence upon translation of *Bacteria* and *Archaea* not having a nucleus.
   bb. Compare and contrast prokaryotic ribosomes and eukaryotic ribosomes.
   cc. Summarize the steps of loading and preparing the tRNA molecule.
   dd. Describe the role of the Shine-Dalgarno sequence in protein synthesis initiation.
   ee. Outline the major reactions for each of three steps in polypeptide chain elongation: aminoacyl-tRNA binding, peptide bond formation, and translocation.
   ff. Explain how polypeptide chain termination is triggered.
   gg. Give examples of covalent modification in nascent polypeptide chains.
   hh. Identify the role of chaperone proteins in the protein folding pathways.

IMPACT OF MICROORGANISMS

3. Microorganisms provide essential models that give us fundamental knowledge about life processes.

   a. Describe an example of a replication process that was discovered and described in bacteria and is widespread among living organisms.

SUPPLEMENTAL MATERIAL

References

- **Beattie TR, Reyes-Lamothe R.** 2015. A replisome's journey through the bacterial chromosome. *Front Microbiol* **6:**562. doi: 10.3389/fmicb.2015.00562
- **Mott ML, Berger JM.** 2007. DNA replication initiation: mechanisms and regulation in bacteria. *Nat Rev Microbiol* **5:**343–354 doi: 10.1038/nrmicro1640
- **McGary K, Nudler E.** 2013. RNA polymerase and the ribosome: the close relationship. *Curr Opin Microbiol* **16:**112–117.(doi: 10.1016/j.mib.2013.01.010

Supplemental Activities

CHECK YOUR UNDERSTANDING

1. It is said that prokaryotic cells are "streamlined" for rapid growth. What characteristics of macromolecule synthesis illustrate this feature of the prokaryotic cell?

2. Some aspects of DNA replication are different in prokaryotes and eukaryotes; others are the same. List the similarities and differences.

3. The average lifetime of mRNA molecules in prokaryotes is a few minutes. This is clearly costly in terms of energy. What might be the selective pressure for such rapid turnover?

4. The prokaryotic chromosome is the site of much activity in a growing cell, including replication and transcription. Are these processes ever in conflict? If so, what is the outcome of the conflict?

5. What prevents ribosomes from initiating protein synthesis at AUG sequences prematurely?

DIG DEEPER

1. The structure of the origin of replication (*oriC*) is more complex than presented in Fig. 8.3, as illustrated in the figure below:

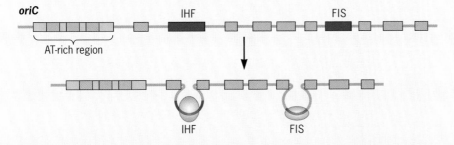

oriC

   a. The DnaA boxes scattered throughout *oriC* do not have the same sequence. As a result, some have more affinity for DnaA, others for DnaA-ADP, and others for DnaA-ATP. Why would a cell contain boxes with reduced affinity for DnaA-ATP, the functional molecule that permits DNA replication?

   b. In addition to DnaA boxes, the origin contains sequences recognized and bound by two DNA-bending proteins: IHF (integration host factor) and FIS (inversion stimulation). Mutants that cannot produce IHF or FIS cannot initiate replication on time for cell division. What role could they have?

2. The three rRNA molecules (5S, 16S, and 23S) are encoded in an operon and, therefore, are transcribed as a polycistronic RNA transcript.

   a. The *rrn* promoter of bacteria such as *E. coli* has the consensus sequence of σ^{70} promoters (Fig. 8.11) and an upstream UP element. Why?

   b. Some prokaryotes have more than one *rrn* operon. *E. coli*, for example, has seven. Other bacteria can have a dozen *rrn* operons. The extraordinarily high demand for the products of these genes is undoubtedly one selective pressure that favors *rrn* redundancy. Can you think of other reasons why a microbe would carry so many *rrn* operons?

   c. Prokaryotic *rrn* operons often encode some tRNAs in addition to the three rRNAs (Fig. 8.13). Which tRNAs would you place in these actively transcribed *rrn* operons?

3. Coupled transcription and translation is a defining characteristic of prokaryotes. Do you think that under exceptional circumstances this could also take place in a eukaryote?

4. If one were to collect all the polyribosomes that were engaged at a given instant in producing the same protein in a bacterial cell, would they be the same size (i.e., each have the same number of ribosomes)? Explain your answer.

5. **Team work:** Prokaryotic and eukaryotic cells take different approaches when synthesizing their DNA, RNA, and proteins. Working in groups:

   a. Explain why DNA replication in eukaryotes and some *Archaea* starts at more than one origin but only in one in *Bacteria*.

   b. Speculate about possible reasons why most eukaryotes do not have operons.

   c. Fathom why the eukaryotic ribosomes are larger than the prokaryotic ones.

Supplemental Resources

VISUAL

- Review the chemistry and structure of DNA and marvel at how the DNA double helix is faithfully replicated by the replisome: https://www.youtube.com/watch?v=JZXT2uOcD2w
- Animation of polyribosomes on an mRNA: https://www.youtube.com/watch?v=ReogkhkoArw
- Protein elongation by the ribosome: https://www.youtube.com/watch?v=3amYZtPGFBc

WRITTEN

- Christoph Weigel tells us about bacteria with more than one origin and no DnaA proteins in the *Small Things Considered* blog post "Adhering to the 'Replicon Model' the Sloppy Way" (http://schaechter.asmblog.org/schaechter/2014/02/adhering-to-the-replicon-model-the-sloppy-way.html).
- Marcia Stone ponders what the ancestral RNAs were like in the *Small Things Considered* blog post "Oddly Microbial: Ribocytes" (http://schaechter.asmblog.org/schaechter/2012/04/oddly-microbial.html).
- Moselio Schaechter remembers the iconic picture that demonstrated the coupling of transcription and translation in bacteria in the *Small Things Considered* blog post. "Pictures Considered #1. Visualizing Coupled Transcription and Translation in *E. coli*" (http://schaechter.asmblog.org/schaechter/2013/03/pictures-considered.html).
- Moselio Schaechter describes the unsuspected functions of ribosomal proteins in the *Small Things Considered* blog post "The World Is Pleiotropic" (http://schaechter.asmblog.org/schaechter/2009/06/the-world-is-pleiotropic.html).
- Merry Youle discusses how epigenetic switches in transcription allow two bacterial cells that are genetically identical to behave very differently in the *Small Things Considered* blog post "Hedging Your Bets" (http://schaechter.asmblog.org/schaechter/2011/01/hedging-your-bets.html).

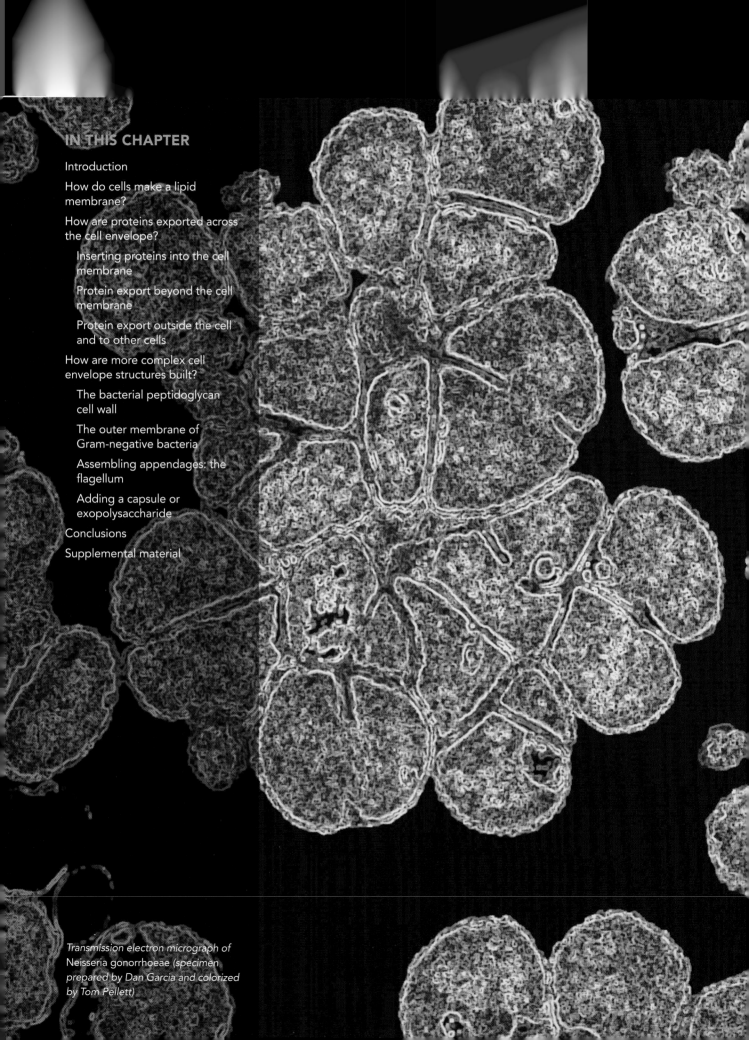

Transmission electron micrograph of Neisseria gonorrhoeae (specimen prepared by Dan Garcia and colorized by Tom Pellett)

CHAPTER NINE

Building the Cell Envelope

KEY CONCEPTS

This chapter covers the following topics in the ASM Fundamental Statements.

EVOLUTION

1. Cells, organelles (e.g., mitochondria, chloroplasts), and all major metabolic pathways evolved from early prokaryotic cells.

CELL STRUCTURE AND FUNCTION

2. *Bacteria* and *Archaea* have specialized structures (e.g., flagella, endospores, pili) that often confer critical capabilities.

INTRODUCTION

The preceding chapters described how the major constituents of a cell are made using energy and substrates harvested from the environment. We now know how microbes synthesize three of their essential macromolecules: DNA to carry the genetic information; RNA as information carriers, catalysts, and regulators; and proteins, which contribute catalytic, structural, and other functions in the cell. The integrity and activity of these macromolecules require a controlled environment, which the cell envelope provides. But consider the challenges that arise from this physical confinement:

- Topologically, all layers of the envelope are closed surfaces and must be physically continuous for cell integrity and viability. Yet, they must expand during cell growth by incorporating newly synthesized material.
- All constituents of the envelope must expand coordinately, with components properly placed.

- Proteins, lipids, and complex polysaccharides must be trafficked to their correct destinations, such as the cell membrane, periplasm, outer membrane, or the exterior of a Gram-negative bacterium.
- Newly assembled envelopes must facilitate cell division.
- As the first sensor of environmental change, the envelope must participate in the cellular response to fluctuations, in many cases by modifying its own structure or composition.

How are the envelope components synthesized and assembled with speed, accuracy, and adaptive flexibility? To date, microbiologists have only a general picture of this process; many interesting questions are yet to be answered. Here, we present a general overview of what is entailed in building a cell with such impressive complexity, yet efficient function.

CASE: **How close are scientists to building a synthetic cell?**

For many years, researchers have been able to replicate biosynthetic processes such as DNA and RNA synthesis *in vitro*. But they have failed at confining these processes within a cell envelope in a manner such that a synthetic cell, self-sustained and able to grow and divide, is formed. As one pragmatic approach, they set a modest goal: to build a simple "protocell," such as the one that evolved into the ancestral cellular forms, which contained a self-replicating molecule within a lipid membrane boundary. Let's assume that they have succeeded.

- What self-replicating molecule could these synthetic protocells have?
- How would the lipid membrane of the protocell permit the right chemicals to permeate so that macromolecular synthesis can be sustained?
- What other features must still be added to convert this minimal cell into a modern, fully functional cell?
- Would you consider the protocell to be "alive"?
- What evidence would you need to designate the synthetic protocell as alive?

HOW DO CELLS MAKE A LIPID MEMBRANE?

LEARNING OUTCOMES:
- Name features common to protocells and the first cellular forms.
- Compare and contrast the structure of phospholipids and *Bacteria* and *Archaea*.
- Describe how a cell builds phospholipids in the cytosol and inserts them into the membrane.
- Summarize why a cell's membrane is not considered a static entity.

Chapter 1 described how, over 4 billion years ago, abiotic reactions could have generated not only the first self-replicating RNAs but also fatty acids, molecules that have both hydrophobic and hydrophilic portions. This amphipathic property of fatty acids promotes their self-assembly as monolayers (micelles) or bilayers (vesicles or liposomes) in which the polar groups are exposed to the aqueous phases (Fig. 9.1). A fortuitous event could have formed the first protocell: a spontaneously forming vesicle engulfing a

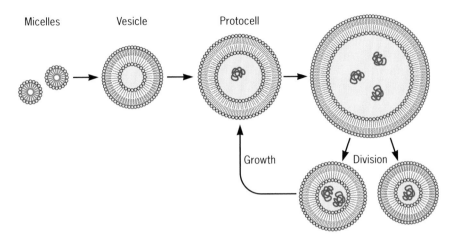

FIGURE 9.1 **Making a protocell with fatty acid micelles.** Formation of fatty acid vesicles from micelles and engulfment of a self-replicating RNA to make the first protocell. Micelles fuse with the protocell membrane, enlarging it until it reaches a size that is too unstable to protect the hydrophobic regions within from the aqueous environment. At that point, the protocell divides into two.

self-replicating RNA (Fig. 9.1). The protocell likely grew by incorporating more micelles, thus expanding its volume to accommodate the RNA molecules being synthesized inside. At some point, the protocell would have become so large that the hydrophobic tails of the fatty acids were no longer shielded from the aqueous environment; ionic forces would split the protocell in two. And so, the three essential features of a living cell—self-replication, growth, and cell division—occurred spontaneously, driven by the natural chemistry of each molecular component. Would it be possible to reproduce similar abiotic reactions in the laboratory to make a synthetic protocell, as suggested in the case study?

The first cellular forms that evolved from protocells inherited the primitive lipid membrane and gradually acquired mechanisms to control its growth and division. And from those cells, new cells evolved. Or as the renowned 19th century German scientist Rudolph Virchow eloquently stated: *Omnis cellula e cellula ("Every cell originates from a cell")*. Note the physical continuity of the cell membrane as cells divide and new generations of cells are formed. The membrane is not synthesized *de novo* in the daughter cells but inherited from the mother cell. Hence, all the membranes in cells are derived from pre-existing ones *(Omnis membrana e membrana?)*. If so, the implications are profound: all the membranes of present day cells are derived from the first successful cell!

Fueling and biosynthesis provide the energy and building blocks a cell needs to expand the lipid membrane and to modify its composition to adapt to the environment. As we learned in chapter 8, the essential steps for making DNA, RNA, and proteins are polymerizing their building blocks into long chains, modifying and folding the products, and placing them in their appropriate cellular locations. In contrast, making a lipid membrane requires more than one type of building block (Fig. 9.2). Fatty acids, sugars, and proteins are the building blocks of lipids and their conjugates, such as lipoproteins and lipopolysaccharides. Moreover, these disparate moieties are not polymerized but rather linked together through a variety of chemical bonds.

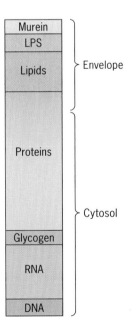

FIGURE 9.2 **Macromolecules needed to build a cell and their preferential target location (envelope or cytosol).**

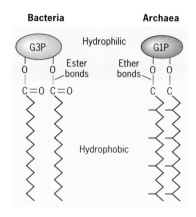

Bacteria   **Archaea**

FIGURE 9.3 **Phospholipids of the bacterial and archaeal cell membrane.** The phospholipids of *Bacteria* and *Archaea* are similar in structure but differ in the type of polar head (glycerol 3-phosphate, G3P, or glycerol 1-phosphate, G1P, respectively), its linkage to the hydrophobic fatty acids (ester or ether, respectively), and type of fatty acid (linear or branched isoprenoid, respectively).

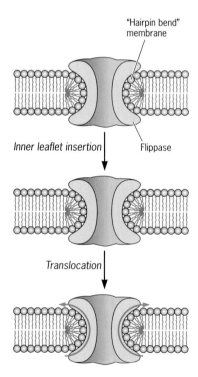

FIGURE 9.4 **Incorporation and translocation of phospholipids into the cell membrane bilayer.** New lipids are shown in red.

As an example, consider the synthesis of glycerol phosphate **phospholipids**, the primary component of all cell membranes. (Although, if phosphate levels are low in the environment, some marine bacteria can substitute them with certain sulfolipids.) As its name indicates, this molecule consists of a glycerol (a sugar alcohol) carrying a phosphate and linked to a hydrophobic lipid tail composed of two fatty acid chains (Fig. 9.3). Recall from chapter 2 that *Bacteria* (and eukaryotes) use *glycerol 3-phosphate* as the phospholipid's polar head and link it to the fatty acid tails with *ester* bonds, whereas *Archaea* use *glycerol 1-phosphate* and *ether* bond linkages, and the hydrophobic tails are isoprenoids (Fig. 9.3). Because of these two distinct chemical compositions, some scientists have suggested that LUCA (the Last Universal Common Ancestor) had a cell membrane composed of a mix of both types of phospholipids. Despite their chemical differences, the amphipathic structure of the molecules is the same in all cells. Their synthesis also shares some features: the glycerol phosphate and the fatty acid moieties are generated separately and then joined via ester or ether links. But consider the challenges of making hydrophobic fatty acids in an aqueous environment. What prevents them from aggregating in the cytosol? To avoid this mess, cells rely on specialized protein carriers, the **acyl carrier proteins,** to engulf the hydrophobic tails and carry them to the polar head. These chaperone-like proteins also carry their newly synthesized glycerol phosphate phospholipids to their destination in a pre-existing membrane. To ensure their correct orientation and position, the phospholipids are inserted into the target membrane with the assistance of special membrane proteins.

Many aspects of membrane growth remain obscure. For example, it is not known with certainty at how many sites the membrane is assembled. We do know that newly formed phospholipid molecules appear first in the inner leaflet of the cell membrane, and some are then transferred to the external leaflet by membrane proteins such as **flippases**. But how these molecular gymnastics are accomplished is not clear. A hypothetical mechanism is shown in Fig. 9.4. We know that lipid membranes must be dynamic to accommodate the cell's needs. For example, phospholipids are continuously recycled to maintain the membrane's stability and integrity. Cells also adjust the fatty acid composition of the lipid membrane to adapt to a fluctuating environment, whose temperature, osmolarity, and pH may change. In some cases, the existing phospholipids are chemically modified while in the membrane; in other situations, new phospholipids are synthesized *de novo* and inserted into the membrane. Not only is the composition of the membrane phospholipids adjustable, cells also control their distribution in the membrane. As a result, different phospholipids are distributed asymmetrically among the leaflets of the membrane bilayer. This, in turn, controls the membrane's physical and chemical properties and, therefore, its functions. So don't be lulled into thinking that lipids are static, stable structural constituents of membranes. The cell membrane undergoes a variety of molecular changes not only to fulfill its critical physiological roles but also to adapt to a fluctuating environment. And all these features would have needed to be considered when designing a synthetic cell, as proposed in the case study.

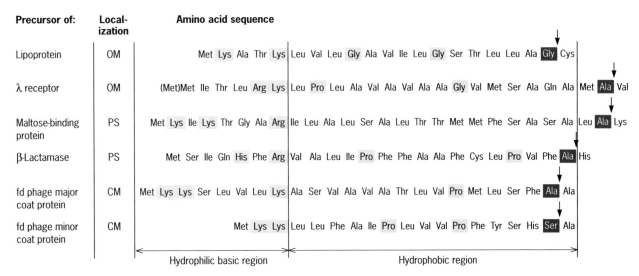

Precursor of: | **Local-ization** | **Amino acid sequence**

FIGURE 9.5 Signal sequences for protein export. Examples of the structure of *E. coli* precursors of proteins targeted to the outer membrane (OM), the periplasmic space (PS), or the cell membrane (CM). The precursors carry an amino-terminal tag (signal peptide) containing hydrophilic and hydrophobic regions, which is recognized and cleaved (arrow) by proteins of the secretion apparatus during export. Basic amino acids of the hydrophilic region are highlighted in green, glycine and proline residues of the hydrophobic region are in orange, and the residues at the cleavage sites (arrows) are in red.

HOW ARE PROTEINS EXPORTED ACROSS THE CELL ENVELOPE?

LEARNING OUTCOMES:
- Describe how a protein is inserted into the cell membrane.
- Identify the role of the Sec system in exporting most periplasmic outer membrane proteins of Gram-negative *Bacteria* and secreted proteins in Gram-positive *Bacteria*.
- Illustrate the challenges faced by Gram-negative *Bacteria* and some solutions regarding inserting proteins in the outer membrane or secreting proteins outside the cell.
- Identify features and structures common to all Gram-negative secretion systems.

Consider the protocell of the case study. The protective lipid membrane provides a controlled internal environment for its molecular machinery. But, how responsive is the protocell to its environment? Can it control what permeates the membrane and what doesn't? Because the lipid membrane is relatively impermeable to many chemicals, it isolates the cell from the surrounding environment. But complete isolation would be akin to a death sentence for the cell. How would it get its nutrients? Or sense stress? How would it interact with other cells? In chapter 2 we learned that microbes expand the function of their lipid membranes by inserting proteins. In fact, membrane proteins are quite abundant, accounting for approximately 70% of the mass of the membrane of many prokaryotes. Furthermore, some of these proteins have the important role of secreting other proteins to specific locations within the cell envelope, such as the periplasmic space or the outer membrane of Gram-negative bacteria, or to the extracellular environment.

Making a protein is one thing, delivering it to a specific location in the cell envelope or outside the cell is another. Yet cells have evolved sophisticated ways to ensure that each protein reaches its cellular target. *How does the cell recognize the appropriate destination of a protein?* And, once a protein needing translocation is recognized, *how is it transported?* Answers to these fundamental questions are beginning to be discovered. But in all cases, *the basic information for inserting a protein in its right place is contained in its own sequence* (Fig. 9.5). As is often the case, most information comes from studies of *E. coli*. Figure 9.6 will be our guide as we learn about some of the mechanisms used to target and export proteins.

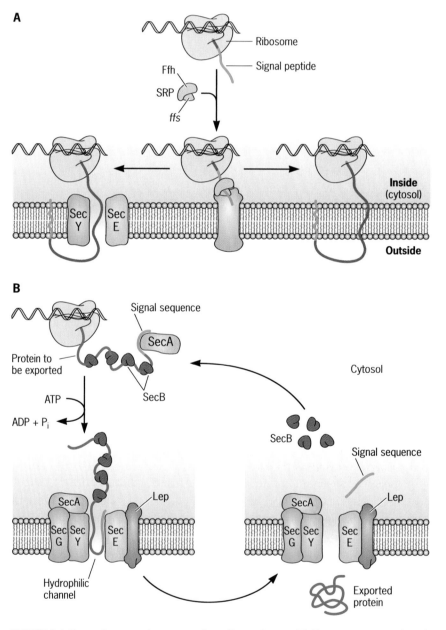

FIGURE 9.6 Exporting proteins across the cell membrane. (A) Shown are two modes of translocating a protein across the cell membrane: direct insertion and insertion using a SRP, the signal recognition particle. (B) The Sec system exports proteins with signal sequences out of the cytoplasm.

Inserting Proteins into the Cell Membrane

Insertion of a protein *into the cell membrane* is the simplest case of protein export (Fig. 9.6A). Are membrane proteins sufficiently hydrophobic to "melt" into the membrane? This simple notion has a kernel of truth. Many cell membrane proteins are indeed strongly hydrophobic, projecting only hydrophilic tails into the cytoplasm or the exterior. Their hydrophobic loops or tails may be small or quite extensive, but they spontaneously insert into the hydrophobic environment of the lipid membrane (Fig. 9.6A). These polypeptides contain all the information needed to slip into the cell membrane—they do not need help.

Other proteins require assistance (Fig. 9.6A). They must first be recognized by a set of dedicated proteins. On the N-terminus of these target proteins is a **signal sequence** of *15 to 30 amino acids*. The signal comes in several varieties (Fig. 9.5), but all will bind to a **signal recognition particle** (SRP) (Fig. 9.6A). SRPs consist of a protein (Ffh) and a small 4.5S RNA molecule (ffs), which together recognize the signal sequence. As the nascent polypeptide is being translated by the ribosomes, SRP binds the signal sequence and *escorts* the protein to dedicated secretory proteins in the cell membrane, such as the SecYE complex (Fig. 9.6A).

Protein Export Beyond the Cell Membrane

What about hydrophilic proteins destined either for the periplasm or for secretion outside the cell? Proteins with a hydrophilic surface cannot readily enter or pass through the hydrophobic phospholipid membranes to reach their proper location. The bacterial cell membrane, for example, contains hundreds of proteins that perform a variety of metabolic functions (chapters 2 and 6). Another hundred or more proteins function exclusively in the periplasm and outer membrane of Gram-negative bacteria. Furthermore, Gram-positive and, to a lesser extent, Gram-negatives secrete enzymes and toxins into their environments. Some bacteria even inject proteins directly into eukaryotic host cells and into *other bacteria*. How? And how are they distinguished from the pool of cytoplasmic proteins, which are also hydrophilic? In fact, bacteria and archaea have evolved at least a half dozen mechanisms for moving this type of protein into and through membranes.

On a journey beyond the cytoplasm, a protein confronts the cell membrane, its first barrier. Here, a machinery called the **Sec system** exports most *periplasmic and outer membrane proteins* of Gram-negative bacteria as well as secreted proteins of Gram-positive bacteria (Fig. 9.6B). Recognition by the Sec system occurs early on, usually while the nascent peptide is still being made on the ribosome. One of the proteins of this widely disseminated Sec system (SecA) recognizes the N-terminal signal sequence, binds to it, and leads it to a cell membrane-spanning channel composed of three other Sec proteins (SecY, SecE, and SecG). While in transit from the ribosome to the channel, polypeptides destined for export are coated with yet another Sec protein (SecB), a dedicated **chaperone** that *prevents the polypeptides from folding*. Kept in an extended form by SecB, hydrophobic and hydrophilic proteins can thread their way through the hydrophilic inner surface of its channel (Fig. 9.7). While passing through the SecYEG channel, a "**signal peptidase**" clips off the protein's signal sequence

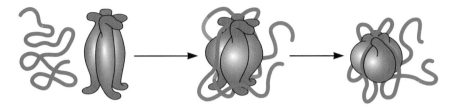

FIGURE 9.7 SecB, the peptide chaperone. The SecB protein serves as a chaperone to keep a peptide from folding upon itself, which would make it difficult to translocate across a membrane. This sketch shows that when SecB interacts with the peptide, its conformation is altered, revealing further binding sites, thus sequestering the protein to a greater extent.

(Fig. 9.6B). Finally, the protein folds in the periplasm or, in the case of Gram-positive bacteria, in the environment. The job is now complete. But passage through the SecYEG channel is not free; translocation requires energy in the form of proton motive force, ATP, or both. SecG and SecA hydrolyze a molecule of ATP, providing sufficient energy to force a length of 20 amino acids through the channel.

Gram-negative bacteria face another challenge: some proteins must be inserted into the outer membrane. Consider the arduous journey of these proteins (Fig. 9.8). First they need to cross the cell membrane. Being membrane proteins, they are hydrophobic; hence, they must depend on protein chaperones to prevent them from aggregating in the periplasm. Finally, they need to be recognized by dedicated outer membrane proteins to insert them into the outer membrane. How are these steps in the export path accomplished? To direct their export across the cell membrane, proteins destined for the outer membrane also carry the N-terminal signal sequence that is recognized and cleaved off by the Sec apparatus (Fig. 9.8). However, while still being exported by the Sec machinery, the nascent proteins are bound by a chaperone protein (Skp) to prevent them from aggregating in the periplasm. The same chaperones then escort the pro-

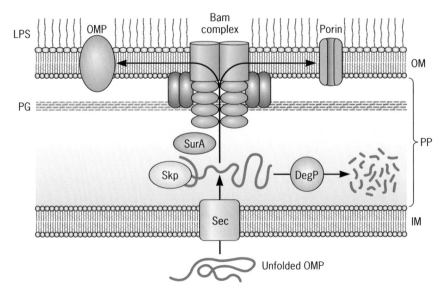

FIGURE 9.8 Exporting proteins to the outer membrane in Gram-negative bacteria.

tein to a dedicated protein complex of the outer membrane (Bam). With the assistance of another chaperone (SurA), the Bam complex folds the protein and inserts it into the outer membrane. If something goes awry, misfolded proteins are degraded in the periplasm by the DegP protease, which prevents them from getting inserted into the outer membrane. However, one step of this journey remains obscure: how the protein manages to cross the peptidoglycan layer. Perhaps the Bam complex is long enough to protrude into the periplasm and to facilitate the protein's passage through the peptidoglycan layer.

Protein Export Outside the Cell and to Other Cells

And what about those proteins the cell must deliver outside or into another cell? Such proteins are hydrophilic, like those exported to the periplasm. In Gram-positives, which do not have the outer membrane barrier, proteins are secreted directly outside the cell once they cross the cell membrane, as just described (Fig. 9.6A). However, in Gram-negative bacteria the proteins must also traverse the outer membrane. In certain cases, the Sec system first exports the cargo to the periplasm, and then a different system takes the protein across the outer membrane. At least six distinct mechanisms have already been characterized, called Type I through Type VI secretion systems, and still others are being discovered (Fig. 9.9, 9.10, Table 9.1). Their secreted cargo contributes to the life of bacteria in many different ways. Some proteins build flagella and pili; others are toxins that affect animal and plant host cells or other bacteria in the neighborhood; or, DNA is delivered to recipient cells. Below we highlight two of these export machines.

Type III secretion systems. The Type III secretion system is a captivating instance of translocation that is neither co-translational nor Sec-dependent. Instead, it evolved from the machinery that builds flagella (discussed shortly and in Box 9.1) and is encoded on a genomic island that has spread among bacterial species by **horizontal gene transfer** (chapter 10). Many pathogenic bacteria (e.g., *Yersinia*, *Salmonella*, *Shigella*, and *Pseudomonas* species) have acquired mobile plasmids that encode Type III secretion systems. Using these needle-like machines, the bacteria can deposit adhesins, toxins, and other virulence factors into the extracellular space or directly into the cytoplasm of animal or plant cells. To maximize efficiency, some of these systems mediate contact-dependent secretion, triggered when the pathogen contacts a host cell. Once injected directly into the cytoplasm, many cargo proteins manipulate the biology of their host cell (Fig. 9.9). Type III secretion requires a large complex of approximately 20 proteins, including chaperones for particular proteins to be secreted and ATP-binding protein to energize the system.

Type VI secretion systems. A Type VI secretory apparatus can translocate effector proteins to other bacteria or eukaryotic cells (and in at least one case, both). These fascinating structures evolved from phages, whose contractive tails inject DNA into their bacterial hosts (Fig. 9.10). Just like the phage tail, the long spike of the Type VI apparatus contracts to inject

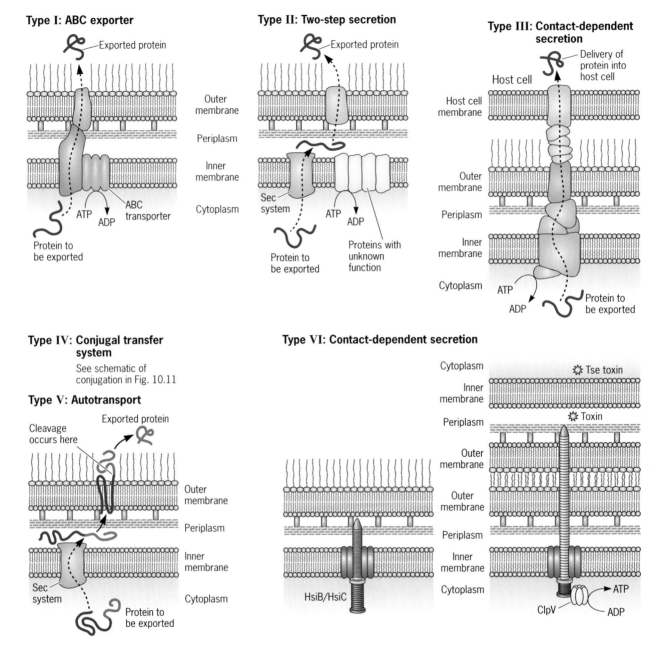

FIGURE 9.9 Several mechanisms for protein export. See Table 9.1 for details.

its cargo. Also like phages, Type VI secretion systems assemble a sheath that promotes contraction. As soon as the target cell is in reach, the Type VI system can puncture it and inject its payload. In many cases, the cargo is a toxic protein that interferes with the target cell's physiology, eventually killing it. A variety of bacteria use these phage-like secretion systems to compete with a related strain of bacteria. Others use these versatile secretory machineries to manipulate the biology of their eukaryotic host cell, either to evade host defenses or to establish a beneficial interaction.

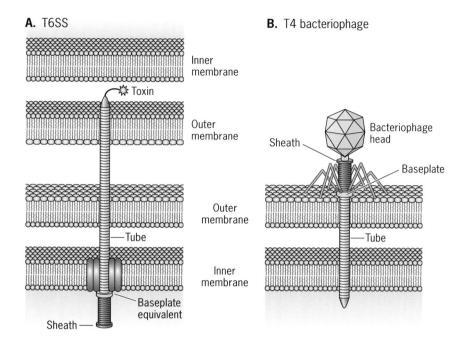

FIGURE 9.10 The Type VI secretion apparatus is strikingly similar to parts of phages.
Note that the long spike of Type VI resembles tail of a phage. In addition, both these structures are contracted by the action of an analogous sheath. In the case of a phage, this results in the passage of DNA into the recipient cell; in Type VI, the entry of proteins into an adjacent cell.

HOW ARE MORE COMPLEX CELL ENVELOPE STRUCTURES BUILT?

LEARNING OUTCOMES:
- Identify how murein precursors travel from the site of assembly, the cytoplasm, to the periplasm.
- Contrast the assembly of new flagellins to flagella with the assembly of new pilin subunits to pili.
- Explain how modular rearrangement and gene duplication have contributed to the evolution of the flagella.
- Describe two strategies used by bacteria to make a capsule or secreted exopolysaccharides.

Most microbes protect their cell membrane with additional layers (chapter 2). Some assemble multiple layers around their cell membrane to confer protection, rigidity, and specific functions. For example, the complex structure of the cell envelope of Gram-negative bacteria contains a periplasmic space above the cell membrane, a peptidoglycan (murein) cell wall, and an outer membrane with a distinct lipopolysaccharide (LPS) in its outer leaflet. As the cell grows, its cell wall and outer membrane must also expand while retaining their integrity and function. Furthermore, the cell must erect appendages such as flagella and pili that also need to be assembled (chapter 2). All of these events are critical to building a fully functional cell and many take place simultaneously. Next we discuss some aspects of how cells add key accessories to their cell envelopes.

TABLE 9.1 **Some protein secretion systems in Gram-negative bacteria**

System	Name	Sec dependency	Description
Type I	ABC exporter	Sec independent	Consists of three proteins, one of which is an ATP-binding cassette; sometimes called the ABC transporter; operates on proteins lacking a signal sequence
Type II	Two-step system	Sec dependent	Moves protein into the periplasm by the Sec system in one step; then, 14 accessory proteins move it across the outer membrane in a second step
Type III	Contact-dependent system	Sec independent	Involves 20 or more proteins (including an ATPase) that span the envelope from cytoplasm to surface; in some cases, activated by contact with a host cell; then injects a virulence protein into the host cell directly
Type IV	Conjugal transfer system	Unknown	Uses same system that transfers DNA from some bacteria to eukaryotic cells
Type V	Autotransport	Sec dependent	Two steps involved, as in Type II, but no helper proteins needed for transfer through the outer membrane
Type VI	One-step system	Sec independent	Assembly of 20 or more proteins (including at ATPase) across the cell envelope to form a phage-related contractile structure for the injection of effector proteins into target cells

The Bacterial Peptidoglycan Cell Wall

Gram-positive and Gram-negative bacteria make a cell wall with a type of peptidoglycan called *murein*, which is assembled in a series of defined steps:

1. First, the sugar precursors, *N*-acetyl glucosamine (NAG) and *N*-acetyl muramic (NAM) are synthesized in the cytoplasm as uridine diphosphate (UDP) nucleotide sugar-linked precursors (UDP-NAG and UDP-NAM), which, in the case of UDP-NAM, also carry a pentapeptide (Fig. 9.11A).

2. The second step takes place at the cell membrane and involves the attachment of the UDP-NAM-pentapeptide precursor to a lipid carrier (a 55-carbon-long alkyl chain with a phosphate at the end; undecaprenylphosphate) (Fig. 9.11A). While bound to the lipid carrier, the NAG is attached to the NAM sugar and the peptide crossbridge is added to the pentapeptide. The lipid carrier then translocates the disaccharide-pentapeptide precursor to the external side of the cell membrane (most probably with help from membrane-bound proteins).

3. Once on the external side of the membrane, the modified disaccharide-pentapeptide is polymerized into nascent peptidoglycan by **penicillin-binding proteins** (PBPs), which covalently attach the repeating subunit into the pre-existing molecular scaffold (Fig. 9.11B). In the case of Gram-positive bacteria, the newly formed peptidoglycan often progresses from its site of synthesis to the periphery of the cell wall, where it is eventually hydrolyzed and sloughed off (Fig. 9.11C).

4. These very same proteins also generate crosslinks between adjacent peptide chains to form a strong meshwork (Fig. 9.12). The transpeptidase activity of these enzymes also links newly polymerized glycan strands to old strands already in place.

In some bacterial species, wall assembly takes place at many sites along the cell surface; in others, such as some Gram-positive cocci, this process is restricted to the equator of the cells. The newly formed peptidoglycan traf-

A

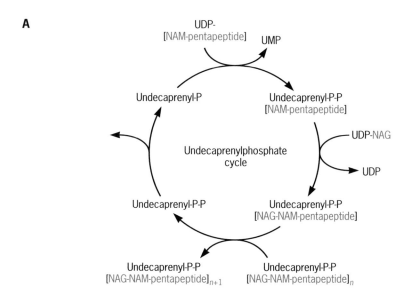

B Gram negative   **C** Gram positive

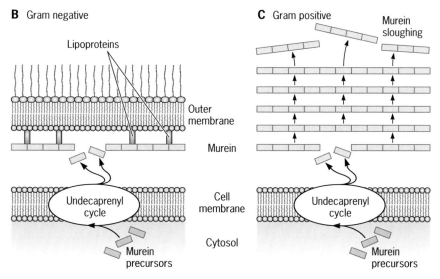

FIGURE 9.11 Synthesis and assembly of the bacterial peptidoglycan (murein) cell wall. (A) The undecaprenyl phosphate cycle synthesizes the peptidoglycan subunit (NAG-NAM-pentapeptide) while bound to the lipid carrier undecaprenylphosphate. P, phosphate; NAM, *N*-acetylmuramic acid; NAG, *N*-acetylglucososamine; UDP, uridine diphosphate; UMP, uridine monophosphate. (B) Wall assembly in Gram-negative bacteria involves the shuttling of the peptidoglycan precursor units (NAG-NAM-pentapeptide) through the cell membrane by the undecaprenylphosphate lipid carrier, its polymerization into nascent glycan in the periplasm, and covalent attachment to the cell wall to make it grow. (C) Wall assembly and turnover in Gram-positive bacteria follow the same steps but, in many species, the newly formed peptidoglycan progresses from its site of synthesis to the periphery of the cell wall, where it is eventually hydrolyzed and sloughed off.

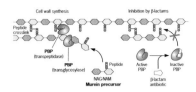

FIGURE 9.12 Peptidoglycan synthesis. Assembly of the peptidoglycan cell wall, its inhibition by β-lactam antibiotics, and mechanisms of resistance to this class of antibiotics are illustrated in this animation (7.5 min). http://www.microbeworld.org/component/jlibrary/?view=article&id=7171.

fics from its site of synthesis to the periphery of the Gram-positive cell, where it is eventually hydrolyzed and sloughed off (Fig. 9.11C). In Gram-negative bacteria, the peptidoglycan cell wall is anchored to the inner leaflet of the outer membrane by lipoproteins (chapter 2). In particular, these polypeptides are modified after their synthesis by the addition of three fatty acids. This hydrophobic anchor within the outer membrane facilitates the covalent bonding of approximately one-third of these lipoprotein molecules to the underlying murein, thereby attaching the wall to the outer membrane (Fig. 9.11B).

To accommodate cell growth, the peptidoglycan synthase enzymes continually remodel the peptidoglycan wall. By encasing the cell membrane, this rigid peptidoglycan grid equips the cell to withstand the turgor pressure generated by its high concentration of cytoplasmic constituents. Collectively, the peptidoglycan synthase enzymes are referred to as penicillin-binding proteins, a designation that emphasizes their vulnerability to this and other β-lactam antibiotics (Fig. 9.12). Binding of the β-lactam ring of these antimicrobials to the enzymes' active site is catastrophic: the cell explodes!

The Outer Membrane of Gram-Negative Bacteria

The outer membrane of Gram-negative bacteria is an extreme example of the asymmetry of the cell membrane: **lipopolysaccharide** (LPS) is present solely in the outer leaflet (chapter 2). The components of LPS, a hallmark Gram-negative molecule, are made by separate pathways that independently synthesize the lipid A core portion and the side chain polysaccharide. Next, these two constituents are joined at the outer membrane. Insertion of phospholipids and proteins into the outer membrane requires energy from the cell membrane's proton motive force. Lipopolysaccharide is thought also to assist in the incorporation of proteins. One hypothesis is that junctions between the outer and cell membrane (Fig. 9.13) lower the

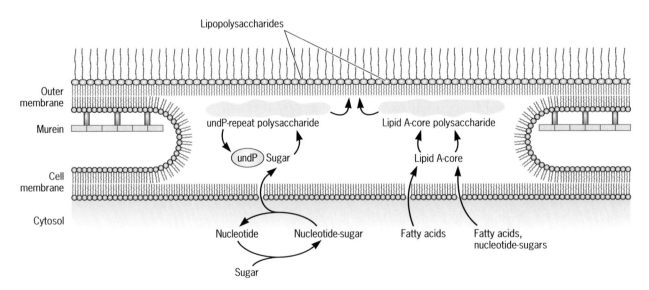

FIGURE 9.13 Lipopolysaccharide synthesis and assembly. Lipopolysaccharide subunits are synthesized in the cell membrane by two parallel processes. One path makes the repeating polysaccharide side chain, built up on undecaprenyl phosphate (undP), the same carrier that functions in murein polymerization. The other makes the core polysaccharide built up on lipid A, which functions as both a primer and the carrier that transports the core across the cell membrane. These two sets of precursors are made on the inner surface of the cell membrane, translocated by proton motive force to the outer surface, and covalently joined there.

energy barrier for growth of these two layers. Unfortunately, firm evidence for such areas of fusion or adhesion has been difficult to obtain.

Assembling Appendages: The Flagellum

Each appendage that is assembled in the cell envelope has a unique mode of synthesis and assembly. Here we consider in some detail the flagella, which consist of several components: basal body, hook, and filament (chapter 2). How are these motility machines assembled? Stepwise, from the bottom up (Fig. 9.14). First, the basal body is assembled and inserted into the cell envelope, with the bottom parts going in first. Then the hook is added. Finally, the flagellar filaments must be made. Given that the filament is quite long and is attached to the dedicated basal structure in the cell envelope, one may guess that new flagellins are added at the bottom of the flagellar filament to increase its length. Not so. Flagella grow by the addition of new flagellin *to the end that protrudes from the cell*. It helps that the flagellar filament is hollow: through its central channel, new flagellin molecules are extruded. When these subunits reach the tip, each molecule spontaneously attaches to its predecessor, and thus the filament elongates (Fig. 9.14). As the flagellum gets longer, the assembly process slows down. This and the inevitable mechanical breakage provide the means to limit flagella length. Note that the assembly of the flagellum is very different from the assembly of other appendages such as pili. These appendages are not hollow and grow by addition of the pilin subunits at the basal end (chapter 2).

How could complex appendages such as the flagellum evolve? Analogous structural components with overlapping functions exist in other cellular appendages, suggesting that their evolution may have been modular. By this reasoning, cells would first acquire a subset of components (module) for a specific function, and then acquire another module with a different function from another organism (Box 9.1). Microbes that put two modules together built a new structure with its own unique function.

Given the complexity of these motility machines, it may not surprise you to learn that the assembly of the bacterial flagellum is a highly regulated process. In fact, the synthesis of flagellin molecules is strictly coupled to their incorporation in the growing filament. Thus, excess flagellins do not accumulate or spontaneously assemble in the cytoplasm. *E. coli*, for example, makes an *inhibitor molecule* (FlgM) that blocks the synthesis of flagellins. However, once assembled on the membrane, *the basal body secretes the FlgM inhibitor* and relieves the repression on flagellin synthesis. New flagellin molecules are then made and secreted by the basal body to assemble them and grow the flagellar filament. Thus, the basal body structure itself cues the cell when to produce the next component needed to build the motility apparatus. What a clever mechanism!

Adding a Capsule or Exopolysaccharide

Many bacteria make their cell envelope even more complex by surrounding it with a slippery coat assembled from a variety of extracellular polysaccharide materials or exopolysaccharides (EPSs; chapter 2). In some cases, the EPS layer remains firmly attached to the cell envelope; it is then called a **capsule** (chapter 2). However, many other bacteria *secrete an*

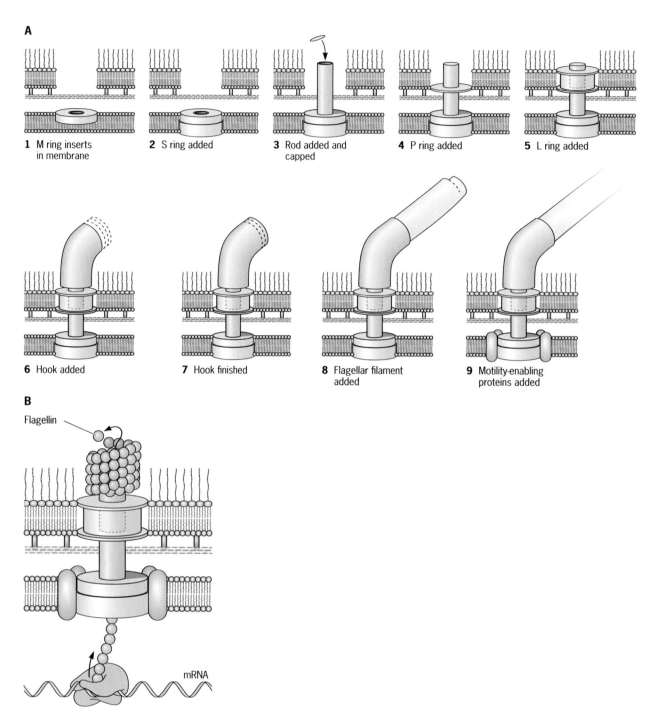

A

1 M ring inserts in membrane

2 S ring added

3 Rod added and capped

4 P ring added

5 L ring added

6 Hook added

7 Hook finished

8 Flagellar filament added

9 Motility-enabling proteins added

B

Flagellin

mRNA

FIGURE 9.14 **How flagella are assembled (the Gram-negative case).** (A) The process begins with the insertion of rings of the basal body into the cell membrane (1 and 2). A rod is added (3), followed by two outer membrane rings (4 and 5). Then, the hook (6 and 7) and the filament (8) are added. The last step is the addition of motility proteins in the cell membrane (9). (B) Growth of the flagellar filament (step 8 in panel A) occurs by extrusion of flagellin subunits through the core of the flagellum.

EPS layer and do not retain it attached to the cell envelope. This type of EPS serves as an adhesive-like material that allows cells to colonize surfaces and promotes the formation of multicellular communities called **biofilms**. The biofilm EPS is part of a complex matrix that also contains other polymeric substances such as nucleic acids and proteins and provides a

BOX 9.1

How Did Bacterial Flagella Evolve?

At a glance, evolution of such a complex structure appears to be a difficult problem because bacterial flagella are made up of some 30 to 60 proteins. If evolution were to take place only by the stepwise accumulation of mutations, making a structure as complex as a flagellum would have been a formidable task. Rather, the evolution of complex structures is likely to have proceeded by the assembly of smaller parts, each acquired for a different purpose. The formation of flagella, then, is likely the result of a **modular rearrangement**, the borrowing and joining of preexisting parts. Because each module has few components, fewer genes are needed to encode for them and their evolution through the accumulation of mutations would have required considerably fewer steps.

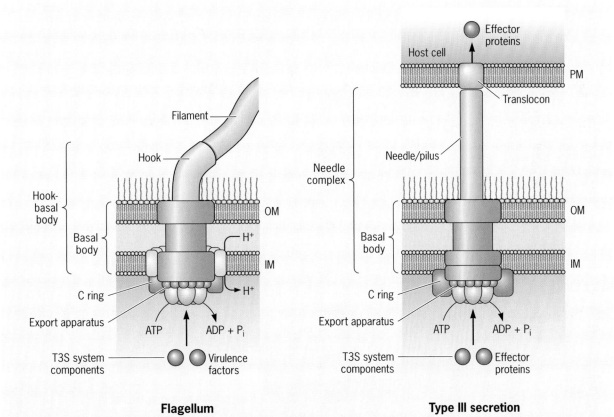

Flagellum   **Type III secretion**

(Modified from Büttner D, *Microbiol Mol Biol Rev* **76**:262–310, 2012.)

Evidence for a modular evolution of bacterial flagella can be seen in the sequence homology and structural similarities between its components and those of other cell constituents. An example is the basal body of the flagellum and the Type III secretion apparatus, a bacterial machinery that exports proteins such as toxins (see above). As the figure illustrates, the two structures look much alike, suggesting that they evolved from the same ancestral module. Likewise, two proteins that constitute the flagellar motor are homologous to outer membrane proteins that mediate uptake of vitamin B_{12} and iron chelates. The ATPase that drives the assembly of the flagellar filament is similar to other ATPases of bacteria and of mitochondria and chloroplasts. In addition, much of the genetic intricacy of the bacterial flagella appears to have arisen by a mechanism common in evolution: gene duplication (chapter 10).

The modular strategy that aided the evolution of complex structures from interchangeable genetic elements is likely to have taken place in many instances, from viruses to protozoa to worms to human beings. To quote Frank Harold: "Both gene duplication and the recruitment of genes from unrelated functions appear to have contributed to the making of flagella. What has been learned does not yet add up to a historical narrative, but it strongly supports the conclusion that flagella are products of evolution, just as wings and legs are. They were not designed by an engineer according to a plan, but cobbled together by a tinkerer."

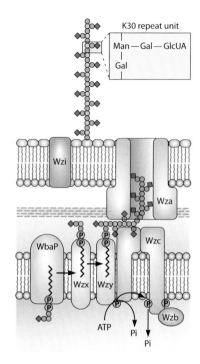

FIGURE 9.15 Secretion of the K30 group I capsule by *E. coli.* The synthesis of the K30 group I capsule by *E. coli* begins with the attachment of the tetraoligosaccharide precursor, the K30 repeating unit, to a lipid carrier and its translocation across the inner membrane. The sugar repeating units, now flipped to face the periplasm, are polymerized into chains and excreted. (Modified from Dong C, et al., *Nature* **444**:226–229, 2006, doi: 10.1038/nature05267, with permission.)

mechanical and selective chemical barrier for the cells. The biofilm matrix thus protects the cells against a number of environmental insults and is responsible for the resilience of biofilms. We will learn more about how cells come together as biofilms to cope with stress in chapter 12.

Bacteria follow one of two strategies to make their capsule or secreted EPS. Either they synthesize the oligosaccharide precursors in the cytoplasm and secrete them as preformed polymers, or they secrete the precursors and polymerize them outside the cell envelope using extracellular enzymes. For this second strategy, the cell assembles a specialized secretory complex that transports the precursors across the cell envelope; the components are polymerized during transport or once secreted outside the cell. For example, some *E. coli* strains produce a capsule called the K30 antigen (or K30 group I capsule) using the Wza multimeric protein translocon, which spans all the layers of the cell envelope (Fig. 9.15). Synthesis begins when the sugar precursor units (an oligosaccharide with four linked sugars) attach to a lipid carrier that facilitates their transport across the cell membrane. Next, other components of this transport apparatus polymerize the precursor units in the periplasm, and then others excrete the full-length polymeric chain outside the outer membrane. Did you notice the similarities between the synthesis of the K30 antigen and peptidoglycan synthesis?

The capacity to make a slippery coat gives many microbes a huge advantage. For example, the Gram-negative bacterium *Pseudomonas aeruginosa* makes at least three different EPSs: alginate, Pel, and Psl. Each EPS has a different chemical composition, and each is secreted by a dedicated translocon. Strains of *P. aeruginosa* isolated from chronic pulmonary infections in cystic fibrosis patients are often mucoid, reflecting the synthesis of large quantities of alginate. But other isolates make small and rugose colonies on laboratory solid growth medium and produce either the Pel or the Psl EPS. All coexist in the cystic fibrosis lungs because each envelope confers unique properties to the cells. By contrast, most environmental isolates do not secrete alginate; instead they produce one of the other two EPS types. The choice of EPS enables the bacterium to colonize a particular environmental niche and compete for the available substrates. A lot of cellular resources are diverted towards the synthesis of the EPS, so clearly the payoff must be great.

CASE REVISITED: **How close are scientists to building a synthetic cell?**

By now it must be clear that a cell is far too complex to build synthetically. However, a more modest goal, building a protocell is attainable. How can a protocell be built from scratch? Let us consider the two main features of a protocell:

- **Compartmentalization**. A protocell is surrounded by a lipid membrane, a physical boundary that separates it from its environment. Since all known cells are surrounded by lipid membranes, a reasonable approach for the experimentalist is to start by making lipid-containing vesicles (Fig. 9.1). Making a simple lipid vesicle is not hard. Indeed, whether monolayered (micelles) or bilayered (liposomes), such vesicles form spontaneously when phospholipids or similar amphipathic molecules are placed in water. In the ancestral protocells, membrane lipids may have consisted of fatty acids, which are simpler and may have been already present in the abiotic environment. Are you skeptical? Realize that fatty acids have been found in meteorites! To be useful, a protocell membrane would have to be selectively permeable to permit the entry of building blocks and sources of energy. In laboratories, membranes with these properties have been created by using fatty acids selected for this purpose. For example, branched fatty acids do not pack together tightly and therefore allow the passage of sugars and even nucleotide molecules. Now, if you could add energy-transducing proteins to the membrane, perhaps a hydrogenase to create a proton motive force and an ATP synthase to make ATP from the energy in the proton gradient (chapter 6), you would be close to making a synthetic cell. Straightforward in principle, but nobody has succeeded at this yet!

- **Self-reproduction.** If one did succeed in making a self-reproducing cell, even this primitive entity would be subject to Darwinian evolution. Simply put, a new cell would win out over what else in its environment, outcompeting other particles that are incapable of duplicating themselves. Inevitably, such a particle would continue to evolve, since any advantageous property would create selective value. So, with such good odds, how far along this synthetic biology path have researchers gone? First, the primitive cell needs a carrier of genetic information that also has the capacity to self-replicate. RNAs are attractive candidates because, in addition to carrying genetic information and some being able to self-replicate, certain RNAs can act as enzymes that may contribute to self-reproduction. Indeed, RNAs incorporated into lipid vesicles have been shown to self-replicate, without any assistance from protein enzymes. The next critical step—to achieve cell division of these vesicles—has not yet come to fruition. Nevertheless, studies of protocells have yielded technical advances that may well be pave the road for achieving the lofty goal of engineering a synthetic cell.

Note that this case study followed a "bottoms-up" approach, starting with very simple chemicals to create a protocell. An alternative, and perhaps not as ambitious, "top-down" line of attack consists of taking constituents from existing cells, such as their chromosomes, to reassemble cells, which would allow experimentalists to introduce desirable modifications. This approach has been successful, as viable cells with "transplanted" chromosomes have been assembled. The organisms receiving new chromosomes have been wall-less mycoplasmas (chapter 2), which are also among the cells with the smallest genomes.

CONCLUSIONS

Building a cell envelope is not trivial: sophisticated metabolic pathways are needed to synthesize precursors, put them together, and assemble them properly without compromising the integrity and essential functions of the cell envelope layers. Core features of today's cell envelopes are limited to the lipid membrane boundary; thus this structure was likely present in LUCA already. The lipid membrane provided a protective enclosure yet permitted the flux of chemicals needed to sustain essential processes such as ribosome-mediated protein synthesis, DNA and RNA synthesis, and energy conservation as ATP. The cell envelope, even its most primitive forms, allows cells to become more efficient and diversify. The descendants were able to acquire increasingly more complex cell envelopes and new functions to better exploit the available resources. This, in turn, required cells to carry larger genomes. The cell envelope, with potentially many layers and remarkable functions, was critical for diversification and helps us appreciate the level of sophistication of cellular life as we know it today.

Representative ASM Fundamental Statements and Learning Outcomes for the Chapter

EVOLUTION

1. Cells, organelles (e.g., mitochondria, chloroplasts), and all major metabolic pathways evolved from early prokaryotic cells.

   a. Name features common to protocells and the first cellular forms.
   b. Explain how modular rearrangement and gene duplication have contributed to the evolution of the flagella.

CELL STRUCTURE AND EVOLUTION

2. *Bacteria* and *Archaea* have specialized structures (e.g., flagella, endospores, pili) that often confer critical capabilities.

   a. Compare and contrast the structure of phospholipids and *Bacteria* and *Archaea*.
   b. Describe how a cell builds phospholipids in the cytosol and inserts them into the membrane.
   c. Summarize why a cell's membrane is not considered a static entity.
   d. Describe how a protein is inserted into the cell membrane.
   e. Identify the role of the Sec system in exporting most periplasmic outer membrane proteins of Gram-negative bacteria and secreted proteins in Gram-positive bacteria.
   f. Illustrate the challenges faced by Gram-negative bacteria and some solutions regarding inserting proteins in the outer membrane or secreting proteins outside the cell.
   g. Identify features and structures common to all Gram-negative secretion systems.
   h. Identify how murein precursors travel from the site of assembly, the cytoplasm, to the periplasm.
   i. Contrast the assembly of new flagellins to flagella with the assembly of new pilin subunits to pili.
   j. Describe two strategies used by bacteria to make a capsule or secreted exopolysaccharides.

SUPPLEMENTAL MATERIAL

References

- **Zhang YM, Rock CO.** 2008. Membrane lipid homeostasis in bacteria. *Nat Rev Microbiol* **6(3):**222–233. doi: 10.1038/nrmicro1839.
- **Whitfield C, Trent MS.** 2014. Biosynthesis and export of bacterial lipopolysaccharides. *Annu Rev Biochem* **83:**99–128.

- **Nikaido H.** 2003. Molecular basis of bacterial outer membrane permeability revisited. *Microbiol Mol Biol Rev* **67(4):**593–656.
- **Schneedwind O, Missiakas D.** 2014. Sec-secretion and sortase-mediated anchoring of proteins in Gram-positive bacteria. *Biochim Biophys Acta* **1843:**1687–1697.
- **Costa TR, Felisberto-Rodrigues C, Meir A, Prevost MS, Redzej A, Trokter M, Waksman G.** 2015. Secretion systems in Gram-negative bacteria: structural and mechanistic insights. *Nat Rev Microbiol* **13(6):**343–359.
- **Erdhardt M, Namba K, Hughes KT.** 2010. Bacterial nanomachines: the flagellum and type III injectisome. *Cold Spring Harb Perspect Biol* **2(11):**a000299.
- **Basler M.** 2015. Type VI secretion system: secretion by a contractile nanomachine. *Philos Trans R Soc Lond B Biol Sci* **370**(1679). pii: 20150021. doi: 10.1098/rstb.2015.0021.
- **Lovering AL, Safadi SS, Strynadka NC.** 2012. Structural perspective of peptidoglycan biosynthesis and assembly. *Annu Rev Biochem* **81:**451–478.
- **Willis LM, Whitfield C.** 2013. Structure, biosynthesis, and function of bacterial capsular polysaccharides synthesized by ABC transporter-dependent pathways. *Carbohydr Res* **378:**35–44.

Supplemental Activities

CHECK YOUR UNDERSTANDING

1. **Unlike the bacterial counterparts, the archaeal phospholipids have:**
   a. Ester bonds.
   b. Ether bonds.
   c. Glycerol 1-phosphate.
   d. Glycerol 3-phosphate.

2. **The polymerization of flagellins during the assembly of the bacterial flagellum . . .**
   a. takes place before the assembly of the basal body and the hook.
   b. takes place after the basal body and the hook are assembled.
   c. takes place at the base of the flagellum filament.
   d. takes place at the tip of the flagellum filament.

3. **The synthesis of lipopolysaccharide (LPS) in Gram-negative bacteria . . .**
   a. requires the preassembly of the LPS components prior to insertion in the outer membrane.
   b. requires energy from the proton motive force.
   c. requires the Sec system.
   d. None of the above.

4. **Which of the following statements is NOT true about the synthesis of the peptidoglycan cell wall in bacteria?**
   a. The sugar and peptide precursors are synthesized in the cytoplasm.
   b. The sugar and peptide precursors are transported independently across the cell membrane by dedicated lipid carriers.
   c. The lipid carriers flip the precursors to the periplasmic side of the membrane.
   d. Penicillin binding proteins (PBPs) catalyze the transglycosylation and transpeptidation reactions that incorporate the precursors into the nascent peptidoglycan.

DIG DEEPER

1. **Team work:** Microbes have acquired multiple mechanisms to export proteins to the environment or into other bacterial or host cells. Research one of these six secretion systems (Types I–VI) to identify a fitness advantage the secretion system and one of its cargo proteins confer to the microbial cell. Develop illustrations or an animation to teach the biology to your classmates. Include in your presentation: the ecology of the microbe; the structure of the apparatus; how secretion activity is regulated; the phenotype of microbes that lack the secretion system; one cargo molecule the secretion system delivers; the impact of that molecule on its target; and the cargo molecule's mechanism of action.

2. Jon Beckwith and colleagues developed a number of genetic methods to deduce the topology of *E. coli* membrane proteins. Read the paper "Genetic analysis of membrane protein topology by a sandwich gene fusion approach" by Erhmann, Boyd, and Beckwith (*PNAS* **87**:7574–7578, 1990, http://www.pnas.org/content/87/19/7574.long) and prepare a presentation with illustrations to teach classmates about one of these genetic tools. Address the following questions:

- Why was alkaline phosphatase chosen as the reporter?
- Why was the coding sequence for alkaline phosphatase inserted into a transposon?
- Why was the MalF protein chosen as the target for this study?
- What are some biological questions that can be addressed with this technique?

3. In their paper "Economical evolution: microbes reduce the synthetic cost of extracellular proteins" (*mBio* **1**(3):e00131-10, 2010, http://mbio.asm.org/content/1/3/e00131-10), Dan Smith and Matthew Chapman calculated the energetic cost of each component of the flagellar apparatus. Study Fig. 2 of their paper (below) to answer the following questions.

- In which compartments do most of the *least* economical proteins reside?
- In which compartments are the majority of the *most* economical proteins?
- FlgM appears to be an outlier. What is its function? Is it an exception to the rule displayed here, or is it an exception that proves the rule? Why?
- What biological process(es) might contribute to the cellular distribution of the least and most economical proteins?
- To test your hypothesis, what other cellular components would you analyze to determine their relative cost?
- Study the other four figures in the Smith and Chapman paper. Do they support your hypothesis?

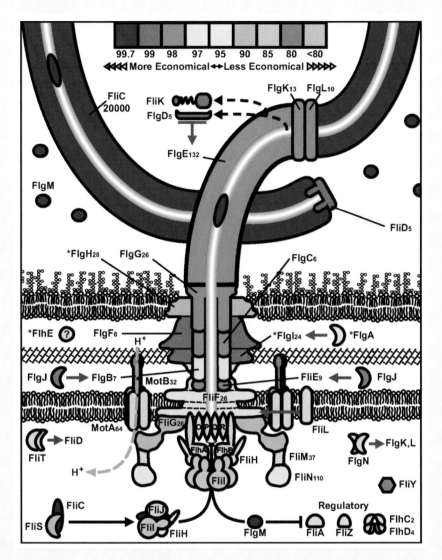

Figure 2. (from Smith and Chapman) Protein abundances and costs of flagellar proteins. Flagellum diagram showing colored economic percentiles of each protein. Proteins with lower Average Synthetic Cost (more economical) have higher percentiles and are dark red. Average Synthetic Cost increases in the following order: dark red to pink, light to dark blue, gray. *, proteins with Sec secretion signal sequences. The number of proteins per flagellum is indicated if known. [From Smith DR, Chapman MR, *mBio* 1(3):e00131-10, doi:10.1128/mBio.00131-10.]

4. In their study "PhoPQ-mediated regulation produces a more robust permeability barrier in the outer membrane of *Salmonella enterica* serovar *typhimurium*" (Murata T et al., *J Bacteriol* **189:**7213–7222, 2007, http://jb.asm.org/content/189/20/7213.long), Murata and colleagues analyzed the impact of modifications to the lipid A moiety of LPS (see Fig. 1 of their paper, below) of *Salmonella,* a microbe that alternates between soil and the gastrointestinal tract of host vertebrates. They compared two strains of regulatory mutants: one that did not modify lipid A (*phoP*) and another that always modified lipid A in broth culture (PhoP^c). Permeability was tested by measuring cellular penetration of a neutral fluorescent dye (Nile red) or an acidic fluorescent dye (eosin Y). What conclusions can be drawn from the data plotted in Fig. 4 of their paper? How might the capacity to alter its lipid A structure increase the fitness of this microbe? What experiments can you design to test your hypothesis?

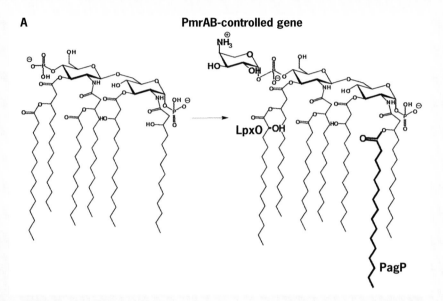

Figure 1. (from Murata et al.) Modification of the lipid A structure by the PhoPQ regulatory system. The modified structures are indicated by bold type and thick lines. (From Murata T et al., *J Bacteriol* **189:**7213–7222, 2007.)

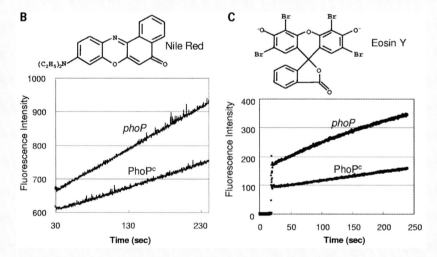

Figure 4. (from Murata et al.) Penetration of a neutral dye and an acidic dye across the OM of *tolC* mutants. Isogenic strains HN1139 (*phoP* null, *tolC*) and HN1140 (PhoP constitutive [PhoPc], *tolC*) were grown and harvested as described in Materials and Methods. For the influx of a neutral dye, Nile red, the cells were resuspended in 50 mM potassium phosphate buffer (pH 7.0), and the dye was added to a final concentration of 2 M at the beginning of the experiment. Fluorescence was monitored at 630 nm with excitation at 540 nm. For the assay of the penetration of an acidic dye, eosin Y, cells were washed and resuspended in 50 mM morpholineethanesulfonic acid (MES)-KOH buffer (pH 5.5) at a concentration of 40 OD 600 units/ml. Eosin Y (Na salt) was added to a concentration of 1 mM, and the mixtures were kept at 37°C for 2 h to allow for passive accumulation of this weakly acidic dye in the cytosol. For the assay, 5l of the mixture was added, at about 20 s, to 2 ml of 50 mM phosphate buffer (pH 7.0), and the fluorescence was recorded at 540 nm with excitation at 520 nm. (From Murata T et al., *J Bacteriol* **189:**7213–7222, 2007.)

Supplemental Resources

VISUAL

- *Exploring Life's Origin* project led by Jack Szostak with funding from the *National Science Foundation* provides educators and the public a series of short animations to illustrate how the first protocells could form. Topics include the need for compartmentalization; formation of fatty acids; formation of vesicles from fatty acid micelles; protocell life cycle (http://exploringorigins.org/resources.html).
- Video of bacterial lysis by penicillin (1 min): https://www.youtube.com/watch?v=UjLmf-cVcMw.
- Movie of flagellar assembly (1:48 min): https://www.youtube.com/watch?v=GnNCaBXL7LY.

SPOKEN

- *iBiology* lecture "The Origin of Life on Earth" by Jack Szostak (55 min): http://www.ibiology.org/ibioseminars/evolution-ecology/jack-szostak-part-1.html.

WRITTEN

- Moselio Schaechter discusses research that characterized the structure of this flagellar motility machine in *The Small Things Considered* blog post "Pictures Considered: The Basal End of Bacterial Flagella" (http://schaechter.asmblog.org/schaechter/2014/08/pictures-considered-19-the-basal-end-of-bacterial-flagella.html).
- Sonja-Verena Albers and Ken F. Jarrell describe what is different about the archaeal flagellum in the *Small Things Considered* blog post "The Archaellum—The Motility Structure of Archaea" (http://schaechter.asmblog.org/schaechter/2015/02/the-archaellum-the-motility-structure-of-archaea.html).
- Merry Youle explains the strategies used by phages to infect Gram-negative cells despite the various cell envelope layers in the *Small Things Considered* blog post "Bridges Across the Periplasmic Moat" (http://schaechter.asmblog.org/schaechter/2014/01/bridges-across-the-periplasmic-moat.html).
- Moselio Schaechter describes the lipid rafts of bacterial cell membranes in the *Small Things Considered* blog post "Rafting Through Time" (http://schaechter.asmblog.org/schaechter/2011/07/rafting-through-time.html).
- Moselio Schaechter discusses how some marine bacteria use sulfolipids for their membranes when phosphate is scarce in the *Small Things Considered* blog post "No Phosphorus? No Problem! (There's More Than One Way to Skin a Phytoplankton)" (http://schaechter.asmblog.org/schaechter/2009/03/no-phosphorus-no-problem-theres-more-than-one-way-to-skin-a-phytoplankton.html).

Schematic of DNA structure

CHAPTER TEN

Inheritance and Information Flow

KEY CONCEPTS

This chapter covers the following topics in the ASM Fundamental Statements.

EVOLUTION

1. Mutations and horizontal gene transfer, with the immense variety of microenvironments, have selected for a huge diversity of microorganisms.
2. The evolutionary relatedness of organisms is best reflected in phylogenetic trees.

INFORMATION FLOW AND GENETICS

3. Genetic variations can impact microbial functions (e.g., in biofilm formation, pathogenicity, and drug resistance).
4. Although the central dogma is universal in all cells, the processes of replication, transcription, and translation differ in *Bacteria, Archaea,* and eukaryotes.
5. Cell genomes can be manipulated to alter cell function.

IMPACT OF MICROORGANISMS

6. Microorganisms provide essential models that give us fundamental knowledge about life processes.

INTRODUCTION

Genetics is remarkably unified: The genetic information of all cellular organisms is encoded the same way, in sequences of bases in deoxyribonucleic acid (DNA); the genetic code is essentially universal; and the mechanisms of replication and expression of genetic material are very similar among all organisms. Thus, genetic studies in model prokaryotes, which are generally much easier to grow and manipulate in the laboratory, provided critical knowledge to establish the fundamentals of modern molecular genetics, including its most central discovery—that DNA is the molecule that encodes the genetic information of all cellular organisms. This is but one of the spectacular examples of how much we have learned about ourselves by studying prokaryotes. Indeed, in the discipline of genetics, we come face to face with the universality of many essential processes that sustain life on Earth. Still, as you will learn in this chapter, there are intriguing differences between prokaryotes and eukaryotes in the ways that genetic information is exchanged between individuals, in the patterns of its storage, and in its susceptibility to change.

CASE: **Tricks to increase the mutation rate**

Dr. E, a seasoned microbiologist, set up an experiment to isolate a mutant of *Escherichia coli* that was resistant to the antibiotic ampicillin. To do this, he plated approximately 500 cells of *E. coli* on each petri dish containing solidified growth medium supplemented with the antibiotic, and incubated the culture until a colony grew that was resistant to the antibiotic. He knew from previous data that the spontaneous mutation rate was 10^{-10}. This means that only one cell in every 10^{10} (or one in 10 billion!) became ampicillin resistant.

- How many plates did Dr. E need to isolate the ampicillin-resistant mutant?
- Why was the spontaneous mutation rate so low?
- Dr. E realized that he would need many, many plates to carry out his experiments unless he figured out a way to increase the mutation rate. How do you think he did it?

WHY DOES GENETIC VARIATION MATTER?

LEARNING OUTCOMES:
- Define "mutation."
- Explain why mutations are important for evolution.

All organisms store information in DNA, transcribe it into RNA, and use mRNA as the template to make proteins. This simple flow of information, from DNA to mRNA to proteins, is commonly known as the *Central Dogma of molecular biology* (Fig. 10.1). Each component has evolved to play a role in inheritance and information flow: DNA stores the information so it can be inherited by the daughter cells and transcribed into ribosomal and transfer RNAs for protein synthesis, regulatory RNAs for coordination, and mRNA templates to make proteins, the major catalysts of the cell. DNA's role as repository of cellular information is no coincidence: being double-stranded it is remarkably stable compared to the self-replicating, single-stranded RNAs that it replaced early on in the evolution of cellular life (chapter 1). Thanks to dedicated proteins, DNA is replicated with exquisite fidelity to preserve the information that is to be passed to the progeny. So important is the encoded information that the cell devotes a great number of resources to prevent and repair damage to DNA. Why, then, do cells risk the comfortable *status quo* by permitting changes to their DNA? The answer is simple: the possibility of a better tomorrow!

Changes in the DNA sequence, if not repaired, can be passed to the RNA and, from there, to proteins (Fig. 10.1). Some changes may be innocuous and have no effect on protein function; others lead to phenotypic variation. Because prokaryotes usually reproduce fast when conditions are favorable, they can accumulate changes in their DNA sequence and rapidly pass them to the daughter cells. Any change in the DNA sequence inherited by the progeny is called a **mutation**. By permitting some genetic variation, prokaryotic cells ensure that the population has some degree of phenotypic variation. In a fluctuating environment, mutations maximize the chances that the progeny, or at least a fraction of it, will grow and survive. In the next section, we will learn more about mutations, whether random or induced, as sources of genetic variation in prokaryotes. Later in the chapter, we will discuss just how promiscuous prokaryotes can get to achieve genetic variation, going as far as exchanging DNA from one cell to another in creative ways. You may already be aware that the capacity of bacterial cells to exchange DNA is the major reason why antibiotic resistance spreads fast. Because DNA is the macromolecule of inheritance, genomes carry the "scars" caused by mutations and DNA acquired from other cells. Technological advances now allow us to sequence, annotate, and analyze whole genomes in a relatively short amount of time. Genomic studies tell a tale of genetic variation that helps us grasp bits and pieces of the incredible evolutionary journey of prokaryotes on Earth. It is the basis of evolution. Without it, life may have emerged, but it would not have lasted.

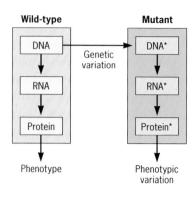

FIGURE 10.1 **Genetic and phenotypic variation.**

WHAT ARE MUTATIONS?

LEARNING OUTCOMES:
- Give an example of how using mutant technology with microbes has elucidated key cellular functions.
- Differentiate auxotrophic mutants from wild-type cells.
- Identify why small mutational changes, such as point mutations, can have variable consequences upon the translated protein, while large mutational changes, such as duplications or inversions, have more dramatic consequences.
- Explain how cells maintain a relatively low mutation rate.
- Name potential mutagens that you have come into contact with.
- Define transposon.
- Identify the function of the two features that all transposable elements share.
- Contrast the effect of a deletion mutation and a transposon insertion on the encoded protein.
- Describe how PCR (polymerase chain reaction) is used together with site-directed mutagenesis in mutant technology.
- Describe how genetic complementation is used to link a gene to a phenotype in mutant technology.

Mutation is a major driver of biological variation and, thereby, evolution. A permanent change in the DNA sequence of a cell will be inherited by a daughter cell (which is now a mutant); as a result, compared to its ancestor, the daughter cell may have different characteristics (i.e., a different *phenotype*; see Fig. 10.1). The power of mutation is easily appreciated in laboratory studies with some model bacteria, which grow rapidly in synthetic media and are amenable to genetic manipulation. Mutants probably constitute the most valuable experimental tool for analyzing cellular functions. How are they used? Simply put, mutants (cells carrying specific mutations) are compared to the strains from which they were derived.

Now a ubiquitous tool of modern experimental biology, mutant technology was developed by G. W. Beadle and E. L. Tatum in 1941 (before genetic information was known with certainty to be encoded in DNA!). The researchers wanted to investigate biochemical pathways of the red bread mold *Neurospora crassa*. They did so by first irradiating the mold's spores with UV to induce mutations in the DNA, and then allowing the spores to germinate and grow into filaments in various types of media. Because *Neurospora* filaments are haploid (i.e., there is only one copy of each chromosome), recessive as well as dominant mutations were expressed. The UV irradiation induced a rich diversity of mutant strains—including **auxotrophic** mutants. Unlike their nonirradiated parents, such mutants were unable to grow on a minimal medium unless it was supplemented with a particular small molecule—an amino acid, a nucleotide base, or a vitamin, for example. The majority of these mutations caused the *loss of function of a particular biosynthetic enzyme*. Using standard genetic crosses and biochemical studies, Beadle and Tatum assigned the sites of the mutations to particular genes. Their results gave birth to the "one gene, one enzyme" theory. Although we now recognize its oversimplification (e.g., some enzymes are composed of multiple subunits, each encoded by a different gene), their experiments established a fundamental principle of biology: most genes encode a protein, and a protein's physiological role can be deduced from the phenotypes of strains mutated in that gene. Mutant technology has been applied successfully to such diverse topics as the pathways of metabolism and the complex cascades of signal transduction.

Kinds of Mutations

Because the term mutation is so broad, it is useful to subdivide it. One way to classify mutations is by their *phenotype*, the observable change in appearance or function of a strain carrying the mutation (Table 10.1). Auxotrophs

TABLE 10.1 **Types of mutant strains**

Designation	Phenotype
Auxotroph	Requires an exogenous building block or growth factor (e.g., an amino acid or vitamin)
Carbon source	Unable to use a particular compound as a source of carbon
Nitrogen source	Unable to use a particular compound as a source of nitrogen
Phosphorus source	Unable to use a particular compound as a source of phosphorus
Sulfur source	Unable to use a particular compound as a source of sulfur
Temperature sensitive	
Heat sensitive	Loses a particular function at a high temperature
Cold sensitive	Loses a particular function at a low temperature
Osmotic sensitive	Loses a particular function at high or low osmolarity
Conditional lethal	Unable to grow in a particular environment (e.g., high temperature) in any medium

TABLE 10.2 **Small mutational changes: point mutations and microlesions**

Kind	Description	Consequences
Base pair	Change in a base pair	Varies from no effect to complete loss of function
Transition	Substitution of one pyrimidine for another or one purine for another in a base pair; e.g., change of AT to GC	Varies from no effect to complete loss of function
Transversion	Substitution of a purine for a pyrimidine and a pyrimidine for a purine in a base pair; e.g., AT to CG	Varies from no effect to complete loss of function
Silent	Change in a base pair leading to a redundant codon (one that encodes the same amino acid as the original codon encodes)	Usually none
Missense	Change in a base pair leading to a codon that encodes a different amino acid	Varies from little effect to complete loss of function
Nonsense	Change in a base pair leading to a stop (nonsense) codon (UAG is called amber, UAA is called ochre, and UGA is called opal)	Usually loss of function
Frameshift	Gain or loss of one or several base pairs (designated +1, −1, +2, −2, etc.) so that the frame of codons is disrupted	Usually loss of function

affect the cell's metabolism, either by failing to make their own building blocks or growth factors or by losing the capacity to use a particular nutrient that is supplied. Some mutants only show a different phenotype under certain conditions, such as the temperature of incubation (temperature sensitive mutants) or osmolarity of the medium (osmolarity sensitive mutants). Some mutations are lethal under certain conditions, but permissive at others (conditional lethal mutants).

Another way to classify mutants is based on the change in DNA that caused the phenotype in the first place; i.e., the underlying *genotype*. Some changes affect only a single base pair (**point mutations**; Table 10.2). Another type, called **frameshift mutation**, adds or removes one or more base pairs and changes the frame of codons and, therefore, the sequence of amino acids (Table 10.2). Large mutations are also possible that rearrange the DNA sequence significantly, such as **deletions**, **insertions**, **translocations**, and **inversions** (Table 10.3). These types of mutations can also have dramatic phenotypic effects. Why might these larger changes have more persistent effects than point mutations? (*Quiz spoiler*: They more rarely reverse spontaneously.)

TABLE 10.3 **Large mutational changes: rearrangements and macrolesions**

Kind	Description	Change	Consequences
Duplication	Formation of a supplemental copy of DNA, usually in tandem	ABC<u>DEF</u>GHI → ABC<u>DEF</u>-DEFGHI[a]	Usually causes loss of function when within a single gene, but no loss of function if the whole gene is duplicated (it increases gene dose)
Deletion	Loss of a segment of DNA	ABC<u>DEF</u>GHI → ABC-GHI	Usually causes a loss of function
Translocation or insertion	Movement of a fragment of DNA from one location to another	ABC<u>DEF</u>GHI → UVW-<u>DEF</u>-XYZ	Usually causes a loss of function at the site of insertion
Inversion	Reversal of the order of a segment of DNA	ABC<u>DEF</u>GHI → ABC-<u>FED</u>-GHI	Frequently causes a loss of function

[a]Hyphen represents the location of an improper or novel junction. Underlining indicates the changed region of DNA.

Sources of Mutations

Spontaneous mutations. Mutations occur spontaneously as DNA is being replicated or when it suffers chemical damage. Although replication is extraordinarily precise, occasionally the wrong base is incorporated in the newly synthesized strand of DNA. Such random mistakes occur at a relatively low frequency, thanks to the exquisite proofreading capacity of the principal DNA polymerase (Pol III; chapter 8). These enzymes sense DNA mismatches and eliminate them using their built-in nuclease activity. During DNA replication, mistakes occur only *once in a million* base pairs. Because mutations are so often associated with replication, by convention the mutation rate is calculated as the number of mutants formed per cell doubling, rather than per hour or day.

Mutations can also result from chemical damage to DNA, such as the spontaneous deamination of cytosine induced by external irradiation or internal reactive oxygen species (including hydrogen peroxide and superoxide). Spontaneous deamination, perhaps the most frequent chemical damage, converts cytosine to uracil; during the next round of replication pairs, the uracil is "read" as thymine and base-paired with adenine, thereby causing a CG-to-TA **transition**. Although these mutations "fool" DNA polymerases, additional surveillance and repair mechanisms correct most of the errors caused by chemical damage and those missed during replication, reducing the error rate to an impressive *one in a billion*. In the case study, Dr. E calculated that the spontaneous emergence of a mutant of *E. coli* resistant to the antibiotic ampicillin was within these ranges. Clearly, the cell has evolved mechanisms to minimize mutation rates so the information in the DNA is not frequently altered. Nevertheless, when microbes are subjected to strong selective pressures, such as the prolonged course of antibiotics needed to treat patients infected by *Mycobacterium tuberculosis* (chapter 24), the mutation rate is sufficient to drive emergence of dangerous antibiotic-resistant microbes. We'll also see that, on a longer time scale, spontaneous mutations contributed to the evolution of the tick-borne *Yersinia pestis,* the cause of plague, from *Yersinia pseudotuberculosis,* a relatively benign gastrointestinal pathogen (chapter 27).

Mutagens

Mutations can be induced by a number of chemical and physical agents (**mutagens**; Table 10.4); some are alarmingly potent. Nitrosoguanidine, for example, can cause multiple mutations in almost every surviving cell of a treated culture. Hence, treating a culture with nitrosoguanidine increases the mutation rate significantly. In the case study, Dr. E needed a way to increase the spontaneous mutation rate of *E. coli* to isolate an ampicillin-resistant mutant more readily. Would you recommend nitrosoguanidine?

Treating a microbial culture with nitrosoguanidine or another chemical or physical mutagen can induce mutations in any gene with nearly equal probability. Thus, experiments are often designed that *enrich* and/or *screen* for the mutant of interest among all others. The ways to enrich and/or screen for mutants are limited only by the investigator's imagination (Fig. 10.2). Two methods can be readily adapted to many schemes:

TABLE 10.4 **Some physical and chemical mutagens**

Agent	Mutagenic action
Physical agents	
X rays, gamma rays	Cause double-strand breaks in DNA, the repair of which leads to macrolesions
UV light	Cause adjacent pyrimidines in DNA to join at positions 4 and 5, forming dimers. Repair of the dimers results mostly in transversions, but also in frameshifts and transitions
Chemical agents	
Base analogs	Become incorporated in DNA and then, owing to their ambiguous pairing on subsequent replication, cause transitions
2-Aminopurine	Can pair with either thymine or cytosine
5-Bromouracil	Can pair with either adenine or guanine
DNA modifiers	
Nitrous acid	Deaminates bases; deamination of cytosine produces uracil and then a CG-to-TA transition
Hydroxylamine	Hydroxylates 6 amino group of cytosine, causing CG-to-TA transition
Alkylating agents (e.g., nitrosoguanidine and ethyl methane sulfonate)	Alkylate DNA bases, distorting DNA structure and resulting in a variety of types of mutations
Intercalating agents (e.g., acridine orange and ethidium bromide)	Intercalate between stacked bases in DNA; replication results in frameshift mutations

1. One *enrichment method* exploits the fact that antibiotics such as penicillin *kill only those cells that are growing* (Fig. 10.2). For example, if you are interested in isolating histidine auxotrophs (His⁻) from a mixture of mutants you grow the culture in a medium lacking that amino acid to stop the growth of the His⁻ yet permit the growth of the others. You then add penicillin. All of the growing cells will incorporate the antibiotic and most will be killed (Fig. 10.3). The mutants of interest (His⁻), which are not growing, will not be killed. Thus, the abundance of this mutant lineage will be enriched in the surviving culture. Once the antibiotic is diluted and histidine is supplied, the surviving mutant cells will grow normally. This penicillin "counterselection" method can be used to enrich and isolate for many other mutants. How would you apply this method to enrich for mutant strains unable to utilize ribose as a carbon source?

2. One *screening method* is **replica plating**, which is as powerful as it is simple. Here is how to use it to identify the histidine auxotrophic mutants (His⁻) you just enriched for (Fig. 10.2). In this experiment the culture containing a mix of mutants is plated on agar to separate the individual cells and allow each of them to grow into a colony. Note that the growth medium in this "master" plate needs to permit the growth of all the cells, whether they are the mutant of interest or not. A petri-dish-size metal or wooden block, covered with sterile fuzzy fabric or filter paper, is then pressed against the surface of the agar plate to create an imprint of each colony. This "imprint" carries cells from each of the colonies in the original position they had on the plate, allowing researchers

Enrichment method

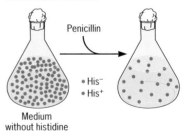

Replica plating

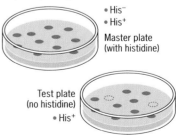

FIGURE 10.2 **Methods to screen for mutants.**

to transfer each clone to new plates and study whether the new media conditions permit their growth. If the "imprint" is transferred to a plate lacking histidine, the mutant of interest (His⁻) will not grow. By comparing the plates that have histidine to those that lack the amino acid, one can identify the His⁻ mutant colonies growing on the master plate. Replica plating can be extended to search for other kinds of mutants; for example, temperature-sensitive mutants. Here, the two replicated dishes contain the same medium, but they are incubated at different temperatures. The possible applications are endless. A modern, high-throughput version of this technique enlists robots that "pick" large numbers of colonies from the master plate and array each of them on the test plates.

Mutations caused by mobile elements. Initially, genetics was confined to studying genes sitting stably on chromosomes. No longer. Major contributors to evolution are the so-called **mobile elements** (or transposable elements), such as **transposons**. Transposons are pieces of DNA that move from one location to another, and they are abundant. All organisms—prokaryotic and eukaryotic—carry such "jumping genes" in their DNA. Most prokaryotes carry several of them, and it appears that about half of our own DNA may consist of transposons. There are different kinds of transposons that "hop" by different mechanisms (Table 10.5), but in all cases their movement is catalyzed by a **transposase** encoded in the trans-

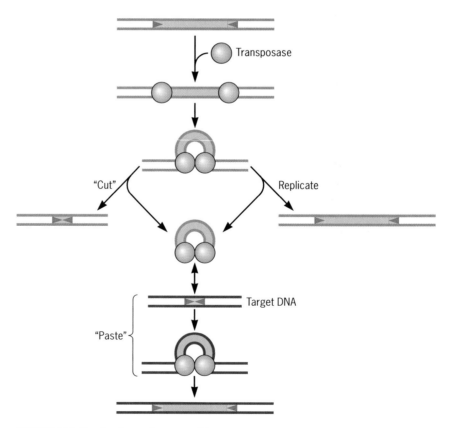

FIGURE 10.3 **Mechanisms of transposition.**

TABLE 10.5 **Types of transposable elements**

Designation	Characteristics
Types of transposable elements	
Insertion sequence (IS) elements	Transposase gene flanked by inverted repeats. Designation: IS followed by an italicized number (e.g., IS1, IS2, IS3)
Noncomposite transposons	Transposase and drug resistance genes flanked by inverted repeats
Composite transposons	Transposase and other genes (unrelated to transposition) flanked by IS elements. Designation: Tn followed by an italicized number
Mechanism of transposition	
Cut and paste	The transposon is cut out of the DNA where it resides and is inserted in a new location.
Replicative	The transposon is replicated; one copy remains at its original location, and the other is located at a new one.

poson. All transposons carry **terminal inverted repeats** that serve as recognition sites for the transposase enzyme. Depending on the type of transposon, the transposase can cut the element and paste it in another location (the **"cut and paste"** mechanism) or make a copy to be inserted elsewhere (the **replicative** mechanism; Fig. 10.3). The insertion site is not random: the transposase *recognizes and cuts specific sequence*s, usually direct repeats, in the target DNA, and integrates the transposon right there.

A closer look at the structure of various transposons allows us to appreciate the conserved elements (transposase and inverted repeats) that are required for transposition (Table 10.5). The simplest and shortest (700–2,000 bp) of all transposable elements is the **insertion sequence** or **IS element**, which contains the gene encoding the transposase flanked by inverted repeats. The transposon can also carry extra genes (most often, antibiotic resistance genes) (**noncomposite transposons**) or even contain two IS elements flanking additional genes (**composite transposons**). These complex transposons are important agents in the spread of antibiotic resistance. Transposon Tn5, for example, carries kanamycin and neomycin resistance cassettes. Thus, insertion of Tn5 in the genome of a recipient strain makes the cell resistant to these antibiotics. Likewise, Tn10 encodes enzymes that pump the antibiotic tetracycline out of cells, which also makes these bacteria resistant to the antibiotic.

Regardless of their structure and mechanism of insertion, a transposable element inserted into a new location in the genome is likely to interrupt a gene. If the sequence where it lands encodes a protein, chances are that such a large insertion will cause complete loss of function, just as a deletion mutation will. Thus, transposons are useful genetic tools in the laboratory. However, keep in mind that some transposons can be excised precisely from their location. Such mutations are not stable and may revert at some frequency.

Site-Directed Mutagenesis

Once a scientist identifies a gene that confers the phenotype of interest, there are multiple ways to probe its function. If the sequence of the gene is already known, specific changes in its DNA sequence can be introduced to investigate how the phenotype of the mutant cell is affected (Fig. 10.4A). Such **site-directed mutagenesis** is a powerful tool of experimental biology. Typically, the target DNA sequence is first amplified *in vitro* using a technique called **polymerase-chain reaction** or **PCR**. The 5′ and 3′ boundaries of the desired locus are established by designing two short sequences of complementary DNA, called primers. Then a reaction mix of template DNA, primers, deoxynucleotides, magnesium, and a heat-tolerant DNA polymerase is placed in a "thermo cycler" machine that alternates the mix between two temperatures. The first (hotter) incubation period denatures the double-stranded DNA; during the second (cooler) incubation, the primers anneal to the resulting single-stranded DNA, and the DNA polymerase extends from the 5′ primer until it collides with the 3′ primer. Then the reaction mixture is returned to the first, hotter, temperature, to denature the nascent and template DNA strands and the primer pairs. After multiple cycles, the DNA sequence defined by the two primers is amplified to generate ample quantities of the desired DNA fragment. How are mutations introduced? Any mutations incorporated into the primer design will be amplified in consequent cycles. By adding a molar excess of primers relative to the template DNA, the altered copies will greatly outnumber the original DNA sequence, making it negligible.

Once the engineered DNA fragment has been PCR-amplified, it can be introduced into the microbial strain of interest taking advantage of the pathways that microbes naturally enlist to exchange DNA, which we describe below. The DNA fragment carrying the mutation can then be integrated into the recipient's chromosome via **recombination** (Fig. 10.4A). If the recombination is successful, the chromosomal locus will be replaced by the mutated one, and a stable mutant strain is generated that can grow and multiply into more mutant cells. The mutant can then be characterized to link the mutation to a phenotypic change.

Genetic Complementation

Linking a particular phenotype to a gene often requires additional confirmation. One way to do this is to introduce a "good" copy of the gene and seeing if this fixes the mutant defect. This process is called **genetic complementation** because the unchanged (wild-type) gene is introduced into the mutant cells. If this restores the wild-type phenotype, it confirms that the gene is indeed responsible for the phenotype (Fig. 10.4B). The wild-type gene is usually reintroduced using extrachromosomal genetic elements or **plasmids** as vectors. Such a vector must have two properties: (i) it needs to be able to express the gene of interest, and (ii) it needs to carry a selectable marker to permit identifying the cells that carry it. A

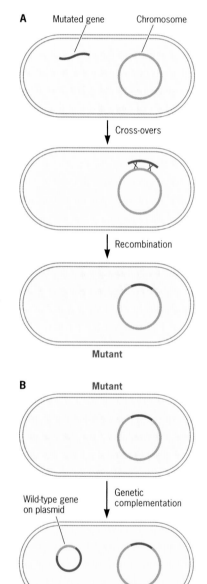

A Mutated gene   Chromosome

Cross-overs

Recombination

Mutant

B **Mutant**

Wild-type gene on plasmid

Genetic complementation

Genetically complemented

FIGURE 10.4 Site-directed mutagenesis via recombination (A) and genetic complementation using a plasmid vector (B).

selectable marker can be an antibiotic resistance gene, such as the ones carried by transposons, which confer on the recipient strain the ability to grow in the presence of the specific antibiotic. A genetic complementation experiment that restores the original phenotype is an elegant demonstration that the phenotype assigned to a mutant was caused by the specific mutation in the gene. But, as it often happens in science, genetic complementation may fail or only partially restore the wild-type phenotype. Can you think of reasons why this could happen?

HOW DO PROKARYOTES EXCHANGE DNA?

LEARNING OUTCOMES:
- Summarize ways in which prokaryotes and eukaryotes differ with respect to the exchange of DNA.
- Discuss how microbes can evolve at a very fast pace without sexual reproduction to rearrange genes.
- Describe how the Griffith natural transformation experiment led to the discovery that genes are made of DNA.
- Identify the role of RecA in transformation.
- Differentiate between transformation, transduction, and conjugation.
- Describe the circumstances and situation of an example of natural transformation.
- Identify the role of artificial transformation in recombinant DNA technology.
- Compare and contrast the mechanisms of generalized transduction and specialized transduction.
- Compare and contrast the outcomes of generalized transduction and specialized transduction.
- Explain what characterizes a cell as an Hfr (high frequency of recombination) cell.
- Compare and contrast the process of conjugation for Gram-negative and Gram-positive bacteria.
- Identify two mechanisms of conjugation unique to some Gram-positive bacteria.

While genetic exchange between eukaryotes occurs by one single mechanism—fusion of gametes—prokaryotes have three totally different mechanisms: **transformation**, **transduction**, and **conjugation** (Fig. 10.5). These mechanisms have been likened to conveying information via a postcard

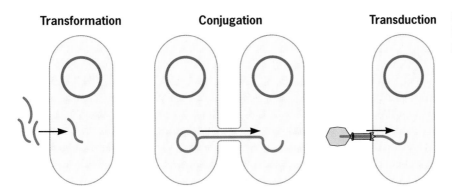

Transformation   **Conjugation**   **Transduction**

FIGURE 10.5 **Mechanisms of horizontal gene transfer in prokaryotes.**

(naked DNA), a letter in an envelope (DNA within a phage), and, if you wish, a birthday present (sexual conjugation). The exchange of chromosomal genes by two of these mechanisms, transduction and conjugation, is apparently an accidental by-product of some other event—infection by, and replication of, a bacteriophage, in the case of transduction and movement of certain kinds of plasmids from one cell to another in the case of conjugation. On the other hand, the third mechanism, transformation, might actually have evolved specifically to mediate genetic exchange.

Before we discuss these three mechanisms, consider how different genetic exchange is in prokaryotes and eukaryotes:

1. Prokaryotic genetic exchange is not dependent on reproduction (formation and fusion of gametes in eukaryotes).

2. In general, prokaryotes only transfer a small bit of DNA. Unless capable of independent reproduction (as in a plasmid), the donated DNA is easily degraded inside the recipient cell and not passed to the daughter cells. But, in some cases, the DNA is recombined into the recipient's DNA and the integrated allele(s) are replicated together with the chromosome and passed to the progeny. Note how different eukaryotic genetic exchange is in this respect: when eukaryotic gametes fuse, their complete set of chromosomes mix. Thus, more DNA is exchanged in events that are seldom futile, as in prokaryotes.

3. Eukaryotic organisms with obligatory sexual reproduction rearrange their genes each generation. Hence, we are different from our parents. Lacking an obligatory sexual stage, microbes instead rearrange their genetic material by exchanging their DNA by other means.

4. Prokaryotes are usually haploid and can directly express their alleles. By contrast, most of the cells of higher eukaryotes are diploid. Thus, even if foreign DNA were introduced in one of the sets of chromosomes, an intact allele would still be present in the other, and the recessive alleles will not be manifested. Why does this matter? The consequences of expressing and not expressing recessive genes are profound. Put in human terms, only about 1 in 3,500 white Americans suffers from the recessive genetic disease, cystic fibrosis; if recessive genes were instead expressed, 1 in 30 people would experience this lung disease.

Although microbes transfer DNA at a low frequency, say, once ever 1,000 cell divisions, the number of microbes is so staggeringly high that the actual number of genetic exchanges in nature is huge. An exact figure is hard to come by, but we estimate that genetic exchange is several orders of magnitude more frequent than the average mutation rate. Such events of genetic exchange, collectively known as **horizontal gene transfer**, allow microbial populations to evolve at a very fast pace—witness the rapid and widespread emergence of resistance to antibiotics. Prokaryotic evolutionary history has been marked by numerous occurrences of horizontal gene transfer, sometimes between distantly related organisms. These lateral gene transfers confound molecular studies of evolution, since we cannot assume that organisms that share particular genes evolved from a common ancestor. Later in the chapter we will learn about prokaryotic genomes and what they tell us about the evolutionary history of the cell.

TABLE 10.6 **Some species of bacteria and archaea known to be capable of natural transformation**

Prokaryotic microbe	Remarks
Gram-positive bacterium	
Streptococcus pneumoniae	Pathogen of various tissues
Bacillus subtilis	Soil-dwelling, mesophilic sporeformer
Bacillus stearothermophilus	Soil-dwelling, thermophilic sporeformer
Enterococcus faecalis	Gut bacterium
Gram-negative bacterium	
Acinetobacter calcoaceticus	Soil dweller, opportunistic pathogen
Moraxella urethralis	Opportunistic pathogen
Psychrobacter spp.	Psychrophiles
Azotobacter agilis	Free-living, soil-dwelling nitrogen fixer
Pseudomonas stutzeri	Soil-dwelling denitrifier
Haemophilus influenzae	Pathogen of various tissues
Neisseria gonorrhoeae	Sexually transmitted pathogen
Campylobacter jejuni	Intestinal pathogen
Helicobacter pylori	Stomach pathogen
Archaea	
Methanococcus voltae	Methanogen
Methanobacterium thermoautotrophicum	Methanogen
Thermococcus kodakarensis	Hhyperthermophile
Pyrococcus furiosus	Hyperthermophile

Transformation

In biology, "transformation" has two distinct meanings: (i) in eukaryotes it refers to the change of a normal healthy cell into a cancerous one; (ii) in prokaryotes, it refers to the genetic change that occurs in a cell after it has taken up soluble DNA from its environment and recombined at least part of it into its genome. The latter applies to both bacteria and archaea and is our focus here.

Cells *capable of taking up DNA* from their environment, and thereby becoming transformed by it, are said to be **competent**. Genetic competence is a transient state. It develops naturally among some species of prokaryotes (mostly bacteria, but also a few archaea; Table 10.6), depending on the cell's physiological state. But even for those species that are not known to be naturally competent, it is often possible to use artificial means to induce them to take up DNA.

Natural transformation. **Natural transformation** holds a special position in the history of genetics because it led to the discovery that genes are made of DNA. In 1928, Frederick Griffith did his elegant, now-famous experiment that pointed the way (Fig. 10.6). Using the pneumococcus (*Streptococcus pneumoniae*), he found that capsule-bearing strains (called S for the "smooth" appearance of their colonies) killed mice, but capsule-free (called R for "rough" or uneven colonies) strains did not. Because controls of heat-killed S cells did not kill, he deduced that the capsule itself was innocuous. On the other hand, a mixture of heat-killed S cells and live R cells was deadly. Yet, the two components did not act synergistically, as he had anticipated. In fact, the dead mice contained only live capsulated cells, which Griffith determined had properties characteristic of the

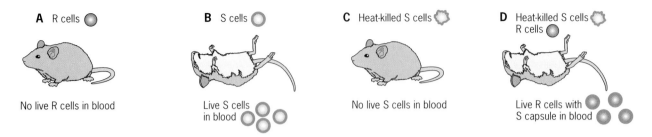

FIGURE 10.6 Griffith's experiments that led initially to the discovery of the transforming principle and subsequently to the discovery that it is composed of DNA. Live R cells, which lack a capsule, are not infective when injected into mice (A), but live S cells, which produce a capsule, are deadly (B) unless heat killed (C). The combination of live R cells and heat-killed S cells becomes deadly (D).

injected R cells. Griffith concluded (rightly) that something, which he called the *transforming principle*, from the dead S cells had "transformed" the R cells, rendering them smooth (capsulated) and deadly (Fig. 10.6). Sixteen years later (1944), O. Avery, C. MacLeod, and M. McCarty purified this transforming principle and showed that it was DNA.

We now know that the "transformed" R cells that Griffith recovered from the blood of sickened mice (injected with heat-killed S cells and live R cells) were **recombinants;** that is, they were R cells that had acquired the capsule genes from the DNA released when the S cells were killed by heat. The DNA was taken up by the R cells and recombined into their chromosome, allowing the recipient cells to produce the S capsule. *S. pneumoniae* is one of the relatively few prokaryotes known to become naturally competent and thus transformable during growth and development (Table 10.6). By taking up soluble DNA from their environment and introducing it by recombination into their genomes, the bacteria can be genetically altered. Usually, natural competence develops in dense cultures. This observation likely reflects natural selection: the probability of successful gene transfer is higher when cells are in close proximity, such as within surface-attached communities or **biofilms**.

The natural transformation systems of *S. pneumoniae*, *Bacillus subtilis*, and *Haemophilus influenzae* have been studied thoroughly; although distinct in some details, they have many common properties. These systems are complex, being encoded by 20 to 30 genes, but they all process the transforming DNA in similar ways (Fig. 10.7). Exogenous double-stranded DNA (dsDNA) binds to proteins of the competence system located on the cell membrane, where it is fragmented into smaller pieces. One strand is then degraded by a nuclease, while the other enters the cell. The strand that is taken up into the cytoplasm becomes coated with a transformation-specific DNA-binding protein of the recombination system (RecA), which helps it survive its hazardous trip through the exonuclease-laden cytoplasm. If the strand is homologous to the cell's resident DNA, RecA facilitates its integration into the chromosome by a mechanism called **homologous recombination**. First, RecA *invades the homologous region* in the double-stranded chromosomal DNA, *displacing one strand* and transiently forming a *stretch of hybrid DNA* or heteroduplex consisting of one strand of resident DNA and one strand of transforming DNA. Next, the invading strand is sealed in place, becoming part of the

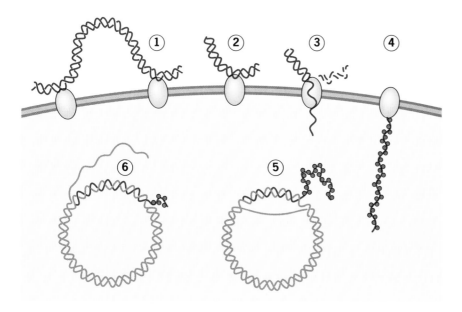

FIGURE 10.7 Principal features of transformation of _S. pneumoniae._ (1) A fragment of double-stranded DNA binds to proteins of the competence system on the cell membrane. (2) Bound DNA is nicked and cut into smaller fragments, and (3) one strand is degraded by a competence-specific nuclease. (4) A single DNA strand enters the cell and becomes coated by RecA proteins. (5 and 6) The entering DNA wrapped in RecA proteins invades a homologous region of DNA on the chromosome and displaces an existing strand, forming a heteroduplex.

recipient's chromosome. When the cell divides, the two strands are replicated to make the chromosomes of the daughter cells, each carrying an old strand and a newly replicated one. Consequently, only one daughter carries the transformed DNA—the one that inherited the "transforming" strand of DNA.

Natural transformation systems evoke several intriguing questions. First, what is the _source of transforming DNA_ in the microbe's natural environment? Bear in mind that only exogenous DNA that is closely related to the recipient's DNA is capable of transforming, because to be maintained it must recombine with homologous regions of the recipient's DNA. Your first thought may be that transforming DNA is released by random lysis of cells in the environment. But if transformation evolved as a means of genetic exchange, this explanation is not satisfying. Free DNA is short-lived in many natural environments because nucleases (DNases) abound. And would evolution generate a sophisticated multicomponent complex that relies on random release of DNA from other cells? Indeed, there is some evidence that donor cells are genetically programmed to release DNA when its probability of being taken up is highest. Thus, natural transformation of _Pseudomonas stutzeri_ occurs when donor and recipient cells touch. Likewise, free DNA can be detected in _S. pneumoniae_ cultures only when populations are dense, governed by a population-dependent form of cell-cell communication called **quorum sensing** (chapter 12).

Second, what is the biological function of natural transformation? Some have proposed that exogenous DNA is taken up as a nutrient. But a counterargument is that some naturally transformable bacteria have evolved mechanisms to recognize and thereby take up DNA *only from closely related strains.* For example, the *H. influenzae* chromosome contains 600 copies of an 11-base-pair recognition site, and its competent cells take up only DNA that contains one of these sites. On the other hand, other naturally transformable species take up DNA indiscriminately. It seems likely that natural transformation evolved as a *means of exchanging DNA among strains of a particular bacterial species,* a substitute for the sexual exchange in eukaryotes. But does that genetic argument adequately explain why natural transformation has been reported for relatively few prokaryotes? At the moment, there are no good answers.

Artificial transformation. Most prokaryotes (probably all, if proper conditions were discovered) can be transformed artificially by treatments that transiently alter the permeability of a cell's membrane, thus facilitating the uptake of soluble DNA. These treatments include such diverse manipulations as exposure to high concentrations of calcium ions on ice, followed by a heat shock, and **electroporation** (making transient holes in the cytoplasmic membrane by exposing cells to millisecond jolts of tens of thousands of volts). Such artificial transformation is an efficient way to introduce plasmids (for example, when attempting the genetic complementation of a mutant gene with the wild-type copy expressed from a plasmid; Fig. 10.4B). Why are plasmids more easily introduced by artificial transformation than linear DNA? Because these circular molecules have no exposed ends, they are resistant to attack by the intracellular *exo*nucleases of most prokaryotes, so they can survive within the transformed recipient. Mutant strains of *Escherichia coli* lacking two such exonucleases can be successfully transformed artificially with linear DNA. Alternatively, if provided in sufficiently high concentrations, some linear fragments of double-stranded DNA will survive inside the cell. And, if the linear fragment encodes regions that are homologous with the chromosome, the foreign DNA may get integrated.

As you can imagine, artificial transformation is a critical step in gene cloning and recombinant DNA technology. Earlier we described how linear fragments of DNA carrying targeted mutations can be engineered by PCR. Now you know it's also possible to introduce this altered piece of DNA into a cell where at some frequency it will naturally recombine with the wild-type gene (Fig. 10.4A). A few recombinants will likely arise. All you need is a way to select for these rare events. Can you think of ways to do this?

Transduction

Prokaryotes also receive bacterial DNA from viruses in a process known as **transduction.** How? Some bacterial viruses, or **bacteriophages** (**phages**; chapter 17) make mistakes when they pack DNA during assembly and can accidentally carry DNA from the bacterial host to a new one. Such "imperfect" virions, which contain host cell DNA in addition to their own genome, are called **transducing particles.** Two different kinds of mistakes can occur, and each is characteristic of a particular type of phage (Table 10.7). During **generalized transduction,** phages randomly pack and, therefore, exchange *any of the host's genes.* However, as its name indicated, **specialized transduction** involves the packing and exchange of *specific genes from the host.*

TABLE 10.7 Differences between generalized and specialized transduction

Property	Generalized transduction	Specialized transduction
Mediator	Virulent or temperate phages	Only temperate pages
DNA in transducing particle	Only host cell DNA	Phage and host cell DNA
Genes transduced	Any host gene	Only genes close to insertion site of prophage

Generalized transduction. During generalized transduction, phages mistakenly and randomly pack genes from the host in their capsids and move them to the next cell they infect. Phage P22, which infects *Salmonella*, is probably the most thoroughly studied and is a good example for all generalized transducing phages. During the infection, the circular phage uses the rolling-circle mechanism, a replicative process described for some plasmids in which one of the strands of the circular DNA is cut and separated to allow the circular strand to replicate; the cut strand is then resealed and replicated. Although rolling-circle replication occurs in two steps, it occurs rapidly because the replication machinery does not have to deal with the challenges of simultaneously replicating two strands of different polarity (as when replicating the chromosome; chapter 8). However, phages carry rolling-circle replication in a slightly different way: rather than making separate genomes one at a time, several rounds of replication occur sequentially, generating a long concatemer of about 10 phage genomes joined end to end, like a toilet paper roll (Fig. 10.8). Each concatemer is then packed into the phage head by what is called a "head-full" mechanism. First, two

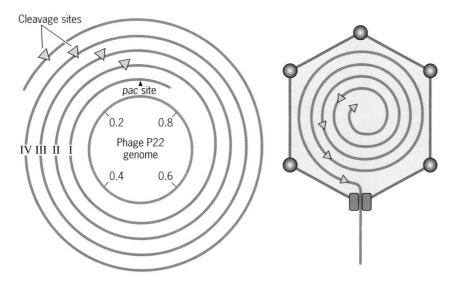

FIGURE 10.8 Packaging DNA by phages that mediate generalized transduction.
(A) Rolling-circle replication of the phage circular genome (inner circle) synthesizes new DNA as a concatemer containing several copies of the genome (outer spiral). (B) Each copy is processed at the *pac* site (triangles) and packed inside the phage head until it is filled with DNA. Successive lengths of DNA (I, II, III, and IV) are each packaged into a separate phage head. Note that the packaged lengths are longer than a phage genome, thus ensuring that each phage particle gets a complete set of genes, no matter where packaging begins.
(B) Structure of the P22 phage showing the concatemers of DNA packed inside the capsid.

phage-encoded nuclease proteins recognize and cut at a frequently occurring 17-bp sequence (called *pac* for *pac*king) in the concatemer; next the enzymes lead the cut end into an empty phage head. Packing proceeds until the head is filled with phage DNA. Then, a second nuclease cuts the exposed part of the concatemer, leaving a neatly filled phage head. What's left of the DNA concatemer is now ready to fill other empty phage heads.

But how can a phage like P22 "mistakenly" pack some of the host's DNA? Some *pac*-like sites also occur in the host cell's chromosome. Thus, the same phage nucleases can inadvertently guide *pac*-containing host DNA to the phage heads for packing. Such generalized transducing particles mistakenly contain only host cell DNA (strange for a virus, don't you think?). It is curious that not all host genes are packed with equal frequency. Why not? (*Quiz spoiler*: A gene's probability of being incorporated depends on its proximity to a *pac*-like site.) However, certain mutant phages have low specificity for *pac* sites and can transduce all host cell genes at nearly equal frequencies; these transducing phage, designated HT for *high frequency of transduction*, are useful experimental tools.

Generalized transducing particles are surprisingly common, about 2% of the progeny in a P22 lysate. However, the DNA from only a fraction of these particles becomes integrated into the genomes of the host cells. The DNA that is not promptly recombined into the host chromosome is usually circularized upon entry and stably maintained in the host's cytoplasm. However, because it cannot replicate independently, these fragments are eventually lost during cell division. Thus, they are not inherited. We call this event *abortive transduction*. Nevertheless, the genes acquired by generalized transduction can be expressed while in the recipient host cell. Some may even confer a temporary growth advantage, until successive rounds of cell division dilute its concentration.

Specialized transduction. Specialized transduction is another, distinct way that phages move host DNA from cell to cell. However, only certain genes are moved (Table 10.7). It is also a consequence of "errors" but, in this case, the mistakes are made by **temperate phages** (discussed in more detail in chapter 18). During an infection, the genomes of these phages become integrated into the host cell's chromosome at a specific site (*att*), a phenomenon called **lysogeny**. Typically, the integrated phage DNA or **prophage** remains "dormant" for some time, replicating with the chromosome and passed to daughter cells after cell division. However, in response to specific environmental cues, the phage can be "awakened" and excised from the host chromosome to initiate a new cycle of virulence. Usually, excision occurs precisely at the two ends of the prophage. But occasionally the register of the cuts is shifted, and some of the host cell DNA is carried along with the phage genome and packed in a viral particle. Because the register might shift in either direction, specialized transducing particles might contain host genes that flank the attachment site on either side. Figure 10.9 shows the normal and abnormal excision of the prophage of the **phage** λ (the Greek letter *lambda*), which has been an indispensable experimental tool to understand specialized transduction and is discussed in more detail in chapter 18. This phage infects *E. coli* cells and integrates specifically at the so-called attachment site (*attB*) on the chromosome.

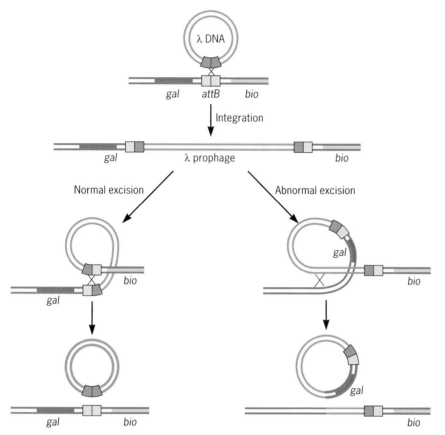

FIGURE 10.9 **Formation of a specialized transductant of phage λ.** Temperate phages like λ integrate their genome into the *attB* site of the host's DNA by site-specific recombination and remain integrated as a prophage until excision is triggered. Normal excision reverses the site-specific recombination process to produce the circular phage DNA again. However, rare excisions may occur during specialized transduction in which one of the neighboring regions is excised, leaving behind some phage DNA in the chromosome and moving host genes to the phage DNA.

This site is flanked by genes encoding biotin synthesis (*bio*) and galactose degradation (*gal*). Hence, either one of these genes can be packed in phage λ's specialized transducing particles. Note that, in either case, a portion of the phage DNA is left behind on the bacterial chromosome.

When a specialized transducing particle infects a new bacterial cell, its DNA can still become integrated into a new host cell's genome (Fig. 10.10). But the game now changes. The integration no longer occurs by *site-specific recombination* at the *att* sites but by *homologous recombination* between the DNA carried over from the old host and its homologous region in the new host. Once the DNA from the specialized transductant becomes integrated, a short diploid region is created that can recombine into the new host DNA

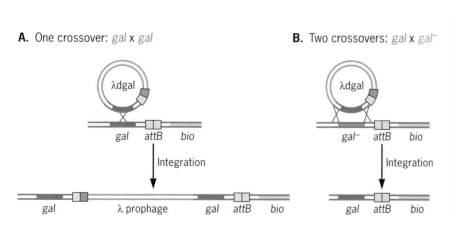

A. One crossover: *gal* x *gal*

B. Two crossovers: *gal* x *gal⁻*

FIGURE 10.10 **Integration of a specialized transductant of phage λ.** A defective phage λdgal can recombine into the DNA of a new host by homologous recombination. If the host cell carries the *gal* gene, a single crossover between the two *gal* copies integrates the phage DNA and generates a *gal* diploid (left). If the new host cell is a *gal* mutant, a double crossover in the regions flanking the mutation allows for the replacement of the mutated gene with the wild-type *gal* copy, thus complementing the mutant phenotype.

at a homologous site. This type of event effectively introduces traits from the old host into the new one. For example, a *gal* gene carried by the phage can be exchanged with a mutated copy (*gal⁻*) in the new host (Fig. 10.10), thereby allowing the new host to grow with galactose as carbon source. In this case, a double crossover replaced the *gal⁻* locus with the wild-type *gal* gene and allowed the *E. coli* mutant cell to grow with galactose. The exchange is then said to have *complemented* the *gal⁻* mutation. Note that this type of *genetic complementation* differs from the one we described earlier, which used plasmid vectors to express the wild-type gene in *trans*, because genetically complementation via specialized transduction produces a haploid, rather than diploid, strain. Furthermore, the copy number of the gene remains the same. Early geneticists capitalized on the ability of specialized transductants to complement mutants to develop genetic tools that greatly advanced our knowledge of microbial genetics and physiology.

Conjugation

Conjugation, the third mechanism of genetic exchange, relies on the remarkable property of certain **plasmids** to transfer from cell to cell. The so-called "broad spectrum" plasmids have developed this skill to a remarkable degree. Some can move between *almost any pair of Gram-negative bacteria* and are even able to transfer to eukaryotes. For example, the Ti plasmid of the plant-pathogenic bacterium, *Agrobacterium*, can transfer genes to plants (chapter 14), and certain plasmids (R751 and F) can transfer bacterial genes to the yeast *Saccharomyces cerevisiae*. Some can also be transferred from bacteria to archaea. Such plasmids may be responsible for much of the horizontal gene transfer that torments students of evolution by scrambling genes between unrelated species. To transfer, the plasmid must carry the genes that mediate this exchange. In addition, some plasmids carry other genes that *encode resistance* to antibiotics or other environmental stresses; historically, these plasmids were called **R factors**. The impact of these plasmids on public health is profound, because they are *responsible for the rapid spread of antibiotic resistance* between populations of bacteria, with serious problems for clinical medicine.

Conjugation among Gram-negative bacteria. Most conjugative plasmids in Gram-negative bacteria carry genes for the formation of a complex multiprotein secretion apparatus for transfer of DNA from the donor cell to the recipient cell called the **type IV secretion system**. The most thoroughly studied conjugative plasmid is the **F** (for fertility) **factor**, which transfers itself between enteric bacteria such as *E. coli*. Its self-transfer apparatus is encoded in an operon containing 13 *tra* genes, for which there is ample room on this relatively large (94.5-kilobase) circular plasmid. One of these genes encodes the protein subunits that make a dedicated pilus, called **sex pilus** (plural, pili). Cells bearing such pili are called F⁺, or donor cells (Fig. 10.11). The pilus extends from the donor cell by polymerizing subunits at its base until its tip attaches to another cell (designated F⁻, or recipient) that lacks an F plasmid. The pilus then retracts by the depolymerization of its subunits, thereby drawing the mating pair together. (Note that the dynamic protrusion and retraction of the sex pilus is

A

B

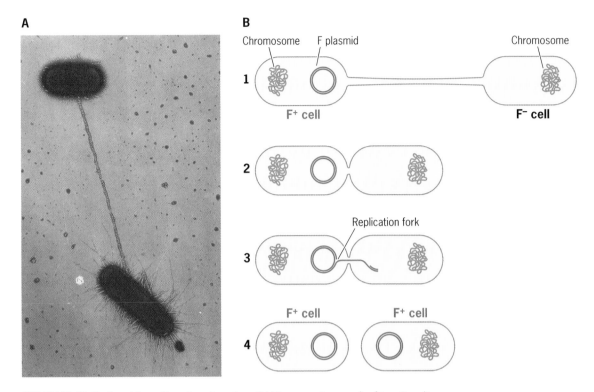

FIGURE 10.11 **F-plasmid-mediated conjugation.** (A) Electron micrograph of two *E. coli* cells held together by a sex pilus (step 1 in panel B). (B) Steps in the process of conjugation. (1) Conjugation begins when a sex pilus on an F⁺ cell attaches to an F⁻ cell. (2) The sex pilus retracts, drawing the cells together. (3) The F plasmid replicates by a rolling-circle mechanism in the donor cell. A nick is made in the dsDNA of the plasmid, and the nicked strand enters the F⁻ cell while the circular plasmid strand is being replicated in the donor cell. (4) In the F⁻ cell, the single strand is duplicated and circularized, converting the F⁻ cell into an F⁺ cell.

analogous to that of type IV pili, which we discussed in chapter 2 for their role in cell attachment and surface motility, but that is as far as similarities go for these two pilus systems.)

The union of the two cells via the sex pilus is stabilized via a pore structure that connects the two cytoplasms. One of the strands of the plasmid is then nicked at a specific site (origin of transfer or *oriT*) and pulled towards the recipient cell, where it is duplicated and circularized. As the single strand is led to the recipient cell, the circular strand in the donor "rolls" while being replicated (the rolling-circle mechanism; Fig. 10.11). As a consequence of the mating process, the F⁻ cell becomes F⁺. When F⁺ and F⁻ cells are mixed, the culture rapidly becomes completely F⁺. "Maleness" in this case is infectious (pun intended).

Rarely (at a frequency of about 1 in 1,000) the F plasmid becomes integrated into the host cell chromosome by a recombination event between *homologous regions* (Fig. 10.12). As the name indicates, this type of recombination can only happen if the homologous regions are present in both F plasmid and chromosome. What regions are these? Some F plasmids carry insertion (IS) elements that match IS elements present on the chromosome (Fig. 10.12). Homology between the plasmid and chromosomal IS elements promotes integration of the F plasmid into the chromosome, generating cell lines designated **Hfr** (for **high frequency of recombination**). Hfr strains carry the F plasmid, just in integrated form, and thus retain the ability to transfer it to a recipient F⁻ cell. However, when the transfer begins the integrated F plasmid *will drag part of the chromosome along*. Thus, chromosomal genes enter the recipient cell along with the F plasmid. Any regions of homology between the incoming chromosomal genes and the recipient's DNA can be used for homologous recombination and integration of the transferred genes. This happens quite often, and it is the reason why these strains are named Hfr.

Mating is a rather slow process, requiring about 100 minutes at 37°C for a complete copy of the *E. coli* chromosome to pass from an Hfr strain to an F⁻ cell. Because DNA enters the recipient cell at a constant rate, matings shorter than 100 minutes will transfer fewer chromosomal genes. In fact, this principle enabled early geneticists to map the positions of genes on the *E. coli* chromosome. For these *interrupted mating* experiments Hfr and F⁻ strains were mixed to promote their conjugation. The cells were then agitated vigorously at different time intervals to break apart the mating pairs. By scoring which genes entered and recombined into the recipient cell based on the mating time allowed, scientists deduced the position of

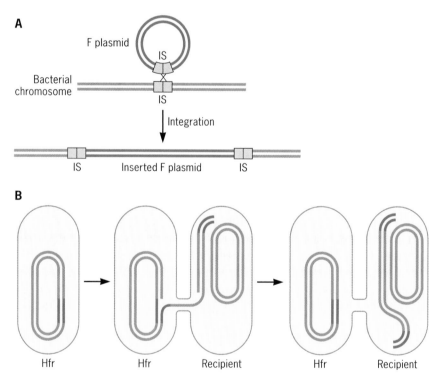

FIGURE 10.12 Integration of F plasmid in host chromosomes (Hfr strain). IS elements in the F plasmid and host chromosome allow for the integration of the plasmid via site-specific recombination in a single crossover event (left). The strain carrying the integrated F plasmid is now designated Hfr and retains the ability to transfer the plasmid DNA to a recipient (F⁻) cell. If the mating lasts long enough, the F plasmid also drags portions of the chromosome to the recipient cell. Once inside the recipient, the incoming genes can be integrated via homologous recombination into the recipient's chromosome at high frequency.

the gene in the chromosome. Now you know why the first maps of the *E. coli* chromosome showed the position of the genes in minutes!

Conjugation among Gram-positive bacteria. Plasmids also mediate conjugation of Gram-positive bacteria. However, instead of mating pairs being attached by a pilus, here mating cells clump together to bring the donor and recipient close. Next, a conjugative type IV secretion system forms a channel across the cell envelopes, interconnecting the cytoplasm of donor and recipient cells. The Gram-positive apparatus contains fewer proteins than the Gram-negative counterparts and is adapted to the unique structure of the Gram-positive cell envelope (cytoplasmic membrane surrounded by a thick peptidoglycan cell wall). Despite these differences, plasmid transfer by Gram-positive and Gram-negative bacteria shares many similarities. Transfer is initiated when one strand of the plasmid is nicked by a plasmid-encoded protein and then led to the recipient cell, where it is replicated and circularized. The circular single strand that remains in the donor cell is then replicated by a rolling-circle mechanism.

Some Gram-positive conjugative plasmids are quite sophisticated, as they are also *pheromone-sensitive*. Consider the Gram-positive gut bacterium *Enterococcus faecalis* (Fig. 10.13). Recipient cells excrete chromosomally encoded **pheromones**, here secreted peptides that are "sensed" only by donor cells of the same species. Much like the secretion of pheromones by female animals to signal their fertility, these pheromones attract bacterial donor cells and signal readiness to mate with the right partner. In the donor, expression of one of the peptides is repressed by plasmid-encoded proteins; consequently, the cell is tuned to that particular pheromone and to the recipient cell with which it can mate. Once the pheromone enters the donor cell, it induces the expression of plasmid-encoded genes needed

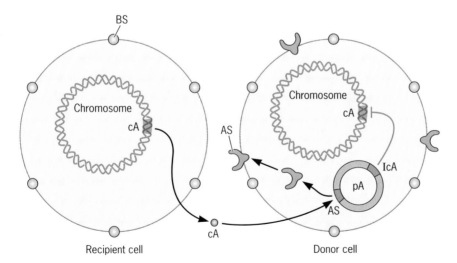

FIGURE 10.13 Conjugation of *Enterococcus faecalis*. The recipient cell produces a chromosomally encoded pheromone (cA) that interacts with a plasmid (pA) in the donor cell, causing it to produce aggregation substance (AS), which migrates to the cell surface. At the surface, AS binds to binding substance (BS) on the surfaces of other cells, forming a clump of cells within which genetic exchange occurs. Although donor cells also carry a gene encoding cA, its expression is repressed by IcA, encoded on pA. Thus, aggregation occurs only among groups of donor and recipient cells.

for transfer. Among those genes are an *aggregation substance*, a protein that is targeted to the cell surface to promote clumping between donor and recipient cells. The mating pair is now ready for genetic exchange. Once the recipient has acquired the pheromone-responsive plasmid, proteins encoded by plasmid genes repress the chromosomal genes that code for that particular pheromone. Nevertheless, the cell continues to produce other pheromones—a constant invitation to mate that reflects the importance of conjugation to these species.

Gram-positive bacteria have yet another creative way to exchange DNA. The conjugative plasmids of the antibiotic-producing bacteria *Streptomyces* encode a single protein (TraB) that takes care of the transfer all by itself. This elegant approach to genetic exchange is remarkably efficient (often 100% efficient!). TraB is structurally and functionally similar to a cell division protein (FtsK) that pumps DNA at the division septum to complete the replication of the bacterial chromosome before two daughter cells finalize their division (chapter 4). Plasmid-encoded TraB polymerizes to form a small ring of six subunits (a hexamer) with an inner hole big enough to accommodate double-stranded DNA. Studies suggest that *the TraB ring does it all*: it makes a pore on the inner membrane and forms a channel across the thick peptidoglycan layer to create a route for the plasmid to travel from donor to recipient.

CAN GENETIC EXCHANGE GET ANY "CRISPER"?

LEARNING OUTCOMES:
- State the known and proposed functions of the CRISPR-Cas system.
- Describe how microbes acquire immunity to foreign DNA with the CRISPR-Cas system.

One of the surprises revealed through genome sequencing was the presence of arrays of DNA repeats in the chromosome of *E. coli* K12. As more sequenced genomes became available, scientists observed that such arrays were common in prokaryotes, both bacteria and archaea. These arrays, designated **CRISPRs**, launched a new era in our understanding of horizontal gene transfer!

Just as vaccines prime our immune system against harmful bacteria and viruses (chapter 22), microbes have also evolved mechanisms of immunity. As we learned earlier, prokaryotic cells can receive foreign DNA via several mechanisms. Why not use it to make a vaccine? Microbes have a battery of DNA processing enzymes collectively known as the *Cas system* that cleave the foreign DNA and "archive" a small fragment within organized clusters or CRISPRs located in the vicinity of the Cas-encoding genes (Fig. 10.14). Their full name, *Clustered Regularly Interspaced Short Palindromic Repeats,* may be a mouthful, but it accurately describes the unique way in which the foreign DNA "souvenirs" are archived in CRISPR arrays: close to each other and flanked by short segments with *inverted (palindromic) repeats*. Think about them as a bar code, each bar a historic token of an encounter with plasmid or viral DNA.

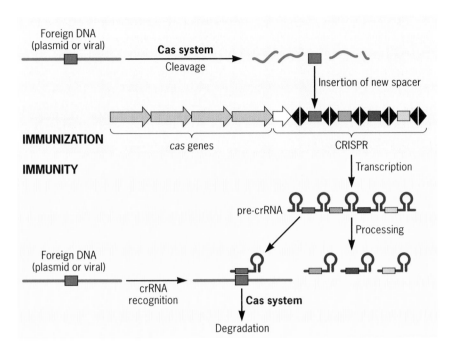

Foreign DNA
(plasmid or viral)

Cas system
Cleavage

Insertion of new spacer

IMMUNIZATION

cas genes

CRISPR

IMMUNITY

Transcription

pre-crRNA

Processing

Foreign DNA
(plasmid or viral)

crRNA
recognition

Cas system

Degradation

FIGURE 10.14 **CRISPR immunization and immunity.**

But how can CRISPR-Cas systems protect the cell from repeat infections? The CRISPR cluster is transcribed as a single, long precursor RNA or pre-crRNA containing the repeats interspersed with spacers that are complementary to the original foreign DNA (called **antisense**) (Fig. 10.14). Because the repeats separating spacers are inverted, they anneal in the single strand of RNA, forming hairpins that "decorate" the pre-crRNA along its length. Dedicated Cas enzymes recognize the hairpins and process the precursor RNA into small CRISPR RNAs (crRNAs), each containing one antisense RNA spacer and all or part of the repeat sequence as a tag. Should old plasmid or viral foes infect again, the cell now has a battery of small antisense RNAs ready to bind the foreign DNA and target it for degradation, thanks to the repeat tag they contain. Mission accomplished! Note also that because the CRISPR system is encoded in the chromosome, it is inherited by the daughter cells. Hence, *the progeny is also immunized.*

Although the primary role of CRISPR-Cas systems was believed to be adaptive immunity against invasive genetic elements (plasmid and virus), this ingenious mechanism of RNA interference is now known to play a number of other roles in bacteria and archaea. Some of the small crRNAs that are derived from the pre-crRNA can also target native RNAs of the cell and interfere with their expression. Thus, cells use their CRISPRs to *silence or downregulate the expression of specific genes.* Why would a cell carry spacers matching its own DNA? It could be coincidental, or adaptive, reflecting the co-evolution of CRISPR sequences and genomes. If the interference helps the cell grow and survive in its environment, it makes sense to keep a spacer that matches one of the native genes, regardless of how the spacer got there in the first place. As for the Cas protein system, there is evidence suggesting that their activities may also *complement the activity of DNA repair proteins* and even *help refold proteins after damage.* If so, the Cas system may have evolved to play a *dual role in adaptive immunity and stress response.*

The discovery of CRISPR-Cas systems has completely transformed our understanding of prokaryotic immunity and regulation via RNA interference. Many applications for these systems have also rapidly emerged. Because immunity to viruses can easily be achieved with CRISPR-Cas systems, industrial microbiologists can now engineer CRISPR arrays to immunize industrial strains of bacteria, such as the cheese-making lactobacilli and streptococci, against viruses. Furthermore, as incoming spacers are incorporated rapidly into the CRISPR array, they can be used to quickly genotype complex populations and investigate their dynamics. The Cas machinery has also captured the imagination of prokaryotic and eukaryotic geneticists, who can now engineer specific Cas proteins and repurpose their activities for genome editing and transcriptional regulation in prokaryotes and eukaryotes, alike. The potential of CRISPR-Cas systems (discovered in microbes!) to revolutionize genetic engineering seems endless. What ethical questions does this technology raise?

WHAT DO GENOMES TELL US ABOUT PROKARYOTES?

LEARNING OUTCOMES:
- Discuss how the advent and rapid evolution of genome sequencing have impacted microbiology.
- Define open reading frame (ORF).
- Describe how researchers can predict putative functional roles of newly discovered ORFs.
- Infer why phylogenetic analysis based on genomes likely better reflects evolutionary relatedness over phylogenetic analysis using single genes.

The discovery of CRISPRs is but one example of how much information can be gained by studying prokaryotic genomes. Traditionally, geneticists focused on identifying phenotypes and correlating these traits with specific genotypes. But the rapid development of technologies to sequence, annotate, and compare genomes has forever changed the scope of genetic studies and deepened our understanding of how cells work and evolve. DNA sequence analysis opened the door to modern **genomics**, the comparative study of whole DNA sequences of organisms.

Early genomic studies focused primarily on the relatedness of microbial species and groups. For example, the *percent of GC* in a group of organisms' genomic DNA could be determined by measuring its density or melting point. If the values for two organisms were similar, they might not necessarily be closely related, but if the values were significantly different, they most certainly were not. The ability of DNAs to *hybridize* proved to be a more sophisticated index of relatedness. *Relative genome size* also gave clues to a cell's potential complexity or array of talents (chapter 3). Although a large number of interesting correlations were obtained, these biochemical approaches also revealed perplexing inconsistencies. For example, the two cyanobacteria *Nostoc* and *Synechocystis* are quite similar in most respects, but the genome of the former (9.1 megabases, or Mb) is 2.5-fold larger than that of the latter (3.6 Mb). Likewise, the genome of the newt (an amphibian; 15,000 Mb) is five times larger than ours (3,000 Mb), and the lily's (100,000 Mb) is over 30 times as large.

The genomics field's next major advance was spurred by methods to determine the sequence of bases in DNA. Genome sequencing made possible totally new ways to study microbes (and other organisms as well). For example, startling information that came from sequencing was the 1977 discovery of the *Archaea* by Carl Woese. So different were their small ribosomal RNA (rRNA) gene sequences from those of other prokaryotes and eukaryotes that he concluded that these microbes constitute a distinct third branch of the biological world (a deduction supported subsequently by a wealth of data). All the information needed for an organism to reproduce and carry out its myriad activities is embedded in the sequence of bases in its DNA. But how do we extract and decipher these molecular patterns?

Initially, DNA was sequenced manually. Slow, laborious, and tedious, the early techniques held only marginal promise of producing sufficient data to sequence any organism's DNA completely, much less to acquire enough data for meaningful comparisons between organisms. A series of technological breakthroughs changed the situation spectacularly. In 1995, the genome of the pathogenic bacterium *H. influenzae* was completely sequenced. More complete sequences were soon announced. Then, in the late 1990s, high-speed sequencing machines entered the scene. What had been one year's work for an individual, machines could accomplish in about an hour. Depending on the sequencing approach, you can now determine short (up to 300-bp long) or long (up to 10 kB) sequences (or reads) in a single reaction (Fig. 10.15). These segments must then be arranged in their natural order in the genome, which could be millions or billions of base pairs long. This task is accomplished with computers and programs that assemble the reads and reconstruct a genome (Fig. 10.15). If, on average, the random bits of DNA sequence cover the genome seven times, a computer program captures enough information from overlapping and redundant individual reads to form a contiguous sequence or **contig**. Still, some manual sequencing is often required to close seemingly inevitable gaps in the sequences. But once this "finishing" is done, a consensus sequence for the genome is generated, and many of the inner secrets of the organism are revealed. Today, the sequences of the genomes of hundreds of organisms (including our own species) are available. We can even sequence genomes from environmental microbes without cultivating them in the laboratory (chapter 19). The problem is no longer how to acquire information; it is how to deal with it.

Annotation

A completely sequenced genome is just a list of A's, G's, T's, and C's, albeit a very long one. The process of converting this raw data into meaningful information is called **annotation** (Fig. 10.16). The first step is to enlist a computer program to search for presumed genes, called **open reading frames** (ORFs), or stretches of DNA *presumed to encode a protein*. To do so, a computer program scans the sequence for stop-and-start sequences separated by gene-like distances. Next, the computer is tasked with comparing sequences of the discovered ORFs to a database (e.g., a big public one called

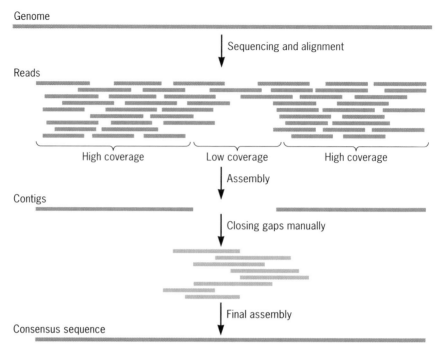

FIGURE 10.15 Sequencing a genome. A genome sequencing project generates fragments of sequenced DNA (reads) that are then aligned to find areas of overlap and reconstruct larger segments of the genome (contigs). Any gaps in the assembly can be closed manually. For example, PCR primers can be designed to amplify the regions between contigs and get the missing sequence. This approach allows the experimenter to close the gaps of the genome and generate a consensus sequence for the whole genome.

GenBank) of the sequences of genes with "known" functions (we shall discuss these quotation marks below). Typically, such annotation *assigns putative functions to about half the genes of most sequenced bacterial genomes*. Using additional computer programs or manual searches, these individual functions can be refined and analyzed to group them into pathways (those of biosynthesis, for example).

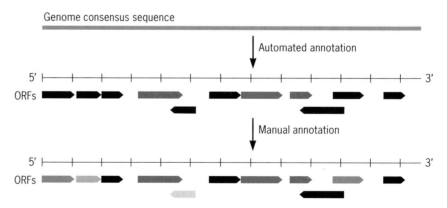

FIGURE 10.16 Genome annotation. Automated genome annotation identifies open reading frames (ORFs) in the consensus sequence of a genome and assigns each to functional categories (indicated here by different colors) based on the similarity of the ORF sequence to other sequences in the databases. The preliminary annotation can be further improved by manually annotating ORFs for which the database matches were only weakly significant.

Even if a deduced protein sequence has no match in existing databases, often it is possible to discover more by searching for particular patterns within the newly discovered ORF sequence. For example, some known biochemical functions are performed by conserved clusters of amino acids, called domains or **motifs**. Certain domains indicate the product protein's ability to bind to particular compounds, such as ATP, nicotinamide adenine dinucleotide (NAD), or DNA. Thus, an ORF with an ATP-binding motif may well be a kinase, and one with an NAD-binding motif is probably a dehydrogenase. Some protein sequences also contain clues to their cellular location. For example, hydrophobic regions may indicate the protein associates with or passes through a membrane. Other proteins might encode recognizable signal sequences at the amino-terminus, indicating that the product protein is probably targeted for secretion to the periplasm or the external environment.

But bear in mind that annotation and protein domain assignments are based on computer comparisons, not experimentation. Hence, annotations reveal *putative* functional roles only. Although a powerful approach to understanding biology, annotation is not without hazards. (It may be comforting to note that we humans still have a critical role to play—both in making biological judgments and in catching mistakes.) But even comparisons to "known" functions determined in the laboratory can be somewhat risky. For example, many proteins carry out multiple functions, some of which may not be obvious under artificial conditions. Thus, while annotation can have great predictive value, hypotheses developed from genome annotation must be tested experimentally.

Relatedness

Genomics has already told us much about the relatedness of various microorganisms, and it promises more. Just as exciting, it tells us about the *relatedness and evolution of proteins*. On the basis of sequence similarity of their genes, the protein products from various sources across the total spectrum of organisms can be grouped into families; families of proteins with lesser similarity can be grouped into **superfamilies**. These groupings have evolutionary significance. Members of the same family, no matter in which organism they are found, presumably evolved from a common ancestral protein, regardless of whether they were spread through vertical or horizontal transmission. Deciphering the relatedness of protein sequence similarity is decidedly more straightforward than deciphering the relatedness of organisms.

Nowadays, computers and the algorithms of some programs are so powerful that it is possible to compare genomes and assess genetic variation based on the accumulation of mutations and horizontal gene transfer. In chapter 19, we will learn how genomics has advanced our knowledge of the immense diversity of microbes inhabiting our planet. In subsequent chapters, we will see how these technical advances identified the genetic mechanisms responsible for the emergence of the multiple antibiotic-resistant MRSA strain from commensal *Staphylococcus aureus* (chapter 23). **Phylogenetic analyses** can also be performed on whole genomes. These methods deciphered the source of an explosive epidemic of cholera in Haiti (chapter 26). They have also illuminated the contribution of horizontal gene transfer to life's evolution. By taking into consideration how genomes have been shaped not only by accumulation of mutations but also by lateral gene transfer, a new view of life phylogeny has emerged. The *Bacteria* and *Archaea* form the primary lineages, branching early on from a common ancestor, and *Eukarya* are a hybrid group that emerged from symbiotic events between *Bacteria* and *Archaea* (chapter 1). Hence, genome-scale data analysis supports a "ring of life" rather than a three-domain tree of life. But here is the amazing part. We have only touched on the wealth of scientific and application benefits that are likely to derive from genomics. Give your imagination free rein.

CASE REVISITED: **Tricks to increase mutation rates**

Dr. E set up an experiment to grow 500 cells of *E. coli* as isolated colonies on the surface of solidified medium contained in a petri dish. When he supplemented the plates with ampicillin, most of the cells died because they were sensitive to the antibiotic. He soon realized that relying on spontaneous mutations was useless. Most mutations in the cell arise from errors in replication and spontaneous chemical damage to the DNA bases. Since the cell corrects most of these mistakes, the rate of spontaneous mutation is just one in 1 billion (notated 10^{-9}). Dr. E calculated that to isolate one spontaneous antibiotic-resistant mutant, he would need to plate 10^{10} cells, which meant he would have to prepare 20 million plates! (Divide 10^{10} by the 500 or so colonies that can be isolated on each plate). Dr. E is a patient man, but not that patient! Hence, to reduce the number of plates he needed to prepare and screen, he used mutagens to increase the mutation rate. Using nitrosoguanidine, he increased the mutation rate to 10^{-4}. What about the number of plates? Well, 10^4 divided by 500 is 20. Thus, he needed just 20 plates to isolate the ampicillin-resistant mutant. Evolution can rely on spontaneous mutations because it acts at time scales of billions of years. Dr. E did not have that luxury and wanted to perform an experiment that he could finish before retirement. He cleverly did so.

CONCLUSIONS

There is a common thread to the genetics of all organisms, but the details of genetic exchange differ profoundly between prokaryotes and eukaryotes. Because of the ease and precision with which prokaryotes can be studied, most of our knowledge of basic molecular biology has come from studies of prokaryotes and the viruses that infect them. Now, with the advent of genomics, studies of genetics have entered a new phase, and the wealth of knowledge and genetic tools generated by studying prokaryotes will be invaluable.

Representative ASM Fundamental Statements and Learning Outcomes for the Chapter

EVOLUTION

1. Mutations and horizontal gene transfer, with the immense variety of microenvironments, have selected for a huge diversity of microorganisms.

   a. Define "mutation."
   b. Explain why mutations are important for evolution.
   c. Explain how cells maintain a relatively low mutation rate.
   d. Discuss how microbes can evolve at a very fast pace without sexual reproduction to rearrange genes.

2. The evolutionary relatedness of organisms is best reflected in phylogenetic trees.

   a. Discuss how the advent and rapid evolution of genome sequencing have impacted microbiology.
   b. Define open reading frame (ORF).
   c. Describe how researchers can predict putative functional roles of newly discovered ORFs.
   d. Infer why phylogenetic analysis based on genomes likely better reflects evolutionary relatedness over phylogenetic analysis using single genes.

INFORMATION FLOW AND GENETICS

3. Genetic variations can impact microbial functions (e.g., in biofilm formation, pathogenicity, and drug resistance).

   a. Identify why small mutational changes, such as point mutations, can have variable consequences upon the translated protein while large mutational changes, such as duplications or inversions, have more dramatic consequences.
   b. Define transposon.
   c. Identify the function of the two features that all transposable elements share.
   d. Contrast the effect of a deletion mutation and a transposon insertion on the encoded protein.

4. Although the Central Dogma is universal in all cells, the processes of replication, transcription, and translation differ in *Bacteria, Archaea,* and eukaryotes.

   a. Summarize five ways in which prokaryotes and eukaryotes differ with respect to exchange of DNA.

5. Cell genomes can be manipulated to alter cell function.

   a. Differentiate auxotrophic mutants from wild-type cells.
   b. Describe how PCR (polymerase chain reaction) is used together with site-directed mutagenesis in mutant technology.
   c. Describe how genetic complementation is used to link a gene to a phenotype in mutant technology.
   d. Identify the role of RecA in transformation.
   e. Differentiate transformation, transduction, and conjugation.
   f. Describe the circumstances and situation of an example of natural transformation.
   g. Identify the role of artificial transformation in recombinant DNA technology.
   h. Compare and contrast the mechanisms of generalized transduction and specialized transduction.
   i. Compare and contrast the outcomes of generalized transduction and specialized transduction.
   j. Explain what characterizes a cell as an Hfr (high frequency of recombination) cell.
   k Compare and contrast the process of conjugation for Gram-negative and Gram-positive bacteria.
   l. State the known and proposed functions of the CRISPR-Cas system.
   m. Describe how microbes acquire immunity to foreign DNA with the CRISPR-Cas system.
   n. Identify two mechanisms of conjugation unique to some Gram-positive bacteria.

IMPACT OF MICROORGANISMS

6. Microorganisms provide essential models that give us fundamental knowledge about life processes.

    a. Give an example of how using mutant technology with microbes has elucidated key cellular functions.

    b. Name potential mutagens that you have come into contact with.

    c. Describe how the Griffith natural transformation experiment led to the discovery that genes are made of DNA.

SUPPLEMENTAL MATERIAL

References

- **Siguier P, Gourbeyre E, Varani A, Ton-Hoang B, Chandler M**. 2015. Everyman's Guide to Bacterial Insertion Sequences. *Microbiol Spectr* **3**:MDNA3-0030-2014. doi: 10.1128/microbiolspec.MDNA3-0030-2014.

- **Johnsborg O, Håvarstein LS**. 2009. Regulation of natural genetic transformation and acquisition of transforming DNA in *Streptococcus pneumoniae*. *FEMS Microbiol Rev* **33**:627–642. https://femsre.oxfordjournals.org/content/33/3/627.long. http://dx.doi.org/10.1111/j.1574-6976.2009.00167.x.

- **Llosa M, Gomis-Rüth FX, Cool M, de la Cruz F**. 2002. Bacterial conjugation: a two-step mechanism for DNA transport. *Mol Microbiol* **45**:1–8. http://onlinelibrary.wiley.com/doi/10.1046/j.1365-2958.2002.03014.x/epdf.

- **Barrangou R, Marraffini LA.** 2014. CRISPR-Cas systems: prokaryotes upgrade to adaptive immunity. *Mol Cell* **54**:234–244. http://www.ncbi.nlm.nih.gov/pmc/articles/PMC4025954/. doi: 10.1016/j.molcel.2014.03.011.

Supplemental Activities

CHECK YOUR UNDERSTANDING

1. **Prokaryotic genetic exchange . . .**

    a. is dependent on reproduction.

    b. occurs at high frequency in nature.

    c. involves several mechanisms.

    d. is called horizontal gene transfer.

    e. often involves the transfer of large segments of DNA.

2. **During transformation . . .**

    a. dsDNA is taken up by the cell.

    b. dsDNA is recombined in the chromosome.

    c. ssDNA is taken up by the cell.

    d. ssDNA is recombined in the chromosome.

3. **Summarize the advantages and disadvantages of using chemical mutagenesis in mutant studies.**

4. **What would be three possible genotypes and phenotypes resulting from a mutation that renders a bacterial cell unable to grow on the same minimal medium that supported the growth of the wild-type ancestor?**

5. **What are the common elements in all transposons, and what role do they play in transposition?**

DIG DEEPER

1. A bacterial strain that is defective in the *recA* gene is unable to undergo homologous recombination.

    a. Would this strain be able to serve as a donor or a recipient strain for generalized transduction? Why?

    b. Would it be able to receive an F plasmid via conjugation if it is F⁻? Why?

    c. Would it be able to take up exogenous DNA and pass it to the progeny? Why?

2. After generalized transducing phages that were grown on a wild-type bacterium were added to a suspension of 10^8 cells of an auxotrophic mutant and the mixture was spread on a petri dish containing a minimal medium (lacking the nutrient that the mutant requires), several dozen colonies developed. How would you determine whether any of these colonies developed from an abortive transductant?

3. Could a frameshift mutation generate a nonsense mutation? How? Illustrate your answer with a schematic.

4. Read the article "Diverse and abundant antibiotic resistance genes in Chinese swine farms" (Zhu Y-G, Johnson TA, Su J-Q, Qiao M, Guo G-X, Stedtfeld RD, Hashsham SA, Tiedje JM. 2013. *Proc Natl Acad Sci USA* **110:**3435–3440. http://www.ncbi.nlm.nih.gov/pmc/articles/PMC3587239/) to answer the following questions:

   a. What human practices correlated with an increased prevalence of genes that confer resistance to antibiotics and metals?
   b. What molecular data did the scientists collect to evaluate the potential for horizontal gene transmission to contribute to the burden of resistance to antibiotics and metals?
   c. Prepare a schematic or animation that summarizes the main findings.
   d. How does this study contribute to public policy discussions to evaluate the benefits and costs of industrial farming practices?

5. In their study "Efficient gene transfer in bacterial cell chains" (*mBio* **2:**e00027-11, 2011; http://mbio.asm.org/content/2/2/e00027-11.full), Babic and colleagues visualized transfer of a conjugative transposon between *Bacillus subtilis* donor and recipient cells by time-lapse fluorescence microscopy (figure). Read the paper to answer the following questions:

   a. How did they engineer red donor and green recipient cells?
   b. What do the green foci indicate?
   c. What mechanisms might explain the appearance of multiple green foci in one cell?
   d. What approach did the scientists use to verify that the transfer occurred by conjugation?

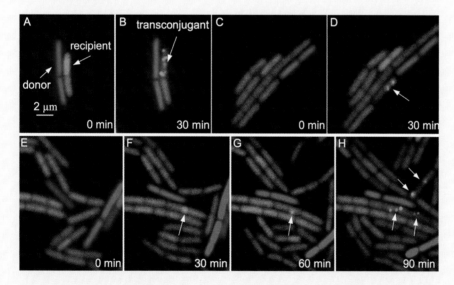

(From Babic A et al., *mBio* **2:**e00027-11, 2011, doi: 10.1128/mBio.00027-11.)

6. **Class debate:** This Week in Parasitology podcast #100 (http://www.microbeworld.org/podcasts/this-week-in-parasitism/archives/2051-twip-100) discusses the application of CRISPR-Cas9 technology to control the mosquito species that spreads the parasitic disease malaria described by Grantz et al. (*PNAS* **112:**E6736-E6743, 2015, http://www.ncbi.nlm.nih.gov/pmc/articles/PMC4679060/)

Research both the promise and the risks of this approach to protecting public health. Organize a mock public hearing to develop policy recommendations. What constituencies should be invited? What other infectious diseases could be managed by this approach?

Supplemental Resources

VISUAL

- Learn about replica plating and other plating methods used in microbiology from video (18 min): http://www.jove.com/video/3064/aseptic-laboratory-techniques-plating-methods
- Artificial transformation made easy in "The Mechanism of Transformation with Competent Cells" (1:41 min): https://www.youtube.com/watch?v=7Ul9RVYG5CM.
- Steps in Cloning a Gene (1:42 min): https://www.youtube.com/watch?v=juP6iHMIYkE
- Replica plating tutorial (1:51 min): https://www.youtube.com/watch?v=zDm9RaOUHcl
- 2013 *FASEB Stand Up for Science* award-winning video (4 min) "Funding Basic Science to Revolutionize Medicine" highlights that studying how microbes protect themselves from viruses has contributed to modern medicine, including diabetes treatment: https://www.youtube.com/watch?v=GmhD-RWNL6c

WRITTEN

- *The Genomics of Disease-Causing Organisms: Mapping a Strategy for Discovery and Defense*, a 2004 colloquium report from the American Academy of Microbiology: http://academy.asm.org/index.php/browse-all-reports/515-the-genomics-of-disease-causing-organism-mapping-a-strategy-for-discovery-and-defense
- *Microbial Evolution*, a 2011 colloquium report from the American Academy of Microbiology: http://academy.asm.org/index.php/browse-all-reports/508-microbial-evolution
- Sean Kearney ponders how frequently microbes exchange genes when living on a host in the *Small Things Considered* blog post "Life on a Pig's Skin" (http://schaechter.asmblog.org/schaechter/2014/11/life-on-a-pigs-skin.html).
- Moselio Schaechter discusses how recombination allows survival at high temperatures in the *Small Things Considered* blog post "Sex to the Rescue" (http://schaechter.asmblog.org/schaechter/2012/03/sex-to-the-rescue.html).
- Jeffrey Morris discusses how genes spread in a microbial population in the *Small Things Considered* blog post "Sex and Evolution, Vibrio Style" (http://schaechter.asmblog.org/schaechter/2015/10/sex-and-evolution-vibrioem-style.html).

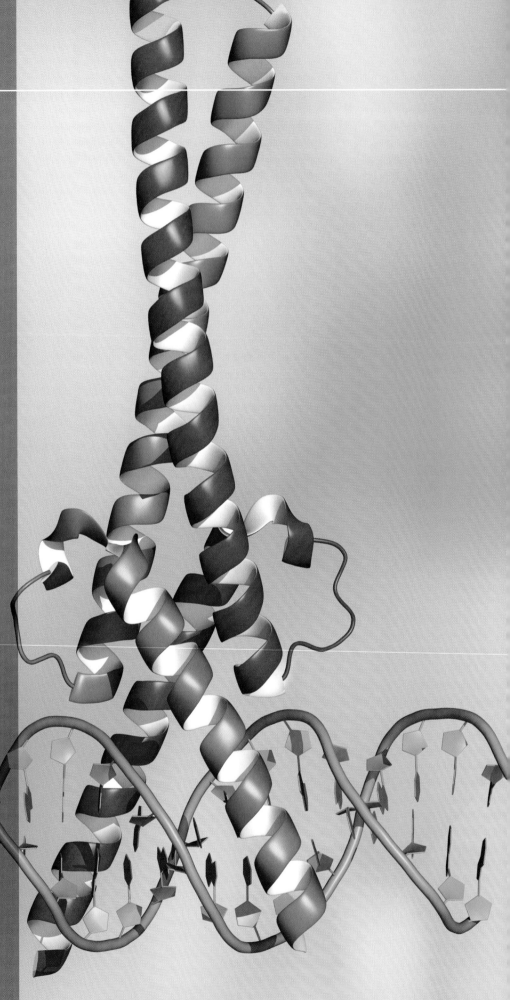

Schematic of transcription factor bound to DNA

CHAPTER ELEVEN

Coordination of Cell Processes

KEY CONCEPTS

This chapter covers the following topics in the ASM Fundamental Statements.

EVOLUTION

1. Cells, organelles (e.g., mitochondria and chloroplasts), and all major metabolic pathways evolved from early prokaryotic cells.

METABOLIC PATHWAYS

2. The survival and growth of any microorganism in a given environment depend on its metabolic characteristics.

INFORMATION FLOW AND GENETICS

3. The regulation of gene expression is influenced by external and internal molecular cues and/or signals.

INTRODUCTION

Our earlier chapters on bacterial metabolism and growth noted that some 2,000 chemical reactions drive the assembly and life of a bacterium. We introduced the notion of phases of metabolism (fueling, biosynthesis, polymerization, and assembly), discussed the organization of metabolic pathways, and described the complex overall sequence of biochemical events that generate a living cell from simple organic or even completely inorganic substrates.

But that is not all it takes to make a living cell. If each of these 2,000 chemical reactions acted independently, they would produce not a cell, but chaos! In this chapter, we will examine how independent chemical reactions are linked into a single, coordinated network—an integrated circuit of enormous complexity and elegance, surpassing the highest achievements of human engineers. Without such control mechanisms, microbes could not respond to changes in the environment appropriately, and many, perhaps all, environments on Earth would never have been colonized.

CASE: How can *Escherichia coli* selectively use sugars from a mixture?

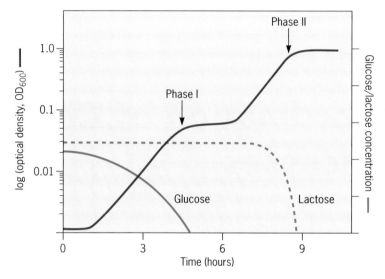

Many heterotrophic bacteria have the remarkable ability to use glucose preferentially over any other sugar. Student JM demonstrated this by growing *E. coli* in mineral medium supplemented with a mix of glucose and lactose (a disaccharide of glucose and galactose). He monitored growth using a spectrophotometer to quantify the increase in optical density of the culture and plotted the data in a semi-log plot. He also measured the concentration of glucose and lactose in the medium throughout the incubation period and plotted the consumption of these sugars on the secondary y axis. He was puzzled when he observed two phases of growth (known as diauxic growth): the first phase correlated with depletion of glucose and the second phase with lactose consumption.

Questions arising:

- How did *E. coli* cells selectively use glucose first?
- How did they switch from growing with glucose to growing with lactose?

WHAT EVIDENCE SHOWS THAT METABOLIC REACTIONS ARE COORDINATED?

LEARNING OUTCOMES:
- Explain the need for coordination of metabolic reactions within a cell.
- Summarize a set of experiments that demonstrates coordination of biosynthesis in a bacterium.
- Speculate about at least one macromolecule (or group of macromolecules) that will be in greater abundance for each phase of bacterial culture growth and undergo coordination of polymerization.

As student JM observed experimentally in the case study, microbes are remarkably attuned to their environment, so much so that they selectively use one sugar over another to support their growth. To accomplish this feat, microbes have mechanisms to sense environmental changes and to respond by rapidly reprogramming their metabolic networks. In the case study, the cells detected glucose and lactose and responded by using glucose first, then lactose. How? Below, we describe a few experiments that provide concrete evidence of the cell's ability to coordinate its working parts and become a single integrated system. Coordination can be demonstrated in each of the major stages of metabolism. For simplicity, we shall begin with biosynthesis.

Coordination in Biosynthesis

Imagine that you prepare a medium consisting of a mixture of inorganic salts and radioactive ^{14}C-labeled glycerol as the sole source of carbon and energy (just enough to support the growth of a bacterial species such as *E. coli* to a density of 10^8 cells per ml). Inoculate the medium with a few cells, and incubate it aerobically. As the culture grows, the cells take up radioactively labeled glycerol, metabolize it via their oxidative reactions, and produce carbon dioxide, which is now labeled. You would also find some radioactive carbon in the cells' building blocks and macromolecules. Indeed, the total protein of the cells would be labeled. Furthermore, if you were to analyze specific amino acid residues in the protein fraction, you would find that they are all labeled as well. Consider the implications of this finding: carbon from the substrate (glycerol) flowed through each metabolic pathway in a coordinated fashion to make amino acids and, from them, proteins. Rather neat and simple.

But here is the interesting part. Repeat this experiment, but now include in the medium not only the labeled glycerol but also an unlabeled amino acid, for example, histidine. You would discover that, as before, the radioactive carbon flowed from the glycerol substrate to carbon dioxide and to the cells' proteins. However, the histidine residues, alone among all the amino acids, *will contain almost no trace of radioactivity.* You can deduce that the cells' histidine derived almost exclusively from the nonradioactive histidine supplied in the medium instead of the radioactive glycerol substrate. In fact, if you were to analyze any of the intermediates of the histidine biosynthetic pathway, you would find that they are also unlabeled. Thus, the cell *turned off the entire pathway* leading to the synthesis of histidine from glycerol. This experiment could be repeated with virtually any of the building blocks, vitamins, or cofactors with the same result. *The presence of any one compound in the medium stops endogenous synthesis of that compound.* Moreover, the same pattern would be observed if the experiment was performed under different temperature, pH, oxygen tension, carbon source, mixture of nutrients, and so on. In all cases, the cells stop making a particular compound when it is readily available in the medium. Remarkable!

Coordination in Fueling

Does coordination exist in fueling reactions as well? The answer is a big yes. Fueling reactions provide the precursor metabolites, reducing power (mainly in the form of NADH), and energy (ATP) needed for the synthesis of building blocks and macromolecules. The cell adjusts the ratios of the products of fueling (precursor metabolites, energy, and reducing power) to satisfy its needs. If precursor metabolites are provided in the medium, the cell coordinates the fueling reactions to generate only the products of fueling that it needs. Let us return to our glycerol experiment, but this time we will grow *E. coli* in a rich medium that supplies most of the amino acids. Would you detect radioactive carbon from glycerol in the protein fraction? Only a little, because the cells no longer need to use the glycerol to make most of the amino acid building blocks. Instead, the cells reroute the glycerol through the fueling pathways to generate an adequate supply of reducing power and ATP to satisfy their biosynthetic needs.

Coordination in Macromolecular Composition

Recall from chapter 4 that, in general, cells maintain their macromolecular composition despite differences in their growth medium. Cells growing on a mixture of amino acids are not richer in total protein; those growing on fatty acids are not richer in lipid content; those growing on nucleosides are not richer in nucleic acids; those growing on sugars are not richer in carbohydrates. On the other hand, the *rate* of growth does affect a cell's composition. Cells growing fast, as when supplied with amino acids, are richer in RNA than the same cells growing more slowly in a minimal medium. Therefore, regulatory circuits must sense the cell's potential to attain a particular growth rate and control the synthesis of each macromolecule.

In summary, all the major fueling and biosynthetic pathways are subject to powerful controls that bring order out of the potential chaos of a system with thousands of individual working parts. These examples demonstrate that metabolic reactions are coordinated. The question now becomes, how?

HOW DO MICROBES REGULATE THEIR METABOLISM?

LEARNING OUTCOMES:
- List four reasons why prokaryotes more often control the amounts of enzymes produced rather than controlling enzymatic activity (as is more often seen in eukaryotes).
- Define "allostery."

If you wanted to design a highly coordinated cell, you might first think about the possible ways the rates of metabolic reactions can be regulated. Consider the following metabolic reaction (where S_1 and S_2 are substrates and P_1 and P_2 are products) catalyzed by an enzyme:

$$\rightarrow S_1 + S_2 \xrightarrow{Enzyme} P_1 + P_2 \rightarrow$$

The rate of the reaction can be controlled by one of three means: (i) varying the activity of the enzyme, (ii) changing the amount of the enzyme, or (iii) varying the intracellular concentration of one or both substrates (or products). Of the three, which would you choose? Microbial cells actually use all these options. However, as reactions in the cell are coupled, the third mechanism—varying the concentration of substrate(s) for a reaction—is

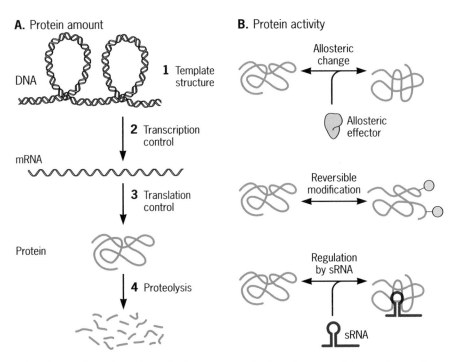

A. Protein amount

DNA

1 Template
structure

2 Transcription
control

mRNA

3 Translation
control

Protein

4 Proteolysis

B. Protein activity

Allosteric
change

Allosteric
effector

Reversible
modification

Regulation
by sRNA

sRNA

FIGURE 11.1 **Overview of metabolic regulatory devices.** Synopsis of the two forms of control that operate to coordinate metabolism: regulation of the amount of a protein in the cell (A) and regulation of its activity (B).

brought about by one of the first two mechanisms. Thus, the primary basis of metabolic coordination is varying either the *amounts* or the *activities* of enzymes (Fig. 11.1).

Controlling Enzyme Amounts

The simplest and neatest trick to control the rates of a metabolic reaction is to modulate the amounts of enzymes that are made. How can this be done? *Any step in the flow of information*, from DNA to mRNA to protein, can be controlled to regulate the rates of synthesis (Fig. 11.1A). For example, the expression of a gene can be controlled by changes in DNA structure such as supercoiling (chapter 3), or it can be controlled at the level of transcription (a major hot spot of regulation for prokaryotes) or translation (chapter 8). Interestingly, controlling the *amounts of enzymes* produced is much more common in prokaryotes than in cells of plants and animals, which often control *enzyme activity*. So, in what ways do prokaryotes differ from cells of higher organisms? Here are four key microbial themes.

- First, microbes, particularly free-living ones, are often *substrate limited*. Microbes cannot afford to make useless or redundant enzymes (remember the energy investment required? [chapter 8]), nor can they survive without producing essential enzymes. To meet both challenges, they rapidly adjust enzyme synthesis.
- Second, coupled transcription and translation allows prokaryotes to *adjust quickly* the expression of genes encoding an enzyme to control its availability. Upregulation can start within seconds and can quickly increase the amount of an enzyme by 1,000-fold or more. Downregula-

tion also starts right away. If transcription of an enzyme-encoding gene is turned off in a growing culture, the amount of that enzyme will halve every time the cell divides. If the generation time is 20 minutes, the enzyme pool can be reduced 8-fold in an hour.

- Third, prokaryotic mRNAs are short-lived. Thus, the pool of mRNA transcripts available for protein synthesis can be completely renewed every few minutes—a marvelous opportunity to control enzyme synthesis rapidly.

- Fourth, prokaryotic genomes organize genes of related functions in multicistronic transcriptional units called **operons** (chapter 8). This facilitates the coordinated control of functionally related enzymes (e.g., enzymes involved in the same metabolic pathway). By adjusting the synthesis of just one long mRNA, the cell can change the rate of synthesis of an entire pathway.

Controlling Enzyme Activity

Controls on protein activity, though less common, are also used by prokaryotes to modulate the rates of a metabolic reaction and to coordinate some cellular processes (Fig. 11.1B). As we discuss in more detail below, activity can be positively or negatively modified in a reversible manner by *covalent modification* (phosphorylation is most common). In other cases, activity is modulated by **allostery**, a process that involves the *reversible association of the protein with another molecule*. Compared to covalent modification, allostery is far more pervasive in microbial regulation.

If reactions can be regulated equally well by controlling the amounts or the activities of enzymes, why do microbes employ both mechanisms? To examine this key question of cell physiology, we need more information about each mode of regulation. Then the benefit of having two modes of regulation will become clear.

HOW IS PROTEIN ACTIVITY MODULATED?

LEARNING OUTCOMES:
- Differentiate covalent modification and allosteric interactions as ways of controlling protein activity.
- Use the histidine biosynthetic pathway as an example to explain how feedback inhibition is an effective means to control biosynthetic pathways.
- Describe how the ratio of the concentration of ATP to those of ADP and AMP is maintained almost constant in bacterial cells over a wide range of energy supply and demand.

Modulation of enzyme activity occurs rapidly, almost instantaneously, adjusting metabolic flow in the microbial cell in a fraction of a second. Using the overview of regulatory devices shown in Fig. 11.1B, we next discuss the two strategies microbes use to regulate protein activity.

Covalent Modification

Both eukaryotes and prokaryotes control enzyme activity using covalent modifications. Some of these (phosphorylation, in particular) are prevalent in bacterial physiology; beautiful examples are motility, chemotaxis, and sensory signal transduction, in which cascades of proteins are activated via phosphorylation (see chapter 12). Other modifications, such as addition of adenylyl, acetyl, methyl, and other residues, are rarer, but the cellular processes they regulate are no less important (for example, adenylation regulates glutamine synthetase, a key enzyme in nitrogen assimilation).

Allosteric Interactions

The most prevalent mode of controlling the activities of enzymes (and other proteins) occurs via **allosteric** interactions. Allostery involves a change in conformation or shape and, therefore, activity upon binding a molecule called an **allosteric effector** (Fig. 11.1B). The effector can be a metabolite (**ligand**) or another protein or RNA (**modulator**), but the result is the same: upon binding the allosteric effector, the protein's conformation is changed and, consequently, so is its activity. Remember two features of allosteric changes: (i) the regulatory site at which the allosteric effector binds is separate from the enzyme's catalytic site, and (ii) the effector need bear no steric resemblance to the substrates (or products) of the enzyme. This is worth emphasizing: allostery provides a means for modifying the activity of an enzyme by substances not even remotely resembling the substrates or products. Therefore, allostery differs fundamentally from competitive inhibition, in which a substance resembling the substrate competes for binding to the active site of the enzymes.

Can you imagine how allostery works? *Allosteric enzymes can exist in at least two conformations*, one with high and another with low activity. The binding of the effector favors one conformation over the other. In some cases, what is changed is the velocity (V_{max}, the maximum rate of reaction) of the enzyme reaction; in others, it is affinity for the substrate (K_m, the Michaelis constant). Allosteric effectors may either increase or decrease the activity of the enzyme, and some enzymes even have both positive and negative allosteric effectors.

Allostery in biosynthesis. Earlier we noted that, given the choice, a bacterium will utilize a building block supplied in the medium rather than make its own (in our example, histidine). Moreover, recall that by tracing the radioactively labeled glycerol, we learned that flow through the histidine biosynthetic pathway ceases just seconds after histidine is supplied. This calls for a rapid mechanism, right? Indeed, the first enzyme in the histidine biosynthetic pathway is allosteric, and histidine is its negative allosteric effector. By this elegant mechanism, exogenous histidine immediately shuts off the entire biosynthetic pathway at its very first step.

In fact, a more dramatic version of this process takes place continuously in the cell to regulate histidine biosynthesis. The internal concentration of histidine, whether made by the cell or imported from the environment, determines the rate of flow from precursor metabolites to end product. And realize that histidine is but one example, not an exception: virtually

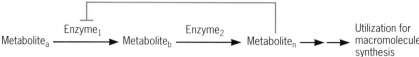

FIGURE 11.2 **Feedback inhibition.** The final product of a series of enzymatic reactions, metabolite$_n$, has the property of binding to the regulatory site of the allosteric protein, enzyme$_1$, and thereby inhibiting it. A scheme of this sort can guarantee that metabolite$_n$ is produced only as rapidly as it is used in some subsequent process, such as macromolecule synthesis.

A Isofunctional enzymes

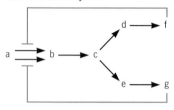

B Cumulative feedback inhibition

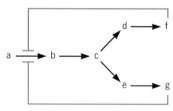

C Sequential feedback inhibition

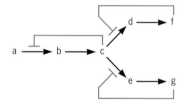

D Inhibition plus activation

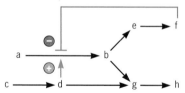

FIGURE 11.3 **Patterns of feedback inhibition found in bacterial biosynthetic pathways.** (A) Isofunctional enzymes for the regulated reaction allow differential feedback effects by the two pathway end products. (B) Cumulative feedback inhibition involves multiple allosteric sites on the regulated enzyme, ensuring that there will be some activity unless all the end products are in excess. (C) Sequential feedback inhibition involves different end products operating separately on the various branches of the pathway. (D) Inhibition plus activation uses both positive (+) and negative (−) allosteric effectors to coordinate complex pathways.

all *building blocks control their own synthesi*s by acting as negative allosteric effectors of the first enzyme in their biosynthetic pathway. This control process is called **feedback inhibition** (or **end-product inhibition**). Since each building block is consumed by macromolecule synthesis, biosynthetic pathways work by *demand feeding*: they produce their end products at the same rate that macromolecular synthesis depletes them. Figure 11.2 summarizes general features of feedback inhibition in biosynthetic pathways. Feedback inhibition can also be adapted to accommodate the branching of biosynthetic pathways and to interact with other pathways (Fig. 11.3).

Allostery in fueling. Allosteric inhibition and activation also regulate the flow of metabolites through fueling pathways (Fig. 11.4). Here, the simple device of end-product control of the first or an early step in a pathway cannot apply. As we learned in chapter 7, the pathways of formation of the 13 precursor metabolites are so interrelated (some pathways are cyclic) that *there are no unique early steps*. A cell metabolizing glucose may produce malate in the tricarboxylic acid (TCA) cycle, one of the last reactions in glucose fueling (Fig. 11.4). However, if malate is provided as a sole carbon and energy source, the same cell may reverse the metabolic flow to generate the precursor metabolites from malate (chapter 7). Thus, controls work internally in each of the main fueling pathways, and both positive and negative allosteric interactions abound (Fig. 11.4).

The rates of formation of ATP and reducing power (NAD[P]H) must also be regulated during fueling. As an example, consider how tightly ATP levels are regulated. Because ATP synthesis and utilization involve a cyclic flow through ADP and AMP, it makes sense that all three adenylates play regulatory roles in fueling reactions (as well as in biosynthetic pathways) (Fig. 11.4). A useful index of a cell's energy status is the **energy charge** of the cell, defined as follows:

$$\text{Energy charge} = \frac{[\text{ATP}] + 1/2[\text{ADP}])}{[\text{ATP}] + [\text{ADP}] + [\text{AMP}]}$$

Mathematically, the energy charge of the cell could vary from 0 (all AMP) to 1 (all ATP), but in fact, the energy charge of prokaryotes under normal conditions is held narrowly between 0.87 and 0.95. Starving a cell for a very long time reduces its energy charge slowly; when it reaches approximately 0.5, the cell is dead. In general, ATP-replenishing pathways (fueling) are inhibited by high levels of energy charge, and ATP-utilizing pathways (biosynthesis and polymerization) are stimulated.

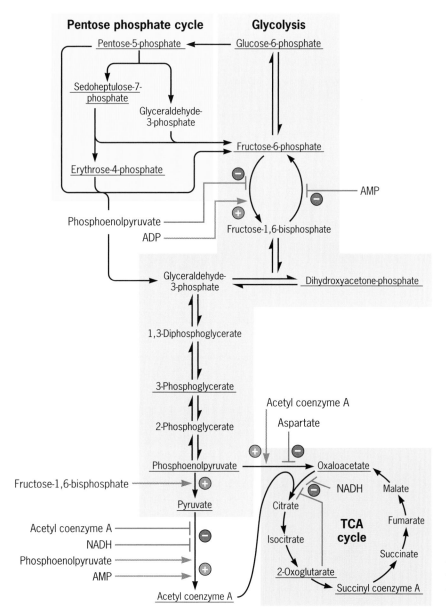

FIGURE 11.4 **Central pathways of fueling reactions showing some of the allosterically controlled steps.** −, negative allosteric effector; +, positive allosteric effector.

Allostery by regulatory RNAs. Prokaryotic cells also use RNA molecules as allosteric effectors to modulate the activity of numerous proteins. Typically small in size, these so-called small RNAs (**sRNA**s) fold into a looped structure containing binding sites for specific proteins. Protein-RNA pairs can inhibit, activate, or modify the protein activity or even tether proteins together to modulate their activities collectively (Fig. 11.5). The 6S RNA of *E. coli*, for example, regulates entry in stationary phase by modifying the activity of the RNA polymerase (RNAP) holoenzyme containing the housekeeping sigma factor σ^{70}. Once bound to 6S RNA, the RNAP-σ^{70} holoenzyme can no longer bind its usual promoters, and genes otherwise activated during exponential phase are repressed. But, as we will discuss later in the chapter, allosteric regulation is but one of the many ways in which sRNAs coordinate processes in the cell.

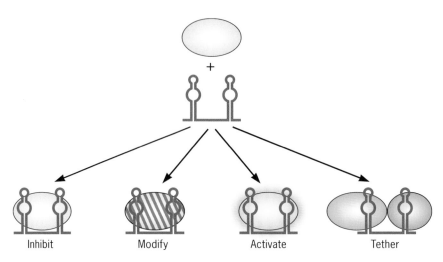

Inhibit   Modify   Activate   Tether

FIGURE 11.5 **Control of protein activity by regulatory sRNAs.** Upon binding the target proteins, regulatory sRNAs can affect their activity in various ways. Some are inhibited (as in the CsrB-CsrA system), some are modified in their activity, and some are activated. In some cases, the same sRNA may bring (tether) two different proteins together, influencing their activities through their association.

HOW ARE PROTEIN AMOUNTS MODULATED?

LEARNING OUTCOMES:
- Describe two ways in which gene expression can be monitored under various growth conditions.
- State how the regulation of gene expression in the *lac* operon is under the influence of external and internal molecular cues/signals.
- Identify the molecules that serve as the repressor and inducer in the *lac* operon.
- List three central conditional properties of operons.
- Speculate on reasons why microbes have such a large diversity of regulatory processes that control the synthesis of proteins.
- Describe six ways in which microbes control the initiation of transcription as a means of regulating gene expression.
- Identify examples of protein-RNA interactions that regulate gene expression in a cell.
- Explain how the structure of the *trp* operon mRNA contributes to controlling gene expression for enzymes involved in biosynthesis of tryptophan through attenuation.
- Give two reasons for the evolution of controls that govern sets of genes or operons together as regulatory units (global regulation).
- Relate operons, regulons, and modulons.
- Use the catabolite repression modulon as an example to identify a reason why some modulons are considered global regulatory systems.
- Explain how an *E. coli* cell uses the catabolite repression system to repress transcription of the *lac* operon in the presence of both glucose and lactose.
- Summarize the role of ppGpp in the stringent response modulon.

Even if all the processes described above were functioning optimally, they would not prevent the synthesis of unneeded proteins. However, such waste does not occur, because *enzyme amounts can be controlled by regulating gene expression*. In the mid-20th century, intensive efforts focused on discovering how bacteria change the rates of synthesis of their enzymes. The initial studies were done with *E. coli* and related enteric bacteria, but all bacteria (and archaea) are extremely adept at turning genes on and off, thereby raising and lowering the levels of enzymes by controlling the expression of the encoding genes.

Evidence that flexibility of gene expression is widespread in both bacteria and archaea can be quickly obtained using two tools in wide use today: *proteomics* and *transcriptomics*. Monitoring the proteome allows one to survey the rates of synthesis and the amounts of most proteins of the cell; transcriptional monitoring tells the profile of all or specific mRNA molecules in the cell. Many different species of prokaryotes radically change their protein and mRNA profiles in response to different growth conditions. In many (but not all) instances, the modulation of protein synthesis occurs at the transcriptional level (that is, by changes in the synthesis of mRNA). But, as we will discuss below, translational control can also be exerted to adjust the amount of proteins that are made.

Gene regulation is an epic tale, best told in two parts. In the first, we ask how individual genes and operons are controlled. In the second part, we examine how sets of genes and operons are coordinately regulated and tied into a cellular regulatory network, a regulatory process that integrates so many gene targets that it is often referred to as **global regulation**.

Regulation of Gene Expression

Much of what we know about the regulation of gene expression comes from pioneering studies of bacterial operons. As we have noted repeatedly, the operon is a hallmark of the prokaryotic cell. It is a unit of transcription: a DNA segment containing a **promoter**, two or more (typically more) genes encoding polypeptides (**cistrons**), and a **transcription terminator**. Thus, all of the genes in an operon are transcribed simultaneously in *a single mRNA molecule*. A historic picture of the regulation of an operon, as first proposed by F. Jacob and J. Monod in 1961 for the genes encoding the enzymes metabolizing lactose in *E. coli*, designated *lac*, is shown in Fig. 11.6. The *lac* operon contains the three genes (*lacZ, lacY,* and *lacA*) encoding all of the proteins needed to metabolize lactose: LacZ (β-galactosidase, the enzyme that hydrolyzes lactose), LacY (a membrane protein, or permease, that transports lactose into the cell), and LacA (a transacetylase of uncertain function). *Upstream* of the operon is the promoter, which is the binding site for RNAP. The promoter overlaps with a control region, or **operator**, which binds an allosteric protein called LacI (the product of the nearby gene *lacI*). LacI is produced constitutively and thus is always available for binding the operator. In the absence of lactose (Fig. 11.6A), LacI binds the operator and prevents RNAP from transcribing the operon (either by preventing it from binding or by blocking its movement). Thus, LacI is a negative regulator, or **repressor**, of the *lac* operon: when bound to its operator, none of the three enzymes encoded by the operon is made. Occasionally, the repressor dissociates from the operator and a few transcript molecules are made, ensuring that the cell will have a low basal level of the three products of the operon.

Why does the cell maintain a small amount of the Lac proteins? Once lactose becomes available, these basal protein levels equip the cell to take

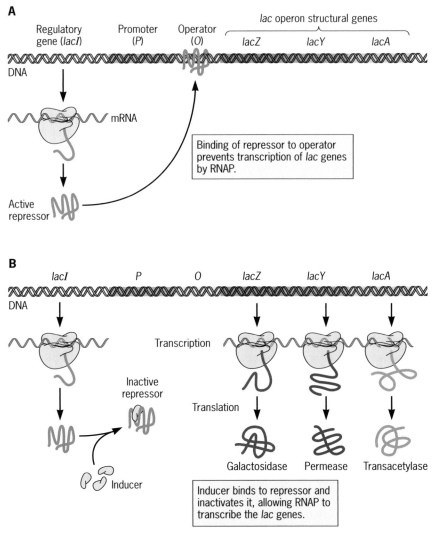

FIGURE 11.6 Original operon model of Jacob and Monod, proposed in 1961 for the regulation of the *lac* genes of *E. coli*. (A) Uninduced state of the *lac* operon in cells growing on substrates other than lactose; (B) induced state during growth on lactose. See the text for an explanation.

up and metabolize some of this sugar, an event that is sufficient to activate the *lac* operon (Fig. 11.6B). Lactose is brought into the cells by LacY permease and is converted into a metabolite called **allolactose** by the β-galactosidase LacZ. This derivative of lactose is the actual **inducer** of the *lac* operon: it *binds the LacI repressor, triggering a conformational change that releases this allosteric regulatory protein from its promoter.* RNAP is now free to make *lacZYA* transcripts. More of the operon products are made, and the cells bring more lactose in. But, you may be wondering, if lactose is all the cells need to make the inducer that removes the repressor of the *lac* operon, why did *E. coli* grow with lactose *only after glucose was consumed* in the case study? Later in the chapter, we will learn the answer: catabolite repression.

Research on the *lac* operon of enteric bacteria generated a number of fundamental concepts of the operon, including the following.

1. Initiation of transcription from the promoter of an operon can be a site of regulation.
2. Initiation of transcription can be controlled by allosteric proteins (i.e., the protein's activity is governed by the binding of specific ligands).
3. Increase in the expression of an operon can be triggered by relief of a negative control.

The *lac* operon is one elegant example of how microbes regulate a metabolic adaptation, but it is not the only mechanism. Microbes exhibit a dazzling diversity of solutions to metabolic and physiologic challenges. In general, operons differ as to (i) the site at which control is exerted, (ii) the mode of control (positive or negative), and (iii) the particular molecular device used to bring about the regulation. Some operons even employ more than one promoter; positioning a second regulatory site within the operon provides greater flexibility and sensitivity of regulation. Every conceivable step that leads from a gene to its finished protein product has been singled out for control in some gene or operon or in some bacterium (Fig. 11.7). Furthermore, some of these regulated steps may involve multiple molecular mechanisms. To examine some tricks that have evolved, we will use a few well-studied examples. To guide the discussion, we will

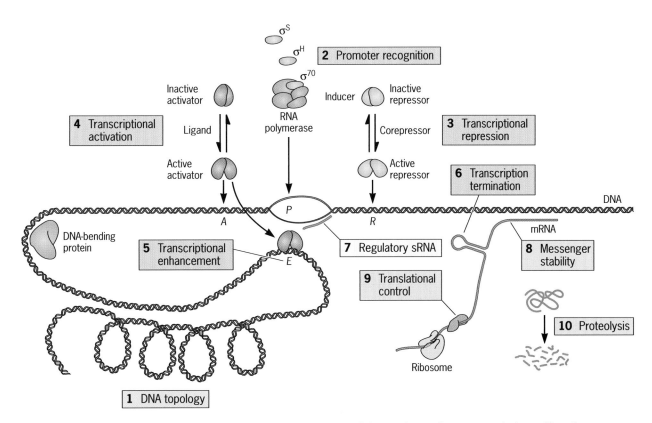

FIGURE 11.7 Some of the many regulatory processes that can control the synthesis of proteins and, thus, affect their amounts in the cell. The boxes show various regulatory processes: those acting on the DNA (1, blue) and during (2 to 6, green) or after (8 to 10, red) transcription. Regulatory sRNAs (7, tan) can modulate transcription or translation.

refer to the various regulatory mechanisms by the numbers (1 to 10) following the order of regulatory events shown in Fig. 11.7.

Transcriptional control. Transcription is a major regulatory hub in prokaryotes and transcription initiation, the most commonly controlled step. By preventing the synthesis of the mRNA transcript, neither the encoding RNA nor the protein is made; thus resources are not wasted. How is this accomplished? Recall that transcription initiation involves the binding of the σ-RNAP holoenzyme to a promoter and the subsequent opening of the DNA in this region (chapter 8). Thus, any regulatory mechanism that controls any of these early events in transcription can be targeted to control the frequency of initiation. And though not as prevalent, transcription termination can also be regulated. Here we examine some of the most relevant steps of control of transcription initiation (steps 1 to 5 in Fig. 11.7) and regulated transcription termination (step 6 in Fig. 11.7). We will also describe regulation of both by sRNAs (step 7 in Fig. 11.7).

1. **DNA topology:** An effective way to control the initiation of transcription is by changing the DNA topology to either prevent or promote σ-RNAP binding to the promoter. Most promoters are, for example, sensitive to the degree to which the DNA is *supercoiled* or *packed* (step 1 in Fig. 11.7). In fact, cells devote many proteins to the control of the topology of their DNA.

2. **Promoter recognition:** A second way to control the initiation of transcription from certain promoters is by interfering with promoter recognition (step 2 in Fig. 11.7). Recall that binding to the σ factor allows RNAP to recognize and bind to the −10 and −35 promoter sequences and locally melt the promoter to initiate transcription (chapter 8). Bacteria control promoter recognition, producing *alternative σ factors,* which when available compete with the housekeeping sigma factor σ^{70} and redirect the σ-RNAP holoenzyme to specific genes in the chromosome. This is an especially effective mechanism to coordinate the expression of numerous genes scattered across the genome in a single step. We will see other examples of global reprogramming later in the chapter and in the next chapters, when discussing the activation of the heat shock response and entry into stationary phase (chapter 12) and spore formation (chapter 13).

3, 4, and 5. **Transcriptional repression, activation, and enhancement:** Initiation of transcription can also be activated, repressed, or enhanced (steps 3, 4, and 5 in Fig. 11.7) in regulatory events involving sequences called **DNA control regions**. Some of these control regions are adjacent to or overlap the −10 and −35 promoter sequences (as the operator in the *lac* operon), but others are dozens of nucleotides upstream of the promoter or downstream of the operon terminator.

In most cases, these control regions are the binding sites for allosteric regulatory proteins. Each binding site is called a **box**, and many are named after the protein they bind (such as the CRP box) or the process they regulate (such as the nitrogen box). As we saw for the *lac* operon, some regulatory proteins called repressors bind to DNA control regions (operators) and decrease the frequency of transcription initiation (step 3 in Fig. 11.7). By contrast, **activators** are positive regulators; that is, they increase the frequency of transcription initiation (step 4 in Fig. 11.7). Regulatory proteins can also bind control regions, called **enhancers**, that are distant from the gene or operon they regulate. To exert their control, the enhancer-binding proteins often loop the DNA, sometimes with help from DNA-bending proteins, to get closer to the promoter region and promote initiation (step 5 in Fig. 11.7).

6. **Transcription termination:** Bacteria also regulate the termination of transcription to control the expression of specific genes. Some bacteria do so by a mechanism called **attenuation**, whereby the majority of transcripts are *prematurely terminated* shortly after initiation. The best-studied example of attenuation (and the system in which it was first discovered) is the *E. coli trp* operon, encoding the enzymes for tryptophan biosynthesis. A 5′ segment of the *trp* mRNA (or leader sequence) that precedes the *trp* operon structural genes can adopt two configurations in the nascent mRNA: one that causes premature transcription termination and another that does not (Fig. 11.8). The speed of translation dictates which configuration this leader mRNA assumes. This, in turn, depends on how fast the ribosomes are at translating two consecutive tryptophan codons in the leader sequence, which serve as "sensing codons" of tryptophan availability.

   - When tryptophan is scarce, the ribosomes stall at these tryptophan codons, waiting for tryptophanyl-tRNAs. The mRNA is then elongated and acquires the antitermination configuration (Fig. 11.8) so the enzymes for making tryptophan are produced.
   - When tryptophan is abundant, the ribosomes translate the sensing codons without delay and prevent the antitermination configuration from forming. The mRNA adopts instead a conformation that pauses the RNAP and then prematurely terminates transcription of the operon.

7. **Regulatory RNAs (sRNAs):** Transcription initiation and termination can also be regulated by sRNA molecules. Recall, for example, the allosteric regulation exerted by the 6S RNA of *E. coli*, which we described earlier. The 6S RNA transcript has a secondary structure that mimics the open conformation of DNA when transcription is initiated (the so-called **open complex**; see chapter 8). As stationary-phase cells produce 6S RNA transcripts, they bind σ^{70}-RNAP holoenzymes, effectively preventing them from interacting with their promoters. As a result, the frequency of initiation of exponential-phase genes is reduced (step 7 in Fig. 11.7). In a few instances, the sRNA molecule base pairs directly with a strand of DNA in the promoter region to prevent transcription from initiating. A more prevalent mode of transcriptional regulation by sRNAs is, however, regulated transcription termination, whereby the sRNA base pairs with a complementary region in the nascent mRNA and acquires a conformation such that a termination loop forms that stops RNAP elongation. As we will discuss later, sRNAs are versatile effector molecules and are also implicated in the control of mRNA stability and translation (steps 8 and 9 in Fig. 11.7).

A

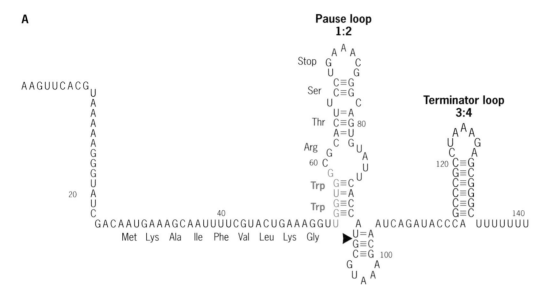

B

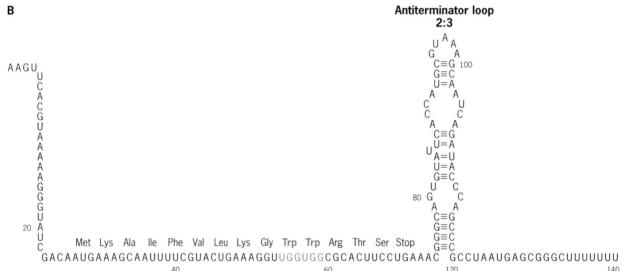

FIGURE 11.8 **Alternative secondary structures of the *trp* leader region of *E. coli*.** The numbers 1, 2, 3, and 4 indicate the RNA segments that form the secondary structures pictured. (A) Termination configuration. The arrowhead indicates the site of transcription pausing. (B) Antitermination configuration.

Posttranscriptional control. Prokaryotic cells also use posttranscriptional mechanisms to control gene expression (steps 8 to 10 in Fig. 11.7), which we highlight next.

8. **mRNA stability:** On average, prokaryotic transcripts are short-lived (chapter 8). By altering the stability of mRNA, a cell can regulate the amount of its protein products (step 8 in Fig. 11.7). For example, upon binding, some sRNAs *trigger the degradation of their target mRNA molecules.*

9. **Translational control:** Initiation of translation can be repressed or activated (step 9 in Fig. 11.7). Noteworthy examples are the operons that encode *ribosomal proteins* (r-proteins). Some of the mRNAs encoding r-proteins contain nucleotide sequences that are recognized and bound

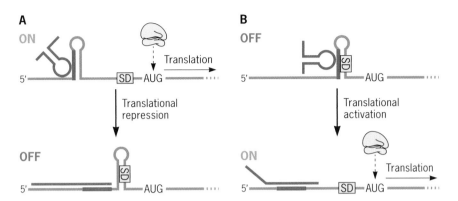

FIGURE 11.9 **Antisense mechanism of translational control by sRNAs.** (A) Repression; (B) activation.

to their own r-protein products when in free form (that is, not yet assembled into ribosomes). Binding of r-proteins to their own mRNAs prevents their translation, resulting in **translational repression**. Consider how clever this mechanism is: the cell balances the rate of r-protein synthesis with the rate of ribosome assembly because the pool of free r-proteins can control the translation of their own mRNAs.

As we mentioned above, translational control can also involve regulatory sRNA molecules. The same base-pairing interactions that direct some sRNAs to bind a nascent mRNA and terminate its transcription also allow them to interfere with mRNAs and affect their translation. For example, binding of an sRNA to its mRNA target can block or expose the **Shine-Dalgarno** sequence in the mRNA to ribosomes, thus preventing or promoting ribosome binding and initiation of translation, respectively (Fig. 11.9). Recognition of target mRNA is often facilitated by **RNA chaperones**, which bind the sRNA molecules, escort them to their targets, and stabilize the sRNA-mRNA interaction. In chapter 12, we will learn more about sRNAs and one well-known chaperone called Hfq.

Regulatory sRNAs are not the only effector molecules that bind mRNAs and control their translation. Metabolites, for example, can also bind target mRNAs with great specificity. Upon binding, the effector molecule changes the structure of the target mRNA to control whether or not they get translated. The binding sequences in the mRNAs, collectively known as **riboswitches**, are commonly found in the 5′ untranslated region of the mRNAs. Translational control is, along with transcriptional termination, the most prevalent regulatory mechanism used by riboswitches to control gene expression. However, some riboswitches can have dual functions, regulating both transcription and translation.

10. **Proteolysis:** Though not high on most prokaryotic cells' lists of regulatory devices, proteolysis plays a definite role (step 10 in Fig. 11.7). Enzyme degradation provides, for example, a means to control a metabolic reaction. More commonly, however, proteolysis is employed to *remove a protein that regulates other proteins.* For example, alternative σ

factors used to reprogram the promoter specificity of the core RNAP under specific conditions are constantly being made and degraded. Proteolysis keeps their cellular level below that of the major σ factor (σ^{70} in *E. coli*) when not needed. When an environmental change calls for a particular σ factor to come into play, *its degradation is stopped*, rapidly raising its concentration. A σ factor can be protected from proteolysis in part by its increased binding to core RNAP. This process is significant in controlling the σ factor that triggers the heat shock response, a topic that we will discuss in the next chapter.

Note that all forms of posttranscriptional regulation we have considered seem wasteful in the sense that mRNA is made but not used. True, but this cost may be offset by other considerations that we know little about. For several reasons, the overall energy budget of using one or another regulatory device is hard to assess. So, the book is not closed on this amazing panoply of regulatory devices. In fact, the broadest question—why there are so many—is wide open (and for you to explore).

Global Regulation

Operons are a sleek way to coordinate the expression of multiple genes in a biosynthetic pathway. But some processes require the coordinate expression of many genes scattered on the genome. We call this global regulation. In the next chapter, we will learn just how important global regulatory networks are for cells to succeed in a fluctuating environment. Here, we will introduce the topic and describe, as an example, two well-known global regulators.

Regulatory units above the operon. Prokaryotes have devised multiple ways to integrate individual genes and operons into *coordinated units*. One simple strategy is to *control more than one transcriptional unit with the same regulatory protein*. How? Place the regulator's DNA control region, or box, in front of each of these genes and operons! Enteric bacteria use this regulatory device in their SOS, oxidation damage, and anaerobic electron transport systems. A second strategy to coordinate multiple genes and operons, which we discussed earlier, is to reprogram RNAP using an *alternative σ factor* that recognizes each of the promoters of member operons. The heat shock (chapter 12) and sporulation systems (chapter 13) of various species employ this regulatory design. But σ factors are not the only modulators of RNAP activity. For one of the largest networks, the **stringent control system**, the member operons are regulated by the nucleotide **guanosine tetraphosphate** (**ppGpp**), which binds and reprograms RNAP to regulate multiple cellular pathways. We discuss this interesting control mechanism below.

How do we name regulatory units above the operon? The group of independent genes and operons *governed by the same regulator* comprise a **regulon** (Fig. 11.10). For example, the *arg* regulon of *E. coli* includes operons scattered around the chromosome that together encode all the enzymes of the arginine biosynthetic pathway. We also use "regulon" to refer to the genes and operons regulated by a particular σ factor (e.g., the RpoS or σ^S regulon). Above regulons are **modulons** (Fig. 11.10). A modulon is a set of different regulons that display similar regulatory responses to a partic-

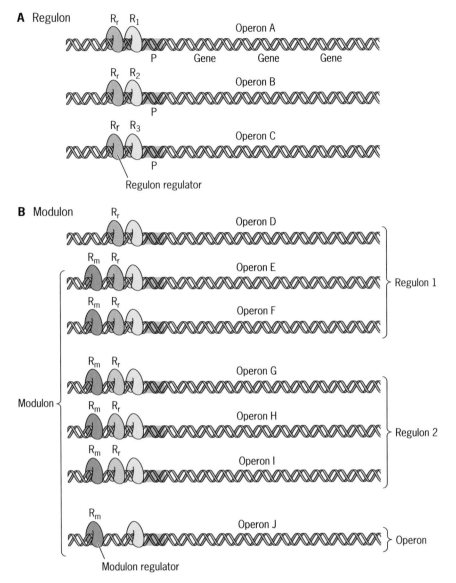

FIGURE 11.10 **Patterns of operon organization into higher (global) regulatory units.**
(A) Regulon organization. Each operon (A, B, or C) is depicted as a segment of DNA
consisting of a control site at which a common protein regulon regulator (R_r) and operon-
specific protein regulators (R_1, R_2, and R_3) act to control transcription of the genes of the
operon. The promoter (P) of each operon is also indicated. (B) Modulon organization. A
modulon consisting of six operons (E to J), controlled by a single modulon regulator (R_m) or
global regulator, is depicted. Operons E and F are part of one regulon (regulon 1), along
with operon D (which is not in the modulon). All three operons (G, H, and I) of regulon 2
are included in the modulon, along with the independent operon J. The two individual
regulon regulators are shown, as are the individual regulators of each operon.

ular physiological condition. Hence, modulons coordinate the expression
of regulons. Below, we describe two examples of modulons.

Examples of global regulatory systems. Some modulons are so pervasive in
their physiological effects that they help us understand the "global" nature
of global regulatory systems. Two modulons that fit this category are the
so-called **catabolite repression** system (which, as you will learn below,

does not really involve a "repressor") and the **stringent response** system. Together, these two systems directly or indirectly control probably *three-fourths of the protein-synthesizing capacity* of the *E. coli* cell.

Catabolite repression. The ability of enteric bacteria to ensure that a premium substrate, glucose, is utilized to the exclusion of lesser substrates is controlled (in part) by a global regulator called the **cyclic AMP receptor protein** (**CRP** or CAP). When this allosteric regulatory protein binds the small nucleotide ligand, cyclic AMP (cAMP), it recognizes and binds CRP-specific boxes located upstream of the transcriptional start site of several operons. Upon binding to the CRP box, cAMP-CRP increases the frequency of transcription initiation, provided the operon has already been induced by the operon's own substrate. (Note that the name "catabolite repression" refers to the cell's response to glucose, which is controlled by a transcriptional *activator*, CRP.)

Catabolite repression has extraordinary importance in the physiology of many bacteria. The operons in its network can be expressed at very high levels when induced by their substrates. Preventing such unnecessary expression of these operons is so critical to bacterial economy that their *control within a modulon* is widespread in the bacterial world, even though the molecular mechanisms of control differ. For example, Gram-positive bacteria lack cAMP yet possess an effective catabolite repression system.

To illustrate how catabolite repression can override other regulatory switches, let's return to student JM, who grew *E. coli* with a mixture of glucose and lactose (case study). Glucose uptake by the cells used the phosphotransferase (Pts) system, a complex transport system that includes proteins that regulate the phosphorylation state and, therefore, activity of the enzyme (**adenylate cyclase**) that makes cAMP from ATP (Fig. 11.11).

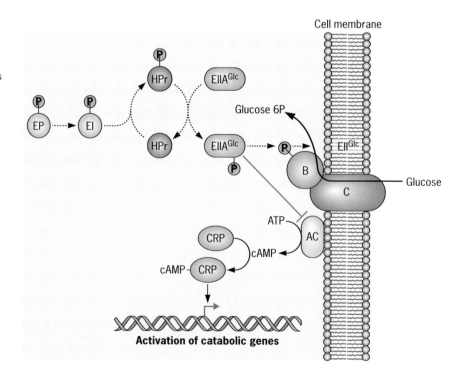

FIGURE 11.11 **Glucose catabolite repression in *E. coli*.** AC, adenylate cyclase; CRP, cyclic AMP receptor protein. The EII^Glc transporter of the phospotransferase system (orange) is regulated by phosphates transferred from enzyme I (pink) to histidine protein (green) to enzyme II (light green).

Activation of catabolic genes

So in JM's culture, when glucose was transported, adenylate cyclase was *inhibited*; consequently, the cellular pool of cAMP declined, and the amount of cAMP-CRP decreased. Without this allosteric activator, RNAP could not transcribe the *lac* operon to high levels even though lactose was also present in the medium to induce the operon. Thus, the cells could not use lactose until the glucose was consumed. Once this happened, the adenylate cyclase was activated and cAMP levels increased, providing the ligand for CRP activation.

Stringent response. The stringent response modulon is a gigantic regulatory network geared to *respond to starvation*, for either energy or building blocks. The response is extensive and affects a large number of operons and regulons with diverse functions (Fig. 11.12). For example, limitation of amino acids *restricts the formation of one or more species of aminoacyl-tRNA*, and this, in turn, stalls ribosomes on the corresponding codons on the mRNA being translated. A sensor on the stalled ribosomes (the enzyme RelA) is then activated to convert GTP into ppGpp, a small molecule appropriately called an "allarmone" for its critical role in signaling the cell that it is starving. The ppGpp alarmone cooperates with other components to *alter RNAP's affinity for many regulatory sites on DNA* and reprogram the cell to respond to starvation (Fig. 11.12). By activating some metabolic processes and repressing others, the stringent response equips the cell to balance the available resources and cope with nutrient restriction. So powerful is the stringent response that bacteria encode multiple elaborate biochemical paths for making, degrading, and responding to ppGpp.

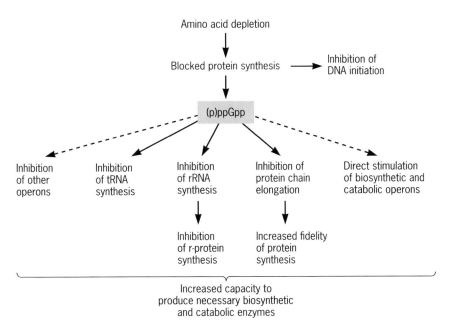

FIGURE 11.12 **Stringent response.** The series of events ensuing after amino acid restriction of bacteria are shown. The solid arrows indicate processes shown experimentally to result from ppGpp accumulation. The dashed arrows depict more speculative relations.

WHY REGULATE BOTH PROTEIN ACTIVITY AND AMOUNTS?

LEARNING OUTCOMES:
- Explain why evolution has primarily favored two modes of metabolic regulation in microbes (controlling either the amount or activity of enzymes) over a single mode, such as allostery.

Why should a cell regulate its metabolism by changing both enzyme activity *and* enzyme amount? *Controlling the activity of an allosteric enzyme is a swifter and more precise mechanism* to adjust flow through the many metabolic pathways. For example, recall the experiment where *E. coli* was grown with glycerol only or with glycerol and histidine; histidine biosynthesis was active in the first culture but repressed in the other. This is a case of end-product inhibition by allosteric enzymes (Fig. 11.2). As soon as histidine was added to the medium, it inhibited the first reaction of the biosynthetic pathway. In fact, a cell could accomplish the task of coordinating metabolic flow in an orderly fashion using allosteric enzymes alone.

If allosteric enzymes can be so effective, why do prokaryotes also go to great lengths to regulate the synthesis of their proteins (Fig. 11.7)? Evidently, in a given situation evolution has pressured microbes to be highly sensitive about what proteins they make and in what quantity. A reasonable theory about the force behind this evolution is the *need for economy and efficiency*. Over half the cell's dry mass is protein, and *making proteins is expensive* (75% of the energy budget is devoted to protein synthesis!). Optimizing the rate of growth in each environment requires that cells not make redundant or irrelevant proteins. By this argument, *metabolic coordination—making order out of chaos*—would seem to be the domain of allosteric control of enzyme activity. On the other hand, competing successfully for food and space would seem to demand the elegant and effective *controls on transcription and translation*. From these considerations, one might speculate that allostery developed at an earlier stage of evolution than did controls on gene expression. What do you think?

CASE REVISITED: **How can *E. coli* selectively use sugars from a mixture?**

Even though lactose was present in the medium and the inducer had removed the LacI repressor, something else was needed to activate the RNAP holoenzyme and stimulate transcription of the *lac* operon. Clearly, the "stimulating factor" was only available when glucose was no longer present. What was it?

As you now know, RNAP needs stimulation by cAMP-CRP. Once bound by cAMP, CRP binds to sequences upstream of the *lac* promoter and interacts with RNAP to help it clear the promoter, allowing the operon to be transcribed. Why was cAMP-CRP not available until all the glucose was exhausted from the medium? Because the protein system that transports glucose (Pts) antagonizes the enzyme (adenylate cyclase) that makes cAMP. As long as glucose was present, the adenylate cyclase was inactivated and the levels of cAMP decreased. But once glucose was used up, cAMP was made, and cAMP-CRP became available. The RNAP received its "stimulating factor" and the *lac* operon was transcribed. The cells were now able to grow on lactose, and the second phase of diauxic growth started. Note that there was a *lag* phase between the two phases. Clearly, it takes time for adenylate cyclase to be activated, for cAMP to be made and bind CRP, and for the cAMP-CRP complex to bind and stimulate the RNAP at the *lac* operon.

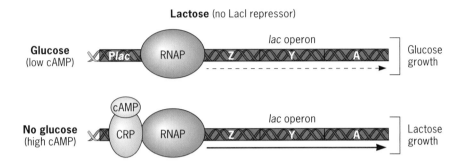

CONCLUSIONS

To avoid chaos, the approximately 2,000 individual chemical reactions a typical prokaryotic cell requires to live and grow must be integrated into a single complex system, with controls on the rate of each reaction. We have seen that specific control devices have evolved to govern *enzyme activity* and *enzyme synthesis*. We have also had a glimpse into the world of feedback control and complex circuitry that integrates the activity of all the cell's parts and processes. As we considered how prokaryotic cells coordinate their metabolism, the question of how they cope with their environment has gradually come to the surface.

In real life, nothing is constant. All niches in the biosphere of Earth are subject to changes in temperature, chemical composition, and all the other parameters important for life. The environment of microbial cells is often subject to catastrophic changes of a sort not usually met by the individual cells of higher plants and animals. Microbes must cope with an environment that continually presents new challenges: changes in nutritive value, temperature, pH, redox potential, barometric pressure, and the presence of toxic factors. Such fluctuations exert an extraordinary burden on the regulatory devices that coordinate the internal processes of the bacterial cell. The story of how metabolic reactions are coordinated cannot be separated from the intricate, clever ways in which microbial cells successfully cope with environmental stresses. Therefore, the next chapter is a continuation of this one.

Representative ASM Fundamental Statements and Learning Outcomes for the Chapter

EVOLUTION

1. Cells, organelles (e.g., mitochondria and chloroplasts), and all major metabolic pathways evolved from early prokaryotic cells.

   a. Give two reasons for the evolution of controls that govern sets of genes or operons together as regulatory units (global regulation).
   b. Explain why evolution has primarily favored two modes of metabolic regulation in microbes (controlling either the amount or activity of enzymes) over a single mode, such as allostery.

METABOLIC PATHWAYS

2. The survival and growth of any microorganism in a given environment depend on its metabolic characteristics.

   a. Explain the need for coordination of metabolic reactions within a cell.
   b. Summarize a set of experiments that demonstrates coordination of biosynthesis in a bacterium.
   c. Speculate about at least one macromolecule (or group of macromolecules) that will be in greater abundance for each phase of bacterial culture growth and undergo coordination of polymerization.
   d. List four reasons why prokaryotes more often control the amounts of enzymes produced rather than controlling enzymatic activity (as is more often seen in eukaryotes).
   e. Define allostery.
   f. Differentiate covalent modification and allosteric interactions as ways of controlling protein activity.
   g. Use the histidine biosynthetic pathway as an example to explain how feedback inhibition is an effective means to control biosynthetic pathways.
   h. Describe how the ratio of the concentration of ATP to those of ADP and AMP is maintained almost constant in bacterial cells over a wide range of energy supply and demand.

GENETICS AND INFORMATION FLOW

3. The regulation of gene expression is influenced by external and internal molecular cues and/or signals.

   a. Describe two ways in which gene expression can be monitored under various growth conditions.
   b. State how the regulation of gene expression in the *lac* operon is under the influence of external and internal molecular cues/signals.
   c. Identify the molecules that serve as the repressor and inducer in the *lac* operon.
   d. List three central conditional properties of operons.
   e. Speculate on reasons why microbes have such a large diversity of regulatory processes that control the synthesis of proteins.
   f. Describe six ways in which microbes control the initiation of transcription as a means of regulating gene expression.
   g. Identify examples of protein-RNA interactions that regulate gene expression in a cell.
   h. Explain how the structure of the *trp* operon mRNA contributes to controlling gene expression for enzymes involved in biosynthesis of tryptophan through attenuation.
   i. Relate operons, regulons, and modulons.
   j. Use the catabolite repression modulon as an example to identify a reason why some modulons are considered a global regulatory system.
   k. Explain how an *E. coli* cell uses the catabolite repression system to repress transcription of the *lac* operon in the presence of both glucose and lactose.
   l. Summarize the role of ppGpp in the stringent response modulon.

SUPPLEMENTAL MATERIAL

References

- **Görke B, Stülke J.** 2008. Carbon catabolite repression in bacteria: many ways to make the most out of nutrients. *Nat Rev Microbiol* **6:**613–624. doi:10.1038/nrmicro1932.
- **Hauryliuk V, Atkinson GC, Murakami KS, Tenson T, Gerdes K.** 2015. Recent functional insights into the role of (p)ppGpp in bacterial physiology. *Nat Rev Microbiol* **13:**298–309. doi:10.1038/nrmicro3448.
- **Lewis M.** 2013. Allostery and the *lac* operon. *J Mol Biol* **425:**2309–2316. doi:10.1016/j.jmb.2013.03.003.
- **Storz G, Vogel J, Wassarman KM.** 2011. Regulation by small RNAs in bacteria: expanding frontiers. *Mol Cell* **43:**880–891. doi:10.1016/j.molcel.2011.08.022.

Supplemental Activities

CHECK YOUR UNDERSTANDING

1. **In the case study, when JM cultured *E. coli* in a medium supplemented with both glucose and lactose, *E. coli* initially utilized glucose rather than lactose. Why?**
   a. A lactose derivative relieves inhibition of the *lac* operon.
   b. The *lac* operon is activated in the presence of lactose and glucose.
   c. In the presence of glucose, cells have reduced amounts of the cAMP-CRP allosteric regulator.
   d. Binding of cAMP to CRP increases transcription of the *lac* operon.
   e. All of the above

2. **Enzyme activity can be modulated by _____.**
   a. allostery by an sRNA molecule
   b. increased expression of the gene encoding the enzyme
   c. phosphorylation
   d. methylation
   e. all of the above

3. **What fitness advantage do bacteria gain by controlling enzyme amount instead of enzyme activity?**
   a. The enzyme is essential for viability.
   b. Cells do not invest energy-synthesizing proteins that are not needed.
   c. The enzyme is available for immediate use.
   d. Less energy is required to synthesize protein and store it.

4. **Of the dozen or so different mechanisms for regulating the synthesis of specific proteins, name two that involve direct participation of ribosomes.**

5. **Explain how control regions located hundreds of nucleotides away from a promoter can control gene expression from that promoter. What name is given to such control regions?**

DIG DEEPER

1. Prokaryotic cells coordinate metabolic pathways by controlling the amount of enzymes or modulating their activity by allosteric interactions.
   a. Which type of control is generally faster?
   b. What advantage would be sacrificed if all metabolic coordination were achieved solely by modulating protein activity?

2. The transcription of bacterial operons is often controlled by regulatory proteins, which bind the operator region and either decrease (repressors) or increase (activators) the frequency of initiation. In the late 1990s, Savageau proposed the "demand theory" of gene regulation (M. A. Savageau, *Genetics,* **149:**1665–1676, 1998), which stated that negative or positive modes of regulation are selected for genes encoding functions in low or high demand, respectively.

a. Based on your knowledge of the regulation of the *lac* operon, would you predict lactose to be often found in environments inhabited by *E. coli*?

b. Arabinose utilization in *E. coli* requires the expression of the *araBAD* genes, which are encoded in an operon that is positively regulated by a regulatory protein (AraC) but only when bound to arabinose. Would you predict arabinose to be abundant in the environment(s) inhabited by *E. coli*?

c. Would you predict operons involved in the synthesis of building blocks to be positively or negatively regulated?

d. Operons encoding biosynthetic enzymes (e.g., synthesis of amino acids) are *generally turned on* unless their building block end product is present in the environment. In these cases, the end-product feedback inhibits an allosteric enzyme early in the biosynthetic pathway. The end product can sometimes perform double duty and activate an allosteric repressor protein to shut off the transcription of the operon. Bacteria in the human gut are constantly exposed to a hefty supply of amino acids derived from dietary protein. Based on the "demand theory," do you expect the human intestinal environment to select for positive or negative modes of control of operons involved in the synthesis of amino acids?

e. Consider now the economic cost if a cell acquires a null mutation in a positive regulator versus a negative regulator. How might this contribute to the pattern that operons encoding products in high demand tend to be positively regulated, whereas those with intermittent or low demand are negatively controlled?

3. Riboswitches are sequence elements typically located in the 5′ untranslated region of mRNA that bind effector molecules with great specificity. Upon binding the effector molecule, the structure of the nascent mRNA changes and its transcription and/or translation are promoted or prevented. Search the Web for examples of riboswitches that

a. terminate transcription of the mRNA prematurely.

b. prevent the translation of the mRNA.

c. use an antisense sRNA as the effector molecule.

d. use a metabolite as an effector molecule.

4. **Team work:** Read the following research article on the *Staphylococcus aureus* sRNA SprD, which regulates bacterial virulence: S. Chabelskaya et al., *PLoS Pathog*, **6:**e1000927, 2010; doi:10.1371/journal.ppat.1000927.

a. What experiments did the authors perform to determine that the sRNA SprD regulates the expression of the Sbi protein at the translational level but not transcriptional level? (Hint: see Fig. 2.)

b. Draw a schematic to illustrate how SprD regulates the expression of the *sbi* mRNA.

c. How did the authors show that SprD RNA enhanced the virulence of *S. aureus* in mice?

d. Discuss possible ways to control virulence of *S. aureus* through SprD.

Supplemental Resources

VISUAL

- Walter Gilbert describes how they designed the experiment that allowed his group to isolate the LacI repressor (1:09 min): https://www.dnalc.org/view/15254-The-experimental -design-for-the-lac-operon-Walter-Gilbert-.html.

- Animation of attenuation at the *trp* operon: http://www.microbelibrary.org/library /biology/2825-regulation-of-biosynthesis-attenuation-of-the-trp-operon.

- Beatrix Suess discusses how synthetic biologists are engineering riboswitches in bacteria to manipulate the expression of genes of interest (22:15 min): https://www.youtube .com/watch?v=Ptsv5ZtnnXw.

WRITTEN

- S. Marvin Friedman tells us how some plant pathogens inject double-stranded sRNAs in their host during an infection to suppress their immune system in the *Small Things Considered* blog post "Plant Pathogen Silences Host's Immune Genes" (http://schaechter .asmblog.org/schaechter/2014/07/plant-pathogen-silences-hosts-immune-genes.html).

- Fred Neidhardt gives a personal account of how capturing cells' protein profiles by two-dimensional gel electrophoresis opened the door to the proteomics field of today in the *Small Things Considered* blog post "How Proteomics Got Started" (http://schaechter .asmblog.org/schaechter/2009/12/how.html).
- Fred Neidhardt discusses how the stringent response coordinates not only bacterial growth but also virulence of a number of pathogens, including *Legionella pneumophila*, in the *Small Things Considered* blog post "Bacterial Physiology and Virulence: The Cultures Converge" (http://schaechter.asmblog.org/schaechter/2010/09/bacterial-physiology-and -virulence-the-cultures-converge.html).

IN THIS CHAPTER

Colorful bacteria at the Grand Prismatic Spring in Yellowstone National Park, USA

CHAPTER TWELVE

Succeeding in the Environment

KEY CONCEPTS

This chapter covers the following topics in the ASM Fundamental Statements.

EVOLUTION

1. Mutations and horizontal gene transfer, with the immense variety of microenvironments, have selected for a huge diversity of microorganisms.

CELL STRUCTURE AND FUNCTION

2. *Bacteria* and *Archaea* have specialized structures (e.g., flagella, endospores, and pili) that often confer critical capabilities.

METABOLIC PATHWAYS

3. The interactions of microorganisms among themselves and with their environment are determined by their metabolic abilities (e.g., quorum sensing, oxygen consumption, and nitrogen transformations).

4. The survival and growth of any microorganism in a given environment depend on its metabolic characteristics.

INFORMATION FLOW AND GENETICS

5. The regulation of gene expression is influenced by external and internal molecular cues and/or signals.

MICROBIAL SYSTEMS

6. Most bacteria in nature live in biofilm communities.

INTRODUCTION

In chapter 4 we learned about the incredible growth potential of pro-karyotes, thanks to their streamlined design and small size. In chapter 11 we noted how regulatory systems optimize the metabolism of the growing cell so that the growth rate is maximized. The numerical consequences of fast, exponential growth go far toward accounting for microbial dominance of Earth. Far, but not far enough! The same features that contribute to their success in colonizing planet Earth—microscopic size and unicellularity—leave microbes exceptionally exposed to environmental forces and biological predators. Disaster looms for the growing cell when its environment lacks nutrients or exposes the microbe to harsh physical or chemical conditions. Toughness matters in the world, but so also does flexibility. Unless the growing cell adapts to adverse conditions, it will die—either from structural damage or from irreversible destruction of its vital components, such as DNA, ribosomes, or the cell membrane. Thus, from the very beginning of life on Earth, selective pressures have led to the evolution of myriad strategies—**adaptive responses**—for coping with adversity. Cells unable to cope with stress were swiftly eliminated. Those with successful strategies survived. In this chapter, we shall introduce some of the prominent means by which microbes face (and conquer!) adversity.

CASE: **How fast can microbes adapt to environmental changes?**

A coyote, on the prowl for rabbits at dusk, pauses to defecate by a desert pool contaminated with metals discharged from a nearby copper mine. As with all vertebrates, the coyote's feces contain lots of bacteria, including *Escherichia coli*. You examine the water in the pool over several days by spreading aliquots on agar media plates and incubating them. Colonies of *E. coli* now appear, whereas none were generated from samples of the water collected before the coyote's deposit. Think about it: cells from the coyote's feces experienced an abrupt shift from their favorable habitat in the gut (dark, warm, nutrient rich, near neutral in pH, anaerobic, and relatively free of toxic metals) to an inhospitable, shallow pool of water (aerobic, nutrient poor, acidic, containing toxic chemicals, and with evening temperatures as low as 10°C and midday temperatures reaching 40°C). How could any *E. coli* cells survive?

(http://l7.alamy.com/zooms/6de61b09-affa-42bd-9687-6ea391f7b532/B9G0GF.jpg)

WHEN ARE MICROBES "STRESSED"?

LEARNING OUTCOMES:
- Identify the defining characteristic of "environmental stress" for a microbe.
- Give examples of microbial stressors.

When describing the microbial world, we use the term **environmental stress** broadly, referring to challenges that range from severe to subtle. The defining feature is a *significant change in the cell's environment*. Notice that one cell's stress may be another's habitat: a hyperthermophilic microbe living steadily at 100°C is not stressed; neither is a halophilic microbe living in salt brine, nor an acidophilic species living in acid mine water. To such organisms, known as **extremophiles** (chapter 4), living under such conditions is just another day at the office. But move a culture of a heat-loving microbe to the laboratory bench, at room temperature, and, in essence, the cell would experience a deep freeze. Dilute the high salt concentration of salt brine, and the halophilic microbes would die. Why? Because these microbes have evolved adaptive mechanisms to grow in the unique conditions of their environment. Hence, what is stressful for any microbe is a *change* in the environment that *makes growth a challenge*.

Consider the *E. coli* cells in the case study that survived their abrupt deposit into an extreme acidic, metal-laden environment. The stress on the cells must have been enormous, particularly when coming from the nutrient-loaded, warm habitat of the coyote's intestines. And the nutritional changes are but a minor problem compared with the acidic conditions and metal toxicity. Yet this scenario is neither uncommon nor exaggerated. In any environment, physical factors and chemical components can vary over a wide range, and thus become a source of stress. And when populations of microbes become large, they can even create their own problems. Not only do cells deplete the local environment's nutrients; like all living things, microbes pollute. Their waste products, such as short-chain fatty acids or hydrogen peroxide, can be quite toxic to themselves and their competitors. How do cells survive such inescapable or sudden environmental stress?

HOW DO MICROBES COPE WITH STRESS?

LEARNING OUTCOMES:
- Speculate about why most of the microbial stress responses and adaptive mechanisms involve the regulation of many genes (frequently >12) at once.
- Outline the general pattern of the signal transduction/two-component stress response.
- Speculate about the type of environment wherein the microbial phosphate response regulon may frequently be turned on.
- Using the *E. coli* phosphate response regulon as an example, describe the nature, role, and location of the primary components of the two-component stress response.
- Identify some effects of heat stress on microbes.
- Describe how bacteria sense a rise in temperature.

- Identify the role of protein DnaK in the heat shock response and feedback control.
- Contrast the signal to regulate gene expression for the two-component stress response with that of the heat shock stress response.

Changing environments exert fierce selective pressures. With their long past, microbes have endured environmental stresses for billions of years. Thus, modern microbes are those that evolved coping strategies. Some of these adaptive responses thwart transient physical or chemical threats; others toughen the cells for long intervals of nongrowth during starvation. Some behaviors enable cells to move toward a food source or away from toxic compounds. Equally sophisticated (and surprising) responses involve actual conversations among cells to plan collective action. Yes, you heard right: *microbes communicate and team up!* Based on such conversations, microbes might aggregate and form floating (pelagic) groups, or attach to a surface and grow as a **biofilm** community. Collectively, the cells gain the advantages of physical protection and community action (two can do more than one!). A particularly extreme response is entering a state of dormancy. For example, some microbes transform into metabolically inert packages, called **spores**, that contain only what is essential for growth at some far-off time when conditions for life improve (chapter 13).

Although microbes exhibit a wide range of adaptive responses, some are quite common among *Bacteria* and *Archaea*. Table 12.1 summarizes some well-studied adaptive systems and provides a small glimpse at their variety. Scan it to appreciate that natural selection has endowed microbes with a battery of sophisticated devices that augment the growth advantage enjoyed by microscopic cells. Some of the responses listed in the table are geared to one specific threat (e.g., heat, UV light, or phosphate starvation), while others have roles in multiple challenges and responses (e.g., entry into stationary phase). We tend to refer to the former as **stress responses** and the latter as **adaptive mechanisms**, but the two terms overlap, and you will encounter both used interchangeably (occasionally even in this textbook). We begin with an analysis of stress responses.

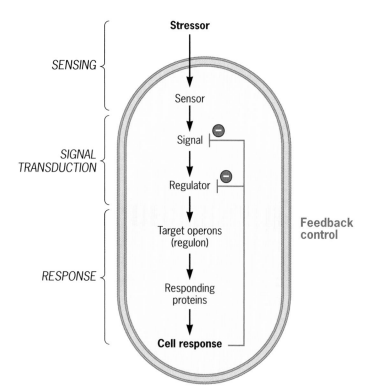

FIGURE 12.1 Generalized signal circuit in a microbial stress response. Microbes cope with stress with responses involving some common steps: sensing of the stressor and production of a cellular signal, which reprograms specific genes (signal transduction) to respond to the stress. A feedback mechanism inactivates the signal and response when the stressor is no longer present.

General Steps in the Stress Response

Although microbes are endowed with numerous stress responses, they have several features in common (Fig. 12.1). At a minimum, a stress response entails (i) *sensing the stressor,* (ii) *doing something about it (response),* and (iii) *turning off the response when no longer needed.* Typically, these three actions are mediated by

TABLE 12.1 **Some major bacterial adaptations to the environment**

Environmental situation	Responding system	Regulated genes (and products)	Type of regulation
Nutrient limitation			
Carbon	Catabolite repression	Genes encoding catabolic enzymes	Activation of CRP by cAMP (see chapter 11)
Amino acid or energy	Stringent response	More than 200 genes for ribosomes and other proteins involved in translation, plus biosynthetic enzymes	(p)ppGpp binding RNAP (see chapter 11)
Ammonia	Nitrogen acquisition from organic sources and low ammonia concentrations	Genes for glutamine synthetase and deaminases	Complex
Phosphate	Acquisition of inorganic phosphate (Pho system)	Alkaline phosphatase, and about 40 other genes involved in utilizing organophosphates	PhoB-PhoR two-component system (see this chapter)
Electron acceptor present			
O_2	Aerobic respiration	Many (>30) genes for aerobic enzymes	Repression of genes of aerobic enzymes upon signal of low O_2
Other electron acceptors	Anaerobic respiration	Genes for nitrate reductase and others	Transcriptional activation
Toxic agents			
UV and other DNA damagers	SOS response	~20 genes for repair of DNA damage by UV or other agents	Transcriptional repression by LexA, relieved upon cleavage of LexA by RecA nucleoprotein
DNA alkylation	Ada system (alkylation response)	Four genes involved in removal of alkylated bases from DNA	Transcriptional activation by Ada
H_2O_2 and similar oxidants	Oxidation response	~12 genes involved in protection from H_2O_2 and other oxidants	Transcriptional repression by OxyR
Heat	Heat shock response	Dozens of genes involved in protein synthesis, processing, and degradation	RNAP programming by σ^H (see this chapter)
Cold	Cold shock response	A dozen genes for macromolecule synthesis	Unknown
High osmolarity	Outer membrane porin response	Genes for OmpF and OmpC porins	OmpR, when phosphorylated, is a negative regulator of *ompF* and a positive regulator of *ompC*
Nongrowth conditions			
Starvation or inhibition	Stationary phase	Hundreds of genes affecting structure and metabolism	Complex (see this chapter)
Starvation or inhibition	Sporulation	Many (>100) genes for structural aspects of spore formation	Complex (see chapter 13)

(i) *protein sensors*, which sense the stressor; (ii) *a transmitted signal that reprograms genes* encoding components of the stress response; and (iii) a *feedback control mechanism* that turns off the response when the "all-clear" has sounded. Thus, cells translate an environmental stress into a biochemical signal, which then coordinates the response to the stress. To emphasize that signals are converted from one form to another, we refer to the components of the stress response collectively as **signal transduction** systems.

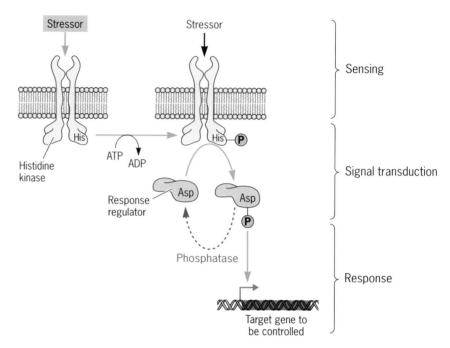

FIGURE 12.2 Generalized scheme of a two-component stress response. In this two-component system, the stressor triggers the autophosphorylation of a histidine residue in a sensory kinase, which then phosphorylates the aspartic acid of a response regulator. The phosphorylated response regulator is now active and binds control regions of target response genes to control their expression. The feedback control, in this case, is achieved by dephosphorylating the response regulator when the stress response is no longer needed.

To cope with many of the stresses that originate in the environment (the stressor), cells employ a common signal transduction design: a sensor protein, often located on the cell surface or in the periplasmic space, senses the stressor and transmits a signal to a **response regulator protein** in the cytoplasm, which triggers the response (Fig. 12.1). The numerous pairs of sensor/response regulator proteins are referred to as two-component systems (though some systems consist of more than two). In many cases, the sensor protein is a **histidine kinase**, a sensory protein that *phosphorylates one of its own histidine residues* when it senses the stressor (Fig. 12.2). The phosphate residue is then transferred to a cognate-response regulator-protein in the cytoplasm to activate it and control the expression (induction or repression) of genes appropriate for that stress response. Feedback mechanisms are necessary to deactivate the system when the response is no longer needed (Fig. 12.1). In the case of two-component systems involving histidine kinase/response regulator pairs, dephosphorylating the response regulator prevents the amplification of the response (Fig. 12.2). If the stressor is no longer present, the response regulator is not phosphorylated again, and any remaining phosphorylated protein will be removed— kind of like unplugging an electronic device.

Examples of Stress Responses
Microbes face myriad threats in their natural habitats, evident from the well-studied stress responses listed in Table 12.1. Given that the *stressors are so varied*, it may not surprise you to learn that microbes have evolved

numerous variations of the general response theme to adapt to fluctuations in their specific environment. Most often, cells respond to stress by reprogramming several genes and operons, collectively known as a **regulon** (chapter 11). To illustrate the ingenuity of microbes and their unique physiological twists, we shall focus on two stress responses. The phosphate response regulon represents the large class of bacterial stress responses mediated by two-component systems. In this case, the stressor is the absence of a nutrient (phosphate). The heat shock response illustrates a near universal, organized response system to temperature shifts that extends throughout the biological world.

The phosphate response regulon. Inorganic phosphate (P_i) is not a stressor—but its absence is. Indeed, phosphate is a major component of all cells, with nucleic acids and ATP being the most notable life-sustaining examples. We can easily see the correlation between cell growth and phosphate availability in lakes and streams affected by the runoff of phosphorus-containing material, whether as laundry detergents or farmland fertilizers. The sudden availability of inorganic phosphate (a process known as **eutrophication**) leads to the massive growth, and then death, of algae, which accumulate as thick mats on the surface waters. Their decomposition consumes oxygen and prevents it from diffusing into the water depths, thus killing aquatic animals and plants. However, in most natural environments, phosphate is present in limited amounts. Not surprisingly, cells are particularly attuned to sense the levels of phosphate in their environment and to scavenge it when it is scarce.

Follow along with Fig. 12.3 to learn how a two-component signal transduction system equips *E. coli* to respond to phosphate limitation. As in

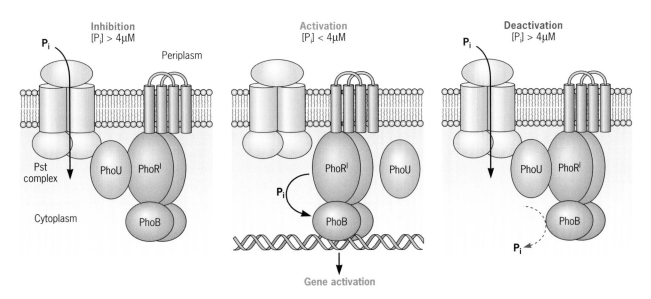

FIGURE 12.3 Operation of the *E. coli* phosphate response regulon. In *E. coli*, the two-component system PhoR-PhoB is physically linked to the protein complex (PstSCAB) that transports inorganic phosphate (P_i) via the chaperone protein PhoU. The structural conformation of the Pst system changes when P_i levels drop below a threshold concentration (4 µM) and these changes are passed to PhoR through PhoU. This triggers the autophosphorylation of PhoR and activation of the signal transduction cascade via phosphorylated PhoB. The phosphate stress response is then activated. Once P_i concentrations increase above the threshold, conformational changes in the Pst complex are reversed.

many other signal transduction systems, the phosphate response is mediated by a sensor and a response regulator. The sensor is a histidine kinase protein called PhoR, and the response regulator is PhoB. When adequate inorganic phosphate is present, the system is inactive: PhoR is not phosphorylated in its histidine residue and thus cannot phosphorylate PhoB. As a consequence, PhoB cannot bind DNA and there is *no activation or repression of genes* of the phosphate response regulon. By contrast, when the inorganic phosphate concentration drops (<4 μM), PhoR autophosphorylates its histidine and passes the phosphate residue to PhoB, which then activates the genes that handle the problem (Fig. 12.3). Thus, this system illustrates the widely used scheme of a histidine kinase/response regulator pair whose phosphorylation cascade is triggered by the stressor signal (in this case, low phosphate levels).

But how does the sensor PhoR "sense" the external concentration of inorganic phosphate? This protein does not bind phosphate directly; instead, it is physically linked via a **"chaperone"** protein (PhoU) to the transport system (the Pst complex) that brings environmental inorganic phosphate inside the cell. As you can see in Fig. 12.3, if phosphate is available in the environment, it is transported by the Pst system, and PhoU remains linked to the sensor PhoR to prevent its autophosphorylation. It makes sense to silence the regulon because the cell is bringing in phosphate in sufficient quantities. Things change, however, when phosphate levels drop below a threshold concentration. Transport through the Pst system is slowed down or interrupted, and the transport protein complex undergoes conformational changes to the chaperone PhoU such that the sensor PhoR can now autophosphorylate. In this way, the signal—low levels of phosphate—is transduced as a phosphorylation cascade between the sensor PhoR and the response regulator PhoB.

Once it is phosphorylated, PhoB can *regulate transcription* of the response genes of the phosphate regulon. Numerous genes are recruited (at least 47 genes; Fig. 12.3 is oversimplified), including enzymes such as alkaline phosphatase and AMP nucleosidase that cleave phosphate residues from organic compounds. The cell not only produces enzymes to utilize other sources of phosphate, it also increases the number of Pst/PhoR/PhoB/PhoU complexes in the membrane to help scavenge phosphate from the environment and tune in to the environment to monitor phosphate availability.

To exert feedback control, the PhoR-PhoS two-component system operates in reverse (Fig. 12.3). As soon as the cell senses sufficient inorganic phosphate levels (>4 μM) in the environment, the Pst transporter conformation changes again and PhoU "sequesters" the sensor PhoR once more to prevent its autophosphorylation. The active Pst complex, influenced by PhoU, also converts PhoR into a phosphatase, which dephosphorylates PhoB. Once dephosphorylated, PhoB is inactive and no longer modulates transcription of any of the response genes. The phosphate response regulon is now off. *Exquisite!*

Using a similar scheme, microbes can respond to a wide range of stressors. However, during the long course of microbial evolution, signal transduction systems have diversified to coordinate a specific and effective response to particular environmental cues.

The heat shock response. Protection against excessive heat is a universal feature of living organisms. Hence, mechanisms that defend against deleterious heat have evolved independently in all forms of life, from microbes to humans. The heat shock response has been particularly well studied in yeast and *E. coli*. We'll look in detail at the latter.

What does a rise in temperature do to *E. coli*? Temperature can have profound effects on the formation and stability of macromolecules. Proper folding is critical to protein function, and misfolded proteins are more than a waste: they aggregate and interfere with normal cellular metabolism. A sudden, heat-induced accumulation of inactive and useless unfolded or misfolded polypeptides poses a serious challenge for the cell, which now must form functional proteins while getting rid of the misfolded ones. Heat can also denature nucleic acids. Think of it: if RNA molecules were to unfold, they would lose their activity, tRNAs would not bring amino acids to the ribosomes, and the ribosomes would unfold. Protein synthesis would come to a halt. The many regulatory roles of small RNAs (sRNAs) would no longer be exerted, bringing chaos to the cell. In chapter 3, we learned about the importance of packing the DNA tightly as a nucleoid. Can you imagine what heat can do to a condensed nucleoid? Clearly, heat can be a big problem for microbes (the word "shock" in "heat shock" is quite appropriate). For this reason, heat stress has been the selective force driving the evolution of the heat shock response in all organisms.

The heat shock response in *E. coli* is mediated by the usual suspects: a sensor, a signal, response genes, and feedback control (Fig. 12.4). To learn how these factors cooperate to produce a fast, effective, and reversible response, read on.

Sensor. Which component is the thermometer? An mRNA encoding a **sigma (σ) factor**! Recall that σ factors bind and direct the RNA polymerase

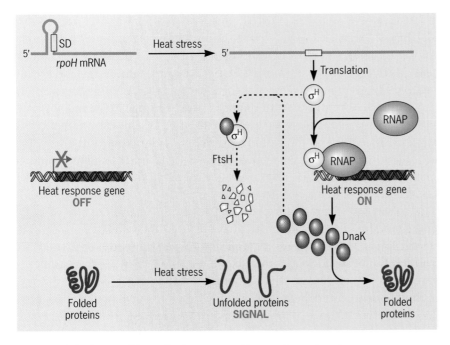

FIGURE 12.4 **The *E. coli* heat shock response.** See text for explanation.

(RNAP) to specific **promoters** from which gene transcription starts (chapter 8). A dedicated sigma factor called σ^H (for "heat") increases the affinity of RNAP for promoters of the heat shock regulon. Normally, *E. coli* makes small quantities of σ^H: at lower temperatures, the Shine-Dalgarno region of its *mRNA* is *sequestered* within the mRNA, preventing ribosomes from efficiently binding and making more σ^H (Fig. 12.4). A rise in temperature *melts the structure and unfolds the mRNA*. With its Shine-Dalgarno sequence no longer "hidden" from the ribosomes, *E. coli* can now make σ^H and express the σ^H regulon. Thus, the σ^H mRNA functions like the sensing element of a thermostat.

Signal. One of the major products of the heat shock regulon is DnaK, a protein chaperone. Recall that these specialized proteins facilitate proper folding and remove fragments of proteins in *E. coli* and most microbes (chapter 8). During growth at constant temperature, the chaperones are maintained at a level just right to do their job. In addition to its housekeeping role, chaperone DnaK binds and targets σ^H for degradation. But what prevents DnaK from prematurely aborting expression of the σ^H heat shock regulon? At elevated temperatures, denatured cellular proteins accumulate, bind DnaK, and *divert it from σ^H*. Thus, in this signal transduction pathway, *denatured proteins are the signal* that triggers the appropriate transcriptional response.

Response genes. The heat shock regulon of *E. coli* has *more than 250 genes* that are activated, either directly or indirectly, by σ^H. Approximately 25% of these encode proteins that change the composition of the cell membrane to preserve its integrity and function at higher temperatures. At the same time, newly expressed protein chaperones help refold the denatured proteins and proteases degrade those beyond repair. The σ^H regulon also induces expression of the DNA repair machinery to repair heat-induced mutations.

Feedback control. Once repair processes are under way, it is advantageous for the cell to *scale back* its continuing production of the many heat shock proteins, some of which are made in large amounts. From what you learned already, can you deduce the feedback mechanism that turns down the heat shock response? Once the protein debris is gone, the many DnaK proteins induced during heat shock are now free to bind and target σ^H to the FtsH protease for destruction (Fig. 12.4). With σ^H out of the way, recruitment of RNAP to the promoters of the heat shock regulon is dampened, and the heat shock response is over. In fact, the response is turned off almost as fast as it was turned on.

Let us return for a moment to the case study, where the coyote's microbes endured transfer from the warm gastrointestinal tract to the desert pool, cool (10°C) in the evening and hot (40°C) during the day. The cells will use the heat shock response to adapt to the hotter, diurnal temperatures but will need a cold shock response to adapt to the cooler evenings. Although different in details, *E. coli* is also equipped to respond to cold shock by inducing expression of an appropriate stress regulon. And so the heat and cold shock responses adapt the cells to grow and survive through daily cycles of temperatures.

IS STATIONARY PHASE A STRESS RESPONSE?

LEARNING OUTCOMES:
- State how a bacterial cell in stationary phase differs from the same cell in exponential growth.
- Identify similarities and differences for all nongrowing cells of a given species for two cells that had different causes for the transition from exponential phase to stationary phase.
- Describe how RpoS mRNA and its translated protein σ^S act as a major regulator of the stationary-phase regulon.
- List examples of genes that are in the stationary-phase regulon and controlled by σ^S.
- Discuss how stationary phase can play a large role in evolution.

You remember stationary phase—the condition generated in microbiology laboratories whenever a flask containing a growth medium is inoculated with a small number of cells, and the culture is then incubated at a chosen temperature for long periods (chapter 4). Growth follows a predictable pattern: after a lag period, growth occurs exponentially for a time, and then slows and ceases—usually in a few hours—when nutrients are used up or toxic products of metabolism accumulate, or both occur. In the environment, stationary phase may well be the *default condition* for most microbes. Whether in the test tube or the environment, growth comes to a halt whenever the conditions no longer support it. No single stress response can deal with this complex situation. Instead, think about stationary-phase physiology as an emergency program of large-scale proportions, in which the cell enters a nongrowth state that may last for a long time. Because it involves so global a change in the cell, entry into stationary phase more closely resembles **differentiation** (chapter 13) than mere adjustment of some of its components.

The Stationary-Phase Cell

Cells in stationary phase differ profoundly from when they were growing. They are tougher and even look different than growing cells (Fig. 12.5). Their metabolism is altered in ways that promote survival rather than

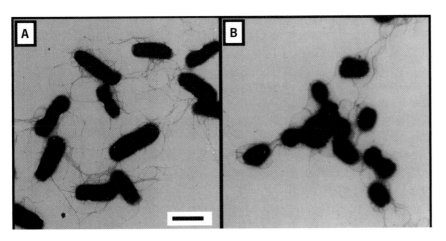

FIGURE 12.5 **Morphological changes in *E. coli* cells during growth (exponential phase) and nongrowth (stationary phase).** Electron micrographs of negatively stained cells of *E. coli* from exponential-phase (A) and stationary-phase (B) cultures showing the size reduction and aggregation of stationary-phase cells. Bar, 20 µm. (From Makinoshima H et al., *J Bacteriol* **185**:1338–1345, 2003.)

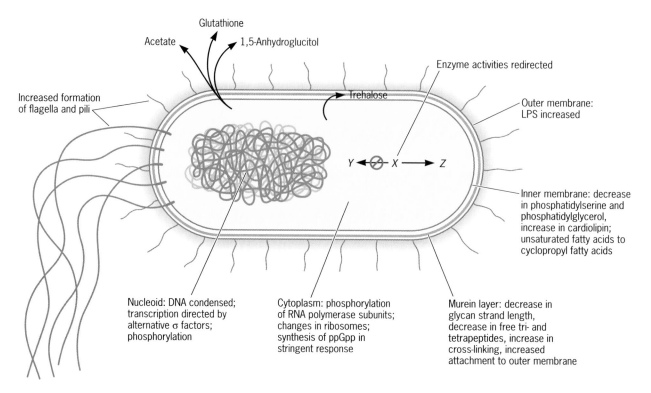

Glutathione

Acetate

1,5-Anhydroglucitol

Enzyme activities redirected

Trehalose

Increased formation of flagella and pili

Outer membrane: LPS increased

Y ← X → Z

Inner membrane: decrease in phosphatidylserine and phosphatidylglycerol, increase in cardiolipin; unsaturated fatty acids to cyclopropyl fatty acids

Nucleoid: DNA condensed; transcription directed by alternative σ factors; phosphorylation

Cytoplasm: phosphorylation of RNA polymerase subunits; changes in ribosomes; synthesis of ppGpp in stringent response

Murein layer: decrease in glycan strand length, decrease in free tri- and tetrapeptides, increase in cross-linking, increased attachment to outer membrane

FIGURE 12.6 **A stationary-phase cell.** The outcome of the differentiation process governed by RpoS (σ^S) is shown by indicating some of the many differences between the stationary-phase cell and one in active growth.

growth. As a result, everything has changed: their overall chemical composition and the chemical and physical nature of each cellular component (Fig. 12.6). In fact, if you were to compare the morphology, transcriptome, or proteome of these cells side by side with actively growing siblings, you might think that you were analyzing a different species. These adaptations contribute to the cell's capacity to survive under conditions that are not favorable for growth.

Cells may be unable to grow for innumerable reasons, including nutrient depletion, accumulation of toxic by-products of their own metabolism, the presence of toxic chemicals, or environmental changes in temperature, osmotic pressure, or pH. But *it matters little what caused the stationary-phase cell to stop growing*. Although each type of stress alters the profile of particular enzymes tailored to that condition, the cells all undergo similar global changes that alter their appearance and increase their toughness. In the next section, we will highlight some of the most notable features of this transition in *E. coli*, but keep in mind that different players operate in some other bacteria.

Transition to Stationary Phase

Differentiation of a growing cell into a nongrowing stationary-phase cell is guided by a sophisticated regulatory circuit. It is remarkable that the many different stressors that can potentially halt cell growth each converge in some common integrator molecules to trigger the production of regulatory molecules that reprogram the cell to enter stationary phase (Fig. 12.7).

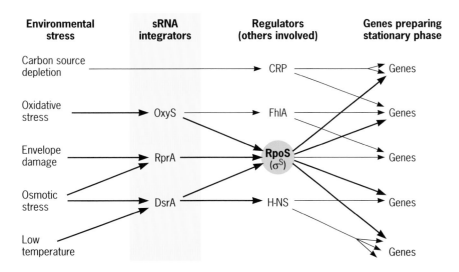

Environmental stress	sRNA integrators	Regulators (others involved)	Genes preparing stationary phase

FIGURE 12.7 **The regulatory cascade governing entry into stationary phase.** sRNA, small RNA.

Note that *the first controllers* are **small RNA** (sRNA) molecules. Rather than encode proteins, sRNAs regulate gene expression mainly during transcription or translation (chapter 11). In the stationary phase, sRNAs are integrators: *they process signals from different stressors and integrate them into a single, coherent action* (Fig. 12.7). These sRNAs stimulate translation of one or more regulators, including the stationary-phase σ factor RpoS (σ^S). As a result, RNAP is directed to the specific operons and genes of the stationary-phase regulon.

Activating the stationary-phase regulon. But how can an sRNA molecule stimulate the synthesis of σ^S? In growing cells, the 5′ region of RpoS mRNA that encodes the Shine-Dalgarno sequence and the AUG start codon is folded on itself, inaccessible to ribosomes. Thus, the mRNA secondary structure prevents translation and synthesis of σ^S. One sRNA integrator called DsrA has three loops or hairpins that share partial complementarity with the mRNA encoding RpoS (Fig. 12.8). In stressed cells, with assistance from the loading protein Hfq, DsrA binds to these complementary regions and partially unfolds the RpoS mRNA. This partial unfolding permits access of RNases to the hybrid DsrA-RpoS double-stranded region so it is cut. Now the Shine-Dalgarno sequence of the RpoS mRNA is accessible to the small subunit of the ribosome, which loads the large subunit and initiates translation of the RpoS mRNA.

Thanks to the DsrA integrator, σ^S is produced and some RNAP enzymes are redirected to the genes of the stationary-phase regulon. Their products lead to the distinctive structures and activities of the stationary-phase cell (Fig. 12.6). Although σ^S is a major player in this reprogramming of gene expression, other global regulators (H-NS, CRP, and FhlA) contribute by modulating and fine-tuning the cell's expression of individual operons of the stationary-phase regulon (Fig. 12.7). A network of this sort, in which *regulators govern regulators, which in turn govern other regulators*, is called a **regulatory cascade**. Complex as it is, this circuitry works with exquisite efficacy to ensure that more than 100 genes are coordinately regulated

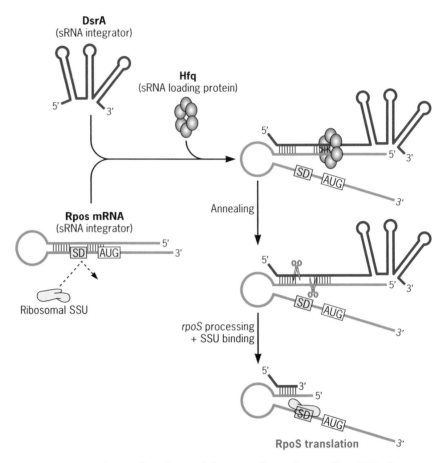

FIGURE 12.8 Regulation of synthesis of σ^S via translational control by sRNA. Once DsrA is expressed in the stress response, it is loaded onto the RpoS mRNA by Hfq to promote base pairing between complementary regions and opening of the RpoS loop that hinders the Shine-Dalgarno (SD) sequence in the mRNA. Much of the hybrid DsrA-RpoS double-stranded RNA is then degraded. The ribosomal small subunit (SSU) can now bind the SD sequence, load the large ribosomal subunit, and initiate translation of the RpoS mRNA.

to transition *E. coli* cells into stationary phase. Indeed, although other microbes display variations on this general theme, all rely on a stationary-phase sigma factor.

The stationary-phase response. The more than 100 genes in the stationary-phase regulon induce multiple changes in the cells, such as decreases in cell volume, changes in morphology, nucleoid compaction, and cell envelope modifications (Fig. 12.6). Many of these genes encode membrane proteins that control the permeability of the cell membrane, cell shape, and division, or lipoproteins of the outer membrane that promote cell-cell aggregation. Other genes increase acid resistance and osmotic stress and even encode an error-prone DNA polymerase that increases the mutation rate. (Why would cells purposely introduce more mutations during DNA replication?) Also in the RpoS regulon there are the more than 30 flagellar proteins that make motility happen. (And why are stressed cells programmed to become motile?) Furthermore, when σ^S competes with the housekeeping sigma factor σ^{70} to bind the core RNAP, it simultaneously switches off growth-related functions. *E. coli* gets two for one!

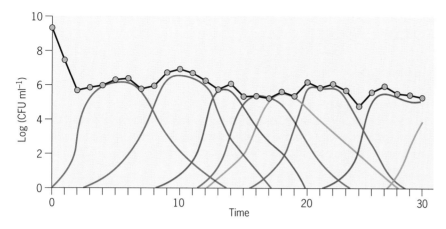

FIGURE 12.9 Sequential growth of mutants in stationary-phase cultures of *E. coli*. As cells enter stationary phase (time 0 on the x axis, in days), they begin to die, and the number of colony-forming units (CFUs) decreases concomitantly until reaching a stable level (indicated by the black plot). Phenotypic characterization of the cells that survive the long-term incubation shows the sequential growth and death of stationary-phase mutants in the culture (colored lines).

The stationary-phase response not only enhances cell survival, it also promotes evolution of bacteria under many circumstances in nature. How? Even during the so-called stationary phase, dynamic processes are at work. In fact, calling stationary-phase cultures "nongrowing" is misleading. Laboratory experiments reveal their under-the-radar behavior (Fig. 12.9). As much as 99% of the *E. coli* population dies within hours of transitioning into stationary phase. Yet about 1% of the cells (typically 50 million of the 5 billion cells in a 1-ml culture) can form colonies on agar plates. Who are these survivors? Cannibals! While most cells are dying and lysing, some cells are slowly growing and dividing using nutrients supplied by their neighbors' cadavers. This cannibalism can go on for months. The growth advantage that such subpopulations of cells have in stationary phase results from mutations introduced by the sloppy DNA polymerase that is induced during the stationary-phase response. The mutant survivors are more fit at feeding on the nutrients from dead cells; they grow and outcompete the others. Different mutant populations rise and die sequentially during the many months that the stationary phase can last (Fig. 12.9). When added as a minority to fresh cultures entering stationary phase, some mutants outcompete the parental strain. Thus, diversity, a key step in evolution, is generated in stationary-phase cultures in the lab. Under what conditions do you think such evolution occurs in nature?

Alternatives to Stationary Phase

Bacteria and *Archaea*, having the advantage (unlike us) of 3 billion to 4 billion years of evolution, have also evolved strategies to permit "suspended animation"; i.e., the ability to come back to a metabolically active state, and to grow, after immensely long periods of quiescence. Forming spores is one strategy practiced by some bacteria, and there are several different variations on this theme. Chapter 13 is devoted to well-studied examples of microbial differentiation and development. For now, it suffices to remember that microbes have evolved many ways to escape adversity. In the next sections, we will explore some other notable tactics.

DO MICROBES USE MOTILITY AND CHEMOTAXIS TO COPE WITH STRESS?

LEARNING OUTCOMES:
- State how flagellar movement is powered.
- Define "chemotaxis."
- Differentiate the distinct roles of the phosphorylation cascade and the methylation system for bacterial chemotaxis based on flagellar motility.
- Explain how accommodation by the methylation system works to prolong runs towards a chemoattractant.
- Identify four ways in which chemotaxis based on flagellar motility can help microbes overcome stress.
- Contrast three forms of motility that bacteria use to move on surfaces.

Microbes often cope with environmental stress by *migrating to a more hospitable location*. The movement of many (but not all) motile bacterial species is achieved by swimming using flagella. Recall from chapter 2 that a flagellum has three parts: a helical filament that functions as a propeller, a hook that works as a universal joint, and the basal body that functions as bushings and bearings to affix it to the envelope. The driving motor lies just internal to the basal body exposed to the cytoplasm. Let us take a closer look at these remarkable machines.

The Runs and Tumbles of Flagellar Motility

How do flagella equip the bacterial cell not just to move, but seemingly to move with purpose? Some clear answers have come from studying *E. coli.* Flagella rotate by turning the flagellar filaments using energy directly via a flow of protons—about 1,000 protons per turn—from the transmembrane electrochemical gradient (the same proton motive force that many microbes use to make ATP; chapter 6). Hence, *a transmembrane ion gradient, rather than ATP, powers the bacterial flagellum.* The filament can be rotated either clockwise or counterclockwise. The flagella of each cell are synchronized to rotate together in the same direction. Counterclockwise rotation causes the flagella to entwine into stiff bundles, producing a propeller-like force that results in a productive directed motion, called a **run** (Fig. 12.10). Clockwise rotation of the flagella dissociates the bundles, causing the cell to **tumble** in place (Fig. 12.10). The flagella can alternate between periods of counterclockwise and clockwise rotation; as a result, motile bacteria can move in brief runs interrupted by periods of tumbling. Each new run occurs in whatever *random direction* the cell happens to be facing after a tumble. Because this type of movement lacks a defined direction (Fig. 12.11A), the cells explore in a random walk rather than heading toward a specific destination.

But purposeful swimming *does* occur. Take, for example, the *E. coli* cells in our case study: they were dumped with the coyote feces into a nutrient-poor pool loaded with toxic metals. Which cells would be most fit: the ones that swim in random orientations or those that purposely steer away from the toxic chemicals and toward some tasty nutrients? The capacity to move toward "better pastures" would clearly be advantageous. This migratory behavioral response to chemical concentration gradients does indeed

A. Polar flagellum

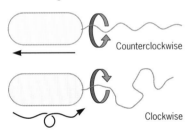
Counterclockwise

Clockwise

B. Many flagella (peritrichous)

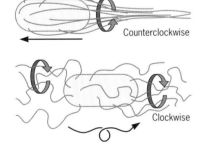
Counterclockwise

Clockwise

FIGURE 12.10 Flagellar behavior and motility. (A) A cell with a single polar flagellum moves from right to left when the flagellum turns counterclockwise and tumbles when the flagellum turns clockwise. (B) In a cell with many flagella all over its surface, counterclockwise rotation produces a coherent flagellar bundle and smooth movement. Clockwise rotation causes the bundle to fly apart; tumbling results.

occur. How important is this process of **chemotaxis**? To date, it has been found in roughly half the bacterial genomes examined.

Chemotaxis directs the cell's movement by controlling the frequency of runs and tumbles. For example, consider cells in a food desert that detect a useful nutrient. Sensing this chemoattractant, individual cells swim rapidly toward the source. A sequence of runs and tumbles directs the cells up the concentration gradient of the chemoattractant. How? Any productive run (i.e., one getting the cell closer to the nutrient source) is prolonged, whereas any neutral or counterproductive run is shortened (Fig. 12.11B). As a result, the cell inexorably approaches the goal through a series of so-called **biased random walks**. Along the way, cells focus their behavior by adjusting the frequency of runs and tumbles: they increase the frequency of tumbling when headed in the wrong direction but suppress the tumbles when they are on the right course. Now, picture the opposite situation: if a swimming cell senses a toxic substance, would it prolong the run or tumbles?

Sensing Where To Go: Chemotaxis

How do bacterial cells sense attractants or repellants? With a "nose"? Well, sort of. Assembled on one pole of bacterial cells are arrays of chemoreceptors called **methyl-accepting chemotaxis proteins** (MCPs) (Fig. 12.12). These chemosensors on the cell membrane continuously scout for chemicals in the environment, detect their concentration, and transmit signals to control flagellar rotation and the length of the runs. Intracellular signals are also transmitted by a two-component system, which integrates

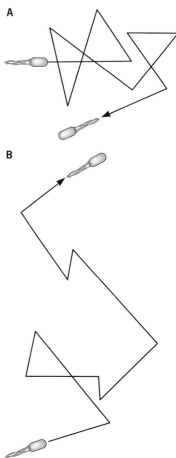

FIGURE 12.11 A biased random walk. (A) Typical random path of a motile bacterium such as *E. coli* in the absence of a chemotactic stimulus. Periods of runs and tumbling alternate by an endogenous schedule. (B) Path taken by this cell when an attractant is present toward the top of the figure. Runs last longer when they are in the direction of the attracting chemical.

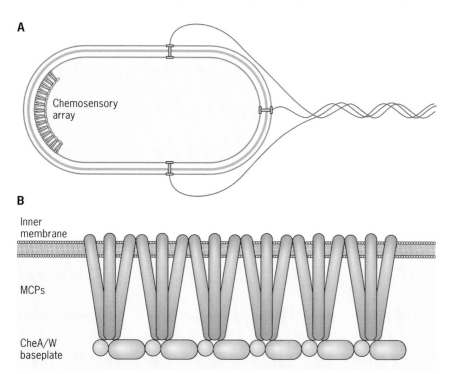

FIGURE 12.12 Organization of chemosensory arrays on the *E. coli* inner membrane. (A) Polar arrangement of the transmembrane chemosensory array of *E. coli*. (B) Side view of the array showing the MCP hexameric complexes (each one is a trimer of dimers, in dark blue) aligned on the membrane and held together by a baseplate of the CheW adapter protein (light blue) and the histidine kinase CheA (red).

information from the many chemoreceptors. This process creates a *chemical memory* that equips the bacteria to swim up a concentration gradient of a chemoattractant or down that of a repellent. Let us examine this circuitry in more detail.

Integration of chemosensory arrays and flagellar motility. Chemicals that diffuse into the periplasm are recognized and bound by a specific MCP transmembrane receptor. A densely packed array contains thousands of these MCPs, each held together by a cytoplasmic baseplate and the sensor CheA (Fig. 12.12). Each of the different MCPs within the array detects a specific attractant and/or repellent, and all of these receptors are integrated with the flagellar apparatus via the CheA sensor (Fig. 12.13).

In the absence of an attractant, CheA autophosphorylates at some frequency and, in turn, transfers its phosphate to the response regulator CheY (Fig. 12.14). Now activated, phosphorylated CheY interacts with switch proteins at the flagellum's base to change the direction of its rotation from its default counterclockwise to clockwise. This sudden reversal of the flagellum's rotation causes the cell to tumble in place. But feedback control is exerted by CheZ, a phosphatase that continuously dephosphorylates CheY-P and returns the flagellar rotation back to the default counterclockwise condition.

Sensing a chemoattractant. Now picture what happens when the cell migrates into an area with a chemoattractant, say, an aspartate that diffuses into the periplasm (Fig. 12.15). Here, together with a cognate periplasmic protein, aspartate complexes to its specific chemoreceptor. Upon binding,

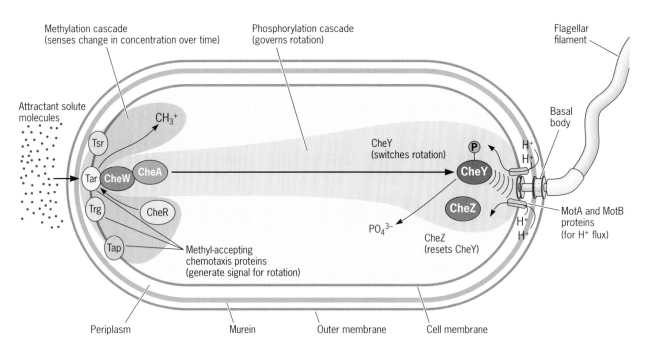

FIGURE 12.13 Integrated response of *E. coli* chemotaxis system and flagellum rotation to an attractant. A diffusible chemoattractant is shown entering the periplasm and complexing with a periplasmic binding protein. In this case, it interacts specifically with Tar, one of the membrane methyl-accepting proteins, triggering a phosphorylation cascade that activates counterclockwise rotation of the flagellum, leading to a run. Methylation of the membrane Tar soon leads to a reversal of this activation, and clockwise rotation causes the cell to tumble in place. The timing of this cycle depends on the new concentration of the attractant that is presented to Tar.

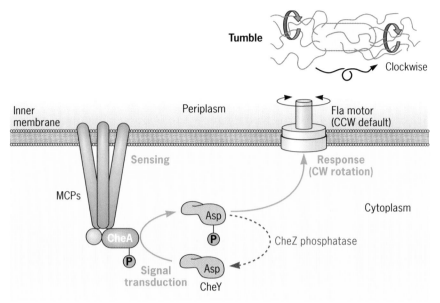

FIGURE 12.14 **Integration of chemosensory arrays and flagellum apparatus in *E. coli* via the CheA-CheY two-component system.** The chemotaxis two-component system of *E. coli* is composed of the CheA sensor of the chemosensory array and the CheY response regulator and uses a phosphorylation cascade as a signal between the two. In the absence of an attractant or a repellent, the CheA sensor autophosphorylates at some frequency and triggers a response that leads to change in flagellar rotation from the default counterclockwise (CCW) to a clockwise (CW) direction. This causes the cell to tumble.

the chemoattractant induces conformational changes in the chemorecep-tor complex that inhibit autophosphorylation of CheA. This inhibition has two major effects. On one hand, CheY does not get phosphorylated and the flagellum rotates in the default counterclockwise direction. Simulta-neously, the same conformational change promotes methylation of the che-moreceptor (Aha! the reason the chemoreceptors are called methyl-accepting proteins!). The protein that adds the methyl group to the chemoreceptor, the CheR methylase, is key to chemical memory. By methylating the chemoreceptor, the methylase resets its conformation so CheA can auto-phosphorylate again *but only when the cell detects a higher concentration of the chemoattractant.*

This process of resetting the sensitivity of the CheA sensor so it can only be activated at higher concentrations of the chemoattractant is called **ac-commodation**. (Accommodation accounts for many properties of human perception, such as our gradual failure to detect an odor after smelling it for an extended time.) Now, only a higher concentration of the attractant

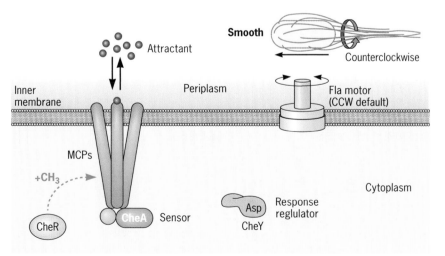

FIGURE 12.15 **Sensing an attractant.**

will prolong the bacterium's run. In effect, accommodation constitutes a *molecular memory* that equips cells to sense a change in attractant concentration and progress toward it. This remarkable molecular sensory system thus possesses many of the characteristics typical of behavioral systems in higher animals. And all this without even one synapse!

Sensing a repellent. Let us consider now the opposite situation, when a repellent, rather than a chemoattractant, binds to a chemoreceptor (Fig. 12.16). (In the chapter's case study, *E. coli* cells found themselves in a pool loaded with toxic metals, which are repellents.) When the repellent binds its cognate chemoreceptor, it causes a conformation change such that the *autophosphorylation of CheA is stimulated* rather than inhibited. This makes more phosphorylated CheY and increases the frequency of the tumbles. The increased activity of CheA also allows it to phosphorylate and activate proteins (CheB) that counteract any methylation so the sensitivity of the chemoreceptors is maintained. This is very important: if desensitized by methylation, the chemoreceptor would only detect higher and potentially lethal concentrations of the repellent. Hence, it is critical that the detection system returns to the pre-stimulus level once a repellent is detected and the signal is transmitted to the flagellum.

Does Chemotaxis Help Microbes Succeed in the Environment?

Chemotaxis is both a *survival* device (for escaping toxic substances and areas with low concentrations of nutrients) and a *growth-promoting* device (for finding food). It can also be a *virulence factor* that promotes colonization of the host by directing a pathogen toward its proper site of attack in the host. For example, the bacterium *Helicobacter pylori* chemotaxes to the stomach wall, its preferred attachment site. Chemotaxis also allows microbial cells to colonize surfaces and grow on them as biofilms, which provide

FIGURE 12.16 Sensing a repellent.

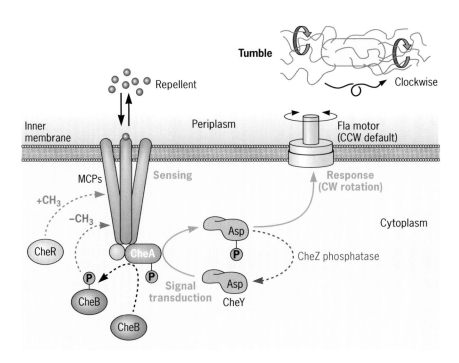

TABLE 12.2 **Bacterial tactic responses**

Tactic response	Description
Chemotaxis	Directed movement in response to chemicals (chemoeffectors), which can be either attractants (positive chemotaxis) or repellents (negative chemotaxis); widespread property of prokaryotes
Aerotaxis	Directed movement in response to oxygen; widespread property
pH taxis	Movement either toward or away from acid or alkaline conditions; *E. coli* moves from either acid or alkali to reach a neutral pH
Magnetotaxis	Directed movement along geomagnetic lines of force; believed to function in guidance up or down rather than north or south and useful in moving toward microaerophilic environments
Thermotaxis	Directed movement toward a range of temperatures usually optimal for the bacterium's growth; widespread property
Phototaxis	Directed movement toward wavelengths of light, usually related to photopigments and their function in the metabolism of the bacterium; property associated with virtually all photosynthetic bacteria

the cells with strength in numbers to cope with environmental stresses. As we shall see below, when we discuss cell-cell communication and social behaviors of microbes, inducing biofilm formation is an important selective advantage of chemotaxis. In fact, this chemosensory system may be a factor in the evolution of multicellularity.

Can properties of the environment other than a food source direct motility? Absolutely. Many bacterial cells can sense gradients of pH, temperature, light, salinity, and oxidation potential and move in a direction that optimizes conditions for that particular bacterial species. Some aquatic bacteria exhibit a particularly arcane version of directed motion called magnetotaxis—motility guided by sensing Earth's magnetic field. In chapter 3, we discussed how these microbes get their magnetic abilities: they accumulate magnetic crystals made of iron inside their cytoplasm, within structures called **magnetosomes**. This talent to orient in a magnetic field has nothing to do with north-south migration but with the ability to know which way is up. How? In most northern latitudes, the magnetic forces point mainly downward toward north, thus guiding cells to move deeper into the mud, a favored habitat for microaerophilic bacteria. Does this mean that magnetotactic bacteria in the Northern and Southern Hemispheres follow magnetic lines in opposite directions? Yes! Other examples of bacterial tactic responses are listed in Table 12.2.

The importance of moving about in the natural world is underscored by the great variety of motility devices bacteria have evolved. The following brief look at some of these will emphasize the lengths to which cells have gone to move in the right direction (so to speak).

Swarming Motility

Some bacteria use flagella for a different kind of motility, one that takes place on surfaces rather than liquids. This **swarming** phenomenon (Fig. 12.17) is a group process; individual cells do not swarm, but "rafts" of rod-shaped cells arranged side by side do. Why such cooperation is required is not yet clear. One interesting hypothesis is that swarms of bacteria can push others along, even transporting otherwise immobile spores of

FIGURE 12.17 **Swarming motility by *Proteus mirabilis*.** Fewer than 10,000 cells of *P. mirabilis* were inoculated on the top spot of the agar plate and allowed to grow for 12 to 16 hours at 32°C. As the cells grew and divided, the colony expanded as terraces due to the swarming motility of the cells. (ASM MicrobeLibrary, courtesy James Shapiro.)

some fungi. Swarming requires a slime layer secreted by the swarmer cells. One component of this slime is surfactin, a lipopeptide with exceptional surface tension-lowering power, synthesized and secreted by several swarming species.

Gliding Motility

A dozen or more genera of bacteria display a form of motility on solid surfaces called **gliding**. This movement occurs *without the use of flagella*. The term may include a number of different mechanisms that share only the fact that they produce flagellaless movement across a solid surface. Gliding has been observed in such diverse groups as cyanobacteria, myxobacteria, and mycoplasmas (chapter 13). In the myxobacterium *Myxococcus xanthus*, for example, the retraction of their pili (chapter 2) powers the movement of cells in groups (called **social motility**), but isolated cells can also glide by yet unknown mechanisms (called **adventurous motility**). Motility clearly matters, whether alone or in groups.

Twitching Motility

Yet another kind of movement is employed by several species of Gram-negative bacteria, including *Pseudomonas aeruginosa*, *Neisseria gonorrhoeae*, and *E. coli*. By **twitching**, these bacteria attach to the surface of host cells or to inanimate surfaces. As the name implies, the cells move in a jerky fashion across solid surfaces (such as an agar medium). Rather than depending on flagella, this type of motility depends on the presence of type IV pili (chapter 2). These remarkable filaments are dynamic, protruding and retracting to adjust their length. By doing so, the structures can extend ahead of the cell until the pilus attaches to the underlying surface (like a ninja grappling hook!). By subsequently retracting the filaments back into the cell, the bacterium translocates across the surface. Twitching motility contributes to biofilm formation (discussed below) and also to the aggregation that triggers differentiation in the myxobacteria (chapter 13).

DO MICROBES COME TOGETHER TO OVERCOME STRESS?

LEARNING OUTCOMES:
- Identify the classes of molecules that Gram-negative and Gram-positive bacteria use for cell-cell communication.
- Define "quorum sensing."
- Use an example of quorum sensing to describe how microbes communicate with each other.
- Describe the general properties of a biofilm.
- List three advantages that a biofilm offers over the planktonic mode of microbial life.
- Use evidence to justify the following statement: "Most bacteria in nature live in biofilm communities."

Biology is rich in examples of major insights arising from the study of seemingly minor, esoteric subjects. A case in point is the realization that microbes are not the lone players once thought but rather are *social* creatures that communicate with each other and exhibit cooperative behavior. Yes, you heard right! Microbes talk to each other and use their unique "language" to join together as a community. And, as you know, there is strength in numbers. So it may not surprise you to learn that microbes respond to many stressful situations by coming together as a community. In many cases, a surface becomes colonized and cells reprogram their activities for communal life within a biofilm. To understand how cells come together, we first need to learn how they communicate.

Cell-Cell Communication

Many microbes coordinate social behaviors with chemical languages by synthesizing and/or recognizing molecules such as **homoserine lactones** in Gram-negative bacteria and peptides in Gram-positive bacteria. To "hear" other cells, a microbe must have *receptors* that bind the chemical *signal* that they make; upon binding, a *response is triggered* to coordinate functions, such as the formation of a biofilm. Next, we examine some notable examples of communication among bacteria.

The LuxI-LuxR system. The revolutionary concept that microbial cells communicate was first appreciated in the 1990s but had its origin many decades earlier in studies of a light-emitting (bioluminescent) bacterium called *Vibrio fischeri*. This bacterium encodes one of the simplest communication circuitries known to date. An enzyme (LuxI) makes a signal or **autoinducer** and a response regulator (LuxR) that binds the signal. Binding to the autoinducer [a small ligand with a big name: N-(3-oxohexanoy)-L-homoserine] activates LuxR and induces expression of the seven-gene *lux* operon, encoding proteins that convert the energy of ATP into light (**luciferase** is one; Fig. 12.18).

The ecology of *V. fischeri* illuminates why population density governs luminescence by these cells (Fig. 12.18). With the first rays of sunlight, these free-living marine organisms transition from the open sea to the light organs of squid. It is within the confined space of the squid's light organ

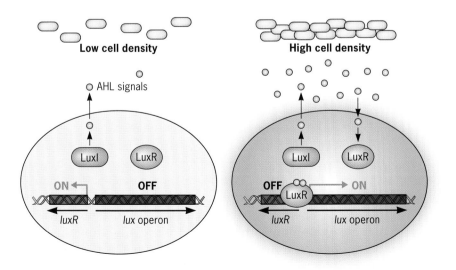

FIGURE 12.18 The LuxI-LuxR quorum sensing system in *Vibrio fischeri*.

that the vibrios grow while secreting the LuxI autoinducer. By nighttime, the population has increased to a high cell density and the autoinducer has become so concentrated that it freely diffuses in the cells. There, the autoinducer binds and activates the LuxR response regulator, which then induces the *lux* operon. *Lights up!* The biochemical reaction that leads to light is catalyzed by the enzyme luciferase, one of the proteins encoded in the *lux* operon, both in vibrios and in fireflies. The system also has a feedback control mediated by LuxR: once bound by the autoinducer, LuxR represses its own expression. The squid also contributes: after the squid feeds each night, the light organ expels vibrios back into the ocean, reducing the cell density, turning off their light. Thus, light emission from the squid *depends on the concentration* of the bacterial population.

How does each partner in this symbiosis benefit? Finding nutrients in the sea is not easy, and predators abound. The squid provides the vibrios with shelter and food. Meanwhile, this night-feeding squid gains a light beam to mask the shadow it would cast on a moonlit night—a camouflage to elude its predators. (In fact, the squid's light organ is equipped with an aperture that modulates its light emission to suit the strength of the moonlight!)

This phenomenon of inducing a response when the cells have reached a certain population number is called **quorum sensing**. The term is apt, because it evokes the idea that it takes a quorum of individuals, that is, a certain number, to take some kinds of actions. What would happen if you confined a single cell of *V. fischeri* to a very tiny space? (*Quiz spoiler:* You would fool it to think it has reached the quorum. In the physical confinement, the single cell would act like many cells in a larger space, producing a high concentration of the autoinducer.)

Quorum sensing is a widespread form of cellular communication that empowers cells to exhibit certain behaviors only when the population exceeds some threshold value. Many parasitic microbes gauge their numbers before taking an action. For example, the lungs of people with cystic fibrosis are often persistently colonized by *Pseudomonas*. To survive in the lungs, these bacteria exploit quorum sensing to establish complex biofilm communities (see below). You may recall from chapter 10 another example of quorum sensing: for the sexual transmission of genetic material by some Gram-positive species, population density is signaled by the secretion of an autoinducer called a **pheromone**. Might this bacterial exchange of molecular signals be the forerunner of the chemical communication so universal among insects? Quorum sensing is widespread in the microbial world; Table 12.3 lists several examples.

Formation of Organized Communities on Surfaces: Biofilms

An advantage of communicating is that, by sensing their own numbers, bacteria can gauge when to aggregate into multicellular communities. Indeed, most bacterial behaviors are communal, not individual. In a variety of environments, multicellular structures offer protection and other advantages. A large proportion of all microbes, perhaps a majority, reside within biofilms. In these communities, the microbes are **sessile** (from the Latin "to sit"). In contrast, most research on the physiology of bacteria has been conducted on cells that are **planktonic** (individually free living) and dra-

TABLE 12.3 **Sampling of bacterial quorum-sensing systems**

Genus	Autoinducer signal molecules	Physiological function(s)
Vibrio	Homoserine lactones	Bioluminescence
Pseudomonas	Homoserine lactones	Pathogenesis
Agrobacterium	Homoserine lactones	Conjugation
Bacillus	Peptides	Competence, development
Enterococcus	Peptides	Conjugation, plasmid maintenance, pathogenesis
Myxococcus	Peptides	Development
Streptococcus	Peptides	Transformation
Staphylococcus	Peptides	Pathogenesis

matically different from sessile cells. Typically, microbial communities form on *surfaces*, hence "biofilm." For example, biofilms form on marine organisms and their debris (called "marine snow"), on rocks in a stream, inside a corroded water pipe, on the hull of a ship, or on prosthetic medical devices implanted in humans. Indeed, biofilms are found everywhere there is a solid surface bathed by water. Obviously, surfaces have physico-chemical properties distinct from those of free solutions. Nutrients tend to adsorb to surfaces, so it makes sense for microbes to associate with them. In particular, dilute organic and inorganic substances are attracted largely by the electrostatic charge of the surface.

A variety of stressors trigger biofilm formation, and each microbe is unique in the signals that it responds to. But in all cases cells follow some common steps—essentially a programmed developmental process—to build a surface-attached community often visible with the naked eye (Fig. 12.19). Attachment is initially reversible, and the cell comes and goes until it sticks to the surface and commits to a sessile existence. Once this happens, the cells secrete a complex matrix of exopolymeric substances (EPS) that becomes the cement they need to build multicellular structures. The **EPS matrix** is more than a structural component of the biofilm: it is the factory where many communal reactions take place. The EPS matrix adsorbs water and nutrients, allows waste products to exit, and provides the cells with a protective armor.

The structure of the biofilm is tailored to the environment. Many microbes build flat biofilms, while others build architecturally complex structures,

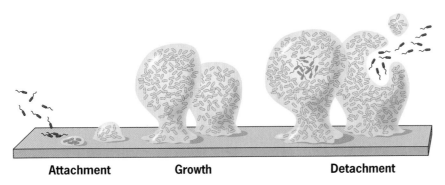

Attachment   **Growth**   **Detachment**

FIGURE 12.19 **Common steps in the formation of a biofilm.**

often made up of pedestals and mushroom-like structures surrounded by open channels of fluid. These architectural details do matter, as the canals distribute nutrients and dissipate waste products. Eventually, some cells disperse to seek new territory; or if more-favorable conditions arise, the whole biofilm community may dissolve and return to planktonic life.

But communal life offers many other advantages. Biofilms typically have enhanced resistance to antibiotics and other toxic substances. In some cases, the inhibitory concentration for an antibiotic is increased as much as 1,000-fold, in part because the cells are encased in a protective shell. In fact, the EPS matrix limits the diffusion of many other substances, retaining them in the top layers of the biofilms. Cells in the top stratum of the biofilm are exposed to toxic concentrations, but the residents underneath are protected. Moreover, within biofilms, individual cells express certain enzymes that confer intrinsic resistance to antibiotics or toxins. Consequently, clinicians trying to eradicate biofilm infections have a major challenge on their hands, and the microbial pathogenesis community has a great interest in biofilm biology.

In nature, biofilms and other communities often consist of several *different microbial species*. For example, the formation of dental plaque (which begins to re-form as you leave the dental hygienist's office) requires that certain "founder" bacterial species, notably *Streptococcus mutans*, stick to the tooth surface first. Bacteria of different species, such as some belonging to a large group called the actinomycetes, stick to these pioneers. Eventually, others stick to the actinomycetes, leading, in time, to the development of a complex scaffold of different kinds of organisms, a multispecies biofilm. We will discuss oral biofilms again in chapter 21, describing in more detail the interactions that allow the resident microbes to have a communal existence.

And microbes do not assemble biofilms solely for defense. As we will describe later, some communities form food chains (chapters 19 and 21). For example, certain marine *Archaea* can oxidize methane, an abundant gas in their environment. However, this process is thermodynamically unfavorable. To drive the reaction, these archaeal species associate with bacteria that metabolically remove the hydrogen generated during methane oxidation. Cooperating in biofilms, the two partners make methane oxidation happen. Since biofilms are ubiquitous, it is likely that such cooperative enterprises are common in the microbial world.

CASE REVISITED: How fast can microbes adapt to environmental changes?

By now you will appreciate that microbes can respond rapidly to stress to avoid irreversible harm. The *E. coli* cells in the case study were suddenly dumped in an inhospitable aquatic environment: a shallow pool of water exposed to intense sunlight during the day, with temperatures dropping from highs of 40°C to 10°C at night. On top of that, the pool was poor in nutrients, acidic, and likely contained toxic concentrations of metals that leached from the copper mine. The fact that some *E. coli* cells survived the ordeal is a testament to their capacity to cope with a range of environmental stressors. What type of stress responses do you think they used in the short term? And what about the long term: would they grow planktoni-cally, assume a resilient nonreplicating state, or form biofilms?

CONCLUSIONS

One stands in awe of the ability of microbes not only to survive conditions that for us are truly extreme but also to adapt to environmental change. The continued existence of a microbial cell is predicated on growth and survival. Both properties are essential. The many means by which mi-crobes adapt to a changing environment are as highly evolved as the pro-cess of growth itself.

Representative ASM Fundamental Statements and Learning Outcomes for the Chapter

EVOLUTION

1. Mutations and horizontal gene transfer, with the immense variety of microenvironments, have selected for a huge diversity of microorganisms.

    a. Discuss how stationary phase can play a large role in evolution.

CELL STRUCTURE AND FUNCTION

2. *Bacteria* and *Archaea* have specialized structures (e.g., flagella, endospores, and pili) that often confer critical capabilities.

    a. State how flagellar movement is powered.
    b. Define "chemotaxis."
    c. Differentiate the distinct roles of the phosphorylation cascade and the methylation system for bacterial chemotaxis based on flagellar motility.
    d. Explain how accommodation by the methylation system works to prolong runs towards a chemoattractant.
    e. Identify four ways in which chemotaxis based on flagellar motility can help microbes overcome stress.
    f. Contrast three forms of motility that bacteria use to move on surfaces.

METABOLIC PATHWAYS

3. The interactions of microorganisms among themselves and with their environment are determined by their metabolic abilities (e.g., quorum sensing, oxygen consumption, and nitrogen transformations).

    a. Identify the classes of molecules that Gram-negative and Gram-positive bacteria use for cell-cell communication.
    b. Define "quorum sensing."
    c. Use an example of quorum sensing to describe how microbes communicate with each other.

4. The survival and growth of any microorganism in a given environment depend on its metabolic characteristics.

    a. Identify the defining characteristic of "environmental stress" for a microbe.
    b. Give examples of microbial stressors.
    c. Identify some effects of heat stress on microbes.
    d. State how a bacterial cell in stationary phase differs from the same cell in exponential growth.

e. Identify similarities and differences for all nongrowing cells of a given species for two cells that had different causes for the transition from exponential phase to stationary phase.

INFORMATION FLOW AND GENETICS

5. The regulation of gene expression is influenced by external and internal molecular cues and/or signals.

   a. Speculate on why most of the microbial stress responses and adaptive mechanisms involve the regulation of many genes (frequently >12) at once.
   b. Outline the general pattern of the signal transduction/two-component stress response.
   c. Speculate about in what environment the microbial phosphate response regulon may be frequently turned on.
   d. Using the *E. coli* phosphate response regulon as an example, describe the nature, role, and location of the primary components of the two-component stress response.
   e. Describe how bacteria sense a rise in temperature.
   f. Identify the role of protein DnaK in the heat shock response and feedback control.
   g. Contrast the signal to regulate gene expression for the two-component stress response with that of the heat shock stress response.
   h. Describe how RpoS mRNA and its translated protein σ^S act as a major regulator of the stationary-phase regulon.
   i. List examples of genes that are in the stationary-phase regulon and controlled by σ^S.

MICROBIAL SYSTEMS

6. Most bacteria in nature live in biofilm communities.

   a. Describe the general properties of a biofilm.
   b. List three advantages that a biofilm offers over the planktonic mode of microbial life.
   c. Use evidence to justify the following statement: "Most bacteria in nature live in biofilm communities."

SUPPLEMENTAL MATERIAL

References

- **Goo E, An JH, Kang Y, Hwang I.** 2015. Control of bacterial metabolism by quorum sensing. *Trends Microbiol* **23:**567–576. doi:10.1016/j.tim.2015.05.007.
- **He K, Bauer CE.** 2014. Chemosensory signaling systems that control bacterial survival. *Trends Microbiol* **22:**389–398. doi:10.1016/j.tim.2014.04.004.
- **Persat A, Nadell CD, Kim MK, Ingremeau F, Siryaporn A, Drescher K, Wingreen NS, Bassler BL, Gitai Z, Stone HA.** 2015. The mechanical world of bacteria. *Cell* **161:**988–997. doi:10.1016/j.cell.2015.05.005.

Supplemental Activities

CHECK YOUR UNDERSTANDING

1. **The physiology of stationary-phase cells _____.**
   a. depends on what caused the cessation of growth
   b. is a general adaptation of cells to the nongrowth state
   c. occurs only when carbon sources become growth limiting
   d. all of the above

2. **Cells within biofilms _____.**
   a. do not replicate
   b. have markedly different physiology than planktonic cells
   c. are more resistant to antibiotics and toxic chemicals
   d. can detach and disperse

3. **Explain how inoculating bacterial cells of an unknown species into a medium at 37°C can lead to either a heat shock or a cold shock response.**

DIG DEEPER

1. Many two-component systems rely on covalent modifications, such as phosphorylation, to modulate protein activity and control the regulation of genes in response to a specific signal. Modulating protein activity via covalent modification would seem to be more costly than doing allosteric binding of a ligand.

   a. What advantage(s) might be offered by covalent modification?
   b. What regulatory mechanisms may benefit from allosteric regulation instead?

2. When an *E. coli* culture is shifted from 30 to 40°C, the growth rate accelerates swiftly but no lag is observed.

   a. Why is no lag phase observed after the temperature shift?
   b. Why do growth rates accelerate?

3. DnaK is a highly conserved bacterial protein chaperone that hydrolyzes ATP to facilitate the folding of nascent proteins and, during the heat shock response, assists in the re-folding of denatured proteins. A psychrophilic bacterium (Icy 1) containing a DnaK homolog was isolated from Antarctic seawater (0°C). However, as the graph shows, the ATPase activity of DnaK purified from strain Icy 1 at various temperatures was markedly different from that of *E. coli* DnaK.

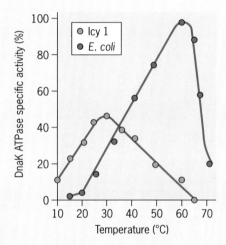

   a. What is the optimal temperature for the ATPase activity of the Icy 1 and *E. coli* DnaK proteins?
   b. Which DnaK protein do you expect to be more efficient at folding proteins at 37°C?
   c. Why did the activity of the Icy 1 DnaK protein decline rapidly at temperatures above 30°C?
   d. *E. coli* grows within the temperature range of 15 to 43°C, but a *dnaK*-null mutant cannot grow at either 15 or 43°C. Do you expect the heterologous expression of Icy 1 DnaK in the *dnaK*-null mutant of *E. coli* to rescue growth at 15°C? At 43°C?

4. In response to growth-limiting conditions, some motile bacteria aggregate in the liquid phase and form cell aggregates, or flocs. Read Alexandre's paper (G. Alexandre, *J Bacteriol*, **197**:3230–3237, 2015; doi:10.1128/JB.00121-15) to learn about cell flocculation and how it may be more similar to biofilm formation than previously acknowledged.

   a. How are flocculated and biofilm-bound cells similar?
   b. How does flagellar motility contribute to flocculation and biofilm formation?
   c. The author describes several examples that illustrate the role of taxis (chemotaxis and phototaxis) in promoting transient aggregation of motile cells. Pick one of the examples and prepare a short (5-minute) presentation describing the tactic behavior of that particular bacterium and the signal transduction system that modulates it. Search the Web to look for additional resources that can help you with the project.

Supplemental Resources

VISUAL

- In the iBiology lecture "Marvels of Bacterial Behavior" (32 min), Howard Berg discusses the physics of *E. coli* motility (http://www.ibiology.org/ibioseminars/biophysics-chemical -biology/howard-berg-part-1.html).
- Biofilm researchers explain biofilms and why they matter in the video "What Are Bacterial Biofilms? A Six Minute Montage" (6 min): https://www.youtube.com/watch?v=lpl4 WCM_9pM.
- Meet Bill Costerton—The "Father" of Biofilms (13:43): https://www.youtube.com/watch ?v=M_DWNFFgHbE.
- In the TED talk "The Secret, Social Lives of Bacteria" (19 min), Bonnie Bassler conveys the wonders of quorum sensing (http://www.microbeworld.org/component/jlibrary/?view =article&id=381).
- In *Microbes after Hours* episode #67, "The Secret Language of Bacteria" (55 min), Bonnie Bassler and Steven Lindow describe for a general audience how quorum sensing contributes to the lives of microbes in the ocean and on plants (http://www.microbeworld.org /component/content/article?id=1347).
- Video of chemotaxis (3:25): https://www.youtube.com/watch?v=h4lv7cBYVug.
- In the video "Dawn of the Cyborg Bacteria" (4:21), Ph.D. students Elizabeth Beattie and Denise Wong describe their research to harness bacterial sensing and motility machinery to create microscopic robots (http://www.microbeworld.org/component/jlibrary/ ?view=article&id=13555).

SPOKEN

- *This Week in Microbiology* podcast episode #104, "Feed Me Polyamines, Biofilm," discusses links between bacterial biofilms and cancer in the colon from the paper "Metabolism links bacterial biofilms and colon carcinogenesis" (Johnson CH et al., *Cell Metab*, **21:**891–897, 2015): http://www.microbeworld.org/podcasts/this-week-in-microbiology /archives/1907-twim-104-feed-me-polyamines-biofilm.
- *This Week in Microbiology* podcast episode #102, "Happiness Is the Spore-Formers in Your Gut," discusses evidence that serotonin synthesis is regulated by spore-forming members of the gut microbiota from the paper "Indigenous bacteria from the gut microbiota regulae host serotonin biosynthesis" (J. M. Yano et al, 2015 Cell 161:264-276): http://www.microbeworld.org/podcasts/this-week-in-microbiology/archives/1889-twim -102-happiness-is-the-spore-formers-in-your-gut.

WRITTEN

- S. Marvin Friedman highlights the role of noncoding DNA exchange in regulatory switching and environmental success in the *Small Things Considered* blog post "Noncoding DNA and Bacterial Evolution" (http://schaechter.asmblog.org/schaechter/2015 /02/noncoding-dna-and-bacterial-evolution.html).
- Gemma Reguera discusses the physics of swarming motility in the *Small Things Considered* blog post "Living on the Edge . . . of the Swarm" (http://schaechter.asmblog.org /schaechter/2012/04/living-on-the-edgeof-the-swarm.html).
- Daniel P. Haeusser discusses the machinery that equips *Mycoplasma motile* to glide across solid surfaces in the *Small Things Considered* blog post "Faster than a Speeding Bolt: *Mycoplasma* Walk This Way" (http://schaechter.asmblog.org/schaechter/2012/10 /fa.html).
- Mark Lyte discusses communication between microbes and the endocrine systems in the *Small Things Considered* blog post "The Road to Microbial Endocrinology" (http:// schaechter.asmblog.org/schaechter/2012/08/the-road-to-microbial-endocrinology.html).
- Gemma Reguera describes how motile cells disperse from biofilms to make holes through the sessile community and increase nutrient flow in the *Small Things Considered* blog post "Holey Biofilm!" (http://schaechter.asmblog.org/schaechter/2013/04/holey -biofilm.html).
- Heidi Arjes tells a story about programmed cell death within biofilms in the *Small Things Considered* blog post "The Slime That Smiles" (http://schaechter.asmblog.org/schaechter /2012/11/the-slime-that-smiles.html).

IN THIS CHAPTER

A subset of photosynthetic vegetative cells of the cyanobacterium Anabaena *(red) express a regulator of differentiation in heterocysts (yellow)*

CHAPTER THIRTEEN

Differentiation and Development

KEY CONCEPTS

This chapter covers the following topics in the ASM Fundamental Statements.

EVOLUTION

1. Cells, organelles (e.g., mitochondria and chloroplasts), and all major metabolic pathways evolved from early prokaryotic cells.
2. Human impact on the environment influences the evolution of microorganisms.

CELL STRUCTURE AND FUNCTION

3. *Bacteria* and *Archaea* have specialized structures (e.g., flagella, endospores, and pili) that often confer critical capabilities.

INFORMATION FLOW AND GENETICS

4. The regulation of gene expression is influenced by external and internal molecular cues and/or signals.

INTRODUCTION

At some point in ancient history, individual bacterial and archaeal cells acquired a new skill: by clumping together, they became better at surviving environmental challenges. In time, these cells developed into the **biofilms** that we described in chapter 12. But that was not all. Prokaryotes acquired other creative ways to form differentiated structures that catch our eye today. Eukaryotic cells then came along and, as we see in today's protists, animals, and plants, advanced this ability to differentiate into multicellular forms of a stupendous variety of shapes and forms. The distinction between the limited differentiation repertoire of the prokaryotes and the range of forms and shapes of the eukaryotes is a fact that clearly distinguishes the two. Why did the prokaryotes not develop into insects, trees, or dinosaurs is an unresolved question, but one that has not been wanting for theoretical considerations. Prokaryotic differentiation may be limited, but it is far from ordinary. We will explore some representative examples in this chapter.

CASE: **Toughening up the prokaryotic way**

Scientists discovered fossilized bees in 25- to 40-million-year-old amber and managed to isolate some of their abdominal tissue. Under the phase-contrast microscope they saw small, bright, round structures. To test if these particles were ancient bacterial endospores, the researchers first soaked some amber pieces in glutaraldehyde and bleach to kill contaminating bacterial cells, extracted the tissue under sterile conditions, and then placed the specimens in tubes containing a rich growth medium. After 2 weeks at 35°C, the culture was turbid with *Bacillus*-like cells.

- Did the *Bacillus* culture originate from the germination of ancient spores? Could it have been a contaminant?
- If indeed the culture had an ancient origin, how could the spores survive for millions of years in amber?
- What further evidence would convince you that the culture was the progeny of ancient *Bacillus* cells?

WHY DO PROKARYOTES DIFFERENTIATE TO BECOME A DIFFERENT VERSION OF THEMSELVES?

LEARNING OUTCOMES:
- Use examples to explain why prokaryotic cells differentiate into structurally and functionally different versions of themselves.
- Contrast typical prokaryotic cell differentiation with plant and animal cell differentiation.
- Identify why domesticated variants of *Bacillus* lost the ability to form biofilms.

In the past two chapters, we considered how microbes adjust their activities to survive and prosper, with respect both to coordinating the processes of growth (chapter 11) and to coping with environmentally imposed stress (chapter 12). We learned how individual cells become structurally and functionally specialized in stationary phase to get much tougher and to endure conditions that, for whatever reason, did not permit continued growth. Some groups of prokaryotes take differentiation to a greater extreme, undergoing progressive changes in form and function in a programmed developmental process (Fig. 13.1). In some cases, a cell differentiates into a different cell type; in others, cells aggregate to form multicellular communities, such as biofilms and other structurally distinct multicellular bodies that play specialized roles in their life cycles. Why would a cell differentiate into a structurally and functionally different version of itself, whether as a single cell or a multicellular entity? The short answer is that a cell transforms to be more fit to thrive and survive in a particular envi-

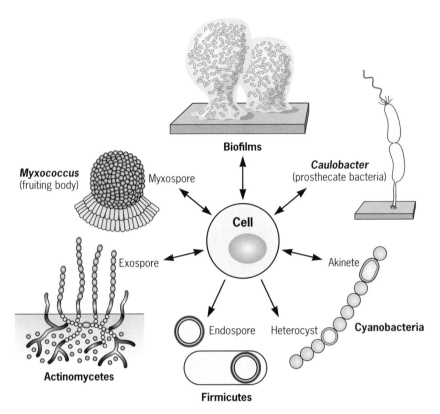

FIGURE 13.1 Examples of prokaryotic differentiation. The prokaryotic world has many examples of cellular differentiation, which are most often reversible (two-way arrows). However, there are also examples of terminal differentiation in which the differentiated type can no longer return to its former self (one-way arrows).

ronment. Hence, differentiation is *an adaptive response of the cells to their environment*. The environmental signals may be diverse, but all converge into the same response: the cell differentiates to acquire structural and functional attributes suitable to the new environment.

One important feature of prokaryotic differentiation is its reversibility. A cell in stationary phase can revert back to its original structure and morphology once environmental conditions favorable for growth return. Cells may come together as a biofilm to cope with stress, but disperse as planktonic cells in response to environmental cues. By contrast, differentiation in plants and animals is genetically programmed and is, for the most part, a one-way journey: specialized cells only rarely go back (dedifferentiate) to become stem cells (undifferentiated cells that are capable of differentiation). However, although prokaryotic cells that differentiate are usually capable of returning to their original cellular form, there are notable exceptions. For example, as we will learn later in the chapter, some filamentous cyanobacteria differentiate into a cell type that can revert back to a vegetative cell (the **akinete**) or that cannot (the **heterocyst**) (Fig. 13.1). Cyanobacterial akinetes are enlarged, dormant cells that can germinate and revert to a vegetative cell when conditions are favorable, whereas heterocysts are terminally differentiated, nitrogen-fixing cells that form when nitrogen sources become scarce. A cyanobacterial cell that becomes a heterocyst undergoes such profound changes necessary to supply fixed nitrogen to neighboring vegetative cells that they can no longer fix carbon via photosynthesis. Hence, they depend on neighboring vegetative cells to supply nutrients derived from the carbon dioxide that they have fixed. Another example of terminal differentiation, discussed in more detail in the next section, is the formation of **endospores** by some bacteria (Fig. 13.1). A cell differentiating into an endospore is called the **mother cell** because it nurtures the developing spore inside its own cytoplasm until its task is completed. Then, the mother cell lyses. This terminal form of spore differentiation contrasts with other examples well known in the bacterial world: spores that are dormant cellular types produced externally to facilitate their dispersal and germination in more hospitable environments (Fig. 13.1).

In addition to *distinct morphological changes*, prokaryotic differentiation is almost always *triggered by environmental cues*. In this respect, it resembles the stress responses discussed in chapter 12. Heterocyst development in cyanobacteria, for example, occurs when the organism is deprived of fixed nitrogen (such as ammonium or nitrate); resistant spores form when a culture runs out of nutrients. Biofilms also form in response to stress, including nutrient limitation. How? In this chapter, we will describe specific examples of bacterial differentiation and development of multicellular structures, noting the underlying themes. Much of our knowledge was generated using model organisms whose genomes have been sequenced and whose genetic systems are well developed. For example, mutants that have lost the ability to differentiate provide valuable insights into the molecular mechanisms that equip cells to differentiate through a series of defined developmental stages. Because these processes often reprogram the cell's physiology profoundly, high-throughput transcriptomic methods have proven instrumental to identify genes that are sequentially repressed or activated. The eventual goal and proof of our understanding of these processes is to construct mathematical models that not only describe the biological events of differentiation, but that can also predict the environmental triggers and internal changes that the cells undergo. Such command of the biology is still a way off, but there is a feeling that it can be achieved.

The capacity of the environment to exert strong selective pressure for microbes to differentiate into other cellular types and into multicellular entities is easily demonstrated in the laboratory. In the absence of natural selective pressures, laboratory strains can become domesticated and lose functions required for differentiation. For decades, for example, microbiologists have relied on growing laboratory bacterial strains in cultures as dispersed planktonic (floating) cells, since these controlled conditions facilitate biochemical, physiological, and genetic manipulations. In nature, however, most bacteria are not planktonic. Instead, they accumulate on solid surfaces, forming biofilms (Fig. 13.1). Many of the strains handled in laboratories "forgot" (genetically speaking) how to respond to environmental cues that otherwise trigger biofilm formation. These are *domesticated variants* of strains that exist in nature. As we learned in chapter 12, biofilms are not merely random associations of cells sticking together. There are precise developmental stages that lead a cell to aggregate in an orderly manner, reprogram its activities, and form multicellular structures with defined architectural features that enable the cells to function effectively as a community. Mutations in genes activated during transition from one developmental step to another would prevent the organism from forming biofilms. This is exactly what happened to laboratory strains of the Gram-positive bacterium *Bacillus subtilis*, a soil-dwelling organism that had been studied for decades as a model system for endospore formation. Biofilm formation had never been observed in a laboratory strain but, once less domesticated strains were studied, scientists were surprised to see the beautiful multicellular structures that developed on the surface of agar plates or at the air-liquid interface. When the domesticated and undomesticated genomes were compared, mutations were identified that allowed scientists to revert the original sequences in the domesticated strain and bring the strain back to its former self (Fig. 13.2).

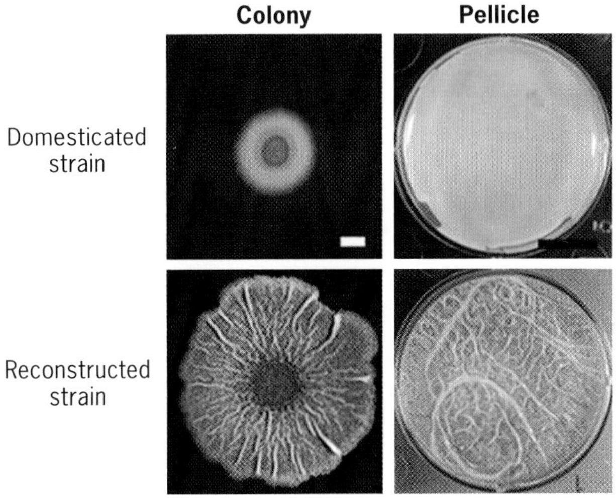

The finding that *B. subtilis* can form biofilms also led to the unexpected discovery that spore formation in this organism, as in many others, is a social event. Cells at the tips of biofilm projections **sporulate**, thereby facilitating spore dispersal. In this chapter, we will also learn about a group of bacteria, the myxobacteria, that come together as a multicellular community with distinct morphology, structure, and function to sporulate. In the case of myxococci, starvation signals the cells to move and congregate on surfaces in a coordinated fashion, aided by a type of surface motility called **gliding motility**. As they move, they prey on other bacteria for sustenance and then differentiate to form an elaborate fruiting body with a distinctive globular head (Fig. 13.1). A fraction of the cells in the globular heads then differentiate into spores called myxospores. Clearly, bacterial cells are equipped to coordinate all aspects of differentiation and developmental behaviors to control their fate. Let us learn more about how some model bacteria accomplish this.

FIGURE 13.2 Domestication of *B. subtilis* and loss of differentiation. The habit of growing laboratory strains of *B. subtilis* in planktonic cultures resulted in the accumulation of mutations and loss of their capacity to grow robust biofilms on the agar surface (colony) or at the air-liquid interface (pellicle). By comparing it to less domesticated or environmental strains, key mutations could be reversed to reconstruct a strain that recovered the ability to form robust biofilms. (Modified from McLoon A et al., *J Bacteriol* **193**:2027–2034, 201.)

WHAT ARE ENDOSPORES?

LEARNING OUTCOMES:
- Compare and contrast bacterial exospores and endospores.
- Identify evidence that the ability to make endospores evolved only once.
- Outline the process of endospore formation.
- Describe structural and biochemical features of endospores that make them vegetative and resistant to unfavorable conditions.
- Identify specific environmental signals that can trigger sporulation.
- Explain how it is possible for several different environmental signals to regulate the single pathway leading to the formation of Spo0A.
- Name the type of proteins that are responsible for the regulation of sporulation genes.

A

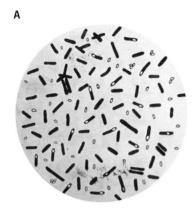

B

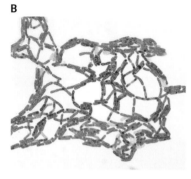

FIGURE 13.3 **Bacterial endospores.**
(A) Micrograph of cells of *Clostridium botulinum* stained with gentian violet; the endospores (oval objects) remain unstained. (B) Micrograph showing cells (red) and spores (green) of a *Bacillus* species stained after treatment with steam to allow a malachite green dye to penetrate the spore's coat. (Panel A courtesy of CDC–Public Health Image Library; panel B courtesy of Larry Stauffer, CDC–Public Health Image Library.)

Many microbial species produce **spores**, a dormant version of a cell that can persist through unfavorable conditions just to germinate when the conditions are right. Like fungi, some filamentous bacteria known as actinomycetes produce spores externally (thus the name **exospores**) by pinching off the tips of the filamentous cells that stick out in the air (Fig. 13.1). Aerial exposure matters here because exospores can *disperse in the air* to initiate new filamentous growth someplace else. Some species of bacteria, however, form differentiated intracellular structures that are also referred to as spores. These are **endospores**, so called because they *develop inside another cell* (Fig. 13.3). Furthermore, endospores *are agents of survival, not dispersal*. In this sense, they are functionally analogous to the stationary-phase cells discussed in chapter 12. Yet they are unique because they are extraordinarily resilient and can survive in a metabolically inert state for extended periods—certainly hundreds of years, perhaps thousands. In the case study we described the bright structures identified in fossilized bees preserved in amber that were millions of years old! Could they have been bacterial endospores that germinated after millions of years of dormancy? Or were they modern-day spores stuck to laboratory surfaces and utensils or floating in the air? Whatever type of spore, ancient or modern, made it into the culture, it germinated into a vegetative cell that grew and reproduced. When appropriate nutrients again become available, endospores do indeed sense the good news and rapidly **germinate** to become vegetative cells (sort of like the awakening of Sleeping Beauty). Can they be woken up after millions of years of dormancy, as in the case study? Endospores are highly resistant to extremes of temperature, radiation, reactive oxygen, acid, and alkali. Indeed, they are the most resistant of all known biological structures. For this reason, endospores readily enter into conversations about extraterrestrial life. Add the extra protection of a block of amber and you have a biological structure that can withstand the passage of time (perhaps millions of years).

In the case study, the researchers used chemical sterilization with glutaraldehyde and bleach to kill any cell in the amber block but the spores. To kill the spores you need harsher treatments. Heat treatment is the standard sterilization approach applied in canning food because it also kills bacterial endospores. Standard sterilization procedures with heat are also traditionally used in microbiology laboratories to sterilize growth medium and vessels within autoclaves. The temperature reached in autoclaves (121°C) kills spores, so you can be quite sure that it will also kill other microbial life forms. All of them? In the last few years, scientists have identified hyperthermophilic archaea that grow at autoclave temperatures. The first to be identified was described in chapter 1 and chapter 4 and was rightly named strain 121. This strain may thrive at temperatures that destroy endospores but it cannot grow at moderate temperatures. Hence, its discovery has not altered the sterilization regimes in use for most purposes. In fact, heat sterilization is still the procedure used to kill endospore-forming bacterial pathogens of humans, among them the agents of anthrax (*Bacillus anthracis*), botulism (*Clostridium botulinum*), tetanus (*Clostridium tetani*), pseudomembranous colitis (*Clostridium difficile*), and gas gangrene (*Clostridium perfringens* and others). For these virulent species, endospores contribute to transmission by their resistance to chemical and physical agents but, once inside the host, they play a limited role in the pathogenesis of these diseases.

Phylogenetic Distribution

In spite of enjoying a prominent status in basic and applied microbiology, endospore formation appears restricted to only one Gram-positive bacterial phylum, the *Firmicutes* (Fig. 13.4), and to several genera, such as *Clostridium* and *Bacillus,* within this phylum (Table 13.1). A notable endospore former within this group is the giant bacterium *Epulopiscium fishelsoni,* described in chapter 1. No archaea are known to make spores. Why? We don't know, but being restricted to one bacterial phylum suggests that making endospores may have *evolved only once*, after the *Firmicutes* branched off from the other bacterial phyla. The evolution of endospore formation also likely occurred once the *Firmicutes* had started to diversify because the phylum also includes many nonspore formers such as *Staphylococcus aureus* (chapter 23) and *Streptococcus pyogenes* among the cocci; nonsporulating rods include the diphtheria bacillus (*Corynebacterium diphtheriae*) and an important agent of food poisoning (*Listeria monocytogenes*, chapter 24). Can you imagine how other bacteria could acquire this skill? (*Quiz spoiler:* possibly by horizontal transfer of the relevant genes, although that is unlikely given their large number.)

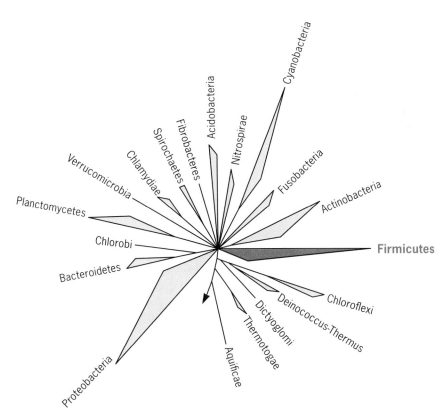

FIGURE 13.4 Phylogenetic tree highlighting only phylum with endospore formers, the *Firmicutes*, among the *Bacteria*.

TABLE 13.1 Genera of endospore-forming bacteria

Genus	Properties
Bacillus	Aerobes and facultative anaerobes; the most thoroughly studied endospores; includes important pathogenic species; e.g., the agent of anthrax
Clostridium	Anaerobes; includes important pathogenic species; e.g., the agents of tetanus, botulism, and gas gangrene
Thermoactinomyces	Thermophilic aerobes; closely related to *Bacillus*
Sporolactobacillus	Endospore-forming lactic acid bacteria
Sporosarcina	Obligate aerobic cocci (the only endospore-forming cocci), with cells arranged in packets of four or eight
Sporotomaculum	Anaerobes that carry out anaerobic respiration, using sulfate as a terminal electron acceptor
Sporomusa	Anaerobic, acetate-forming (acetogenic) bacteria
Sporohalobacter	Anaerobic, salt-resistant bacteria from the Dead Sea

Endospore Formation, Step by Step

The complex sequence of events that transforms a vegetative cell into an endospore has been most intensively studied using the Gram-positive aerobe *B. subtilis* as a model organism (Fig. 13.5). In all cases, the process starts with the typical step of replicating the chromosome, but then takes a different turn when the cell begins to divide asymmetrically to form a small cell with a copy of the genome that will eventually differentiate into a spore. The process of sporulation is triggered by *near depletion of any of several nutrients* (carbon, nitrogen, or phosphorus) in a relatively dense culture. The timing of this trigger is extremely critical: nutrients must be close to depletion but some must still remain to help the mother cell nurture the endospore forming within. This is why cells approaching complete starvation fail to form endospores. But when the right conditions are sensed, the cell commits to sporulating and a sequence of events is triggered that culminates in endospore formation.

Although there are subtle variations, endospore formation in all species progresses through essentially the same series of events that occur in *B. subtilis* (Fig. 13.5). Note that the stepwise process is divided by convention into stages that are designated 0, before any indication of sporulation is detectable, through VII, when the mature endospore is released. As cells grow and have completed or are close to completing replication of their chromosomes (stage 0), the nucleoid lengthens, becoming a structure called an **axial filament** that spans the length of the cell and positions the origins of replication of the newly replicated chromosomes closer to each of the cell poles. This is the stage (stage I) that signals the beginning of sporulation. A sporulation septum then forms at a point about a quarter of a cell length, closer to one of the poles rather than in the typical mid-cell position. For this reason, the sporulation septum is called the **polar septum**. The larger portion of the cell that remains after the polar septum becomes the mother cell, and its role is to nourish the developing spore. The smaller portion of the dividing cell is the **forespore**, which eventually becomes the actual spore. As the septum begins to form, only about 30% of the

A

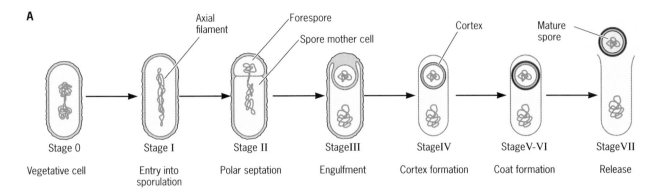

B

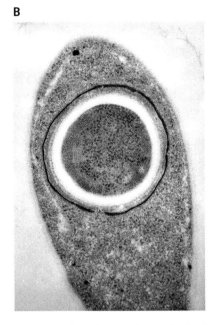

FIGURE 13.5 **Endospore formation in the model organism *B. subtilis.*** (A) Stages in the process of endospore formation. (B) Electron micrograph of a thin section of an almost completely developed spore of *Clostridium tetani* (stages V and VI). The core is almost completely dehydrated and packed with ribosomes. The cortex and spore coat are clearly seen.

chromosome that will eventually be in the spore is on the forespore side. What to do? One of the copies of the chromosome is literally pumped into the forespore by a mechanism similar to the transfer of DNA that occurs during conjugation (chapter 10).

Until this point, sporulation resembles a cell division event, differing only in that the division is asymmetric. But here is where it takes a different turn: the mother cell *behaves as a phagocyte*, sending out pseudopod-like extensions that engulf the forespore (this may be the best, perhaps the only reliably, known example of phagocytosis in prokaryotes) (stage III; Fig. 13.5). When the edges of the mother's extensions meet at the other side of the forespore, their cytoplasmic membranes fuse, engulfing the forespore completely. The forespore (as is the case with phagocytized cells) is then surrounded by two cytoplasmic membranes: its own and the mother cell's. The outer surfaces of the two membranes face one other. Now is the time for the forespore to layer up and get tough (stages IV to VI; Fig. 13.5). To do this, a thick murein layer, called the cortex, is laid down between the two

membranes. The murein of the cortex differs in several chemical properties from that of the cell wall of vegetative cells. Outside these membranes
(from the viewpoint of the forespore), a protein spore coat forms that
might be surrounded by a membranous layer called the **exosporium**. As
it becomes surrounded by layers of protective coverings, the forespore
changes internally. It synthesizes large quantities (up to 15% of the spore's
dry weight) of a spore-specific, low-molecular-weight compound, **dipicolinic acid**. When complexed with calcium, this compound binds water,
causing the spore to become drier and compact. Eventually, at stage VII of
sporulation (Fig. 13.5), the mother cell lyses, releasing the mature spore.

These carefully programmed steps transform a cell division into a process that differentiated a large, metabolically active vegetative cell into a
small, compact version of itself that has no detectable metabolism (Fig. 13.5B).
The mature endospore appears to be completely inert. It contains no measurable ATP or reduced pyridine nucleotides, the reducing power of a metabolically buzzing cell. Its interior (called the core) is 1.5 to 3 times drier
than the vegetative cell and takes up water only upon germination, which
accounts in part for its remarkable resistance to the moist heat of pressurized steam at temperatures exceeding the boiling point. The "Sleeping
Beauty" now awaits the moment in which it can be woken up, perhaps
millions of years later as suggested in the case study.

Reprogramming a Vegetative Cell To Make an Endospore

The stages of sporulation we have just described from stage 0, before any
indication of sporulation is detectable, through VII, when the mature endospore is released (Fig. 13.5), are genetically programmed. Indeed, it is
possible to interrupt each step by mutating the genes that control it. This
genetic approach has identified many genes essential for sporulation,
which are designated by the particular stage at which the process is ar-
rested. For example, mutations in the gene *spo0A* arrest spore formation at
stage 0. This gene also happens to encode the protein, that once activated,
sends the cell into the stages of sporulation I to VII. Hence, the sporulation
process can be divided into a "before" and an "after": once Spo0A is activated and the concentration of activated Spo0A reaches a threshold, this
regulator triggers the cell's reprogramming and committing the cell to make
an endospore. This is clearly an oversimplification of an intricate regulatory network. (It is so complex that the process has not been fully elucidated even after being studied intensely for the past 50 years!) But this split
will conveniently help us dissect the genetic control of sporulation into two
parts: (i) *how Spo0A is activated* and (ii) *what activated SpoA does*.

Activation of Spo0A. Spo0A is activated once it is phosphorylated
(Spo0A~P) as part of a cascade collectively called the **Spo0A phosphorelay
system** (Fig. 13.6). The Spo0A phosphorelay is a relatively straightforward
phosphorylation cascade, much like the two-component systems described
in chapter 12, involving a kinase, a response regulator, and a phosphorylation signal between the two. There are, however, several notable differences. The cascade is started, for example, by not one but five sensor kinases.
Each kinase autophosphorylates in response to environmental triggers and
transfers the phosphoryl group to Spo0F. Then, through the mediation of

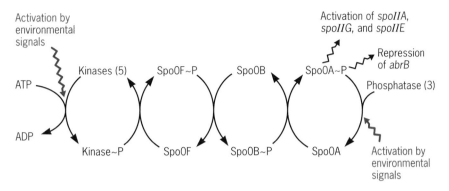

FIGURE 13.6 **The phosphorelay system that regulates sporulation and sporulation-associated events in** *B. subtilis.* Sites of regulation are shown in dark red.

a **phosphotransferase** (Spo0B), Spo0F~P donates its phosphate group to Spo0A. Note that, although appearing as a linear phosphorelay pathway, five sensor kinases serve as input proteins to transduce several environmental signals. The overall system is further regulated and fine-tuned by three **phosphatases** that remove phosphate groups from the end product (Spo0A~P) or from one of the intermediates. Thus, both phosphorylation and dephosphorylation collaborate to set the optimal internal concentration of Spo0A~P. In fact, *both kinases and phosphatases are activated by a particular signal*; thus, these enzymes respond to a large repertoire of environmental cues. Sporulation, then, can be set in motion by many different changes in environmental circumstances.

Two of the major environmental triggers of sporulation are the near depletion of nutrients and sufficient cell density. Each of the five kinases is, for example, activated independently by different signals, including the near depletion of one of several nutrients. Cell density is also a trigger. It is determined by **quorum sensing** (chapter 12) using certain peptides that accumulate to a critical threshold before they are sensed by surface receptors. This binding triggers the phosphorylation of two of the kinases and the intermediate proteins Spo0F and Spo0B, thus setting in motion the cascade that leads to the accumulation of Spo0A~P.

The response of the cell to nutrient depletion and cell density, measured as the concentration of Spo0A~P in the cell, is gradual. At low levels, Spo0A~P cannot prevent the synthesis of a **repressor** (AbrB) that turns off the expression of several genes required for sporulation. The cell is not ready yet to commit to dormancy. It is sensing the first signs of nutrient depletion and a crowded environment, but it might still be able to obtain additional nutrients and thereby postpone the need to sporulate. Hence, at this stage, the cell takes several actions: it *synthesizes antibiotics and toxins* as possible chemical mediators of predation (a prey is food!), *induces competence* to increase the chances of getting foreign DNA for genetic variation (desirable here) and perhaps as food (chapter 10), *turns on motility and chemotaxis* to move to new areas in pursuit of nutrients (chapter 12), and *expresses transport systems and catabolic pathways* to utilize alternative sources of nutrients that might be present. The cell works hard to maintain growth at all costs. But should these emergency responses fail and sporulation signals continue to be sensed, the concentration of Spo0A~P will increase to levels

TABLE 13.2 **Sigma factors that direct expression of sporulation genes**

Sigma factor	Encoding gene	Location	Stage at which active
σ^H	*sigH*	Predivision sporangium	0
σ^F	*sigF*	Forespore	II
σ^E	*sigE*	Mother cell	II
σ^G	*sigG*	Engulfed forespore	IV
σ^K	*sigK*	Mother cell	IV

that repress the synthesis of AbrB. Without the AbrB repressor, genes coding for proteins needed for sporulation are expressed. The cell is now committed to sporulation, and differentiation will continue until an endospore is formed and released.

Activity of Spo0A~P. Perhaps not surprisingly, the regulation of sporulation is more complicated than controlling the expression of the starvation-associated reactions. Sporulation depends on genes being expressed at the proper time—some immediately, to start the process, and others later, at particular stages of spore development. In addition, genes must be expressed in the proper location, some within the mother cell, some within the forespore. This differential expression occurs in spite of the fact that the mother cell and forespore are genetically identical.

Although our understanding of all the genes that regulate sporulation and how they interact is far from complete, the general outlines of the process are taking shape. The central actors are **sigma factors** (proteins that bind to RNA polymerase and thereby confer their ability to interact specifically with certain promoters [chapters 8 and 11]). During vegetative growth, a sigma factor designated sigma-A (σ^A) predominates. Then, during sporulation, five other sigma factors are synthesized sequentially that cause various spore genes to be expressed at specific times during sporulation and in specific locations (mother cell or forespore) (Table 13.2). This prominent role of sigma factors is reminiscent of the major role played by sigma-S (σ^S) in preparing growing cells to enter stationary phase (see chapter 12). But during sporulation the interplay between sigma factors also matters.

HOW DOES *CAULOBACTER CRESCENTUS* MORPH INTO TWO DIFFERENT CELLS?

LEARNING OUTCOMES:
- Speculate advantages for *Caulobacter crescentus* from differentiating into two different cells, a swarmer cell incapable of reproducing and a sessile cell.
- Describe two ways in which *Caulobacter crescentus* cells control development and differentiation.

The bacterium *Caulobacter crescentus* has its own unique differentiation style: it can divide into two cells that are nothing like each other (Fig. 13.7). The cycle starts with a motile, rod-shaped cell called the swarmer cell, which has a polar **flagellum** and **pili**. When a swarmer cell divides, the two prod-

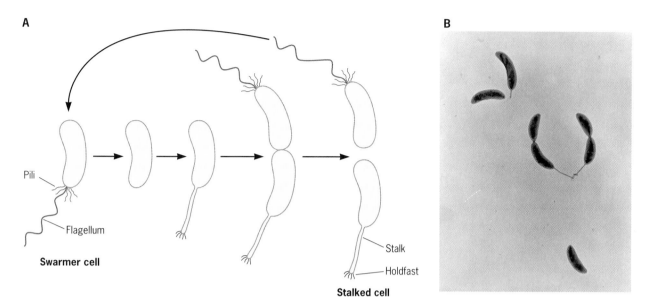

FIGURE 13.7 **Differentiation in *Caulobacter crescentus*.** (Panel B courtesy of Yves Brun.)

ucts of division differ in both appearance and behavior. One of the cells is a swarmer, like the mother cell, but the other is nonmotile, or **sessile**. The nonmotile cell bears a **prostheca** in its old pole (in distinction to the "new" pole formed by division). The prostheca is a cellular projection containing a filamentous part or stalk and a terminal organ called a holdfast, which anchors the cell to a solid surface. For *Caulobacter* to attach, almost any surface will do; e.g., a rock in a pond or the faucet of a sink.

The *Caulobacter* Cell Cycle

Producing a differentiated cell that can attach to an environmental surface sets in motion a developmental cycle that is essential to the reproduction of *C. crescentus* and other prosthecate bacteria (Fig. 13.7). A stalked cell continues to divide, each time producing a new swarmer, which actively swims away, powered by its polar flagellum. Swarmers, which are chemotactic, are clearly agents of dispersal: they leave their tethered, stalked mother cell seeking to colonize new areas with supplies of nutrients.

Interestingly, the newly formed swarmer cannot replicate its DNA or divide—not until it differentiates to become a stalked cell, which, like its mother cell, is capable of dividing and producing more swarmer cells. Each swarmer cell goes through a highly structured developmental progression before it can reproduce; altogether, at 30°C the process takes about 90 to 180 minutes (depending on the medium) (Fig. 13.7). For about the first 15 minutes, a swarmer swims chemotactically. Then, it dismantles its distinctive swarmer cell features: it sheds its flagellum, retracts its pili, synthesizes a holdfast at the same pole, and degrades its chemotaxis apparatus (which is also polar). Next, it replicates its DNA and synthesizes a stalk at the site bearing a holdfast. At 110 minutes, it starts to divide to produce a new swarmer. Considering how elaborate the swarmer's developmental progression is both morphologically and biochemically, it may not surprise you to learn that many genes are turned on and off throughout the process.

How many? About a fifth of *Caulobacter*'s genes are activated or repressed at discrete times during the cell cycle.

Control of Asymmetric Development

The cell cycle of *Caulobacter* has two particularly important features. One is the powerful influence of *cellular polarity*. Bacterial cells are inherently asymmetric because when they divide by binary fission they produce daughter cells with an "old" and a "new" pole, the latter being the one synthesized in the most recent cell division. Bacteria like *Caulobacter* capitalize on this distinction between new and old pole to take asymmetry to the extreme. For example, most of the significant events in *Caulobacter* development—formation of flagella, pili, chemotactic apparatus, and holdfast-bearing stalk—occur at the cell's old pole. The cell seems to be compartmentalized, although it has no internal membranes to define separate compartments. However, protein-labeling methods (such as **green fluorescent protein** tagging) show that certain constituents, including regulatory proteins, accumulate at particular intracellular locations. Indeed, localization of regulatory molecules is essential for such spatially differentiated development. The detailed basis for such localization remains the subject of active research. For some proteins, a "diffuse and capture" method operates: the molecules move rapidly through the cell by diffusion and are captured by a high-affinity interaction with some cellular component at the pole.

The other important feature is the *precise reprogramming* that turns on and off the biosynthesis of cell components during cell division. *Caulobacter* cells appear to follow a "just-in-time" policy of gene expression (like what we saw above in the case of sporulation). Genes encoding components of particular cellular structures and functions—for example, flagellar biosynthesis and associated chemotaxis systems—are expressed just before their products are used. Likewise, when no longer needed, these systems and their components are quickly destroyed by specific proteases that are also synthesized just in time.

Several phosphotransferase cascades, of the sort we encountered in spore formation, seem to be dominant controllers of the *Caulobacter* cell cycle, as well. Each appears to control a modular set of proteins that cooperate to carry out a particular cellular function. Regulatory linkages "talk" to the various modules and synchronize them. One particular response regulator protein, CtrA, is the master regulator, playing a role analogous to that of Spo0A in sporulation. Directly or indirectly, it controls about a quarter of the genes that are regulated in the cell cycle.

Lastly, we introduce here a multitalented molecule that participates not only in the control of asymmetric development in *Caulobacter* (Fig. 13.8), but also in many other differentiation pathways. It is **cyclic di-GMP**, the so-called **universal second messenger** (Fig. 13.8A). It works as a signal molecule that affects the transcriptional regulation of a large number of genes. When a *C. crescentus* cell divides asymmetrically, cyclic-di-GMP levels are over five times higher in the stalked cell than in the flagellated swarmer cell (Fig. 13.8B). The asymmetric distribution of the second messenger reprograms the swarmer and motile cells differently so each can follow a distinct path of differentiation. The scope of cyclic di-GMP's power is truly

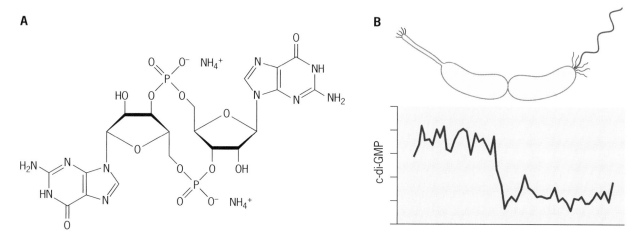

FIGURE 13.8 **Asymmetric distribution of cyclic di-GMP in *Caulobacter crescentus.***

amazing. This molecule interacts with many different regulatory proteins to control a large number of phenomena besides *Caulobacter* differentiation, notably sporulation, biofilm formation (chapter 12), adhesion, motility, surviving nutritional stress, and virulence, among many others. Being a widespread but purely bacterial constituent, it will not surprise you to learn that some hosts recognize cyclic di-GMP as a bacterial "calling card" that alerts the immune system to unwanted invaders.

HOW DO MYXOBACTERIA FORM FRUITING BODIES AND SPORULATE?

LEARNING OUTCOMES:
- Identify selective advantages conferred by cellular differentiation (i.e., aggregates and fruiting bodies) in myxobacteria.
- Identify selective advantages conferred by the ability of myxobacteria to produce myxospores.
- Describe how myxobacteria regulate the cycle of development.

The myxobacteria are a group of soil-dwelling *Proteobacteria* (Fig. 13.4), in the class *Deltaproteobacteria*, that undergo an elaborate process of differentiation involving multicellularity and sporulation as an integral part of their life cycle (Fig. 13.9). Individual cells gather into swarms that move by gliding motility over solid surfaces, killing and consuming other microbial cells in their path. No wonder they have been called wolf packs. The swarms are not selective: they devour filamentous fungi, yeasts, and protozoa, as well as other bacteria. Being prokaryotes, myxobacteria are *incapable of ingesting their prey*. Instead, they secrete antibiotics and enzymes that kill and lyse cells, along with other enzymes that break down the released macromolecules. They then take up the resulting small molecules, primarily amino acids. This mode of killing and consuming microbial cells depends on group action: only by pooling their resources can myxobacterial cells create a sufficient concentration of lytic enzymes to mediate the process. For example, dilute cultures of myxobacteria are incapable of metabolizing

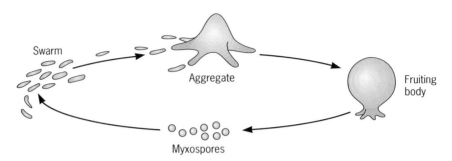

FIGURE 13.9 **Life cycle of *Myxococcus xanthus*.**

the protein casein, whereas denser cultures can readily do so to utilize the amino acids that are released.

When prey becomes scarce (as signaled by near exhaustion of amino acids in their environment), the developmental cycle begins (Fig. 13.9). Individual cells switch from "adventurous" (A) to "social" (S) motility, which now has a component of pilus-driven twitching motility (chapter 2). Cells become more closely packed in "aggregation centers." Then, they pile on top of one another to form an elaborate multicellular structure called a fruiting body, which in some species can be nearly a millimeter high. Fruiting bodies are not haphazard piles of cells; they have a *definite form and size*, characteristic of the particular species that produces them (Fig. 13.10). At the tips of these fruiting bodies, globular spore-containing structures or **sporangioles** form. About half the cells within these structures lyse, releasing nutrients that are used by the survivors to differentiate into resistant resting cells called **myxospores** (Fig. 13.9). Although not nearly as hardy as endospores, myxospores survive considerably more heat, UV irradiation, and desiccation than the vegetative cells can handle. In soil stored at room temperature, myxospores can survive for at least 15 years! When conditions become favorable, myxospores germinate, initiating a new growth cycle.

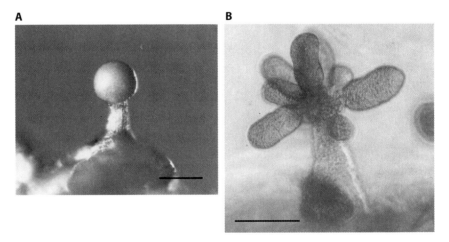

FIGURE 13.10 **Fruiting bodies of myxobacteria.** Fruiting bodies of myxobacteria vary in complexity from the relatively simple one of *Myxococcus stipitatus* (A) to the elaborate structure produced by *Stigmatella aurantiaca* (B). Bars, 50 μm. (Reprinted from Reichenbach H, *in* Dworkin M, Kaiser D, ed, *Myxobacteria II*, American Society for Microbiology, 1993.)

The fruiting bodies of various myxobacteria are quite diverse in shape (Fig. 13.10). Those of *Myxococcus xanthus*, the most thoroughly studied myxobacterium, are quite simple: they consist of a single dome-shaped sporangiole. In contrast, those of *Stigmatella* spp., which are usually found on the bark of rotting wood, are elaborately shaped and richly colored. These bright orange structures consist of several sporangioles, each of which sits at the tip of a branched stalk. The elevated position of the sporangioles in the fruiting bodies probably facilitates the dispersal of the myxospores. Worms or insects moving toward a more nutrient-rich environment might carry the whole sporangiole along to a new location that has not been depleted of microbial prey. The concentration of myxospores within the sporangiole ensures that *germination will give rise to a community of cells*; this is important because, as we have seen, myxococcal feeding is a group effort.

Regulation of Development

Because the developmental cycle of myxobacteria relies heavily on interaction between groups of cells rather than on differentiation of individuals, we probably should not be surprised to learn that cell-to-cell signaling directs the process. At appropriate times during development, cells of *M. xanthus* produce at least five molecular signals (designated A, B, C, D, and E) that direct and coordinate the process. Some, but not all, of these intercellular signals have been identified chemically. For example, the A signal, which is produced when cells sense starvation, is a mixture of amino acids and peptides. The C signal is a mixture of two proteins. These molecular signals and evidence for the existence of those that have not been characterized chemically were discovered by genetic strategies. For example, a mutant strain that by itself fails to initiate or complete the developmental cycle can do so when mixed with wild-type cells (or cells of certain other developmentally defective mutants).

HOW DO FILAMENTOUS CYANOBACTERIA UNDERGO DIFFERENTIATION AND DEVELOPMENT?

LEARNING OUTCOMES:
- Distinguish structure, function, and metabolic pathways for the three major forms of cellular differentiation in *Nostoc punctiforme*: heterocysts, akinetes, and hormogonia.

Many other bacteria differentiate and undergo development. Some of the most thoroughly studied are listed in Table 13.3. Here we will mention some notable features about differentiation in filamentous cyanobacteria, as they include both reversible and terminal differentiated cell types (Fig. 13.1). Possibly, *Nostoc punctiforme* and related filamentous cyanobacteria win the differentiation record for prokaryotes (Fig. 13.11). Here are the kinds of differentiation *N. punctiforme* undergoes:

1. **Heterocysts:** Nitrogen limitation induces some cells within a filament to terminally differentiate and become nitrogen-fixing specialists called heterocysts. Differentiation in this case includes the synthesis of the

TABLE 13.3 **Some well-studied examples of bacterial differentiation and development**

Group	Example	Differentiation and development
Actinobacteria	*Streptomyces coelicolor*	Forms both aerial and substrate mycelia. Forms exospores at tips of aerial mycelium.
Cyanobacteria	*Nostoc punctiforme*	Forms nitrogen-fixing heterocysts. Forms resting cells called akinetes. Produces migrating groups of cells called hormogonia. Forms nitrogen-fixation symbioses with certain plants, ferns, and bryophytes.
Firmicutes	*Bacillus subtilis*	Forms endospores largely in the tips of protuberances that extend from pellicles on liquid surfaces.
Myxobacteria	*Myxococcus xanthus*	Individual motile, feeding cells aggregate and develop into fruiting bodies that form myxospores within themselves that can germinate to form feeding cells.
Stalked bacteria	*Caulobacter crescentus*	Stalked cells divide, producing a stalked cell and a swarmer, which develops to become a stalked cell.
Many kinds of bacteria	*Proteus mirabilis*	Groups of cells perform swarming migration when individual cells differentiate into elongated (20- to 80-μm-long) multiflagellate (up to 50) cells that move together; then, they dedifferentiate into the original cell form.
Chlamydiae	*Chlamydia trachomatis*	Undergoes a mandatory developmental cycle in which an infectious, metabolically inactive elementary body enters a host cell by phagocytosis; within the phagocytic vacuole, it differentiates into a noninfectious but metabolically active reticulate body, which multiplies to fill the cell. The reticulate bodies dedifferentiate, becoming elementary bodies that are released as the host cell lyses.
Endosymbiotic nitrogen fixers	*Sinorhizobium meliloti*	Free-living bacterial cells enter the root hairs of leguminous plants through an infection tube to differentiate into nitrogen-fixing bacteroids within a nodule that the plant forms; bacteroids are capable of dedifferentiating to become free-living bacterial cells.

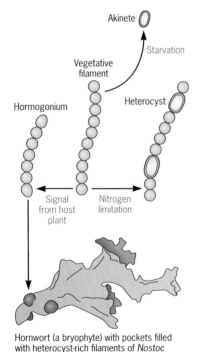

Hornwort (a bryophyte) with pockets filled with heterocyst-rich filaments of *Nostoc*

FIGURE 13.11 *Nostoc punctiforme* patterns of differentiation.

nitrogenase enzymes, degradation of **photosystem** II, and formation of thick external layers that minimize oxygen diffusion. In addition, proteins are expressed inside the heterocyst that scavenge any traces of oxygen that could have made it through. The goal is to protect the oxygen-sensitive nitrogenase so it can perform its function despite the oxygen generated by vegetative cells during photosynthesis. The heterocysts are still connected to the neighboring vegetative cells by channels, which they use to exchange nutrients. The heterocysts feed the combined nitrogen they make through nitrogen fixation to the vegetative cells and receive the nutrients produced by photosynthesis. Division of labor at its best!

2. **Akinetes:** Nutritional limitation for carbon and phosphorous, as well as temperature stress, triggers some cells in the filament to differentiate into enlarged, resistant, resting cells called akinetes. Differentiation of cells into akinetes also involves the synthesis of new cell wall layers and nutrient reserves that can sustain cell growth once the resting cell germinates. They are not as resistant as endospores, but they serve the purpose: whereas vegetative cells remain viable for about 2 weeks in a dark, dry state, akinetes can survive for 5 years.

3. The cyanobacterial filament can also differentiate to form agents of short-distance dispersal called **hormogonia** in response to a wide range of environmental changes, which may be positive or negative for growth such as nutrient limitation, salt concentration, and interactions with

potential symbiotic hosts. Each hormogonium is a short filament that transiently becomes motile by differentiating smaller, gliding cells and terminal cells with characteristic pointed ends. In some respects, these differentiated filaments are analogous to the swarmer cells of *Caulobacter* because they do not grow or reproduce and their role is simply dispersal. Also, like swarmer cells, the differentiated state is transient and the cells can revert back to their vegetative state once a new niche is colonized. This niche can be a plant host that secretes chemicals to attract the hormogonium and establish symbiotic interactions. This also triggers the next and last of the differentiation events that *Nostoc* and related cyanobacteria undergo.

4. Upon receiving an appropriate chemical signal from certain plants, ferns, or bryophytes, hormogonia migrate toward them, enter pockets that the host forms, and differentiate to form a mass of nitrogen-fixing heterocysts. Nitrogen fixation, as it happens in other symbiotic associations between plants and bacteria (chapter 21), supplies the plant with much-needed sources of combined nitrogen to grow, while the plant feeds the heterocysts all of the other nutrients it requires.

Clearly, it takes complex regulatory networks to sense specific signals and transduce them in ways that send vegetative cells down one specific path of differentiation out of the many that filamentous cyanobacteria undergo. It would take a whole book to describe what is known about this fascinating biology, and even then we would be covering only a fraction of the inner workings of such complex differentiation processes. Certainly this elaborate system of differentiation will challenge microbiologists for some time.

CASE REVISITED: Toughening up the prokaryotic way

Is it possible for ancient bacterial spores to remain viable after millions of years in the abdominal tissues of fossil bees preserved in amber? The experiment described in the case study certainly suggests so. But was their turbid culture sufficient proof that the bright, microscopic structures were indeed endospores that had germinated and grew into *Bacillus* cells? Many scientists are skeptical, arguing that it is unlikely that spores could remain viable after such a long time. Absolute proof for this report would be hard to come by. A more conclusive proof could be produced by PCR-amplifying specific genes, such as 16S rRNA, from DNA extracted from the spores and matching it to the 16S rRNA sequence obtained from the grown *Bacillus* cultures. But recovering intact DNA directly from spores is difficult because harsh procedures such as mechanical shearing with glass beads are needed to break their tough coat and cortex and release their DNA, procedures that would degrade the DNA. It has been reported that small (microscale) chambers can be used to deposit spores within a drop of liquid where they can be superheated to release DNA of PCR quality. However, these approaches have not yet made it into mainstream laboratories. Hence, conclusive evidence is still not readily available.

However, if the amber protected the abdominal tissues of the fossilized bee, which are considerably more sensitive to environmental insults and the passage of time than bacterial spores, it is plausible that the spores remained viable throughout this time. It is also well known that *Bacillus* species establish symbiotic associations with bees, whereby the bacterial metabolism contributes to the preservation and health of the larva. Furthermore, DNA sequences matching *Bacillus* species have been amplified from the abdominal tissues of bees similarly preserved in amber. Hence, symbiotic associations between spore-forming bacteria and bees are likely ancient and dating back millions of years. If so, perhaps the spores of these now extinct *Bacillus* symbionts of bees remained viable in the protective amber environment, germinating millions of years later in a laboratory test tube.

CONCLUSIONS

Certainly bacteria are not just single-celled creatures that enlarge and divide. Many have remarkable capacities to differentiate and undergo development, and the overwhelming majority of them live together in multicellular groups as biofilms. The heterogeneous prokaryote domain is rich with biological properties beyond those described here; we shall survey this remarkably diverse group of organisms in the next chapter.

Representative ASM Fundamental Statements and Learning Outcomes for the Chapter

EVOLUTION

1. Cells, organelles (e.g., mitochondria and chloroplasts), and all major metabolic pathways evolved from early prokaryotic cells.

    a. Identify evidence that the ability to make endospores evolved only once.

2. Human impact on the environment influences the evolution of microorganisms.

    a. Identify why domesticated variants of *Bacillus* lost the ability to form biofilms.

CELL STRUCTURE AND FUNCTION

3. *Bacteria* and *Archaea* have specialized structures (e.g., flagella, endospores, and pili) that often confer critical capabilities.

    a. Use examples to show the benefits of the differentiation of prokaryotic cells into structurally and functionally different versions of themselves.
    b. Contrast typical prokaryotic cell differentiation with plant and animal cell differentiation.
    c. Compare and contrast bacterial exospores and endospores.
    d. Outline the process of endospore formation.
    e. Describe structural and biochemical features of endospores that make them vegetative and resistant to unfavorable conditions.
    f. Speculate advantages for *Caulobacter crescentus* from differentiating into two different cells, a swarmer cell incapable of reproducing and a sessile cell.
    g. Identify selective advantages conferred by cellular differentiation (i.e., aggregates and fruiting bodies) in myxobacteria.
    h. Identify selective advantages conferred by the ability of myxobacteria to produce myxospores.
    i. Distinguish structure, function, and metabolic pathways for the three major forms of cellular differentiation in *Nostoc punctiforme*: heterocysts, akinetes, and hormogonia.

INFORMATION FLOW AND GENETICS

4. The regulation of gene expression is influenced by external and internal molecular cues and/or signals.

a. Identify specific environmental signals that can trigger sporulation.

b. Explain how it is possible for several different environmental signals to regulate the single pathway leading to the formation of Spo0A.

c. Name the type of proteins that are responsible for the regulation of sporulation genes.

d. Describe two ways in which *Caulobacter crescentus* cells control development and differentiation.

e. Describe how myxobacteria regulate the cycle of development.

SUPPLEMENTAL MATERIAL

References

- **Hilbert DW, Piggot PJ**. 2004. Compartmentalization of gene expression during *Bacillus subtilis* spore formation. *Microbiol Mol Biol Rev* **68**:234–262. http://dx.doi.org/10.1128/MMBR.68.2.234-262.2004.

- **Higgins D, Dworkin J**. 2011. Recent progress in *Bacillus subtilis* sporulation. *FEMS Microbiol Rev* **36**:131–148. http://dx.doi.org/10.1111/j.1574-6976.2011.00310.x.

- **Sarwar Z, Garza AG**. 2015. Two-component signal transduction systems that regulate the temporal and spatial expression of *Myxococcus xanthus* sporulation genes. *J Bacteriol* **198**:377–85. http://dx.doi.org/10.1128/JB.00474-15.

- **Curtis PD, Brun YV**. 2010. Getting in the loop: regulation of development in *Caulobacter crescentus*. *Microbiol Mol Biol Rev* **74**:13–41. http://dx.doi.org/10.1128/MMBR.00040-09.

Supplemental Activities

CHECK YOUR UNDERSTANDING

1. **Prokaryotic differentiation is**

   a. unique to *Firmicutes*

   b. irreversible

   c. an adaptive response to the environment

   d. a costly process with little evolutionary benefit

   e. coordinated by pheromones

2. **What TWO mechanisms equip *Bacillus subtilis* to sporulate in response to environmental cues?**

   a. allostery

   b. phosphorelay system

   c. quorum sensing

   d. methylation

   e. pheromones

3. **Compared to vegetative cells, endospores are more resistant to**

   a. UV light

   b. heat

   c. acid

   d. desiccation

   e. all of the above

4. **What characteristic do heterocysts (in cyanobacteria) and spore mother cells (in *Bacillus*) share?**

5. **Is there evidence for functional compartmentalization within a stalked *Caulobacter* cell? What is it?**

DIG DEEPER

1. You have three strains of a domesticated *Bacillus* species in your laboratory, all exhibiting varying abilities to form biofilms. Strain A has lost the ability to form biofilms, strain B forms biofilms normally in response to changes of nutrients and culture conditions,

and strain C forms biofilms spontaneously regardless of the culture condition or availability of nutrients.

   a. Design an experimental strategy to identify genetic changes that contribute to the ability of *Bacillus* to form biofilms.

   b. What genetic change(s) may account for the spontaneous biofilm formation by strain C? Design an experiment to test your hypothesis.

2. Read the research article on *Clostridium difficile* by Garneau et al. in *BMC Infect Dis* **14:**29 2014, (http://www.ncbi.nlm.nih.gov/pmc/articles/PMC3897887/pdf/1471-2334-14 -29.pdf), to answer the following questions.

   a. Why is *C. difficile* spore formation a risk to humans?

   b. What method did the scientists use to quantify sporulation?

   c. Despite an extended period of antibiotic treatment that is known to kill *C. difficile*, some patients still suffer recurrent *C. difficile* infections. Design an experimental strategy to determine if sporulation contributes to recurrent infections in these patients. What positive and negative controls would you include? What clinical data would you obtain?

3. Spo0A is part of the Spo0A phosphorelay and two-component regulatory system that coordinates sporulation of *Bacillus subtilis*.

   a. Spo0A initiates sporulation when *B. subtilis* senses high cell density. Draw a schematic to illustrate the key steps.

   b. What mutations could you engineer to block activation but not expression of Spo0A?

   c. What mutations could you engineer to generate a strain that initiates sporulation at a cell density higher than normal? Lower?

4. **Team work:** Working in groups, research one of the following endospore-forming bacteria. Develop visual aids to present your findings to the class.

Bacillus anthracis
Clostridium botulinum
Clostridium difficile
Clostridium tetani
Clostridium perfringens

   a. What is the life cycle of the bacterium?

   b. Is there anything unique about how the bacterium forms endospores?

   c. Why is the bacterium a risk to humans?

   d. How could you reduce human exposure to this pathogen? What are the pros and cons of each approach?

5. **Team work:** Dental plaque is a biofilm that attaches to the surface of the tooth, especially on the gum line. Working in groups, research human dental plaque to answer each of the following questions. Develop visual aids to present your findings to the class.

   a. Identify three environment factors that promote biofilm formation in the mouth.

   b. Why is the ability to form a biofilm on the surface of the tooth an advantage for oral bacteria?

   c. Identify three oral bacterial species that are commonly found in dental plaque.

   d. How does plaque contribute to dental disease?

   e. Do oral biofilms contribute to other diseases in humans?

   f. How can biofilm formation by oral bacteria be prevented or reduced?

Supplemental Resources

VISUAL

- *iBiology* lecture "Developmental Biology of a Simple Organism" by Richard Losick (29 min): http://www.ibiology.org/ibioseminars/microbiology/richard-losick-part-1.html
- Bacterial Spore Formation Animation Video (1.3 min): https://www.youtube.com/watch ?v=NAcowliknPs

SPOKEN

- *BacterioFiles 209* audio "*Myxococcus* Menace Motivates Megastructures" discusses research on this microbial predator in the soil (9.5 min).

WRITTEN

- Moselio Schaechter discusses bacteria that move small objects along surfaces, including fungal spores, in the post "The *Paenibacillus* Moving Company" (http://schaechter .asmblog.org/schaechter/2012/01/the-paenibacillus-moving-company.html).
- Alan Derman reviews the research article by Werner et al. *(PNAS* **106:**7858–7863, 2009), which describes the first bacterial "localisome," a profile of the intracellular location of a large number of the proteins of *Caulobacter*, in the post "Location, Location, Location" (http://schaechter.asmblog.org/schaechter/2009/08/location-location-location.html).
- A Microbe Word article discussing some interesting facts about microbial spore formation: http://www.microbeworld.org/interesting-facts/how-do-they-do-that/microbial-spore -formation

Microbial Diversity

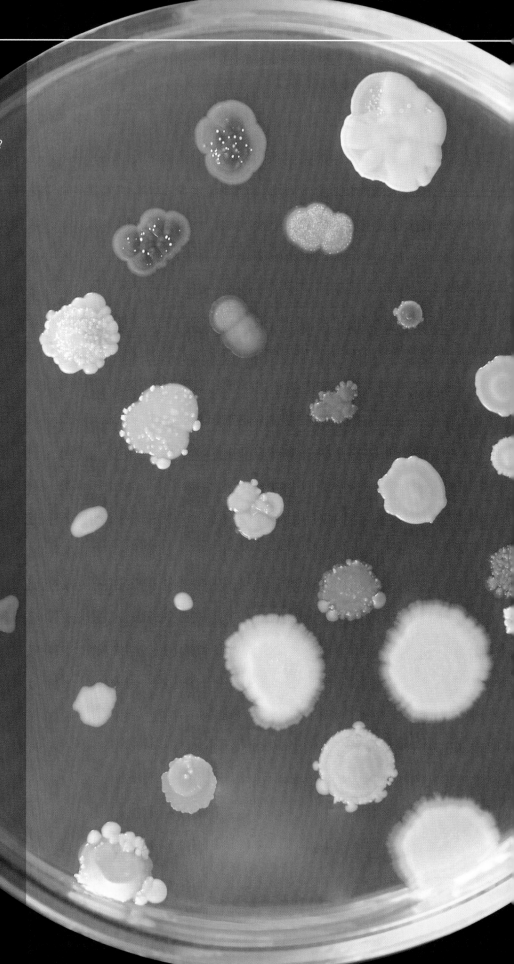

Colonies of a variety of bacterial species

CHAPTER FOURTEEN

Bacteria and *Archaea*

KEY CONCEPTS

This chapter covers the following topics in the ASM Fundamental Statements.

EVOLUTION

1. Cells, organelles (e.g., mitochondria, chloroplasts), and all major metabolic pathways evolved from early prokaryotic cells.
2. The traditional concept of species is not readily applicable to microbes due to asexual reproduction and the frequent occurrence of horizontal gene transfer.

CELL STRUCTURE AND FUNCTION

3. *Bacteria* and *Archaea* have specialized structures (e.g., flagella, endospores, pili) that often confer critical capabilities.

METABOLIC PATHWAYS

4. The survival and growth of any microorganism in a given environment depend on its metabolic characteristics.

MICROBIAL SYSTEMS

5. Microorganisms are ubiquitous and live in diverse and dynamic ecosystems.

IMPACT OF MICROORGANISMS

6. Microbes are essential for life as we know it and the processes that support life (e.g., in biogeochemical cycles and plant and/or animal microbiota).
7. Microorganisms provide essential models that give us fundamental knowledge about life processes.
8. Humans utilize and harness microorganisms and their products.

INTRODUCTION

I n Part I, we dealt with microbial physiology, a study that by its nature seeks generalizations that apply across the microbial world; its goal is unity. In the next part of the book, we make a dramatic shift to focus on diversity in the microbial world. We begin this chapter describing the diversity of prokaryotes (*Bacteria* and *Archaea*), leaving the eukaryotic microbes (fungi and **protists**) and the viruses for subsequent chapters. As we shall see, there is no doubt that the greatest microbial diversity, even if not yet fully cataloged, is found among prokaryotes.

CASE: Is it a bacterium or an archaeon?

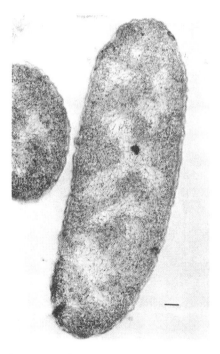

(From Forsberg CE et al., *J Bacteriol* **104**:1338–1353, 1970.)

Student CJ was asked to examine a thin section of a microbial cell under an electron microscope (scale bar, 100 μm) and answer an apparently simple question: "Can you tell if it is a bacterium or an archaeon?"

• What would your answer be?
• Why?

HOW DID *BACTERIA* AND *ARCHAEA* EVOLVE FROM A COMMON ANCESTOR?

LEARNING OUTCOMES:
• Identify specific key characteristics of the Last Universal Common Ancestor (LUCA).
• Explain why the endosymbiotic acquisition of mitochondria and chloroplasts by ancestral eukaryotes was an essential event that led to the diversification and explosive evolution of eukaryotes.

- Describe how mutations can be maintained in a cell and thus build the capacity for natural selection.
- Discuss how the Luria and Delbrück experiments with *E. coli* lent support for natural selection over the Lamarckian view of evolution.

The conviction that life began only once leads to its own quandaries. The best data now available indicate that the two prokaryotic lineages of present life—*Bacteria* and *Archaea*—separated early in their evolutionary development (Fig. 14.1) and subsequently diversified. At some point in its evolution, an archaeal host "ingested" a bacterium, which later became the mitochondrial organelle. This seemingly inconspicuous event had tremendous repercussions: the emergence of the third lineage of life, the *Eukarya*. Looking at the phylogenetic tree shown in Fig. 14.1, it is easy to appreciate that the *Eukarya* may have emerged as a hybrid branch of the *Bacteria* and *Archaea*. Hence, the branch point of the *Bacteria* and *Archaea* represents the shared ancestry of all extant organisms. As we learned in chapter 1, this common ancestor is named LUCA.

What was the ancestor of the earliest *Bacteria* and the *Archaea* like? Logic dictates that LUCA had to possess features the two lineages share, including the genetic code that serves as the repository for the cell's genetic information, ATP as the primary energy currency, ribosome-mediated protein synthesis, and a cell membrane. To have all of these properties, the universal ancestor must have been highly sophisticated, at least metabolically. Furthermore, considering that life may have emerged 4 billion years ago and the oldest microfossils are around 3.5 billion years old, LUCA must have evolved all of these cellular features in the intervening half billion years. This is certainly a long time in human timescales, but, in evolutionary terms, it is a relatively short period. How could such a complex organism evolve so quickly?

Some believe that LUCA arose among populations of simple cellular forms or "cellular progenotes" (chapter 1). Such progenotes likely had a primitive metabolism and slow growth. Certain flaws even offered advantages. The mechanism to copy their genetic material was probably sloppy and introduced many mutations, but a high mutation rate would

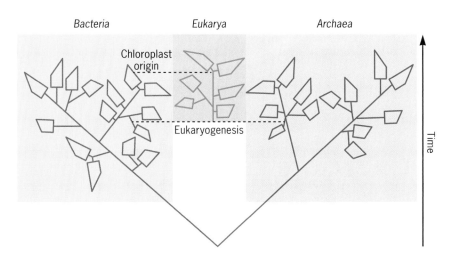

FIGURE 14.1 Taxonomic classification of prokaryotes.

BOX 14.1

Explosive evolution and the origin of eukaryotes

Perhaps the most fundamental evolutionary jumps of all were the endosymbioses that led to the acquisition of the ancestors of the first organelles (mitochondria and chloroplasts) and emergence of the first eukaryotic cells. Many aspects of early microbial evolution are just speculation, but the origins of these two organelles in eukaryotes are well established. The ancestral mitochondrion evolved from a bacterium that was ingested by an archaeal ancestral cell (orange dashed line in figure). A second endosymbiotic event between such a cell and a cyanobacterium led to the evolution of chloroplasts and photosynthetic eukaryotes (green dashed line). Phylogenetic analyses of DNA sequences of mitochondrial and chloroplast origin tell the tale with outstanding precision: mitochondria originated from a bacterium in the *Proteobacteria*, whereas chloroplasts are derived from a captured cyanobacterium. As we will learn in chapter 21, endosymbioses, such as the ones that drove the evolution of eukaryotic cells, are not uncommon in today's microbes. In fact, the eukaryotic world, from protists to animals and plants, is filled with examples of bacterial endosymbionts. What made the endosymbiotic events that led to the evolution of mitochondria and chloroplasts so explosive? Some say it could have been energy. Mitochondria and chloroplasts are organelles devoted to energy transduction: mitochondria for ATP production from **aerobic respiration** and chloroplasts for **oxygenic photosynthesis**. Once these ancestral eukaryotes could respire oxygen and/or transduce energy from light, they were able to occupy new niches and diversify, in-

vesting the excess energy they were generating to become more complex and acquire novel functions. To increase their complexity and evolve rapidly, eukaryotes increased their genome size (they needed to synthesize more components).

To understand how genome size and energy are interrelated, consider the energy budget of a cell. On average, a growing prokaryotic cell devotes 2% of its energy budget to DNA replication and 75% to protein synthesis. But if your average *Escherichia coli* bacterium, whose genome is about 4.6 million base pairs (Mbp), had to replicate the much (1,000 times) larger genome of corn (*Zea mays*), the prokaryotic cell would need about 20 times more energy. Imagine the expense of protein synthesis! It has been proposed that the ancestral eukaryotes overcame this energetic barrier by acquiring more membrane surface, the site of energy transduction. Mitochondria provided numerous membranous extensions for aerobic respiration whereas chloroplasts did the same for energy transduction during photosynthesis. By expanding their bioenergetics membranes, the cells amplified their capacity to form transmembrane ion gradients and thus increased opportunities to harvest the energy of the electrochemical gradient as ATP via ATP synthases (chapter 6). The cell's energy budget could have skyrocketed, enabling the ancestral eukaryote to expand its gene content up to 200,000 times! With more components and modules in their evolutionary toolbox, cells could have become more complex, acquired new functions, and colonized new niches. Indeed explosive.

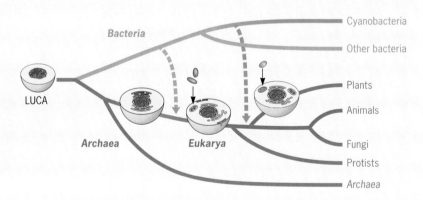

promote rapid evolution. As the more competent cells outcompeted the more inept ones, populations of cellular progenotes evolved rapidly toward efficiency until the most proficient organism (LUCA) rose above all others. According to this theory, with a sudden evolutionary jump, a cellular progenote, LUCA, became more fit than any other to colonize new or yet unoccupied ecological niches and to diversify. Similar explosive evolutionary events could have occurred, for example, when an archaeal host acquired a bacterial endosymbiont that later became the mitochondrion (Box 14.1). This pattern of rapid emergence fits what we know about the evolution of prokaryotes even to this day. Recall, for example, the modular steps of evolution that gave rise to the bacterial flagellum (chapter 9). Once cells became motile, they were able to swim to new (and better) environments (chapter 12). New traits, new life.

How did the *Bacteria* and *Archaea* diversify to such a degree through the course of evolution? In chapter 10, we learned about one major force of genetic variation in prokaryotes: **mutations**. Luria and Delbrück, followed by the Lederbergs (Box 14.2), demonstrated in their laboratories

BOX 14.2

Mutations first, then selection

In 1943, Salvador Luria and Max Delbrück obtained elegant experimental support for Darwin's proposal for natural selection of preexisting variations using *E. coli*. As a selective force, they used phages, the viruses that infect bacteria (chapter 10). These microbiologists knew that if they mixed large numbers of susceptible bacteria with virulent phages and plated them on an agar-containing medium, a small number of phage-resistant colonies would always arise. Luria and Delbrück designed an experiment to distinguish between two leading theories. Would exposure to phages induce mutations in some bacterial cells that would make them phage-resistant, as Lamarck would have argued? Or, had mutations been accumulating naturally, and exposure to the phage would merely select for the preexistent phage-resistant cells, as Darwin would have predicted? Luria and Delbrück showed that the latter was the case. They grew a large number of separate cultures of susceptible bacteria, each from an inoculum too small to likely contain phage-resistant cells. When the cultures were fully grown, a sample from each was mixed with phage and plated. The number of resistant colonies that developed was *highly variable*—much more variable than mul-

tiple samples taken from a single culture. These results convincingly fit a mathematical model (the Poisson distribution) that predicts the highly variable number of mutants that arise randomly in the individual cultures. The cultures where mutations occurred early in their growth would have large numbers of resistant progeny cells (a jackpot!), whereas mutations occurring later would have fewer resistant progeny (hence, "fluctuation"). If phage resistance were the result of exposure to the phages, the variation among samples from different cultures and from the same culture would be similar ("no fluctuation").

Nine years later, Joshua Lederberg and Esther Lederberg used their newly devised technique of replica plating (chapter 10) to do a simpler, but possibly even more convincing, experiment to prove the same point. They replicated a master plate of bacterial colonies to fresh plates, some containing an antibiotic. A number of antibiotic-resistant colonies developed. They then went back to the master plate and picked each corresponding colony (which had never seen the antibiotic!). These progenitors were also antibiotic resistant, convincingly demonstrating that resistance was preexistent and not induced by exposure to the antibiotic.

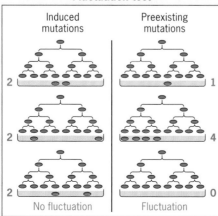

Fluctuation test

Induced mutations | Preexisting mutations

No fluctuation | Fluctuation

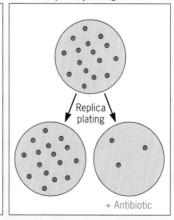

Replica plating test

Replica plating

+ Antibiotic

that mutations occur randomly and that the environment selects for favorable ones, as proposed in Darwin's theory of natural selection. In addition, *gene duplication* occurs with some frequency in prokaryotes; thus, many mutations that arise by chance will occur in one copy or the other. Because extra copies of a gene generate more of its product for the cell, this advantage alone could favor maintenance of the gene duplicates. Given time, one of the gene copies may acquire mutations that confer a new function, thus diversifying the genomic capacity of the cell. In other words, *redundancy fosters the evolution of innovation*.

Being small and reproducing relatively quickly, the early prokaryotes accumulated mutations rapidly. Moreover, they acquired mechanisms to exchange genetic material from one cell to another (**horizontal gene transfer**). Thus, here are the two major forces of genetic variability: *mutations and horizontal gene transfer* (chapter 10). The planet provided fluctuating environmental conditions that exerted selective pressure, so only those variants most fit for life in a particular environment would grow and survive. *Bacteria* and *Archaea* branched early in life's evolution and diversified quickly, even before the major endosymbiotic events that led to the emergence of the first eukaryotes (Fig. 14.1). For a long time, *Bacteria* and *Archaea* were the only forms of life: from them, all other organisms evolved.

HOW DIVERSE ARE TODAY'S PROKARYOTES?

LEARNING OUTCOMES:
- Explain why the traditional classification of a species based on morphology for plants and animals is not appropriate for defining microbial species.
- Discuss how using the term "operational taxonomic unit" is better suited than the term "species" when describing relationships among microbes.

How can one extract coherence and a sense of order out of a sea of information as vast as the myriad prokaryotes? Certainly, since Aristotle, and probably well before, humans have taken the same general approach to sorting out a seemingly bewildering diversity: identifying groups of individuals that resemble one another and collecting them into larger assemblages. In this way, we simplify the picture by focusing on a more manageable number of groups rather than on overwhelming numbers of individuals. This approach works only partially for microbes. Microscopy, for example, gives us a glimpse at the diversity of microorganisms, but it barely touches the surface. Usually defined as creatures invisible to the unaided eye (i.e., smaller than about 100 µm; chapter 1), microbes occur in all three biological domains—*Eukarya*, *Bacteria*, and *Archaea*—although the two prokaryotic ones predominate. Eukaryotic cells are easily distinguished under the microscope by the presence of a nucleus. But can we tell *Bacteria* and *Archaea* apart with a simple microscopic observation? CJ, our case study student, was asked to do this. He probably concluded that it was a prokaryotic cell because there was not a defined nucleus. Instead, he saw a defined central area, the **nucleoid** (chapter 3). That's about it. Morphology—so critical for the traditional classification of plants and animals—is of little use for the classification of most prokaryotes. Furthermore, many prokaryotes change their shape and other morphological features in response to chemical and physical environmental cues (chapter 13). The early microbiologists, who relied on microscopy for most of their studies, misinterpreted the various shapes and cell forms they observed as being stages of an individual microbe's life cycle, a phenomenon they called polymorphism. At the time, some biologists even concluded that all prokaryotes belonged to a single species! Now, wouldn't that be convenient when studying for a microbiology exam?

Luckily for us, we now have a wide array of powerful methods to estimate how diverse prokaryotes are, even if we know little or nothing about most of them. For example, we know of about half a million insect species, yet only about 12,000 prokaryotic species have been validated. This number clearly underestimates the existing prokaryotic diversity, which some microbiologists estimate to include 10 million species. Such figures must be taken with a grain of salt, however, because even the definition of "prokaryotic species" remains controversial. (The species concept is one of the thorniest issues of the microbial world, one that has thwarted even the sharpest minds.)

A widely used approach to classify prokaryotic species depends on genetic relatedness, typically based on handy genealogy-specific genes, such as the 16S rRNA gene (here is an idea for our case study student, CJ!). For example, the sole criterion often used to designate a new prokaryote species is a 16S rRNA sequence less than 97% similar to that of any other known species. However, this approach to classifying prokaryotes species is hampered by the observation that some organisms sharing more than 97% homology in their 16S rRNA genes differ greatly phenotypically, more than would be expected for organisms from the same species. To bypass this limitation, the relatedness of the entire genome, rather than of a single gene, can be used as the criterion to delineate a species. This method helps to some extent, but it still does not fully resolve the issue. Indeed, DNA analysis has led to the realization that the distance between genomes, measured as sequence homology, is enormously greater in prokaryotes than in eukaryotes. Think about it: if we were to rely solely on genomic relatedness to classify eukaryotic species, most primates would be a single species!

What to do about the vagaries of the species concept? It is undoubtedly useful, perhaps essential, in considering evolution and relationships between organisms (remember the title of Darwin's seminal book, "*The Origin of Species*"). One practical way of handling this problem has been to replace "species" with another term that can be used when only sequence data are available. By general agreement, scientists use the **operational taxonomic unit** or **OTU**. An OTU is separated from others by somewhat arbitrary, but agreed upon, thresholds of homology, typically, in ribosomal RNA (Table 14.1). Whether the sequence belongs to a cultivated or an uncultivated microbe is of no importance in this case; sequence homology alone is used to classify the organism phylogenetically. So, how many bacterial and archaeal OTUs are there? We don't really know, but the figure is probably in the millions.

TABLE 14.1 **Taxonomic classification based on 16S rRNA gene % identity**

Shown are the major taxonomic categories of prokaryotes and the threshold sequence homology (%) used for their classification. Numbers in brackets are confidence intervals.

Category	Threshold
Species	98.7
Genus	94.5
Family	86.5
Order	82.0
Class	78.5
Phylum	75.0

WHY IS IT SO DIFFICULT TO ASSESS THE DIVERSITY OF PROKARYOTES?

LEARNING OUTCOMES:
- State reasons why it is generally difficult to assess the diversity of prokaryotes, especially for environmental microbes.
- Describe how cultivation-independent techniques can be used to determine the presence of *Bacteria* and *Archaea* in a sample.

Given the vast number of prokaryotes, why has only a small fraction of them been recognized and studied? There are many reasons. Consider for example *E. coli*. Taxonomically, we can use genotypic and phenotypic traits to assign it to a domain (*Bacteria*), phylum (*Proteobacteria*), class (*Gammaproteobacteria*), order (*Enterobacteriales*), family (*Enterobacteriaceae*), genus (*Escherichia*), and species (*E. coli*) (Fig. 14.2). However, within this species are strains that vary all over the map both in genotype and phenotype. Some have a genome that is 20% larger than others. Some strains are harmless. (In fact, the laboratory strain K-12 has been drunk by volunteers without untoward effects—but, please, do not attempt to reproduce this!) Other strains cause urinary tract infections and severe hemorrhagic disease. And yet, despite their differences, they are all called *E. coli*, in good part because they have a "core genome" in common. If such great diversity existed in animals, they would belong to distinct orders or even different phyla. And here the *E. coli* are, all one "species." (As a practical solution, microbiologists use the term "pathotype" to describe different phenotypic groups of *E. coli*.)

Other factors exacerbate classification of microbes. Just looking at a prokaryote, with few exceptions, is not sufficient to identify it, given that so many look alike. (In the case study, CJ realized this too soon.) Cultivating most prokaryotes, the *sine qua non* of characterizing and studying them, can be quite difficult. In many environments, today we are able to cultivate less than 1% or so of the number of apparently viable cells visible by microscopy. (An uncultivated species can be assigned only a tentative name preceded by the term "*Candidatus*.") However, scientists are avidly working to develop procedures to grow more microbes in the laboratory, and much progress is being reported. Indeed, many bacteria and archaea previously believed to be "uncultivable" have been coaxed into making colonies on agar plates or growing in liquid media (chapter 19). The secret

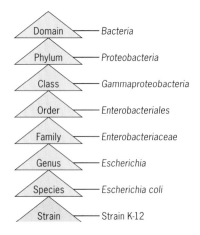

FIGURE 14.2 **Levels of taxonomic classification.**

- Domain —— *Bacteria*
- Phylum —— *Proteobacteria*
- Class —— *Gammaproteobacteria*
- Order —— *Enterobacteriales*
- Family —— *Enterobacteriaceae*
- Genus —— *Escherichia*
- Species —— *Escherichia coli*
- Strain —— Strain K-12

seems to be attention to the nutritional conditions and patience—it can take weeks or months for cells to grow! Microbiologists who have been most successful in culturing recalcitrant strains attribute most of their achievements to supplying substrates at extremely low concentrations and keeping in mind that, in nature, many microbes grow very slowly. Such approaches have markedly increased the number of cultivable microbes from certain environments, in some cases to about 7%. Still, culturing the vast majority of the microbes that exist in nature remains a challenging, but important, goal. Curiously, most pathogens that infect humans, animals, and plants can be grown in the lab with relative ease (although there are dramatic exceptions, e.g., the leprosy bacillus, *Mycobacterium leprae*). One explanation for pathogens being cultivable is that it may be easier for scientists to mimic the environment of the human body. Another is that while infecting their host many pathogens go at it alone whereas many microbes in natural environments grow in communities (chapter 19).

But even microbes that cannot yet be cultivated can be detected using molecular techniques. Take a sample of soil, water, or clinical material, and extract the DNA. Even though you usually cannot figure out the genomic sequence of each organism present, you can use PCR primers (chapter 10) to amplify distinct marker sequences. Again, 16S rRNA is a favorite. Using this "cultivation-independent" technique, you can both count and identify the OTUs present in the sample to profile the diversity of organisms in the community. It is also possible to bypass the PCR step and directly sequence DNA extracted from the sample using *high-throughput sequencing* methods. In combination with *modern bioinformatics*, an astounding amount of information can be gleaned from such studies. Such indirect approaches have made available a large number of 16S rRNA sequences. Most of these sequences correspond to not-yet-cultivated microbes, the so-called "*microbial dark matter.*" We may not know much about them yet, but we know they are there and are extremely diverse. We further discuss the details, advantages, and limitations of cultivation-dependent and cultivation-independent approaches in chapter 19.

HOW TO MAKE ORDER OF THE INCREDIBLE PROKARYOTIC DIVERSITY?

LEARNING OUTCOMES:
- Identify a consequence upon Earth of the emergence and expansion of cyanobacteria 2.5 billion years ago.
- Summarize how *Prochlorococcus* is well adapted to the nutrient-poor ocean.
- Describe how the alphaproteobacterium *Agrobacterium tumefaciens* induces its plant host to provide nutrients for the bacterium.
- Identify how the *A. tumefaciens* Ti plasmid can be harnessed to benefit humans.
- Name a microbe from each of the six classes of *Proteobacteria*.
- List three ways in which *Actinobacteria* are beneficial for the environment and/or to humans.

- Speculate why scientists do not yet agree upon the reasons why *Streptomyces* naturally produces antibiotics.
- Evaluate whether the term "prokaryote" is still a useful description of *Bacteria* and *Archaea*, given the three-domain tree of life and the many differences between the domains.
- Identify distinguishing features of archaeal membranes.
- List biochemical differences between *Archaea* and *Bacteria*.
- Infer how unique structures in *Archaea* contribute to their prevalence in extreme environments.
- Identify reasons why *Archaea* were undiscovered until the advent of molecular biology and why the diversity of extant *Archaea* is still being discovered.
- Describe the role of methanogens in the carbon biogeochemical cycle.
- Use extreme halophiles as an example to justify the statement: "the survival and growth of any microorganism in a given environment depend on its metabolic characteristics."

Until significant amounts of information about DNA and protein sequences became available in the 1970s, there was no rational basis for assigning prokaryotes to phylogeny-based taxa. Now, with the wealth of available sequence information, this can be done. Conservative estimates are that the number of yet-uncultivated *bacterial phyla* may be greater than 1,300. The analysis of microbial mats from a single site in Mexico alone produced 43 bacterial novel phyla (although the great majority of them are infrequent). Environmental surveys indicate that archaea are also metabolically diverse and, consequently, ubiquitous and abundant. Some scientists estimate that the number of uncultivated archaeal phyla may be as high as that for bacteria.

How then do we make order of this incredible prokaryotic diversity? As a guide, we will use the taxonomic classification shown in Fig. 14.2, which starts with the upper domain level (*Bacteria* or *Archaea*) and then narrows the evolutionary relatedness of prokaryotes down to the species levels. So great is the diversity even at the phyla level that we will focus on selected phylum-level clades only.

Bacteria

The domain *Bacteria* was long believed to be the largest and most complex of the two prokaryotic domains. However, the perceived difference might be greater than the actual one because, as we will discuss later, new species and groups of *Archaea* continue to be discovered. The domain is further divided into several phyla (Fig. 14.3), but many more are believed to exist. The size of each bacterial phylum varies enormously. The smallest contains only a single recognized species. The largest, the **Proteobacteria,** is huge, with six different classes (alpha through zeta) and at least 1,300 species in all. In spite of such diversity, knowing the phylum to which a bacterium belongs can provide valuable information. Indeed, the phylum designation has become the accepted shorthand method—in publications, as well as conversation—for microbiologists to communicate something about an unfamiliar bacterium: "It belongs to this or that phylum." Try it! If I tell you that a bacterium belongs to the **Cyanobacteria** phylum, you

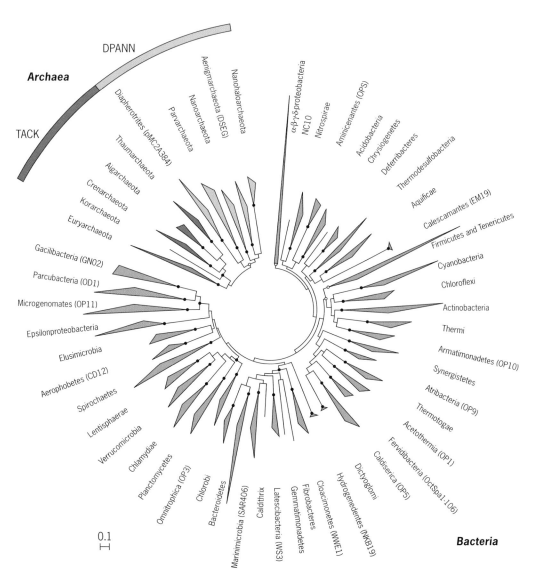

FIGURE 14.3 Phyla in the *Bacteria* and *Archaea* domains. Unrooted phylogenetic trees showing most of the phyla identified to date in the *Archaea* and *Bacteria* domains. The two archaeal superphyla (TACK and DPANN) are highlighted in dark and light blue, respectively. The scale bar represents 10% estimated sequence divergence. (Modified from a figure in Rinke et al., *Nature* **499**:431–437, 2013.)

will immediately know that it carries out oxygenic photosynthesis (chapter 6). Fear not, we will not even try to discuss all the bacterial phyla—a formidable task! Instead, we focus here on four of the largest and perhaps best known (Fig. 14.4), keeping in mind that other phyla are discussed elsewhere in this book.

The **Cyanobacteria** *phylum.* This group is featured throughout the book, and we will continue to highlight its critical contribution to life's diversification. The phylum *Cyanobacteria* contains the oxygenic photosynthesizers of the prokaryotic world and the ancestors of modern chloroplasts, the eukaryotic organelles of plants and algae (Box 14.1). Cyanobacteria emerged on Earth some 2.8 to 2.5 billion years ago and diversified explosively, thanks to their unique energy transduction mechanism to harvest

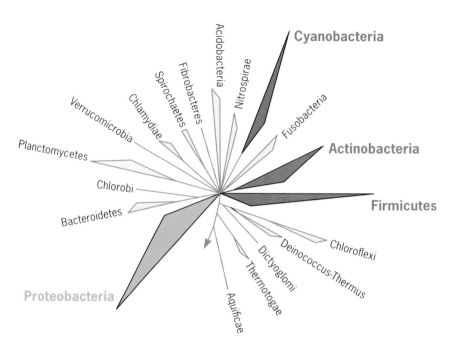

FIGURE 14.4 Rooted phylogenetic tree showing representative phyla in the *Bacteria.* Phyla described in the text (*Cyanobacteria, Proteobacteria, Firmicutes,* and *Actinobacteria*) are highlighted in colors.

energy from light while generating oxygen as a byproduct (chapter 6). Not surprisingly, they are a highly diverse group and occupy numerous ecological niches on Earth. Many cyanobacterial cells are large as bacteria go—up to 50 μm in diameter. By devoting considerable space to their photosynthetic membranes, they have increased their energy budget, gained more genes, and increased in complexity (Box 14.2). Their morphology also varies, from unicellular to filamentous forms. Many cyanobacteria are also highly differentiated. For example, recall the various differentiated cells that *Nostoc* species form (chapter 13). One of them, the **heterocyst**, is a terminally differentiated cell uniquely specialized to fix nitrogen and can easily be distinguished within a filament composed of mostly vegetative cells (Fig. 14.5A).

The ecological success of cyanobacteria is unmatched. Consider *Prochlorococcus marinus*, a member of the marine phytoplankton and arguably the most abundant photosynthetic organism on our planet. It was discovered not that long ago, in 1986, and probably only then because it is unlike most other cyanobacteria. Its oval cells are small, around 0.6 μm in diameter, which also makes it among the smallest known photosynthetic organisms. We will return to this champion in chapter 19 as an example of how challenging some microbes are to cultivate in the laboratory. The breakthrough came by providing the bacterium with a most minimalistic diet: it needs only light, carbon dioxide, and mineral salts to grow. Consistent with its reduced metabolic capabilities, *Prochlorococcus* also has a small genome—1.7 million base pairs—containing a paltry 1,716 genes, which may be the minimal genome for a photoautotroph.

A slightly larger (about 0.9 μm in diameter) cyanobacterium, *Synechococcus*, is only a bit less abundant. Looking at a cell of *Synechococcus*,

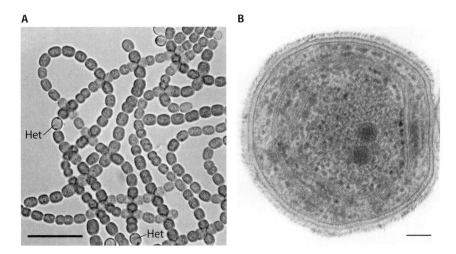

A

B

Het

Het

FIGURE 14.5 **Cyanobacteria.** (A) A chain of cells of the cyanobacterium *Nostoc paludosum.* Het indicates heterocysts, differentiated cells that carry out nitrogen fixation. Bar, 20 mm. (B) Electron micrograph of a thin section of a *Synechococcus* sp. This marine cyanobacterium contains abundant membranous structures, typical of many photosynthetic and chemosynthetic organisms. Bar, 100 nm. (Panel A, courtesy of J. E. Frias; panel B, reprinted from McCarren J, *J Bacteriol* **187:**224–230, 2005.) (Reprinted from Ingraham JL, Ingraham CA, *Introduction to Microbiology*, 2nd ed., Brooks/Cole, Pacific Grove, CA, 2000; source: D. A. Glave, Biological Photo Service.)

with its intricate membrane pattern for light energy transduction, helps us appreciate how the cell has evolved an architecture that maximizes its photosynthetic efficiency (Fig. 14.5B). This morphology in turn translates into ecological success. Indeed, cyanobacteria were the original producers of atmospheric oxygen, and they still contribute to about half of all the oxygen released to the atmosphere. The other half comes from plants and algae, also photoautotrophs. But, then again, their oxygen-producing capacity resides in chloroplasts derived from anciently captured cyanobacteria. So oxygen production on the planet follows the cyanobacterial way.

The **Proteobacteria** *phylum.* The highly diverse plylum *Proteobacteria*, the largest among the bacterial phyla (Fig. 14.4), is subdivided into six classes (*Alpha* through *Zeta*) of Gram-negative bacteria and includes some of the bacterial groups we are most familiar with. Here we find well-known pathogens, such as the causative agent of whooping cough (*Bordetella pertussis*, a *Beta*), salmonellosis (*Salmonella enterica*, a *Gamma*), and ulcers (*Helicobacter pylori*, an *Epsilon*). Let us examine some of the notable representatives and their defining characteristics.

In the *Alphaproteobacteria* class we find genera that associate with plants and either cause disease or benefit the host. *Agrobacterium tumefaciens*, for example, is a soil bacterium that invades, and causes tumors in, broadleaf (dicot) plants. This ordinary-sized (about 1 by 3 μm), Gram-negative bacterium does something quite extraordinary: *it shares some of its genes with its plant host.* You could say it is a genetic invader! *A. tumefaciens* attaches to a plant at the site of a wound and uses a Type IV secretion system (chapter 9) to insert some of its genes into the genomes of surrounding plant cells, thereby subverting them to become "better" hosts. The hosts submit to the pathogen by forming a protective niche. The infection is visible to us

FIGURE 14.6 **A gall on an oak tree caused by** *Agrobacterium tumefaciens*, **an alphaproteobacterium.**

as an ugly tumor-like growth on the plant, usually near the crown (the connection of stem and root; Fig. 14.6). These **crown galls** do not kill plants but reduce the value of nursery stock, which is why considerable effort has been expended to understand the mechanism of infection and to control the disease (Box 14.3). We will learn of other examples of bacteria-plant associations in chapter 21, including the beneficial association between legume plants and another *Alpha, Rhizobium*, which we also mention briefly in Box 14.3.

The *Proteobacteria* phylum includes five other classes, *Beta* through *Zeta*, some of which are huge (e.g., the *Beta*- and *Gammaproteobacteria*). All include members that have a great impact on human affairs and planetary processes. So important are they that you will encounter them repeatedly throughout this book. Here we simply describe some notable representatives of the five classes.

The **Betaproteobacteria** contain two important genera of pathogens. *Bordetella pertussis* causes whooping cough, largely a childhood disease. *Neisseria gonorrhoeae* causes gonorrhea, and *Neisseria meningitidis* causes meningococcal meningitis (chapter 22).

The **Gammaproteobacteria** are even larger and contain 13 orders, including the enterobacteria, vibrios, pasteurellas, and pseudomonads. The enterobacteria include human pathogens, most of which infect the digestive system.

BOX 14.3

The intimate affairs of *Alphaproteobacteria* and plants

The relationship between *A. tumefaciens* and the plant it infects is surprisingly intimate, even cooperative. The interaction begins when the wounded plant releases (among other compounds) a phenolic compound, acetosyringone, to which *A. tumefaciens* responds using functions encoded in a large plasmid called Ti (for *tumor-inducing*). The high concentration of acetosyringone closer to the wound induces the expression of *A. tumefaciens* virulence factors, such as an endonuclease that cuts a piece of DNA from the Ti plasmid called *transfer DNA* (T-DNA). The excised T-DNA then leaves *A. tumefaciens*, via a Type IV secretion system (chapter 9) encoded on the plasmid, to enter a plant cell and integrate into a chromosome. From there it directs the plant cell's metabolism for the benefit of *A. tumefaciens*: it causes the plant to produce proliferation-inducing plant hormones, which form the gall, and to synthesize an unusual set of compounds called opines (amino acids) and agrocinopines (phosphorylated sugar derivatives), which *A. tumefaciens* uniquely uses as nutrients. In other words, by means of a genetic invasion, *A. tumefaciens* creates a near-ideal habitat for itself at the plant's expense.

How can we protect the plants from this infectious bacterium? Antibiotics work, but such treatments are too costly.

But here is a useful fact: *A. tumefaciens* is innocuous once it loses the Ti plasmid. Thus, scientists use a Ti-lacking strain as a biological agent to protect the plants from *A. tumefaciens*. They simply dip seeds, seedlings, or cuttings in a suspension of these plasmidless, and therefore avirulent, bacteria. The microbes act by an unusual and highly specific mechanism: they produce agrocin 84 (a toxic analogue of adenine) that *A. tumefaciens* will incorporate as if it were adenine itself. Once inside, this toxin kills the invader, protecting the plant from infection. Hence, the plasmidless strain provides an effective preventative treatment.

Other good things are worth mentioning about the *Alphaproteobacteria*. Incorporating foreign DNA into the T-DNA of the Ti plasmid of *A. tumefaciens* has become the *favored route for genetically engineering plants*. In this manner, the *A. tumefaciens* Ti plasmid has been used to make transgenic plants resistant to insects and herbicides. Furthermore, as Ti is a self-transmissible plasmid, it can be passed to other related *Alphaproteobacteria* via conjugation (chapter 10). When transferred to the close relative *Rhizobium*, the Ti plasmid allows *Rhizobium* to infect plants and form galls. But, in this case, there is big bonus for the plant: nitrogen fixed by the bacteria!

One of the *Salmonella enterica* strains causes typhoid fever, while others are responsible for common types of food poisoning; *Shigella* species cause shigellosis, a form of dysentery; *Yersinia pestis* causes bubonic plague (chapter 27); and the well-studied *E. coli* has a few pathogenic strains (remember "pathotypes"?). The vibrios include *Vibrio cholerae*, the species that causes cholera (chapter 26). The pasteurellas include two genera of devastating pathogens, *Pasteurella pestis*, the agent of the plague, and *Haemophilus influenzae*, which causes pneumonia, meningitis, and a number of other infections. The pseudomonads include a species, *Pseudomonas aeruginosa*, that often infects weakened hosts, especially burn victims and cystic fibrosis patients. The pseudomonads also include species equipped with the remarkable metabolic ability to eat many nutrients, including pollutants. Hence, they show promise as bioremediation agents to restore contaminated environments.

The ***Deltaproteobacteria*** include Gram-negative bacteria with unique respiratory talent. *Desulfovibrio* respires by reducing sulfur compounds, and *Geobacter* species respire using iron. Hence, these organisms make critical contributions to the cycling of these elements (chapter 20). Also in this phylum are the myxobacteria, a group of soil bacteria that glide on surfaces, forming packs of cells or swarms (chapter 13). When nutrients are scarce, their social motility equips the bacteria to aggregate and to form sophisticated fruiting bodies, where cells differentiate into dormant, spore-like forms (myxospores) that ensure prolonged survival until conditions are favorable for growth.

The ***Epsilonbacteria*** include **microaerophilic** species such as *Helicobacter pylori*, the causative agent of stomach ulcers. A few genera live in the digestive track of animals, including humans, and can be opportunistic pathogens. Also in this class are **chemolithotrophs** from high-temperature habitats, such as hydrothermal vents and cold seeps.

The ***Zetaproteobacteria*** include a single species, the iron-oxidizer *Mariprofundus ferrooxydans*. Analysis of environmental DNA sequences from yet uncultivated members from this class suggests they promote iron cycling at oxic-anoxic interfaces in deep sea and near-shore environments.

The* Firmicutes *phylum. The unifying characteristic of this highly diverse bacterial phylum is the *low GC content* (often less than 50%) of their members' genomes. They are also classified as a Gram-positive phylum, though members of one group, the mycoplasmas, are wall-less (chapter 2). This is also the only bacterial phylum that includes members able to form **endospores** (chapter 13). Endospore formation is restricted to two of the three classes in the phylum, *Bacilli* and *Clostridia*. In the *Clostridia* we also find the endospore former *Epulopiscium fishelsoni*, the giant fish endosymbiont discussed in chapter 1. The phylum also includes non-spore formers. In the *Bacilli* class we encounter, for example, some notable human pathogens such as the cocci *Staphylococcus aureus* (chapter 23) and *Streptococcus pyogenes* and non-spore-forming rods such as the diphtheria bacillus (*Corynebacterium diphtheriae*) and *Listeria monocytogenes*, an important agent of food poisoning (chapter 24). But the same class includes *Lactobacillus* bacteria, which we use to make yogurt and cheese.

A

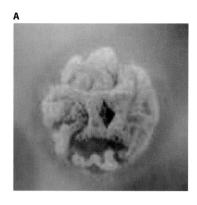

B

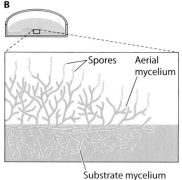

FIGURE 14.7 **The actinobacterial genus *Streptomyces*.** (A) Pastel-colored colonies on an agar plate; (B) drawing of a cross section of a colony. (Panel A courtesy of K. J. McDowall and K. Jolly.)

Genetic exchange is widespread in the *Firmicutes* and not restricted to members of the same phylum or even the *Bacteria*. Approximately one-quarter of all of the fully sequenced bacterial plasmids, for example, belong to the *Firmicutes*, and many contribute to horizontal transmission of antibiotic resistance markers. High genetic variation is also reflected in the variety of environments that these bacteria colonize. Thus, we encounter aerobic or facultative aerobic bacteria (*Bacilli* class) and strict anaerobes (*Clostridia* class) as well as species adapted to extremes of temperature (cold and hot—up to 70°C) and salt (as in the genus *Halobacillus*). Some members are adapted to living in association with mammalian hosts, as we will learn in chapter 19. The *Firmicutes* are, for example, the first colonizers of the intestine of human newborns and retain their prominent status throughout the adult life of the individual. True specialists, these bacteria are only found in the microbial community of the gut, a symbiotic association that is critical to our digestion and overall health.

The Actinobacteria *phylum*. The ***Actinobacteria*** consitute a phylum, as well as a class, of Gram-positive bacteria characterized by an unusual *high guanine-cytosine (GC) content* of their genomes and include species of ecological, industrial, medical, and scientific importance. Perhaps the most famous members of the phylum are those in the genus *Streptomyces*, within the order *Actinomycetes*. They are abundant in soil, accounting for much of the breakdown of organic matter that occurs there. In fact, the pleasant odor of freshly turned soil comes from the volatile compounds (geosmins) produced by these bacteria. When a sample of soil is streaked on a simple inorganic medium, many of the colonies that develop are species of *Streptomyces*. These microbes are readily identified by their striking pastel colors: various shades of violet, blue, orange, yellow, and red (Fig. 14.7). The cultures also have the characteristic soil odor. You can be almost certain that a colony is a species of *Streptomyces* by touching it with an inoculating needle. The colony sticks together: you can only pick up all of it, not just part of it. The integrity of a *Streptomyces* colony reflects its structure (Fig. 14.7). *Streptomyces* is filamentous, and it grows into a fungus-like mycelium (hence the "-myces" in the name), although in cross section, the filament looks like a perfectly ordinary Gram-positive bacterium. Part of the mycelium (the substrate mycelium) grows into the agar, anchoring it there. On top, the aerial mycelium extends upward, and at the tips of its filaments, long chains of **spores** form. The intertwining of the filaments is what makes the colony stick together (chapter 13).

Possibly the greatest impact that *Streptomyces* species have on our lives comes from their ability to *produce antibiotics*. The majority of all antibiotics in use today are made by various species of *Streptomyces*. (Exceptions are the penicillins, which are made by fungi, and some other antibiotics made by species of *Bacillus*.) Each commercially relevant antibiotic is made by a particular species, perhaps reflecting patent protection more than biological distinction (Table 14.2). Species of this genus are also used to produce antiparasitic agents, herbicides, immunosuppressants, antitumor drugs, and enzymes used in the food and other industries.

Antibiotics are products of the secondary metabolism typical of *Streptomyces*, as well as of many fungi and plants. Like other **secondary me-**

TABLE 14.2 **Some *Streptomyces* species and the antibiotics they produce**

Species	Antibiotic
S. aureofaciens	Tetracycline
S. erythreus	Erythromycin
S. fradiae	Neomycin
S. griseus	Streptomycin
S. lincolnensis	Clindamycin
S. noursei	Nystatin

tabolites, antibiotics are synthesized when nutrients become growth-limiting. Why? One thought is that *Streptomyces* spp. produce antibiotics in a desperate attempt to eliminate competition for scarce nutrients. Some microbiologists argue that this cannot be the case because antibiotics are likely to be ineffective against competitors in a natural setting. Some—streptomycin, for example—are tightly adsorbed and inactivated by clay minerals once released in the soils. Also, there is the intriguing coincidence that the *major antibiotic producers—Streptomyces, Bacillus,* and fungi—*all produce spores* of one sort or another. Instead, might antibiotics be signaling agents? Go figure! Because of their biotechnological potential, many *Streptomyces* genomes have been fully sequenced, and draft assemblies are available for many more. They tend to be large genomes; in fact, some are among the largest bacterial genomes yet sequenced. The genome of *Streptomyces coelicolor*, for example, has 8.5 million base pairs and 7,825 predicted genes. This is probably not a surprising result in view of the complex developmental cycle of *Streptomyces*, their abundant capacity for producing secondary metabolites, and their fluctuating environment.

Archaea

The realization that a group of prokaryotes, now called *Archaea*, differs profoundly from all other prokaryotes (the *Bacteria*) occurred in the mid-1970s, when Carl Woese began to sequence 16S rRNA from various prokaryotes. The nucleotide sequences of 16S rRNA from some were so different from the rest that Woese and his colleagues concluded they must constitute a distinct group of prokaryotes, which he called *Archaebacteria* to suggest their suspected ancient origins. Now a preponderance of biologists agree that these microbes are no more closely related to *Bacteria* than they are to eukaryotes! In the classical three-domain tree of life, *Archaea* constitute the third major division (domain) of living things, in addition to *Bacteria* and *Eukarya* (eukaryotes). Whole-genome phylogenetic analyses contest this tripartite division of the biological world and show the *Eukarya* as a hybrid group that emerged after the bacterial and archaeal lineages started to diversify (Fig. 14.1). Regardless of the phylogenetic approach, the *Archaea* appear clearly to be an independent line of descent that separated early on from the *Bacteria* as the first prokaryotic cells evolved from a common ancestor.

Acceptance of the distinct evolutionary origin of *Bacteria* and *Archaea* relegates the term prokaryote to a *mere description of the particular cell structure* that two domains (*Bacteria* and *Archaea*) happen to share. But the *Archaea* and *Bacteria* have other properties in common. Most members of both domains are only about 1/10 the length of eukaryotic cells (an average 1 μm long compared to the typical eukaryotic length of 10 μm). Their 70S ribosomes (differing from the 80S ribosomes of eukaryotes) contain distinct prokaryote rRNAs: 16S in the small subunit and 5S and 23S in the large (chapter 8). When examined with an electron microscope, as CJ from the case study was asked to do, it is apparent that prokaryotes lack the nucleus and the complex, intracellular, membrane-bound organelles (e.g., the Golgi apparatus and mitochondria) typical of eukaryotes. Also, neither *Bacteria* nor *Archaea* have evolved along the majestic lines of the eukaryotes. With some exceptions, they have remained unicellular and, if you wish, simple. It is worth noting that most of the defining properties of prokaryotic cells are negative—what they lack rather than what they have—and hence do not imply relatedness. Still, the term prokaryote is useful if for no other reason than that it links small-celled organisms, which, as a consequence of their size, share many similarities of cell physiology. In fact, it is often impossible to tell a bacterial from an archaeal cell on morphological grounds alone.

In certain biochemical respects, however, *Archaea* resemble eukaryotes. Their DNA and RNA polymerase enzymes, for example, are similar to the eukaryotic ones. Also, unlike *Bacteria*, *Archaea* do not use an *N*-formyl methionine residue as the amino acid incorporated during translation (chapter 8). Other biochemical differences include resistance to some antibacterial antibiotics. In yet other biochemical respects, *Archaea* differ from *both* eukaryotes and *Bacteria*. Most remarkably, their cytoplasmic membranes contain glycerol *ethers* attached to **isoprenoid** groups, rather than fatty acids attached by *ester* bonds to a molecule of glycerol (**phospholipids**). Furthermore, archaeal membranes can be *bilayers* of glycerol diethers, much as phospholipid bilayers, or *monolayers* composed of diglycerol tetraethers (chapter 2). It seems likely that their unique biochemical properties represent adaptations to the extreme environments in which many *Archaea* reside. In addition, some *Archaea* (those that make methane) produce coenzymes found nowhere else in nature (Box 14.4).

Yet other biochemical properties of the *Archaea* distinguish them clearly from their fellow prokaryotes, *Bacteria*. The walls of almost all bacteria are composed of the **peptidoglycan** called **murein** (chapter 2). *No known archaeal cells contain murein* in their walls. A few make a different peptidoglycan called **pseudomurein**, which lacks the *N*-acetylmuramic acid and D-amino acids found in bacterial murein (chapter 2). Most have walls made of protein.

Another curious distinction of the known *Archaea* is that, even though some are found as commensals in animals, apparently none are pathogens of either plants or animals. One can only speculate on the reasons for this. Possibly the major groups of *Archaea* evolved and became largely fixed in their lifestyles before eukaryotic hosts became available for them to exploit, or perhaps *Archaea* evolved in environments that were devoid of eukaryotic microbes.

Archaeal taxonomy: an evolving view of archaeal diversity. The taxonomy of the *Archaea* was, for a long time, simple and restricted to two recognized phyla, the ***Euryarchaeota*** and ***Crenarchaeota,*** for which cultivated repre-

BOX 14.4

Making methane the archaeal way

Methane formation can be viewed as a form of anaerobic respiration (chapter 6). In the simplest case (mediated by many methanogens), hydrogen gas (H_2) is the electron donor and carbon dioxide gas (CO_2) is the terminal electron acceptor:

$$4\,H_2 + CO_2 \rightarrow CH_4 + 2\,H_2O$$

Eight electrons are transferred from four molecules of hydrogen to a molecule of carbon dioxide in order to make one molecule of methane (CH_4). In the process, eight protons are transferred across the membrane to create a **proton motive force** that can be used to generate ATP or to drive other energy-utilizing cellular processes. This unique energy metabolism requires specialized enzymes with dedicated cofactors. In fact, methanogens are biochemically unique in that they produce cofactors not found elsewhere in nature. The stepwise reduction of carbon dioxide occurs while attached to these cofactors.

Among the various species of methanogens, there are three variations on this general theme:

1. The electron donor, hydrogen, can be replaced by an organic compound (formate or one of several alcohols; for example, ethanol, 2-propanol, 2-butanol, or 2-pentanol).
2. The electron acceptor, carbon dioxide, can be replaced by a methyl-group-containing compound (e.g., methanol, trimethylamine, or dimethyl sulfide).
3. Acetate can serve as both a donor and an acceptor. The acetate molecule is cleaved. Its methyl group, which serves as an electron donor, is oxidized to carbon dioxide, and its carboxyl group, which serves as an electron acceptor, is reduced to methane.

Many of the various substrates for methanogens (carbon dioxide, hydrogen, formate, acetate, and short-chain alcohols) are abundantly produced from higher-molecular-weight organic compounds by fermentative bacteria that are present in anaerobic environments. By converting these products of fermentation to a gaseous product (methane), methanogens provide a route by which organic materials escape from anaerobic environments. As methane rises into an aerobic environment, some of it is oxidized by **methylotrophs** (methane-oxidizing bacteria) to carbon dioxide. However, much of the methane produced by methanogens escapes into the atmosphere, where it acts as a greenhouse gas.

---

sentatives were available (Fig. 14.8). Among them were the **methanogens**, a group of *Euryarchaeota* that reduces carbon dioxide (CO_2) to methane (CH_4). The rest of the known archaeal species belonged to one of these two phyla and all were extremophiles, i.e., they thrived in extremely hostile environments. Some grew well at temperatures higher than the boiling point of water or even at autoclave temperatures (121°C; chapters 1 and 4). Other *Archaea* set biological records for tolerating high concentrations of salts and high acidity. Because such environments were probably more common during Earth's earliest development, these abilities of *Archaea* led to the proposal that some of our earliest ancestors might have been members of the *Archaea*. We now know that some bacteria thrive in extreme environments, and some archaea grow under quite ordinary conditions. But growth in extreme environments still remains a general characteristic of the *Archaea*.

The simple view of *Archaea* as restricted to two phyla was challenged when the first environmental DNA sequencing studies began to identify sequences corresponding to new, candidate archaeal phyla called the *Thaumarchaeota*, *Aigarchaeota*, and *Korarchaeota*, collectively designated, along with the *Crenarchaeota*, as the TACK superphylum (Fig. 14.3 and 14.8). Environmental genomics studies and single-cell genomics further filled the gaps at the base of the archaeal branch, identifying several new candidate phyla for ultra-small *Archaea* that defied cultivation and which were grouped collectively in the Dynamic Programming meets Artificial Neural Networks

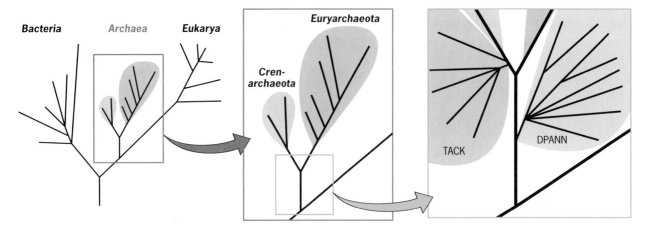

FIGURE 14.8 Diversity of phyla in the domain *Archaea*. The classical tree of life, with the three domains (*Bacteria*, *Archaea*, and *Eukarya*), traditionally included only two archeal phyla, the *Crenarchaeota* and *Euryarcheota*. Environmental surveys and single-cell genomics have however challenged this simplistic view and have identified numerous other phyla, most of them with yet uncultivated archaeal members, that can be grouped in the TACK and DPANN superphyla. Some scientists have even proposed that the *Eukarya* branched from the TACK superphylum. (Modified from a figure in Eme L, Doolittle WF, *Curr Biol* **16**:R227–30, 2015, doi: 10.1016/j.cub.2014.12.044.)

(DPANN) superphylum (Fig. 14.8). The reconstruction of the genomes of some of these candidate phyla also revealed some surprising facts. They all have small genomes, lack many core metabolic pathways, and are predicted to have fermentative and possibly symbiotic lifestyles based on carbon and hydrogen cycling. The deep-branching position of the DPANN group also led to the proposal that the root of the archaeal tree lies within this group. Clearly, this is an evolving field that continues to fill the gaps of knowledge about the *Archaea* and their evolution and, by extent, the evolution of *Eukarya* from an ancestral archaeal host.

Even with incomplete knowledge of the actual diversity of archaeal species and the physiology of these yet to be cultivated *Archaea*, the cultivation-independent sequence data and analyses provide novel insights into the metabolic diversity and ecology of the *Archaea*. Key points are:

1) The *Archaea* are phylogenetically diverse, perhaps as diverse as the *Bacteria*.
2) They are not restricted to extreme environments. In fact, *Archaea* are extremely abundant in moderate environments such as soils and sediments, as well as marine and freshwater aquatic habitats.
3) Methanogenesis likely appeared only once and early on in the evolution of the *Euryarchaeota*, as it is not represented in any other archaeal phyla (or anywhere else in biology for that matter). This finding also suggests that it is unlikely that the archaeal ancestor was a methanogen, as had been proposed in the past.
4) The deep branching of the DPANN superphylum, which contains ultrasmall *Archaea* with very small genomes and limited metabolic capabilities, suggests that the archaeal ancestor may have also been small and may have had a simple physiology based on the anaerobic cycling of carbon and hydrogen.

5) One of the candidate phyla, the "Lokiarchaeota," is particularly puzzling because its members are phylogenetically closer to eukaryotes than to prokaryotes and their genomes possess many eukaryotic-like genes. Thus they are thought to be "on the way" to eukaryotes. No doubt, there is yet a lot to be learned about the *Archaea*.

***Some notable* Archaea.** Taxonomy aside, some members of the *Archaea* are so unique at the physiological level that they merit special mention. We will highlight the two phyla (*Crenarchaeota* and *Euryarchaeota*) for which cultivated representatives are available, focusing on the most familiar groups.

Crenarchaeota: *thermoacidophiles and other notable members of the phylum.* The **thermoacidophiles,** as the name suggests, consist of extreme **thermophiles** (growing at more than 70°C temperatures) that are also **acidophiles** (the optimal pH of some being as low as 2.0). Most of these organisms are found in highly acidic hot springs. This is the group to which the extreme thermophile *Pyrolobus fumarii* belongs. Most *Crenarchaeota* metabolize elemental sulfur in some way as part of their fueling reactions. Some are aerobic chemolithotrophs that oxidize sulfur. Others are chemolithotrophs that reduce it to sulfide (H_2S) while oxidizing hydrogen (H_2).

Surprisingly, large numbers of *Crenarchaeota* have been detected (by molecular methods) in the open ocean, particularly at depths below a few hundred meters. Their abundance suggests they play an important, but still unknown, ecological role in the marine environment. Their metabolism must differ markedly from those of other *Crenarchaeota*: they are not thermophiles or acidophiles and are unlikely to metabolize sulfur because the open ocean contains only minuscule amounts of this element. Regardless of what they do, the discovery of **mesophilic** *Crenarchaeota* challenged the traditional view that *Archaea* were either methanogens or extremophiles.

Euryarchaeota: *methanogens and extreme halophiles.* The *Euryarchaeota* make up the larger group of *Archaea* and comprise several physiological types, including the methanogens (methane producers) and extreme halophiles (those that grow in the presence of high concentrations of salts). As their name indicates, methanogens produce methane, whereas extreme halophiles are adapted to growing at high (4 M) concentrations of salts. As we will soon learn, they clearly represent two extremes of physiology.

METHANOGENS

The methanogens are a specialized group characterized by their ability to generate methane as a byproduct of their metabolism, in many cases using hydrogen to reduce carbon dioxide (Box 14.4). Despite their very specialized metabolism, they are ecologically diverse: they prosper in all carbon-rich anaerobic environments, from those in the **psychrophilic** (cold) to those in the hyperthermophilic (very hot) temperature ranges of growth. Bubbles that rise from most quiet ponds contain methane produced by methanogens in the organically rich, anaerobic mud at the bottom, a fact that can be easily verified by collecting the gas and noting that it burns with a blue flame. Gas released when cattle and other ruminants belch also contains methane produced by methanogens in the animals' rumens. Collectively, ruminants produce huge amounts of methane, which account for 10 to 20% of all the methane in Earth's atmosphere (we will revisit this methane belching phenomenon in chapter 21). About a third of us humans also produce methane as an intestinal gas due to the metabolic activities of methanogens that live in our intestines. (In case you are interested in knowing more about the role of microbes in flatulence, here are two facts: individuals who do produce methane when they pass gas do so for life; the rest produce hydrogen, one of the substrates for methane production in methanogen-colonized humans.) In many other environments, including landfills and anaerobic sewage digesters, methanogens produce large quantities of methane, which can be collected. This application of methanogens is discussed in chapter 28. Methane (called natural gas by the petroleum industry) is a valuable fuel. But it is also a powerful greenhouse gas, thereby constituting an ecological hazard by contributing to global warming. This concern is exacerbated by the fact that the total global production of methane is continuously increasing, currently at the rate of about 1% per year.

EXTREME HALOPHILES

The extremely halophilic *Euryarchaeota* (the so-called **haloarchaea**) are a metabolically diverse group that shares the common feature not only of growing in near-saturated brine (4 M), but also of *depending on high salt concentrations to grow*. If their environment is diluted even to 1 M saline (still a lot of salt), their protein walls weaken, and the cells lyse. Haloarchaea flourish in salt-evaporation ponds, which turn bright red from the carotenoid pigments these microbes produce. From an airplane flying over San Francisco Bay, evaporation ponds appear as bright-red checkerboard squares (Fig. 14.9). Since biblical times, humans have used the appearance of this red color as an index for when to drain the "bitterns" (brine high in magnesium and sulfate) from the crystallized salt that sinks to the bottom of the pond, thereby producing a better-tasting product.

FIGURE 14.9 **Aerial photograph of evaporation ponds in San Francisco Bay.** As the salt concentration rises, halophilic archaea proliferate, turning the ponds red. (Courtesy of the Image Science and Analysis Laboratory, NASA Johnson Space Center.)

Extreme halophiles live in environments so "salty" that many essential processes in other organisms would grind to a halt. Unlike moderately halophilic bacteria, which exclude salt from their intracellular environment, haloarchaea avoid plasmolysis by concentrating KCl intracellularly—up to 5 M—when growing in near-saturated brine. Necessarily, their proteins are resistant to the high concentrations of salt that would denature proteins of most organisms. These archaea face yet one more challenge to grow in brine: being strict aerobes, haloarchaea are precariously dependent on a supply of oxygen, a gas with low solubility in their high-salt environment. Some haloarchaea, including the well-studied *Halobacterium halobium*, possess an emergency alternative to aerobic respiration, namely, a light-driven proton pump (**bacteriorhodopsin**) that generates sufficient ATP when oxygen becomes limiting (Fig. 14.10). Low concentrations of ambient oxygen signal the cell to synthesize specialized regions in its cytoplasmic membrane called **purple membranes**, where bacteriorhodopsins are housed. The light *induces the protein to pump protons*, creating a transmembrane proton gradient capable of generating ATP. A similar protein, rhodopsin, located in the retinas of animals is the light receptor necessary for vision.

CASE REVISITED: Is it a bacterium or an archaeon?

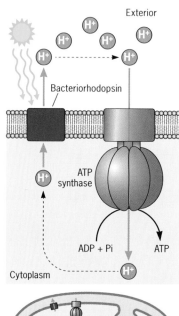

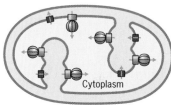

FIGURE 14.10 Bacteriorhodopsin, the light-driven proton pump of haloarchaea. (Modified from source: "Bacteriorhodopsin chemiosmosis" by Darekk2—own work. https://commons .wikimedia.org/wiki/File:Bacterior hodopsin_chemiosmosis.gif#/media /File:Bacteriorhodopsin_chemiosmosis .gif)

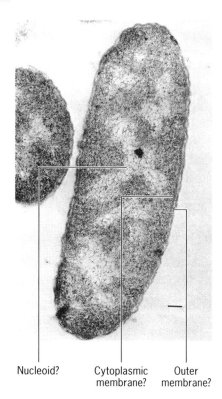

Nucleoid?   Cytoplasmic   Outer
              membrane?   membrane?

Student CJ examined a thin section of a microbial cell under an electron microscope and had to figure out whether it was a bacterium or an archaeon. He knew that it was indeed a prokaryote because, wisely, he noticed that there was no nucleus or membrane-bound organelles, a hallmark of the structure of a eukaryotic cell. He observed a less dense mass inside the cell that was likely the nucleoid, surrounded by a darker area (the cytoplasm) with small dark dots consistent with being ribosomes. He also observed two darker boundaries in the periphery, perhaps the cell membrane and the outer membrane of a Gram-negative cell envelope, but he could not distinguish any more detail. In summary, the cell structure was clearly prokaryotic, and he suspected it was a Gram-negative bacterium, but he could not tell with certainty. Why not? Bacterial and archaeal cells may look structurally similar under the microscope. However, they followed distinct evolutionary paths from a common universal ancestor, and their differences are manifested at the genetic and biochemical level. How could CJ prove this? He could extract DNA from the microbe and sequence the 16S rRNA. He could then compare the sequence against those available in the databases to infer phylogenetic relationships. This approach would unequivocally elucidate if it was a bacterium or an archaeon. Furthermore, based on the closest relatives, he could even assign the sequence to another taxonomic unit such as a phylum, a class, and a genus. Ancestry is to be found in the DNA sequence.

CONCLUSIONS

The prokaryotic domains, the *Bacteria* and *Archaea*, stand out for their remarkable diversity, reflected in their many metabolic capabilities and widespread presence in the environment. Studies of cultivated species hint at the potential diversity of these two groups but, only now, with the advent of more powerful cultivation-independent methods, are we beginning to grasp the true diversity of these two domains. Their long journey on Earth, together with their ability to grow and generate genetic variation fast, has propelled their diversification. At some point in the evolution of life on Earth, representatives of both groups came together and established symbiotic associations that led to the emergence of eukaryotes and their subsequent diversification. It is a world dominated and influenced by prokaryotes. And what we know is just the tip of the iceberg!

Representative ASM Fundamental Statements and Learning Outcomes for the Chapter

EVOLUTION

1. Cells, organelles (e.g., mitochondria, chloroplasts), and all major metabolic pathways evolved from early prokaryotic cells.

   a. Identify specific key characteristics of LUCA.
   b. Explain why the endosymbiotic acquisition of mitochondria and chloroplasts by ancestral eukaryotes was an essential event that led to the diversification and explosive evolution of eukaryotes.
   c. Describe how mutations can be maintained in a bacterial cell and thus build the capacity for natural selection.
   d. Evaluate whether the term "prokaryote" is still a useful description of *Bacteria* and *Archaea*, given the three-domain tree of life and many differences between the domains.

2. The traditional concept of species is not readily applicable to microbes due to asexual reproduction and the frequent occurrence of horizontal gene transfer.

    a. Explain why the traditional classification of a species based on morphology for plants and animals is not appropriate for defining microbial species.
    b. Discuss how using the term "operational taxonomic unit" is better suited than the term "species" when describing relationships among microbes.

CELL STRUCTURE AND FUNCTION

3. *Bacteria* and *Archaea* have specialized structures (e.g., flagella, endospores, pili) that often confer critical capabilities.

    a. Describe how the α-proteobacterium *Agrobacterium tumefaciens* induces its plant host to provide nutrients for the bacterium.
    b. Identify distinguishing features of archaeal membranes.
    c. List biochemical differences between *Archaea* and *Bacteria*.
    d. Infer how unique structures in *Archaea* contribute to their prevalence in extreme environments.

METABOLIC PATHWAYS

4. The survival and growth of any microorganism in a given environment depend on its metabolic characteristics.

    a. Summarize how *Prochlorococcus* is well adapted to the nutrient-poor ocean.
    b. Speculate why scientists do not yet agree upon the reasons why *Streptomyces* naturally produces antibiotics.
    c. Use extreme halophiles as an example to justify the statement: "the survival and growth of any microorganism in a given environment depend on its metabolic characteristics."

MICROBIAL SYSTEMS

5. Microorganisms are ubiquitous and live in diverse and dynamic ecosystems.

    a. State reasons why it is generally difficult to assess the diversity of prokaryotes, especially for environmental microbes.
    b. Describe how cultivation-independent techniques can be used to determine the presence of *Bacteria* and *Archaea* in a sample.
    c. Name a microbe from each of the six classes of *Proteobacteria*.
    d. Identify reasons why *Archaea* were undiscovered until the advent of molecular biology and why the diversity of extant *Archaea* is still being discovered.

IMPACT OF MICROORGANISMS

6. Microbes are essential for life as we know it and the processes that support life (e.g., in biogeochemical cycles and plant and/or animal microbiota).

    a. Identify a consequence upon Earth of the emergence and expansion of cyanobacteria 2.5 billion years ago.
    b. Describe the role of methanogens in the carbon biogeochemical cycle.

7. Microorganisms provide essential models that give us fundamental knowledge about life processes.

    a. Discuss how the Luria and Delbrück experiments with *E. coli* lent support for natural selection over the Lamarckian view of evolution.

8. Humans utilize and harness microorganisms and their products.

    a. Identify how the *A. tumefaciens* Ti plasmid can be harnessed to benefit humans.
    b. List three ways in which *Actinobacteria* are beneficial for the environment and/or to humans.

SUPPLEMENTAL MATERIAL

References

- **Eme L, Doolittle WF.** 2015. Microbial diversity: a bonanza of phyla. *Curr Biol* **25**:R227–30. doi: 10.1016/j.cub.2014.12.044.
- **Lane N, Martin W.** 2010. The energetics of genome complexity. *Nature* **467**:929–934. doi: 10.1038/nature09486.

- **Rosselló-Móra R, Amann R**. 2015. Past and future species definitions for *Bacteria* and *Archaea*. *Syst Appl Microbiol* **38**:209–216. doi: 10.1016/j.syapm.2015.02.001.
- **Spang A, Saw JH, Jørgensen SL, Zaremba-Niedzwiedzka K, Martijn J, Lind AE, van Eijk R, Schleper C, Guy L, Ettema TJ**. 2015. Complex archaea that bridge the gap between prokaryotes and eukaryotes. *Nature* **521**:173–179. doi: 10.1038/nature14447.

Supplemental Activities

CHECK YOUR UNDERSTANDING

1. **What differentiates *Eukarya* from *Bacteria* and *Archaea*?**
   a. Cell wall
   b. Nucleus
   c. Different ancestor
   d. Cellular size

2. **Which of the following are exclusive of prokaryotes?**
   a. Oxygenic photosynthesis
   b. Methanogenesis
   c. Respiration
   d. Endospore formation
   e. Peptidoglycan cell wall

3. **List the two major forces of genetic variation behind the diversification of *Bacteria* and *Archaea*.**

4. **Cultivation-independent methods have revealed a great diversity of yet-to-be-cultivated archaeal species. Summarize what novel insights have been gained from these studies about the phylogenetic and metabolic diversity and ecology of the *Archaea*.**

DIG DEEPER

1. The *Firmicutes* is the only bacterial phylum that includes endospore formers.

   a. No *Archaea* are known to make spores. Why?

   b. The following paper describes an approach to study the relative abundance and distribution of genes required for endospore formation among the *Firmicutes*: **Bueche M, Wunderlin T, Roussel-Delif L, Junier T, Sauvain L, Jeanneret N, Junier P**. 2013. Quantification of endospore-forming *Firmicutes* by quantitative PCR with the functional gene *spo0A*. *Appl Environ Microbiol* **79**:5302–5312)

      i. What genes were used as markers of spore formation? Why?

      ii. What gene was most frequently found in the major genera of endospore-forming *Firmicutes*?

      iii. Based on the phylogenetic tree shown in Fig. 1 and the frequency of sporulation genes in each clade, formulate a hypothesis that explains how many times and at what point did endospore formation emerge during the diversification of this phylum.

      iv. Would it be possible for other bacteria or even archaea to acquire the capacity to form endospores from the *Firmicutes*? How likely could this be?

2. *Actinobacteria* is an ancient phylum that comprises high-GC Gram-positive bacteria and is integrated by members with various morphologies and lifestyles. Use the information in the following article to learn more about this diverse phylum and answer the questions below: **Ventura M, Canchaya C, Tauch A, Chandra G, Fitzgerald GF, Chater KF, van Sinderen D**. 2007. Genomics of Actinobacteria: tracing the evolutionary history of an ancient phylum. *Microbiol Mol Biol Rev* **71**(3):495-548.

   a. The authors describe examples of specific genetic events (gene duplication, horizontal gene transfer, gene loss, and chromosomal rearrangements) that can explain the diversification of specific actinobacterial taxa. Which ones?

b. Name three actinobacterial taxa that are capable of mycelial growth.

c. Working in groups, select one of the following actinobacterial groups and identify their most salient features. Present your summary as a 5-minute presentation to the class.

      i. *Bifidobacterium*

     ii. *Tropheryma*

    iii. *Propionibacterium*

    iv. *Mycobacterium*

     v. *Nocardia*

    vi. *Corynebacterium*

   vii. *Leifsonia*

  viii. *Streptomyces*

    ix. *Frankia*

     x. *Thermobifida*

3. **Team work:** Several properties of the candidate archaeal phylum "Lokiarchaeota" support the notion that eukaryotes emerged from an archaeal ancestor. Working in groups, search the Web to gather information about the properties of this yet uncultivated archaeal phylum that lend support to this hypothesis. Summarize your findings in a 5-minute presentation to your class.

Supplemental Resources

VISUAL

- Video showing the mycelial growth of some *Actinomycetes*, in the *Actinobacteria* phylum (2:05 min): https://www.youtube.com/watch?v=Sk0ll39g3y4
- iBiology lecture about "Methanogens" by Dianne Newman (Cal Tech/HHMI) (2:36 min): https://www.youtube.com/watch?v=TyM2TKmI2pl
- Psychrophilic arctic methanogens make A LOT of methane! Check it by yourself in this video "Dr. Katey Walter and the Flaming Arctic Methane" (0:16 min): https://www.youtube.com/watch?v=HvhhsGnC1-c

SPOKEN

- How to pronounce *Firmicutes* (0:15 min): https://www.youtube.com/watch?v=SejEm43Znaw
- Short lecture about the *Actinobacteria* explaining their most notable features (4:01 min): https://www.youtube.com/watch?v=6gezol4QFgM

WRITTEN

- *Marine Microbial Diversity: The Key to Earth's Habitability*, a 2005 colloquium report from the American Academy of Microbiology: http://academy.asm.org/index.php/browse-all-reports/522-marine-microbial-diversity-the-key-to-earths-habitability.
- *The Rare Biosphere*, a 2011 colloquium report from the American Academy of Microbiology: http://academy.asm.org/index.php/browse-all-reports/501-rare-biosphere.
- *Microbial Evolution*, a 2011 colloquium report from the American Academy of Microbiology: http://academy.asm.org/index.php/browse-all-reports/508-microbial-evolution.
- Moselio Schaechter reminds us that endosymbiosis is a common occurrence in extant microbes in the *Small Things Considered* blog post "Pictures Considered #31. An Endosymbiont On Its Way To Becoming an Organelle?" (http://schaechter.asmblog.org/schaechter/2015/11/pictures-considered-31-an-endosymbiont-on-its-way-to-becoming-an-organelle-.html).
- Gemma Reguera describes the untapped diversity of metabolically active bacteria that inhabit the coldest places of the Earth in the *Small Things Considered* blog post "The Cold Side of Microbial Life" (http://schaechter.asmblog.org/schaechter/2014/03/the-cold-side-of-microbial-life.html).
- Merry Youle ponders "Do prokaryotes, being small, make small proteins?" in the *Small Things Considered* blog post "Going to Great Lengths" (http://schaechter.asmblog.org/schaechter/2011/05/going-to-great-lengths.html).
- Moselio Schaechter discusses small endosymbiotic bacteria within small insects in the *Small Things Considered* blog post "The Bacterium That Doesn't Know How To Tie Its Own Shoelaces" (http://schaechter.asmblog.org/schaechter/2009/05/the.html).

- Merry Youle discusses the diversity of the microbial dark matter in the *Small Things Considered* blog post "Fine Reading: Exploring the Microbial Dark Matter" (http://schaechter .asmblog.org/schaechter/2013/11/fine-reading-exploring-the-microbial-dark-matter .html).

Macro X-ray image of a mold

CHAPTER FIFTEEN

The Fungi

KEY CONCEPTS

This chapter covers the following topics in the ASM Fundamental Statements.

CELL STRUCTURE AND FUNCTION

1. While microscopic eukaryotes (for example, fungi, protists, and small algae) carry out some of the same processes as bacteria, many of their cellular properties are fundamentally different.

INFORMATION FLOW AND GENETICS

2. Although the central dogma is universal, applicable to all cells, the processes of replication, transcription, and translation differ in *Bacteria*, *Archaea*, and *Eukarya*.
3. The regulation of gene expression is influenced by external and internal molecular cues and/or signals.

IMPACT OF MICROORGANISMS

4. Microbes are essential for life as we know it and the processes that support life (e.g., biogeochemical cycles and plant and/or animal microflora).
5. Microorganisms provide essential models that give us fundamental knowledge about life processes.
6. Humans utilize and harness microorganisms and their products.

INTRODUCTION

Stepping across the great cellular divide of the biological world, we shall now consider the other large group of microscopic organisms, the eukaryotic microbes (Table 15.1). They constitute a hugely important portion of the living world, challenging the prokaryotes (*Bacteria* and *Archaea*) in total mass and diversity. Eukaryotic microbes are usually larger than most prokaryotes, but they vary enormously in size and shape. They can be found in a wide variety of environments: in bodies of water, in soil, and on and in plants and animals, including both healthy and diseased humans.

The taxonomy of eukaryotic microbes is particularly complex, befitting the large number of organisms in this group. We will not discuss taxonomy in detail but will consider the eukaryotic microbes as belonging to two large groups: the fungi in this chapter and the protists in chapter 16. Together, these organisms shape our global ecosystem, but the fungi stand out among them for being responsible for most of the recycling of the Earth's organic material. In this chapter we discuss the ubiquity of the fungi and their essential role as decomposers of plant material. We will also learn that many fungi are symbiotic or parasitic, some causing serious diseases in humans and plants. Fungi also make a large number of compounds of industrial and pharmaceutical importance. The yeasts, in particular, find numerous uses in daily food preparation, such as during the making of bread and alcohol beverages. And, as we will learn, yeasts are also very useful model systems in biological research because many of them are easy to manipulate and to study in the lab. Be prepared! The fungal world is an amazing one.

CASE: Making bread and wine

You have two projects in mind, to bake a loaf of bread and to make your own wine. For this purpose, you use two different but related strains of yeast. For the bread, you mix flour, water, and the yeast and knead them patiently and lovingly together. You do this until, according to your mother, the mass has "the consistency of your ear lobe." Next you let the dough "rise" until it's about double in size (figure) before baking it in an oven.

TABLE 15.1 **Main groups of eukaryotic microbes**

Group	Examples of types	Characteristic cell envelope and constituents
Fungi[a]	Yeast, molds	Cell wall containing glycoproteins (e.g., mannoproteins), polysaccharides (e.g., chitin, glucans), other
Protozoa	*Paramecium*, amebas, *Giardia*, *Plasmodium* (agents of malaria), *Tetrahymena*	Pellicle, interior to plasma membrane; consists of interconnected protein molecules; gives shape to the cell
Algae[a]	*Euglena*, *Chlorella*, diatoms, dinoflagellates	Chrorophyll, cell wall containing cellulose, silica in some, calcium in some

[a]Not all are microscopic; e.g., mushrooms (fungi) or seaweed (algae)

For the wine, you crush the grapes and add the yeast. You store this liquid in a nearly airtight container and wait for the fermentation to proceed. Then you bottle the wine and wait for it to age properly.

- Why is kneading the dough necessary for making bread?
- What kind of biochemical activities are the yeasts carrying out at the various steps in both bread and wine making?
- What metabolic energy-related reason accounts for the differences in exposure to air?
- What would have happened had you stored the grape juice under aerobic conditions, and, conversely, had you kept the bread dough under anaerobic conditions?

Bread rising. (From Photobucket, Steve4him.)

WHAT ARE THE FUNGI?

LEARNING OUTCOMES:
- List three characteristics of fungi that make them different from other eukaryotic microbes, plants, and animals.
- Identify three ways in which fungi have made an impact on your life.
- Explain how fungi can help plants grow.
- Justify the statement: "Without fungi, we would be denied our very existence."

Say "fungi" and you should be thinking of delectable dishes made with wild mushrooms and beer, wine, and bread, which are made with yeasts, as in the case study. Fungi may evoke less pleasant scenarios: athlete's foot and other infections, moldy food forgotten in the refrigerator, or mildew on old leather shoes. Fungi do all these things and many more. The fungi are, in fact, as diverse in size and shape as these examples suggest. They range from simple unicellular organisms, such as the yeasts used to make bread and wine in the case study, to gigantic bracket fungi in old forests. Indeed, one of the largest specimens known is over 1 meter (*yes, one meter!*) across: *Bridgeoporus nobilissimus*, the "most noble polypore." It is a large, shaggy, shelf fungus growing from old tree stumps with "the upper surface reminiscent of a green pizza with a crew cut," according to a mushroom hunter (Fig. 15.1). Alas, it is not edible. (Let us share with you that one of us is an avid wild mushroom hunter and has found this hobby to be both rewarding and challenging.)

Most fungi are filamentous (yeasts are the exception). Thus, specific terms are used to describe their distinctive morphological features. A fungal filament, for example, is called a **hypha**, and a collection of hyphae is called a **mycelium**. Commonly, the diameters of both fungal filaments and yeast cells are similar to those of typical vertebrate cells, about 4 to 10 micrometers (μm).

Fungi are also plentiful. A square meter of forest land typically contains about 100 grams of fungi, making them by far the most abundant of all microbes in forest sites. This figure translates into a guesstimate of well over two tons of fungi for every human being on Earth! In the air of rooms in occupied buildings, such as your classrooms, fungal spores are about as

FIGURE 15.1 One of the largest known mushrooms, the bracket fungus *Bridgeoporus nobilissimus*, growing on a dead grand fir tree near Redwood Creek, CA (with a hungry mycologist). (Photo courtesy of Kamyar Aram and Ashley Hawkins.)

abundant as bacteria (Fig.15.2). Thus, we inhale them in large amounts, usually with no untoward effects.

What are all of these fungi doing? Their major activity is the recycling of vegetable matter in the planet. Inconspicuous though they may be, fungi are indeed the great decomposers. They can break down many complex organic compounds, including cellulose and lignin, the main components of wood (Fig. 15.3). Termites and other wood-eating insects depend on fungi and bacteria in their guts to break down these otherwise indigestible compounds. Were it not for fungal activity, dead plants and trees would accumulate to great depths and lead to a colossal accumulation of carbon. Without fungal activity, there *would not be enough carbon dioxide* to sustain plant and microbial photosynthesis. Thus, all animals, which ultimately depend on plant life, would perish. Without fungi, we would be denied not only porcini risotto, but our very existence.

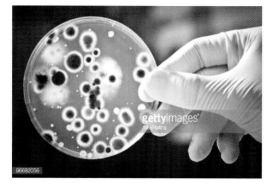

FIGURE 15.2 **An agar-containing petri dish that had been exposed to the air and then incubated for a day.** The larger fuzzy colonies are of molds, the smaller ones of yeasts and bacteria. (Panel A, credit Alexander Raths/iStock/Getty Images; panel B, Microbe-World, Wistreich Collection.)

Fungi play another major environmental role as partners with plant roots; these symbiotic microbes are called **mycorrhizae** (for "fungus-roots"). Mycorrhizae supply the plants with water and certain minerals, such as phosphorus. The fungal filaments are intimately associated with the roots, thus effectively extending the plant's reach over large distances (Fig. 15.4). The hyphae here function as enormously long drinking straws. In addition, mycorrhizae serve as communication conduits between individual plants, allowing the transmission of signals that alert neighboring plants to the presence of invading insects. In these ways, mycorrhizal fungi foster the life of plants. Without them, the growth of both natural and cultivated plants would be significantly reduced.

Being endowed with varied biochemical abilities, it is not surprising that fungi make a large number of compounds that are useful to humans. Indeed, fungi provide many antibiotics, including the first useful one, penicillin, made by a mold called *Penicillium*. Other industrial products made by fungi include the statins used to reduce cholesterol levels, the immunosuppressant cephalosporin, a variety of food additives, such as vitamin B2 and lactase (aka β-galactosidase, used to make milk inoffensive for people who are lactose intolerant), and several industrial enzymes, even those used to make "stone-washed" jeans (Table 15.2).

Fungi also cause disease. Table 15.3 lists some of the principal fungal infections of humans. Note that these include some rather common ones that many of us are acquainted with, such as athlete's foot and vaginal "yeast" infections. Other so-called "systemic" diseases affect deeper organs of the body, often with severe consequences. Some of these, such as coccidioidomycosis ("valley fever") and histoplasmosis, resemble tuberculosis (keep this in mind when you visit TB in chapter 24). The agents that cause these infections are always found in the environment. Coccidioidomycosis, in fact, is typically found among people living in the Western desert of the United States, because the spores of the causative fungus, *Coccidioides immitis*, survive well in the dust. In such places, archeologists and their students are well advised to wear masks on digs. Characteristic of these diseases is their greater severity in immunocompromised people. In fact, such diseases are often the cause of death in terminal patients who have undergone cancer chemotherapy or those with untreated AIDS. This tells us that a functional immune system is effective in keeping such agents at bay, even if we encounter them with some frequency. In addition, encounter with fungal spores causes allergies and asthma.

Lastly, fungi and their fungus-like distant cousins, the water molds or oomycetes, are serious pathogens of plants (Table 15.4). They cause the majority of infections affecting plants, many with lethal consequences, including the Irish potato famine and the chestnut blight. In addition,

FIGURE 15.3 **Decay of a tree trunk showing mushrooms, the fruiting bodies of a fungus that has caused the wood rot.** (Detail from "View from Mount Holyoke, Northampton, Massachusetts, after a Thunderstorm" [1836] by Thomas Cole.)

TABLE 15.2 **Some useful things made by fungi**

Foodstuffs (beer, wine, bread, cheeses, vitamins)
Antibiotics
Statins to lower cholesterol
Immunosuppressants for organ transplantation
Industrial enzymes (various uses; e.g., for laundry
　detergents)

FIGURE 15.4 **Mycorrhizal fungal filaments attached to plant roots.** (Courtesy Sylvie Herrmann.)

TABLE 15.3 **Some major fungal pathogens of humans**

Name of fungus	Main disease
Candida	"Yeast infection" (vaginitis, thrush, others)
Aspergillus	Chronic lung infections, sinusitis, others
Cryptococcus	Meningitis
Histoplasma	Respiratory infections, others
Coccidioides	"Valley fever," respiratory infections, others
Pneumocystis	Pneumonia (almost only in the immunocompromised)
Trichophyton, Microsporum	Ringworm, athlete's foot

fungi cause diseases that are known and feared by farmers, including blights, rusts, wilts, rice blast, and corn smut. Many of these diseases can be controlled by breeding resistant host varieties and by fungicides.

ARE YEASTS A TYPE OF FUNGI?

LEARNING OUTCOMES:
- Explain why yeasts are an excellent model system for eukaryotic cell biology.
- Differentiate the conditions under which *S. cerevisiae* makes wine in comparison to conditions under which *S. cerevisiae* makes bread.

Yeasts are *definitely* fungi! They also happen to be the best-studied members of the fungi, and, in recent years, they have also become one of the best models for eukaryotic cell biology. Studies carried out with yeasts have provided us with a detailed understanding of such processes as the

TABLE 15.4 **Some major fungal plant pathogens**

Name of fungus	Taxonomic group	Examples of disease
Magnaporthe oryzae	Ascomycete	Rice blast
Puccinia spp.	Basidiomycete	Rust (many hosts)
Fusarium graminearum	Ascomycete	Blights (many hosts)
Fusarium oxysporum	Ascomycete	Wilts (many plants)
Blumeria graminis	Ascomycete	Powdery mildews (grasses)
Mycosphaerella graminicola	Ascomycete	Blotches of wheat
Colletotrichum spp.	Asexual	"Anthracnose spots" and blights (many hosts)
Ustilago maydis	Basidiomycete	Corn smut (edible as "huitlacoche"

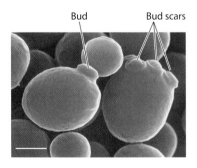

Bud   Bud scars

FIGURE 15.5 **Scanning electron micrograph of yeast cells showing a bud and several bud scars.** (Courtesy of A. Wheals.)

control of the cell cycle and cell division and the regulation of gene expression under diverse circumstances.

By "yeast" most people mean the common baking or brewing yeast, *Saccharomyces cerevisiae*, used to make bread and wine in the case study. Like most yeasts, *S. cerevisiae* replicates by budding (a **bud** is the bubble-like extrusion of a progeny cell; Fig. 15.5). We will discuss this interesting form of cell division in more detail in the next section. *S. cerevisiae* is also the species favored by most fungal researchers. However, there are other kinds of yeasts that replicate the way bacteria do, by binary fission; i.e., splitting in half. Keep in mind that many "non-yeast" fungi can grow as single cells (thus, in the yeast form), as well as in filaments, depending on the circumstances. The reasons for this **dimorphism** are not fully understood, but this property most likely extends the organisms' biological repertoire, enabling them to thrive in different environments.

One advantage of working with *S. cerevisiae* is that it can grow in simple laboratory media containing only a single carbon source, such as glucose or acetate, plus mineral salts. Use rich media, and the yeast cells will double every 2 hours. Yeast also grows on solid medium. Colonies become visible within 2 days, which makes this yeast easy to use for scientific research. In addition, *S. cerevisiae* can grow either aerobically or anaerobically. In the presence of oxygen, it *respires;* i.e., oxidizes carbon sources to CO_2 and water. For the case study, did you hypothesize that CO_2 gas production by the yeast made the bread dough rise? You were right! Kneading the dough is important because it supplies a lot of oxygen to allow the yeast to grow and generate CO_2 gas as a by-product. The gas forms bubbles, or tiny pockets, that expand the volume of the dough and make the bread fluffy and palatable after baking. Can you predict how fluffy and palatable the bread would be if you did not knead the dough? (Hint: When growing anaerobically, yeast ferments sugars to more ethanol and less CO_2.)

Yeast *fermentation*, on the other hand, is critical during wine making (case study). In chapter 6, we learned that there is a significant difference in the energy yield between respiration and fermentation. For this reason, under anaerobic conditions, the yeast needs to metabolize more of the sugar to gain energy for growth. This is what happens when the yeast is growing anaerobically using sugars provided in the grape juice. Because the yeast is fermenting the grape sugars, it produces more alcohol and less CO_2. And this is not restricted to wine making. Brewers, because they are interested in boosting alcohol production, use anaerobic conditions. Bakers, who need to maximize CO_2 production, use aerobic ones. Which condition would you use to make yeast for sale? (*Quiz spoiler:* Aerobiosis because, compared to fermentation, respiration yields greater energy and, in turn, generates more yeast cells.)

HOW ARE THE LIFESTYLES OF YEASTS DIFFERENT FROM THOSE OF OTHER FUNGI?

LEARNING OUTCOMES:
- Compare and contrast the life cycle of an animal to the life cycle of a yeast.
- Describe how fungi cells know when to mate.
- Compare and contrast meiosis in an animal to meiosis in a yeast.
- Explain how haploid fungal cells can spontaneously convert to the other mating type to ensure colony self-sufficiency.

Budding, the characteristic cell division of most yeasts, is a somewhat unusual form of cell division. Unlike binary fission, which most prokaryotic cells follow, budding is asymmetrical. The bud starts out as a protrusion from the "mother cell" and gradually enlarges until it is nearly the size of the mother. At this point, the newly formed cell separates from the mother.

Budding has a traumatic-sounding consequence: at the site where the bud separates from the mother cell, a little differentiated plate, called a **bud scar**, is formed (Fig. 15.5). Therefore, the number of bud scars indicates the history of a yeast cell: the more prolific, the more scars. But there's a catch: no new buds can form at the site of a bud scar. The mother cell can only accommodate a couple of dozens of these birthmarks, after which it cannot divide any more: It has become "old." On the bright side, this property of yeasts has made it into a valuable model for the study of *senescence*.

Unlike prototypical "higher" organisms that alternate their life cycles between haploid (one set of chromosomes) and diploid (two sets), many yeasts can grow in either state for prolonged periods. Some yeasts, such as *Candida albicans* (the agent of thrush, vaginal infections, and others), are mostly diploid; other species are forever haploid. A variation on the theme is found in other fungi: the mushrooms. Their filaments, which arise by the joining of two gametes, have two nuclei sharing the same cytoplasm. In such cells, called **dikaryons** (or heterokaryons), the two nuclei coexist without fusing, like detached roommates rather than a committed couple. The cells of a mushroom remain in the dikaryon stage until they are ready to make spores. At that time, the two nuclei fuse and undergo meiosis right away, leading to the production of haploid spores (Fig. 15.6). Note that these organisms have a remarkably short diploid stage, at least from the point of view of a "higher" organism.

Fungi have no obligatory germ line and do not make specialized sperm cells and/or eggs. Rather, their cells can become **gametes** directly. As in higher organisms, fungal gametes make zygotes by mating only with cells of another gender. Cells of one gender (also called a **mating type**) go about their metabolic business and do not become involved in sex with cells of the same mating type. One striking feature of some fungi, especially among those that make mushrooms, is that they have more than two genders,

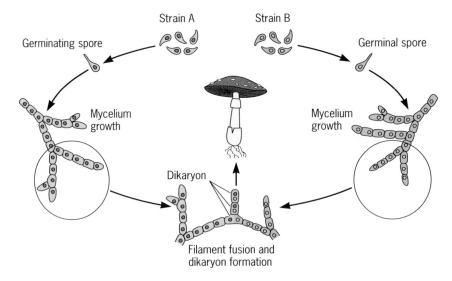

FIGURE 15.6 **How to make a mushroom.** Spores of different mating types (shown as having red or tan nuclei) germinate and form filaments (hyphae). The hyphae fuse to make a new filament with both kinds of nuclei, called a dikaryon. The dikaryon eventually differentiates to make a fruiting body, which we call a mushroom.

often thousands. The reason is that their sex-determining genes come in many varieties. Since mating can occur only between strains with different mating types, this diversity enhances the chances of encountering a compatible partner in nature.

Fungal cells differentiate into gametes only when cells of the opposite mating type are present. What is the "mating call" among the fungi? The signals are diffusible molecules called **pheromones** (secreted chemicals that affect behavior). The mating pheromones of yeasts, for example, are *small peptide molecules* that diffuse through the environment, with each mating type secreting its own kind of pheromone. When a signal from the opposite mating type is perceived, a cell becomes a gamete, i.e., it becomes competent to mate. This is a reciprocal behavior: cells of each mating type are equally alerted to the possibility that mating is at hand. When yeast cells become gametes, they stop dividing and elongate toward the pheromone source, becoming pear-shaped (Fig. 15.7). Such receptive cells are called "shmoos," a term evocative for fans of the old "Li'l Abner" comics. These elongated blobs fuse to become a structure that resembles two pears joined at their small ends. The two haploid nuclei then fuse into a diploid nucleus, becoming a **zygote**.

After mating, yeast cells may proliferate in the diploid state for a long time (unlike the mushrooms). However, once nutritional conditions become unfavorable, the diploid cells differentiate into **haploid spores**. Unlike bacterial endospores, which are extraordinarily resistant to heat and chemicals (chapter 13), fungal spores are only partly able to withstand harsh conditions. However, the spores endure starvation and, therefore, can survive in nutritionally fallow environments.

How are spores made? The diploid cells must undergo meiosis to make haploid spores. The signal for the onset of meiosis is, as was just mentioned, nutritionally poor conditions. Meiosis leads to the formation of four cells, which in yeasts are encased in a common envelope, looking something like four potatoes in a bag. In other ascomycetes, these cells undergo another division, resulting in a sac with eight spores (Fig 15.8). This structure, called an **ascus** (plural, asci), gives this group of fungi their taxonomic name, the Ascomycetes. This is the fungal group that includes most yeasts and molds, many human and animal pathogens (such as the agents of candidiasis, valley fever, and histoplasmosis), truffles, and morel mushrooms. Other fungi, including most mushrooms, are called the Basidiomycetes. As their name indicates, these fungi make a cell called the **basidium**, from whose surface haploid spores emerge by budding (Fig. 15.8).

Let us explore in more detail how Ascomycetes, the group that yeasts belong to, make their spores (called ascospores) and how these spores germinate. In a nutritionally suitable environment, the ascospores germinate and undergo vegetative growth as haploid cells. Following Mendelian rules, two of the cells in each ascus will be of one mating type, and the other two will be of the opposite mating type. If a single germinated spore were separated from the others and kept apart, its progeny would seem to be deprived of sex forever. Fate is not so unforgiving: cells of each mating type *spontaneously convert* to the other mating type. This change in identity occurs by a physical rearrangement of genes in a manner that broadly resembles the generation of antibody diversity (see chapter 22). Here is how

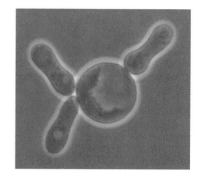

FIGURE 15.7 Yeast cells mating. Three hourglass-shaped mating yeast cells are shown together with a round diploid cell. (Courtesy of K. Johnson.)

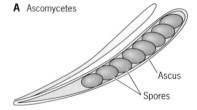

A Ascomycetes

Ascus

Spores

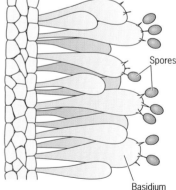

B Basidiomycetes

Spores

Basidium

FIGURE 15.8 Two ways that fungi produce sexual spores. (A) In the Ascomycetes, which include most yeast, molds, and a few mushrooms, the spores are encased in a sac called an ascus. (B). In the Basidiomycetes, which include most mushrooms, the spores bud from a cell called a basidium.

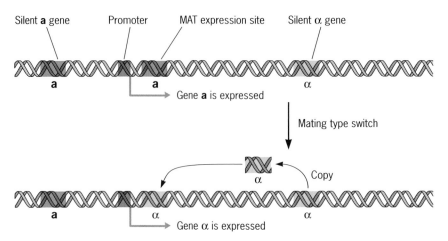

FIGURE 15.9 Switching of mating types in yeast. There are two mating types, called a and α, only one of which is expressed at a time. The genes for both are silent, unless a copy of one of them is transferred to a site, MAT, where transcription can take place. With some frequency, DNA recombination enzymes exchange the resident expressed gene with a copy of a silent gene, resulting in a switch of mating types.

this genetic trick works. Each yeast cell has two genes, one (called the "a" gene) encoding one mating type and the second (called the α gene) encoding the other (Fig. 15.9). But both of these mating type loci are *silent*, due to being packaged in chromatin structures that exclude the transcription machinery. However, each of the silent genes can be transferred into a new region on the chromosome that promotes its expression. In a way, each gene serves as a "CD" that can be physically inserted into a site (the "CD player") to express its content. Switching occurs at some random frequency when the mating gene in the "player" site is replaced by the other mating gene via **homologous recombination**. Thus, during growth, mating-type switches occur at some frequency. As a result, a colony contains cells of both mating types, even though the population arose from a haploid spore of one type only.

WHY IS YEAST SUCH A POPULAR GENETIC TOOL?

LEARNING OUTCOMES:
- Discuss the reasons why geneticists use yeasts as a model system for eukaryotic genetics.
- Describe the process by which yeasts make proteins from foreign DNA.

The lifestyle of yeasts makes them relatively easy to work with in the laboratory, which is one reason for their wide use among researchers. Mating can be induced at will by mixing cultures of different mating types. Each of the four products of a meiosis is enclosed in a single ascus that can be separated by using a micromanipulator viewed with a microscope. Few other eukaryotes can be genetically analyzed so rapidly, conveniently, and precisely.

There are other advantages to using yeasts for genetic research: they have a relatively small genome (14 million bases), only about three times the size of that of *Escherichia coli*. The genome is divided into 16 tiny chromosomes, each averaging only 800 kilobases in size, which makes them attractive models for the study of eukaryotic DNA replication and chromosome behavior. Yeasts can be thought of as stripped-down eukaryotic cells that do all of a cell's essential business without the fancy stuff (such as differentiating into a neuron or producing antibodies).

The haploid/diploid states that yeast undergo are also advantageous. Having stable haploid and diploid phases allows deleterious mutations to be maintained in the diploid cells but analyzed in the haploid cells. The effect of mutations can be examined immediately after sporulation; lethal mutations can be inferred by applying Mendel's rules to each ascus but saved in the heterozygous diploid form for further analysis.

Tools are also available to introduce foreign DNA into a yeast cell by transformation (see chapter 10). Although the cell wall of yeasts is very tough, it can be breached by electric shock (electroporation) or by treatment with certain salts, such as lithium acetate, to allow the introduction of the foreign DNA into the cell. Once inside, the DNA integrates with high efficiency into the targeted chromosomal location via homologous recombination, thus making it easy to engineer any gene at will.

Constructing **plasmids** that replicate in yeast is also relatively easy. Some plasmids even reproduce in *E. coli* as well, allowing researchers to use this faster-growing bacterium host to make copies of the plasmid, clone any gene of interest, and then transfer the plasmid back to the yeast cell for expression. Such "shuttle vectors" take advantage of both genetic systems, those of yeast and of *E. coli*. It is also possible to construct and propagate plasmids containing remarkably large amounts of foreign DNA (sometimes up to 800 kilobases). These gigantic plasmids are called **yeast artificial chromosomes** (YACs). It would be difficult to introduce such large plasmids into a bacterium. But yeasts take up the plasmids and replicate them. These properties make yeasts convenient vehicles for the study of DNAs from many sources, including humans (the characterization of eukaryotic genomes usually depends on cloning large chromosomal fragments).

Yeasts are also attractive systems for the production of recombinant proteins. Being eukaryotic, they allow most eukaryotic proteins to become properly folded and post-translationally modified, e.g., by glycosylation (chapter 8). The first genetically engineered human vaccine, hepatitis B core antigen, and the first genetically engineered enzyme used in food production, rennin, were produced in yeasts. Can you think of other proteins that could be mass-produced in yeasts?

CASE REVISITED: Making bread and wine

Making bread and wine relies on the activity of related yeast strains, but the final product could not be more different. Why? The difference is best revealed by the fact that bread making is an aerobic process, wine making an anaerobic one. This is one reason bread dough is kneaded: it is a good way to oxygenate the yeast. By contrast, grape juice is covered to prevent oxygenation.

While making bread, the yeast will respire, that is, use oxygen as the terminal electron acceptor. As a result, each sugar molecule in the dough will be converted into six molecules of CO_2 and H_2O. So much CO_2 is produced that bubbles of the gas form and make the bread rise. If deprived of air, the yeast would ferment, thus making much less CO_2 (two molecules per molecule of sugar) and producing ethanol as a by-product. The result would be a fairly unpalatable flat bread.

However, yeast fermentation is desirable for wine making. Here, the desired product is the ethanol, which is generated during anerobic growth. Had the grape juice been aerated, the yeast would have respired, producing a nonalcoholic bubbly grape juice. Because less energy (ATP) is derived from fermentation than respiration, wine making is intrisically slower than bread making. Of course, by aging the wine after the fermentation has essentially ceased, proper flavors are imparted to the wine. Thus, bakers work faster than winemakers.

CONCLUSIONS

Fungi make up a large portion of the total biomass. They are essential for life on earth because they decompose most plant material. Fungi are found in a large number of habitats, including soils, the surface of plants and inanimate objects, and on and in bodies of animals. Many fungi live in symbiotic or parasitic relationships with animals, including humans and plants. Fungi make a large number of compounds of industrial and pharmaceutical importance. They are enormously useful in biological research because many of them, yeasts especially, are easy to manipulate and to study.

Representative ASM Fundamental Statements and Learning Outcomes for the Chapter

CELL STRUCTURE AND FUNCTION

1. While microscopic eukaryotes (for example, fungi, protozoa, and algae) carry out some of the same processes as bacteria, many of the cellular properties are fundamentally different.

    a. List three characteristics of fungi that make them different from other eukaryotic microbes, plants, and animals.

    b. Compare and contrast the life cycle of an animal to the life cycle of a yeast.

    c. Compare and contrast meiosis in an animal to meiosis in a yeast.

INFORMATION FLOW AND GENETICS

2. Although the central dogma is universal in all cells, the processes of replication, transcription, and translation differ in *Bacteria*, *Archaea*, and *Eukarya*.

    a. Explain how haploid fungal cells can spontaneously convert to the other mating type to ensure colony self-sufficiency.

    b. Describe the process by which yeast makes proteins from foreign DNA.

3. The regulation of gene expression is influenced by external and internal molecular cues and/or signals.

    a. Describe how fungi cells know when to mate.

IMPACT OF MICROORGANISMS

4. Microbes are essential for life as we know it and the processes that support life (e.g., biogeochemical cycles and plant and/or animal microflora).

    a. Explain how fungi can help plants grow.

    b. Justify the statement: "Without fungi, we would be denied our very existence."

5. Microorganisms provide essential models that give us fundamental knowledge about life processes.

   a. Discuss the reasons why geneticists use yeasts as a model system for eukaryotic genetics.

   b. Explain why yeasts are an excellent model system for eukaryotic cell biology.

6. Humans utilize and harness microorganisms and their products.

   a. Identify three ways in which fungi have made an impact on your life.

   b. Differentiate the conditions under which *S. cerevisiae* makes wine in comparison to conditions under which *S. cerevisiae* makes bread.

SUPPLEMENTAL MATERIAL

References

- **van der Wal A, Geydan TD, Kuyper TW, de Boer W**. 2013. A thready affair: linking fungal diversity and community dynamics to terrestrial decomposition processes. *FEMS Microbiol Rev* **37**(4):477–494. doi: 10.1111/1574-6976.12001.
- **Bonfante P, Genre A**. 2015. Arbuscular mycorrhizal dialogues: do you speak 'plantish' or 'fungish'? *Trends Plant Sci* **20**(3):150–154. doi: 10.1016/j.tplants.2014.12.002.
- **Kaeberlein M**. 2010. Lessons on longevity from budding yeast. *Nature* **464**:513–519. doi: 10.1038/nature08981.
- **Liti G**. 2015. The fascinating and secret wild life of the budding yeast *S. cerevisiae*. *eLife* **25**:4. doi: 10.7554/eLife.05835.

Supplemental Activities

CHECK YOUR UNDERSTANDING

1. **What differentiates fungi from *Bacteria* and *Archaea*?**
   a. Cell wall
   b. Nucleus
   c. Fermentation
   d. Cellular size

2. **Which process does NOT involve fungi?**
   a. Decomposing wood
   b. Delivering water to plant roots
   c. Making yogurt
   d. Signaling between plants
   e. Making champagne

3. **How does replication of budding yeast differ from fission yeast?**

DIG DEEPER

1. Apply what you learned in the case study to illustrate with schematics the critical biochemical steps in wine or bread making.
   a. How are the two processes similar?
   b. What are the major differences?

2. Mycorrhizal fungi associate with plant roots and extend their mycelia far away from the root system to promote the absorption of essential nutrients that are dispersed in the soils and otherwise difficult for the plant root system to adsorb.

   Read the following *Small Things Considered* blog posts: "Mycorrhizal Fungi: The World's Biggest Drinking Straws and Largest Unseen Communication System" (http://schaechter .asmblog.org/schaechter/2013/08/mycorrhizal-fungi-the-worlds-biggest-drinking -straws-and-largest-unseen-communication-system.html) and "The Power of Fungal Genetics—Cassava for Food Security and Sustainability in Colombia" (http://schaechter .asmblog.org/schaechter/2015/03/the-power-of-fungal-genetics.html). Work in groups to answer the following questions:

    a. How do the fungi benefit from the association?

    b. Intense cropping practices use chemical products, such as fertilizers and pesticides, to stimulate plant growth, but these chemical treatments also reduce the content of mycorrhizal fungi in agricultural soils and the water demand of the plants. Why?

    c. The mycorrhizal fungi also connect neighboring plants to each other through the underground mycelia network. Breaking this connection makes the plants more sensitive to infestation by aphids. Why?

    d. Based on the plant growth-promoting effects of mycorrhizae, propose an experiment to demonstrate the applicability of using these fungi as natural fertilizers.

3. Despite their reputations, termites are not very good at digesting plant material (cellulose) even with the aid of the many microbes they carry. For example, African termites only partially digest the plant material they ingest. After carrying their meal in their stomachs back to their nest, they then regurgitate the meal into small piles. The termites also transport a fungus that grows on these piles. Why? Termites happen to be the first fungus farmers on Earth!

    Read the resource "What agriculture can learn from termites and fungi" (http://phys.org/news/2015-04-agriculture-termites-fungi.html) to answer these questions:

    a. Does this arrangement benefit the fungus?

    b. Does it or harm the termite? How?

4. **Team work:** Work with classmates to generate at least five reasons why *Saccharomyces cerevisiae* is an excellent laboratory research tool.

5. **Team work:** Considering that fungi have coexisted with prokaryotes (*Bacteria* and *Archaea*) for approximately a billion years, it is not surprising that scientists have identified many examples of interspecies communication between the two groups.

    a. Working in groups, use PubMed to identify in the primary literature an example of interspecies fungus-prokaryote interactions.

    b. Identify if a signal(s) is produced by the participating microbes. If so, what does that signal convey about the producer's physiological state? What impact does the signal have on the recipient cell? Is the interaction beneficial? For whom?

    c. Develop schematics or animations to teach others about the relationships between the microbes.

    Some examples:

- *Microbial interference in quorum sensing*
- *Acquisition of xenosiderophores*
- *Cross-feeding in the gut or oral microbiomes*
- *Metabolite sharing between pathogenic or commensal species*
- *Impact of quorum sensing molecules on the growth of other microbes*
- *Peptidoglycan stimulating bacterial differentiation*
- *Induction of antibiotic resistance*
- *Induction of antibiotic production*
- *Biofilm formation in response to sublethal doses of antibiotics*

Supplemental Resources

VISUAL

- A pair of talks on sex in yeast by Andrew Murray:
  Yeast Sex: An Introduction (27 min): http://www.ibiology.org/ibioseminars/andrew-murray-part-1.html.
  How to Shmoo and Find a Mate (52 min): http://www.ibiology.org/ibioseminars/andrew-murray-part-2.html.

SPOKEN

- Meet the Scientist Episode 12: Nancy Keller, "*Aspergillus* and the Fungal Toxin Problem" (21 min): http://www.microbeworld.org/careers/audio-interviews/255-mts12-nancy-keller-aspergillus-and-the-fungal-toxin-problem.

- Meet the Scientist Episode 61: Charles Bamforth, "Beer: Eight Thousand Years of Bio-technology" (52 min): http://www.microbeworld.org/careers/audio-interviews/804-mts61-charles-bamforth-beer-eight-thousand-years-of-biotechnology.
- *This Week in Microbiology* podcast episode #112 "Mushroom Pickers and Mushroom Kickers" discusses why definitions in biology often change, and why the small molecule terrain is important for the growth of a soil fungus (1 hour): http://www.microbeworld.org/podcasts/this-week-in-microbiology/archives/1994-twim-112.
- Science Friday podcast: Two Oregon brew experts discuss the science of craft brewing, including how yeast, hops, malt, and water come together to create the perfect pint (16:30) min: http://www.sciencefriday.com/segment/06/20/2014/beer-science-crafting-the-perfect-pint.html.

WRITTEN

- Contributions of yeast to early civilizations and their early study: "Did Leeuwenhoek Observe Yeast Cells in 1680?" http://schaechter.asmblog.org/schaechter/2010/04/did-van-leeuwenhoek-observe-yeast-cells-in-1680.html.
- Check out just how creative fungi are when it is time to disperse their spores:
    "Riding the Spore Wind." http://schaechter.asmblog.org/schaechter/2010/10/riding-the-spore-wind.html.
    "Fungi that Spit Like Baseball Players." http://schaechter.asmblog.org/schaechter/2011/08/fungi-that-spit-like-baseball-players.html.
- *Small Things Considered* blog post "Mycorrhizal Fungi: The World's Biggest Drinking Straws and Largest Unseen Communication System" by Moselio Schaechter discusses an amazing feat of mycorrhizae. http://schaechter.asmblog.org/schaechter/2013/08/mycorrhizal-fungi-the-worlds-biggest-drinking-straws-and-largest-unseen-communication-system.html.
- *Small Things Considered* blog post "Little Known Glomalin, a Key Protein in Soils" by Moselio Schaechter discusses how one of the most abundant proteins in soils is fungal in origin. http://schaechter.asmblog.org/schaechter/2013/05/little-known-glomalin-a-key-protein-in-soils.html.
- *Small Things Considered* blog post "An Evolutionary Tale of Zombie Ants and Fungal Villains & Knights" by Gemma Reguera discusses how genes spread among bacterial populaitons and their consequences. http://schaechter.asmblog.org/schaechter/2012/07/an-evolutionary-tale-of-zombie-ants-and-fungal-villains-knights.html.
- *Small Things Considered* blog post "Plant Pathogen Silences Host's Immune Genes" by Marvin Friedman discusses fungi interactions with their host. http://schaechter.asmblog.org/schaechter/2014/07/plant-pathogen-silences-hosts-immune-genes.html.
- American Academy of Microbiology report "If the Yeast Ain't Happy, Ain't Nobody Happy: The Microbiology of Beer." http://academy.asm.org/index.php/faq-series/435-beer.
- *The Fungal Kingdom: Diverse and Essential Roles in the Earth's Ecosystem,* a 2008 colloquium report from the American Academy of Microbiology. http://academy.asm.org/index.php/browse-all-reports/174-the-fungal-kingdom-diverse-and-essential-roles-in-earths-ecosystem.

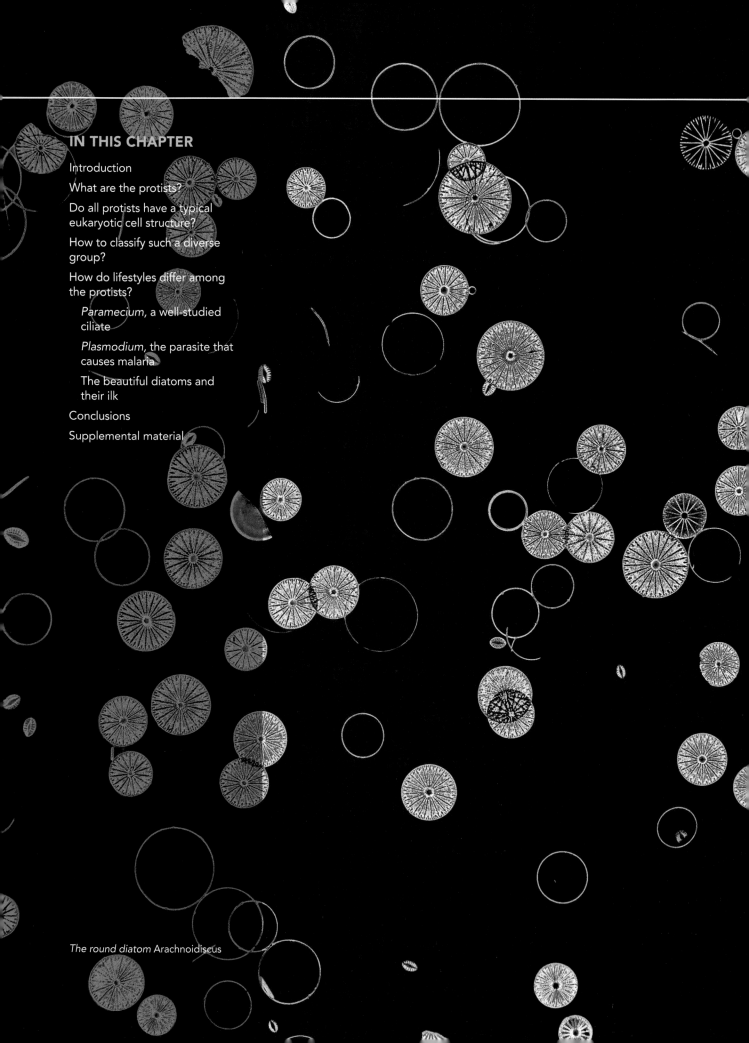

IN THIS CHAPTER

The round diatom Arachnoidiscus

CHAPTER SIXTEEN

Protists

KEY CONCEPTS

This chapter covers the following topics in the ASM Fundamental Statements.

EVOLUTION

1. Cells, organelles (e.g., mitochondria and chloroplasts), and all major metabolic pathways evolved from early prokaryotic cells.
2. Mutations and horizontal gene transfer, with the immense variety of microenvironments, have selected for a huge diversity of microorganisms.
3. Human impact on the environment influences the evolution of microorganisms (e.g., emerging diseases and the selection of antibiotic resistance).

CELL STRUCTURE AND FUNCTION

4. The structure and function of microorganisms have been revealed by the use of microscopy (including bright field, phase contrast, fluorescent, and electron).
5. While microscopic eukaryotes (for example, fungi, protozoa, and algae) carry out some of the same processes as bacteria, many of the cellular properties are fundamentally different.

INFORMATION FLOW AND GENETICS

6. Although the central dogma is universal in all cells, the processes of replication, transcription, and translation differ in *Bacteria*, *Archaea*, and eukaryotes.

MICROBIAL SYSTEMS

7. Microorganisms can interact with both human and nonhuman hosts in beneficial, neutral, or detrimental ways.
8. Microbes are essential for life as we know it and the processes that support life (e.g., in biogeochemical cycles and plant and/or animal microbiota).
9. Microorganisms provide essential models that give us fundamental knowledge about life processes.

IMPACT OF MICROORGANISMS

10. Humans utilize and harness microbes and their products.

INTRODUCTION

I n chapter 14 we learned about the extraordinary diversity of extant prokaryotes (*Bacteria* and *Archaea*) and also discussed the endosymbiotic events between the two prokaryotic groups that led to the emergence of eukaryotes. In chapter 15, we began to explore the diversity of eukaryotes, looking in more detail at the fungi. Now, we move to the study of the **protists**. This group comprises mostly unicellular eukaryotes of varied morphology and size and great phylogenetic diversity (some estimate there are perhaps 50,000 different species of protists!). Not surprisingly, protists are metabolically diverse and abundant, particularly in aquatic environments. The group also includes some that cause disease in humans and animals. Furthermore, among the protists we find numerous examples of endosymbiotic relationships with bacteria, which give us a glimpse at how eukaryotic organelles, such as mitochrondria and chloroplasts, could have originated from bacterial endosymbionts over a billion years ago. As fascinating as these microbes are, we are just beginning to understand their true diversity and the impact that their activities have in the functioning of many ecosystems. Prepare yourself for an amazing ride!

CASE: Is it a protist?

Ms. G, a 22-year-old female, traveled to East Africa (Kenya and Tanzania) with her parents to celebrate her graduation from college in the northeastern U.S. Her trip included a photo safari, where she was bitten by mosquitoes. On her flight home, 9 days after the safari adventure, she developed flu-like symptoms with headache, muscle aches, and a 40°C temperature. On her return, she consulted a physician who, given her recent travels and symptoms, suspected an infection by a mosquito-borne parasite. The doctor immediately took a blood sample and sent it to the microbiology laboratory for microscopic examination. Upon looking at the Giemsa-stained blood cells (see figure), the doctor immediately prescribed antimalarial drugs. The next day Ms. G felt worse and developed a short intense chill that was followed by a high fever lasting 6 hours. Two days later, these symptoms recurred.

• Why did the doctor suspect malaria as the culprit of Ms. G's symptoms?
• Why did the symptoms recur while on medication?

WHAT ARE THE PROTISTS?

LEARNING OUTCOMES:
• List three ways that protists influence nutrient cycles or food webs.
• List two ways that humans use protists or their products.
• Describe the distinguishing characteristics of coccolithophores.
• Identify three ways in which protists have shaped the Earth.

You may know them as the **protozoa** and the microscopic algae. The term protist may bring to mind paramecia, amebas, or mosquito-borne para-

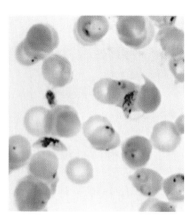

Plasmodium blood smear

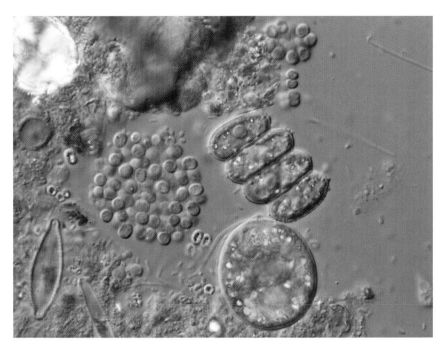

FIGURE 16.1 **Micrograph showing abundant unicellular eukaryotes (protists) in a drop of pond water scum (40× magnification).**

sites that cause malaria or sleeping sickness. Protists are a hugely diverse group of some 50,000 mostly unicellular species, all with a eukaryotic body plan. These organisms have evolved as separate, distinct groups but often with perplexing kinships. For example, certain photosynthetic groups (e.g., *Euglena*) are more closely related to certain protozoa (e.g., trypanosomes) than they are to algae or diatoms.

It may not surprise you to learn that protists, diverse and abundant as they are, play a major role in the shaping of our planet. Together with photosynthetic cyanobacteria, protists carry out perhaps half of all photosynthesis on Earth, mainly in the oceans. They are particularly abundant in aquatic ecosystems (Fig. 16.1). So abundant are they that you can count one million flagellates, one of the many groups of protists, in 1 ml of the water from some lakes! Other protists live by eating smaller microbes, thus being big players in determining how organic matter is recycled. In fact, the protists' appetite for aquatic bacteria is the major cause of mortality among prokaryotes, which contributes to keeping the number of bacteria in check. Predatory protists also matter to us because they prevent cyanobacterial blooms, which would otherwise release toxins in the lakes and poison potable water. In fact, the presence of some protists is used as an indicator of a "healthy" aquatic ecosystem.

Certain protists also contribute to the geology of our planet by making insoluble shells of silica (the diatoms) or calcium carbonate (other algae) that accumulate in huge deposits. The White Cliffs of Dover in England are the cemetery of untold numbers of shells of a little alga called a coccolithophore (*Emiliana huxleyi*). These organisms fix carbon dioxide (CO_2) via photosynthesis and divert substantial amounts of the carbon to make

FIGURE 16.2 A coccolithophore.
Shown is "Ehux," short for *Emiliania huxleyi*, the most abundant of all coccolithophores. Under favorable conditions, it makes huge marine blooms, outnumbering all other members of the phytoplankton by as much as 10:1. The source of their beauty is the coccoliths, or platelets of calcium carbonate that cover the cells and give the organisms their name. The blooms are highly reflective, causing more light and heat to be reflected into space instead of heating the ocean. The construction of huge numbers of coccoliths and their sinking to the ocean floor make a difference in how CO_2 can be stored in the atmosphere to contribute to the greenhouse effect. However, they not only influence climate; eventually, they form enormous chalk and limestone rocks.

their beautiful carbonaceous shells (Fig. 16.2). They make "blooms" that can become enormous, covering as much as 100,000 km² of ocean surface (almost the size of England or the state of Ohio). They grow their shells to accommodate the increases in cell size and can divide into motile cells, leaving the shells behind. Predation by viruses can also kill the cells confined within their shells. The empty shells then sink to the bottom of the oceans, where they become buried in the sediments for posterity. The mineralization activities of the coccolithophores are impressive: 80% of all the inorganic carbon in the oceans sinks to the deep-sea sediments as coccoliths. Despite their enormous importance in the cycle of matter, coccolithophores are little known and have not gotten their "day in the sun." It would not be unusual if you had never heard of them.

For another example of the contribution of protists to geological processes, consider the limestone blocks of the pyramids of Egypt, which consist of the shells of a type of amebae called the Foraminifera. These organisms live in the oceans and rain down to the deep-sea sediments due to the weight of their shells. How did they make the limestone mountains of Egypt, which is where the pyramids' limestone blocks came from? Some 30 to 50 million years ago, most of northern Africa, including Egypt, was covered by an ocean. In the bottom regions of this ocean the Foraminifera shells piled up, eventually becoming rocks of calcium carbonate or limestone. As the ocean waters receded, the bottom rocks became exposed and provided the Egyptians with the rock material to make the pyramids.

In addition to their many roles in natural systems, protists are also beneficial in man-made processes (Table 16.1). A key step in the treatment of wastewater is its aeration in large ponds so aerobic bacteria can degrade the organic matter. Protozoa team up with the bacteria to degrade the organic matter, but once the nutrients are gone, protists eat the bacteria and then each other. This natural process clears the water and eliminates the threat of pathogens. And what about the filters we use to clean the water in aquariums, swimming pools, or hot tubs? Most are made with diatom shells. The pores of the shells are so small (as small as 1 µm) that they trap even the smallest particles. When ground to a fine powder, diatom shells have the appearance of talcum powder, but if inhaled or swallowed by insects, they are lethal. The microscopic silica particles are so abrasive that they rupture the soft internal organs of the insect. The particles can also attach and damage the cuticle of the insect, promoting the efflux of water from the tissues, causing the insect's rapid death from dehydration. Hence, diatom powder works as a natural pesticide. The same abrasive properties also make

TABLE 16.1 **Useful things that protists do**

- Carry out a lot of photosynthesis, thus sequestering much CO_2 and releasing oxygen
- Consume great amounts of bacteria, including those dangerous to animals and humans
- Help decompose plant matter (i.e., are part of the major food webs)
- Help purify wastewater
- Provide a source of food to aquatic animals
- Make industrial products, e.g., diatomaceous earth.

diatom shell powder a favorite ingredient in cleaning powders and metal polish formulations. In some instances, diatom shell powder is even added to toothpastes and soaps. Next time you take a bath, thank the diatoms!

DO ALL PROTISTS HAVE A TYPICAL EUKARYOTIC CELL STRUCTURE?

LEARNING OUTCOMES:
- List two examples of organelles found in protists besides mitochondria and plastids.
- Explain how a cell can have mitochondrial genes, but no mitochondria.
- Present evidence that mitosomes are "mitochondrial remnants."
- Compare and contrast the function of hydrogenosomes with mitochondria.

Most protists have a typical eukaryotic organization—nuclei, mitochondria, and, in those that photosynthesize, plastids. However, some take nucleation to new heights and carry many more than one nucleus. One group of protozoa, the radiolaria, have members with 100 nuclei or more per cell (Fig. 16.3). As we will learn later in the chapter, the malaria parasite *Plasmodium* infects blood cells and then undergoes a complex life cycle, which includes a stage (schizont) in which the parasite makes tens of thousands of nuclei. (Do you think the doctor from the case study saw multinuclei within the stained blood smear?) On the simpler side, *Giardia*, the agent of hiker's diarrhea, and its relatives have no mitochondria. You may be tempted to think that they represent descendants of a primitive line of eukaryotic cells that had not yet acquired the endosymbiotic bacterium destined to become a mitochondrion. Perhaps not: the genomes of these organisms contain what are probably bacterial genes. Why is this a clue that *Giardia* evolved from an ancestor with a premitochondrial endosymbiont? (*Quiz spoiler*: Typically, as an endosymbiont becomes an organelle, many of its bacterial genes are transferred to the nucleus.) And *Giardia* is not alone: other protists, such as the flagellate *Trichomonas* and the ameba *Entamoeba*, also lack mitochondria, yet all contain genes of mitochondrial ancestry in their genomes. These chromosomal genes tell an interesting story about the evolutionary paths that led to the diversification of the early eukaryotes and support the notion that the ancestor of all extant eukaryotes had mitochondria (Box 16.1).

You may wonder how these microbes function without a mitochondrion, the energy factory of the eukaryotic cell. *Giardia* and *Entamoeba* and some microsporidia have **mitosomes**. These are also organelles surrounded by a double membrane, but, unlike mitochondria, they do not make ATP. This actually makes sense: mitosomes have been found only in parasitic protists that live under anaerobic or **microaerophilic** conditions inside their host. Oxygen is limited in these environments and is not available to fuel the mitochondrial ATP-generating reactions. Yet mitosomes, like mitochondria, contain proteins involved in the assembly and maturation of iron-sulfur proteins. This suggests that these organelles are, in fact, mitochondrial remnants. This finding is consistent with the phylogenetic prediction that the ancestor of all extant eukaryotes had a mitochondrion

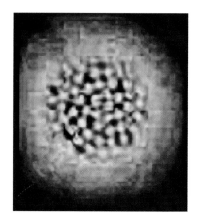

FIGURE 16.3 **How multinucleated can you get?** The many nuclei of the radiolarian *Collosphaera huxleyi*. Stained with DAPI, a dye that stains DNA. (Modified from Suzuki N et al., Bull Natl Mus Nat Sci Ser B, 35:169–182, 2009.)

BOX 16.1

Amitochondriate protists?

Giardia is not the only group that lacks the typical mitochondrion. Other protists, such as *Trichomonas and Entamoeba*, anaerobic flagellates found in hindguts of termites (e.g., *Trichonympha*), and the microsporidia fungi also lack these membrane-bound organelles. Yet, despite their amitochondriate status, they carry genes in their genomes of mitochondrial origin. One such gene is *cpn60*, which encodes a heat-shock protein that helps proteins fold properly. Researchers analyzed the mutations that have accumulated in the *cpn60* gene of the amitochondriate and mitochondriate eukaryotes, including the human version of the gene (see *Homo sapiens* branch in the phylogenetic tree), and found that all the gene variants derived from a common ancestor (the root of the phylogenetic tree). Researchers also observed that the amitochondriate groups branched early from the mitochondriate groups, including those represented by protists. This suggests that the common ancestor of all extant eukaryotes had a mitochondrion. But was it shed at an early stage in the evolution of eukaryotes to give rise to the amitochondriate protists? As we discuss in this chapter, these "amitochondriate" protists do have mitochondria-like organelles, which may be relics of the ancestral mitochondrion.

(Box 16.1). However, this ancestral mitochondrion was not just shed in the course of evolution. Rather, it was reduced as its function adapted to the need of the microbe.

Certain protozoa, such as the sexually transmitted pathogen *Trichomonas vaginalis*, have yet another variation on the mitochondrial theme. These organisms contain **hydrogenosomes**, membrane-bound organelles that produce ATP like mitochondria via reactions that generate hydrogen as a by-product (thence their name "hydrogenosomes"). Unlike mitochondria, typical hydrogenosomes contain no DNA. However, the nuclear genes that encode these reactions have bacterial counterparts, suggesting that these organelles, too, may have surrendered all their genes to the host cell's nucleus. There is an exception that seems to confirm this inference: a protozoan (*Nyctotherus ovalis*), found in the hindgut of cockroaches, has hydrogenosomes that do contain DNA and, apparently, ribosomes as well. Furthermore, the membranes of the *Nyctotherus* hydrogenosomes contain cardiolipin, a lipid of the mitochondrial (and bacterial) membrane. Hence, like mitosomes, hydrogenosomes also appear to be remnants of an ancestral mitochondrion. So, in the end, it appears that all eukaryotes, even the so-called "amitochondriate" protists, carry derivatives of the ancestral mitochondrion. They never got rid of it. The organelle just evolved for functions suitable to their host's particular mode of life.

HOW TO CLASSIFY SUCH A DIVERSE GROUP?

LEARNING OUTCOMES:
- Compare and contrast mechanisms of motility in eukaryotic microbes and bacterial cells.
- Speculate why there is such high diversity among protists.

TABLE 16.2 **Some of the major groups of protists**

Group	Examples	Means of locomotion	Main mode of nutrition
Flagellates	*Trypanosoma* (African: sleeping sickness; American: Chagas' disease), *Giardia* (hiker's diarrhea), *Trichomonas* (sexually transmitted infection)	Flagella	Absorption (uptake of soluble food)
Amoeboids	*Entamoeba histolytica* (dysentery, abscesses), *Naegleria fowleri* (encephalitis)	Pseudopodia	Phagocytosis (uptake of particulate food)
Ciliates	*Paramecium*, *Balantidium* (human intestinal infection)	Cilia	Ingestion (of particulate food via a mouth-like organ)
Apicomplexa (so called because they produce a structure called an apicoplast)	*Plasmodium* (malaria), *Toxoplasma* (toxoplasmosis)	Nonmotile (except for some stages in the life cycle)	Absorption (uptake of soluble food)

Discussing the classification of protists is like opening Pandora's box: many evil things can escape! So we will try to keep it simple. One of the first things you observe when you examine a drop of pond water under a light microscope is that it contains many protists (Fig. 16.1) and that some of the cells swim actively in all directions while others simply float around. This simple observation provides an easy way to classify protists as motile or nonmotile. As you increase the resolution (for example, using an electron microscope), you begin to appreciate morphological details in all of them. The motile cells may display long, thin filaments for locomotion (flagella), others may extend a part of their body (pseudopods), and others may have many short hair-like appendages (cilia) to move around. You just managed to classify the motile protists into flagellates, ameboids, and ciliates, respectively (Table 16.2). Note that all the members of the nonmotile group (apicomplexa) are parasites of animals. Except for some stages in their life cycle (e.g., when cells need to move from one animal host to another), locomotion is really not needed. You will learn about this later in the chapter, as we discuss the lifestyle of the malaria parasite, *Plasmodium*. The motile groups, on the other hand, are often free-living and need locomotion to access nutrients and avoid predation. *Paramecium*, a well-studied ciliate that we will discuss in detail later in the chapter, moves its many cilia to stir the water around and trap nutrients, including other microbes. It also uses cilia to escape from predators, although that strategy does not always work (Fig. 16.4).

FIGURE 16.4 Predation among ciliates. Scanning electron micrographs of a ciliate protozoan, *Didinium*, swallowing another ciliate, *Paramecium*. (Left) Side view of the initial stages; (right) top view showing the prey almost completely ingested. (Courtesy of G. Grimes and S. L'Hernault.)

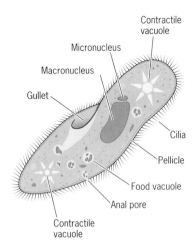

FIGURE 16.5 A paramecium. This complex single-celled organism has structures involved in food uptake (gullet), food digestion (food vacuoles, anal pore), secretion (contractile vacuoles), locomotion (cilia), safeguarding of the genome (micronucleus), and using genetic information (macronucleus).

Do not be fooled by this relatively simple classification. Protists are *extraordinarily diverse*, with members in each subgroup belonging to *distant lineages*. Morover, protist cell structures can also be enormously complex for unicellular organisms. Some are gigantic in size, reaching centimeters in length, although most are measured in tenths of millimeters. The ciliates are stunning in their cell complexity, possessing structural analogs that resemble, at least vaguely, a mouth, a stomach, a bladder, and an anus (Fig. 16.5). Their structures can be even more intricate: Some dinoflagellates (including the agents of red tides) possess an eye-like organ, the ocelloid, with a lens, a chamber, and a retina-like structure (Fig. 16.6). All of this in a single cell! One wonders why dinoflagellates have not become multicellular, with distinct functions assigned to different cells. Nobody has yet given a convincing argument for their refusal to adopt the more popular solution by divvying up some of their biological functions among different cells. Can you think of one?

The protists also have very different metabolisms and lifestyles. Because of this variety, to classify them scientists have resorted to complementary approaches that rely on morphological and metabolic features and phylogeny. Even with these many criteria, protists have been assigned to six "supergroups." The name "supergroup" is already cluing us into the complexity with in each group. This outstanding diversity should not surprise us though. As we learned in chapter 1, the unicellular eukaryotes emerged on Earth nearly 2 billion years ago, whereas the first multicellular eukaryotes emerged 750 million years ago (Fig. 16.7). Fungi did not evolve until much later. Two billion years is a long time to evolve and diversify!

HOW DO LIFESTYLES DIFFER AMONG THE PROTISTS?

As you can imagine, being so diverse, protists can have very different lifestyles. In general, protozoa are unicellular and free-living organisms, though some, like the ameba *Dictyostelium*, can attach to, and move on, surfaces and make complex multicellular structures called fruiting bodies (Fig. 16.8). But even those protists that remain unicellular throughout their life cycle can undergo complex cellular transitions. Below, we concentrate on the lifestyles of two widely known protozoa: *Paramecium*, a well-studied model organism, and *Plasmodium*, the parasite that causes malaria and that exemplifies how intricate a protist's lifestyle can be (and which will bring us back to our case study). We will also introduce you to the diatoms, which are microscopic algae and among the most abundant organisms in the phytoplankton of lakes and oceans.

Paramecium, a Well-Studied Ciliate

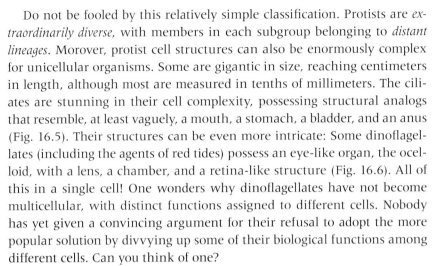

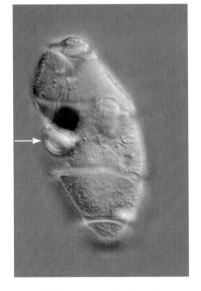

FIGURE 16.6 Some dinoflagellates possess an eye-like structure (the ocelloid) with a lens, a chamber, and a retina-like structure. The arrow points to the bright lens. (Courtesy Franz Neidl, Photomacrography.)

LEARNING OUTCOMES:
- List two functions of cilia in ciliates.
- Compare and contrast the structures and functions of the macronucleus and micronucleus in paramecia.
- Speculate on the benefits to each partner in the relationship between paramecia and the toxic bacterial cells.

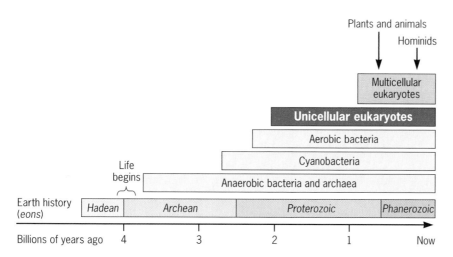

FIGURE 16.7 **The emergence of unicellular eukaryotes on Earth occurred some 2 billion years ago, after the aerobic bacteria arose.**

- Describe how paramecia avoid senescence.
- Define cortical inheritance.
- Discuss three major contributions to the study of biology, beyond knowledge about protists, that have been made as a result of studying protists.

To stir up your interest in science, your high school teachers may have asked you to look at a drop of pond water or scum under the microscope (Fig. 16.1). Slithering by were all sorts of creatures, going to and fro in apparently haphazard motion. Among the larger ones are particularly active creatures that look like hairy slippers, the paramecia. These creatures fascinate not only young students but also professional biologists: paramecia are among the largest and most complex single-celled organisms, and they carry out their physiological and genetic transactions in intriguing ways. You could argue about what constitutes large size in cells. After all, an ostrich egg is a single cell. However, it does not move about like a *Paramecium*, and it is programmed for a single mission.

Paramecia and their ciliate allies are well adapted to a free life. They inhabit not only pond water but also oceans, lakes, rivers, and soils. Competition in these environments is harsh, and paramecia themselves are prey to other protists, becoming part of the food chain (Fig. 16.4). However, paramecia are, in turn, predators that make a living as "grazers," ingesting any nutritious particles they encounter. They are not picky eaters; they eat all sorts of organic particles and also bacteria, yeast, and other protists. Their complex cell structure helps in this endeavor (Fig. 16.5). As their name indicates, ciliates have cilia. These appendages are aligned in neat rows and move like little whips to propel the cell and to drag nutrients, including other microbes, toward them. The sway of the cilia sweeps the food particles directly into a groove on the cell (the "gullet"), which functions as a mouth. The food is then digested in a special compartment called a "food vacuole," and the nutrients released in the process are then excreted to the cytoplasm of the cell. Any undigested food material is excreted through the anal pore.

FIGURE 16.8 **Fruiting bodies of a slime mold, *Dictyostelium discoideum*.** http://www.the-scientist.com/?articles.view/articleNo/30520/title/Micro-Farmers/.

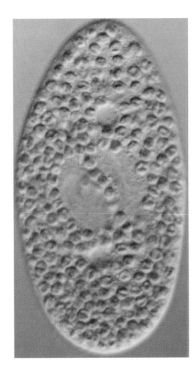

FIGURE 16.9 **A paramecium fully loaded with endosymbiotic algae.** (From Fujishima M, Kodama Y, *Eur J Protistol*, **48**:124–137, 2012, with permission.)

Predation and hospitality. Paramecia possess other interesting features. They swallow any particle in their vicinity, including other paramecia. But interestingly, some of their intended prey have evolved mechanisms to fight back. Some bacteria, for example, carry in their genomes **prophages** that encode a toxin. These bacteria make the toxin and then, if they are ingested by a paramecium, they secrete the toxin to kill the would-be predator. Some paramecia, in turn, have evolved immunity to the toxin and can ingest and carry the killer bacteria inside. The bacteria find shelter and food within the paramecium, whereas the paramecium host becomes itself toxic to any nonimmune paramecium that attempts to ingest it. After all, it is loaded with toxin!

But not all is predation in the paramecia. Hospitality is, in fact, a fashionable way of life for these protists. Their social registry of endosymbiotic guests includes a number of bacteria other than the toxin-producing ones. Some of these have lost many of their genes and cannot live freely outside their host. What do you think? Are they on their way of becoming organelles? They are found in many different locations within the host cell, including in both kinds of nuclei, and a number of them coexist in the same host. Paramecia also associate with algae and carry them in astounding numbers (Fig. 16.9). Here, both host and endosymbiont can live freely without their partner but come together in nutrient-limited environments. This dance makes sense, because photosynthesis and autotrophy allow the algae to use light as an energy source to fix carbon dioxide and make the nutrients that the host cannot acquire from the environment; the paramecium cells, on the other hand, protect the algae and use motility to reach the surface waters where light exposure is optimal and concentrations of carbon dioxide highest.

The challenges of cell division in paramecia. Paramecia are so big (although not especially so for a ciliate) that one could expect them, hippopotamus-like, to go about their business in a deliberate and unhurried manner. Not so. Some species can divide every 2 hours. For these large and complex cells, rapid growth represents special challenges; i.e., it requires that a lot of things fall into place. The cell must double every one of its organelle-like components in the time it takes to divide. The machinery involved in duplicating components must now work at high efficiency, synthesizing a lot of the enzymes that make nucleic acids, proteins, carbohydrates, and lipids. Incidentally, RNA enzymes, or **ribozymes**—the exception to the rule that enzymes are proteins—were discovered in a relative of *Paramecium* called *Tetrahymena*. And so was RNA splicing, as well as telomeres, the enzyme telomerase, and their role in aging! If Nobel prizes were given to the organisms rather than to the scientists who studied them, *Tetrahymena* would certainly qualify.

But let's turn back to the challenges of dividing such a large and sophisticated cell. Asexual reproduction involves dividing a cell into two cells, each carrying a set of all the nuclei and organelles (Fig. 16.10). The problem then becomes: how does a paramecium synthesize and make copies of all the cellular components before division? This reconstruction requires a lot of mRNA, but there is a limit to how rapidly mRNA can be made from DNA. Ciliates solve this problem in a way unique in biology: they have

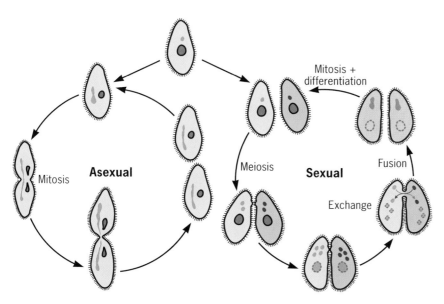

FIGURE 16.10 Reproduction in paramecia. Asexual reproduction (left) allows paramecia to divide by binary fission and uses mitosis to divide the micronucleus in two (the macronucleus is roughly split in half). After several cycles of asexual division, the cells age and will die unless they engage in sex. During sexual reproduction (right), two cells attach to each other through their gullets and connect their cytoplasm (conjugation). The macronucleus disintegrates, and the micronucleus divides by meiosis twice to make four nuclei. Three of them disintegrate, and the remaining one divides by mitosis. Each conjugant now has two haploid nuclei, and they exchange one of them (the "migratory" or "male" nucleus). After the exchange, the two nuclei fuse and form the "zygote" nucleus, which divides by mitosis to generate eight nuclei. Half of the nuclei enlarge, while the other half remain small. Only one of each will remain functional to become the macronucleus and micronucleus of the cell. The cells are now rejuvenated and can start a new cycle of asexual reproduction.

two nuclei, one containing DNA with copies of all genes (the **micronucleus**) and another, much larger, containing many copies of selected genes (the **macronucleus**) (Fig. 16.5). The micronucleus serves as an ordinary nucleus; i.e., the repository of the genetic information of the cell. The macronucleus, on the other hand, is a bag of repeated copies of selected genes that are made from the "real" nucleus (the micronucleus). Depending on the species, the macronucleus contains between 40 and 1,000 copies mainly of the genes needed for growth. As you can imagine, the macronucleus is a busy place, with molecular copying machines buzzing. Forming a macronucleus from a micronucleus is a complex process involving a selective reduction in the cell's genetic inventory to 15% or less of its gene content. This is challenging because the genes to be included in the macronucleus are not contiguous in the micronuclear genome, so they must be reshuffled. As the keeper of information for future generations, the micronucleus, like any respectable nucleus, must duplicate with extreme precision by mitosis. Not so with the macronucleus. It has so many copies of each gene that the micronucleus need not be partitioned precisely; it does not need a precise mitotic apparatus, as does the micronucleus. It simply divides in half, like a piece of chewing gum being pulled apart.

One more interesting fact about paramecia and certain other protists is that they do not divide indefinitely: they die after a number of cell divisions. This phenomenon of senescence is reminiscent of what happens in animal cells. In vertebrates, senescence is caused by the shortening of telomeres (the ends of linear chromosomes) after repeated cell divisions. Something different causes protist cells to get old and die, but what that is remains a mystery. However, paramecia have figured a way out of the cycle of senescence. They can reset their biological clock by undergoing sexual reproduction (Fig. 16.10). To do so, two paramecia cells, using a sticky substance, attach to each other through their gullets and make a bridge that connects their cytoplasms. Several things happen then: the macronuclei of both cells disappear, and the micronuclei undergo meiosis, forming four haploid micronuclei. Out of the four, only one haploid micronucleus retains the capacity to divide via mitosis. After this organelle divides in two, one (called the "male" nucleus) travels to the opposite cell. Now each cell has one functional micronucleus and at least one nonfunctional (called "stationary") one. The two micronuclei then fuse to make a diploid micronucleus in each cell. The cells are then ready to separate. With brand new micronuclei, courtesy of sex, the cells make more copies of the organelle, transition through a multinucleated state, and then differentiate to have the standard micronucleus and macronucleus. In many ways, it is as if they have been reborn: they can start a new cycle of binary divisions (asexual reproduction). Thus, these species can avoid senescence and endure as long as they occasionally engage in sex. Much of how this works is still a mystery. But then again, so is all sex.

Yet another contribution of ciliates to our understanding of genetics and evolution comes from the phenomenon called **cortical inheritance**. This refers to the inheritance of surface (cortex!) properties without the participation of genes. Mating in paramecia involves cell fusion, and, later, separation of the conjugating cells. The separation is not always gentle: occasionally one of the partners will pick up a patch of the cortex ("skin") from the other. The cortex carries cilia that are oriented in a given way. The ones in the new patch will be oriented in the opposite direction (Fig. 16.11). Cilia orientation is important, because cilia allow a paramecium to swim and feed. If enough cilia are facing the wrong orientation, swimming will be aberrant, and the cell may starve. This natural occurrence can be reproduced experimentally by surgically reversing part of the cortex. The point here is that the new pattern of cilia, and therefore the altered swimming behavior, is inherited without involving any change in genes; it is dependent entirely on the local geometry of the cortical patch involved. We call such a phenomenon **epigenetics** (see chapter 18 for more on epigenetics involving prions and chapter 25 on herpesvirus latency).

Plasmodium, the Parasite That Causes Malaria

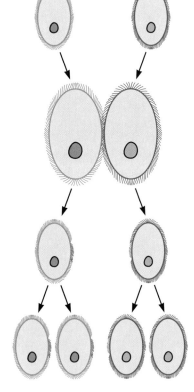

FIGURE 16.11 Cortical inheritance in a ciliate. The two sexually compatible cells happen to have cilia facing in the opposite direction. Upon mating (middle) and separation, some of the "skin" is transferred to the other cell. The progeny will have a patch of cilia facing the "wrong" orientation. This change is inherited for several generations.

LEARNING OUTCOMES:
- Outline the life cycle of *Plasmodium*, including gametocytes, gametes, and zygotes.
- Describe how *Plasmodium* infects red blood cells.
- Explain the method by which the malaria parasite, *Plasmodium*, harms its hosts.
- Predict how some virulent species of *Plasmodium* have become resistant to many antimalarial drugs.

Malaria has been, and continues to be, one of the great scourges of humankind. This disease affects some 300 million people, especially in tropical areas of the world, and causes 1 to 1.5 million deaths per year; it ranks among the major deadly infectious diseases. The pathogen is transmitted by biting mosquitoes that are found mainly, but not solely, in tropical regions. These mosquitoes do not range much further than 2 miles, an important consideration when attempting their control. To a large extent, the disease can be controlled by sanitary measures, such as the use of insecticides, drainage of pools of water where mosquitoes develop, and the use of mosquito netting. Unfortunately, such measures are expensive on a large scale and beyond the means of some countries in the developing world. Several drugs can prevent and treat malaria. Some, such as quinine, have been used for centuries. However, some of the most virulent species of the parasite have become resistant to many antimalarial drugs, and insecticides can become ineffective because mosquitoes can also develop resistance to them.

Human malaria is caused by four species of the genus *Plasmodium*, the most virulent one being *P. falciparum*. *Plasmodium* parasites alternate between two obligatory life stages, one in the mosquito—their only vector—and the other in their vertebrate host. To complete its life cycle, *Plasmodium* goes through an extraordinarily intricate choreography, including a mandatory sexual phase. The parasite has a dozen or so distinctive stages, each of which has been given a name. We will discuss only a few of these.

Life cycle inside the mosquito. As with most bloodsucking insects, only female mosquitoes feast on vertebrate hosts. (Unlike the males, females need a rich source of protein for making their eggs.) When feeding on a malaria-infected host, the mosquito ingests parasitized red blood cells (i.e., blood cells carrying the *Plasmodium*). The parasites then proliferate in the mosquito, which bites another host and delivers an infective load of parasites. However, matters are not so simple. The parasites must undergo a complex series of developmental changes before they reach an infective load sufficient to cause disease (Fig. 16.12).

The blood meal that the mosquito acquires from the infected host contains parasites in several stages of development, but only the so-called **gametocytes** differentiate into gametes that can invade the insect. Once in the mosquito's gut, the gametes exit the red blood cells that carried them to the mosquito, and the male gametes give rise to eight wildly motile, sperm-like cells that seek out female gametes. Within 30 minutes of entering the mosquito, mating is completed, and diploid **zygotes** are formed. Sex again. The zygotes then differentiate into motile cells that invade the mosquito gut. During this process, meiosis takes place to yield haploid cells. These divide to produce tens of thousands of offspring that are released into the hemolymph (blood) of the mosquito. In this form, the parasites invade the salivary glands. The mosquito is now loaded with parasites and ready to inject them into the bloodstream of the next person or animal that it bites!

Life cycle inside the human. In humans, *Plasmodium* rapidly travels through the bloodstream to the liver (Fig. 16.12). Depending on a number of factors,

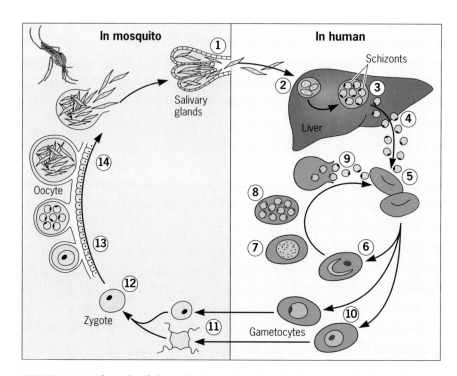

FIGURE 16.12 Life cycle of the malaria parasite. Parasites (sporozoites) released from the salivary gland of a female mosquito are injected into a human (1). They travel through the bloodstream and enter the liver (2). There, they mature into cells called tissue **schizonts** (3). They are later released back into the bloodstream (**merozoites**; 4), where they invade red blood cells (5). However, some parasites remain dormant in the liver (2). In some forms of malaria (due to *Plasmodium vivax* and *Plasmodium ovale*), these cause relapsing fever and chills. In the red blood cells, the parasites mature into ring-shaped forms (6) and other asexual stages (**trophozoites** and schizonts; 7 and 8). When fully mature, the parasites lyse the red blood cells and are released to invade uninfected red blood cells (9). Within the red blood cells, some parasites differentiate into sexual forms (**gametocytes;** 10). When taken up by a mosquito, the gametes further differentiate (11) and mate to produce a **zygote** (12), which penetrates the mosquito's gut (13) and develops into a structure called an **oocyte**. In time, oocytes produce cells (14) that migrate to the mosquito's salivary glands (1) and repeat the cycle.

including the immune state of the host, *Plasmodium* either multiplies rapidly within liver cells or remains dormant but able to reactivate to cause relapsing bouts of malaria at a later time. (Could this be the reason why Ms. G, from the case study, relapsed?) When parasites are released from the liver, they enter the circulation and infect red blood cells. The detour through the liver may have evolved so the parasites can persist in the body.

One of the best-studied aspects of the *Plasmodium* life cycle is its penetration into red blood cells. These host cells are not phagocytic (they do not "ingest" particles); they must be coaxed into allowing *Plasmodium* to enter them. The initial interaction between parasite and red blood cell consists of binding mediated by ligand-receptor interactions. Next, the parasite reorients, so that its pointed "apical" end is in contact with the host cell surface. This end of the cell has its distinctive shape because it carries a DNA-containing organelle called the **apicoplast**. You may recall from our discussions about the classification of protists (Table 16.2) that *Plasmodium* species belong to a nonmotile group of protists, the Apicomplexa. They carry this name because they all possess an apicoplast. Although its function is not fully understood, this organelle is essential for the life of the parasite. Its structure is derived from plastids, which goes to show that a human parasite had a photosynthetic ancestor!

Once inside the blood cell, the parasite matures into a ring-shaped form and other asexual stages (trophozoites and schizonts) (Fig. 16.12). Such stages of differentiation inside the red blood cell are easily distinguished in blood smears of malaria patients after staining with certain dyes, as in the case study. The differentiation of the parasite within the blood cells leads to its multiplication. Once fully mature, the parasites lyse the red blood cells and infect healthy red blood cells. It is like a chain reaction. The parasites also prepare for the trip back to the mosquito: some parasites differentiate into sexual forms (gametocytes) inside the blood cells (Fig. 16.12). When a mosquito bites an individual and sucks his/her blood, it aspirates some of the gametes, which can now further differentiate and mate to produce a diploid zygote. The latter stage is the form of the parasite that penetrates the mosquito's gut, differentiates, and eventually migrates to the mosquito's salivary glands, ready to repeat a new cycle of infection.

An interesting fact about the infective process is that, once inside the red blood cells, the parasite begins to control the host's cellular properties to make it the ideal vehicle of infection. For example, infected red blood cells become stiff. The spleen recognizes them as being "old" and destroys them. However, the parasites have evolved a protective mechanism to avoid being destroyed along with the cell. They induce knob-like structures on the red blood cells that make them adhere to the surface of the blood vessels and keep them from ending up in the spleen. Despite this, many red blood cells are lysed. Uninfected ones are lysed along with the infected ones, which suggests that infection by *Plasmodium* may trigger some autoimmune mechanism that ruptures red blood cells and causes the symptoms of malaria. (Immunity in malaria and the possibility of creating a vaccine are intricate subjects, and we will spare you here.) But some infected blood cells survive the attack by the immune system, allowing the parasite to undergo a life cycle inside the human cells and multiply.

Coinciding with the lysis of blood cells are the typical symptoms of malaria, namely chills that can be so severe as to lead to violent shaking and loud chattering of teeth, even in hot weather. This is usually followed by a feeling of great weakness, throbbing headaches, vomiting, and high fever. It is thought that red blood cell lysis releases parasite-produced molecules that trigger the production of fever-inducing cytokines. The patient then breaks out in a drenching sweat until the fever subsides. Note that Ms. G from the case study presented with many of these symptoms. This, together with the fact that she had traveled to African countries where malaria is endemic, prompted the doctor to suspect malaria as the culprit. This clinical picture is repeated, sometimes quite regularly, every few days (depending on the species of *Plasmodium*), because the multiplication and release of the parasites take place with a certain degree of synchrony. (Could this be the reason why Ms. G relapsed while on medication?) The destruction of red blood cells can be so great that vast amounts of hemoglobin are excreted, giving the urine a deep blackish-red color (hence the name of this condition, "blackwater fever").

What evolutionary forces have selected for such an intricate lifestyle? We do not really know. Unlike most other human protozoan parasites, plasmodia have a sex life; they invade two kinds of human cells (those of the liver and the red blood cells) and go through a complex life cycle for survival and transmission. Other parasites, including relatives of *Plasmodium*, do not follow such a complicated design. The trypanosomes that cause sleeping sickness in Africa, for example, do not undergo marked morphological changes in the insect or in the mammalian host. Other trypanosomes that cause Chagas' disease in South America and the more distantly related *Leishmania* species (the agents of leishmaniasis, a disease that affects both deep tissues and the skin) have only two morphologically distinct forms, one in the insect and one in the human. Why such disparate lifestyles have evolved evades a simple explanation. However, only a few malarial plasmodia enter the bloodstream after the insect bites, whereas trypanosomes and *Leishmania* cause a local lesion at the site of the insect bite, proliferate, and then enter the circulation in larger numbers. So perhaps plasmodium needs an intricate life cycle because it has not evolved mechanisms to multiply proficiently and, in this way, gain strength in numbers. One bright side is that this pathogen's complex life cycle provides researchers many different targets to interfere with the protist's proliferation and transmission, so there are more opportunities to develop drugs or vaccines. However, although much progress has been made in the past years, it has not as yet been sufficient to control malaria, which plagues many parts of the world.

The Beautiful Diatoms and Their Ilk

LEARNING OUTCOMES:
- Describe three distinguishing characteristics of diatoms.

The oceans are full of exciting forms of life that cannot be seen with the naked eye. Collect the small particles that float in the sea and observe them under the microscope. In addition to many prokaryotes, you would see an array of eukaryotic microbes of great diversity and endless beauty. Many display a body structure remarkable for its eye-pleasing geometric arrangement. Some of the organisms are round, others oblong, and yet others are joined in chains. These microbes make up the plankton, an essential element of the oceans' food chain. Many of these organisms are diatoms (Fig. 16.13).

So who are these microbial beauties? Diatoms are one-celled algae. They are extraordinarily abundant both in the plankton and in the sediments of marine and freshwater ecosystems. Individual diatoms range in size from 2 mm to several millimeters, although very few species are larger than 200 mm. They are highly varied: over 50,000 species exist today. A striking feature of the diatoms is their cell walls, which are made of silicon dioxide (in other words, a kind of glass) enveloped by organic material. (Yes, they live in glass houses, which they themselves build!) Silicon is one of the most abundant elements on the planet, so it makes sense that microbes have figured out ways to use this abundant resource. Good chem-

FIGURE 16.13 **Examples of diatoms.** Scanning electron micrographs of differently shaped diatoms. (Courtesy of R. Edgar and the Center for Diatom Informatics.)

ists as they are, diatoms learned soon that once oxidized, silicon becomes silicon dioxide or silica, a hard material that we humans use to make glass. These silica cell walls have complex geometric patterns and are some of the loveliest forms you will find in nature (Fig. 16.13). People who have observed them tend to wax eloquent and call them by such names as "nature's gems" or "plants with a touch of glass."

Why silica? The design and material of the diatom cell wall make for a very tough structure that is resistant to mechanical breakage. However, the silica in diatom shells is far from inert. It is thought to speed up photosynthesis, possibly by acting as a buffer to keep the pH within an optimal range. The ornate designs in the shells, full of pores and indentations, provide a tough armor but also leave large areas of the cell surface exposed to water and carbon dioxide to facilitate photosynthesis. Diatoms make further use of silica: it activates a variety of genes, including the one coding for a diatom's DNA polymerase. The shells of certain diatoms are made of two unequal parts that fit into one another, an arrangement that, in round species, resembles a pillbox or a petri dish. This architecture results in an unusual mode of replication, where each half serves as a template for a new shell (Fig. 16.14). When the shell size becomes too small, the cell leaves it behind, undergoes meiosis, and becomes gametes (male and female). The two types of gametes can fuse to make a zygote, which grows a new shell and starts the process again.

Diatoms are an important element of the food web. They are photosynthetic and are thought to be responsible for 20 to 25% of all organic carbon fixation on the planet. Because they are so plentiful, these protists are both an important food source for marine organisms and a major producer of oxygen for our atmosphere. Not all diatoms float freely, though; many cling to surfaces, such as those of aquatic plants, mollusks, crustaceans, and even turtles. Some whales carry dense growths of diatoms on their skin. Many protists and small plankton consume the smaller diatoms whole, but some invade the large diatoms and eat them out of their shells. Even though diatoms get eaten, their shells are almost indestructible and have accumulated over geological time, going back at least to the Cretaceous period, to make enormous deposits. White chalky rocks consisting almost entirely of fossil diatom shells are known as diatomite or diatomaceous earth. These deposits are mined commercially to make

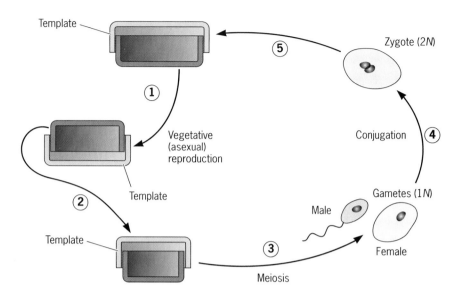

FIGURE 16.14 **Life cycle of petri dish-shaped diatoms.** Upon asexual reproduction (1), the "top lid" shell of a cell serves as a template for the "bottom." In turn (2), this bottom becomes a top and serves as the template for another bottom. With each division, the cells become encased in a smaller shell (blue). When the shell size becomes too small, the cells exit, undergo meiosis, and become gametes (3). These can fuse to make a zygote (4), which can grow a shell to start the process again (5). For convenience, the alternating half shells are drawn as light and dark, and only a few of the divisions are shown.

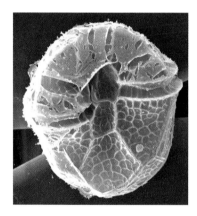

FIGURE 16.15 **A freshwater dinoflagellate, *Peridinium*, seen under a scanning electron microscope.** Note the armor-plated appearance due to a cellulose-containing cell wall and a characteristic groove known as the cingulum. Most dinoflagellates are marine, but some are also found in freshwater lakes and rivers. About half of them are photosynthesizers; the others are heterotrophs. When present in large numbers, they produce red tides that are poisonous to vertebrates. (Courtesy of S. Carty.)

abrasives, cleansers, and paints and for filtering agents for various liquids, including wine. As we mentioned earlier, brushing one's teeth may put one in contact with diatoms because some types of toothpaste contain processed diatomaceous earth. Because individual diatom species thrive under different climatic conditions, the analysis of fossil diatoms provides paleontologists with information on past environments.

Silica shells are but one way to make a protective armor. Cousins of the diatoms make a rigid cell wall made of cellulose or other organic polymers rather than minerals (Fig. 16.15). The protists that encase themselves in a polymeric cell wall are the dinoflagellates. Some of them produce red-colored photosynthetic pigments that give the algal blooms their distinctive red color and their name, "red tides" (Fig. 16.16). Dinoflagellates make an interesting spectacle at night: they are bioluminescent and glow in a oxygen-dependent manner; for this reason, the presence of these organisms in the coastal waters is revealed by flashes of light on the crest of the breaking surf or in the wake of a ship. But do not be fooled by these beautiful protists. Some of them produce potent neurotoxins that can be transmitted through the food chain to affect, and sometimes kill, shellfish, fish, birds, marine mammals, and even humans.

There are many questions peculiar to the diatoms, regarding their unusual use of silicates, their varied shapes, their unique modes of reproduction, and more. Another group of protists, the dinoflagellates, is equally interesting. Perhaps readers of this book may someday become motivated to answer some of these questions. *Hint:* Diatoms can be cultured in the laboratory. Keep us posted!

FIGURE 16.16 **A visible bloom, or "red tide," of the dinoflagellate** *Lingulodinium polyedrum* **along the coast of La Jolla, California.** (Photo courtesy of Kai Schumman.)

CASE REVISITED: **Is it a protist?**

Remember Ms. G, who traveled to East Africa for a photo safari and returned with flu-like symptoms, headaches, high fever, and chills? Did you conclude that she had been bitten by a *Plasmodium*-infected insect and that, as the doctor suspected, she had malaria? The doctor used a smear of her blood to show that some of Ms. G's red blood cells contained structures consistent with some of the developmental stages that the parasite transitions through during asexual replication. The blood smear also showed a *Plasmodium* cell outside the blood cell. In this case, the doctor was right to treat her with antimalarial drugs. Did you figure out why she relapsed? Could it be that Ms. G had latent parasites in the liver? Is it possible that she experienced sequential bursts of lysis of the red blood cells? Could she have been infected with a form of *Plasmodium* that was resistant to the antimalarial drugs?

CONCLUSIONS

Protists comprise a huge number of life forms that vary greatly in morphology and size. Many play essential roles in the cycles of nature; some cause disease in humans and animals, and some serve as models for basic research. Certain ciliates have greatly amplified our understanding of molecular biology, permitting the discovery of such unexpected phenomena as RNA splicing and RNAs functioning as enzymes. We can expect more from the study of these fascinating organisms. We are only beginning to understand their roles in the major biogeochemical cycles and the recycling of carbon. In the next chapter, we continue with the theme of biological diversity and consider another central ingredient of the biological world, the viruses.

Representative ASM Fundamental Statements and Learning Outcomes for the Chapter

EVOLUTION

1. Cells, organelles (e.g., mitochondria and chloroplasts), and all major metabolic pathways evolved from early prokaryotic cells.

   a. Explain how a cell can have mitochondrial genes, but no mitochondria.
   b. Present evidence that mitosomes are "mitochondrial remnants."
   c. Compare and contrast the function of hydrogenosomes with mitochondria.

2. Mutations and horizontal gene transfer, with the immense variety of microenvironments, have selected for a huge diversity of microorganisms.

   a. Speculate why there is such high diversity among protists.

3. Human impact on the environment influences the evolution of microorganisms (e.g., emerging diseases and the selection of antibiotic resistance).

   a. Predict how some virulent species of *Plasmodium* have become resistant to many antimalarial drugs.

CELL STRUCTURE AND FUNCTION

4. The structure and function of microorganisms have been revealed by the use of microscopy (including bright-field, phase-contrast, fluorescent, and electron).

   a. Describe distinguishing characteristics of coccolithophores.
   b. Describe three distinguishing characteristics of diatoms.

5. While microscopic eukaryotes (for example, fungi, protozoa, and algae) carry out some of the same processes as bacteria, many of the cellular properties are fundamentally different.

   a. List two examples of organelles found in protists besides mitochondria and plastids.
   b. Compare and contrast mechanisms of motility in eukaryotic microbes and bacterial cells.
   c. Describe how paramecia avoid senescence.
   d. Define cortical inheritance.
   e. List two functions of cilia in ciliates.

INFORMATION FLOW AND GENETICS

6. Although the central dogma is universal in all cells, the processes of replication, transcription, and translation differ in *Bacteria*, *Archaea*, and eukaryotes.

   a. Compare and contrast the structures and functions of the macronucleus and micronucleus in paramecia.

MICROBIAL SYSTEMS

7. Microorganisms can interact with both human and nonhuman hosts in beneficial, neutral or detrimental ways.

   a. Speculate on the benefits to each partner in the relationship between paramecia and the toxic bacterial cells.
   b. Outline the life cycle of *Plasmodium*, including gametocytes, gametes, and zygotes.
   c. Describe how *Plasmodium* infects red blood cells.
   d. Explain the method by which the malaria parasite, *Plasmodium*, harms its host.

8. Microbes are essential for life as we know it and the processes that support life (e.g., in biogeochemical cycles and plant and/or animal microbiota).

   a. Identify three ways in which protists have shaped the Earth.
   b. List three ways that protists influence nutrient cycles or food webs.

9. Microorganisms provide essential models that give us fundamental knowledge about life processes.

   a. Discuss three major contributions to the study of biology, beyond knowledge about protists, that have been made as a result of studying protists.

IMPACT OF MICROORGANISMS

10. Humans utilize and harness microbes and their products.

   a. List two ways that humans use protists or their products.

SUPPLEMENTAL MATERIAL

References

- **Adl SM, Simpson AG, Farmer MA, Andersen RA, Anderson OR, Barta JR, Bowser SS, Brugerolle G, Fensome RA, Fredericq S, James TY, Karpov S, Kugrens P, Krug J, Lane CE, Lewis LA, Lodge J, Lynn DH, Mann DG, McCourt RM, Mendoza L, Moestrup O, Mozley-Standridge SE, Nerad TA, Shearer CA, Smirnov AV, Spiegel FW, Taylor MF**. 2005. The new higher level classification of eukaryotes with emphasis on the taxonomy of protists. *J Eukaryot Microbiol* **52**(5):399–451.
- **Shiflett AM, Johnson PJ**. 2010. Mitochondrion-related organelles in eukaryotic protists. *Annu Rev Microbiol* **64**:409–429. doi: 10.1146/annurev.micro.62.081307.162826.
- **Gast RJ, Sanders RW, Caron DA**. 2009. Ecological strategies of protists and their symbiotic relationships with prokaryotic microbes. *Trends Microbiol* **17**(12):563–9. doi: 10.1016/j.tim.2009.09.001.

Supplemental Activities

CHECK YOUR UNDERSTANDING

1. **Which eukaryotic structure(s) can take an unusual form in protists?**
   a. Nucleus
   b. Cell envelope
   c. Mitochondria
   d. All of the above
   e. A and C

2. **What is one way that protists are NOT classified?**
   a. Motility
   b. Reproduction
   c. Source of energy
   d. Cellular structures
   e. None of the above

3. **Why can protists be used as an indicator of a "healthy" aquatic ecosystem?**

4. **Compare the DNA content of the micronucleus to that of macronucleus of paramecia. How does each nucleus contribute to the life cycle of paramecia?**

DIG DEEPER

1. In addition to contributing to the burial of carbon in deep-sea sediments, coccolithophores also influence the cycle of sulfur and climate because these protists use sulfur compounds from the oceans to make dimethylsulfoniopropionate (DMSP), which marine bacteria then convert into the volatile dimethyl sulfide (DMS). Once released to the atmosphere, DMS promotes cloud formation.

   Read the *Small Things Considered* blog post "Ehux: The Little Eukaryote with a Big History" (http://schaechter.asmblog.org/schaechter/2012/08/ehux-the-little-eukaryote-with-a-big-history.html). Work in groups to answer the following questions:

   a. How would cloud formation affect the growth of the coccolithophores? And carbon sequestration in the ocean sediments?
   b. What effect would massive blooms of coccolithophores have on climate?
   c. Viruses have also been linked to the decimation of coccolithophore blooms; why?

2. Using what you learned about malaria (case study) and the life cycle of *Plasmodium* in humans, plot the body temperature (°C) of a patient sick with malaria (y axis) as a func-

tion of the life cycle stage, using the numbers illustrated in Fig. 16.12 (*x* axis). Illustrate multiple days of infection.

3. Using the PubMed search engine, identify an article evaluating a potential new treatment for malaria. Write a one-page summary that describes the treatment.

   a. What part of the malaria life cycle does it target?
   b. What is its mechanism of action?
   c. Are there advantages and disadvantages to the new treatment?
   d. Are there potential ethical issues with the new treatment?

4. **Team work:** Speculate why some anaerobic protists have hydrogenosomes instead of mitochondria. To get started, consider the following hypotheses:

   • *Hydrogenosomes generate ATP without consuming oxygen.*
   • *Hydrogenosomes help anaerobic eukaryotes maintain the redox state of the cell since they do not produce as much NADH.*
   • *The production of hydrogen by anaerobic protists has been selected for in specific ecological niches.*

   a. Propose an experiment to test each of these hypotheses. What independent and dependent variables would you measure and how? What would be appropriate experimental controls?
   b. What are the possible outcomes of the proposed studies?
   c. What conclusion would you draw from each outcome?

5. Learn more about ocelloids, eye-like structures of single-celled protists, by listening to *This Week in Microbiology* podcast episode #103 (http://www.microbeworld.org /component/content/article?id=1896) and reading the accompanying paper by Hayakawa et al. (*PLoS One* 10(3):e0118415, 2015, http://journals.plos.org/plosone/article?id =10.1371/journal.pone.0118415). Write a newspaper-style article to describe this fascinating science for the general public.

6. Deposits of diatoms are mined commercially for a range of purposes. Research one of the products below, and share your findings with the class using visual aids.
   • abrasives
   • cleansers
   • paints
   • filters
   • toothpaste
   • insecticides

Supplemental Resources

VISUAL

• Powerful microscopes reveal England's white cliffs to be made up of billions of tiny coccolithophore fossils, BBC (1:17 min): http://www.bbc.co.uk/nature/life/Coccolithophore #p00bxk9c.
• *Didinium* eats *Paramecium* (10:19 min): https://www.youtube.com/watch?v=rZ7wv2LhynM.
• *Paramecium* eating pigmented yeast (1:59 min): https://www.youtube.com/watch?v =I9ymaSzcsdY
• Malaria caught on camera breaking and entering red blood cell (video) (http://www .newscientist.com/blogs/nstv/2011/01/malaria-caught-breaking-and-entering-red-blood-cell.html)

WRITTEN

• The smallest free-living eukaryote: "Pico Who?": http://schaechter.asmblog.org/schaechter /2008/07/pico-who.html
• The complex genomic make-up of a ciliate: "The Genes, The Whole Genes, & Nothing But The Genes": http://schaechter.asmblog.org/schaechter/2009/08/the-genes-the -whole-genes-nothing-but- the-genes.html
• The bacterial endosymbionts of the *Paramecium* macronucleus: "Life in a Big Mac": http:// schaechter.asmblog.org/schaechter/2009/07/l.html

- The incredibly complex forms made by protists: "Houses Made by Protists": http://schaechter.asmblog.org/schaechter/2012/02/fine-reading-houses-made-by-protists.html
- How do diatoms make their beautiful glass shells?: "Hard Biology http://schaechter.asmblog.org/schaechter/2011/05/hard-biology.html
- Slime molds "cultivate" bacteria as foodstuff: "Farmer Joe Dictyostelium": http://schaechter.asmblog.org/schaechter/2011/02/farmer-joe-dictyostelium.html
- Global impact of climate change on coccolithophore ecology, effects at trapping atmospheric carbon in the deep ocean: "The heartbreak of Emiliania huxleyi" *Watts Up With That* blog post on climate change (http://wattsupwiththat.com/2011/10/14/the-heartbreak-of-emiliania-huxleyi/).
- How did foraminifera shells end up in the pyramids? "Pyramids, Forams, and Red Sea Reefs: Field notes from Lorraine Casazza" from the University of California Museum of Paleontology (http://www.ucmp.berkeley.edu/science/fieldnotes/casazza_0711.php).

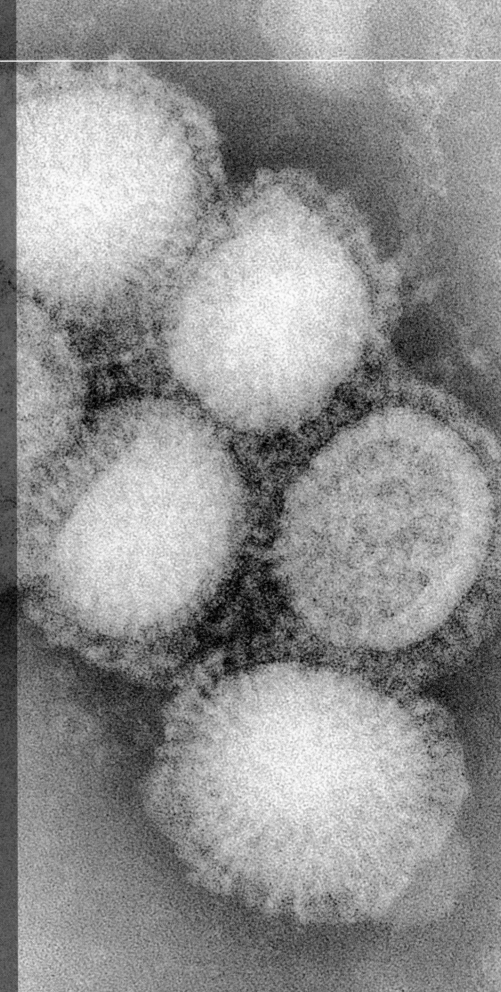

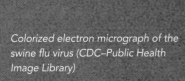

Colorized electron micrograph of the swine flu virus (CDC–Public Health Image Library)

CHAPTER SEVENTEEN

The Viruses

KEY CONCEPTS

This chapter covers the following topics in the ASM Fundamental Statements.

EVOLUTION

1. Mutations and horizontal gene transfer, with the immense variety of micro-environments, have selected for a huge diversity of microorganisms.

CELL STRUCTURE AND FUNCTION

2. The structure and function of microorganisms have been revealed by the use of microscopy (including bright field, phase contrast, fluorescent, and electron).
3. The replication cycles of viruses (lytic and lysogenic) differ among viruses and are determined by their unique structures and genomes.

METABOLIC PATHWAYS

4. The growth of microorganisms can be controlled by physical, chemical, mechanical, or biological means.

GENETICS

5. The synthesis of viral genetic material and proteins is dependent on host cells.

IMPACT OF MICROORGANISMS

6. Microorganisms provide essential models that give us fundamental knowledge about life processes.
7. Because the true diversity of microbial life is largely unknown, its effects and potential benefits have not been fully explored.

INTRODUCTION

What exactly are viruses? They possess some of the qualities of living things, such as the ability to replicate, mutate, and recombine. But they are not cells. They are inert particles that depend on the cells they infect to reproduce. To multiply, they first lose their identity as a particle and deliver their genome into the host cell. Yet while commandeering cells for their own reproductive advantage they cause profound changes in the cell's metabolic and genetic activities and the structure of the microbial communities they live with. Viruses are present in very large numbers in the environment and are spectacularly varied in shape and size. Take a minute and count the names of viruses you know anything about. You may be surprised at the number.

Viruses influence all forms of cellular life. They have shaped evolution, are constantly involved in ecological relationships between living organisms, and, of course, cause disease. Because of all of this, the study of viruses is central to biology.

In this chapter we will emphasize bacterial viruses or **phages**, which serve as the paradigm for the rest and share fundamental characteristics with viruses that infect eukaryotic cells. In chapter 18, we deal with "latent viruses," those that have long-term interactions with their hosts. Then, in chapter 19, we consider the ecology of viruses. Other viruses that infect humans are discussed in chapter 25.

CASE 1:   The tale of a phage that infected an *Escherichia coli* cell

Some human feces ended up in a pond where an *E. coli* cell it contained encountered an infective viral particle. The virus attached to the surface of the bacterium and, within a few seconds, introduced its nucleic acid (in this instance, DNA) into the bacterium. Less than 1 hour later, the bacterium broke open and released several hundred virus particles, which went on to infect other *E. coli* cells. One of the neighboring cells was simultaneously infected by many (let us say 100) of these viral particles, yet it did not die. It underwent a different fate. Instead of lysing, the bacterium remained alive for many generations. The infected progeny continued to grow until one of the cells became exposed to sunlight, with dire consequences. Within 2 hours, this bacterium broke open and, as in the first instance, released hundreds of progeny virus particles. A diluted sample of the pond water was mixed with a laboratory strain of *E. coli*. After incubation, the plate showed holes (called "plaques") in the growth of the bacteria (see Box 17.1). Upon further study, the virus turned out to belong to a family of "temperate viruses" related to a well-known model virus (phage) called lambda.

- Where did the original phage particles come from?
- How did the phage particles stick to the *E. coli*?
- How did the DNA penetrate the cell?
- What happened next?
- Why did the second infection lead to such a different result?
- What did sunlight do to the infected bacteria?

CASE 2:   When viruses mind your business

CJ, a 21-year-old male student, returned from a spring vacation where he had oral and vaginal sex with a single partner. Within a week, he felt painful sores on his penis and a sore throat. He went to see a doctor who, based on the history and the symptoms, diagnosed him as having herpes. This conclusion was confirmed by a PCR for herpesvirus DNA carried out on a specimen from the lesions on the penis. Given this history, it is unlikely that the female partner had visible lesions. CJ was treated with the drug acyclovir, with a favorable outcome.

- If the female partner had no lesions, where did the virus that infected CJ come from?
- What was going on within CJ in the week between exposure and symptoms?
- Why did the lab perform PCR instead of a simpler test, such as culturing the sample on a blood agar plate?
- Why did the doctor not prescribe an antibiotic such as penicillin?

WHO ARE THE VIRUSES AND WHY DO THEY MATTER?

LEARNING OUTCOMES:
- Distinguish viruses from cellular forms of life.
- Compare and contrast lytic and lysogenic viral cycles.
- Identify examples of important discoveries in biology that were made possible by using viruses as experimental tools.

If viruses did not exist, could anyone have conjured them up? They are inert particles, but they infect cells and take control over their biosynthetic machinery to make the components needed to assemble new virus particles, sometimes in the hundreds. Could one imagine small infectious particles that multiply by losing their physical integrity; that is, by first coming apart inside a host cell and then reassembling into progeny? Could anyone have predicted that they would influence all forms of life, that they would be spectacularly varied in shape and size, and that they would be present in huge numbers in the environment? In fact, viral genomes (the **virome**) vastly outnumber those of all other biological cells.

The abundance of viruses in nature is astonishing. A reasonable estimate is that there are 10^{31} virus particles in waters, soils, and associated with the bodies of animals, plants, fungi, bacteria, and archaea. In chapter 19, we will learn just how abundant, diverse, and widespread viruses are. Their rate of replication is commensurate with such numbers. It has been calculated that over *10^{23} viral genomes replicate every second* and that 10^{22} of them mutate every second as well. Consider the single virus that infected an *E. coli* cell in case study 1. The bacterial virus (or phage) injected its DNA in the cell, and 20 minutes later hundreds of viral particles, each one carrying a copy of the original genome, had been manufactured and released.

It is *in the virosphere that most of evolution takes place*. Hence, viruses play a central role in biology: they have shaped evolution and are commonly involved in ecological relationships between living organisms.

Strikingly, viruses do not appear in the conventional depictions of the "tree of life," which are commonly based on comparisons of the small subunit ribosomal RNAs (16S and 18S) of prokaryotic and eukaryotic cells, respectively (Fig. 1.9, chapter 1). Why is that? Viruses lack ribosomes and so have no 16S and 18S RNAs. Without the ability to synthesize proteins, viruses depend on shifting the host's translation machinery to make the proteins that will be assembled into the viral progeny. Hence, even though viruses possess some of the qualities of living things, such as the ability to replicate, mutate, and recombine, they cannot reproduce independently, and therefore *they are not cells.*

The example presented in the first case study also helps us understand the two life cycles that viruses can have (Fig. 17.1). The first phage infected the cell, and 20 minutes later, the bacterium burst open to release several hundred viral particles or **virions**. Such an infective cycle that leads to the generation of the viral progeny and death of the host cell is typical of **lytic** viruses. However, another *E. coli* cell did not die upon infection. Rather, it continued to live and reproduce normally for several generations until exposure to sunlight triggered a new viral lytic cycle. This cell had been infected simultaneously by several virions that successfully introduced their genomes into it. But rather than taking over the cell machinery to make new virions, the viral genome became integrated into host chromosome (it is now called a **prophage**). There, it remained latent for several cell generations until being "awakened" (induced) by the sunlight. This phage is a temperate or **lysogenic virus**. It maintains the full infective potential but "hitched a ride" on the cell's chromosome before going on attack mode. Why would a virus remain dormant, hiding in the host's chromosome?

It may not surprise you to learn that most viruses infect *Bacteria* and *Archaea* only, thus are innocuous to humans, though student CJ in case study 2 may think otherwise. He came back from his spring vacation with

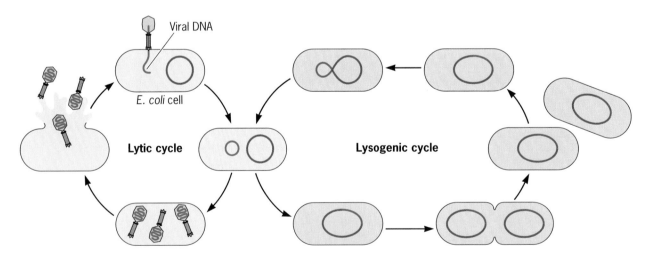

FIGURE 17.1 **Lytic and lysogenic cycles of viruses.** Viruses attach to the host cell and introduce their nucleic acid. Here a T4 phage is shown infecting an *E. coli* cell and introducing its linear dsDNA genome. Once inside the host cell, the viral genome circularizes and is replicated and transcribed to make the components of the viral progeny, which is released by bursting the host cell (lytic cycle). Alternatively, the circular dsDNA genome of the virus can integrate into the host chromosome as a prophage (lysogenic cycle). The prophage is replicated with the host chromosome and passed to the daughter cells until environmental stimuli induce its excision and a new lytic cycle begins.

a sexually transmitted disease (STD) caused by a herpesvirus. CJ's sexual partner transmitted the virus but did not manifest genital lesions. (Where was the virus hiding in the sexual partner?) Nevertheless, the fact is that only a small fraction of human viruses cause disease. The intimate relationship between humans and viruses is reflected by the fact that viruses are also part of our genetic makeup. Remarkably, some 8% of our DNA consists of genes that are derived from retroviruses, the group of agents to which HIV belongs. Just what they do for us is still being unraveled, but, clearly, we have an intimate, possibly largely beneficial, relationship with them. This contrasts with the common view of human viruses as pathogenic agents that cause serious diseases. For example, smallpox and measles viruses have killed untold numbers of people in history, and polio has incapacitated many others.

The study of viruses has also been a major playing field for many notable discoveries in biology. Major contributions include the role of DNA in inheritance, the elucidation of the genetic code, the discovery of mRNA splicing, and many others. These discoveries are not surprising because viruses, especially phages, are superb experimental tools—they are relatively easy to manipulate genetically and biochemically, and they multiply and evolve quickly. In chapter 10, we learned about the ability of phages to accidentally pack DNA from their host and take it to another cell, sometimes packing random pieces of DNA (general **transduction**) and others carrying along with the viral genome specific genes of the host DNA (specialized transduction). Scientists capitalize on these phenomena to use phages as vehicles to deliver genetic material from one cell to another. You may be familiar with bacterial viruses or phages used in classic studies of transduction such as the lambda phages.

HOW DIVERSE ARE VIRUSES?

LEARNING OUTCOMES:
- Name structural components common to all viruses.
- State an important difference between icosahedral and filamentous viruses, other than shape.
- Identify how scientists have learned details of the molecular organization of viruses, given their relatively small sizes.
- Speculate what would be required for larger viruses to become cellular entities and replicate independently.

Our best guess is that all living organisms are infected by viruses. With such a large range of hosts to infect, perhaps it is not surprising that viruses are very diverse, varying greatly in all their properties, such as size, shape, and genetic makeup. Indeed, viral particles range in diameter from the nanometer to the micrometer range (Fig. 17.2). The smallest of viruses is about 20 nm in diameter, about the size of a prokaryotic ribosome. At the other end, the largest known virus is larger than the smallest bacteria (it is about 1 µm in length and 0.5 µm in width) and is even bigger than the smallest eukaryotic cells. This Gargantua of a virus has hundreds of thousands of times the volume of the smallest virus!

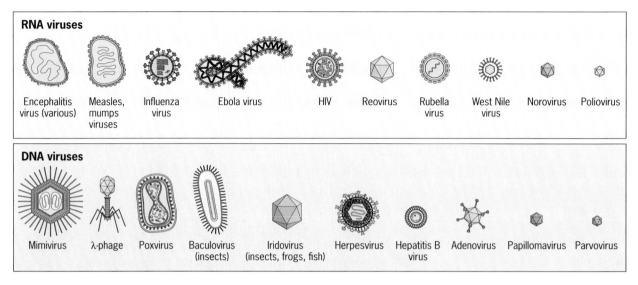

FIGURE 17.2 **Examples of shapes and sizes of viruses.** Note the wide range of shapes and relative sizes. (Viruses are not drawn to scale.)

Viruses also display many different shapes (Fig. 17.2). Some viruses look like sticks, some like geodesic domes, and others like lunar landers. But all these shapes have a common structural basis; namely, they carry their *nucleic acid within a protein shell*. The nucleic acid can be either DNA or RNA (rarely both, with some notable exceptions), and either single-stranded or double-stranded. The protein shell or **capsid** is made up of structural subunits, known as **capsomers**, that can be made up of the same or different proteins. In the simplest and smallest viruses, capsomers are arranged to form one of two structures: an icosahedron, made up of 20 equilateral triangular faces, or a helical filament, made up of capsomers arranged as a spiral with a hollow core (Fig. 17.3). Both shapes are determined by built-in properties of the capsomers: isolated capsomers spontaneously

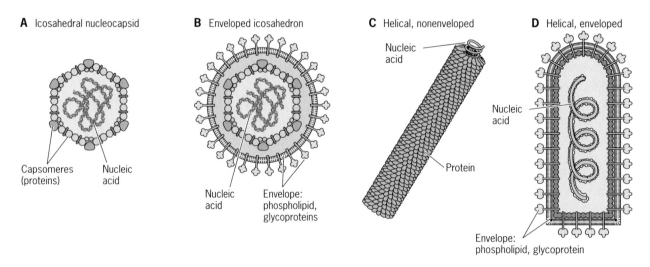

FIGURE 17.3 **Basic viral forms.**

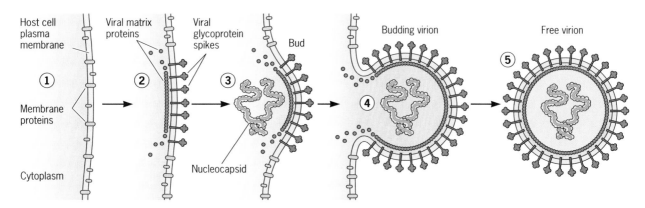

FIGURE 17.4 Viral budding through the plasma membrane. (1) The host cell membrane before or early in the infection. (2) Virally encoded matrix-protein molecules become associated with the plasma membrane. The viral glycoprotein spikes become incorporated into the membrane. (3) The viral nucleic acid and proteins (nucleocapsid) are assembled near the membrane and budding begins. (4) Budding continues as more viral spikes are inserted into the membrane. (5) Budding is complete, and a mature virion is released.

self-assemble (crystallize) into the shape of the virion even in the absence of nucleic acids. The size of icosahedral viruses is determined by this self-assembly property alone.

A simple icosahedral virion is made of 60 capsomers, 3 in each of its 20 faces. Larger icosahedral viruses are made of many more capsomers, but, because capsomers are asymmetric, the number of them that can assemble to form a symmetric equilateral triangular face follows strict rules. Thus, icosahedral virions do not contain just any number of capsomers. The next step after 60 capsomers is 180, then 240, 540, 960, 1,500, and more. Here, biology simply obeys geometry. The length of filamentous viruses, on the other hand, is determined by the length of the nucleic acids they contain. If a filamentous virus acquires foreign nucleic acid, the virions simply become longer (a useful property if you want to make proteins by genetic engineering).

Some viruses, such as human immunodeficiency virus (HIV), have a *mixed morphology*. HIV has a filamentous nucleic acid core surrounded by an icosahedral capsid (we warned you about exceptions to rules), although it does follow many of the same geometric principles as icosahedral capsids. Some viruses, such as the one that infected CJ in the second case study, are more complex, because the capsid is surrounded by a lipid-carbohydrate-protein **envelope**, or viral membrane. The proteins that make up this structure are encoded by the viral genome, but its lipids and carbohydrates are derived from one of the host cellular membranes (e.g., the nuclear membrane, endoplasmic reticulum, Golgi, or plasma membrane). These cellular components are picked up as the virus extrudes (buds) through a host cell membrane in its process of maturation (Fig. 17.4). In addition, some viruses have specialized structures for attachment to the host cell, such as the *spikes* of the influenza viruses or the lunar lander-like assembly of some bacterial viruses (Fig. 17.5). Although virions are metabolically inert, they are not biochemically helpless. Virally encoded proteins are used for attachment to host cells, replication, and even modification of their nucleic acid (to be discussed in detail below). Some of the

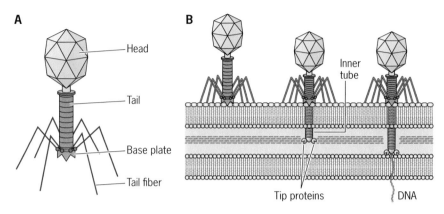

A

- Head
- Tail
- Base plate
- Tail fiber

B

- Inner tube
- Tip proteins
- DNA

FIGURE 17.5 Structure of *E. coli* phage T4. (A) Structure of the T4 virion. (B) During an infection, the virion attaches to the host cell by interactions between the tail fibers and base plate with receptors (LPS and outer membrane porins) on the bacterial surface. These interactions cause the release of the inner tube of the virion, whose tip proteins degrade the peptidoglycan and interact with the inner membrane. The DNA contained in the phage head can now pass into the host cell.

larger viruses, such as the one that causes smallpox or a huge virus found in some amebas, called Mimivirus, are very complex and invite the question: what would it take for them to become cellular entities that are able to replicate independently?

Modern imaging techniques have revealed details of viral structures that are nothing short of stunning. By using a combination of two of them, **cryo-electron tomography** (cryo-EM) and X-ray crystallography (see chapter 2), scientists can now "look" inside single virus particles and make out morphological detail at the level of atoms (Fig. 17.6). In cryo-EM, biological material is frozen at temperatures so low that molecules are stopped in their tracks and ice crystals are not formed. Still frozen, the samples are examined under an electron microscope that functions like a CAT scanner; that is, it makes *optical sections* through the object. These can be used to reconstruct the whole virus. But since each particle will be frozen in a different orientation, a 3-D reconstruction of the physical shape of the virus requires some fancy computer work. Putting this information together with that obtained through X-ray analysis results in stunning images of the outside, as well as the inside, of the tiniest of viruses.

What has been learned from such advanced studies are details of the molecular organization of viruses, including the ways they attach to their host cells, eject their genome, assemble progeny particles, and exit the host cells. Such information not only helps us understand how viruses carry out their business but also sheds light on the general properties of biological constituents. Chief among these properties is the magic of self-assembly. Most of the structures that go into making a virion are assembled by the interaction of their macromolecular parts. In many cases, no enzymes or energy are needed because the ability to self-assemble is built into the macromolecules themselves, although assembly of some viruses requires the action of proteases or the input of energy. Viruses may be inert particles, yet their unique ways are still amazing. And this is just the beginning . . .

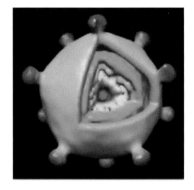

FIGURE 17.6 The detailed structure of an archaeal virus revealed by cryo-electron tomography.
Source: (From Fu CY et al., *Structure* **18:**1579–1586, 2010, with permission.)

HOW ARE VIRUSES CLASSIFIED?

LEARNING OUTCOMES:
- Explain why the true diversity of viruses and many important ecological and evolutionary relationships remains to be discovered.
- Identify 6 ways to classify viruses.

Consider the variety of viruses that infect humans (Fig. 17.7). Indeed, hundreds of viruses are known to cause disease (Table 17.1). Even "simple" organisms, such as *E. coli*, can be infected by hundreds of distinct phages. Moreover, each kind of virus consists of different strains that differ in virulence and antigenic properties (**serotypes**). At present, thousands of viruses have been characterized, and we know from culture-independent

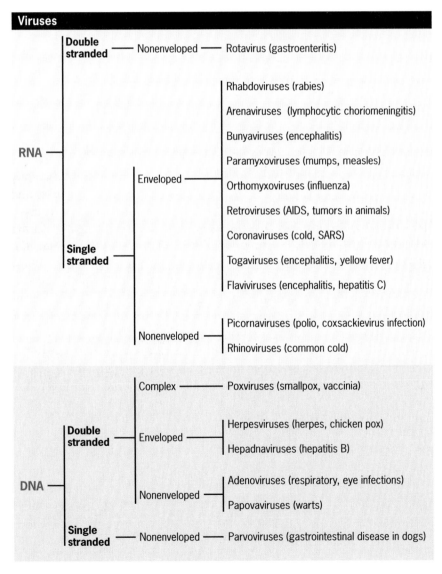

FIGURE 17.7 **Main groups of human viruses.** This is not a representation of viral phylogeny and is shown for practical purposes only.

TABLE 17.1 **One host, many viruses: major human diseases caused by viruses**[a]

RNA viruses

Influenza
Common cold (caused by over 100 kinds of viruses)
SARS[b]
West Nile encephalitis
Hantavirus pulmonary syndrome
Rabies
Mumps
Measles
Rubella
Polio
Gastrointestinal disorders (several different kinds of viruses; e.g.,
　　Norwalk agent, rotavirus)
Ebola hemorrhagic fever
HIV infection, AIDS

DNA viruses

Fever blisters (herpes simplex)
Genital herpes
Hepatitis B
Smallpox
Chickenpox, shingles (varicella, zoster)
Infectious mononucleosis (Epstein-Barr virus)
Cytomegalovirus infection
Adenovirus infection
Papillomatosis (warts)

[a]Several hundred viruses are known to infect humans. Some of the diseases are specific for humans (e.g., smallpox); others (e.g., influenza) are not. Some viruses cross kingdom barriers and can affect both animals and plants.
[b]SARS, severe acute respiratory syndrome.

methods (chapter 19) that this is but a small fraction of what is out there. How can we classify these many viruses?

Viral classification defies simple criteria. The major virus groups have relatively little in common. Their variety is, in fact, astonishing when compared to that of the cellular forms of life. The major virus groups share little structure or nucleic acid and protein sequences. That is, they lack some class of molecules that unites them (e.g., no equivalent of bacterial 16S or 18S ribosomal RNA), making it more difficult to determine their kinship. Because of this, viruses do not appear in phylogenetic trees that rely on the analyses of conserved and universal genes of all cellular organisms. The patterns of viral evolution are therefore elusive and the topic of much discussion. Included in such debates may be the seemingly eternal and controversial question: "Are viruses alive?" For reasons of self-preservation, we'll abstain from trying to provide an answer.

To bypass these phylogenetic limitations, viruses are often classified by their *host* (bacterial, plant, animal, etc.), *size and shape* (Fig. 17.2), presence or absence of an *envelope*, and the *nucleic acid* they contain (Table 17.2). The first criterion is the plainest: when we think of viruses we immediately think of the hosts they infect. Often, viruses are quite specific in their choice of hosts, and this helps us classify them. Thus, the lambda-like phage that infected *E. coli* in the pond (first case study) infects only enterobacteria. And the specific herpesvirus that infected CJ (second case study) is not known to infect any other animal species outside of the laboratory. However, junctions exist: some relatives of CJ's

TABLE 17.2 **Some animal viruses**

Group	Nucleic acid polarity[a]	Examples of disease[b]
RNA viruses		
Single stranded		
Poliovirus	Positive	Polio
Coronavirus	Positive	Colds, SARS
Hepatitis A virus	Positive	Hepatitis A
HIV	Positive	AIDS
Rabies virus	Negative	Rabies
Measles virus	Negative	Measles
Influenza virus	Negative	Influenza
Double stranded		
Rotavirus		Gastroenteritis
DNA viruses		
Single stranded		Rash (humans), GI disease (dogs)
Parvovirus		
Double stranded		
Adenovirus		Colds, eye infections
Herpesvirus		Herpes, encephalitis, infectious mononucleosis, chickenpox
Poxvirus		Smallpox, vaccinia
Hepatitis B virus[c]		Hepatitis B

[a]Positive indicates that the virion's RNA can serve directly as mRNA. Negative indicates that the virion's RNA must first be copied into a complementary strand, which then serves as mRNA.
[b]SARS, severe acute respiratory syndrome; GI, gastrointestinal.
[c]Hepatitis B virus DNA has a single-stranded stretch.

herpesvirus naturally infect other mammals, yet herpesviruses rarely cross over and infect humans (causing a zoonosis, chapter 27). Other viruses are frequently zoonotic: the influenza virus affects pigs and fowl as well as humans. Still, an animal virus is not able to infect a bacterium and vice versa.

Viruses are also grouped based on their genome; i.e., whether it contains DNA or RNA, single-stranded or double-stranded, linear or circular. Each group can be further classified according to their structure and shape. Figure 17.7 illustrates, for example, a classification of human viruses based on their nucleic acid (DNA or RNA first, then single- or double-stranded) and whether they are enveloped or nonenveloped. Note that the poxviruses, such as the one that causes smallpox, are classified here as "complex" because of the intricacy of their envelope, which includes a core envelope surrounded by additional envelope layers decorated with surface tubules.

Classifying viruses based on their host, nucleic acid, and structure works for viruses we know about. But how can we classify the many environmental viruses about which we know little? For example, the oceans contain a huge number of viruses, upward of 10 million particles per ml of seawater, as determined by counting them under the microscope (chapter 19). Most of these viruses are believed to be phages (viruses that infect bacteria) because most of the microbial cells in this environment, and therefore potential hosts, are bacteria. About half of the bacteria in seawater carry prophages, which can be induced (for example, by ultraviolet [UV] irradiation, as in the first case study). Because most of the bacteria in seawater cannot yet be cultivated, we can use only indirect cultivation-independent methods to extract their nucleic acids, sequence them, and use computational analyses to figure out which bacteria are hosts to viruses. As we will discuss in chapter 19, progress has been made toward directly sequencing the viral genomes or **virome** in the population, but the methodologies are not as mature as those used for their hosts. This ignorance surely obscures exceedingly important ecological and evolutionary relationships.

HOW DO VIRUSES INFECT THEIR HOSTS?

LEARNING OUTCOMES:
- Name the five major steps of viral replication.
- Discuss how the structure of viruses enables virions to attach and enter host cells.
- Describe how phages are capable of penetrating the thick cell envelope of Gram-positive and Gram-negative bacteria.
- Name three kinds of enzymes carried by virions.
- Contrast genome replication strategies for positive-polarity and negative-polarity RNA viruses.
- Propose a reason why reverse transcriptase is a good target for HIV antiretroviral drugs.
- Describe how the nucleic acids of retroviruses replicate.
- Discuss how viral proteins are synthesized during the viral replication cycle.
- Give examples of two replication strategies, used by viruses, that do not involve lysing host cells.
- Identify the origin of the envelope of enveloped viruses.
- Explain how to visualize and quantify viral growth using viral-sensitive bacterial and animal cells in culture.

For a virus to infect a specific host-cell type, it must first attach to it, recognize it as a susceptible host cell, and penetrate the membrane and gain access to the host macromolecular-synthetizing machinery. When a virus encounters a susceptible host cell, a lytic infection proceeds as follows: (1) the virion attaches to the cell, (2) its nucleic acid enters, (3) (for DNA viruses) their DNA is transcribed by the host RNA polymerase into proteins that direct the cell to make viral components, (4) the viral components are assembled into progeny particles, and (5) these progeny particles are released by lysis of the cell. For RNA viruses, step 3 is different (see below). Each of these steps is intricate and distinctive for each kind of virus, but the general theme applies to all.

Attachment and Penetration

Viruses bump into their host cells at random, and once in a while, perhaps every 10^3 to 10^4 collisions, they stick. This step, called **adsorption**, does not require energy, but it does require specific ionic conditions and the right pH. Once a virion has been adsorbed, **ligands** on its surface bind specifically and tightly to **receptors** on the host cell. That step equips viruses to recognize a specific susceptible host cell. The more complex viruses (e.g., tailed phages, influenza virus, and other enveloped viruses) have dedicated binding structures, such as spokes that jut from the surface (Fig. 17.2). The host's receptors are **glycolipids** or proteins (often **glycoproteins**). In the case of the flu virus, for example, there is a tidy jigsaw puzzle fit between the ligand protein on the virus and the cell receptor.

In the first case study, for example, an *E. coli* cell was initially infected by a lambda-like virion (see Fig. 17.5 for T4, another well-known phage similar to lambda in general organization). The phage adsorbed onto the cell randomly but, once there, its long tail fibers were bound to their specific receptors, the lipopolysaccharides (LPS) and a porin of the cell's Gram-negative outer membrane. Next the tail of the virion contracts, and the base plate attaches firmly to the bacterial surface. This, in turn, causes structural changes that release an inner tube with such force that it punctures the host outer membrane. The inner tube also contains proteins on its tip that degrade the peptidoglycan barrier in the periplasm and interact with the inner membrane to permit the transfer of the viral DNA into the host cell.

Not all viruses penetrate the cell envelope like lambda and its relatives. In a unique way, each has adapted to the type of host they infect. In the second case study, for example, CJ was infected by a herpesvirus, a type of virus possessing a lipid membrane envelope (Fig. 17.2). To enter the host cell, the viral membrane first has to *fuse with the host cell membrane*, a feat carried out by the proteins of the envelope that recognize specific receptor of the host cell membrane and induce the fusion of the viral envelope (Fig. 17.8). Not all enveloped viruses use this entry mechanism. Others, such as influenza virus, are first taken up by **endocytosis**, thereby trapping the virion in an endocytic vesicle. Then the viral envelope fuses to the endocytic membrane, which allows the virus to escape into the cell's cytoplasm. Influenza virus is a well-understood model for the fusion process: When the host vesicle acidifies to pH 5, the amino-terminal end (called the fusion peptide) of one of the two envelope subunits flips outward and becomes exposed to the aqueous environment. Because the fusion peptide is highly hydrophobic, it becomes embedded in the vesicle membrane. The protein then folds over, pulling the vesicle membrane up against the virus membrane, a maneuver that allows the two membranes to fuse and create a pore in the process. The way is now clear for the viral capsid to enter the cytoplasm.

These viruses face one additional challenge: their nucleic acid is still contained within the capsid. To be infective, the virus needs to release its genome. This "uncoating" process can take place during or after the virus enters the cell. Bacterial viruses, such as the lambda phage described earlier, uncoat *while* entering the host cell. But enveloped viruses need to uncoat *after* they enter. Exactly where and how uncoating occurs can be difficult to determine because the capsid assembly is not easy to manipulate experimentally. In HIV, the capsid helps transport the viral genome to the nucleus and aids in keeping the viral polymerase (**reverse transcriptase**) near the viral genome, so uncoating here doesn't occur until a few hours after entry.

These examples highlight different modes of virus penetration. Though varied, the modes generally share one central aspect: they lead to the *dissociation of the viral nucleic acid from its capsid*. It is this process of disassembling, in order to replicate, that most clearly distinguishes viruses from

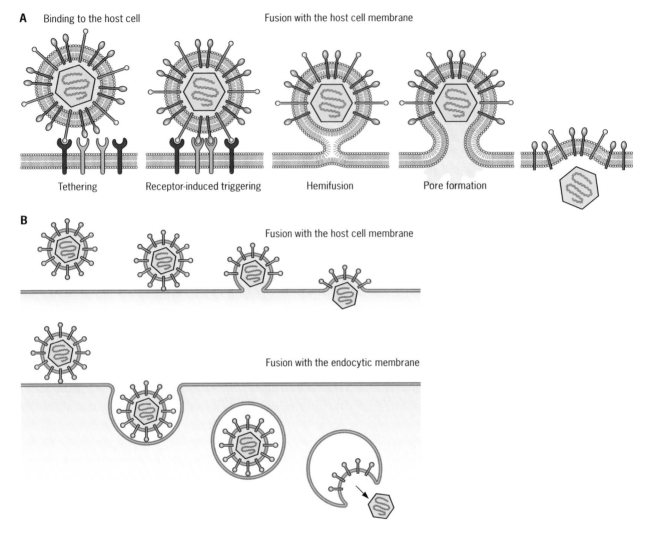

A Binding to the host cell

Fusion with the host cell membrane

Tethering

Receptor-induced triggering

Hemifusion

Pore formation

B

Fusion with the host cell membrane

Fusion with the endocytic membrane

FIGURE 17.8 **Entry mechanisms of enveloped viruses.** (A) Enveloped viruses, such as the herpesvirus, use proteins in their lipid envelope to bind to specific receptors on susceptible cells and trigger the fusion of the viral envelope with the cell membrane. (B) Other enveloped viruses directly fuse their envelope with the plasma membrane or enter the cell by endocytosis and then fuse their envelope with the vesicle membrane.

cellular forms of life. *Cells (except for spermatozoa) retain their bodily integrity; viruses do not.*

Replication of the Viral Nucleic Acid

Regardless of the nucleic acid they carry, all viruses must find a way to make mRNA transcripts that the host ribosomes can recognize and translate into proteins. With the variety of nucleic acids that viruses carry, it comes as no surprise to learn that viruses have evolved different ways to generate these transcripts (Fig. 17.9 and 17.10). Here are some examples.

DNA viruses. Viral double-stranded DNA (dsDNA) genomes have the advantage that they can blend right into the host cell's machinery. Their structure (dsDNA, double helix) can be recognized and transcribed by the host RNA polymerase. To aid in their replication, some linear dsDNA viral genomes circularize. Consider the lambda phage in the first case study.

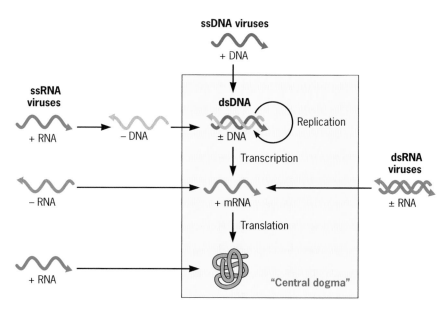

FIGURE 17.9 **Viral strategies to replicate, transcribe, and translate their genomes inside the host cell.**

The linear genome circularizes rapidly and begins replicating in a fashion similar to that of some circular plasmids (**rolling circle replication**; see chapter 10). The herpesvirus that affected CJ in the second case study (and other human viruses, such as the papillomaviruses that cause warts) also carry linear dsDNA genomes, so they can be transcribed and translated right away, much like our own chromosomal DNA. To be replicated, the linear herpesvirus genome first circularizes, just like the lambda phage genome does.

Other viruses have more original ways of doing business (Figs. 17.9 and 17.10). Those that contain *single-stranded DNA* (ssDNA) acquire the other strand of DNA, which is then replicated by the host DNA polymerases to make dsDNA that can be transcribed and translated by the cell's machinery, but first one additional step is required: the viruses need to make ssDNA genomes of the correct polarity. Therefore, the steps are:

ssDNA (infecting virus) → dsDNA (replicating form) → ssDNA (progeny)

RNA viruses. RNA viruses have either ssRNA of positive or negative polarity or dsRNA genomes. Some use the RNA genomes as templates to make dsDNA replicas whereas others bypass this step all together and make coding RNAs (**positive sense**) that the host ribosomes can translate into proteins. Consider viruses with ssRNA of *positive polarity* (the so-called positive-stranded RNA viruses). They carry a genome that is itself the *RNA sense strand* that the host ribosomes recognize and translate upon entry. Here the viral RNA plays a *double role*: it is both the genome of the virus and its mRNA. This group includes such notorious viruses as the agents of polio, hepatitis C, and the common cold. Replication by these viruses requires a special viral enzyme, an **RNA replicase** (an RNA-dependent RNA polymerase), that can *make RNA from RNA*. The viruses first make a complementary copy of the

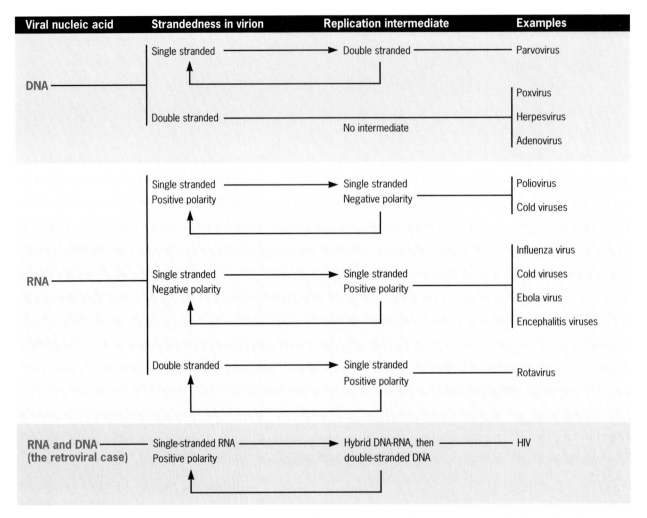

Viral nucleic acid	Strandedness in virion	Replication intermediate	Examples
DNA	Single stranded	Double stranded	Parvovirus
	Double stranded	No intermediate	Poxvirus Herpesvirus Adenovirus
RNA	Single stranded Positive polarity	Single stranded Negative polarity	Poliovirus Cold viruses
	Single stranded Negative polarity	Single stranded Positive polarity	Influenza virus Cold viruses Ebola virus Encephalitis viruses
	Double stranded	Single stranded Positive polarity	Rotavirus
RNA and DNA (the retroviral case)	Single-stranded RNA Positive polarity	Hybrid DNA-RNA, then double-stranded DNA	HIV

FIGURE 17.10 Main viral replication strategies. The arrows indicate replication resulting in the nucleic acid of the virion. Positive polarity means that an RNA molecule can serve directly as mRNA. Negative polarity means that it cannot serve as mRNA but must first be copied into its complementary nucleic acid.

RNA (negative polarity) and then use it as a template to make the positive RNA strands needed for the progeny, as follows:

(+) ssRNA (infecting virus and mRNA) → (−) ssRNA (replicating form) → (+) ssRNA (progeny)

Yet another alternative is seen in **retroviruses**, such as HIV, the virus that causes acquired immunodeficiency syndrome (AIDS). Its genome is also ssRNA of positive polarity, but, oddly perhaps, rather than using it as mRNA, these viruses follow the dsDNA intermediate step. The positive ssRNA genome is copied into a ssDNA copy of opposite polarity by the viral reverse transcriptase enzyme, which is packed in the virion with the ssRNA genome. When the ssRNA genome is introduced into the host cell, the reverse transcriptase uses it as a template to synthesize the two DNA strands. In its double-stranded form, the DNA is *integrated into the* *host genome* by the activity of an **integrase**, another of the class of enzymes that are carried in virions. The host RNA polymerase then transcribes the integrated form of viral dsDNA into viral RNA transcripts, which become the ssRNA genomes of the progeny. Replication of the RNA genome thus proceeds via a dsDNA intermediate by the following steps:

(+) ssRNA (infecting virus) → (–) ssDNA → dsDNA (integrated into host genome) → (+) ssRNA (progeny)

Note that the retroviral reverse transcriptase enzyme that converts the viral RNA into DNA *is unique to these viruses* and a few others and is therefore a good target for antiretroviral drugs. So is a protease carried in the virion that is needed for the maturation of viral proteins (see "HIV Infection and AIDS" in chapter 18. Another interesting fact about retroviruses, such as HIV, is that their genome also exists as an integrated DNA form. For this reason, retroviruses have both RNA and DNA and belong to a category of their own. There is clearly a selective advantage for this complex replication cycle: forming the DNA intermediate allows the virus to integrate into the genome of the host and thereby be carried with it indefinitely, undetected by the immune system.

Matters are not so simple for negative-stranded RNA viruses. Their RNA genomes have negative sense; i.e, they are the noncoding complementary strand and cannot be translated. To make mRNA transcripts, the virion carries its own *preformed RNA replicase*. Soon after entering the cell, the replicase starts working and copies the viral RNA into an mRNA complement (positive polarity), which can now be translated. This group of viruses includes the agents of influenza, mumps, Ebola, and many plant diseases. All of them package their replicase in the virion. These viruses face one additional complication: the negative-stranded RNA must be fully accessible to the enzyme that copies it, and thus, it must be devoid of possible structural impediments, such as double-stranded regions. To ensure ready access, the RNA's sugar-phosphate backbone is sometimes "impaled" onto a protein scaffold; thus, its nucleotide bases become directly exposed to the replicase. The steps here are:

(–) ssRNA (infecting virus) → (+) ssRNA (replicating form and mRNA) → (–) ssRNA (progeny)

To consider every possible known replication strategy, add to the list the double-stranded RNA viruses (Fig. 17.9) and a few others with mixed strategies (Fig. 17.10). In summary, there is remarkable latitude in the ways the viral genetic information is copied and how it flows from the genome to proteins. This clearly defies a precept of the **Central Dogma**, wherein DNA flows to RNA and then to proteins. Here, the flow of information is from RNA to DNA. Note, however, that the information never flows backward from proteins to nucleic acids, which so far remains an immutable tenet of the Central Dogma. What brought about so many variations on this central theme? No one knows for sure.

Making viral proteins. The viral cycle begins with infection of the host and ends with the assembly and release of progeny virions. In between, many events take place. The viral nucleic acid is first transcribed, translated, and replicated. This is called the **eclipse period** because intact viral particles cannot be found. However, this is not an idle time; a lot is going on within the infected cell.

Making viral nucleic acids and proteins is not a haphazard business. Rather, it is a symphony of well-timed and interrelated movements coupled to an amplification process that leads from one to hundreds of viral genomes (the tempo used in music *"andante,* going on *prestissimo"* seems appropriate here: to go from slow to very fast). To make viral proteins, the first thing that must be synthesized is their mRNA (unless we're talking of positive-stranded RNA viruses, whose genome is its own mRNA). Once mRNA is available, viral proteins are made by the host cell's protein-synthesis machinery. But viruses do more than mobilize the host cell to make the viral constituents. Many *subvert the host cell's metabolism* for the production of a lot of virions. For example, certain phages shut down multiple host functions, freeing their machinery to focus on viral components.

In a typical lytic infection by a phage, there are three main waves of transcriptional activity, designated early, middle, and late (Fig. 17.11). The first wave of transcription is initiated by the host RNA polymerase, which recognizes viral genes with a consensus-promoter sequence (chapter 8). This step leads to the expression of proteins that co-opt the host's biosynthetic machinery. Next, transcriptional regulators redirect the RNA polymerase to the middle genes. During this second wave, genes for viral DNA replication and host interference are transcribed. Now cellular functions are under total viral control. One of the middle genes encodes a viral sigma factor that binds RNA polymerase and redirects the enzyme to the late genes. These genes encode the proteins for making the structural components of the virions. Finally, viral proteins break the cell membrane and cell wall to cause cell lysis. With many phages, 100 or more virions will have been assembled and released.

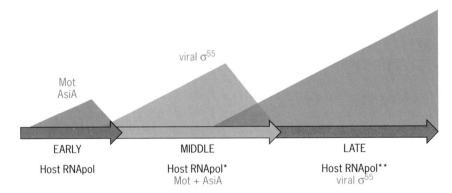

FIGURE 17.11 Temporal transcription of T4 genes upon infection. Upon entry in *E. coli*, a linear dsDNA genome, such as that of phage T4, circularizes, and the viral early genes are transcribed by the host RNA polymerase (RNApol). The proteins made in this phase (Mot, AsiA) begin DNA replication and modify the RNApol (indicated with stars) to redirect it toward the middle phase of transcription. Then, a viral sigma factor (σ^{55}) redirects the modified RNApol to make the late-phase genes, which encode the structural components of the virions and proteins needed for assembly.

In lambda and other phages, viral subversion reaches the heights of impudence. Soon after infection (in the early phase), bacterial DNA is broken down and its fragments cannibalized to support viral DNA synthesis. Any trace of cellular DNA that survives cannot be transcribed because RNA polymerase, taken hostage by the virus, is redirected to transcribe the viral DNA. Meanwhile, other virus-encoded proteins inhibit translation of bacterial mRNA. Thus, by clobbering almost every aspect of the host's machinery, viruses are extremely efficient biosynthesis parasites. Many phages generate hundreds of progeny virions in 20 minutes or less. Of course, for this scheme to work, the phage must redirect host functions quickly; thus, host subversion factors are synthesized early in infection. This sequence is not restricted to phages. Figure 17.12 shows, as an example, the timed expression of proteins of an adenovirus (an agent of respiratory and other infections). Although slower than for phages, the expression speed is still impressive. Less than 4 hours after entry, the first viral proteins (early proteins) begin to accumulate. The first complete infective virions are detected after 12 hours, and their numbers continue to increase until the cell lyses and viral progeny are released. No doubt about it, viral replication is a carefully planned and precisely executed affair.

Although many viruses follow this general scheme, not all do. Recall that many viruses invade their host cells, but, instead of lysing them, they incorporate their genomes into that of the host. This life cycle, called **lysogeny**, is the focus of the next chapter. Other viruses hedge their bet. For instance, some viral genomes persist for long periods within their host cells without replicating. This viral subversion strategy ensures that their host cells survive for a long time and thus become long-lived virus-making machines. For example, the virus that causes infectious mononucleosis (Epstein-Barr virus) enters epithelial and lymphoid cells, which normally have short life spans. But, by inhibiting **apoptosis** (programmed cell death), the virus "immortalizes" its infected host cell. Immortalized cells, which sometimes produce cancers (see chapter 18), can generate virus for longer periods or harbor dormant (latent) viral genomes for extended periods (chapter 25). Other viruses that use this strategy include the human papilloma virus and hepatitis C.

Virion assembly and release from the host cell. Once enough viral nucleic acid and proteins are made, the virions can be assembled. For icosahedral viruses, first the capsid is made as the viral capsomers self-assemble into crystal-like 3-D arrays in the general shape of the virion. The capsids are then filled with the viral nucleic acid. For nonenveloped viruses, this is the final step to producing viable virions. Host-cell lysis,

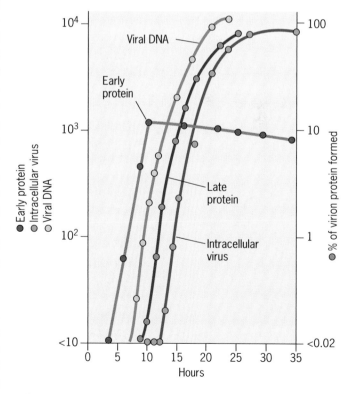

FIGURE 17.12 **A typical "one-step growth curve" of an animal virus.** Adenovirus replication in cell culture. First, viral early proteins are made; they direct the host cell to make viral DNA and late proteins (including virion structural proteins). After some time, complete infective virions are formed within the host cell.

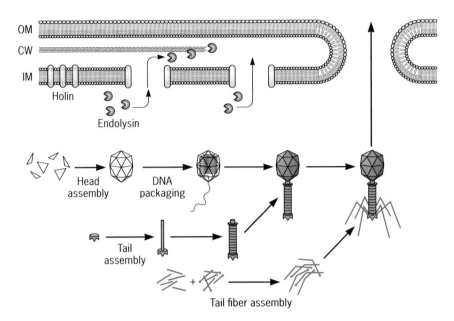

FIGURE 17.13 Virion assembly and release from the host cell. The T4 phage stepwise assembly process in which individual parts are assembled and then put together. Assembly is timed to the formation of porins (viral holin protein) on the inner membrane (IM) and release of the viral endolysin, which degrades the cell wall (CW). The outer membrane (OM) then breaks open, and the mature virions are released.

induced by viral proteins, releases the infectious virions. For example, in the first case study, the lambda phage that infected *E. coli* begins plotting its exit already in the middle wave of transcription of viral genes (Fig. 17.11). A viral protein (holin) assembles on the inner membrane where it makes pores (Fig. 17.13). Another viral protein accumulates in the cytoplasm (endolysin) and gains access to the **periplasm** once the holin pores form. This protein has **lysozyme** activity; i.e., it degrades the **peptidoglycan cell wall** of the cell. Without this fortifying layer, the outer membrane bursts, and the cell lyses. The assembly process is choreographed, first making the individual parts (DNA, head, tail, and fibers) and then putting them together (Fig. 17.13). The spread of the many infective virions released from a single cell can be catastrophic for the bacterial population. The ripple effect of the infection spreading from one cell to its neighbors can be visualized in the laboratory as macroscopic plaques of cell lysis (Box 17.1).

Virions of enveloped viruses have a more complicated birth. They are released from the host cell by budding, thus becoming surrounded by a segment of the host cell's membrane. Before budding begins, surface proteins encoded by the virus are incorporated into discrete patches of the host cell's membrane. Next the capsids bind to these patches and become extruded, coated by an envelope. (Note that this is a reversal of the entry mechanism, where the viral envelope fused with the host cell membrane [Fig. 17.8].) This viral release process does not directly damage the cell, which remains intact. This nonlytic exit mechanism is another strategy to extend the life of its host, which thereby produces progeny viruses for a long period of time.

BOX 17.1

Visualizing and quantitating viral growth

Imagine mixing a small number of bacterial or animal virions with a large number of sensitive bacteria or animal cells in culture and placing (plating) them on an agar plate. After a suitable period of incubation (a few hours or less for bacterial viruses; a day or more for animal viruses), the plate will be filled with uninfected host cells. However, here and there, there will be circular areas that look like holes in the "lawn" of host cells. In the example shown in the figure, replicating *E. coli* cells formed a lawn on the surface of agar-solidified

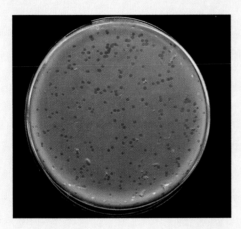

medium. The clearings of the cell lawn are areas where the cells were lysed after being infected by a lytic phage. These areas, called **plaques**, are the results of a microepidemic. Each plaque started from a single infected host cell that eventually lysed. The released virus particles attached to adjacent uninfected host cells, and the process repeated itself, leading to a centrifugal spreading of the infection. The process stops when the cells stop growing because viral reproduction requires that the host cells be metabolically active. There are variations on this theme. In the case of plant viruses, plaque counts can also be obtained by spreading viruses on an abraded surface of a leaf of a susceptible host. The number of lesions that develop is proportional to the number of virus particles in the preparation. In the early days of virology, this procedure was used with the tobacco mosaic virus.

It is a simple matter to use plaque counts to quantify the number of virus particles or infected cells in a preparation. As in the case of colony counts, one multiplies the number of plaques on a plate by the dilution factor of the preparation. But this tells you only the number of virions capable of proliferation. To determine the *total number of virus particles*, the sample can be examined under an electron microscope and the number of particles in a given volume counted.

HOW DO ANTIVIRAL THERAPIES WORK?

LEARNING OUTCOMES:
- Identify a reason why many potential antiviral therapies are toxic to the host, even though this is generally not a problem for antibacterial therapies.
- Give examples of the mechanisms of antiviral therapies.

Can you figure out why treating viral infections is intrinsically more problematic than treating diseases caused by bacteria? Bacteria differ biochemically from their host in many ways, and each of these presents a possible specific target for a drug. This is not the case with viruses, which depend on their host cell for many of their metabolic transactions. Consequently, many of the compounds that inhibit virus replication are toxic to us—as in cancer chemotherapy. But there are exceptions, such as when some step in virus replication is sufficiently different that it can be selectively inhibited. Mind you, the antiviral armament is not nearly as impressive as the antibacterial one, but some antiviral drugs are quite effective. For instance, take acyclovir, which inhibits herpesviruses and was given to CJ, the patient in the second case study. This drug is a synthetic guanosine analog that inhibits a virally encoded DNA polymerase but not the host's enzyme.

How this comes about is interesting. To become active, acyclovir must first be phosphorylated to a monophosphate, which a viral kinase does better than the host's kinase, and next to acyclovir triphosphate, which can then be incorporated into DNA by the virus-encoded DNA polymerase. Thus, the drug is efficiently activated in infected cells but not uninfected ones. Once incorporated into the nascent DNA strand, the phosphorylated acyclovir blocks elongation of the emergent DNA molecule. Acyclovir is used to treat several kinds of herpesvirus infections, including genital herpes.

You have surely heard that drugs are available to control HIV infection. Each targets a different step in the life cycle of the virus (Fig. 17.14). Such drugs have saved the lives of millions of people infected with this virus and have successfully impeded progression of the disease to AIDS. This arsenal is impressively varied, which permits the administration of several drugs at once. Why are multiple drugs needed? Because the virus rapidly mutates to become resistant to any one drug, but rarely to the combination. You can figure out why, right? (*Quiz spoiler:* The rate of mutation for resistance to *all* drugs is calculated by multiplying the rate for each single drug.) For HIV, we have drugs that inhibit multiple viral factors:

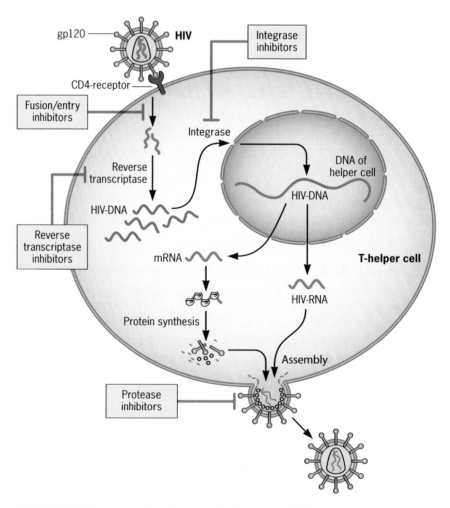

FIGURE 17.14 **Mode of action of drugs effective against HIV.**

- integrase enzyme that incorporates the viral DNA into the genome (integrase inhibitors).
- membrane glycoprotein that mediates entry (fusion/entry inhibitors).
- protease that cleaves a long protein made from the virus RNA (protease inhibitors).
- reverse transcriptase that makes HIV DNA from RNA (reverse transcriptase inhibitors).

In principle, the treatment of most HIV patients could control the disease and seriously limit, if not eliminate, its spread. In practice, the price (and the prolonged administration period required for these drugs) is beyond what low-income people and under-resourced countries can afford.

One last point. Viruses do not always replicate precisely. In fact, mutants arise with great frequency in the process, a concern for the clinician because some of these strains are resistant to antiviral drugs. RNA viruses mutate much more frequently than DNA viruses, since most RNA polymerases lack the "editing function" of DNA polymerases (see chapter 8) and thus cannot correct mistakes made during replication.

CASE 1 REVISITED: A tale of a phage that infected an *E. coli* cell

The pond water surely contained phage particles released during lysis of an infected bacterium. Tail fibers equip the phage to attach to a host cell's surface receptors. Penetration of the DNA into the cell takes place by a structural rearrangement of the virion, whereby the tail contracts and the DNA is ejected. Next, the phage DNA commandeers the biosynthetic machinery of its host and redirects it to make viral components instead of its own. In the second infection, it is likely that the phage established a lysogenic infection. (Chapter 18 covers lysogeny.) Later, UV irradiation from the sun induced the integrated phage to undergo a lytic cycle. This phenomenon, called **prophage induction**, activates the dormant phage, allowing its replication and the eventual lysis of the host cell. The decision to lyse or to remain lysogenic depends, among other things, on the number of phages infecting one cell.

CASE 2 REVISITED: When viruses mind your business

Most likely, CJ's partner was an asymptomatic carrier of herpesvirus. Without revealing its presence, enough virus was present for transmission to someone else. In the week before CJ had any symptoms, the virus was actively replicating at or near the site of entry, eventually causing the inflammation that led to the visible lesions. (We'll learn about mechanisms of herpes pathogenesis in chapter 25.) The clinical laboratory performed a PCR reaction because, unlike most pathogenic bacteria, viruses do not grow on agar plates. Tests to identify virus in a sample are usually directed toward detecting their nucleic acid. An alternative diagnostic strategy is to quantify, in blood, the number of antibodies specific to the virus, an indication that the virus has stimulated the immune system. The doctor did not prescribe an antibacterial therapeutic, such as penicillin, because such drugs are totally ineffective against viruses. Such misuse of antibiotics has increased the population of pathogenic bacteria that are resistant to many and, in some cases most, antibiotics.

CONCLUSIONS

Let us agree that the world of virology is far from monotonous. Viruses vary greatly in shape, structure, and mode of replication. They encompass every version of nucleic acid: single- or double-strands, linear or circular, RNA or DNA. One thing is clear: viruses are not cells. Nor do they operate in a vacuum. Viruses are totally dependent on cells to provide them with energy and the machinery for synthesizing proteins. Some viruses encode the enzymes needed to make their own nucleic acid and, in some cases, even carry an enzyme in their capsid. If such viruses were endowed with a mechanism to obtain energy and to make proteins, would they become capable of unaided reproduction? In other words, would they become cells? This question seems particularly relevant to some of the large enveloped viruses, such as the giant viruses and those that cause smallpox, which are endowed with a significant complement of nucleic acid-synthesizing enzymes.

Representative ASM Fundamental Statements and Learning Outcomes for the Chapter

EVOLUTION

1. Mutations and horizontal gene transfer, with the immense variety of microenvironments, have selected for a huge diversity of microorganisms.

   a. Identify six ways to classify viruses.

CELL STRUCTURE AND FUNCTION

2. The structure and function of microorganisms have been revealed by the use of microscopy (including bright-field, phase-contrast, fluorescent, and electron).

   a. Name structural components common to all viruses.
   b. State an important difference between icosahedral and filamentous viruses, other than shape.
   c. Identify how scientists have learned details of the molecular organization of viruses, given their relatively small size.

3. The replication cycles of viruses (lytic and lysogenic) differ among viruses and are determined by their unique structures and genomes.

   a. Compare and contrast lytic and lysogenic viral cycles.
   b. Speculate what would be required for larger viruses to become cellular entities and replicate independently.
   c. Name the five major steps of viral replication.
   d. Discuss how the structure of viruses enables virions to attach and enter host cells.
   e. Describe how bacteriophages are capable of penetrating the thick cell envelope of Gram-positive and Gram-negative bacteria.
   g. Identify the origin of the envelope of enveloped viruses.

METABOLIC PATHWAYS

4. The growth of microorganisms can be controlled by physical, chemical, mechanical, or biological means.

   a. Identify a reason why reverse transcriptase is a good target for HIV antiretroviral drugs.
   b. Explain how to visualize and quantify viral growth using viral-sensitive bacterial and animal cells in culture.
   c. Give examples of the mechanisms of antiviral therapies.

INFORMATION FLOW AND GENETICS

5. The synthesis of viral genetic material and proteins is dependent on host cells.

   a. Identify three kinds of enzymes that virions carry.
   b. Contrast genome replication for positive-polarity RNA viruses and negative polarity RNA viruses.
   c. Describe how the nucleic acids of retroviruses replicate.
   d. Discuss how viral proteins are synthesized during the viral replication cycle.
   e. Give examples of two replication strategies used by viruses that do not involve lysing host cells.
   f. Identify a reason why many potential antiviral therapies are toxic to the host, whereas this is generally not a problem for antibacterial therapies.

IMPACT OF MICROORGANISMS

6. Microorganisms provide essential models that give us fundamental knowledge about life processes.

   a. Distinguish viruses from cellular forms of life.
   b. Identify examples of important discoveries in biology that were made possible by using viruses as experimental tools.

7. Because the true diversity of microbial life is largely unknown, its effects and potential benefits have not been fully explored.

   a. Explain why the true diversity of viruses and many important ecological and evolutionary relationships remains to be discovered.

SUPPLEMENTAL MATERIAL

References

- **Viruses**, on *Microbe World*, a resource from the American Society for Microbiology. http://www.microbeworld.org/types-of-microbes/viruses
- **Wessner DR**. 2010. The origins of viruses. *Nature Education* **3**(9):37.
- *Viruses Throughout Life & Time: Friends, Foes, Change Agents*, a 2013 colloquium report from the American Academy of Microbiology. http://academy.asm.org/index.php /browse-all-reports/5180-viruses-throughout-life-time-friends-foes-change-agents
- *The BIG Picture Book of Viruses*. Maintained by Dr. Robert F. Garry, Tulane University. http:// www.virology.net/Big_Virology/BVHomePage.html
- *All the Virology on the WWW*: Maintained by Dr. Robert F. Garry, Tulane University. http:// www.virology.net/garryfavweb.html.

Supplemental Activities

CHECK YOUR UNDERSTANDING

1. **Which of the following are possible origins of the lipids and carbohydrates in the envelope of enveloped virus?**
   a. Virus capsid
   b. Viral genome
   c. Host ribosomes
   d. Host Golgi membrane
   e. Host nuclear membrane

2. **What does a negative-polarity RNA virus carry to replicate its genome?**
   a. Viral RNA that encodes an RNA replicase
   b. A preformed RNA replicase
   c. A host RNA polymerase
   d. A host DNA polymerase

3. **Which part of the Central Dogma do some viruses violate?**
   a. DNA is the template for RNA.
   b. RNA is the template for proteins.

c. Protein is the template for proteins.

d. Proteins make DNA.

DIG DEEPER

1. Gina has prepared a fresh suspension of virus for infection studies in the lab. Before she can use it, she needs to calculate the concentration of live virus, or determine the "titer." To do so, she plates 5×10^6 host cells in each of three plates, adds 5 μl, 25 μl, or 125 μl of virus stock, diluted in 5 ml of tissue culture medium, to each monolayer of host cells, and then incubates the cultures at 37°C.

   • If, for the three dilutions, she counts 55, 310, and 1,290 plaques, respectively, what is the viral titer?

   • Multiplicity of infection (MOI) is the average number of virus particles infecting each cell. What volume of the virus stock will she need to infect 2×10^6 cells at an MOI of 1?

2. Consider the data for the phage growth experiment graphed below (http://aem.asm .org/content/73/8/2532.figures-only). Initially the culture was infected with approximately 5 phage particles per bacterium. PFU means plaque-forming units, or infectious phage virions. The red line indicates the untreated culture; the blue line indicates samples treated with chloroform to kill the bacteria and induce their release of intracellular virions (which are not sensitive to chloroform).

   • Why do PFU rise earlier in the chloroform-treated samples?

   • What happened in the treated and untreated cells before 15 or 30 minutes, respectively?

   • Why does the number of PFU level off?

   • Why does the number of phages increase 100-fold in such a short time?

   • At what time point would you expect the bacterial cell envelopes to start being degraded?

   • Label the eclipse and the latent periods.

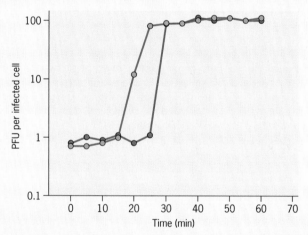

3. National Center for Case Study Teaching in Science "Between the Living and the Dead" by Kari Mergenhagen explores the differences between viruses and bacteria while teaching about the basic components and "life" cycle of a T-even bacteriophage. http:// sciencecases.lib.buffalo.edu/cs/collection/detail.asp?case_id=378&id=378

4. Teach the life cycle of one of the following types of viruses to classmates using drawings or animations.

   • Single-stranded DNA virus.

   • Negative-polarity single-stranded RNA virus.

   • Positive-polarity single-stranded RNA virus.

   • Retroviruses.

5. *Case it!* project from University of Wisconsin-River Falls funded by National Science Foundation, "Human Influenza": molecular biology simulations augment clinical and lab scenarios. http://www.caseitproject.org/influenza-mobile/.

6. Smallpox is a viral disease that killed hundreds of thousands of people over the centuries. It is the only viral disease that has been truly eradicated, after successful mass vaccination programs were initiated by WHO. Research and write a newspaper-style article to educate the general public about the current debate on smallpox research. Describe:

   - Causative agent of smallpox: type of virus, cellular receptors used for entry, type of cells infected, life cycle.
   - Symptoms and course of the disease.
   - Virus strain used for the immunization program and the logistics of the immunization program. What is the current status of smallpox immunity worldwide?
   - Why is smallpox virus research regulated? Why are individual viral genes available for research but the complete viral genome is not?
   - Argue why we should or should not revive this virus for research purposes.

7. **Team work:** Unraveling the mysteries of viruses has led to technical advances well beyond the field of virology. Listed below are some of the technologies that have been created thanks to viruses. Work in groups to develop a 10-minute presentation to explain the contribution a virus has made and how that advance may affect society.

   - Immortalized cell lines
   - Bacterial transduction
   - Gene therapy
   - Cancer prevention and control
   - Biological pest control
   - Bacteriophages as antibiotics
   - Protein expression systems

Supplemental Resources

VISUAL

- *Microbes After Hours* lecture "Return of Influenza" from the American Society for Microbiology to educate the general public (1 h 7 min). https://www.youtube.com/watch?v=4geOQWeeMg&feature=c4-overview-vl&list=PLD2Sv5NskYtwPnpiWnu63hANr0Bo0Xzu0.
- Picornaviruses lecture by Vincent Racaniello, with slide set (1 h 15 min). http://www.microbeworld.org/component/jlibrary/?view=article&id=9749.
- *Virology* blog 2011 keynote lecture "The World of Viruses" by Vincent Racaniello at the XXII meeting of the Brazilian Virology Society discusses 10 seminal virologists and 10 compelling virology stories of recent years (57 min): https://www.youtube.com/watch?v=8u43UDu41pg.
- *iBiology* lecture on viral entry by Ari Helenius. http://www.ibiology.org/ibioseminars/microbiology/ari-helenius-part-1.html.
- *iBiology* lecture on virus structures by Stephen Harrison. http://www.ibiology.org/ibioseminars/microbiology/stephen-c-harrison-part-1.html.
- *iBiology* lecture "Bacteriophages: Genes and Genomes" by Graham Hatfull. http://www.ibiology.org/ibioseminars/microbiology/graham-hatfull-part-1.html.
- *iBiology* lecture "Plant Viruses and Crops" by Roger Beachy (28 min). http://www.ibiology.org/ibioseminars/plant-biology/roger-beachy-part-1.html.
- *iBiology* lecture "Genome Sequencing for Pathogen Discovery" by Joseph DeRisi (27 min). http://www.ibiology.org/ibioeducation/joseph-derisi-genome-sequencing-for-pathogen-discovery.html.
- National Science Foundation video by Richard Lenski demonstrates how lambda evolves new mechanism to invade *E. coli*: http://www.nsf.gov/news/newsvideos.jsp?cntnid=122949&mediaid=71835&org=NSF.
- National Public Radio video "Flu Attack! How a Virus Invades Our Body" on *MicrobeWorld*. http://www.microbeworld.org/component/jlibrary/?view=article&id=10035.
- *iBiology* lecture "Danger from the Wild: HIV, Can We Conquer It?" by David Baltimore (35 min). http://www.ibiology.org/ibioseminars/microbiology/david-baltimore-part-1.html.

- Animations of productive life cycle of animal viruses from ASM MicrobeLibrary. http://www.microbelibrary.org/library/virus/3205-productive-life-cycle-of-animal-viruses-animations.

SPOKEN

- *This Week in Virology* podcast about viruses: an informal yet informative conversation about viruses accessible to a broad audience. http://www.microbe.tv/twiv/.

WRITTEN

- *Virology* blog post "Detecting Viruses: The Plaque Assay" by Vincent Racaniello teaches plaque-forming units. http://www.virology.ws/2009/07/06/detecting-viruses-the-plaque-assay/.
- *Virology* blog post "Viruses and Viral Disease" by Vincent Racaniello on the origin of segmented RNA viruses. http://www.virology.ws/2014/06/12/origin-of-segmented-rna-virus-genomes/.
- FAQ *Influenza*, a colloquium report from the American Academy of Microbiology to educate the general public. http://academy. asm.org/index.php/faq-series/793-faq-west-nile-virus-july-2013.
- *Virology* blog post "Why Do Viruses Cause Disease?" by Vincent Racaniello ponders how viruses benefit from the diseases they cause. http://www.virology.ws/2014/02/07/why-do-viruses-cause-disease/.
- *Small Things Considered* blog post "Oddly Microbial: Giant Viruses" by Marcia Stone discusses viruses that are bigger than some bacteria! http://schaechter.asmblog.org/schaechter/2013/11/oddly-microbial-giant-viruses-.html.
- *Small Things Considered* blog post "Five Questions about Filoviruses" by Jamie Henzy discusses Ebola virus and others. http://schaechter.asmblog.org/schaechter/2014/09/five-questions-about-filoviruses.html.

*Leukemia cells (blue) that contain
Epstein-Barr virus (green) (CDC–Public
Health Image Library)*

CHAPTER EIGHTEEN

Viral Latency

KEY CONCEPTS

This chapter covers the following topics in the ASM Fundamental Statements.

EVOLUTION

1. Mutations and horizontal gene transfer, with the immense variety of microenvironments, have selected for a huge diversity of microorganisms.

CELL STRUCTURE AND FUNCTION

2. The replication cycles of viruses (lytic and lysogenic) differ among viruses and are determined by their unique structures and genomes.

INFORMATION FLOW AND GENETICS

3. The synthesis of viral genetic material and proteins is dependent upon host cells.

MICROBIAL SYSTEMS

4. Microorganisms, cellular and viral, can interact with both human and non-human hosts in beneficial, neutral, or detrimental ways.

INTRODUCTION

When it comes to surprises, viruses never cease to deliver them. They are not just our foes but also our friends. In chapter 17, we learned about viruses mainly as destructive agents that damage their host cells and cause disease. This is a hugely imperfect view of what viruses really do because many of them *coexist with their host cells* for long periods, remaining latent until induced. Such associations often have momentous consequences, as they can lead to important *phenotypic or genetic changes* within the host cell and to the *exchange of genetic material* between cells. As a consequence, viruses play a key role in evolution, a fact that is often underappreciated. In this chapter we will examine the mechanisms that allow viruses to infect an animal or a prokaryotic cell and remain latent and the genetic and evolutionary consequences of viral latency in the host.

CASE: A case of shingles

Mrs. K, a 75-year-old grandmother, had felt tired and lethargic for a week and also complained of a painful itch on the right side of her back and chest. For a day or so, parts of her chest became covered with painful raised blisters (figure). She described the pain as that of being "pierced with needles." The location of the rash as a band on one side of the trunk made the doctor suspect shingles, a disease caused by an infection by varicella-zoster virus, a herpesvirus. The doctor who examined her also learned that Mrs. K had had chickenpox as a child, which is caused by the same virus, but had complained of no other relevant symptoms since then. The physician prescribed the antiviral drug acyclovir and painkillers, in the hope that they would reduce her pain but without expecting that the drugs would cure her. Mrs. K endured several months of pain and discomfort, eventually feeling better. She was told that a vaccine for shingles was available and that it was appropriate for her, even though she had had the disease.

- Why did the doctor suspect a connection between shingles and the patient's childhood bout of chickenpox?
- Where did the virus hide all those years?
- Why would a vaccine against shingles be recommended for patients who have already had the disease?

WHAT IS VIRAL LATENCY?

LEARNING OUTCOMES:
- Describe two ways that viruses establish latency.

Why would a virus remain latent inside the host cell rather than use it to make more infective viral particles? Viruses depend on their hosts for reproduction. Hence, it should not surprise us to learn that some viruses have evolved to become less infectious but remain in the host without immediately causing a damaging infection. In the study case above, Mrs. K had chickenpox as a child, a disease caused by one of the herpesviruses (the *varicella-zoster virus* or VZV). The virus proliferated and probably spread to other human hosts (chickenpox is very contagious). But the virus never completely disappeared from her body. Somehow, the viral DNA remained in some of the host cells after the symptoms of chickenpox had subsided but then got reactivated when Mrs. K was 75 years old. How did the virus remain hidden for so long without manifesting its presence?

As in Mrs. K's case, some viruses can persist in a host cell without proliferating and causing damage. This state is known as **latency**, which can occur both in prokaryotic and eukaryotic hosts. There are two general ways to establish latency. One is for the virus to become an **episome,** that is, an *extrachromosomal element* that replicates alongside the host cell, like a plasmid (chapter 10). Often, the viral element does not make its presence felt. As will be discussed in chapter 25, herpesviruses, including the one that infected Mrs. K in the study case, do just that. The other way to establish latency is for the genome of the virus to *integrate into the host genome*. This form is called a **provirus** for animal viruses and a **prophage** for bacterial ones. Retroviruses such as HIV and many phages integrate into chromosomal DNA. This more intimate relationship between virus and host requires extra steps. Now think: are viruses capable of this lifestyle more likely to be DNA or RNA viruses? (*Quiz spoiler:* RNA is not known to integrate into the host genome unless it is copied into a DNA molecule. Plus RNA would have a hard time becoming an extrachromosomal element. Why? RNA, which is usually single stranded, is unstable and susceptible to cleavage by nucleases.)

HOW DO ANIMAL VIRUSES BECOME AND REMAIN LATENT?

LEARNING OUTCOMES:
- Explain the differences between true latency and clinical latency.
- List three characteristics of host-viral interactions that lead to latency.
- Describe how latent viruses can evade the cell-mediated immune response.
- Explain the role of reactivation in disease progression.

Becoming Latent

Latent viruses cause infections that are often characterized by symptomatic bouts appearing *long after the virus is acquired*. In humans, latency is the hallmark of some long-term infections (Fig. 18.1). Take our case of

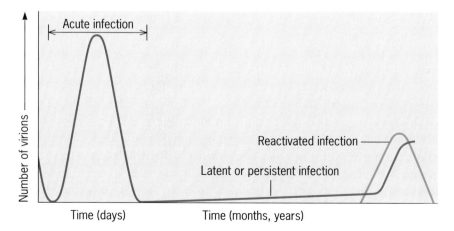

FIGURE 18.1 **The temporal patterns of viral infection.**

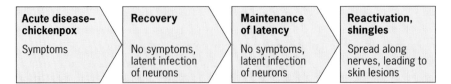

FIGURE 18.2 **The time course of shingles.**

shingles (Fig. 18.2). This is a nasty inflammation of peripheral nerves that affects people of all ages but is more common among the elderly. In most cases, the patient was infected with the virus as a child, when it caused chickenpox. The virus then goes into hiding as an extrachromosomal element or episome in certain neurons, only to emerge many decades later as the result of an usually unknown stimulus that could have been a diminished immune system or some medication. Although this infectious agent, VZV, is a member of the herpesvirus family, it is not closely related to the one that causes genital and other lesions (remember CJ from Case 2 in chapter 17?).

There are several distinct outcomes of long-term viral-host interactions, latency being only one of them, persistence and cancer being others (Fig. 18.3). Some viruses become latent as episomes, others by integrating their genomes into the host's chromosome. Practically all of us carry a number of latent viruses in blissful ignorance since they rarely reactivate to cause symptoms. Latent infections are also caused by other herpesviruses, including Epstein-Barr virus, the agent of infectious mononucleosis or "kissing disease" (chapter 25). Among the viruses that persist for long periods of time, HIV is unique for its distinctive timetable. In chapter 17, we learned how the RNA genome of the HIV virus is copied into a DNA molecule by the viral **reverse transcriptase** enzyme so it can get integrated into the host's chromosome. Once there, the HIV genome lies dormant for many years and, if untreated, inexorably proceeds to cause AIDS. Although this pattern resembles latency, the virus is only *clinically* latent: HIV is actually engaged in a standoff with the host immune system, replicating at a rate the immune system can just tolerate. In a highly predictable course of events, in the absence of antiviral therapy, the immune system eventually wears down, and the virus increases its replication. Symptoms then emerge, ending clinical latency and leading to the grim infections that characterize untreated AIDS.

Remaining Latent

Obviously, the cells that carry latent viruses must be long lived. This is true for the neurons that carried Mrs. K's virus since the childhood bout of chickenpox (study case). And just as obviously, the latent viruses must not kill the host cells that carry them. Many viral infections elicit a form of the adaptive immune response called **cell-mediated immunity** (chapter 22). When activated, immune cells specifically seek out *infected cells* in the body and destroy them. Why is this mechanism (instead of antibodies) particularly relevant in combatting viral infections? (*Quiz spoiler:* Because viruses spend much of their existence intracellularly, they are shielded from circulating defenses such as antibodies.) Why did cell-

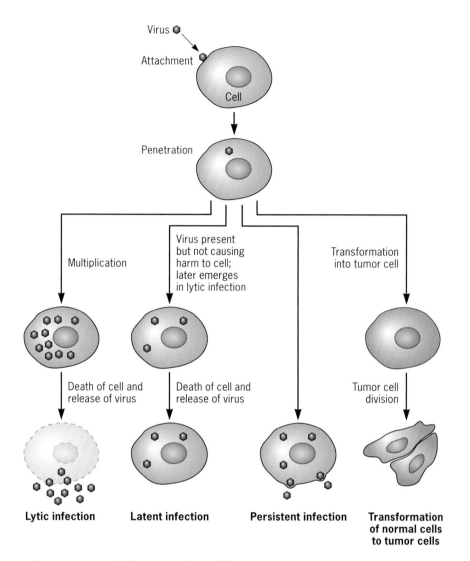

Virus

Attachment

Cell

Penetration

Multiplication

Virus present but not causing harm to cell; later emerges in lytic infection

Transformation into tumor cell

Death of cell and release of virus

Death of cell and release of virus

Tumor cell division

Lytic infection

Latent infection

Persistent infection

Transformation of normal cells to tumor cells

FIGURE 18. 3 **Viral infections can have different outcomes.**

mediated immunity fail to clear Mrs. K's body of her virus study case? Some latent viruses (e.g., Mrs. K's VZV, other herpesviruses, and retroviruses) make proteins that inhibit or divert this branch of the immune system, rendering it ineffective. In addition, some of these viruses actively inhibit **apoptosis** (programmed cell death), thus ensuring prolonged survival of their host cell.

Having solved the problem of host cell survival, a latent virus must keep the expression of its genes under control. Generally speaking, it must *shut off genes involved in reproduction* to avoid unbridled multiplication and cell lysis. At the same time, it must *keep "on" those genes required for prolonged maintenance in the host cell.* How latent viruses shift to persistence mode is best known for lysogenic phages, as we'll discuss later, but there are telling examples among animal viruses. In chapter 25, we'll learn that the herpesvirus that infected CJ in chapter 17's case study makes a set of RNAs called latency-associated transcripts that suppress the expression of the viral genes involved in the lytic cycle.

The concept of viral latency includes one other key aspect: **reactivation**. That is, under some conditions, the latent viruses once again become virulent and multiply. These events most likely happened to Mrs. K from the study case. In many instances the precipitating event is not known but generally speaking involves stress, trauma, physiological change such as menstruation, fever, or various kinds of immunosuppression (e.g., age related or drug induced). Could age have triggered the reactivation of the VZV latent virus in Mrs. K?

HOW DOES VIRAL LATENCY AFFECT EVOLUTION, HEALTH, AND DISEASE IN ANIMAL HOSTS?

LEARNING OUTCOMES:
- List two ways that viruses help human hosts.
- Differentiate animal host cell transformation from bacterial cell transformation.
- Describe how latent viruses can influence oncogenes.

Why do we carry latent viruses? We discussed the idea of viruses taming their infectivity to preserve a sufficient host population to ensure that the virus can continue to spread and replicate in the long term. And, counter-intuitive as this may be, some latent viruses can actually be good for you! There are reports of such viruses conferring protection against bacterial infections. For example, mice infected with relatives of the human Epstein-Barr virus become resistant to certain pathogenic bacteria such as foodborne *Listeria* (chapter 24) and the plague bacillus (chapter 27). There are also claims that mice infected with Epstein-Barr virus have increased resistance to grafts of tumors. Latent viruses also protect us from infection by similar viruses, either by inducing an immune response or blocking the specific entry routes used by the virus to enter your cells.

Of course, certain latent viruses can be very bad for you. We have already encountered such infections in the case of Mrs. K and with HIV. Latent viruses have also been blamed for a number of chronic diseases whose causes seem mysterious, including certain forms of cancer, chronic fatigue, or diabetes. If in fact they are involved, are these viruses the direct cause of these diseases, do they contribute to our predisposition to the disease, or are they indirect by-products of the disease state? These questions are being investigated, but definitive proof seems to be hard to come by. (A large prize is in the waiting, should you establish conclusively a viral cause of such diseases!)

Viral latency has indeed been linked to certain forms of cancer (Table 18.1). This process is called **cell transformation**, a term not to be confused with the bacterial process of transformation whereby naked DNA is taken up and recombined into the microbial chromosome (chapter 10). In eukaryotes, transformation refers to the cellular changes that result from infection by certain latent viruses, especially when the changes result in cancerous cells. Genes called **oncogenes** subvert mechanisms that control cell replication, causing cell transformation. Oncogenes are

TABLE 18.1 **Some virus-induced human cancers**

Virus	Genome	Cancer type or site
Epstein-Barr virus (EBV)	dsDNA	Stomach, lymphoma (Hodgkin and non-Hodgkin), nasopharynx
Kaposi sarcoma	dsDNA	Kaposi sarcoma
Human papillomavirus (HPV)	dsDNA	Cervix, anus
Hepatitis B	dsRNA	Liver
Hepatitis C	ssRNA	Liver
HIV	Retroviral	Kaposi sarcoma, non-Hodgkin lymphoma
Human T-cell lymphotropic virus	Retroviral	T-cell leukemia, lymphoma

altered forms of normal regulatory genes that now disrupt signal transduction pathways responsible for controlling cellular gene expression. In some cases, oncogenes are carried by the virus. Alternatively, cell transformation can occur when insertion of a viral genome alters either the expression or the coding sequence of a host (onco)gene. Viruses capable of transformation include herpesviruses, papillomaviruses, and retroviruses, such as Rous sarcoma virus.

You might be wondering if latent viruses contribute to the evolution of eukaryotic organisms. A most plausible thought, and an important one. The full impact over time of such viruses on the host species remains to be elucidated, but we already have some striking facts at hand. Surprisingly, humans and other animals carry in their genomes a profusion of so-called **endogenous retrovirus sequences**. These sequences make up about 8% of our genome! As we've learned, retroviruses integrate a copy of the viral genome into the host genome. In the integrated state, the viral genome is carried along with that of the host and replicated at each cell division. When integration occurs in cells of the germ line, namely sperm or egg cells, the genes of the virus become *part of the host's genetic makeup.* They are then inherited by offspring and are susceptible to the same evolutionary forces as the host genes.

So, what do endogenous retroviruses actually do? Residence in the genome allows retroviruses to act as transcriptional elements, that is, to *supply the function of promoters* from which downstream genes may be transcribed. But do these viral sequences code directly for proteins themselves? There is one well-characterized instance of two animal proteins that originated from endogenous retroviruses: **syncytins**, which contribute to the development of the placenta. Syncytins are in fact retroviral envelope proteins that were originally part of a provirus but ended up being adopted by the host thanks to their ability to promote the development of the placenta in some mammals (Fig. 18.4). Humans have syncytins as well. In other words, none of us would be here without these viruses! However, we do not all carry the same viruses, a fact that can be expected to affect our individual genetic identity and, at a population level, drive the diversification of species. Despite the formidable presence of endogenous retroviruses (8% of our DNA), we still know rather little about the viral residents in our genomes.

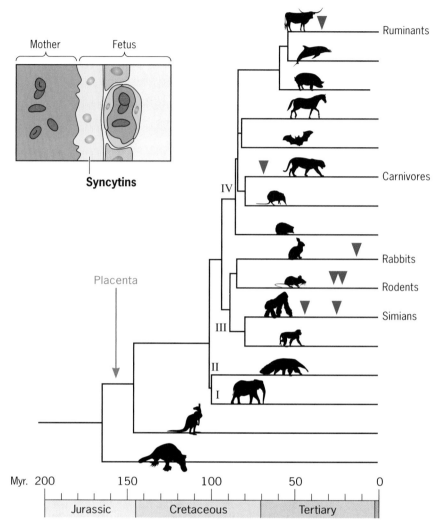

FIGURE 18.4 Capture of retroviral genes and the origin of placental syncytins.
Different lineages of placental mammals captured various retroviral *env* genes encoding syncytins independently (purple triangles represent time of capture events). The timescale unit is million years.

HOW DO PHAGES BECOME AND REMAIN LATENT?

LEARNING OUTCOMES:
- Differentiate temperate or lysogenic phage from prophages.
- Distinguish between virulent and temperate viruses.
- Define viral induction and give an example of how prophages are induced.

Before turning to the prokaryotic viruses, let's review a few definitions introduced in chapter 17. Prokaryotic viruses that can integrate into the host genome are said to be **temperate** or **lysogenic** (the term "latent virus" is usually used in the eukaryotic world, proof that these two worlds don't always speak the same language). Because most of what is known about prokaryotic viruses comes from studies of bacterial viruses or phages, the integrated form is called a **prophage**. The bacteria that carry prophages are then said to be lysogenized, and the phenomenon is referred to as **lysogeny**. Try building a sentence with these terms (*ha!*).

A simplified version of the lysogenic cycle is shown in Fig. 18.5. As with animal viruses, here too there are variations on the basic theme. In prokaryotes, the viral genome may circularize and remain as an autonomous extrachromosomal element rather than integrating into the host genome. With bacteria, the favored term is not episome but **plasmid** (chapter 10). Like other plasmids, the phage DNA may replicate in synchrony with the bacterial chromosome and be transmitted to the progeny. Thus, integration into the chromosome is not an essential feature of lysogeny. The key point is that when the phage pursues the temperate lifestyle, it replicates in step with its host cell and no more.

Which would better foster the survival of the virus over time, being virulent and making a large number of progeny immediately or being temperate and preserving its genome within the host? Unbridled replication of virulent viruses could lead to the destruction of a large proportion of host cells and possibly to a dead end (i.e., no transmission to a new host). On the other hand, being exclusively temperate would mean perpetual confinement. A compromise would be to alternate between virulence and temperance; that is, make particles on occasion but keep the genome in a safe repository the rest of the time. That is just what lysogenic phages do.

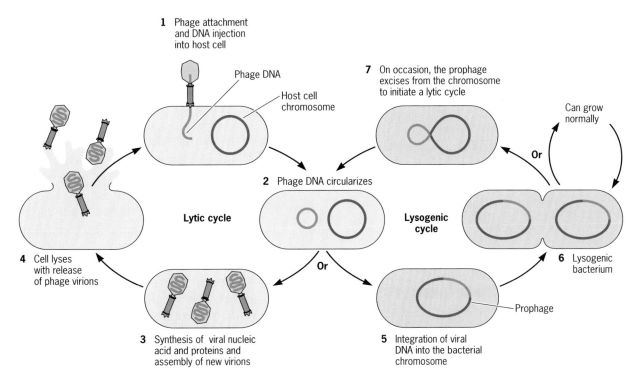

FIGURE 18.5 Two lifestyles of a temperate phage: lysogenic and lytic. Phage attachment to a sensitive host bacterium (1) is followed by the injection of viral DNA into the host (2). In typical cases, the viral DNA circularizes and either replicates (3), leading to the production of progeny phage and cell lysis (4), or integrates into the host chromosome to become a prophage (5). In this state, the host cell has become a lysogen that can replicate for many generations (6). On rare occasions, the prophage excises spontaneously from the bacterial chromosome (7), a process that can be augmented by induction (e.g., treatment with UV light or mutagens). Upon excision, the phage DNA can initiate a lytic cycle (3 and 4).

The stable association between a lysogenic phage and its host bacterium can be interrupted, generally by conditions that threaten the life of the host cell. For example, if a lysogenic bacterium is irradiated with ultraviolet (UV) light, its prophage will excise itself and replicate freely (Case 1 in chapter 17). This phenomenon, called viral induction, is a "bailing out" mechanism that allows temperate viruses to survive even if their host is killed. How can you tell that a cell's genome contains a prophage? (*Quiz spoiler:* By exposing the cells to an inducer, e.g., UV light, and seeing if viruses are produced.) Even without such an experiment, bioinformatics often reveals the presence of viral genes. In fact, well over half the sequenced bacterial genomes include phage sequences.

How Does the Lambda Phage Become and Remain Latent?

LEARNING OUTCOMES:
- Outline the integration cycle of lambda.
- Define the functions of DNA ligase, integrase, and repressor protein and explain the role of each in lambda's integration cycle.
- Differentiate between site-specific and homologous recombination.
- Describe two classes of mutations that can convert lambda from a temperate phage to a virulent one.

How do lysogenic phages manage to balance the virulent and temperate lifestyles? The most studied temperate phage is an *Escherichia coli* phage called lambda, which has become a model for our understanding of the phenomenon of lysogeny. The description below details the process for lambda, but most of the steps are similar for other temperate phages.

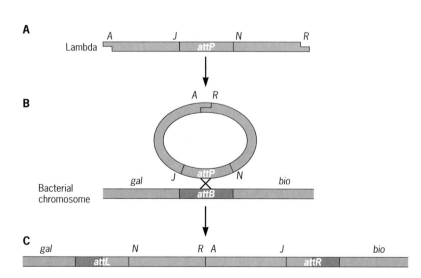

FIGURE 18.6 Mechanism of integration of a temperate-phage genome into the host cell chromosome. The ends of the linear lambda phage genome (A) come together to form a circle (B). The phage attachment site, *attP*, recombines with a site on the bacterial genome, *attB*. As the result of a crossover between these two sites, the phage genome is integrated into that of the host (C). In the process, the attachment sites have been changed to *attL* and *attR*.

Integrating the genome into the host's chromosome. To go latent, the lambda phage integrates its genome into the bacterial host's chromosome (Fig. 18.6). Be prepared for a somewhat complex, yet impressive, story. In the lambda virion, the DNA is a *double-stranded linear* molecule. Shortly after being injected into an *E. coli* cell, it becomes *circular*. Let us think for a moment, how does a linear DNA molecule circularize? Protruding from the ends of these DNA molecules are *single-stranded complementary sequences*. These are "sticky ends" whose bases on one end can pair with those of the other, thus forming a circle. The two single-strand nicks where the pairing occurred are closed by **DNA ligase** to form a *complete covalently closed circle*. This can now integrate into the host

DNA circle by a single crossover event. Why is making DNA circles so popular among most (but not all) phages and bacteria? (*Quiz spoiler:* Bacterial cells have powerful **exonucleases** that degrade linear DNA from the ends unless the ends are protected or absent altogether. A circular chromosome is an easy way to prevent this from happening.)

If viruses were to integrate into their hosts' genomes willy-nilly, they would risk inactivating genes essential for the host's life. To avoid this danger, most phages integrate into *specific locations* on the chromosome. In the case of lambda phage, recombination between its DNA and that of the chromosome takes place on each molecule only at one specific site (*att*) (Fig. 18.6). Other phages integrate at more than one site (Fig. 18.7). Note that these *viral and host sites do not share homologous sequences*. In this regard this process of **site-specific recombination** differs from the more general **homologous recombination** process (chapter 10). Homologous recombination can take place anywhere along the DNA as long as, per its name, two sequences are homologous. Therefore, the two processes are quite different.

Site-specific recombination between phage and chromosomal DNAs depends on the activity of a special viral protein called **integrase**, which recognizes the two dissimilar sites, one on the bacterial chromosome and one on the viral DNA. The integrase recognizes these sequences simultaneously and catalyzes their recombination. Hence, if the phage expresses the integrase while infecting the host cell, the viral genome will be integrated and the virus will become a prophage. As you probably figured out already, the decision to go lysogenic or lytic is going to depend on a series of transcriptional cascades that either induce expression of the integrase or prevent it, respectively.

Remaining latent. A prophage remains a prophage and does not generate virus particles because *most of its genes are silenced*. Silencing is required for maintaining the lysogenic state and is mediated by a special protein, a **repressor** called CI. This protein keeps the viral genes that would coordinate infection from being expressed, *except for the repressor* itself (and a few others). Like many repressors used by bacteria to block gene transcription (chapter 10), the lambda repressor binds to operator sequences in the promoter regions of the viral genes and prevents the RNA polymerase enzyme from transcribing them. The operators happen to flank the CI repressor gene. As more repressor binds the operator regions, they dimerize and loop the prophage, keeping it silent (Fig. 18.8). As long as active repressor is present, the prophage genes needed to excise the phage genome from the host genome and then replicate remain silent. When the repressor is no longer present, the prophage is excised and new phage particles are made. The eventual result is lysis of the host cell and the release of progeny phages.

With some thought, you may recognize that it should be possible to convert lambda from being a temperate virus into a virulent one by mutation. Two classes of mutants do just that: in one, mutations *inactivate the repressor* protein; in the other, they *alter the operator site* so repressor can no longer bind. These lambda mutants are analogous to *constitutive mutants* in the *lac* operon that are always "on" because mutations in the operator regions

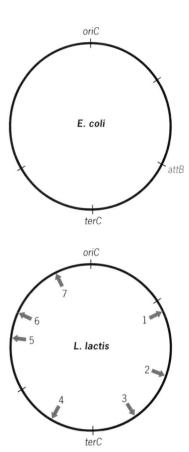

FIGURE 18.7 **Sites of integration of P355 prophages in the *Lactococcus lactis* circular chromosome.** Shown is the minute map of the chromosome and the position of the origin (*oriC*) and termination (*terC*) of replication. P355 prophages integrate at multiple sites in the circular chromosome of *Lactococcus lactis*, an important bacterium in cheese making. (Modified from Kelly W, *Front Microbiol* 30 August 2013, http://dx .doi.org/10.3389/fmicb.2013.00257.)

prevent the repressor from binding (chapter 10). In fact, the two systems of regulation, repression in lambda and regulation of β-galactosidase, were discovered almost at the same time by two interactive groups of scientists working at the Pasteur Institute in Paris during the 1950s.

It is clearly advantageous for the virus to have an optimal level of repressor. Too little, and the phage would replicate and kill the cell. Too much, and viral induction would be difficult to achieve. The "just so" concentration of repressor is maintained because the repressor protein *regulates the expression of its own gene* (Fig. 18.8). Such **autoregulation** is a feedback loop. Or, if the analogy works better for you, a device like a thermostat: too much heat, and the circuit is off; too little, and it is on. In the case of lambda: too much repressor, the transcription of its gene is off; too little, and it is on.

The importance of the repressor is illustrated by two simple experiments:

- If an *E. coli* culture is infected with a small number of virus particles, only a few enter each cell and little repressor is made. Consequently, the phages multiply, and most cells lyse.
- If the culture is infected with a large number of virus particles, many enter each cell, quickly synthesizing a high level of repressor, and most cells become lysogenized.

We introduced these concepts in the chapter 17 case study of human feces contaminating a pond; now we discuss the lytic-versus-lysogenic decision in more detail below.

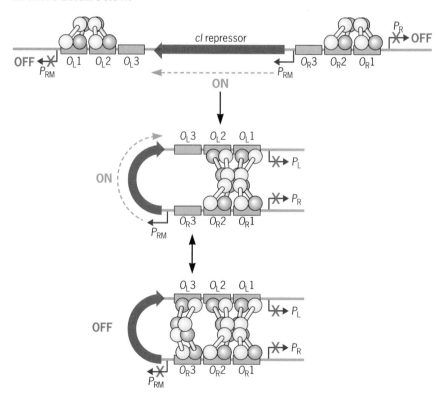

FIGURE 18.8 Stable maintenance of the lambda prophage. The lambda prophage expresses the CI repressor, which dimerizes and binds the left and right operator regions to prevent the transcription of the other viral genes. The repressor dimers on the left and right operators bind to each other, forming tetramers, which loop the DNA and control the expression of the *cI* gene as a function of CI repressor levels.

Lambda Prophage Induction

LEARNING OUTCOMES:
- Describe the series of events that allow deleterious environmental conditions to initiate the lytic cycle.
- List three factors that influence whether lambda will be lytic or lysogenic.
- Explain how the lambda repressor influences the outcomes of infection.
- Predict the outcome of growing *E. coli* in a high-glucose medium if the cells did or did not carry the lambda prophage.

Environmental conditions deleterious to bacteria are cues for the phage to escape. In particular, anything that inactivates the phage repressor induces the prophage and initiates a lytic cycle. One inducing agent (Case 1 in chapter 17) is UV light, which damages the host cell's DNA and provokes a cascade of molecular events. In the process of repairing the damaged DNA, cells accumulate short stretches of single-stranded DNA, which are bound by the recombination protein **RecA**. The ssDNA-RecA complex has the function of *degrading a host repressor* called LexA, which keeps its **SOS response** off (chapter 12). Interestingly, LexA and the lambda repressor CI are structurally similar, and both are recognized and degraded by the ssDNA-RecA nucleoproteins. Hence, as the SOS response is turned on, the ssDNA-RecA also *cleaves the lambda repressor*, thereby reactivating the prophage. Free of CI repressor, the other phage genes are expressed. One of these phage genes encodes an enzyme called **excisionase**, which cooperates with the phage integrase to carry out the reverse genetic crossover required for prophage excision. *Beautiful!*

If no induction took place, could we determine if a prophage is lurking in a bacterial genome? UV induction may not work if, for instance, the prophage is defective in a step in lysis. What else can one do (other than a bioinformatic scanning of the bacterial genome)? One cannot usually just examine a culture or a colony and tell whether the bacteria are lysogenic, but a simple test may work. If after infection of a bacterial cell with the same phage, lysis occurs, the culture is not lysogenic; if it doesn't lyse, the culture is lysogenic. But think about it: you can only do this if you know which phage you are dealing with.

Deciding between lysogeny and lysis. Being an alert reader, you may notice a gap in this story: What determines if a culture will go the lysogenic or the lytic route? Be ready, because this gets a bit tangled. Let us go back to the beginning, the infection of a virgin nonlysogenic culture. We painted the picture of how the viral genome becomes integrated and the cells become lysogenic. The fact is that the virus always goes down the lytic path first, starting to replicate its circular DNA while slowly building up proteins that, if conditions are right, will help the CI repressor get expressed. The actual outcome—lysogenization or lysis—depends on several factors. We have encountered one already, the *number of infecting particles* per cell. But also in play are the *environment and the physiological state* of the host. For example, when an *E. coli* cell grows in a rich nutritional environment (e.g., in the presence of high glucose concentrations), it tends to be lysed upon infection with lambda. In a poor medium, with low glucose concentrations, it tends to become lysogenized. So, how is the decision made to "go lytic" instead of lysogenic? Read on.

What ultimately governs the outcome of infection are properties of the lambda repressor. Lysogeny requires the expression of sufficient quantities of CI to induce the expression of the integrase and silence the other genes. Consider this as an obstacle race, where two helper proteins, CII and CIII, also need to accumulate to help make the repressor (Fig. 18.9). CII induces the expression of the repressor protein CI but is *unstable* because *it can be cleaved by the cell proteases in the cell*. The virus further makes another protein (CIII) to protect CII, but even with this "bodyguard" CII levels are kept low. The repressor CI is also unstable and easily cleaved by the proteases of the host cell. (We have already seen that one such protein, the RecA nucleoprotein, comes into play when the cell's DNA is damaged.) There is more: another protease called Hfl can also cleave the lambda repressor. This protease is *sensitive to nutrient conditions*: its activity is influenced by the cellular level of cyclic AMP (cAMP). From chapter 11, you may remember that the level of cAMP in *E. coli* varies with the amount of glucose in the medium. If the cell has lots of glucose, levels of cAMP are low; the reverse is true for cells in glucose-deficient conditions. The *amount of the Hfl protease is inversely proportional to the concentration of cAMP*, so more protease is made when glucose is abundant than when the sugar is scarce. Consequently, when glucose is plentiful, Hfl more frequently cleaves the lambda repressor, and the lytic cycle is set in motion. Thus, a complex choreography in the host cell dictates whether CI levels rise sufficiently for the phage to take the lysogenic route and, therefore, whether the lambda phage stays or goes.

With that background as a guide, let us return to the test to determine if an *E. coli* culture is lysogenic. Are you ready? Do the experiment in a high-glucose medium, and the level of cAMP in the cells will be low, Hfl protease activity high, and the amount of lambda repressor low. Consequently, the phages would lyse the bacteria. On the other hand, if the recipient cells were already lysogenized, they would contain enough lambda repressor to inhibit the lytic cycle. *Yes!*

The switch between a lytic and a lysogenic cycle is not just a laboratory phenomenon and probably has selective value for virus and host. *E. coli* alternates between life in a nutrition-rich environment (the intestines

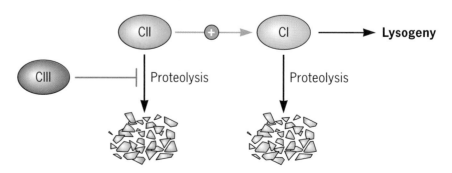

FIGURE 18.9 **Lambda protein encounters the cell's proteases on the way to lysogeny.**

of animals) and in nutritionally poor ones (waters, soils, and sediments). In fact, natural isolates of *E. coli* are often lysogenic for phage lambda or its relatives. It seems plausible that switching between two lifestyles helps both phage and host bacteria survive. In chapter 19, we will learn that this is not unique to the lambda phage. Viral populations in the environment also coordinate their lysogenic and lytic cycles to nutrient availability and the prokaryotic cell population that they infect.

WHAT ARE THE GENETIC AND EVOLUTIONARY CONSEQUENCES OF LYSOGENY?

LEARNING OUTCOMES:
- Compare and contrast generalized transduction from specialized transduction.
- Describe two examples in which lysogeny can alter bacterial host cell function.

Genetic Consequences

Being lysogenic means that the host is carrying extra genes. Obviously, these are viral in origin, but is that all? Surprisingly, prophages may also carry *genes that are not viral*. Where do these come from? Guess that they originated during an earlier infection and you'd be right. How so? In chapter 10, we learned that when a prophage is excised from the host genome, *recombination does not always take place precisely at the ends of the prophage*. Rarely, excision occurs a little to the right or to the left. In some cases, the resulting phage genome includes *neighboring bacterial genes*. A subsequent infection introduces these genes into another cell. The resulting phenomenon, called **specialized transduction**, is discussed in chapter 10. Contrast this with the related but different phenomenon, **generalized transduction**, where a phage capsid picks up random pieces of DNA from its host during the process of viral maturation. The resulting particles, called **transducing particles**, are not true virions as they lack a full phage genome. They are, however, capable of attaching to a new host cell, introducing their DNA, and thus delivering new genes to it.

Acquiring new genes by lysogeny has profound consequences for the physiology of host bacteria and their ability to survive and cause disease. A surprising number of temperate phages carry genes that *contribute to bacterial virulence*. These genes render otherwise innocuous bacteria pathogenic. Examples include the bacteria that cause diphtheria, cholera, scarlet fever, and staphylococcal food poisoning. In many instances, prophage genes code for powerful toxins; in others, they code for new antigenic types. In addition, because these genes are carried on mobile elements (chapter 10), they may be transferred from one bacterium to another and lead to the emergence of new strains (as in epidemic cholera, chapter 26).

Lysogeny can also alter the host genetically even if the prophage carries no bacterial genes. Lambda and many other temperate phages integrate at precise sites on the host chromosome, but other phages are less meticulous. Some of them (e.g., a phage called Mu) can integrate at *random sites on the chromosome*. The gene at the site will be inactivated by this integration, thus producing what we call an **insertion mutation**. Note that this is equivalent both to a retrovirus that generates an oncogene and to the insertion of a transposon (chapter 10). Indeed, Mu has been called "a transposon masquerading as a phage."

Evolutionary Consequences

Phages have strongly affected bacterial evolution, possibly on a large scale. It is estimated that at least 50% of all phages are temperate and thus capable of lysogenizing their hosts. Of course, virulent phages also affect evolution, especially by pressuring the host to evolve phage-resistant survivors. In chapter 10, we learned about the CRISPR strategy of capturing a tiny piece of the infective phage in the chromosomal DNA to make antisense RNAs ready to target a new infective DNA and degrade it. The cell therefore gets immunized against new infections. But temperate phages play an even more important role beyond mutation and selection in shaping their hosts' evolution. Their capacity to transfer *whole packets of genes simultaneously* leads to genetic changes that may be more pervasive than single gene mutations. To expand the picture even further, genomic analysis tells us that many bacterial genomes contain sequences characteristic of phages but not the complete phage genomes. These genetic remnants indicate that a lysogenic phage was integrated in the past and subsequently lost some of its genes; thus, the count of viable prophages underestimates the frequency of lysogeny. The proportion of bacteria found to be lysogenic varies greatly between habitats, but at least some lysogens can be found in practically all of them, terrestrial and aquatic. These facts alone strongly suggest that lysogeny played an important role in bacterial evolution and continues to do so to this day.

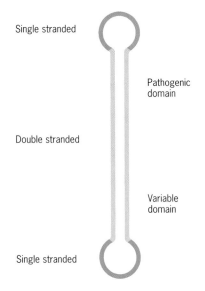

Single stranded

Pathogenic domain

Double stranded

Variable domain

Single stranded

FIGURE 18.10 Structure of a typical viroid.

WHO ARE THE VIROIDS?

LEARNING OUTCOMES:
- Compare and contrast the structure of viroids with that of viruses.
- Describe the viroid replication cycle.
- Describe the replication cycle of hepatitis D.

We now enter a world of virus-like entities that should continue to defy your imagination. As we have seen, viruses are varied enough, but in addition, there are elements that do not fit the classical definition. It is as if evolution has been particularly playful here. We are talking about the **viroids**. These are minimalist entities that will require you to consider even more new information. That's biology for you—alternatives, sometimes contradictory ones, may be tiresomely frequent but at least they are often fascinating!

Viroids are an alternative to "regular" viruses that cause diseases mainly in plants. Viroids are simply *small circular RNA molecules* curled upon themselves to create an extensive *double-stranded segment* (Fig. 18.10).

Hence, they are *not enclosed in a protein shell*. Moreover, they *do not encode even a single protein*. Their impressive double-strandedness makes viroids so resistant to destruction by ribonucleases that they don't need to be protected by a capsid! Viroids cause diseases with curious names, such as avocado sunblotch, peach latent mosaic, and coconut cadang-cadang. As their names may suggest, some of these diseases cause serious economic problems. In fact, viroid diseases have seriously threatened coconut cultivation in the Philippines and chrysanthemum growing in the United States.

How do viroids cause disease? The answer isn't yet clear, but one possibility that has gained favor involves **RNA silencing**. Some small RNAs of viroids may target host mRNA or DNA to alter the host's gene expression. Interestingly, one of their striking characteristics is their small size. The spindle tuber viroid that infects potatoes contains only 359 nucleotides! Yet, these small circles of RNA elicit diseases reminiscent of those caused by viruses many times their size.

How do viroids replicate? Since they have dispensed with protective capsid proteins, they are not burdened to carry genes to encode them. However, they do face the same problem as the RNA viruses: how to make RNA from an RNA template (chapter 17). A second challenge is how to circularize their genome. The way viroids accomplish these tasks is quite intricate. A host RNA polymerase that can *make RNA using an RNA template* replicates the RNA of infecting viroids. Such enzymes are *common in plants but have not been identified in animals or bacteria*. Replication is carried out from a circular RNA template; the polymerase makes complementary copies of it by going around and around the circle. Remember, this so-called **rolling-circle replication** mechanism is used by certain DNA viruses and plasmids (chapters 10 and 17). The result is not a single copy of the template but a long string of copies repeated in tandem, like a roll of toilet paper. To make viroids, this long molecule must be **cleaved** into segments of the proper size. This trimming can be done in one of two ways. In some viroids, the RNA itself has enzymatic activity; this **ribozyme** cuts the long RNA string into single copies. For other viroids, cleavage is done by a host **endonuclease**. In both cases, the resulting pieces of RNA must be ligated into circles that then assume the viroid's molecular structure.

The only known viroid-like human agent is called (wrongly) hepatitis D *virus*; it is actually something between a virus and a viroid. Although it also consists of RNA, the hepatitis D "satellite virus" differs from plant viroids in that it does assemble a capsid. This agent also encodes proteins, but, oddly perhaps, not those needed to make a capsid. How does it make its covering? Actually, it doesn't. Instead, it parasitizes another virus to hijack *its* capsid, sort of like a hermit crab using an empty shell made by a sea snail. The capsid it seizes is the one made by the hepatitis B virus. Some of the hepatitis D-encoded proteins appear to help usher the RNA into the borrowed capsid. The disease hepatitis D therefore occurs only if the person is also simultaneously infected with the hepatitis B virus. This "two-for-the-price-of-one" infection results in a more severe disease than hepatitis B virus infection alone.

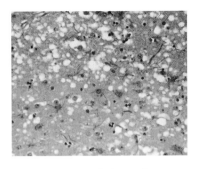

FIGURE 18.11 Pathology of prion disease. A section of a brain with spongiform encephalopathy revealing numerous small cavities in the tissue (light areas). (Courtesy Al Jenny, Public Health Image Library, APHIS.)

"Normal"

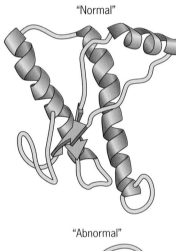

"Abnormal"

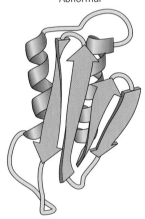

FIGURE 18.12 A prion in its "normal" and "abnormal" state. The structure of the prion protein in its "normal" and "abnormal" state. The normal protein is rich in alpha-helices (red) while the abnormal one has many beta-sheets (blue).

WHAT ARE THE PRIONS?

LEARNING OUTCOMES:
* Compare and contrast prions and viruses.
* Describe how prions differ from their "normal" cell precursors.

Certain infectious diseases cause degenerative damage to tissues. Those that afflict the central nervous system are particularly serious, often having lethal consequences for humans and animals. Perhaps the most notorious of these conditions is mad cow disease; others with similar manifestations are Creutzfeldt-Jakob disease in humans and scrapie in sheep. Collectively, these diseases are called **spongiform encephalopathies** because they generate holes in the brain, making it appear sponge-like (Fig. 18.11). An important clue to the infectious nature of one of these human diseases, **kuru**, was realizing that the affected people had consumed human brains as part of their funeral rituals. Transmission of some of these diseases was confirmed using laboratory animals. Researchers suspected that the agent might be a virus, but they were unable to find any viruses, bacteria, or any other agents from known groups. In time, it was established that the agent, then named a "**prion**," was highly unusual in that it did not contain nucleic acids; *it was composed entirely of a single protein* (Fig. 18.12). Such a claim was startling because all other known infectious agents, indeed all other self-replicating entities, possess nucleic acid-based genomes. Now for a confession: The topic of prions does not belong in a chapter on viruses but, frankly, we could not think of a better place to put it.

The mystery of spongiform encephalopathies was solved in part when it was discovered that prion proteins have unusual properties. Like other proteins, they can fold into two different three-dimensional configurations (Fig. 18.12). In one configuration, they are normal constituents of the cells of the central nervous system and play important roles in cell biology. In another configuration, they gain the unusual ability to *act as templates* that induce their normally folded counterparts to *refold into prions*. Prions and their normal precursors have the same amino acid sequence and are encoded by the same genes: *they differ only in how they are folded*. The precursor proteins are rich in α-helices and have little in the way of β-sheets; prions are the converse. When misfolded into prions, the proteins become highly resistant to proteases, harsh chemicals, and high temperatures. Prion proteins also aggregate into waxy translucent fibrils called **amyloids**. Disinfecting prion-containing material requires extreme measures. For example, contaminated objects must be soaked for 1 hour in 1 N sodium hydroxide to destroy prion proteins.

How do prions impose their alternative folding on normal molecules? It is not known whether prions act on proteins that are still in the process of folding or alter those that have already folded into their natural configuration. The function of the normal form of the prion protein is also not entirely known. Mutant mice that are defective in this protein do not display symptoms. One can speculate that if prions affected some essential protein, the manifestations of disease would be more immediate.

Prion diseases are slow in developing. When enough prions accumulate in the brain and spread to adjacent cells, brain function is impaired. In cows, "mad" refers to insanity, not to anger. In sheep, the disease is called *scrapie* because sick animals rub themselves intensely against any surface, scraping off their wool and skin. As clusters of brain cells die, the diseased tissue resembles a Swiss cheese full of holes (Fig. 18.11). Some human neurological degenerative diseases, including Alzheimer's, Huntington's, and Parkinson's, involve protein misfolding, although they do not seem to be transmitted between individuals and thus are not considered infectious. It is important to note that prions are not limited to vertebrates but have also been found in fungi such as yeast. Here, they do not cause "mad yeast disease," but just what they do is not well established. Both beneficial and pathogenic effects have been proposed, so the matter needs further resolution.

In general terms, the activity of prions falls into the realm of **epigenetics**, heritable changes that *do not result from altered genome sequences*. The two forms of the prion proteins are encoded by the same sequence; thus their virulence does not depend on genetic changes. Examples of epigenetic inheritance in complex organisms are commonplace—think of the ability of, say, liver cells to reproduce their own kind even though they have the same genome as all other cells. In some instances, modifications of DNA by methylation account for the change in the pattern of gene expression that characterizes a given cell type (as in herpesvirus infection, chapter 25). Prions may be involved in other epigenetic phenomena, such as long-term memory or even somatic cell differentiation. In unicellular organisms, epigenetic mechanisms appear to be more limited. However, examples emerge, such as yeast proteins that act like prions. It remains to be seen how widespread the prion mechanism is in other organisms.

CASE REVISITED:　A case of shingles

Mrs. K suffered from a condition common in the elderly, caused by the reactivation of the chickenpox or VZV that she had acquired in childhood. The original infection stimulated an immune response to the virus, but this defense waned with time. This virus has a remarkable capacity for surviving in neurons without causing damage or eliciting an immune response. In the unlucky persons who develop shingles, the virus has become reactivated, usually for unknown reasons. However, the weakening in the immune response associated with age may have contributed to Mrs. K's condition. Note that the same virus can cause two distinct conditions, chickenpox, which is usually mild, and shingles, which can be extraordinarily painful.

Chickenpox affects almost everybody who is not vaccinated; over a lifetime, the odds of coming down with shingles are about 30%. Shingles involves peripheral nerves, but it is not known to cause cancer, which distinguishes it from other chronic viral diseases such as infection by some retroviruses and papillomaviruses, among others.

Why do you think the doctor advised Mrs. K to get vaccinated? The thought is that the vaccine would evoke a stronger immune response than did her infection. Indeed, the vaccine, which contains an attenuated virus, has proved to reduce the cases of shingles by about half and to lessen the severity of illness in the others. Given the pain caused by shingles, vaccination is used widely. Meanwhile, new vaccines are being developed, in part because the live virus, even though attenuated, may cause severe disease in immunocompromised persons.

CONCLUSIONS

Latent viruses are major players of the virus world. They contribute to evolution, health, and disease and are encountered frequently in both animal hosts and bacteria. Whether in integrated form or as independent extrachromosomal elements (episomes or plasmids), viruses can remain latent for long periods of time until reactivated by environmental and cellular factors. Viruses, together with the minimalistic viroids and prions, reveal the great plasticity of the biological world, extending our basic concepts from the cell as the basis for life to chemical messengers of genetic information. We learn from these entities that certain phenomena central to our concept of what is living, such as replication and mutation, are not the province of cells alone.

This wonderland invites further speculation. How did viruses arise? Are they "reduced cells," the result of loss of properties required for independent existence, or did they evolve as novel entities? Or did they follow other evolutionary pathways? What are we to make of their immense variety, numbering in the hundreds even for a single host? Unlike bacteria, which evolved independently, viruses did so in association with their hosts. And, on another topic, are prions more widespread than presently known, and do they function in basic biological phenomena, such as differentiation? These and many other questions are challenging topics and fertile areas for further investigation.

Representative ASM Fundamental Statements and Learning Outcomes for the Chapter

EVOLUTION

1. Mutations and horizontal gene transfer, with the immense variety of microenvironments, have selected for a huge diversity of microorganisms.

   a. Compare and contrast generalized transduction from specialized transduction.

CELL STRUCTURE AND FUNCTION

2. The replication cycles of viruses (lytic and lysogenic) differ among viruses and are determined by their unique structures and genomes.

   a. Describe two ways that viruses establish latency.
   b. Describe how latent viruses can evade the cell-mediated immune response.
   c. Differentiate temperate or lysogenic phage from prophages.
   d. Distinguish between virulent and temperate viruses.
   e. Define viral induction and give an example of how prophages are induced.
   f. Outline the integration cycle of lambda.
   g. Describe the series of events that allow deleterious environmental conditions to initiate the lytic cycle.
   h. List three factors that influence whether lambda will be lytic or lysogenic.
   i. Describe two examples in which lysogeny can alter bacterial host cell function.
   j. Compare and contrast the structure of viroids with that of viruses.

INFORMATION FLOW AND GENETICS

3. The synthesis of viral genetic material and proteins is dependent on host cells.

    a. Describe how latent viruses can influence oncogenes.

    b. Define the functions of DNA ligase, integrase, and repressor protein, and explain the role of each in lambda's integration cycle.

    c. Differentiate between site-specific and homologous recombination.

    d. Describe two classes of mutations that can convert lambda from a temperate phage to a virulent one.

    e. List three factors that influence whether lambda will be lytic or lysogenic.

    f. Predict the outcome of growing *E. coli* in a high-glucose medium if the cells did or did not carry the lambda prophage.

MICROBIAL SYSTEMS

4. Microorganisms, cellular and viral, can interact with both human and nonhuman hosts in beneficial, neutral, or detrimental ways.

    a. Explain the differences between true latency and clinical latency.

    b. List three characteristics of host-viral interactions that lead to latency.

    c. Explain the role of reactivation in disease progression.

    d. List two ways that viruses help human hosts.

    e. Differentiate animal host cell transformation from bacterial cell transformation.

    f. Describe the viroid replication cycle.

    g. Describe the replication cycle of hepatitis D.

    h. Compare and contrast prions and viruses.

    i. Describe how prions differ from their "normal" cell precursors.

SUPPLEMENTAL MATERIAL

References

- **Villarreal LP.** 2005. *Viruses and the Evolution of Life.* ASM Press, Washington, DC.
- **Buzdin A.** 2007. Human-specificendogenous retroviruses. *Sci World J* **7:**1848–1868.
- **Traylen CM, Patel HR, Fondaw W, Mahatme S, Williams JF, Walker LR, Dyson OF, Arce S, Akula SM.** 2011. Virus reactivation: a panoramic view in human infections. *Future Virol* **6:**451–463.
- **Virgin HW.** 2014. The virome in mammalian physiology and disease. *Cell* **157:**142–150.

Supplemental Activities

CHECK YOUR UNDERSTANDING

1. **A clinically latent virus is NOT:**
   a. actively replicating
   b. detectable by PCR
   c. causing symptoms of disease
   d. cleared by the host immune system

2. **Episomes, proviruses, and prophages are three viral forms that contribute to:**
   a. infection
   b. replication
   c. transmission
   d. latency

3. **Viral nucleocapsids acquire envelopes during:**
   a. entry
   b. replication
   c. assembly
   d. release
   e. all of the above

DIG DEEPER

4. **Team work:** Human papillomaviruses (HPV) have been implicated in a form of cervical cancer. Working in groups, research HPV to answer the following questions:

   a. What human cells does HPV typically infect?
   b. Does HPV replicate within cells through a lytic or lysogenic replication cycle?
   c. HPV encodes oncogenes that are associated with cervical cancer. What impact do these oncogenes have on host cells?
   d. Are HPV oncogenes activated during a lytic or lysogenic replication cycle?
   e. How do oncogenes increase the fitness of HPV?

5. Read about antiretroviral therapy (ART) for HIV treatment on the AIDS.gov website: https://www.aids.gov/hiv-aids-basics/just-diagnosed-with-hiv-aids/treatment-options /overview-of-hiv-treatments/. HIV is treated using a combination of drugs, which can be grouped into six classes:

   - Non-nucleoside reverse transcriptase inhibitors (NNRTIs)
   - Nucleoside reverse transcriptase inhibitors (NRTIs)
   - Protease inhibitors (PIs)
   - Fusion inhibitors
   - CCR5 antagonists (CCR5s)

   Working in groups, research one of the drug classes to answer the following questions:

   a. What cellular function does this drug class inhibit?
   b. Does this drug class directly inhibit HIV replication?
   c. Why is this single drug class not sufficient to treat HIV?

6. **Team work:** Working in groups, select a disease caused by a virus, using Table 18.1 or the literature. Research the strategies the virus uses to cause the disease. Develop a way to teach your findings to the general public using visual aids, such as schematics, animations, or theatre. Include an annotated bibliography to guide audience members who want to learn more.

7. Scientists have enlisted the lamda bacteriophage to generate mutations in bacteria and yeast using only PCR products, saving the cloning steps.

   a. Use the PubMed search engine to locate articles and to learn about this technology.
   b. Draw a schematic to illustrate how lambda phage can be used to delete genomic sequences of interest. Cite the articles you used to generate your diagram.
   c. How can this technology advance scientific research?
   d. How can it advance agriculture, technology, or human health?

8. **Class debate:** There is public debate about the use of the human papillomavirus vaccine.
   a. Working in groups, research

   - HPV virus
   - Human diseases caused by HPV
   - Epidemiology of HPV infections
   - How the vaccine works
   - How the vaccine program is being implemented

   b. In class, organize a mock hearing on whether a local public school system should require vaccination. Constituencies to represent may include:

   - School nurses
   - Local pediatricians
   - Patients who became ill with HPV
   - Parents who are in favor of vaccinating their children with HPV vaccine
   - Parents who oppose the HPV vaccine
   - Parents who oppose all vaccines

   c. As a class, develop recommendations for a national policy on HPV vaccination.

9. A mixture of bacteria and their phages was placed on the surface of an agar plate and incubated overnight. The size of the plaques resulting from phage infection was quite uniform, but some were clear while others were turbid. (Figure)

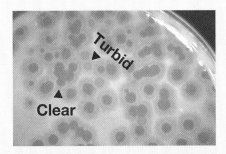

SUPPLEMENTAL QUESTION 9 A phage infection that generates clear and turbid plaques.

- What would cause a turbid plaque?
- What is the significance of the observation that the turbid plaques are similar in size?
- Generate at least two possible genetic mechanisms.

10. Identifying cells with latent viral infection is challenging even in tissue culture models, making it difficult to study the mechanisms of latency. Read the report of a model system for identifying latent and actively infected cells using flow cytometry by Dahabieh et al. 2013 (A doubly fluorescent HIV-1 reporter shows that the majority of integrated HIV-1 is latent shortly after infection. *J Virol* **87**:4716-4727, 2013, http://www.ncbi.nlm.nih .gov/pmc/articles/PMC3624398/). Answer the following questions:

    a. After infection of cultured cells with the HIV-1 vector RGH, how do the scientists identify uninfected, latently infected, and actively infected cells? Use Fig. 1 as a guide (below).
    b. Why did the scientists use a very low multiplicity of infection (MOI=0.2)?
    c. Below is a table of the percentages of different fluorescent-positive cells at different days postinfection of Jurkat cells with the HIV-1 vector RGH. Using Fig. 1B (below) as a guide, plot a graph and use it to answer the following questions:
        i. Were a majority of the infected cells in a latent or active state early in the infection period (day 1-2)?
        ii. Late in infection (day 3-7)?
    d. The incubation period was extended. At day 10 postinfection the cultures were treated with HAART drugs, and the data tabulated below were obtained. Interpret the data and draw conclusions why the particular pattern is seen.
    e. Why were "Green only" cells seen only at day 1 postinfection but not later? Do you expect to see any "Green only" cells at all?

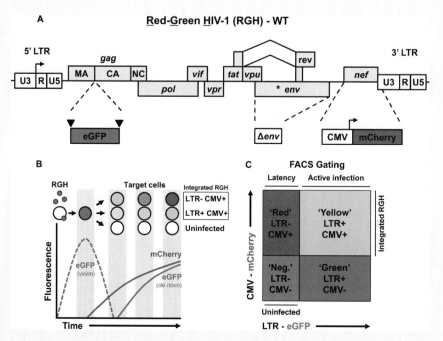

SUPPLEMENTAL QUESTION 10 Schematic representation of the Red-Green-HIV-1 (RGH) vector. (A) HIV-1 B-LAI Δenv is labeled with eGFP as an in-frame *gag* fusion flanked by HIV-1 protease cleavage sites (inverted triangles), and by a CMV$_{IE}$-driven mCherry cassette located in place of *nef*. WT, wild type. (B) Schematic depiction of RGH infection of target cells and the resultant fluorescent protein profiles over time. (C) Schematic depiction of HIV-1 RGH-infected cell populations detected by FACS analysis.

Day postinfection	% Green only cells	% Red only cells	% Yellow cells (Red+Green expression)
1	12	0.7	0.6
2	0.06	5.1	0.8
3	0.05	4.0	1.0
4	0.06	3.7	1.8
5	0.08	3.4	2
6	0.04	2.3	1.7
7	0.05	2.2	1.8
10	0.03	2.0	1.7
15	0.05	1.8	1.2
20	0.02	1.6	0.6
40	0.03	1.8	0.5
60	0.03	1.7	0.4

Supplemental Resources

VISUAL

- Lysogeny animation by the Harvard College Interactive Molecular Biology program. http://sites.fas.harvard.edu/~biotext/animations/lysogeny.html
- HIV life cycle animation with introduction by Ann Brokaw for the Howard Hughes Medical Institute BioInteractive program: http://www.hhmi.org/biointeractive/hiv-life-cycle
- *iBiology* lecture on prions by Susan Lindquist (46 min): http://www.ibiology.org/ibioseminars /cell-biology/susan-lindquist-part-1.html
- Virology lecture series by Vincent Racaniello: http://www.virology.ws/course/

SPOKEN

- *This Week in Virology* podcast episode #275 "Virocentricity" discusses the role of viruses in evolution with Eugene Koonin. http://www.twiv.tv/2014/03/09/twiv-275-virocentricity-with -eugene-koonin/

PART III

Microbial Ecology

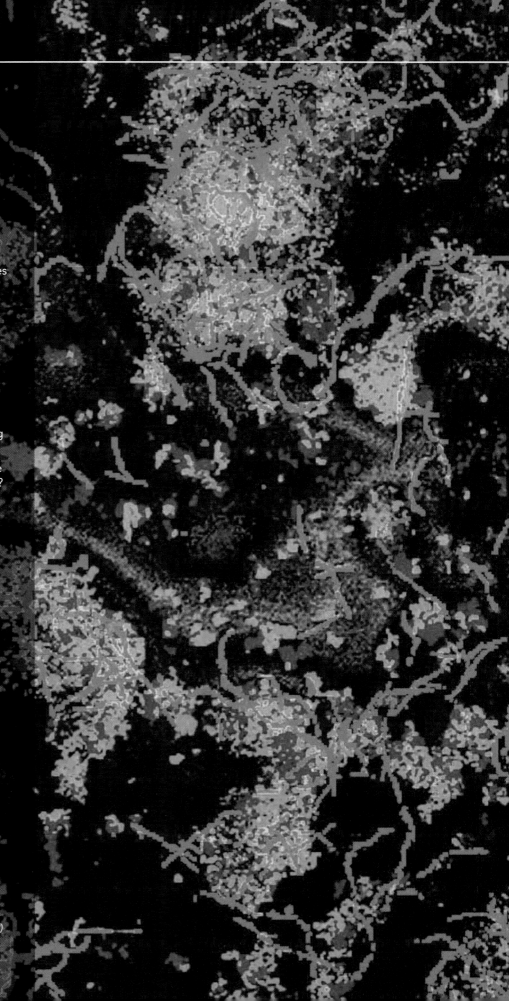

IN THIS CHAPTER

Confocal micrograph of a three-species biofilm community of the human oral bacteria Streptococcus oralis *(green),* Veillonella *sp. (blue), and* Fusobacterium nucleatum *(red)*

CHAPTER NINETEEN

Microbial Communities

KEY CONCEPTS

This chapter covers the following topics from the ASM Fundamental Statements.

EVOLUTION

1. The traditional concept of species is not readily applicable to microbes due to asexual reproduction and the frequent occurrence of horizontal gene transfer.

METABOLIC PATHWAYS

2. *Bacteria and Archaea* exhibit extensive, and often unique, metabolic diversity.
3. The survival and growth of any microorganism in a given environment depend on its metabolic characteristics.

MICROBIAL SYSTEMS

4. Microorganisms are ubiquitous and live in diverse and dynamic ecosystems.
5. Microorganisms and their environment interact and modify each other.

IMPACT OF MICROORGANISMS

6. Microbes are essential for life as we know it and the processes that support life (e.g., biogeochemical cycles and plant and/or animal microbiota).

INTRODUCTION

This chapter starts a new section of the book that focuses on microbial ecology. Ecology is the study of interactions between organisms and their environment—how the chemical, physical, and biological environments affect particular kinds of organisms and how the organisms affect their environments. You might ask, haven't we have been discussing this topic throughout the book? True enough, but the next three chapters will put greater emphasis on the grander scheme of the activities of groups of microbes rather than on those of individuals. We will pay particular attention to how microbes have molded Earth's environment and how they maintain our planet in a balanced, life-sustaining condition. But first we need to know how we study microbes in the environment and what these studies tell us about who is out there and what they are doing.

CASE: What makes sauerkraut so funky?

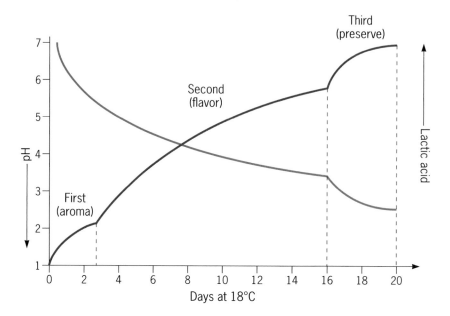

Consider sauerkraut. To make it, dice the cabbage, add salt, shield it from the air in a sealed jar, and presto, a few weeks later, the final product is ready for consumption. Adding salt to the cabbage cells releases a large amount of soluble substances—sugars, amino acids, and vitamins—that jump-start microbial growth in the jar. But here is the amazing part: The process occurs every time with nearly unfailing precision regardless of the cabbage used, who makes it, or where they make it. How is that possible? Sauerkraut always develops in three stages, as indicated on the graph above. Note how the concentration of lactic acid and the pH in the sauerkraut vessel fluctuate in a stepwise manner over the first 20 days after the salt is added to the cabbage and the jar is sealed.

- Why are there three distinct stages?
- Is the same microbe responsible for the whole process?
- Can you form a hypothesis about why cabbage tastes different after this 20-day treatment?
- Where do the microbes that make sauerkraut come from?

WHO IS THERE AND WHAT ARE THEY DOING?

LEARNING OUTCOMES:
- Describe two ways that *Bacteria* and *Archaea* use and survive on low-energy-yielding biochemical reactions.
- Support the following statement using specific examples: "Microbes are ubiquitous and live in diverse and dynamic ecosystems."

The fundamental themes of microbial ecology rest on three characteristics of microbes: (i) *ubiquity*—they are present just about everywhere that liquid water exists; (ii) *abundance*—they occur in huge numbers; and (iii) *metabolic power*—they are extremely active, and in astoundingly diverse ways. Plus, *they are seldom alone but frequently in communities.* Through extensive interactions, communities of microbes perform functions that individual cells could never achieve. In the case study we considered a relatively simple environment, a vessel filled with cabbage and salt and sealed to shield it from air. Yet the stepwise accumulation of lactate and concomitant reduction of the pH already hint at three microbial activities acting sequentially to convert cabbage into sauerkraut. Try to picture now the staggering number of microbes that would be needed to recycle the many substrates available in natural environments. How many microbes are there? A 2010 global census of the most abundant marine microbes detected a stupendous *phylogenetic diversity* of prokaryotic ribosomal RNA (16S rRNA) sequences, particularly bacterial, compared to eukaryotic 18S rRNA sequences (Fig. 19.1). Indeed, a single liter of seawater contains on average about 1 billion bacteria representing about 20,000 different "species." (Recall from chapter 14 that the definition of microbial "species" remains a thorny topic; a bacterial "species" may, in fact, contain many strains with substantial genomic and metabolic differences.) And this is just in seawater. Estimates are that there are over 10^{30} *Bacteria* and *Archaea* on Earth. They are found everywhere there is water, even in environments under the most extreme conditions of temperature, hydrostatic pressure, and salt concentration (chapter 4). And many microbes are associated with some of our body parts. Our intestines alone house more bacterial cells than there are human cells in our bodies. We are also outnumbered at the global scale: a cubic milliliter of the contents of our large intestine or about 10 g of soil contains more microbes than the total number of humans that have ever lived on Earth!

Not surprisingly, the metabolic potential of such a diverse array of microbes is immense and contributes to the ubiquity of microbial life. If the components of a thermodynamically feasible reaction exist in a particular environment, a microbe will almost certainly exploit it to harness energy to grow and reproduce. Of course, there must be a lower limit to how much energy a reaction must yield for a microbe to use it successfully. (Recall that we measure energy yield as the change in Gibbs free energy, or ΔG, chapter 6). But it is remarkable how low this value can be: prokaryotes seem able to squeeze every bit of energy out of chemical compounds. They do so in one of two ways:

1. Although small bits of energy released in low-yield reactions may not be enough to make ATP, they can be used to pump protons across the

Bacteria

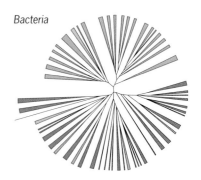

Archaea

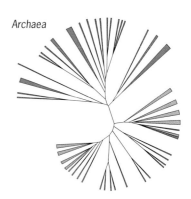

Eukarya

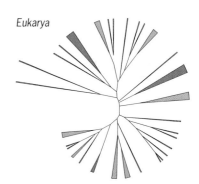

FIGURE 19.1 Census of microbial diversity in the oceans illustrates the outstanding diversity of prokaryotic (bacterial and archaeal) groups compared to eukaryotes. Radial dendrograms of the most abundant bacterial, archaeal, and eukaryotic sequences (small ribosomal subunits) obtained in environmental surveys that studied the global distribution of marine microbes in the planet's oceans. Color branches indicate the location of the samples (benthic in brown, water-column in blue, and sponge- or coral-associated in orange). (Modified from http://comlmaps.org/mcintyre/ch12.)

cell membrane. Eventually, the proton gradient so generated provides sufficient energy to make ATP. This strategy of **transmembrane ion gradients** allows many microbes to harvest energy during respiration and photosynthesis (chapter 6).

2. Microbes team up to increase the amount of energy that can be squeezed out of a reaction. Often times, one microbe catalyzes one reaction and another eats its reaction product to prevent **feedback inhibition** (chapter 11) of its partner's metabolism. Such scavenging shifts the reaction's equilibrium and renders it able to support growth. For example, the anaerobic degradation of hydrocarbons generates butyrate, which some bacteria use as carbon source. If the butyrate were to accumulate, hydrocarbon degradation would stop. Interactions in which two microbes *feed together* are called **syntrophy** and will be studied in more detail in chapter 21.

To get a glimpse of the vast variety of microbes that exist in nature, their diverse forms of metabolism, and their ability to grow almost anywhere, let's consider the reactions that take place just below ground in an open field (Fig. 19.2). Residing in the top, well-aerated layers are the type of **chemoheterotrophs** that utilize organic compounds as carbon and energy sources for **aerobic respiration**. Living in the small anaerobic pockets that exist within this otherwise aerobic region are other chemoheterotrophs that *ferment* carbohydrates. As we go deeper and enter the realm of shallow groundwater, a largely anoxic zone, we find microbes that live by various forms of **anaerobic respiration**. These microbes respire a variety of terminal electron acceptors, such as nitrate, ferric iron, and sulfate often while oxidizing organic compounds, such as acetate, that trickle down from the fermenters living in the soil above. They can also oxidize

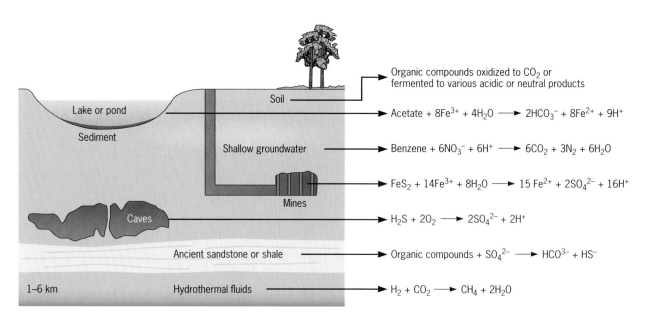

FIGURE 19.2 Metabolic conversions performed by microbes encountered at various depths in the Earth. As we move deeper into the ground, from the aerated top layer of the soils to the ancient sandstones (shales) and hydrothermal fluids, organic matter is degraded by microbes to release gases (e.g., CO_2 and CH_4) to the atmosphere. And these are just but a few examples of the myriad metabolic reactions that microbes carry out in the planet.

pollutants, such as benzene, trichloroethylene, and various pesticides. We also find **autotrophs** that derive energy from anaerobic respiration to fix carbon dioxide (CO_2).

Descending further into the ground, sometimes to several kilometers deep, you might be surprised to continue to encounter microbial communities (Fig. 19.2). Indeed, microbes can be recovered from the salty water that leaks into deep mines from deep regions of the Earth's crust. How can microbes grow and reproduce at such great depths? These extremely hot environments are home to microbes classified as **hyperthermophiles**. Many of them live at the expense of hydrogen gas (H_2) formed in the Earth's magma by the geochemical reduction of water in that high-temperature, high-pressure environment. Microbes couple the oxidization of hydrogen to the reduction of metal ions or carbon dioxide. **Methanogens**, for example, use hydrogen to reduce carbon dioxide and produce methane (CH_4). Some of the methane is oxidized by other resident microbes, and the rest rises to upper regions of the soil to feed specialized microbes called **methylotrophs**. Any remaining methane is released to the atmosphere. As you can see, the microbial diet is extraordinarily diverse—almost nothing goes to waste.

The effect of the collective activities of microbial communities is staggering. Microbes recycle carbon, nitrogen, sulfur, and phosphorus compounds, among others, and release to the atmosphere life-sustaining gases such as oxygen, nitrogen, and carbon dioxide (chapter 20). They transform minerals in aquatic and terrestrial environments, catalyzing redox reactions that contribute to the formation of metal ores, which we humans then exploit to sustain our industrial activities. Being abundant, ubiquitous, and metabolically diverse clearly pays off. Though these traits also create challenges for microbial scientists, methods have been developed that allow us to pursue two broad questions: who is there, and what are they doing? We may not have exact answers yet, but what we do know highlights just how creative and influential microbial life is.

HOW DO WE KNOW WHO IS THERE?

LEARNING OUTCOMES:
- Explain the "Great Plate Count Anomaly."
- Discriminate between enrichment cultures and pure cultures.
- Identify benefits and drawbacks of using enrichment cultures.
- Summarize three reasons so many environmental microbes have eluded cultivation in the laboratory.
- Identify benefits and drawbacks of using microbial pure cultures to understand the microbial ecology of natural environments.
- Describe the process of amplicon-based community fingerprinting.
- Differentiate the terms species, operational taxonomic unit, and phylotype.
- Identify benefits and drawbacks of fluorescence *in situ* hybridization (FISH).

The aim of all ecology is to determine which organisms are present in a particular environment and what they are doing there. For higher organisms, considerable ecological information can be gained by visual observation; not so for microbes—although considerable progress is being made. Unlike higher organisms, morphology reveals little about the identity of a microbe. Furthermore, because microbial activities are largely chemical, only rarely can we see what microbes are doing. In the next section we will learn about the classical and state-of-the-art approaches microbial ecologists use to characterize these largely invisible but no less fascinating communities.

Cultivation-Dependent Methods

For over a century, microbiologists have relied on **cultivation** to identify the microbes in a particular environment. The idea is to mimic in the lab the microbes' natural habitat. If the environment is rich in plant matter, for example, scientists would set up cultures with plant-derived substrates such as cellulose to selectively grow microbes with the metabolic capacity to utilize it. Incubating cultures aerobically or anaerobically or at different temperatures would select for *distinct groups* of cellulose degraders. The pH and the nutritional composition of the medium could also be modified to better match the environmental conditions and promote the growth of more microbes. But even with careful attention to the cultivation parameters, only a fraction of the environmental microbes can be recovered in pure culture. However, ingenious but painstaking methods designed to mimic natural environments, such as slowly feeding nutrients at minuscule concentrations or incubating cultures for prolonged periods, are steadily decreasing the gap between the numbers of microbes observed and the numbers cultivated.

Enrichment cultures. Cultivation techniques routinely rely on an enrichment step, which, as the name suggests, selectively favors the growth of the desired microbes by adjusting the conditions of culture. For example, to enrich for anaerobic photosynthesizers in the phylum *Chloroflexi*, you will need to add a small amount of soil to an oxygen-free medium supplemented with sulfur compounds that these organisms use as electron donor for photosynthesis (Fig. 19.3). You will also need to fill the headspace of the culture vessel with carbon dioxide gas to provide the carbon source (and remove any oxygen) and incubate the cultures in the presence of light. As a consequence, the culture becomes enriched in these microbes. The same principle can be applied to enrich for other classes of microbes (Table 19.1).

Enrichment cultures are a powerful tool for studying microbial ecology. If one suspects that a particular kind of microbe is carrying out a particular transformation in a particular environment, a first step is to inoculate an appropriate medium with material from that environment. If the suspected microbe becomes enriched, it most assuredly is present in that environment. This approach is the reason *we know more about aerobic chemoheterotrophs* than most other microbes: these microbes are easiest to cultivate. We are also motivated to study them because most known pathogenic microbes fall into this class. Some aerobic chemoheterotrophs do play important roles in the environment, but this group probably has a *smaller ecological impact than do anaerobes and autotrophs.*

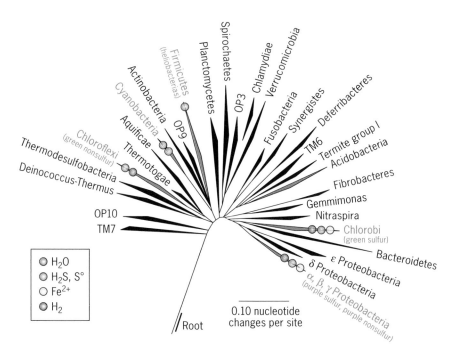

FIGURE 19.3 Distribution of photosynthetic bacteria among bacterial phyla. Bacterial 16S rRNA gene phylogenetic tree, showing phyla with photosynthetic representatives highlighted in green (groups within the phylum are indicated in parentheses). The electron donors used for photosynthesis in each phylum are indicated by colored dots. (Modified from Macalady JL et al., *Philos Trans R Soc B* 2013, http://rstb.royalsocietypublishing.org/content/368/1622/20120383.)

This example highlights the experimental biases that enrichment approaches often introduce. First, microbes that are easier to cultivate tend to be enriched more often. Second, it is not always possible to culture the microbes that exert a particular ecological impact. Finally, as we will see throughout this chapter, it is naïve to assume that microbes act in nature as they do in culture.

Is it them or us? Why is it difficult to cultivate many of the microbes that flourish in nature? *Interdependent nutritional consortia* of prokaryotes are

TABLE 19.1 **Some examples of enrichment culture**

Microbial class	Critical cultural condition	Rationale
Thermophiles	Incubate at a temperature in the thermophilic range, e.g., 55°C.	Only thermophiles can grow at such temperatures.
Endospore formers	Boil soil inoculum before adding to culture medium to kill vegetative cells.	Very few vegetative microbial cells can withstand being boiled; endospores can.
Nitrogen-fixing cyanobacteria	Incubate aerobically in the light in a mineral salts medium lacking fixed nitrogen.	Only certain cyanobacteria can grow aerobically and phototrophically and fix nitrogen.
Sulfate-reducing bacteria	Incubate in the dark, anaerobically, with a nonfermentable carbon source and sulfate ion.	Under such conditions, sulfate-reducing bacteria can derive energy by anaerobic respiration; photosynthesis, aerobic respiration, and fermentation are not possible.
Microbes able to degrade a particular pesticide	Incubate aerobically in the dark in a mineral salts medium with the pesticide as the only source of carbon and energy.	Under such conditions, capability to degrade the pesticide is essential for growth.

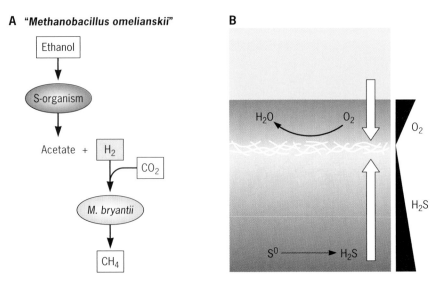

FIGURE 19.4 Why are some microbes challenging to cultivate? (A) The consortium *"Methanobacillus omelianskii"* is composed of a bacterium (S-organism, top) that ferments ethanol to acetate and H_2 and an archaeon *(Methanobacterium bryantii*, bottom) that removes the H_2 to reduce CO_2 and make methane (CH_4). (B) *Beggiatoa* cells grow as thick filamentous mats at the oxic-sulfide interface, where concentrations of O_2 and hydrogen sulfide (H_2S) gases are at the lowest. In both cases, it is difficult to reproduce the exact conditions needed to support the growth of each microbe in the lab.

common in nature. In such groups, one microbe may feed on its neighbor's products to prevent their accumulation. Consider the consortium called *"Methanobacillus omelianskii,"* once thought to be a single species that could grow in pure culture by converting ethanol and carbon dioxide to methane (Fig. 19.4). In fact, *"M. omelianskii"* is a combination of a bacterium and an archaeon, neither of which can grow singly on a mixture of ethanol and carbon dioxide. The bacterium (S-organism, for lack of a better name) ferments ethanol to acetate and hydrogen but only if the hydrogen is removed by the archaeon *Methanobacterium bryantii*, which uses it to make methane. With this information in hand, could you design enrichment cultures to isolate each species separately?

In other cases, a unique culture environment is required to establish growth-sustaining concentrations of seemingly incompatible metabolites. Oxygen and sulfide (H_2S) fall into this category: Oxygen is poisonous to strict anaerobes that reduce sulfur (S^0) and sulfur-containing compounds to sulfide; at the same time, sulfide poisons the electron transport chain of aerobic organisms. To minimize interference, these two metabolic processes are spatially separated in sediments, oxygen respiration above and sulfide production below (Fig. 19.4). Yet species of the bacterium *Beggiatoa* grow by metabolizing these two incompatible gases. To do so *Beggiatoa* cells position themselves at the interface at which the two gases meet: sulfide diffusing from below and oxygen from above (Fig. 19.4). Such conditions can be mimicked in the laboratory, but not easily.

Another challenge to cultivation methods is that we simply do not know what conditions certain organisms need to grow. For example, an environmental survey carried out in the late 1980s identified very abundant 16S rRNA bacterial sequences in the Sargasso Sea that had no cultivated relatives. The group, collectively known as the SAR11 clade, was later found to be extraordinarily abundant in all of the planet's oceans, sometimes accounting for half the prokaryotes in the surface waters and a quarter of all in the deeper regions. (Oceans are estimated to harbor some 10^{28} of these bacterial cells!) Yet, the enigmatic SAR11 clade defied cultivation for over a decade. How did microbiologists solve this biological puzzle? By discarding the standard practice of using rich media! The first SAR11 representative, *Pelagibacter ubique*, was enriched in a growth medium made from sterilized seawater with very low nutrient concentrations that best mimic the natural marine habitat. Under these restricted conditions, cells grew very slowly, doubling every 25 to 50 hours. But *grow they did* and once recovered in pure culture, their secret was revealed: their metabolism is particularly efficient under low nutrient conditions, especially when phosphorus is limited. These bacteria may have adopted an exceedingly bland diet, but in return they got a life with few competitors and predators. This strategy surely paid off: *P. ubique* became the most abundant bacteria in the planet's oceans. (Language aficionados, can you decipher their name?)

Why cultivation matters. Pure cultures offer many opportunities to study a microbe and what it does. Its genome can, for example, be sequenced and its metabolism and physiology investigated under controlled conditions. This information can also provide clues to what similar but not-yet-cultivated microbes are doing in the environment. For example, if you detect cyanobacteria in abundance in a particular location, you can predict that oxygenic photosynthesis will be a major driver of that ecosystem's function, since cyanobacteria use light as energy source to fix carbon dioxide and release oxygen as a by-product (Fig. 19.3). Furthermore, you can also predict that the environment will be largely aerobic, and that aerobic chemoheterotrophs will likely live there as well. Or, if an environmental survey detected abundant *Chloroflexi*, cultivated anaerobes known to carry out anoxygenic photosynthesis, we could predict the environment to be rich in H_2, H_2S, and/or S^0, as these are electron donors that can replenish electrons to the photosystem of these bacteria after each cycle of light excitation (chapter 6). Similar predictions could be made for uncultivated members of other phototrophic groups (Fig. 19.3), but this is not easily done for most other microbes.

Unfortunately, a long-recognized and unpleasant fact of microbial science is that most microbes in nature have defeated cultivation attempts. If you were to *count under the microscope* the number of microbial cells present in a soil or water sample for example, you will find that the total count exceeds by several orders of magnitude the number of microbes that can be cultivated from the same sample. Estimates are that only one out of 100 microbes observed under the microscope can be grown in standard solid media in the laboratory. This discrepancy is known as "The Great Plate Count Anomaly." As we mentioned earlier, in nature microbes often function as members of communities, not as distinct individuals. Thus certain consortia of bacteria accomplish chemical conversions not possible for each member in pure cultures (Fig. 19.4A). This codependence makes the recovery of microbes in pure culture especially difficult. But there are limitations for cultivating microbes in the laboratory as well. Domestication carries a price: some pathogens lose their virulence, others replace their antigens, and yet others alter their metabolic activities.

Cultivation-Independent Methods

Given the challenges of isolating microbes in pure culture, how can we determine who is there? **Cultivation-independent** approaches have been developed that indirectly assess the abundance and taxonomic affiliation of microbes in environment samples. Thus, we can evaluate the presence and relative abundance of a microbe by detecting microbe-specific genes such as the one encoding the rRNA small subunit (16S for prokaryotes and 18S for eukaryotes) (Fig. 19.1). Sometimes the identity of whole microbial communities can be revealed with such indirect methods. Indeed, high-throughput methods to sequence environmental DNA and RNA and to characterize proteins and metabolites directly in environmental samples are revolutionizing our understanding of microbial life. Next we discuss a few notable approaches to answer the question "who is there?"

Amplicon-based community fingerprinting. Profiling a microbial community typically starts with extracting DNA from the sample to sequence regions encoding specific microbial genes, such as the small subunit ribosomal RNA (SSU rRNA) locus (Fig. 19.5). For this **amplicon**-based

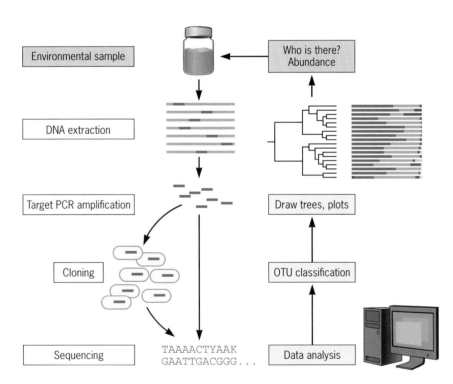

FIGURE 19.5 Community fingerprinting based on targeted amplicons. A widespread technique to profile a microbial community is to extract DNA from an environmental sample and PCR-amplify a target gene (e.g., SSU rDNA) with primers conserved among specific clades. Whether sequenced directly or after cloning the genes in bacterial hosts, the sequence data can be analyzed to group related sequences in operational taxonomic units (OTUs). The sequences are then compared to sequences in databases to infer relatedness, which we can visualize in phylogenetic trees or various plots.

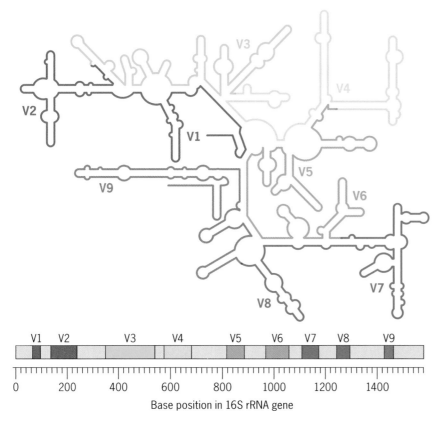

V1 V2 V3 V4 V5 V6 V7 V8 V9

0   200   400   600   800   1000   1200   1400

Base position in 16S rRNA gene

FIGURE 19.6 **The 16S rRNA of *E. coli* and its nine hypervariable regions (V1-9).**

technique, the target sequences are first amplified from the extracted DNA sample using the polymerase chain reaction (PCR) (chapter 10). Because PCR amplification requires the use of primers specific to the target DNA sequence, microbial ecologists are restricted to targeting highly conserved genes. The 16S rRNA of prokaryotes, for example, contains both conserved and variable regions (Fig. 19.6). Primers (the so-called **"universal" primers**) have been designed against the most conserved regions that hybridize to *most* prokaryotic (bacterial and archaeal) 16S rRNA genes. But primers can also be designed to target the variable regions of the 16S rRNA gene, which are only conserved among members of specific taxonomic groups. The stretches of DNA amplified with these primers (amplicons) can then be analyzed in various ways. Traditionally, gel electrophoresis was used to identify the amplicons based on their size, melting temperature, or restriction enzyme sites. Alternatively, amplicons can be cloned in plasmid vectors, amplified in a bacterial host, and then sequenced. But with the advent of modern sequencing technologies, scientists can bypass the more demanding cloning step and sequence the amplicons directly. This sequence is then compared against sequences in gene databases to identify homologous sequences (Fig. 19.5). Based on the degree of homology with known genes, phylogenetic trees can be made to infer the relatedness of the sequence to those from known taxa.

To classify the microbial community, its amplicon sequences are lumped into **operational taxonomic units** (OTUs). Microbial ecologists use specific thresholds of homology to classify the OTUs at the each level (species, 98.7%; genus, 94.5%; family, 86.5%; order, 82.0%; class, 78.5%; and phylum, 75.0%). A profile of a sample's OTUs can answer a broad range of questions. For example, is a particular representative of a phylum or even a domain present? This method also reveals the *relative abundance* of each OTU in the native community. It can also retrieve sequences not previously encountered. Microbes that are *identified solely by sequence similarity* are called **phylotypes**.

Amplicon-based community fingerprinting does have several limitations. The so-called "universal" primers used to amplify 16S rRNA sequences from environmental DNA are not as "universal" as their name implies. Further, as primers are designed based on sequences available in databases, they do not fully capture the diversity present in the environment. Another bias is introduced by the PCR step, because some sequences amplify more readily than others. Furthermore, the least abundant microbes are often under-sampled and, sometimes, missed all together. Despite these limitations, amplicon-based approaches are powerful tools in microbial ecology and widely used, particularly in combination with other cultivation-independent methods. For example, some techniques allow scientists to retrieve large regions of sequence and even reconstruct whole genomes from environmental samples *without* the use of primers. This technique, known as **metagenomics**, was introduced briefly in chapter 14 and is discussed in detail next. The volume of data generated by these studies is impressive. As of 2015, there were almost 4 million entries for individual 16S rRNA sequences in public databases, and more than 80% of them were generated using environmental surveys. Most of these sequences have no cultivated representatives, thereby revealing the so-called *"microbial dark matter."* And this is just for 16S rRNA sequences!

FISHing for microbes. The same PCR primers used to generate amplicons for community profiling (Fig. 19.5) can also be used to make DNA probes that *specifically hybridize* with the DNA within intact cells from a particular group of microbes. First developed in the 1980s, this **fluorescence *in situ* hybridization** (FISH) technique was adopted in the 2000s as a valuable cultivation-independent microbial ecology method. FISH DNA probes are labeled with fluorescent dyes, hybridized to complementary sequences in the genomes of cells in a sample, and then visualized with a fluorescence microscope. For example, FISH probes that distinguish between the 16S rRNA gene of *Bacteria* and *Archaea* can be used to tag members of these two *domains* selectively (Fig. 19.7). Alternatively, probes can be designed to tag specific *families* of microbes or even specific *genera* or *species*.

In a particular environment, FISH not only reveals which microbes are present but also their associations (Fig. 19.7). As we have seen repeatedly, if two microbes are found to be closely associated in nature, they probably interact metabolically and might even be dependent on one another for growth. As an example, Fig. 19.7 shows a methane-oxidizing aggregate of an archaeon and two bacteria, whose proximity allows the consortium to oxidize methane while reducing sulfate. Interactions among microbes can also be inferred using nanoscale secondary ion mass spectrometry (Nano-SIMS). This impressive technique images the chemical composition of a

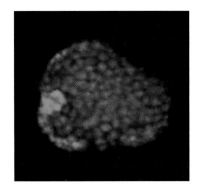

FIGURE 19.7 FISH probes to visualize a consortium of methane-degrading and sulfate-reducing prokaryotes mediating the anaerobic oxidation of methane. Confocal micrograph showing a cell aggregate approximately 10 μm in diameter stained with FISH probes targeting the rRNAs of a group of sulfate-reducing bacteria (*Desulfosarcina* and *Desulfococcus*, in green) and a group of methane-oxidizing archaea called ANME-2 (in red). The close proximity of the two cell groups allows for metabolic cooperation to enable the oxidation of methane coupled to the reduction of sulfate. (Reprinted from Boetius A et al., *Nature* **407**:623–626, 2000, with permission.)

cell or group of cells with nanoscale resolution using a specialized Nano-SIMS instrument. Figure 19.8 displays a micrograph of a filamentous cyanobacterium (*Anabaena*) with vegetative, photosynthetic cells and a differentiated N_2-fixing cell or **heterocyst** (chapter 13). The cyanobacterial heterocyst is colonized by *Alphaproteobacteria* belonging to the *Rhizobium* genus, which can be visualized using a FISH probe specific for *Alphaproteobacteria*. After the coculture is incubated with ^{13}C-bicarbonate and ^{15}N-dinitrogen, the NanoSIMS device maps the carbon and nitrogen **isotopes** on each cell. The images show that nitrogen, derived from the labeled dinitrogen gas that was fixed by the heterocysts, spreads to both the vegetative cells and the *Rhizobium* bacteria. The bacteria also take up carbon fixed by the vegetative cells via photosynthesis. The metabolic exchange from the cyanobacteria to the bacterial epibionts is beautifully revealed!

There are other ways to "fish" for microbes. **Fluorescent antibody probes** can also identify microbes in nature. This highly specific method

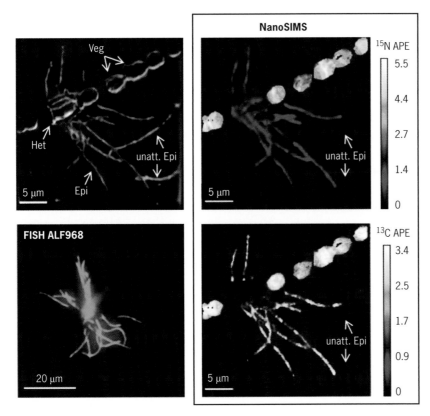

FIGURE 19.8 Combined FISH and NanoSIMS to link taxonomy to metabolic activity.
(Top left) A cyanobacterial filament of *Anabaena* is shown with round vegetative cells (Veg) and a heterocyst (Het) surrounded by attached and unattached epibionts (Epi and unatt. Epi, respectively). (Bottom left) A FISH probe (ALF968) specific for the alphaproteobacterial *Rhizobium* genus reveals the taxonomic classification of the epibionts. (Top right) NanoSIMS reveals the distribution of ^{15}N-nitrogen fixed by the heterocysts into the vegetative cells and, to a lesser extent, in the attached bacterial epibionts. (Bottom right) ^{13}C-carbon enrichment, which results from the fixation of CO_2 by the vegetative cells during photosynthesis, is also seen in the epibionts. (Modified from Behrens S et al., *Appl Environ Microbiol* **74:**3143–3150, 2008, with permission.)

will identify particular species or strains but not classes of microbes (as FISH can). Thus, it cannot be used to answer broad questions regarding the general types of microbes present in a particular environment, but it can provide specific information about one particular type.

The Holy Grail of methods to "fish" for microbes in the environment is, without a doubt, **single-cell genomics**. The goal of this technique is to separate and isolate a single cell directly from the environmental sample, amplify its whole genome, and sequence it. Sounds like science fiction, right? Success may be closer than you think. Cells are first separated and sorted using flow cytometers (chapter 4) or microfluidic devices. Once a single cell is isolated, it is lysed gently to release its DNA, which then must be amplified with PCR-based techniques before sequencing. Once sequenced the amplicons are assembled to reconstruct the genome. The technique is costly and demanding: it requires specialized instruments and personnel. And it does not always work. Yet the technology is advancing at such a fast pace that the time may soon come when sequencing a genome from a single cell becomes a routine approach to fish out even the most stubborn microbes and unveil their inner secrets. But do you recognize a drawback? To access and sequence the new genome, the cell was first destroyed. How can you get it back? The genome may provide critical information to "fish" for the microbe in the environment, but this task can be daunting, like looking for a needle in a haystack.

HOW DO WE KNOW WHAT MICROBES ARE DOING IN THE ENVIRONMENT?

LEARNING OUTCOMES:
- Describe what is measured in each of the following techniques: metagenomics, metatranscriptomics, metaproteomics, and metabolomics.
- Discriminate between "genome" analysis and "metagenome" analysis.
- Identify how metatranscriptomics has better capability to determine what the microbes in a natural community are doing over metagenomics.
- Identify benefits and drawbacks of using a metaproteome to answer the question "what are they doing?"
- Identify benefits and drawbacks of using a metabolome to answer the question "what are they doing?"

Some of the techniques described above, alone or in combination with other methods, can also provide information about what microbes are doing in a particular environment. Predictions can be made by analyzing large sets of gene sequences (or even whole genomes), transcripts, proteins, and metabolites retrieved from environmental samples (Fig. 19.9). Metagenomics can provide valuable information about the potential of a microbe to perform a function. Additional information can be generated by analyzing the sequences of the mRNAs transcripts (**metatranscriptomics**), the proteins that are expressed (**metaproteomics**), or the metabolites that are synthesized (**metabolomics**) (Fig. 19.9). These methods rely heavily on high-throughput technologies for the separation and identification of the target molecules. They also generate data sets so large that their analysis becomes computationally demanding. Furthermore, each technique introduces its own unique biases, both in the experimental and computational parts of the work. Yet, even with their limitations, the methods have proved highly informative, revealing functional attributes of environmental communities that we could not have predicted by studying a few cultivated representatives.

Sequence-Based Functional Profiling of Microbial Communities: Metagenomics and Metatranscriptomics

Metagenomics. We learned that retrieving DNA sequence information from a single SSU rRNA gene (Fig. 19.5) provides taxonomic information about specific groups and their relative abundance. But what are these microbes *doing*? And how do they contribute to the community? One way to address such questions is to obtain sequence information for more than one gene, particularly genes that can be linked to specific microbial functions (Table 19.2). Metagenomics accomplishes just that. By going literally "beyond the genome," metagenomics attempts to retrieve DNA sequences from all the genomes represented in a microbial community (the **metagenome**). First, DNA extracted from the environmental sample is sheared into small fragments (sometimes millions) for sequencing (Fig. 19.10). Computer programs then analyze the sequences, identify regions of overlap among the fragments, and assemble them into larger contiguous fragments or **contigs**. The contigs then undergo a quality screening to discard various artifacts that creep in. For example, artifacts often arise at repetitive sequences. Those contigs that pass the control step are scanned for *signatures of transcriptional units* such as promoters and transcription terminators, *regions of homology* to sequences in databases, and *conserved protein domains* or *motifs*. (Note that annotation of metagenomes follows the same approach used to annotate genomes [chapter 10].) The longer the contig, the more information can be retrieved from metagenomics data.

The quality of the metagenome data depends on the amount of DNA retrieved from every member of the microbial community in the sample. In fact, sampling is the major limitation of the metagenomic approach: sufficient amounts of starting DNA from every member of the microbial community are required to produce high-quality contigs. Consequently,

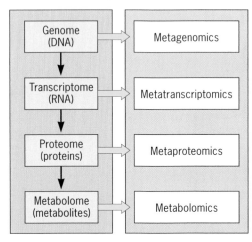

FIGURE 19.9 Cultivation-independent methods that assess what microbes are doing in the environment.

TABLE 19.2 **Examples of microbial functional traits useful in microbial ecology studies**

Microbial functional trait	Marker
Growth rate	Ribosomal copy number
Metabolic potential	Genome size, gene content
Dormancy	Sporulation-specific genes
Stress resistance	Stress response sigma factors
Osmoregulation	Trehalose synthesis
Cold tolerance	Cold-shock proteins, trehalose synthesis
Motility	Chemotaxis and flagellar genes
O_2 tolerance	Catalase and peroxidase genes
Metal resistance	Heavy metal efflux pumps
Antibiotic resistance	Efflux pumps, ribosomal protection, enzymatic inactivators

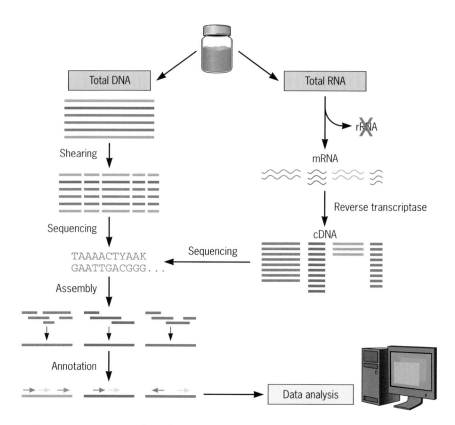

FIGURE 19.10 Sequence-based "meta-omics" (metagenomics and metatranscriptomics). DNA or RNA extracted from a sample can be used to retrieve metagenomic and metatranscriptomic data, respectively, and make functional predictions about the microbial community. In metagenomics the total DNA is sheared, sequenced, and assembled into contigs to reconstruct large stretches of the community genomes. In metatranscriptomics, the total RNA is treated to remove the more abundant rRNA, and the mRNA that is enriched in this step is reversely transcribed into cDNA prior to sequencing. Transcript data are then assembled and annotated against a reference metagenome.

metagenomics surveys typically retrieve data from the most abundant microbes in the sample, undersampling the low abundant ones. To bypass this limitation, metagenomics studies are often combined with amplicon-specific (e.g., SSU rRNA) profiling methods (Fig. 19.5) to retrieve specific taxonomic information about the microbes in the community and their relative abundance. As we will explain later, approaches to sequence the total RNA of the community (the **metatranscriptome**) can also complement metagenomic studies to enhance the predictive value of the data retrieved.

The first metagenomics survey was carried out on samples taken from the Sargasso Sea in the North Atlantic Ocean off Bermuda, where the very abundant SAR11 bacterial clade was identified. The success of this environmental survey relied on a handy step: the samples were first filtered to concentrate the prokaryotic cells, allowing the extraction of sufficient DNA for sequencing. The experimental design paid off: the survey resulted in the sequencing of over a billion base pairs. Analysis of these voluminous data yielded a wealth of information. An estimated 1,800 microbial species were represented in the samples, and of these, 148 were previously unknown prokaryotic phylotypes. Over 1.2 million previously unknown genes were identified, providing a measure of the degree of microbial diversity in the environment sampled.

Such massive information about numbers and kinds of unknown microbes and genes in the environment is humbling: How much more we have to learn! Keep in mind though that such data offer only a glance into what these microbes are doing in nature. Yet informed guesses about a newly discovered microbe can be generated by determining its kinship to known, well-studied microbes and then making the assumption that it is doing something similar. Table 19.2 shows examples of known genes that can serve as proxy for microbial functions and are, therefore, useful in metagenomics studies. Genes encoding proteins for specific metabolic pathways can be used to infer the metabolic potential of specific groups. For example, in a complex community, autotrophs are predicted from the presence of genes encoding proteins of carbon dioxide fixation pathways, some of which can be unique to specific groups of microbes (chapter 7). Other metabolic capabilities such as degradation of recalcitrant organic substrates, metal tolerance, and antibiotic resistance can be inferred from marker genes linked to these functions. Growth rate is a telling characteristic, but because of the many different activities in the cell that contribute to it, it cannot be linked to a single gene. However, the potential of a cell to grow fast may be predicted from other attributes, such as multiple copies of the rRNA operon, abundance of outer membrane proteins (for Gram-negative bacteria), and motility genes. Again, there are limitations. Most metagenomes can only predict functions for 30 to 50% of the reads they generate. How can we improve the predictive value of the metagenomics data? One approach is to learn how to cultivate the new organisms so that physiological and genetic studies of pure cultures can be launched. As you know, this is quite challenging since the nutrition of microbes within communities is interdependent (Fig. 19.4), and we have few clues about the physiology of these newly discovered microbes. Another approach is to reconstruct genomes from metagenomics data and/or sequence genomic DNA from single cells. Lastly, we need to develop more powerful computational tools to accommodate the large data sets that can now be retrieved from the environment, single cell genomes, and cultivated representatives. This is clearly a team effort for microbial ecologists, physiologists, and computational scientists.

Metatranscriptomics. From metagenomics data we gain insight into the *potential* metabolism and physiology of the organisms in the community, but which genes are actually expressed in the environment? One option is to focus only on the living cells, which can be separated from dead cells using dyes that differentially stain the two, followed by sorting via flow cytometry (chapter 4). Alternatively, actively metabolizing organisms can be labeled using carbon sources containing the heavy carbon isotope ^{13}C, which will make their DNA heavier; density gradient centrifugation then separates the "active" DNA fraction (labeled with ^{13}C) from the "inactive" one (unlabeled). The DNA isolated from the active cells can then be used for community fingerprinting (Fig. 19.5). Isotope labeling experiments such as this are laborious and usually require a large number of cells. Instead, why not sequence all of the transcripts in the community?

To identify what genes are expressed in a community and to what levels (*relative transcript abundance*), all the mRNAs from an environmental sample can be isolated and sequenced (Fig. 19.10). As with culture-based transcriptomics methods, metatranscriptomics requires the mRNAs to be enriched by first removing the very abundant rRNAs in the sample. The mRNAs are then copied into DNA molecules (cDNA) with a **reverse transcriptase enzyme**, and then the cDNA is sequenced by standard methods. Finally, transcript sequences are assembled and annotated. A good quality metagenome can greatly help with the assembly and annotation, but short mRNAs are still difficult to retrieve. (The longer the transcript, the more sequence it generates, and the easier it is to assemble it and annotate it.) As with metagenomics data, annotation is also limited to sequences for which genes of known function are already available in the database. But even with this limitation, metatranscriptomics data can provide valuable insights into what microbes are doing in the environment and even help design cultivation approaches to recover novel organisms in pure culture (Box 19.1).

Metatranscriptomics data provide valuable taxonomic and functional information about the microbial community, allowing the researcher to profile the community structure, function, and diversity simultaneously. Furthermore, because each gene can generate many RNA copies (Fig. 19.10), this amplification by transcription may permit the detection of gene sequences from low abundance organisms, which are often undersampled in metagenomics surveys. An *integrated multiomics approach* that combines both metagenomics and metatranscriptomics becomes even more powerful because biases introduced by one technique are compensated by the other. For example, technology is available to differentiate the DNA extracted for metagenomic analyses from the cDNA generated for metatranscriptomics by *barcoding with distinguishing tags*. Both DNA samples can then be pooled together and sequenced simultaneously. Computer programs can then detect the distinctive tags to sort the DNA sequences from the cDNA information. This allows the researcher to simultaneously retrieve information about the metagenome and metatranscriptome of the community from the same batch of cells.

Beyond DNA and RNA: Metaproteomics and Metabolomics

Genes and RNA transcript sequences retrieved in metagenomics and metatranscriptomics data, respectively, are proxies for protein expression but are not always accurate. For example, a significant fraction of genes in any cultivated microbe (sometimes half of them) are *controlled at the translational level*. Hence, even if a gene is transcribed, it is risky to assume the encoded protein is made. To bypass this limitation, microbial ecologists can profile all the proteins present in a microbial community (the **metaproteome**). Meta-omics methods have also been developed to isolate and characterize the metabolites that are produced by the activity of all those proteins (the **metabolome**). The two methods, called metaproteomics and metabolomics, respectively (Fig. 19.9), profile the activities of microbial communities with an unprecedented level of detail.

BOX 19.1

The "unculturable" leech symbiont that is no more

The nutritional interdependence of many microbes in environmental communities has long hindered the successful cultivation of individual members. How can we isolate one member if we do not know which essential nutrients its neighbors provide? Consider the microbial population that inhabits the digestive tract of the medicinal leech *Hirudo verbena*. This bloodsucking worm attaches to the skin of animals, including humans, using its posterior sucker. On the other end of the body are three powerful jaws equipped with sharp teeth, which the leech uses to pierce the host's skin and make it bleed. The leech's saliva contains anticoagulants that prevent the wound from healing and anesthetics that make the bite painless. These compounds equip the leech to suck out the blood at leisure, sometimes ingesting volumes many times its body weight in a single feeding. Once the meal reaches the enlarged cavity of the digestive system, called the crop, the blood is condensed as a viscous liquid by removal of water and osmolytes. This fluid can be stored in the crop for months at a time. Slowly it passes to the intestinum region, where nutrient assimilation occurs.

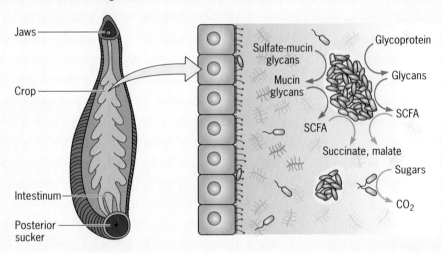

Residing in the crop are microbes that are essential for the leech's survival. The crop microbiome is dominated by only two microbes: *Aeromonas veronii* (a gammaproteobacterium) and a *Rikenella*-like bacterium (a member of the phylum *Bacteroidetes*). These microbes process the blood-derived fluid to extract nutrients for themselves. But the nutrients are also adsorbed by the intestinal epithelium, simultaneously feeding the leech. What exactly do the two microbial symbionts do in the crop? To learn, scientists successfully cultivated *A. veronii* and sequenced its genome. The *Rikenella*-like bacterium resisted cultivation, although researchers did reconstruct its genome using cultivation-independent methods. Still, the genomic information provided little information about the role of the symbionts in the leech nutrition. Next, the researchers isolated the total RNA from the blood-derived fluid stored in the crops of four leeches for metatranscriptomics analysis (Fig. 19.10). By mapping the cDNA reads against the draft genomes of the two symbionts, the scientists deduced that the *Rikenella*-like bacterium removes sulfate from the sulfated-mucin glycans that line the leech's intestine, and then ferments the glycan-derived sugars to acetate and other short-chain fatty acids (SCFA). The bacterium also extracts glycan sugars from the glycoproteins of the envelope of the ingested red blood cells. Meanwhile, *A. veronii* ferments the glycan sugars to carbon dioxide and metabolizes the SCFA produced by the other bacterial partner into succinate and malate.

One thing was clear from the metatranscriptomics data: mucin glycans are the main carbon and energy source that support growth of the symbionts in the crop. To verify this interpretation experimentally, the researchers designed a mucin-rich growth medium and successfully isolated colonies for each of the symbionts. Some colonies contained the two symbionts, consistent with the nutritional interdependence of the two predicted by the metatranscriptome data: the *Rikenella*-like bacterium fermented mucins into acetate, which *A. vernii* used to make succinate and malate. The *Rikenella*-like bacterium that had long defied cultivation attempts was finally available in pure culture, thanks to metatranscriptomics. Because of its appetite for mucins, it was appropriately named *Mucinivorans hirudinis*.

Metaproteomics. As with the other meta-omics methods, metaproteomics technologies follow steps to separate, quantify, and analyze proteins extracted from an environmental sample in a high-throughout fashion (Fig. 19.11). After extraction, the proteins can be separated in two-dimensional (2D) gels based on their mass (first dimension) and then charge (second dimension); each spot is then extracted from the gel to elute the protein. Alternatively, this demanding gel separation step can be bypassed using chromatography columns (such as in high performance liquid chromatography or HPLC) to separate all of the proteins in the mix. Characterization of each of the proteins is usually done by peptide mass fingerprinting, a technique that uses the enzyme trypsin to cleave the proteins at specific amino acid sequences. After such treatment, each protein's distinct pattern of peptide sizes can be used to match the protein to other proteins available in databases. Piecing together each protein based on the pattern of peptides generated after a tryptic digest is a challenging

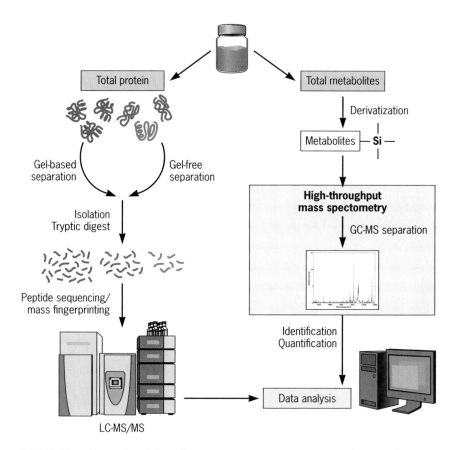

FIGURE 19.11 Beyond nucleic acids sequencing: metaproteomics and metabolomics. Proteins or metabolites extracted from microbes in a sample can be used to retrieve metaproteomic and metabolomic data, respectively, to make functional predictions about the microbial community. In metaproteomics the total protein in the sample is separated and each protein is digested into smaller peptides, whose mass and sequence can be used to identify homologous proteins available in protein databases. In metabolomics, the metabolites of the cells are extracted and derivatized with chemical tags to enable their separation, identification, and quantification using high-throughput spectrometry methods. Computational programs are then used to reconstruct metabolic networks based on the metabolites identified.

task, performed using computer programs. Alternatively, the amino acid sequence of each peptide can be determined using mass spectrometry methods to identify conserved protein domains that are required for specific functions. If available, a good quality metagenome for the sample can be used as a reference to annotate the metaproteomics data. For a complex microbial community, think about the amount of DNA and protein data that coupled metagenomics and metaproteomics studies need to retrieve and compare! Often times, the computational load is simply too high to be handled. In fact, we are generating more data than we can analyze computationally.

Despite its limitations, metaproteomics is particularly useful to identify spatial and temporal patterns of protein expression and infer how microbial communities adapt to their environment. For example, we now know that peptides associated with the most abundant taxa of marine bacteria (SAR11, *Prochlorococcus*, and *Synechococcus*) are enriched in oligotrophic (low nutrient) oceanic surface waters. As we described earlier, these are conditions that SAR11 cells have adapted to grow in. In fact, these bacteria scavenge sparse nutrients very efficiently. Indeed, the SAR11 metaproteome is rich in proteins involved in phosphate and amino acid uptake. On the other hand, *Prochlorococcus* and *Synechococcus* peptides retrieved from metaproteomic data match proteins involved in autotrophic growth (carbon dioxide fixation) and photosynthesis. Metaproteomic surveys therefore explain why these three groups are abundant and widely distributed in the oceans.

Metabolomics. Because microbial activities are largely chemical, their metabolites allow us to infer what microbes are doing. We can make some predictions based on "who is there" and what genes are transcribed into RNA or expressed into proteins. But metabolomics takes this one step further: it identifies and quantifies a large number of small molecules (usually less than 1 kDa in mass, though larger molecules can also be resolved) in a mixture. Since these small molecules are generated as intermediates and products of a cell's metabolism, they serve as chemical fingerprints of specific metabolic processes. For example, the precursor metabolites of the central pathways of fueling as well as the building blocks they generate (e.g., amino acids) can be used as proxies for the activity of those biosynthetic pathways and for growth (chapter 7). Similarly, the identification and abundance of metabolites generated in carbon dioxide fixation pathways such as the Calvin cycle (chapter 7) are indirect indicators of the activity of autotrophic metabolisms. Hence, metabolomics data can provide insight into the activities of groups of microbes within microbial communities.

As with metaproteomics, the steps involved in metabolomic surveys include an experimental phase (collection, extraction, separation, identification, and quantification of the target molecules, in this case metabolites) and computational analysis of the metadata (Fig. 19.11). One advantage of the metabolomics approach is that it requires smaller samples than do metaproteomics surveys. Moreover, the extraction of metabolites is compatible with the extraction of nucleic acids. Hence, metabolomics surveys can be carried out alongside nucleic acid sequence-based omics analyses of the same sample. Sample collection is perhaps the most critical step in a metabolomics experiment because, if not collected fast, some short-lived metabolites will be degraded. Special chemicals can be added to the sample to prevent the cells from lysing and leaking their metabolites. The metabolites are then chemically extracted; if needed, extracellular and intracellular fractions can be extracted separately. Whereas the intracellular metabolites indicate only the metabolic activities of the cells, the extracellular metabolome may reveal how cells interact with each other and the environment.

Once the metabolites are extracted from the sample, they are chemically processed (derivatized) to add chemical tags (usually alkyl or silicon groups) that facilitate their separation in high-resolution mass spectrometry instruments (typically gas chromatography mass spectrophotometer or GC-MS) or by nuclear magnetic resonance (NMR) (Fig. 19.11). Each metabolite elutes as a separate peak based on its mass and can be identified by comparing its elution time against a dataset of known metabolites. Metabolite abundance in the sample can also be inferred from the area of the peak. Computational programs normalize the metabolomics data, reconstruct the cellular pathways that may have generated each metabolite, perform statistical analyses, and also compare several data sets. Such comparisons are particularly useful for studying temporal and spatial variations of metabolic activities at the community level. Furthermore, by identifying and quantifying extracellular small molecules, metabolomics data can provide information about interactions among community members. For example, autoinducers and antibiotics serve as a proxy for cell density-dependent coordinated and antagonistic behaviors, respectively. In practice, it is very *difficult to extract metabolites in sufficient quantities* from environmental samples, particularly when one considers the small spaces that microbes inhabit. A cubic centimeter of soil can contain many spatially separated microbial communities and processes, yet is insufficient for this type of biochemical analysis. And how much antibiotic can be isolated from a cubic centimeter of seawater? Not enough to detect. Yet, trace concentrations may be sensed by microbes living there.

The small molecules that are secreted *in situ* by a microbial community can be identified by another technique—one with a formidable name, *nanospray desorption electrospray ionization* (nanoDESI). Here is how it's done. An extremely thin jet of a solvent is aimed at the target, say a bacterial colony on an agar surface, then the chemicals that are released are captured and analyzed in a mass spectrometer. Hundreds of compounds, many previously not known in this context, can be identified. Using this technique, it is possible to obtain the metabolic profile of living bacteria directly from a petri dish. By placing colonies of different organisms adjacent to each other, the chemistry of their interactions can be determined. One of the beauties of this technique is that no lysis, extraction, or other sample preparation is required!

IS EVERYTHING EVERYWHERE?

LEARNING OUTCOMES:
* Defend the "everything is everywhere" hypothesis with specific examples that you have observed.
* Explain how trait-based microbial biogeography can predict how a microbial community will respond to changes in the environment.
* Indicate how the combination of culture-based studies and cultivation-independent methods can work synergistically to reveal how microbes survive and interact with their environment.

In the dawn of the 20th century, a famed microbiologist, Martinus Beijerinck, proposed that most microbes are cosmopolitan, i.e., distributed everywhere. He noted that particular groups of microbes could be enriched in a culture by providing the specific conditions that promoted their growth. Further, the microbes were always there, regardless of the sample, but their presence only became apparent when right conditions for their growth were provided. This concept was later summarized in a now famed quote: "Everything is everywhere, but the environment selects." How could microbes be everywhere? Studying dispersal, Beijerinck observed that microbes could scatter in innumerable ways—in dust particles spread by wind, by water droplets, by animals, and by plant dispersal structures such as seeds, just to name a few.

The notion that microbes are everywhere is indeed provocative. It would explain why sauerkraut can consistently be made from any cabbage, with any vessel, and no matter who prepares it and where (case study). The microbes are there, and the environment selects for them. But do the same microbes make sauerkraut here and elsewhere? Or, are the microbes taxonomically distinct, but share the same metabolic traits that are selected for in the sauerkraut jar?

Early studies about the cosmopolitan nature of microbes set the foundation of a new field in microbial ecology, so-called **microbial biogeography**, which studies the distribution of microbes in space, time, and across environmental gradients. To test the "everything is everywhere" hypothesis, microbial biogeographers first determine "who is there" and "what they are doing." They then compare how the community composition and activity change geographically, over time, and in response to fluctuations in the environment. Consider the technical challenges that these ambitious studies impose: information needs to be collected about the structure and function of many different communities and analyzed to determine who and what changes spatially, temporarily, and in response to specific environmental conditions!

The biogeography field initially relied on taxonomic studies that identified groups of microbes by detecting taxon-specific genes such as the SSU rRNA (Fig. 19.5). These studies provided little information about the function of these organisms in the environment because, as we know, the microbial species concept is ill defined and even strains within the same "species" can be metabolically diverse. To expand the scope of the studies, sequences that could be assigned to functional traits of microbes were included in the surveys (Table 19.2). Genomic and metagenomic data became particularly useful in *trait-based biogeography studies*. As we mentioned earlier, maximum growth rates can sometimes be inferred from the ribosomal copy number (Table 19.2). The ribosomal gene content can also be used to predict the response time of a microbe to nutritional cues from its environment, such as a sudden increase in nutrients. The idea is intuitive: a microbe with more ribosomal operons makes more ribosomes and, therefore, can synthesize more proteins on short notice. Having a higher capacity for protein synthesis, these microbes can respond to nutrient increases rapidly, thus growing faster and outcompeting microbes carrying only one ribosomal operon. However, multiple ribosomal operons can be an energetic burden when nutrients become scarce, because even downregulating the expression of the ribosomal genes, more ribosomes may be made than needed to satisfy the cell's protein synthesis demands. Indeed, nutrient-rich environments are often dominated by microbes with multiple ribosomal operons; as nutrients become depleted, we instead see communities dominated by microbes with one ribosomal operon. (Realize that ribosomal copy number can sometimes be misleading. For example, the genome of dormant cells may contain multiple ribosomal operons yet the cells are not growing.)

The combination of culture-based studies and indirect methods can also help assess the distribution of traits, rather than the microbes themselves. A concrete example of this concept is the discovery of **proteorhodopsin**. Some cultivated **haloarchaea** (salt-resistant archaea) boost their energy budget with a membrane protein called **bacteriorhodopsin** (chapter 14). This protein uses light energy to pump protons across the membrane and generate sufficient proton motive force to make ATP (Fig. 19.12). Metagenomics surveys identified a homologous sequence, which was termed proteorhodopsin, in the genome of an uncultivated marine bacterium.

Looking for proteorhodopsin-like sequences among metagenomic data collected from many oceans confirmed their abundance and global distribution. This study cemented the idea that proteorhodopsin-based phototrophy is a global strategy for energy transduction and a significant microbial process in the oceans, allowing marine microbes to harvest energy for growth even in oligotrophic (low nutrient) aquatic environments.

Patterns of distribution of microbes and their specific activities can also be studied with metaproteomics and metabolomics methods (Fig. 19.11). Here we consider the example of thick **biofilms** that form in the highly acidic waters that drain from some abandoned mines. Since iron (ferrous, Fe^{2+}) is abundant in these environments, iron-oxidizing bacteria are often isolated from such biofilms. Bacteria cultivated from these biofilms have an electron transport chain with heme-containing proteins called cytochromes, which oxidize

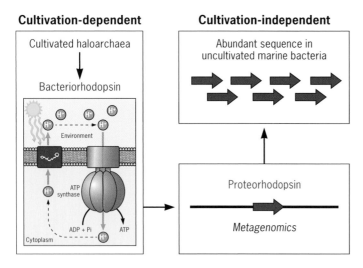

FIGURE 19.12 Combined cultivation-dependent and independent methods identify a globally distributed proteorhodopsin in marine bacteria. Information about the bacteriorhodopsin proton pump that cultivated archaea use to make ATP allowed for the identification of a related sequence (proteorhodopsin) in the genome of an uncultivated marine bacterium, which had been reconstructed from metagenomic data. Metagenomics surveys then confirmed that proteorhodopsin-like sequences are abundant and globally distributed in the oceans.

Fe^{2+} to gain energy for growth. A metaproteomics survey confirmed the prevalence of proteins involved in heme synthesis in the biofilms; in addition, it identified a novel cytochrome, Cyt_{579}, which had the unusual property of absorbing light in the visible region of the spectrum at 579 nm. Subsequent studies with cultivated biofilm isolates demonstrated that Cyt_{579} functions as an accessory protein in the electron transport system that oxidizes iron.

But is everything really everywhere? Microbial biogeography studies are providing strong evidence supporting the notion that the *structure of communities is based on functional traits* rather than on taxonomic relatedness. In the case study the function that lactic acid bacteria contribute is fermenting sugars into lactic acid. As lactic acid accumulates and the pH drops, groups of lactic acid bacteria capable of fermenting optimally at that pH will be selected for and will soon outnumber the rest. Hence, the environment (the vessel with fermenting cabbage) selects for organisms that have the functional traits suited to exploit the available resources (sugars released when salt is added to the cabbage). As the resources (type of sugars) and the environmental parameters (pH) change, organisms better fitted to grow under these conditions are selected for and outgrow the others. The taxonomic classification of these bacteria may not always be the same, depending on where, who, and how the sauerkraut is made. But the *process* is consistently the same. Similarly, microbes with comparable functional traits, whether residing here or elsewhere in the planet, will be selected for if the environmental conditions are right. Hence, the immense diversity of microbes and ability to reproduce and evolve new traits rapidly may ensure that functional traits are always represented in every niche. Then, the environment will exert selective pressures that sort each group and ensure that their relative abundance and interactions make the community perform more efficiently with what resources are available.

WHAT ARE ALL THOSE VIRUSES DOING OUT THERE?

LEARNING OUTCOMES:
- Describe why environmental viruses eluded scientific discovery for so long.
- Give examples of environments where viruses greatly outnumber *Bacteria* and *Archaea*.
- Summarize three ecological roles of viruses in the environment.

Though viruses have long been a powerful tool in bacterial genetics (chapter 10), they did not catch the eye of microbial ecologists until relatively recently. Because of their small size compared to most microbial cells, viruses went undetected in many of the early environmental surveys that depended on retrieving material by filtration. Viruses are also highly diverse, making it difficult to study them at the community level. Furthermore, unlike cells, they lack ubiquitous marker genes such as rRNAs that could be used to assess their diversity in the environment (chapter 17). Yet, viruses are incredibly abundant. Stain a drop of seawater with fluorescent dyes specific for nucleic acids and you will soon realize that most of the stained particles are too small to be microbial cells but small enough

A

B

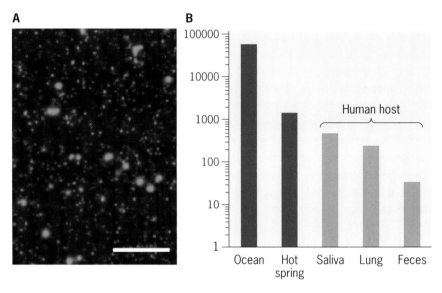

FIGURE 19.13 **Viral abundance and diversity.** (A) Abundance of virus-like particles (VLPs, small objects) in a sample of seawater compared to microbial cells (larger objects). A small volume of seawater (1 to 2 ml) was filtered (0.02-μm pore size), and the nucleic acids of any particles retained on the filter were stained with a green fluorescent dye. (From Patel A et al., *Nature Prot* **2:**269–276, 2007, with permission.) (B) Average species richness of phages in the oceans, Yellowstone hot springs, and various human samples, which are environments with abundant and diverse bacterial hosts. (Modified from de Paepe M et al., *Front Cell Infect Microbiol*, 2014, http://dx.doi.org/10.3389/fcimb.2014.00039.)

to be viruses (Fig. 19.13A). There are approximately 10 times more viruses than cells in seawater (Table 19.3). The total viral count is actually a whopping 10 million particles per milliliter. Most of them are bacterial viruses (phages), since bacteria are more abundant than other microbes in this environment. Half of the bacteria in these communities, if not more, carry **prophages**. Archaea are equally abundant in the oceans, particularly in the seabed sediments, which are dominated by—guess what?—archaeal viruses. Estimates are that there are 10^{30} viruses in the oceans, making them the most abundant biological entities in marine environments.

Indeed, by also counting "<u>v</u>irus-<u>l</u>ike <u>p</u>articles" (VLPs) by fluorescence microscopy in other environments, we discover that viruses are abundant wherever there are microbial cells to infect, often outnumbering their hosts by several orders of magnitude. Numbers in soils and marine sediments are even higher than in ocean waters, consistent with the higher abundance of potential prokaryotic hosts in these environments. Each gram of sediment collected from the bottom of the ocean contains 10^{10} (10 billion!) viral particles. And so do soils: agricultural fields, in particular,

TABLE 19.3 **Abundances of organisms in 1 ml of seawater**

Type of organism	No./ml of seawater
Zooplankton	≪1
Phytoplankton	
Algae	3,000
Protozoa	4,000
Photosynthetic bacteria	100,000
Heterotrophic bacteria	1,000,000
Viruses, including phages	10,000,000

beat all other environments in viral abundance, typically harboring thousands of viral particles per bacterial cell, a ratio at least 10 times higher than in aquatic environments.

The protocol to profile the viral component of an environmental microbial community (or the **virome**) starts with the separation of the VLPs from microbial cells using filters with a pore size (usually < 0.22 μm), which retain cells but allow the viral particles to pass through. This simple protocol also reduces contamination from the abundant microbial DNA and extracellular nucleic acids. It does, however, discriminate against the very large viruses, which are retained by such filters. By enriching for VLPs, one can extract enough nucleic acid material from both DNA and RNA viruses for metagenomics sequencing. DNA and RNA viral genomes can be handled separately to study the genomic diversity of both types, using methods described for the sequencing of DNA and RNA in metagenomics and metatranscriptomics studies, respectively (Fig. 19.10). RNA needs to be copied into cDNA using a reverse transcriptase step prior to sequencing. Then, the cDNA samples are sequenced individually or simultaneously after barcoding each sample differentially. The first metagenomics surveys of viral communities were performed in marine environments (the marine virome) and yielded an enormous diversity of viral reads that, for the most part, had no match in the available databases (Fig. 19.13).

Following the maxim that wherever there are microbial cells there are viruses, places inside our body inhabited by microbial communities (mainly bacteria) also contain abundant and diverse populations of viruses (Fig. 19.13). Bacterial populations are particularly dense in the terminal portion of the vertebrate gut and, as a result, many are shed in feces. A single gram of human feces carries along with the indigenous gut microbes a whopping 10^8 to 10^9 VLPs! What do we know about the gut virome? Approximately half of the human fecal virome contains viral **integrase** sequences. Recall that integrases are viral proteins that promote the integration of viral DNA into the bacterial host's chromosome (**lysogeny**; chapters 10 and 18). The integrated viral DNAs or prophages remain latent until reactivated by environmental cues to start a new cycle of infection. Hence, at least 50% of the human virome detected in the feces may be temperate phages.

Although lysogeny is also deduced from viromes retrieved from other natural environments, its incidence varies both in space and time. This observation emphasizes that, like the microbial communities they are part of, viral populations are dynamic and their abundance is subject to spatial and temporary variations. Coastal waters often contain more viruses and have a lower frequency of lysogeny than offshore waters, possibly reflecting the higher levels of nutrients available in this environment, which can support the growth of more microbial hosts. Particle counts and infectivity also show seasonal variability, being higher in the warmer summer months than in winter, just like the microbial population that serves as hosts. Clearly, the two are intimately connected.

What are all those viruses doing out there? They play ecological roles that are critical for all processes taking place in the oceans, soils, and sediments (Fig. 19.14). Their most obvious role is predation, a type of interaction that we will explore more in chapters 21 and 26. By infecting

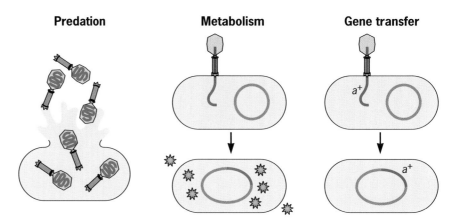

FIGURE 19.14 **Ecological role of viruses.**

and lysing microbial cells, viruses keep cell numbers at bay, preventing blooms that would disrupt the microbial community structure and the careful ecosystem balance needed to cycle nutrients and exchange gases with the atmosphere. Together with protists (chapter 16), they are the most important bacterial predators in the oceans. They prey on phytoplankton, particularly during phytoplankton blooms. This is exactly what happens to the vast ocean blooms of the unicellular alga *Emiliana huxleyi*, which collapse abruptly when infected by viruses (chapter 16). Viral predation is particularly high in groups of microbes that grow and increase in numbers fast, as we'll see in explosive epidemics of the bacterial disease cholera (chapter 26). By "killing the winner," viruses stabilize the community structure that is so critical for microbes to function in the environment.

The ability of some viruses to integrate their DNA into the host's chromosome as a prophage (lysogeny) also has ecological implications. As we learned in chapter 10, once integrated as a prophage into the host chromosomes, temperate viruses can influence the metabolism of the host cells (Fig. 19.14). Is lysogeny widespread? It appears so. Half of the bacteria cultivated from marine environments lyse and release viral particles when treated with chemical or physical agents that activate integrated viral DNA sequences (prophages). Indeed, half of all the sequenced genomes of marine bacteria carry prophage-like sequences. Evidence also suggests that lysogeny is more prevalent when temperatures are lower and microbial cell numbers (thus potential hosts) are lower as well. But once cell host numbers increase, for example in the summer months, lysis prevails over lysogeny. Hence, the viral life cycle adjusts dynamically to the host's physiology and environmental conditions.

Last, but not least, viruses are major vehicles for **horizontal gene transfer** (Fig. 19.14). In **generalized transduction** (chapters 10 and 18), phages mistakenly pack host DNA in viral particles and transduce it to a new host, where it can recombine into the chromosome. The lysis of host cells by viruses also releases substantial amounts of DNA that survivors can take up by **transformation** (chapter 10). Some *Alphaproteobacteria* inhabiting the oceans also produce VLPs called **gene transfer agents** (GTAs). These particles are similar in size and morphology to tailed phages but, unlike them, they indiscriminately pack a large amount of DNA (~ 4.5 kb) and transfer it to another host cell. GTAs do not lyse the host cells; they are mere vehicles for DNA exchange from cell to cell. Moreover, they transduce at a high frequency, much higher than transducing viral particles, and are responsible for numerous events of horizontal gene transfer in marine environments. Clearly, viral ecology is integral to microbial ecology, and we are beginning to understand just how much viruses contribute.

THE HUMAN MICROBIOME: ARE WE JUST LIKE ANY OTHER ENVIRONMENT?

LEARNING OUTCOMES:
- Explain how advances in scientific technology advanced our knowledge of the human microbiome.
- Identify the major groups of microbes that inhabit microbe-rich parts of the body (such as the gut).
- Use examples to provide evidence for the importance of the human microbiome in promoting human health.
- Give examples that exemplify how the human microbiota are well adapted to their microenvironment.
- Describe how deviations from normal interactions between human hosts and the microbiome can lead to disease, such as in the case of Crohn's disease.

Humans provide a terrific environment for microbes. Many of our body parts are warm, cozy, and nutritious. What microbe would not love us? So important are the environments we provide for microbes that many have made parts of our body their permanent residence (Fig. 19.15). There they form communities that mature with us as we age. And they drive the evolution of our immune defenses (chapter 22). The stability and functions of our microbial communities, collectively known as the human **microbiota**, are critical to our health. We also house certain microbes sporadically. These come and go, but some take advantage of the opportunity to colonize otherwise sterile parts of our body and cause disease. Because of this, the bacteria (and relatively few archaea) that make up our microbial complement have become the object of intense studies, and a lopsided amount of information is available for the human microbiota compared to environmental communities. Here we will consider just a few interesting facts to highlight that, to microbes, we are just like any other environment.

Who Is In or On Us?
There are perhaps 10 times more microbial cells in our body than human cells in our tissues. The microbial gene count or **microbiome** compared to that in the human genome is even more impressive (at least 100 to 1). The traditional approach to study our microbial companions relied on their cultivation and/or direct microscopic examination. But, as we know, these approaches are very limited in scope. Probably reflecting the relatively abundant nutrient environment of most body sites, growing in the lab

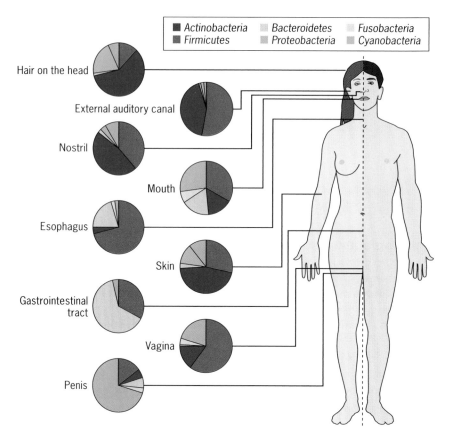

FIGURE 19.15 The human microbiome. (Modified from Spor et al., *Nat Rev Microbiol* **9**:279–290, 2011, with permission.)

many of our microbes requires rich media (containing amino acids, purines and pyrimidines, and vitamins, among other components). A notable exception is *Escherichia coli*, which can also grow on "minimal" media having only a few nutrients. Can you fathom why this may be? (*Quiz spoiler: E. coli* strains are also found in contaminated waters and other nutrient-poor environments, thus are adapted to nutritionally varied habitats.) In the early days of microbiology, the simplest method to isolate *E. coli* from human feces was to incubate inoculated plates aerobically. Consequently, it was thought that *E. coli* was the most abundant bacterium in human feces. It turns out that it is the most abundant **facultative aerobe** (i.e., it grows in the presence or absence of oxygen; chapter 6). Once anaerobic techniques were put to use and the abundant anaerobes cultivated, *E. coli* was demoted to become about a puny 0.1% of the fecal microbes. We now know that many of the bacteria associated with humans are **strict anaerobes**, and some are *facultative aerobes*.

Cultivation-independent methods have greatly expanded our knowledge of the microbes we carry (Fig. 19.15). Diversity and abundance vary not only with the part of the body they claim as their habitat but also with the individual host. In most body sites, especially in the large intestine and the nose and mouth, bacteria are in close proximity with other kinds. Thus, they have evolved to interact with their neighbors, a topic that we will develop further in chapter 21. In a few parts of the body, e.g., the stomach or the vagina, the diversity is restricted (Fig. 19.15). Typically, the vagina is colonized primarily by a group of Gram-positive bacteria called the lactobacilli, which by producing copious amounts of lactic acid and other antibacterial compounds become gatekeepers against invading species. (Note that lactic acid bacteria are also responsible for dropping the pH during the making of sauerkraut in the case study. Could they play a similar role?) Yet generalizations are not always possible. Some women, for example, have no lactobacilli in their vagina. By contrast, the terminal portion of our intestines (the colon) harbors trillions of bacteria that grow and proliferate in this protective and nutritious environment. In return for our hospitality, they digest complex carbohydrates for us, boost our immune system, and fend off pathogens (although they do not always win). Treat an individual with an oral antibiotic and you risk killing many of the indigenous gut microbes, reducing the diversity and functionality of the microbial community, and exposing the host to opportunistic pathogens.

So important are the microbial communities that live with us that they dynamically adjust their structure and function as we age and make lifestyle changes. The gut microbiome, for example, is assembled in a stepwise manner, increasing in richness as infants move from a milk-based diet during the breast-feeding stage to an adult diet in school-age children (Fig. 19.16A). At birth, newborns carry some gut bacteria inherited from their mothers, but these are quickly displaced and/or outnumbered with the infants' own bacteria in the phylum *Firmicutes*. *Proteobacteria* are often detected soon after, followed by the *Actinobacteria*. Once solid foods are introduced, *Bacteroidetes* get established and displace most of the bacteria from the other groups, except for the *Firmicutes*. By the time children reach school age, the composition of the gut microbiome is essentially adult-like and dominated by *Firmicutes* and *Bacteroidetes*, though the relative abundance of these two phyla varies from person to person. These two groups remain the two dominant phyla as the host ages, though the microbiome undergoes a maturation process (just like the human host), whereby diversity and stability increase over time (Fig. 19.16B). Indeed, species richness is not high in school-age children, but increases gradually as the person grows into adulthood and continues to age (Fig. 19.16B). With greater diversity may come stability; that is, the gut microbiome may become more resilient to perturbations as it ages and matures with the

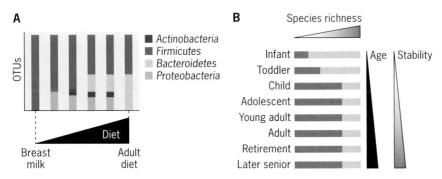

FIGURE 19.16 **Assembly (A) and stabilization (B) of the human gut microbiome.**

host. Otherwise, our health may be compromised. Thus, the gut communities of individuals with obesity, insulin resistance, and/or inflammatory bowel conditions are less diverse than those of healthy individuals. Although correlation does not always mean causation, these studies add to the growing body of evidence suggesting that the composition and diversity of our gut microbiota impact our health.

What Are These Microbes Doing?

Cultivation-independent methods support the notion that we are "raising" true specialists to better serve us. First, many microbes that live in our gut are *rarely detected in any other environment*. Our unique profile suggests that humans act like an enrichment culture, selecting for microbes that function better in our distinctive gut microenvironment. So specialized are these microbes that it is possible to use stools from healthy individuals to restore "perturbed" gut microbiomes. Perturbations occur, for example, after a course of oral antibiotics. When the normal, protective flora does not recover promptly, some patients are prone to recurrent infections from the opportunistic intestinal pathogen *Clostridium difficile*. Indeed, fecal transplantations with stools from healthy individuals have been used successfully to reestablish the "healthy" gut microbiome in affected patients and prevent *C. difficile* disease.

Integrated meta-omics approaches also reveal functional traits of the gut microbes that confirm their adaptation to the intestinal microenvironment. Genes encoding essential building blocks, such as some amino acids, are only expressed at low levels by bacteria residing in our intestinal tract. This is because our intestinal epithelium makes amino acids and supplies them to the microbes. Furthermore, although many of these microbes (e.g., members of the *Firmicutes*) form spores when conditions are unfavorable for growth, sporulation genes are strongly repressed in the gut microbiomes of healthy individuals. Clearly, there is no advantage in going dormant for microbes living in a protective environment with plenty of nutrients and little perturbation. By contrast, *Methanobrevibacter smithii*, the predominant archaeon in the human gut, expresses methanogenic genes at a high level. This archaeon removes hydrogen (H_2) gas produced during the fermentation of complex organic substrates by *Firmicutes* and *Bacteroidetes*. If the hydrogen were to accumulate in the gut, it would slow down the activities of the fermentative bacteria, as well as the digestion of carbohydrates and other fermentable substrates. Better to have methanogenic archaea in our guts to improve digestion, don't you think?

All microbiomes from a particular environment share a set of common loci, or **core genes**. For example, the metagenome of stools from healthy individuals, whether dominated by *Firmicutes* or *Bacteroidetes*, are consistently enriched in genes *encoding functions for carbohydrate degradation*. This particular core metagenome is unique to the gut environment. By contrast, the core metagenome of the oral microbiome includes genes with functions in polysaccharide biosynthesis, particularly dextran, which is abundant in the oral cavity and involved in tooth decay. And the core metagenome of the vaginal microbiome carries genes the contribute to degradation of glycogen and peptidoglycan, consistent with the limited availability of nutrients in this part of the body and the need to scavenge carbon either from internal storage compounds (glycogen) or by preying on other bacteria (peptidoglycan from the bacterial cell wall). This level of specialization highlights how adapted our microbes are to the environments we provide.

The gut metaproteome of healthy individuals also reveals notable functional traits and interactions between the microbial community and the human host. Approximately one-third of the proteins in the gut metaproteome are from microbes. Half of them are involved in housekeeping functions such as protein synthesis and energy metabolism, consistent with an active microbial population. Proteins involved in fatty acid metabolism are also represented. Many gut microbes make fatty acids that the colonic epithelium then adsorbs. Metaproteomics data also reveal that fatty acids are consumed by some of the gut microbes to prevent their accumulation to levels that could trigger the inflammation of the colonic epithelium. By contrast, these proteins are found at low levels or are not detected in metaproteomes from patients with inflammatory bowel disease (Crohn's disease). Host proteins are also present in the stool metaproteome of healthy individuals, and can account for up to two-thirds of all of the proteins in the samples. Many of these host proteins play roles in the maintenance of the integrity and function of the intestinal epithelium, thus serving as markers for colonic health. These same proteins are detected at lower levels in patients with Crohn's disease, who also have abnormal intestinal epithelia. Clearly, our association with microbes is mutualistic (chapter 21), i.e., we both benefit from the interactions. We feed them and protect them; in return, they perform functions that contribute greatly to our health. We are indeed evolving together, as in a true symbiosis.

CASE REVISITED: What makes sauerkraut so funky?

It may not be your favorite food, but sauerkraut illustrates a linked chain of microbial events as they happen in the environment—three fermentations to be exact. This process occurs thanks to the many different microbes contained in the cabbage itself and the selective pressures exerted by the environment. One at a time, microbial activities are linked sequentially as if in an assembly line (an ecological succession). Let us track down the critical steps. We need to add salt to start the sequence: no salt, no sauerkraut. Why? Salt induces plasmolysis; that is, cabbage cells release fluids containing a large amount of soluble substances—sugars, amino acids, and vitamins. These substances provide nutrients for some of the bacteria and yeasts in the cabbage. Right away, microbes start to grow in the sealed jar: some make enzymes to soften the cabbage and change its texture; others ferment the sugars and make organic acids (mainly lactic, acetic, and propionic acids) as well as mannitol, a sugar alcohol. These fermentations also give the cabbage the distinctive aroma of sauerkraut and lower the pH to levels unfavorable for the first group of microbes.

Stage 1 is completed. The pH in the jar is now lower but optimal for a new group of lactic acid bacteria, which take over and ferment any residual substrates. Soon the sauerkraut acquires its distinctive flavor and has so much lactic acid that the pH drops even more (to about 3.5). Such an acidic pH kills most bacteria, except for a group of lactic acid bacteria, which begin the third and last round of fermentations and produce more lactic acid. The pH now drops to about 2.0, stopping further microbial growth and preserving the sauerkraut for a long time, provided that air is excluded.

Three distinct bacterial populations have been at work, the first to give aroma and prepare the food for the next, the second to impart the flavor, and the third to make it into a preserved food. It does not matter what cabbage or vessel is used, who makes the sauerkraut, or where they make it: sauerkraut results every time because capable microbes are always there. Then, the environment selects for the activities of one group at a time so the metabolic products of one microbial population promote the activities of another in a sequential manner. On a small scale, this example exemplifies how microbes link their activities in the environment to carry out processes that the individuals cannot do alone. Perhaps one day you will so be inclined to probe "who is there" and "what they are doing" as cabbage is converted into sauerkraut? (How would you do it?) Now consider the challenge that microbial ecologists face: the microbial processes that make sauerkraut are but an insignificant fraction of the integrated activities of millions and millions of microbes in environmental communities. Collectively, these microbial communities sustain life in our planet.

CONCLUSIONS

Microbes are ubiquitous, abundant, and metabolically diverse. Yet, with a tool set that integrates cultivation-dependent and independent approaches, microbial ecologists are unraveling the diverse strategies that microbes use to grow in communities in the environment, whether in a pond, an agricultural soil, or our gut. The community composition matters because the function of each microbe is integrated in the community to perform functions that the individual cannot perform on its own. And as a well-oiled machine, the community adjusts its structure and activities to the surrounding environment. In the next chapter, we will learn just how pervasive the activities of microbial communities are on the global scale. Then, in chapter 21, we will look at the type of interactions that influence the collective actions of microbes and their ability to colonize any place on Earth where there is water and exploit the resources available there. In chapter 22, we will focus on one environment in particular, the human host, to learn about our sometimes conflicting relationship with microbes.

Representative ASM Fundamental Statements and Learning Outcomes for the Chapter

EVOLUTION

1. The traditional concept of species is not readily applicable to microbes due to asexual reproduction and the frequent occurrence of horizontal gene transfer.

   a. Differentiate the terms species, operational taxonomic units, and phylotypes.

METABOLIC PATHWAYS

2. *Bacteria* and *Archaea* exhibit extensive, and often unique, metabolic diversity.

   a. Describe the process of amplicon-based community fingerprinting.
   b. Identify benefits and drawbacks of fluorescence *in situ* hybridization (FISH).
   c. Describe what is measured in each of the following techniques: metagenomics, metatranscriptomics, metaproteomics, and metabolomics.
   d. Discriminate between "genome" analysis and "metagenome" analysis.
   e. Identify how metatranscriptomics has better capability to determine what the microbes in a natural community are doing over metagenomics.
   f. Identify benefits and drawbacks of using a metaproteome to answer the question "What are they doing?"
   g. Identify benefits and drawbacks of using a metabolome to answer the question "What are they doing?"

3. The survival and growth of any microorganism in a given environment depend on its metabolic characteristics.

   a. Describe two ways that *Bacteria* and *Archaea* use to survive on low-energy-yielding biochemical reactions.
   b. Explain the "Great Plate Count Anomaly."
   c. Discriminate between enrichment cultures and pure cultures.
   d. Identify benefits and drawbacks of using enrichment cultures.
   e. Summarize three reasons why so many environmental microbes have eluded cultivation in the laboratory.
   f. Identify benefits and drawbacks of using microbial pure cultures to understand the microbial ecology of natural environments.
   g. Give examples that exemplify how the human microbiota are well adapted to their microenvironment.

MICROBIAL SYSTEMS

4. Microorganisms are ubiquitous and live in diverse and dynamic ecosystems.

   a. Support the following statement using specific examples: "Microbes are ubiquitous and live in diverse and dynamic ecosystems."
   b. Defend the "everything is everywhere" hypothesis with specific examples that you have observed.
   c. Describe why environmental viruses eluded scientific discovery for so long.
   d. Give examples of environments where viruses greatly outnumber *Bacteria* and *Archaea*.
   e. Summarize three ecological roles of viruses in the environment.

5. Microorganisms and their environment interact and modify each other.

   a. Explain how trait-based microbial biogeography can predict how a microbial community will respond to changes in the environment.
   b. Indicate how the combination of culture-based studies and cultivation-independent methods can work synergistically to reveal how microbes survive and interact with their environment.
   c. Describe how deviations from normal interactions between human hosts and the microbiome can lead to disease, such as in the case of Crohn's disease.

IMPACT OF MICROORGANISMS

6. Microbes are essential for life as we know it and the processes that support life (e.g., biogeochemical cycles and plant and/or animal microbiota).

   a. Explain how advances in scientific technology advanced our knowledge of the human microbiome.

   b. Identify the major groups of microbes that inhabit microbe-rich parts of the body (such as the gut).

   c. Use examples to provide evidence for the importance of the human microbiome to promoting human health.

SUPPLEMENTAL MATERIAL

References

- **Franzosa EA, Hsu T, Sirota-Madi A, Shafquat A, Abu-Ali G, Morgan XC, Huttenhower C.** 2015. Sequencing and beyond: integrating molecular 'omics' for microbial community profiling. *Nat Rev Microbiol* 13(6):360–372. doi: 10.1038/nrmicro3451

- **Martiny JB, Bohannan BJ, Brown JH, Colwell RK, Fuhrman JA, Green JL, Horner-Devine MC, Kane M, Krumins JA, Kuske CR, Morin PJ, Naeem S, Ovreås L, Reysenbach AL, Smith VH, Staley JT.** 2006. Microbial biogeography: putting microorganisms on the map. *Nat Rev Microbiol* 4(2):102–112. doi:10.1038/nrmicro1341

- **Sankara SA, Lagiera J-C, Pontarottib P, Raoulta D, Fourniera P-E.** 2015. The human gut microbiome, a taxonomic conundrum. *Syst Appl Microbiol* 38(4):276–286. doi: 10.1016/j.syapm.2015.03.004

Supplemental Activities

CHECK YOUR UNDERSTANDING

1. **What leads certain microbes to live as consortia?**
   a. Nutrient exchange
   b. Synergistic utilization of nutrients
   c. To maximize energy yields
   d. All of the above

2. **An enrichment culture . . .**
   a. promotes the growth of the most abundant microbe in a sample.
   b. promotes the growth of a particular taxonomic group.
   c. promotes the growth of microbes with a particular metabolic trait.
   d. is a pure culture.

3. **Culture-independent methods can be used**
   a. to assess the microbial diversity in a particular environment.
   b. to predict the metabolic activities of a microbial community and its members.
   c. to infer interactions between microbes in a complex community.
   d. All of the above.

4. **Contrast and compare amplicon-based community fingerprinting to metagenomics. What are the advantages and limitations of each technique?**

5. **Why do microbial ecologists often take an *integrated multi-omics approach* that combines both metagenomics and metatranscriptomics to survey microbial communities?**

6. **Contrast and compare the metatranscriptomics, metaproteomics, and metabolomics methods used to understand the activities of microbial communities in the environment.**

7. **How is the human microbiota any different from any other microbial community in nature?**

DIG DEEPER

1. Some anaerobic bacteria convert butyrate into acetate and hydrogen, a reaction with a positive free energy change ($\Delta G^{\circ\prime} = 48.4$ kJ/mol). However, removal of the hydrogen product makes the free energy change of the butyrate-to-acetate conversion change to -31.2 kJ/mol.
   a. Do you expect a monoculture of the bacterium to grow with butyrate?
   b. How do you expect generation times to change as hydrogen accumulates in the culture? Use the graph provided to plot the correlation between generation time (y axis) and hydrogen concentration (x axis). Define the units you will use for the generation time and hydrogen concentration.

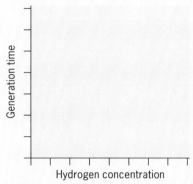

   c. Would you expect the bacterium to grow better or worse in coculture with a methanogen?
   d. Based on the thermodynamics of the reactions, can you predict the composition of the microbial communities that metabolize butyrate to acetate in nature?

2. **Team work:** The famed Russian microbiologist Sergei Winogradsky is regarded as the first microbial ecologist. In the 1880s he designed a column (the Winogradsky column) that replicated the microbial communities living in pond sediments so he could study their metabolic activities. By adding different nutrients to the column, he was able to enrich for communities that cycled different elements such as carbon, nitrogen, and sulfur. Search the web for information about how to make a Winogradsky column. Then work in groups to answer the following questions:
   a. The bottom of the column can be packed with pond mud mixed with shredded paper, calcium carbonate, and egg shells (source of calcium sulfate). The rest of the column is then filled with mud and a layer of pond water and sealed to prevent evaporation. What microbial communities are enriched at the bottom of the column with the paper, calcium carbonate, and calcium sulfate?
   b. Winogradsky columns are often incubated with one side exposed to light. What microbial communities will be enriched on the side of the column exposed to the light?
   c. After several weeks of incubation, a whitish/yellowish community of microbes grows near the top of the sediment surface. Who are these microbes and what supports their metabolism?

3. **Team work:** Work in groups to design an experiment to isolate a nitrogen-fixing microbe from a soil sample. You may find the following videos useful:
   - "How to Set Up an Anaerobic Jar" (https://www.youtube.com/watch?v=oRyj5vWlMe0)
   - Tricks to make anaerobic media and transfer cultures anaerobically: "Video 3: Transferring Cultures" (https://www.youtube.com/watch?v=UOnVyZxZ1ys) and "Preparation of Anaerobic Media" (https://www.youtube.com/watch?v=jTRS7OundUw)
   a. How would you collect the sample?
   b. How would you enrich for the growth of nitrogen-fixers from your soil sample?
   c. How would you isolate individual microbes from your enrichment culture?
   d. How can you confirm that your isolate is indeed fixing dinitrogen gas?
   e. How would you determine if the isolate is a bacterium or an archaeon? What phylum?

f. Design a cultivation-independent approach to assess the relative abundance of your isolate in the original soil that it was isolated from.

g. Design an experiment to test if your isolate is fixing nitrogen in the original soil environment.

4. Cultivation-independent methods can provide novel insights into the structure and functioning of novel microbial communities. The following article uses 16S rRNA amplicon sequencing and metagenomics to gain insights into the role of microbial communities in biogeochemical transformations in oil reservoirs. Read the article and use the information to answer the questions below. Summarize the answers and present them to the class.

Hu P, Tom L, Singh A, Thomas BC, Baker BJ, Piceno YM, Andersen GL, Banfield JF. 2016. Genome-resolved metagenomic analysis reveals roles for candidate phyla and other microbial community members in biogeochemical transformations in oil reservoirs. *mBio* **7:**1. doi:10.1128 http://mbio.asm.org/content/7/1/e01669-15

a. What hypothesis is the team testing?

b. What cultivation-independent methods did the authors used? (Consider the pros and cons of the methods used.)

c. What are the major findings of the work?

d. What type of methods would have been useful to investigate the metabolic activities of the microbial communities under investigation?

Supplemental Resources

VISUAL

- Video from S. Sanschagrin et al. showing how 16S rRNA-amplicon sequencing is performed (10:24 min): http://www.jove.com/video/51709/next-generation-sequencing-of -16s-ribosomal-rna-gene-amplicons
- Video from E. Lopez-Medina et al. showing how to extract high quality RNA from the mouse gut microbiome (8:52 min): http://www.jove.com/video/3293/rna-isolation-pseudo monas-aeruginosa-colonizing-murine
- Video discussing how your dog may be sharing good microbes and helping you stay healthy: "The Microbiome Project: Dogs" (2:02 min): https://www.youtube.com/watch?v =5EXXcyDcTEo

SPOKEN

- iBiology talk by Richard Losick from Harvard University: "Are We More Microbial than Human?" (21:55 min): https://www.youtube.com/watch?v=FMmN2qlhNIl
- *iBiology* lecture "Termites and Their Symbiotic Gut Microbes" by Jared Leadbetter (37 min): http://www.ibiology.org/ibioseminars/jared-leadbetter-part-1.html
- *Microbes After Hours* talk by Martin Blaser: "Missing Microbes" (1 hour) discusses evidence that modern lifestyles impact the microbial communities within the human population. http://www.microbeworld.org/podcasts/asm-after-hours/2046-speak-about -his-book-missing-microbes-how-the-overuse-of-antibiotics-is-fueling-our-modern -plagues
- *Microbes After Hours* talk by Rachel Dutton and Mateo Kehnler: "The Microbiology of Cheese" (69 min) discusses the microbial ecosystems that convert milk to cheese. http://www.microbeworld.org/podcasts/asm-after-hours/1698-the-microbiology-of -beer-the-microbes-after-hours-series-6-8-pm-thursday-october-10-2014the -microbiology-of-cheese-live-june-10-at-asm-headquarters

WRITTEN

- Learn how cultivation-independent methods have provided novel insights into rare taxa, which individually may not be abundant, yet collectively can contribute up to two-thirds of the biomass of some microbial communities: *The Rare Biosphere*, a 2011 colloquium report from the American Academy of Microbiology. http://academy.asm.org/index .php/browse-all-reports/501-rare-biosphere
- Merry Youle and Gemma Reguera discuss the discovery of the smallest microbes using cultivation-independent methods and their isolation in pure culture later on in the *Small*

Things Considered blog post "The Most Abundant Small Things Considered (SAR 11, *Pelagibacter ubique*)." http://schaechter.asmblog.org/schaechter/2015/02/the-most -abundant-small-things-considered.html

- Gemma Reguera discusses the microbial communities that inhabit the coldest places of Earth in the *Small Things Considered* blog post "The Cold Side of Microbial Life": *http:// schaechter.asmblog.org/schaechter/2014/03/the-cold-side-of-microbial-life.html*

- Susan S. Golden, with help from students of The University of California–San Diego, discusses how metagenomics and transcriptomics revealed the circadian clocks of marine microbes in the *Small Things Considered* blog post "The Transcriptomic Motion Picture of Marine Microbes": http://schaechter.asmblog.org/schaechter/2015/09/the -transcriptomic-motion-picture-of-marine-microbes.html

- Moselio Schaechter discusses the composition and activities of the microbes inhabiting the largest underwater microbial mats ever discovered in the *Small Things Considered* blog post "Commuting to Work." *http://schaechter.asmblog.org/schaechter/2010/09 /commuting-to-work.html*

- Gemma Reguera and Moselio Schaechter discuss the differences between the microbiomes of Western and non-Western people in the *Small Things Considered* blog post "There Is No 'Healthy' Microbiome." http://schaechter.asmblog.org/schaechter/2014 /11/fine-reading-there-is-no-healthy-microbiome-.html

- FAQ *Microbes Make the Cheese*, a colloquium report from the American Academy of Microbiology to educate the general public. http://academy.asm.org/index.php/faq -series/5282-faq-microbes-make-the-cheese

- FAQ *Human Microbiome*, a colloquium report from the American Academy of Microbiology to educate the general public. http://academy.asm.org/index.php/faq-series/5122 -humanmicrobiome

The many colors of the water in the Rio Tinto of Spain reflect the diversity of microbes that cycle the elements in this environment

CHAPTER TWENTY

Cycles of Elements

KEY CONCEPTS

This chapter covers the following topics from the ASM Fundamental Statements.

CELL STRUCTURE AND FUNCTION

1. *Bacteria* and *Archaea* have specialized structures that often confer critical capabilities.

METABOLIC PATHWAYS

2. The interactions of microorganisms among themselves and with their environment are determined by their metabolic abilities.

MICROBIAL SYSTEMS

3. Microorganisms are ubiquitous and live in diverse and dynamic ecosystems.

IMPACT OF MICROORGANISMS

4. Microbes are essential for life as we know it and the processes that support life.
5. Humans utilize and harness microorganisms and their products.

INTRODUCTION

Microbes may be small, but they do things on a grand scale. They have a major impact on forming and setting the concentrations of the major gases in the atmosphere—nitrogen, oxygen, and carbon dioxide. They play critical roles in degrading remains of plants and animals that would otherwise accumulate and eventually sequester enough carbon and other bioelements to make life impossible. Microbes are also responsible for a long list of other less readily apparent transformations that have profound effects on our environment. Their recycling activities modify the geochemical signatures of soils, sediments, rivers, lakes, and oceans. They also catalyze redox reactions of a variety of metal salts, thereby contributing to the formation of Earth's huge ore deposits and causing rocks to weather. They even alter the climate. In this chapter we will learn about some of their most notable contributions to the cycle of some essential elements and their global impact on Earth.

CASE: Microbes with a taste for plastic?

(Courtesy of Heal the Bay.)

Throughout the book we have emphasized that microbes, thanks to their long tenure on planet Earth, have mastered an astounding array of metabolic activities to utilize just about any substrate that can provide any of the essential elements (such as carbon, nitrogen, sulfur, and phosphorus) and energy for growth. Manufactured plastics, however, are notoriously recalcitrant to microbial action. And the environment pays the price: only 5% of all the plastic waste is recycled; the rest is discarded after a single use, taking hundreds of years to decay. In 2010, roughly 8 million tons of plastic waste ended up in the oceans, killing hundreds of thousands of birds and marine animals every year, leaching contaminants into our waters, and polluting our beaches. Microbes cannot help us. Or can they? In 2015, a group of scientists reported that waxworms (also known as Indian mealmoths, the larvae of *Plodia interpunctella*) could feed on a diet of polyethylene, the plastic that makes the Styrofoam™ material of hot beverage cups and insulation materials. Were the microbes in the worm's gut consuming the ingested plastic? Polyethylene consists, after all, of interconnected strings of bonded carbons and hydrogens and packs a lot of energy.

- Could the worms digest the plastic without microbes?
- Could microbes associate with the worms and degrade it?
- Design an experiment to test the hypothesis that the worm's gut microbes feed on the plastic and also provide nutrients to the worm.

WHAT IS THE CONTRIBUTION OF MICROBES TO BIOGEOCHEMICAL CYCLES?

LEARNING OUTCOMES:
- Identify the broad role of microbes in biogeochemical cycles on Earth.
- Name the general type of reaction in microbially mediated biogeochemical cycles.

Metabolic diversity—particularly with respect to fueling reactions—is the hallmark of prokaryotes (see chapters 6 and 7). These metabolic innovators are capable of an impressive list of chemical conversions, some of which are essential for the continued existence of all other forms of life (Table 20.1). In the case study, we asked you to consider whether microbes could also feed on plastic, a manufactured material introduced in the environment only in recent years. Most microbes grow and reproduce (chapter 4) and genetically diversify (chapter 10) rapidly. Thus, the presence of an abundant potential source of carbon and energy source such as plastic may have exerted selective pressure on microbes that had evolved metabolisms to feed on what was has been for long considered an unpalatable substrate. The human gut is, after all, colonized by specialized microbes that have diversified their metabolism so together, as a team, we can all gain sustenance from a broader range of foods (chapter 19). Could an analogous symbiosis allow worms and microbes to feed on plastic? In chapter 19 we described how to set up enrichment cultures to isolate microbes based on their specific metabolic activities. How would you go about trying to culture microbes from the guts of the plastic-eating worms?

The interconversions of matter in the biosphere are brought about to a large degree by microbes, and some of these crucial changes are mediated only by prokaryotes (Table 20.1). A useful route to comprehending and summarizing all these highly interrelated conversions is to sort them out according to the chemical changes that the various major bioelements—carbon, oxygen, nitrogen, sulfur, and phosphorus—undergo in nature. These interconversions are in approximate balance, that is, the sum of the reactions that use one particular form of a bioelement is approximately matched by the sum of those that replenish it. Thus, it is possible to describe the various interconversions of bioelements as cycles of utilizations and replenishments. These are called **biogeochemical cycles**.

In the following sections, we will emphasize the roles that microbes play overwhelmingly and uniquely and we'll touch on the considerable human impacts on these cycles. The recalcitrance of plastics to degradation,

TABLE 20.1 **Some metabolic functions mediated only by prokaryotes**

Name	Conversion
Nitrogen fixation	N_2 to ammonia
Denitrification	Nitrate to N_2
Anammox	Anaerobic conversion of ammonia and nitrite to N_2
Nitrification	Ammonia to nitrite and nitrite to nitrate
Commamox	Ammonia to nitrate
TMAO reduction	Trimethyl amine *N*-oxide (TMAO) to trimethyl amine
Sulfate reduction	Sulfate to sulfide
Arsenate reduction	Arsenate to arsenite

which we presented in the case study, shows how much carbon is sequestered by human activities. The introduction of nitrogen-loaded fertilizers in agriculture has similarly impacted the cycling of nitrogen. Each cycle has its own, unique steps and microbial contributions. Yet a generalization applies to almost all them: their steps are *oxidation or reduction reactions.* In most cases, the oxidation state of the bioelement changes throughout the cycle. For example, in the carbon cycle, organic matter is oxidized to carbon dioxide (CO_2), which is then reduced again to organic matter. Why is that? Because, as we learned in chapter 6, at the most fundamental level life harnesses energy and assimilates elements in *reactions that transfer "restless" electrons.* The same principles govern biogeochemical cycles.

HOW DO MICROBES CYCLE CARBON?

LEARNING OUTCOMES:
- Define the phrase "microbial infallibility."
- Describe the importance of autotrophic microbes to the carbon cycle.
- Explain how microbial consortia and interspecies electron transfer contribute to carbon cycling.

Carbon accounts for approximately half of the dry weight of a microbial cell, making it the *most abundant cellular element.* Not surprisingly, microbes have diversified their metabolisms to process just about any source of carbon, organic or inorganic (perhaps even plastic, as shown in the case study). Indeed, microbes are major players in the cycling of carbon on Earth. Figure 20.1 summarizes the essence of the carbon cycle—a cycling between atmospheric carbon dioxide and fixed carbon, either organic or inorganic. Although simple in outline, this cycle encompasses all the complexities of the metabolic and phylogenetic diversity of the biological world, mainly the

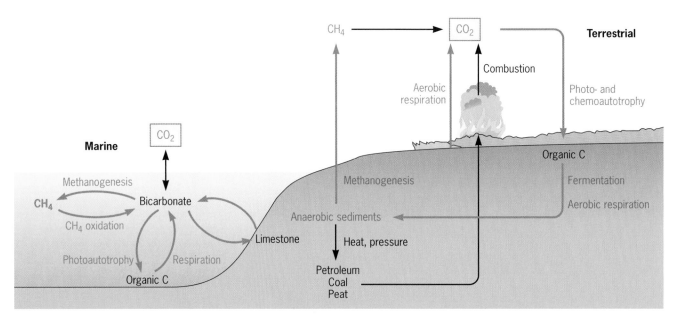

FIGURE 20.1 The carbon cycle. The microbial contributions to the cycling of carbon (C) are indicated in brown.

ability of microbes to fix atmospheric carbon dioxide and slowly return it to the atmosphere through the collective activities of microbial consortia co-operating to degrade the entrapped organic carbon in the environment.

We humans have intervened massively in this cycle with consequences both unintended and not fully understood. By burning fossil fuels at an ever-increasing rate, we have markedly increased the concentration of carbon dioxide in the Earth's atmosphere (Fig. 20.2), contributing to the global warming we are now experiencing. Humans have also added a qualitatively new complexity to the cycle. *Naturally* occurring organic carbon molecules, without exception, are susceptible to microbial attack. Some, such as humus (the black, lignin-rich plant-derived material that accumulates in soils and sediments), are degraded at quite low rates, but there are no naturally occurring organic compounds that cannot be broken down, however slowly, by some microbes (a doctrine sometimes called *microbial infallibility*). However, humans have accomplished what nature could not. Many plastics seem to be completely resistant to microbial attack, and other ecologically dangerous compounds are attacked only extremely slowly. They form a dead-end sump in the carbon cycle. If discarded in nature, they will remain there for a very long time. Though, as we described in the case study, some microbes may have already acquired the ability to degrade them. Perhaps there is hope, after all.

Figure 20.1 shows a simplified view of how the carbon cycle works. Autotrophic microbes, both **phototrophs** and **chemotrophs**, fix the atmospheric carbon dioxide (CO_2) to provide carbon substrates for a number of other microbes in both terrestrial and aquatic environments. This pool of organic carbon, which is supplemented with additional carbon fixed by plants during photosynthesis, supports the growth of aerobic and anaerobic microbes with respiratory and fermentative metabolisms (chapter 6). The cycling of organic compounds trapped in anaerobic environments also generates carbon dioxide, which methanogenic archaea convert into

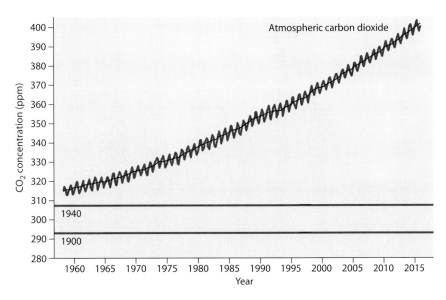

FIGURE 20.2 Concentration of CO_2 in the atmosphere, measured at Mauna Loa, Hawaii. The value at the end of 2015 was over 400 parts per million.

methane (CH_4) to return the carbon to the atmosphere (for more about the **methanogens**, see chapter 14).

Methanogenesis is a critical link in Earth's carbon cycle because the product, methane, being gaseous and sparingly soluble in water, is released into the atmosphere (Fig. 20.1) Worldwide, methanogens produce prodigious quantities of methane from organic material trapped in various anoxic environments—including silt under oceans and lakes and even from the intestinal tracts of ruminant animals and termites (in the next chapter we will discuss just how much methane is produced by microbes in the rumen and how critical this is). This plays a critical role in intertwining the carbon and oxygen cycles (Box 20.1). In addition, huge quantities

BOX 20.1

Cycling carbon and oxygen: an intimate affair

The evolution of **oxygenic photosynthesis** by ancestral **cyanobacteria** some 2 billion years ago oxygenated the planet and exerted tremendous selective pressure on microbes to evolve ways to tolerate it. Some microbes also learned to intertwine the carbon and oxygen cycles. In doing so, microbes diversified their metabolisms in myriad ways and transformed the Earth. The figure illustrates the many ways in which microbes cycle oxygen gas (O_2). Oxygenic photosynthesis by microbes (cyanobacteria and algae) and plants releases oxygen as a by-product and is often coupled to carbon dioxide (CO_2) fixation. This metabolism maintains stable levels of oxygen in the atmosphere and produces organic carbon to sustain heterotrophic **respiration** and **fermentation**. As we learned in chapter 6, oxygen can serve as electron acceptor for respiration. It also oxidizes the crust of the planet and generates oxidized compounds such as metal oxides, which some anaerobic microbes use as alternative electron acceptors for respiration.

Organic carbon is metabolized by fermentative and respiratory microbes to produce carbon dioxide, which methanogens convert into methane (CH_4). The methane supports the growth of methane oxidizers, which cycle the carbon back to the atmosphere as carbon dioxide. Methanogenesis also produces water molecules as a by-product, thus cycling the oxygen atoms and providing the electron donor (H_2O) for oxygenic photosynthesis.

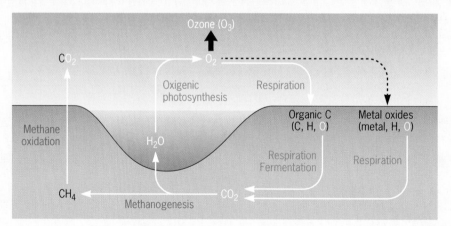

Note an important fact: ultraviolet (UV) light and, to a lesser extent, electrical discharges convert some of the atmospheric oxygen into ozone (O_3), which accumulates in the stratosphere. The ozone layer works as a shield that absorbs much of the incident UV radiation and protects the biosphere from its harmful effects. It also heats the upper layer of the atmosphere, thus inducing a greenhouse effect that is critical for life on Earth. Humans have influenced the oxygen cycle significantly in the past decades by using oxygen for combustion and releasing significant quantities of greenhouse gases such as carbon dioxide and methane to the atmosphere. They have also generated great quantities of ozone-depleting organic chemicals that contain a combination of chlorine, fluorine, and bromine. The uncontrolled release of these chemicals has resulted in the depletion of more than half of the ozone layer over some parts of Antarctica (the so-called Antarctic ozone hole). Can you think of the consequences of this on life on Earth?

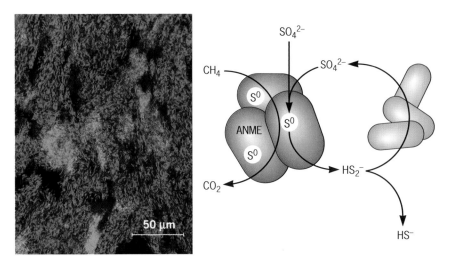

FIGURE 20.3 Cooperative oxidation of methane to carbon dioxide. FISH probes targeting the rRNAs of a group of sulfate-reducing bacteria in green and a group of methane-oxidizing archaea in red were used to reveal the association of the two groups in a cross section of a mat from the Black Sea seen with a confocal microscope (left). The close proximity of the archaeal and bacterial cells has been proposed to promote the exchange of sulfate and its reduced product, hydrogen sulfide (H_2S), between the two to enable methane oxidation (right). (Panel A from Michaelis W et al., *Science* **297**:1013–1015, 2002, with permission.)

of methane from biological and nonbiological sources are buried beneath the ocean floor as solid-phase gas hydrates and as free gas. Some of the atmospheric methane is converted into carbon dioxide but most of it accumulates and contributes greatly to the greenhouse effect that is responsible for the global warming of the planet. (Methane in our atmosphere increased from 1,620 parts per billion in 1984 to 1,842 parts per billion in 2014 but now seems to be leveling off.)

We might ask what happens to all the methane released by methanogens. Until relatively recently, the only known fate of methane, other than being burned by humans, was oxidation by aerobic bacteria called **methylotrophs** or escape into the atmosphere, where it acts as a powerful greenhouse gas. It was thought that methane must escape from its anoxic reservoir before it could be degraded. In the 1970s, however, geochemical evidence suggested that large quantities of methane are degraded within anoxic regions of the oceans. In the late 1990s, a remarkable route of disappearance was discovered that proved to be biological. Using culture-independent methods, such as those described in chapter 19, scientists demonstrated that a two-member microbial consortium consisting of an archaeon and a bacterium were teaming up to anaerobically oxidize methane to carbon dioxide while reducing sulfate to hydrogen sulfide (Fig. 20.3). The metabolic reaction carried out by the consortium was analogous to that of methanogens, but in reverse, as follows:

$$CH_4 + 2H_2O \rightarrow CO_2 + 4H_2$$

In order for this energetically unfavorable reaction to become feasible and therefore able to support microbial growth, sulfate (SO_4^{2-}) is reduced to hydrogen sulfide (H_2S) concomitantly. (Note that these microbial activities also intertwine the carbon and sulfur cycles.) The reaction looks like this:

$$5H_2 + SO_4^{2-} \rightarrow H_2S + 4H_2O$$

It is now well accepted that methane-oxidizing archaea can carry out the two reactions, methane oxidation and sulfate reduction, on their own. Still, they often form microbial consortia with sulfate-reducing bacteria. Why the two consortium partners associate is not well understood. It has been suggested that the sulfide generated by the methane-oxidizing archaea is scavenged and reoxidized by the bacterial partners (Fig. 20.3). This interaction keeps sulfide concentrations sufficiently low to "pull" the reaction of methane oxidation to the right and also replenishes sulfate as an electron acceptor for the archaeon. More recently, a model has been proposed in which the archaeal cells directly transfer electrons to the bacteria to couple their metabolisms. This process, called **interspecies electron transfer**, will be discussed in more detail in the next chapter.

Here are some interesting facts about methane oxidizers. They consume 80% of the methane produced in marine environments. However, none of them has yet been cultivated. Everything we know about them and their bacterial consortia was elucidated solely using cultivation-independent approaches. For example, confocal micrographs of methane-oxidizing mats retrieved from the Black Sea sediments and stained with archaeal and bacterial FISH probes (chapter 19) reveal the close proximity of the two partners (Fig. 20.3). The microbial mats where these samples were taken from form over methane vents in the floor of the sea and are huge (some as high as 4 m and as wide as 1 m). Carbon dioxide released by the consortia reacts with alkaline seawater locally and forms calcium carbonate, which structurally stabilizes the vents. A lot has been learned about these organisms in spite of their never having been cultivated. Perhaps you will be the microbiologist that isolates the first methane oxidizer.

WHY ARE MICROBES CRITICAL TO THE CYCLING OF NITROGEN ON EARTH?

LEARNING OUTCOMES:
- Give examples of how cultivation-independent techniques have been crucial to elucidating the Earth's nitrogen cycle.
- Identify the importance of nitrogen-fixing microbes in the Earth's nitrogen cycle.
- Differentiate assimilatory from dissimilatory nitrate reduction.
- Identify three nitrification pathways.
- Give evidence for the presence and importance of comammox in the nitrogen cycle.
- Compare and contrast denitrification and anammox.

Nitrogen is an essential element for all cells, accounting for approximately 13% of the dry weight of, for example, *Escherichia coli*. Not surprisingly, microbes are particularly versatile at processing just about any nitrogen source in the planet, contributing greatly to its global cycling (Fig. 20.4). The cycle starts with a **nitrogen fixation** step, whereby atmospheric dinitrogen (N_2) is converted into a fixed nitrogen compound. In chapter 7, we learned that some groups of bacteria and archaea catalyze the biological reduction of dinitrogen to ammonia or nitrate. This is not an easy feat. The nitrogen-nitrogen triple bond (N≡N) must first be broken, a process that requires a very high activation energy. Prokaryotes are the only organisms capable of breaking this bond and, thereby, of fixing nitrogen. Small amounts of dinitrogen are also produced in electrical storms and through volcanic activity. Thus, for much of the Earth's history, nitrogen fixation was the near-exclusive province of prokaryotes, until the early 20th century, when the German chemist Fritz Haber developed a chemical method (called the Haber method) to convert dinitrogen to ammonia. Now, about half the world's supply of fixed nitrogen is produced by the fertilizer industry from dinitrogen gas using chemical methods.

Whether from microbial or industrial activities, all the fixed nitrogen that supplies the nutritional needs of life on the planet is derived from the huge reservoir of atmospheric dinitrogen (80% of Earth's atmosphere), and all is returned to it (Fig. 20.4). In geological terms, the cycling of Earth's nitrogen supply through atmospheric dinitrogen is rapid. Its half-life is only about 20 million years, which is fast, considering how much nitrogen gas accumulates in the atmosphere. More startling is the estimate that were the microbes of Earth to cease fixing nitrogen, the soil would be depleted of fixed nitrogen in 1 week. The consequences of this would be

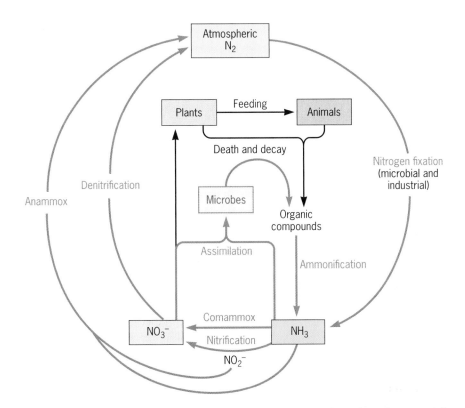

FIGURE 20.4 The nitrogen cycle. The microbial contributions to the cycling of nitrogen (N) are highlighted in brown.

catastrophic. Plants, for example, would stop growing and eventually die. Indeed, the "Green Revolution," a system of improved agriculture that has had a major impact on food production worldwide and decreased starvation, was based principally on providing more nitrogen to crops, mainly in the form of ammonia or nitrate fertilizers.

As seen in Fig. 20.4, microbes cycle nitrogen in two completely different ways. *All microbes use it as a nutrient* to make cellular constituents such as proteins, nucleic acids, and phospholipids, a process known as **assimilation**, which we discussed in chapter 7. *Some microbes* also use the nitrogen sources *as a source of energy* (chapter 6) and the term **dissimilation** is used instead. For example, some autotrophs derive energy by oxidizing ammonia or nitrite to make nitrate. Some **autotrophs** and **heterotrophs**, on the other hand, derive energy by anaerobic respirations that use oxidized nitrogen compounds, such as nitrate (NO_3^-), nitrite (NO_2^-), nitric oxide (NO), and nitrous oxide (N_2O), as a *terminal electron acceptor*. Some nitrogen compounds can fulfill both roles. Nitrate, for example, can be reduced to ammonia to serve as a nutrient (*assimilatory* nitrate reduction) or to gain energy during anaerobic respiration (*dissimilatory* nitrate reduction).

Given its importance as both a nutrient and energy source, it should not surprise you to learn that microbes have evolved creative ways to scavenge for any source of fixed nitrogen, organic or inorganic. This metabolic versatility is illustrated by the fact that some of the steps in the assimilation of fixed nitrogen sources are the exclusive provinces of prokaryotes (Fig. 20.4 and Table 20.1). They include two ways to achieve the complete oxidation of ammonia to nitrate (called **nitrification** and **comammox**, respectively) and two routes by which fixed nitrogen is returned to gaseous form (**denitrification** and **anammox**). We discuss each of them in more detail below.

Nitrification and Comammox

The *oxidation of ammonia* to *nitrate via nitrite* is called *nitrification* (Fig. 20.5). In the first step of the process, ammonia-oxidizing *Bacteria* or *Archaea* (such as the bacterial genus *Nitrosomonas*) oxidize ammonia to nitrite while reducing oxygen. In the second step, nitrite-oxidizing bacteria typified by the genus *Nitrobacter* oxidize nitrite to nitrate, also reducing oxygen in the process. Thus, nitrifiers are often regarded as *specialized chemolithoautotrophs* that gain energy for growth by oxidizing either ammonia or nitrite under *aerobic conditions*. Nitrite being the intermediate, the microbes that catalyze each step of nitrification, ammonia and nitrite oxidizers, are often found in close physical proximity in order to promote the exchange of the nitrite by-product and couple their metabolisms efficiently.

Curiously, the complete oxidation of ammonia to nitrate is not only feasible, but it is thermodynamically more favorable than each reaction individually (Fig. 20.5). Thus, scientists speculated for long that a complete nitrifier could exist in nature that would catalyze the complete oxidation of ammonia to nitrate in a process known as *comammox* (for <u>com</u>plete <u>am</u>monia <u>ox</u>idation).

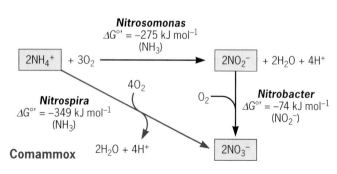

FIGURE 20.5 **Nitrification and comammox.** See text for details.

It was not until 2015 that the existence of a comammox microbe was demonstrated. Several species of bacteria in the genus *Nitrospira* were isolated that carried out the complete oxidation of ammonium to nitrate. Some low levels of nitrite accumulated transiently in the cultures but they were eventually oxidized to nitrate by the cells. In addition, the microbes also grew in medium with nitrite. The sequenced genome of one representative comammox microbe revealed pathways both for ammonia and nitrite oxidation and the genes in each pathway were activated at the same time in cultures to oxidize ammonia to nitrate. Thus, the same microbe was able to carry out the two reactions (ammonium and nitrite oxidation) concomitantly. This goes to show you what turns microbiologists on!

Interestingly, the genes encoding the enzymes (ammonia monooxygenase and hydroxylamine dehydrogenase) that allow some *Nitrospira* species to oxidize ammonia all the way to nitrate are phylogenetically distinct and thus unique to the comammox microbes. The gene sequences also happened to be abundant in *metagenomic sequences* retrieved from many environments. This suggests that comammox is an important microbial process in the nitrogen cycle. Nitrifying prokaryotes are widespread in nature and highly active. As a result, ammonia (and nitrite) is short lived in aerobic environments, such as the upper layers of soil. When anhydrous ammonia gas is added as fertilizer to soil (a common practice in modern agriculture), it reacts with water to form ammonium ion, which binds tightly to clay soils and is therefore immobile. Nitrifying and comammox bacteria come to the rescue, quickly converting the ammonium ion to nitrate ion, a form that moves freely through soil and is readily used by plants.

Denitrification and Anammox

The final step of the nitrogen cycle returns fixed nitrogen to the atmosphere as dinitrogen gas (Fig. 20.4). This step is *almost exclusively biological* and, *for the most part, anaerobic* and can use nitrate (denitrification) or ammonium (anammox) as substrate. Below we discuss each process individually.

From nitrate to N_2: denitrification. Until recently, denitrification was the only known microbial process to return dinitrogen back to the atmosphere. The process relies on a *cascade of anaerobic respirations* through which nitrate (NO_3^-) is successively reduced to nitrogen gas (N_2) via intermediate nitrogenous oxides such as nitrite (NO_2^-), nitric oxide (NO), and nitrous oxide (N_2O), as follows:

$$NO_3^- \rightarrow NO_2^- \rightarrow NO + N_2O \text{ (gas)} \rightarrow N_2 \text{ (gas)}$$

A wide range of prokaryotic and eukaryotic microbes, *both autotrophs and heterotrophs*, denitrify, though heterotrophic denitrification is more common. Autotrophic denitrifiers include, for example, *Thiobacillus denitrificans*, an obligate chemolithoautotrophic bacterium that can denitrify using some seemingly unpalatable electron donors such as sulfur compounds, ferrous iron (Fe[II]) and tetravalent uranium (U[IV]). Among the most abundant heterotrophic denitrifiers are *Pseudomonas* species, a metabolically versatile group of bacteria known for its ability to metabolize a broad spectrum of carbon sources. But, often times, microbes specialize in carrying out only a subset of the reactions involved in denitrification and work as a team in multispecies assemblages to complete the denitrification process all the way to dinitrogen.

Whether catalyzed by a single microbe or by the concerted action of several, the sequential reactions that reduce nitrate to dinitrogen create the proton motive force needed for ATP production during respiration, as shown in the complete denitrification reaction:

$$2NO_3^- + 10e^- + 12H^+ \rightarrow N_2 + 6H_2O$$

Note that one of the intermediates formed in this reaction, nitrous oxide (N_2O), is also a gas and, if not removed as soon as it is produced, it can escape to the atmosphere. This is a very potent greenhouse gas that destroys the protective ozone layer shield (Box 20.1). Thus, there is a lot of interest in studying the microbial activities that contribute or prevent nitrous oxide emissions in order to develop approaches that mitigate its atmospheric release. This is particularly critical in rivers and streams with high nitrate loadings such as those polluted with agricultural runoffs, which carry high amounts of nitrate from fertilizers.

One interesting thing about denitrification is that the biological reactions preferentially use lighter (^{14}N) rather than heavier (^{15}N) **isotopes** of nitrogen present in the environment. Such isotopic fractionation allows scientists to study the contribution of denitrification to the cycling of nitrogen in the environment. This approach revealed, for example, that in marine environments, denitrification is a major contributor to nitrogen losses to the atmosphere. Many areas of the oceans (the so-called oceanic oxygen minimum zones, typically at depths of 200 to 1,000 meters) have concentrations of oxygen low enough to promote the anaerobic cycling of nitrogen. These zones contribute to the release to the atmosphere of 35% of all the oceanic nitrogen. The Arabian Sea alone contributes about half of this, mainly through the activities of denitrifiers. Globally, denitrification accounts for about two-thirds of all of the loss of nitrogen in the oceans.

Nitrate is also abundant in sewage and municipal wastewater. Wastewater treatment plants first reduce the load of carbon in the wastewater in aeration tanks that stimulate organic carbon decomposition by aerobic metabolisms; the wastewater is then transferred to an anaerobic tank to stimulate denitrification and reduce the nitrate to dinitrogen. During the denitrification step the carbon load is further reduced, as the carbon substrates remaining in the wastewater are used by the heterotrophic denitrifiers as carbon and energy sources. In some treatment plants, however, organic carbon in the form of methanol, ethanol, acetate, and glycerol is added to boost the growth of denitrifiers and improve the efficiency of nitrate removal. Think about denitrifiers next time you drink tap water!

From ammonia to N₂: anammox. Until recently, it was thought that denitrification was the only microbial process that recycled fixed nitrogen to return it as dinitrogen back to the atmosphere. Then in the mid 1990s, a new group of bacteria that can produce dinitrogen gas from fixed nitrogen was discovered. These bacteria mediate an _anaerobic ammonia oxidation_ called *anammox*—a route now recognized as being a major way by which fixed nitrogen (in the form of ammonia and nitrite) is recycled to atmospheric dinitrogen gas (Fig. 20.4). This route was missed because the microbes that mediate the process could not be cultivated and besides, conventional denitrification seemed an adequate explanation for the fixed nitrogen-to-dinitrogen step of the nitrogen cycle. However, the contribution of anammox has been shown to be significant. Estimates are that at least one-third, if not more, of dinitrogen production in the oceans is derived from anammox.

It is instructive to consider just how anammox was discovered. Two Dutch microbiologists were faced with the following dilemma: ammonia, but not nitrate, was disappearing from an aerobic reactor in a wastewater treatment plant they were studying. Why one but not the other? Nitrite was also disappearing; thus, they suspected a reaction in which ammonia was oxidized by nitrite, yielding a dinitrogen molecule containing one nitrogen atom from ammonia and another, from nitrite, i.e.:

$$NH_4^+ + NO_2^- \rightarrow N_2 + 2H_2O$$

They proved their hypothesis by adding isotopically labeled ammonia ($^{15}NH_4^+$) and nitrite ($^{14}NO_2^-$) to the reactor and showing that dinitrogen was produced that contained one atom of ^{15}N and one of ^{14}N. Then they showed that the process was almost certainly biological because it was stopped by heat, gamma irradiation, and compounds that are known to disrupt the cell's proton gradient and, therefore, ATP generation via **chemiosmosis** (see chapter 6).

In spite of discovering the anammox reaction, the investigators could not culture the organism that was mediating it, but they could enrich for it in a flow-through reactor fed ammonia, nitrite, and bicarbonate. A mixed culture, 80% of which was made up of similar-appearing coccoid cells, developed. By density gradient centrifugation, these coccoid cells were concentrated to more than 90% purity. The cell preparation could convert ammonia and nitrite into dinitrogen while fixing carbon dioxide. (Note that the anammox bacterium was an *autotroph*; by contrast, denitrifiers are most often *heterotrophs*, that is, they use organic sources of carbon.) Still, the cells could not be recovered in pure culture. Sequencing their 16S rRNA gene showed that they were bacteria belonging to the phylum *Planctomycetes* (which we met in chapter 3 when describing the distinctive invaginations that their cell membrane makes inside the cytoplasm and around the nucleoid, which fooled scientists into thinking it was a nuclear membrane). The putative species in the enriched culture was named "*Candidatus* Brocadia anammoxidans" ("*Candidatus*" indicates that it hasn't been cultivated). But the bacterium still has defeated cultivation attempts. How can these bacteria be enriched to about 90% purity yet defeat attempts to recovery in pure culture? This is most likely due to the fact that anammox bacteria grow very slowly under these conditions (the approximate doubling time is 11 days). Still, the discovery of anammox microbes illustrates how much can be learned about the activities of microbes in nature, even if they cannot be cultivated. It also illustrates how difficult it is to cultivate certain microbes, even when you know a lot about them.

THE MICROBIAL CYCLING OF SULFUR: WHY DOES IT MATTER?

LEARNING OUTCOMES:
- Describe how specialized morphological features may allow *Desulfobulbaceae* to oxidize hydrogen sulfide for energy.
- Describe the role of chemoautotrophic bacteria in supporting life at deep-sea hydrothermal vents.

The sulfur cycle (Fig. 20.6) resembles the nitrogen cycle in that sulfur, like nitrogen, serves two distinct metabolic roles for microbes: for all organisms, it is an essential nutrient (although sulfur is less abundant in the cell than nitrogen), and for some, it enters into both oxidative and reductive pathways that generate ATP. In the latter role, it is cycled in huge amounts.

The terrestrial reservoirs of sulfur are enormous and provide microbes with sources of the essential element for assimilation and to support their growth metabolism. Massive deposits of sulfur in the form of gypsum, for example, occur in the beds of ancient lakes and are being formed in some lakes today. Gypsum is a soft calcium sulfate mineral ($CaSO_4 \cdot 2H_2O$) that some bacteria exploit as a nutrient. The process occurs in two steps of the sulfur cycle: an *anaerobic* step in which sulfate-reducing bacteria convert the sulfate (SO_4^{2-}) into hydrogen sulfide (H_2S), and a second step whereby phototrophic bacteria oxidize hydrogen sulfide to elemental sulfur (S^0). The sulfur is used by a wide range of microbes as an electron acceptor for anaerobic respiration, which generates hydrogen sulfide as well.

Hydrogen sulfide is also oxidized by a variety of autotrophic bacteria, mostly through *aerobic respiration* but in a few cases through *anaerobic respiration*. This conversion occurs abundantly in just about any place hydrogen sulfide is produced, often by sulfate-reducing bacteria in an anaerobic region—mudflats, for example. Because hydrogen sulfide spontaneously oxidizes in air, chemoautotrophic sulfide oxidizers locate themselves at

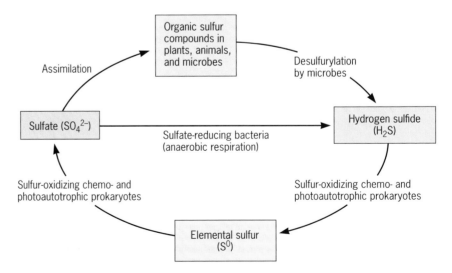

FIGURE 20.6 **The sulfur cycle.** See text for details.

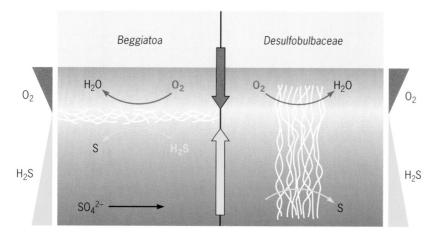

FIGURE 20.7 **The *Beggiatoa* and *Desulfobulbaceae* approaches to coupling the oxidation of sulfide to oxygen reduction in sediments.** See text for details.

the interface where rising hydrogen sulfide comes in contact with oxygen—the only place they can have access to both nutrients (Fig. 20.7). *Beggiatoa* bacteria, for example, position themselves at an interface at which the two gases meet, hydrogen sulfide diffusing upwards from the bottom layers of the sediments and oxygen diffusing from the top (Fig. 20.7). As we discussed in chapter 19, trying to replicate such conditions in the laboratory is not easy, one reason it is so difficult to cultivate some microbes in the laboratory.

Bacteria in the family *Desulfobulbaceae* take a completely different approach to grow with both hydrogen sulfide and oxygen (Fig. 20.7). These bacteria grow abundantly in sulfidic marine sediments, forming long (up to 1.5 cm) multicellular filaments that spatially connect the oxic and sulfide layers. The microbes couple the oxidation of the hydrogen sulfide at the bottom of the sediment with the reduction of oxygen in the top layer. This coupled metabolic process happens so fast that it can only be explained if the microbes move the electrons generated from hydrogen sulfide oxidation as electrical currents along the filaments. For this reason, the process is referred to as *electrogenic* sulfide oxidation.

The mechanism that could allow the bacterial filaments to transport electrical currents, as if they were bacterial power cords, is not known. But some features of their cell envelope provide some clues into how this may be accomplished. These bacteria are Gram negative and thus have a periplasmic space and an outer membrane (chapter 2). However, the periplasms of neighboring *Desulfobulbaceae* cells are interconnected and house tubular structures that span the length of the multicellular filament, as if they were intercellular conduits. The presence of these periplasmic tubes gives the outer membrane a characteristic wrinkled appearance. Studies also show that the filaments' joined periplasm is highly charged, which could allow them to carry electrical currents. How this is possible is not known. Regardless of how they do it, it is clear that these microbes have evolved a creative way to couple the oxidation of the toxic hydrogen sulfide to the respiration of oxygen, two seemingly incompatible processes. In doing so, they couple spatially separated biogeochemical reactions and link the cycles of sulfur and oxygen.

In most locations, sulfide oxidizers utilize hydrogen sulfide that is supplied by the metabolism of other microbes, such as during the reduction of sulfate, but there are spectacular exceptions. Geochemically produced hydrogen sulfide spews out of hydrothermal vents in the mid-ocean floors, where Earth's tectonic plates separate and new crust is constantly being formed. A complex community consisting of hundreds of species, both prokaryotic and eukaryotic, thrives there. Unlike the rest of the biosphere, which depends on the primary productivity of photosynthesis, the community in this sunless region depends entirely on chemoautotrophic microbes that oxidize hydrogen sulfide from the vents at the expense of oxygen diffusing down from the ocean surface. In some cases, this dependence on hydrogen sulfide-oxidizing microbes is remarkably intimate. For example, the huge tubeworms that grow in such environments lack an intestinal system. Instead, they are filled with a spongy tissue, the **trophosome**, that is densely colonized by sulfide-oxidizing bacteria which supply the worm with nutrients.

The sulfur cycle also has an important atmospheric component. Large amounts of volatile sulfur-containing compounds, chiefly **dimethyl sulfide**, enter the atmosphere and are changed there by sunlight into other forms that influence the weather. This aspect of the sulfur cycle is discussed later in the chapter.

HOW DO MICROBES CONTRIBUTE TO THE CYCLING OF PHOSPHORUS?

LEARNING OUTCOMES:
- Identify the important role of microbes in the phosphorus biogeochemical cycle.

The phosphorus cycle (Fig. 20.8) is the simplest of the biogeochemical conversions because, for the most part, this element stays in the same oxidation state, +5 (that is, as **phosphate**). Phosphate, either organic (as esters, amides, or anhydrides) or inorganic, is the principal form of phosphorus in biological systems. Worldwide cycling of phosphate is extremely slow, because *this cycle has no significant gaseous intermediates*. Most forms of

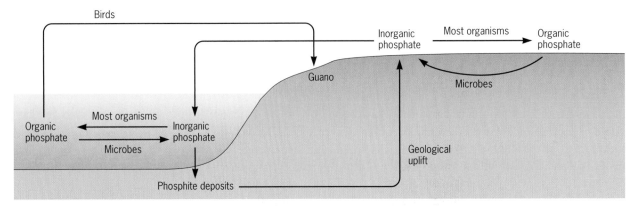

FIGURE 20.8 **The phosphorus cycle.** See text for details.

phosphate, being soluble, are leached from soil and eventually taken to the oceans. Return of phosphate from the oceans to land masses occurs to a minor extent when marine birds that feed on marine animals deposit their feces on land. (These feces, collectively known as guano, are also rich in nitrogen and are increasingly sought after as a plant fertilizer in organic farming.) However, most phosphorus from the seas returns to Earth's landmasses only by geological uplift of the ocean floors—a very slow process. The products of these uplifts are phosphate rock, which is mined largely (about 90%) for use in fertilizer.

To a lesser extent, nonphosphate forms of phosphorus do participate in the phosphate cycle. A gaseous form, phosphine (H_3P), which spontaneously ignites in air, has long fascinated humans as a possible cause of "the will o' the wisp," the ghostly pale blue light sometimes seen hovering over swamps. Phosphine does exist in nature. It occurs in minuscule quantities (nanograms to micrograms per cubic meter) in the lower atmosphere, but it is not clear whether any of it has a biological origin. Other reduced forms of phosphorus, including phosphites (PO_3^{3-}) and hypophosphites (PO_2^{3-}), are also found in nature, and some bacteria are capable of oxidizing them for use as sources of phosphate. In some environments, significant quantities of phosphonates ($R\text{-}PO_2^{2-}$) are encountered. These are of clear biologic origin, made by prokaryotes and eukaryotes (phosphonate antibiotics are made by some streptomycetes; chapter 14). Although the C-P bond of phosphonates is quite stable, certain bacteria can break it and utilize phosphonates by converting them to phosphate.

WHAT ELSE CAN MICROBES RECYCLE?

LEARNING OUTCOMES:
- Explain how the specialized structures of *Geobacter* allow it to reduce metal oxides.
- Identify how *Geobacter* can be used in bioremediation.

Only dissolved nutrients can enter the cells of prokaryotes. Eukaryotes can ingest particulate nutrients by phagocytosis, but prokaryotes apparently lack this ability. Nevertheless, prokaryotes can utilize a variety of insoluble nutrients, including starch, cellulose, and even agar. They do so by secreting enzymes that break down these insoluble polymers into soluble subunits, which then enter the cell. Thus, prokaryotic digestion occurs in the cell's immediate external environment. They have an "external stomach."

In recent years, startling exceptions to this generalization have been discovered: insoluble materials, such as iron (Fe[III]) and manganese (Mn[IV]) oxides, that cannot be broken down into soluble subunits can nevertheless be exploited as nutrients by certain bacteria. A genus of bacteria, *Geobacter*, which we will revisit again in the next chapter and in chapter 28, is capable of anaerobic respiration using one or another of these solid oxides as the terminal electron acceptor (Fig. 20.9). It does so without taking these minerals into the cell. In the case of iron, most of the ferric iron (Fe[III]) contained in the solid rust-colored mineral particles is reduced (i.e., electrons are transferred from the cell to them) during

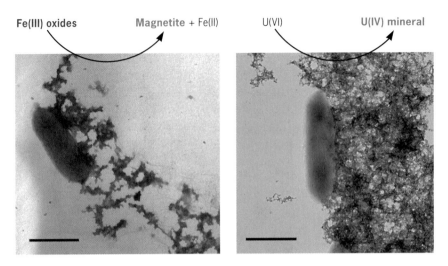

Fe(III) oxides   Magnetite + Fe(II)   U(VI)   U(IV) mineral

FIGURE 20.9 *Geobacter* bacteria use pili as nanowires to transfer respiratory electrons to iron oxides (left) and to bind and mineralize uranium (right). Scale bar, 1 μm. (Courtesy of Gemma Reguera.)

respiration. Some of the ferrous ion (Fe[II]) that is so produced during respiration is solubilized and used by Fe(II)-oxidizing bacteria as the electron donor for their metabolism. Not all, however; some ferrous iron remains trapped in the mineral particles, which become a magnetic mineral called magnetite.

How can *Geobacter* bacteria "breathe" insoluble mineral particles? They have evolved an unusual cell envelope to do so. The cells assemble a single, lateral flagellum that allows them to swim chemotactically towards soluble metal ions (Fe[II] and Mn[II]) produced by the reduction of iron and manganese oxides, respectively. Hence, the motile cells can "sense" where the minerals are located by following the trail of the reduced soluble metal. The cell envelope of *Geobacter* bacteria is also loaded with metalloproteins called *c*-cytochromes, which help transport across the cell envelope electrons that are generated during respiration. The side of the cell where the flagellum assembles also has numerous pili (see chapter 2). The pili enable other bacteria to attach to surfaces, but in *Geobacter* they are conductive and transport electrons between the cell and the minerals dispersed in the soils and sedimentary environments. Thus electrons, or an electric current if you like, flow from organic nutrients in the cell through an electron transport chain and out of the cell through the pili to the chunks of iron oxides in the external environment (Fig. 20.9).

A fascinating consequence of *Geobacter*'s ability to donate electrons to a solid surface is its capacity to generate a *usable electric curren*t in a rig where an electrode serves as the only electron acceptor. In chapter 28 we will discuss how this unique physiology can be harnessed for biotechnological applications. Here we will mention just one more reaction catalyzed by *Geobacter*, one that contributes to the biogeochemical cycling of *uranium*. In the environment, this radionuclide is often immobilized in uranium ores, but some can be present in soluble form as the toxic uranyl divalent cation (UO_2^{2+}). The intense mining of uranium ores during the Cold War did in fact generate vast amounts of subsurface waters, soils, and sediments contaminated with this toxic form of uranium. Being soluble, it spreads fast once it reaches water bodies. This also makes it more bioavailable (i.e., easier to ingest by animals and to assimilate by all living cells). How can this radionuclide be contained? It turns out that the pili of *Geobacter* bind the uranium *with gusto* and use it as an electron acceptor for respiration (Fig. 20.9). The soluble uranyl cation, U(VI), becomes a U(IV) mineral upon reduction by the pili. Thus, the pili effectively mineralize the uranium contaminant, a process that shows promise for the bioremediation of uranium-contaminated environments. Note that the bacterium is not completely altruistic here. By binding the uranium extracellularly, the pili prevent it from permeating inside the cell envelope, where it can bind proteins and phospholipids and interfere with its essential functions. Thus, the conductive pili provide an efficient protective mechanism that immobilizes the toxic molecule far away from the cell envelope. In the process it dumps some respiratory electrons and gains energy for growth. *Clever!*

WHAT ARE THE GLOBAL IMPACTS OF MICROBIAL RECYCLING ACTIVITIES?

LEARNING OUTCOMES:
- List microbes that have a large ecological impact on the global surface ocean through primary production.
- Identify the primary source of organic carbon for aerobic heterotrophs in deep ocean waters.
- Speculate why the concentration of benthic microbes exceeds that of pelagic microbes by five orders of magnitude.
- Describe structural adaptations of ocean floor microbes, such as *Thioploca* and *Thiomargarita*.
- State reasons why microbes should affect climate.
- Identify the role of microbes in the negative feedback loop of cloud formation.

To assess the global impacts of the microbe-mediated transformations that we've just discussed, we may ask where they take place. The simple answer is, just about everywhere. Few places on the planet are too hot, too cold, too acid, too alkaline, too salty, or at too high a hydrostatic pressure for microbes to thrive. Wherever there are microbes, the essential elements are recycled. We present a few of the major habitats.

Soil

The upper layers of soil are an especially active microbial ecosystem in which many steps of the biogeochemical cycles occur. They teem with microbes. A gram of typical soil contains a million to a billion bacterial cells, 10 to 100 m of fungal hyphae, thousands of algal cells, and countless protozoa. As we discussed at the beginning of this chapter, microbial growth is not limited to the upper layers. Although the number of microbes decreases in lower layers, microbial activity continues down to the extraordinary depths of 5 or 6 km below the surface of the Earth (Fig. 20.1). Everything is recycled!

Oceans

The upper regions of the oceans, which cover 71% of Earth's surface, are the other globally important microbial ecosystem. It is estimated that about half of the photosynthesis (the carbon dioxide-fixing and oxygen-producing arm of the carbon cycle) that occurs on the planet is carried out by **phytoplankton** (floating phototrophic microbes) living in the upper layers of the ocean where enough light penetrates to support their growth. In addition to being major modulators of the proportions of gases in Earth's atmosphere, phytoplankton constitute the beginning of the food chain for life in the seas. Other marine organisms, from krill (small shrimp-like creatures) through crustaceans and fish to whales, depend directly or indirectly and almost exclusively on the primary productivity of the phytoplankton. The most abundant members of the phytoplankton are *single-celled cyanobacteria*, belonging principally to just two genera, *Synechococcus* and *Prochlorococcus*. In chapter 19, we introduced these marine cyanobacteria and the relatively recent realization of their critical contribution to the ocean's and to the world's carbon cycle. The abundance of tiny (for a cyanobacterium) *Synechococcus* (its cells are about 0.9 µm in diameter) was recognized in 1979. The even smaller *Prochlorococcus* (0.5 to 0.7 µm) was discovered in 1988. Eukaryotic phytoplankton, such as diatoms and algae, are more than 10 times larger but, being more abundant and metabolically active, the cyanobacterial phytoplankton exerts a greater ecological impact.

Like those in soil, microbes in the ocean are not restricted to the upper regions. Microbes are found all the way down to the regions of darkness, cold, and crushing hydrostatic pressures that exist at the bottom of the seas (which have a median depth of 4 km). We already discussed the sulfide-oxidizing bacteria that flourish on the ocean floor near hydrothermal vents. Although distinctly layered, so that the water column presents different environments at different depths, the seas are subject to stirring, both by the action of winds and tides on the surface and by circulating currents. As a result, the water column becomes partly mixed but not entirely.

Keeping in mind that the oceans of the world vary enormously, what can we say about the nutritional environment at various depths? In addition to dissolved nutrients, the oceans contain, among other things, seaweed, crabs, and fish and are greatly affected by their bottom sediments, as well as the solid earth at the coasts. In the open seas, the concentration of available organic carbon material is low but measurable. There is enough organic carbon to sustain a considerable population of heterotrophic microbes. In fact, the concentration of bacteria in surface seawater is about 1 million/ml, a figure that is relatively constant throughout the world's oceans.

Light is plentiful on the surface, but only about 1% of sunlight reaches a depth of 100 m. Thus, the zone where *phytoplankton* flourish is quite narrow. Some bacteria ensure that they remain at this favorable location by changing their buoyant density via gas vacuoles (see chapter 3). In contrast, the zone favorable to aerobic heterotrophs is much broader. Oxygen, which like sunlight, enters the ocean at the top, either from that in air or from that produced by phytoplankton, extends to the bottom. Its concentration at depths of 4 km is still half as great as it is on the surface. Phosphorus is scarce on the surface and more abundant below about 1 km.

Consequently, considerable microbial life is found in deep waters. Of course, aerobic heterotrophs also need organic nutrients. These are largely supplied by the constant "rain" of organic detritus that slowly settles through the water column. This material, known as **marine snow**, consists of polysaccharide matrices readily visible with the naked eye and in which are embedded living or dead animals and plants. Sometimes, marine snow assumes near blizzard proportions and can limit the visibility of ocean divers to a few feet. Marine snow sediments at rates in the range of 30 to 70 feet per day and, when it hits the bottom, becomes available for the microbial and nonmicrobial residents of the depths. By far the greatest number of **pelagic** microbes (those in the water column) are attached to this particulate matter, which supplies them with nutrients.

The sea floor is a different environment than the water column. The organic nutrients there are steadily replenished by the falling organic detritus. The concentration of **benthic** microbes (those in the sediment on the bottom) exceeds that of pelagic microbes by up to 5 orders of magnitude. Oxygen is quickly depleted in this environment. This is the home of microbes that use nitrate, sulfate, or ferric iron as their terminal electron acceptor for respiration (see chapter 6). Those that utilize sulfate live deeper in the sediments and generate hydrogen sulfide, which other anaerobes can oxidize to sulfur using nitrate as their terminal electron acceptor. Nitrate is fairly abundant in seawater, but it is depleted in the sulfide-rich sediment where these bacteria proliferate. How do such bacteria acquire their nutrients, which are separated in space, one (nitrate) in the water column and the other (sulfide) in the sediment? Earlier we described the vertical growth of *Desulfobulbaceae* to spatially bridge the oxygen and hydrogen sulfide layers of marine sediments (Fig. 20.7). Coupling sulfide oxidation to nitrate reduction uses equally creative ways. Cells of *Thioploca*, for example, become encased in a long mucous sheath, which serves as a transport route between the underlying sediment and the overlying water (an example of a sheathed bacterium is shown in Fig. 20.10). *Thioploca* cells travel up and down these sheaths, acquiring sulfide from below and the nitrate needed to oxidize it from above. These organisms take an elevator to work! *Thiomargarita* solves the problem in a different way. Its cells are also encased in a mucous sheath but remain relatively stationary and evenly separated in the mucous strands. Instead of commuting between food and oxidant, they sit and wait, accumulating nitrate while waiting for a puff of food in the form of hydrogen sulfide to pass by. They get so fat that each cell in the filament can be seen with the naked eye. Consider, for example, *Thiomargarita namibiensis*, whose cells are between 300 and 700 µm in diameter, the largest among all bacteria (Fig. 20.11). Most of the cell volume is occupied by a vacuole where the nitrate is stored and concentrated at levels up to 10,000 times higher than in seawater. The internal vacuole is so large that the cytoplasm of the cell is displaced as an outer shell only 1 to 2 µm thick. The researchers who found this organism in the waters of Namibia named it *T. namibiensis*, which means sulfur pearl of Namibia. This is because, besides nitrate, *T. namibiensis* also stores elemental sulfur, which shines as an opalescent blue-green whiteness, making the cell filament resemble a strand of pearls (Fig. 20.11).

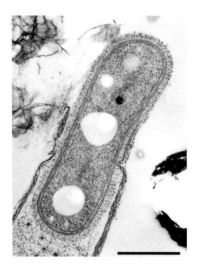

FIGURE 20.10 **A thin section of a sheathed bacterium (*Leptothrix discophora*) examined under an electron microscope.** Bar, 1 µm. (Courtesy of T. J. Beveridge.)

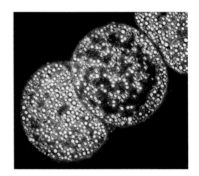

FIGURE 20.11 ***Thiomargarita namibiensis* cells.** The white bodies are sulfur granules made by the oxidation of hydrogen sulfide. Most of the cell consists of an internal vacuole filled with nitrate. (Courtesy of H. Schulz.)

Microbes, Climate, and Weather

That microbes should affect the climate of our planet is not surprising, because they play critical roles in the turnover of the main atmospheric gases such as nitrogen, oxygen, and carbon dioxide. The atmospheric reservoirs of oxygen and nitrogen are so vast that their turnover times are in the millions of years. The turnover time of carbon dioxide is considerably shorter, only a few years. As we have discussed, atmospheric carbon dioxide is the greenhouse gas that exerts major effects on Earth's long-term climate. It is now becoming clear that microbes also affect *local weather* by causing atmospheric changes that take place over a period of just days.

How can microbes influence the weather locally? The story is about cloud formation. It is still an unfolding story, but the general plot is already known. Figure 20.12 illustrates the process. Clouds form when certain particles or compounds in the air act as nuclei for water vapor to condense on, thereby forming fine droplets. Bacteria and fungal spores, swept upwards by winds, act as such nuclei and contribute to rain making. Some of the nucleating compounds contain sulfur. Where do these come from? A large fraction of them are from nonbiological sources, such as volcanic emissions and the burning of coal and high-sulfur petroleum. These sources produce sulfur dioxide (SO_2), which is oxidized by sunlight to sulfur trioxide (SO_3) in the atmosphere. Then, upon hydration, the sulfur trioxide compound becomes sulfuric acid (H_2SO_4), the principal nucleating form. Soon, in a matter of days, this sulfuric acid is returned to Earth in

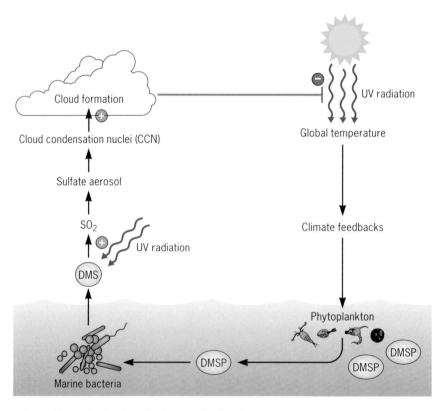

FIGURE 20.12 The DMS cycle. See text for details.

water droplets (the so-called acid rain) or on the surfaces of particulate matter.

As it turns out, large amounts of nucleating sulfur compounds have a biological origin. This part of the story begins in the top layers of the oceans, where the phytoplankton, including eukaryotic members such as the coccolithophores (see chapter 16), are found in prodigious quantities (Fig. 20.12). On cloudless days, the sun's intense ultraviolet irradiation stresses them. In response, these eukaryotic microbes make huge amounts of a protective compound, dimethylsulfoniopropionate (DMSP). During the periods of algal blooms, DMSP from dying algal cells can reach concentrations in the millimolar range in the surrounding seawater. A variety of bacteria in the water metabolize DMSP, breaking it down to a volatile compound, dimethyl sulfide (DMS), which escapes into the atmosphere (Fig. 20.12). In the atmosphere, DMS reacts with oxygen to create various sulfur compounds, including SO_2, that act as nuclei of the water droplets that form clouds. Massive amounts are made: the total annual flux of biogenic DMS to the atmosphere approaches 50 million tons of sulfur per year.

Notice that the process acts as a negative feedback loop (Fig. 20.12). As the intensity of sunshine increases, algal stress increases, causing more DMSP to be made; the sequence of reactions from DMSP to nucleating compounds leads to the formation of more clouds, which decrease the intensity of sunshine that reaches the surface waters. Then, algae become less stressed, they make less DSMP, and fewer clouds form. The cycle repeats. Do these cycles occur over land? The answer is not yet clear, but it is well established that microbes do make DMS wherever they grow. Some microbes produce DMS from dimethyl sulfoxide (DMSO), an industrial waste commonly found in sewage. Indeed, the main "rotten-cabbage" smell in wastewater treatment facilities is due to DMS. Some is also produced in our intestines and, together with other gases we expel (mainly hydrogen sulfide), gives our flatus its distinctive unpleasant smell. It is well known, in fact, that the smell of flatus increases significantly in individuals who have diets rich in sulfur-containing amino acids.

CASE REVISITED: Microbes with a taste for plastic?

Did you hypothesize that microbes associated with the worms provided nutrients for the worms by degrading the plastic? Then you reached the same conclusion as the scientists who carried out this study. They were able to isolate two bacterial strains from the worms' gut in enrichment cultures (see chapter 19) containing polyethylene films as the sole carbon and energy source. After about a month of incubation, they noticed that bacterial **biofilms** coated the plastic film and degraded its surface. They isolated two strains from the biofilms, *Enterobacter asburiae* YT1 and *Bacillus* sp. YP1. Each of the strains was able to partially degrade the plastic and use it as a sole source of carbon and energy, although slowly. After 2 months, each of the strains degraded only about a tenth of the plastic material. The scientists also detected a dozen soluble products derived from the plastic, suggesting that the microbes were solubilizing the films and releasing potential nutrients for the worms. It appears that persistent plastic pollution in the environment has indeed exerted selective pressure on some organisms to use it as a nutrient. In this example, a symbiotic association between bacteria and worms was necessary. Why is this? By chewing the plastic the worm provides a more digestible meal for the gut microbes. The gut microbes likely work synergistically to solubilize the plastic meal and provide the worm with nutrients. But this is purely speculative and based on similar symbiotic associations between gut bacteria and their hosts, including humans. In the next chapter, we will learn just how important microbial interactions of this type are to sustain life as we know it.

CONCLUSIONS

The commanding fact that the overwhelming majority of prokaryotes in most environments have not yet been cultivated announces clearly that microbial ecology remains a work in progress. However, the far greater challenge is discovering the metabolic roles of microbes that resist cultivation. We learned about their global impacts in the cycling of elements such as carbon, oxygen, nitrogen, phosphorus, and sulfur. But microbes contribute to many more reactions than we can describe in this book. Certainly, we cannot fully predict what uncultivated microbes are capable of doing. But based on the little we know we can predict that many elemental reactions on Earth are influenced whether directly or indirectly by microbial activities.

Representative ASM Fundamental Statements and Learning Outcomes for the Chapter

CELL STRUCTURE AND FUNCTION
1. *Bacteria* and *Archaea* have specialized structures that often confer critical capabilities.

   a. Describe how specialized morphological features may allow *Desulfobulbaceae* to oxidize hydrogen sulfide for energy.
   b. Explain how the specialized structures of *Geobacter* allow it to reduce metal oxides.
   c. Describe structural adaptations of ocean floor microbes, such as *Thioploca* and *Thiomargarita.*

METABOLIC PATHWAYS
2. The interactions of microorganisms among themselves and with their environment are determined by their metabolic abilities.

   a. Name the general type of reaction in microbially mediated biogeochemical cycles.
   b. Give examples of how cultivation-independent techniques have been crucial to elucidating the Earth's nitrogen cycle.
   c. Differentiate assimilatory from dissimilatory nitrate reduction.
   d. Identify three nitrification pathways.
   e. Give evidence for the presence and importance of comammox in the nitrogen cycle.
   f. Compare and contrast denitrification and anammox.
   g. Identify the primary source of organic carbon for aerobic heterotrophs in deep ocean waters.
   h. Speculate why the concentration of benthic microbes exceeds that of pelagic microbes by 5 orders of magnitude.

MICROBIAL SYSTEMS

3. Microorganisms are ubiquitous and live in diverse and dynamic ecosystems.

    a. Define the phrase "microbial infallibility."

IMPACT OF MICROORGANISMS

4. Microbes are essential for life as we know it and the processes that support life.

    a. Identify the broad role of microbes in biogeochemical cycles on Earth.
    b. Describe the importance of autotrophic microbes to the carbon cycle.
    c. Explain how microbial consortia and interspecies electron transfer contribute to carbon cycling.
    d. Identify the importance of nitrogen-fixing microbes in the Earth's nitrogen cycle.
    e. Describe the role of chemoautotrophic bacteria in supporting life at deep-sea hydrothermal vents.
    f. Identify the important role of microbes in the phosphorus biogeochemical cycle.
    g. List microbes that have a large ecological impact on the global surface ocean through primary production.
    h. State reasons why microbes should affect climate.
    i. Identify the role of microbes in the negative feedback loop of cloud formation.

5. Humans utilize and harness microorganisms and their products.

    a. Identify how *Geobacter* can be used in bioremediation.

SUPPLEMENTAL MATERIAL

References

- **Madsen EL.** 2005. Identifying microorganisms responsible for ecologically significant biogeochemical processes. *Nat Rev Microbiol* **3(5):**439–436. doi:10.1038/nrmicro1151
- **Offre P, Spang A, Schleper C.** 2013. Archaea in biogeochemical cycles. *Ann Rev Microbiol* **67:**437–157. doi: 10.1146/annurev-micro-092412-155614
- **van Kessel MA, Speth DR, Albertsen M, Nielsen PH, Op den Camp HJ, Kartal B, Jetten MS, Lücker S.** 2015. Complete nitrification by a single microorganism. *Nature* **528(7583):**555–559. doi: 10.1038/nature16459
- **Daims H, Lebedeva EV, Pjevac P, Han P, Herbold C, Albertsen M, Jehmlich N, Palatinszky M, Vierheilig J, Bulaev A, Kirkegaard RH, von Bergen M, Rattei T, Bendinger B, Nielsen PH, Wagner M.** 2015. Complete nitrification by *Nitrospira* bacteria. *Nature* **528(7583):**504–509. doi: 10.1038/nature16461

Supplemental Activities

CHECK YOUR UNDERSTANDING

1. **List three chemical transformations of matter carried out by microbes that are essential for the continuation of life on Earth, and explain why each is essential.**

2. **What microbial processes could have been stimulated by the introduction of nitrogen-loaded fertilizers? How would this affect the concentration of dinitrogen in the atmosphere?**

3. **How were naturally occurring deposits of elemental sulfur formed?**

DIG DEEPER

1. The accumulation of oxygen in the atmosphere when the ancestral cyanobacteria emerged over 2 billion years ago oxidized the Earth's crust and generated large quantities of nitrate and sulfate.

    a. What microbial reactions of the nitrogen cycle do you expect to have been stimulated?
    b. What microbial reactions of the sulfur cycle do you expect to have been stimulated?
    c. Do you expect these conditions to affect the cycling of carbon? How?

2. Because of the British blockade of Europe during the early 19th century, manure piles were used extensively to supply Napoleon's armies with the nitrate needed to make black powder, a mixture of nitrate, sulfur, and charcoal then used as gunpowder. The manure was first mixed with soil and spread on the ground. To maximize nitrate production, the manure piles had to be aerated by turning them repeatedly. Such piles are miniature replicas of one arm of the global nitrogen cycle.

   a. What microbially mediated processes made nitrate in the manure?
   b. Why was the manure mixed with soil first?
   c. How does turning the pile speed the conversion?
   d. What microbial reactions of the nitrogen cycle did not take place within the manure piles, and why?

3. **Team work**: Methane-oxidizing archaea couple methane oxidation to the reduction of sulfate, a reaction that generates sulfide. However, more energy for growth could be derived if nitrate, rather than sulfate, were used as an electron acceptor for methane oxidation. For this reason, scientists speculate that microbes exist in nature that can couple the anaerobic oxidation of methane with denitrification. There are indications that increased fertilization of soils with nitrogen causes a decrease in carbon sequestration

   a. How could you use cultivation-independent approaches to demonstrate that such a reaction exists in nature?
   b. Propose a cultivation approach to isolate a methane oxidizer that has such metabolic capability.

4. **Team work:** Work in groups to design a strategy to stabilize the levels of methane and carbon dioxide in the atmosphere. Consider the pros and cons of your strategy. Present your proposal and arguments to the class.

5. **Team work:** In the case study, we discussed the isolation of worm symbionts that degraded films of polyethylene terephthalate (PET), a plastic commonly used in packaging and polyester fabric. In 2016, a team of Japanese scientists reported in the journal *Science* the recovery in pure culture of a free-living bacterium from enrichment cultures using PET films as sole carbon and energy source (Yoshida S et al., *Science* **351**(6278): 1196–1199, 2016, doi: 10.1126/science.aad6359). Read the article and the corresponding news coverage. Prepare an oral or poster presentation of the key findings, potential, and current challenges for this research effort.

Supplemental Resources

VISUAL

- The phosphorus cycle is all about minerals and fertilizers (3:12 min): https://www.youtube.com/watch?v=_IBx0zpNoEM
- Your aquarium health and the nitrogen cycle (15 sec): https://www.youtube.com/watch?v=1XC7xT0mIbY
- How scientists measure nitrous oxide emissions from agricultural soils (1:09 min): https://www.youtube.com/watch?v=fAHkpOJ4cwA

SPOKEN

- This Week in Microbiology episode #117, "Finding the Comammox," discusses two 2015 *Nature* papers on a single species of *Nitrospira* bacteria that oxidizes ammonia via nitrate to nitrate. http://www.microbeworld.org/podcasts/this-week-in-microbiology/archives/2042-twim-117-finding-the-comammox

WRITTEN

- Gemma Reguera ponders how the melting of the planet's frozen soils (the so-called permafrost) could stimulate microbial activities and carbon cycling? in the *Small Things Considered* blog post "The Cold Side of Microbial Life" (http://schaechter.asmblog.org/schaechter/2014/03/the-cold-side-of-microbial-life.html).
- Gemma Reguera discusses the discovery of electrogenic sulfide oxidation by *Desulfobulbaceae* multicellular filaments in the *Small Things Considered* blog post "Living Wires of

the Ocean Floor" (http://schaechter.asmblog.org/schaechter/2013/01/living-wires-of-the
-ocean-floor.html).

- Keep up to date with current carbon dioxide concentrations in the atmosphere: http://
  co2now.org
- Understanding the Global Warming Potential (GWP) of greenhouse gases: http://www3
  .epa.gov/climatechange/ghgemissions/gwps.html

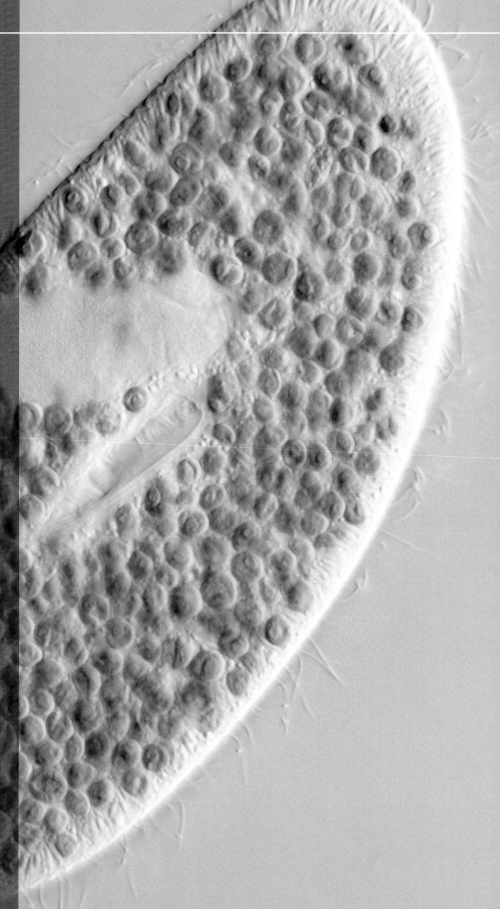

IN THIS CHAPTER

Paramecium bursaria with symbiotic algae (green)

CHAPTER TWENTY-ONE

Microbial Interactions

KEY CONCEPTS

This chapter covers the following topics in the ASM Fundamental Statements.

EVOLUTION

1. Cells, organelles (e.g., mitochondria and chloroplasts), and all major metabolic pathways evolved from early prokaryotic cells.
2. The evolutionary relatedness of organisms is best reflected in phylogenetic trees.

CELL STRUCTURE AND FUNCTION

3. *Bacteria* and *Archaea* have specialized structures (e.g., flagella, endospores, and pili) that often confer critical capabilities.

METABOLIC PATHWAYS

4. The interactions of microorganisms among themselves and with their environment are determined by their metabolic abilities (e.g., quorum sensing, oxygen consumption, nitrogen transformations).

MICROBIAL SYSTEMS

5. Microorganisms, cellular and viral, can interact with both human and non-human hosts in beneficial, neutral, or detrimental ways.

IMPACT OF MICROORGANISMS

6. Microbes are essential for life as we know it and the processes that support life (e.g., in biogeochemical cycles and plant/animal microbiota).

INTRODUCTION

Knowing how complex and diverse microbial communities are in the environment (chapter 19) and the impact that their collective activities have on our planet (chapter 20), it comes as no surprise to learn that microbes are highly interactive. Their interactions are as diverse as the microbes themselves and involve any living component of their ecosystem such as other microbes, plants, and animals. Throughout the book we have discussed how microbes communicate and how they interact to live as surface-attached communities or biofilms. But microbial interactions are far more diverse than what we have described thus far. No partner is off limits; microbes interact with other microbes of the same or different species, and with plants and animals. Sometimes microbes cooperate, other times they antagonize or even kill their partner. In fact, a single microbe may cooperate one minute, just to turn against its partner next. Why? Because microbes are constantly sensing and responding to chemical, physical, and biological cues from their environment, all of which fluctuate rapidly. Their ability to tune their responses to other cells, microbial or nonmicrobial, is key to their survival. Whether through cooperation or conflict, microbial interactions shaped the path of life's evolution and, in doing so, microbes transformed the Earth into the living planet it is today.

CASE: Living together or apart?

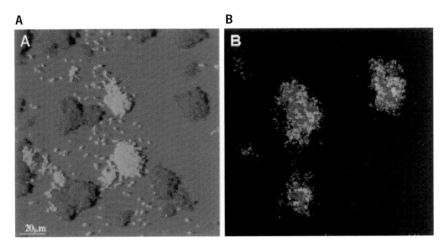

(From Nielsen AT et al., *Environ Microbiol* **2**:59–68, 2000, with permission.)

Dr. T isolated a consortium of two soil bacteria (*Burkholderia* sp. and *Pseudomonas* sp.) that were able to form **biofilms** on plastic surfaces when the medium was supplemented with citrate (panel A) or the industrial pollutant 3-chlorobiphenyl (panel B) as sole carbon and energy source. FISH probes that targeted the 16S rRNA of each bacterium (red probe for *Burkholderia* sp. and green probe for *Pseudomonas* sp.) revealed that each bacterium formed separate biofilm pillars with citrate but grew in close proximity to each other with 3-chlorobiphenyl.

Questions arising:

- Can you form a hypothesis about why the two bacteria separated to grow on citrate but came together to degrade 3-chlorobiphenyl?
- How would you test your hypothesis?

COOPERATION OR CONFLICT?

LEARNING OUTCOMES:
- Identify the most common reason for microbes to interact, either in a positive or negative way.
- State reasons many microbial interactions are transient and change over time.
- Speculate why neutralism, or living solely by oneself, is rare in the microbial world.

The spectrum of associations involving microbes ranges from the beneficial, or mutualistic, to the damaging, or parasitic (Table 21.1). The boundaries are not always precise, and definitions can be elusive because this is a shifting landscape, biased by the vast reactivity of all living things. Most often, interactions are driven by the pressure to acquire nutrients needed to grow and survive in a particular environment. Competition for food is widespread in nature (let the best competitor win!). Taken to the extreme, one microbe can eat another (**predation**) or feast on its resources (**parasitism**). However, microbes also cooperate, sometimes to the benefit of both partners (as in **mutualism**) or only one (**commensalism**).

Protists, which we discussed in chapter 16, provide numerous examples of cooperative and conflictive behaviors in the microbial world, which help illustrate the diversity and the often transient nature of microbial interactions. We learned, for example, that some species of *Paramecium*, a ciliate protist, prey on bacteria but themselves become prey to other ciliates (Fig. 16.4). In both situations, the interaction is one of predation; what changes is the predator and who is the prey. Some paramecia engulf other microbes but do not kill them; rather, they keep them inside and benefit from their metabolic activities. We described, for example, the case of a

TABLE 21.1 **Types of microbial interactions or symbioses**

Type of interaction	Description	Outcome	
		Cell 1	Cell 2
Neutralism	No interaction	No effect	No effect
Commensalism	One benefits, the other is unaffected	Benefit	No effect
Mutualism	Mutual benefit	Benefit	Benefit
Syntrophy	Cross-feeding of nutrients, often with mutual Benefit	Benefit	Benefit
Predation	One consumes the other	Benefit	Harm
Parasitism	One consumes nutrients from the other	Benefit	Harm
Antagonism (amensalism)	Products from one harm the other	Benefit No effect	Harm
Competition	Cells compete for growth-limiting nutrient	Harm	Harm

Paramecium that harbors a great number of unicellular algae in its interior (Fig. 16.9). The algae live in a protective environment inside the host and do not need to move around to gain sufficient light exposure for photosynthesis. The photoautotrophic metabolism of the algae provides the nutrients the host needs and allows it to thrive in nutrient-limited environments. Such environments also limit the growth of predators, so paramecia get the additional benefit of avoiding predation. The association is symbiotic because the two partners live together, in this case to mutual benefit. Yet the relationship is reversible; the host and the endosymbiont no longer need each other once organic carbon becomes abundant so the two live freely until nutrient scarcity hits again. It is a marriage of convenience. In some paramecia, however, the endosymbionts have adapted so much to their host that they have lost unnecessary genes and can no longer live outside and on their own. (Could they be on their way to becoming organelles?) And then there is the paramecium that carries a toxin-producing bacterium so it can kill other paramecia and thus avoid being eaten.

Diversity is also found in the *type of interacting partners*. Microbes interact with other microbes of the same or different species as well as with plants and animals. Most interactions of microbes with plants and animals are transient. But some partnerships have coevolved to the point that the two cannot survive without each other. This is the case of some insects and their endosymbionts, which we will describe in more detail later on. As we learned in chapter 19, humans also carry a great number of microbes, mainly bacteria, in various body parts. Indeed, the human gut is colonized soon after birth and a microbial community establishes there that grows and matures with the individual. The microbes and the human host benefit each other in many different ways. However, not all of the microbes are beneficial. For human beings, today's benevolent companion may become tomorrow's assailant. A person suffering from AIDS becomes the victim of infections by otherwise innocuous members of the body's normal **microbiota**. The balance between a helpful and a destructive relationship is therefore delicate and can be readily altered by genetic or induced changes in one of the partners. The two share common steps in the way the association is established: *encounter, association, and multiplication*. The two partners then enter into intense back-and-forth negotiations whose outcome determines the kind of interaction that will prevail.

The transient nature of many microbial interactions in response to environmental cues is easily appreciated in the case study. The same two bacteria grow separately or in intimate contact in biofilms, depending on the carbon source present in the medium. Could they compete in one instance and cooperate in another? One would think that physical proximity would promote cooperation. But what benefit do the microbes get? And, would both partners benefit from the interaction (as in mutualism) or only one (as in commensalism)? The example in the case study also highlights the prominent role that nutritional cues (in this case, the type of carbon source) have in the establishment of positive or negative interactions. The quest for nutrients and energy exerts a tremendous pressure on microbes to interact with others or with plants and animals. The motto "if you can't beat them, join them" seems quite appropriate here. In the next sections, we examine some notable examples of just how diverse microbial interactions are and how they impact life on our planet.

WHAT IS SYMBIOSIS?

LEARNING OUTCOMES:

- Define symbiosis.
- Use evidence to support the endosymbiotic theory of mitochondria and chloroplasts.
- Use examples to differentiate organelles, endosymbionts, and intracellular parasites.
- Identify differences between modern-day "ex-microbe" organelles, such as mitochondria and chloroplasts, and their free-living ancestral bacteria.
- Describe the function of organelles other than mitochondria and chloroplasts that are suspected to have an endosymbiotic origin.
- Speculate an advantage to hosting endosymbionts in specialized bacteriocyte cells.
- Describe the nutritional mutualism of *Buchnera* and aphids.
- Explain the evolutionary consequences of the unique mode of *Buchnera* symbiont transmission to host cells.
- Identify how *Wolbachia* increases its genetic fitness through its mode of being passed onto the progeny of insects.
- Describe how interactions between nitrogen-fixing bacteria and legumes lead to the formation of root nodules.
- Identify how legumes accommodate the physiological needs of their N_2-fixing endosymbionts.
- Name physiological or structural features of ruminants that make the symbiosis with rumen microbes possible.
- Explain how microbe-microbe interactions make it possible for ruminants, such as cows, to subsist on a cellulose-heavy plant diet.

In biology, the term **symbiosis** is broadly used to describe organisms that *live together*, typically with benefits to both partners, though this is not always the case. Hence, although symbiotic associations are typically *mutualistic* (Table 21.1), exceptions do occur. Sometimes two partners live together although one may not gain much, or may even be harmed from the association (Table 21.1). The sacrifice nevertheless pays off in the long term: as we have learned studying organelles that were once microbes such as mitochondria and chloroplasts (chapter 1), a symbiotic event can lead to the evolution of new organisms (in this case, eukaryotic cells). Clearly, symbiosis is a pervasive force in evolution. Through it, organisms develop novel ways to occupy environmental niches, produce energy, acquire nutrients, or defend themselves from predation. Examples are all around us. Some are easily noticed, such as the lichens (fungus-alga [sometimes fungus-cyanobacterium] partnerships) that adorn rocks and trees or the root nodules of legumes (whereby bacteria supply plants with a nitrogen source). Other symbiotic relationships are not so readily visible. One would have to travel to the depths of the ocean to appreciate the association of sulfur-oxidizing bacteria with hydrothermal vent tubeworms (chapter 20). Sometimes the effect of symbiosis is only revealed when the connection between the partners is severed. Thus, the importance of the human intestinal bacteria becomes apparent when they are reduced in number by treatment with antibiotics. People so treated become susceptible to diarrhea caused by yeast or other opportunistic pathogens, suggesting that, at a minimum, intestinal bacteria play a role in keeping out unwanted intruders.

Mitochondria, Chloroplasts, and Other "Ex-Microbes"

Mitochondria and chloroplasts. The most wide-ranging and perhaps most enduring symbiotic venture involving microbes has been the acquisition of mitochondria by animals and plants and of chloroplasts by plants. These organelles were once bacteria that found their way inside another cell and evolved together (the **endosymbiotic theory**). The origin of these symbioses, which happened a billion or so years ago, led to a momentous event, the *development of the eukaryotic cell*. Although organelles such as mitochondria and chloroplasts hardly resemble their ancestors (Fig. 21.1), they reveal their bacterial origin in various ways. First, both divide by binary fission. Some organelles even retain proteins that are typically required for bacterial division. Indeed, chloroplasts and certain mitochondria use an *FtsZ ring for division* (Fig. 21.2) and, in some instances, proteins of the bacterial Min system (see chapter 4). Second, protein synthesis in mitochondria and chloroplasts starts with formylated methionine rather than with methionine (as in eukaryotes) and uses ribosomes resembling those of prokaryotes (they are smaller than eukaryotic ribosomes), which are also sensitive to antibiotics that inhibit prokaryotic protein synthesis. Furthermore, in general, they have circular chromosomes and no histones (although some chloroplasts may have linear chromosomes and contain histone-like proteins).

The evolution of endosymbiotic bacteria into modern mitochondria and chloroplasts also involved the *transfer of many of the ancestral bacterial genes to the host cell nucleus* and the *loss of many others*. Such genes serve as molecular markers of the symbiotic event and provide valuable information into the evolutionary paths that the ancestral endosymbionts followed to become organelles. Chloroplast genomes, for example, have been reduced to 110 to

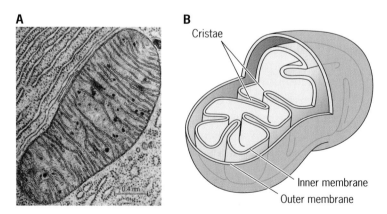

A

B

Cristae

Inner membrane

Outer membrane

FIGURE 21.1 **An electron micrograph (A) and a schematic drawing (B) of a mitochondrion.** (Panel A credit Keith R. Porter, The Free Encyclopedia.)

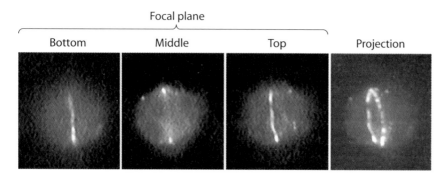

FIGURE 21.2 **The FtsZ protein in a chloroplast of the plant *Arabidopsis*.** The protein was visualized by using fluorescent anti-FtsZ antibodies. The three panels on the left are optical sections through the bottom, middle, and top of a chloroplast. In the far right panel, labeled Projection, the images were stacked and rotated 30% to show the FtsZ ring. (From Vitha S et al., *J Cell Biol* **153**:111–119, 2001, with permission.)

200 kb in size (compared to the 1.6 to 11.6 Mb range of modern cyanobacterial genomes). The reduced genome mostly retains genes encoding proteins required for their main cellular function, energy harvesting through photosynthesis, and genes involved in protein synthesis (ribosomal RNAs and proteins as well as transfer RNAs). Mitochondrial genomes, on the other hand, are 10 times more reduced and carry only a few dozen genes, coding mainly for ribosomes and transfer RNAs. Most other organelle genes have been acquired by the nucleus, including most of the genes needed for aerobic respiration, the organelle's main function. Hence, most respiratory proteins must be transported across the cytoplasm and into the mitochondria. Why not keep these genes in the mitochondria? The mutation rate of the mitochondrial DNA is quite high, possibly due to the lack of a proofreading function in its DNA replication machinery (chapter 8). In fact, several congenital diseases have been attributed to *mutations in the mitochondrial genome*. Thus, sending genes important for mitochondrial function to the nucleus may protect them from mutagenesis. In any case, the medical importance of such mutations is relatively small because most of these mutations take place in somatic cells and are not inherited. And since there is usually more than one mitochondrion per cell, a mutation in one may not have a large genetic effect. Mitochondrial genetics is colored by the fact that these organelles are inherited only from the mother—those of the sperm cells are degraded after fertilization.

Phylogenetic analyses of organelle-like genes residing in the nucleus have also provided insights into the evolutionary journey of the ancestral mitochondrion, from which modern mitochondria evolved. In chapter 16 we learned that some protists do not carry the typical mitochondria. Instead, they have membrane-bound organelles called **mitosomes**, which do not make ATP, or **hydrogenosomes**, which make hydrogen gas (H_2). These organelles are descendants of an ancestral mitochondrion that lost a lot of its original genes. But not all. Some were transferred to the host genome, and there they remain (although they have accumulated numerous mutations throughout their evolutionary journey within the host's genome). Mitochondria and mitochondria-like organelles have become so reduced genetically from their original free-living status that are far from being capable of independent existence. Yet the mitochondrial DNA shares

BOX 21.1

Organelle or parasite, which road to take?

Animals, including humans, can also house microbes inside their cells. In some cases the relationship is not beneficial but parasitic. Why don't these microbes become beneficial endosymbionts? Consider for example rickettsiae, a group of intracellular pathogens of eukaryotic cells. They cannot live outside their host and, to this date, scientists have not been able to grow them independently in their host in the laboratory. Are they organelles in the making? Not likely. Rickettsiae carry genes (e.g., succinic dehydrogenase genes) that are phylogenetically related to modern mitochondria. Like mitochondria, they have lost part of their genome, although not nearly as much. Rickettsiae have, for example, a limited ability to produce ATP and need to obtain it from their environment, the host cell's cytoplasm. Rather than becoming an organelle, rickettsiae acquired genes not found in other bacteria that encode transport systems for ATP and ADP. So what may have started as a symbiosis took a different turn when these bacteria figured out a way to "steal" ATP from the host. This need unbalanced the relationship between the microbes and the host and made them parasites. Interestingly, rickettsiae have not lost genes required for the demanding task of transferring from one host cell to another. Somehow this evolutionary journey made sense for them because it ensured their transmission and survival. And so a parasite, rather than an organelle, was born.

To make things more exciting, there is a group of bacteria phylogenetically related to rickettsiae and mitochondria that *includes endosymbionts*. These organisms, often referred to as **rickettsiae-like endosymbionts** (or RLE), are believed to

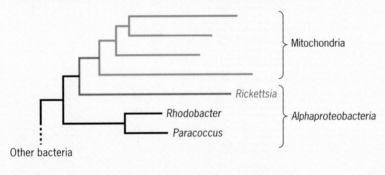

(Modified from Searcy DG, *Cell Res* **13**:229–238, 2003.)

have emerged before the ancestors of present-day rickettsiae. This finding strengthens the idea that a mutualistic interaction was first established between an ancestral bacterium and a eukaryotic host from which the RLE, present-day rickettsiae, and mitochondria evolved in independent events. Each took very different evolutionary journeys: RLE remained endosymbionts, rickettsiae turned into intracellular pathogens, and mitochondria became integral organelles of the eukaryotic cell.

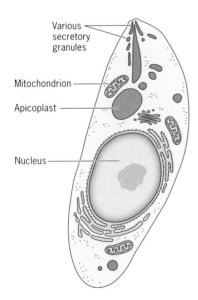

FIGURE 21.3 **A schematic drawing of a protozoon showing the apicoplast.**

homology to DNA from rickettsiae, infamous members of the *Alphaproteobacteria* that cause typhus and other diseases. Interestingly, rickettsiae remain obligate intracellular parasites to this day, although they are a long way from becoming organelles. This suggests that mitochondria and rickettsiae share a common ancestor but followed different evolutionary journeys inside their host (Box 21.1).

Other "Ex-Microbes." Mitochondria and chloroplasts are not the only "ex-microbes" to inhabit eukaryotic cells. In chapter 16, we learned that certain protists, such as *Plasmodium* species (the parasites that cause malaria), have structures called **apicoplasts** (Fig. 21.3). These organelles are essential, and their hosts cannot survive without them. They contain just enough DNA to encode about 35 genes. This DNA is related to that of chloroplasts, although apicoplasts reside in nonphotosynthetic organisms. Apicoplasts, then, have functions other than for photosynthesis, including synthesis of fatty acids and repair, replication, and transcription of DNA. Apicoplasts seem to be the result of an ancient event in which a eukaryotic ancestor acquired a chloroplast but converted it from a photosynthetic factory to one dealing with other biochemical activities.

DNA-containing organelles have further surprises. And how! Certain protozoa, e.g., trypanosomes (the group that contains the agents of sleeping sickness and Chagas' disease), have a highly specialized mitochondrion located at the base of the cilium called the **kinetoplast** (Fig. 21.4). The DNA contained in the kinetoplasts (so-called kDNA) is organized as a mesh of interwoven DNA circles that resembles the chainmail of a medieval armor (Fig. 21.5). The circles occur in two sizes, *maxi* (20 to 40 kb) and *mini* (0.5 to 1 kb). There are 20 to 50 maxicircles per kinetoplast, each encoding ribosomal RNAs and proteins involved in energy transduction. Minicircles are more abundant (5,000 to 10,000 minicircles per kinetoplast), are also heterogeneous in sequence, and do not encode proteins. To date, the only known function of the minicircles is to code for RNAs called *guide RNAs* or gRNAs, whose role is to *edit the maxicircle RNA transcripts* so they can be translated. Kinetoplastid RNA editing appears to be an ancient process but its evolutionary origin remains obscure. Could it be a relic of some other ancestral microbial endosymbiont?

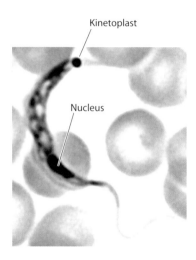

FIGURE 21.4 **A blood smear reveals a *Leishmania* cell with its distinctive apical kinetoplast.** (Courtesy of M. F. Wiser.)

Bacterial Endosymbionts of Insects: Organelles in the Making?

Why do eukaryotic cells have only a few kinds of microbe-derived organelles? Why not more? Go figure! The advantages of establishing this type of symbiosis clearly benefited the host: in a single step, the recipient cell acquired all the genetic material needed for functions as complex as respiration or photosynthesis and thus became equipped to undertake bold new ventures. And in so doing, it changed the course of evolution: the ancestor of modern eukaryotic cells was born. Obviously, the microbe that

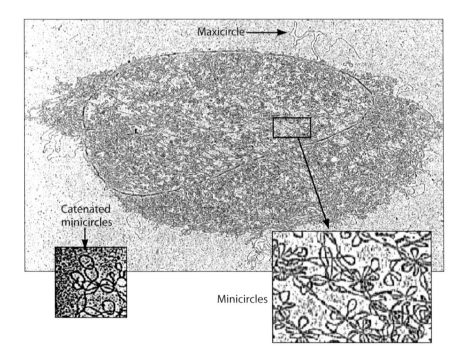

FIGURE 21.5 **Kinetoplast DNA from *Leishmania*.** Kinetoplast DNA consists of a giant network of interconnected (catenated) minicircles and maxicircles. There are approximately 10,000 minicircles and 50 maxicircles per network. This DNA makes one of the most unusual structures known. (Courtesy of M. F. Wiser.)

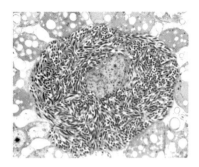

FIGURE 21.6 **A bacteriocyte, an insect cell filled with endosymbiotic bacteria.** This electron microscope thin section shows how tightly endosymbionts can be packed within a host cell, here, one in the fat body of a German cockroach. The bacteriocyte is about 100 mm in diameter. (Courtesy of L. Sacchi.)

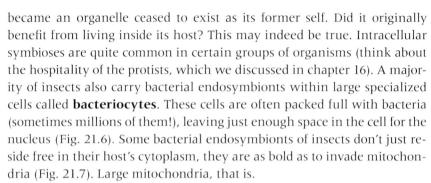

became an organelle ceased to exist as its former self. Did it originally benefit from living inside its host? This may indeed be true. Intracellular symbioses are quite common in certain groups of organisms (think about the hospitality of the protists, which we discussed in chapter 16). A majority of insects also carry bacterial endosymbionts within large specialized cells called **bacteriocytes**. These cells are often packed full with bacteria (sometimes millions of them!), leaving just enough space in the cell for the nucleus (Fig. 21.6). Some bacterial endosymbionts of insects don't just reside free in their host's cytoplasm, they are as bold as to invade mitochondria (Fig. 21.7). Large mitochondria, that is.

The associations between bacteria and insect hosts are often based on *nutritional interdependence*, that is, their survival depends on the other partner providing essential nutrients—a class of mutualism called syntrophic interaction (Table 21.1). In other cases, however, the relationship *affects the sex life of the host* and leads to changes in speciation and lifestyle. In all cases, the symbiont-host relationship is not as intimate as that of organelles. Genes from the symbiont have not been transferred to the host's nucleus but remain in the genome of the symbiont. Yet they may be on the road to become organelles because both depend on each other for survival and, as a result, their evolutionary journey is intertwined.

Nutritional symbiosis between bacteria and insects. A particularly well-studied example of nutritional symbiotic association is that between aphids, small soft-bodied insects that are common pests of plants, and Gram-negative bacteria called *Buchnera*. Aphids have two big problems: they cannot produce 10 of the amino acids needed to make proteins (similar to humans) and they are sap suckers. Why is feeding on sap a problem? Plant sap is poor in proteins and cannot supply all their needed amino acids. Hence, aphids could feed on sap for days on end and they would still starve because of the lack of essential amino acids. So they turn to bacteria for help, harboring millions of bacterial endosymbionts in dedicated symbiotic cells similar to bacteriocytes (Fig. 21.6) and using them as amino-acid-making factories.

So interdependent are the two partners that the evolution of *Buchnera* seems to have proceeded synchronously with that of the host aphids (Fig. 21.8), an example of coevolution. Considering the biochemical complexities of making precursors such as amino acids (chapter 7), the exact complementary relationship between the host and the symbiont must have required a good number of evolutionary steps. Because aphids left a fossil record, this concordance provides a clock for bacterial evolution. Modern *Buchnera* bacteria, like many other bacterial endosymbionts, have small genomes, about one-seventh that of *Escherichia coli*. Their genomes appear to have lost most of the genes of their ancestors (who were probably distant relatives of *E. coli*) and kept only those required for existence within the host. In addition, *Buchnera* carry some of the biosynthetic genes (e.g., those for tryptophan) that are essential to the symbiosis in high-copy plasmids (chapter 10). This provides multiple copies the genes and allows the endosymbionts to make the great quantities of the amino acids that the host needs. This clearly benefits the host. What benefit do the bacteria get? Conveniently, the bacteria cannot make the other set of

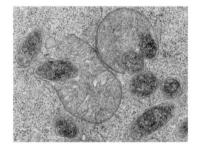

FIGURE 21.7 **Symbiotic bacteria in the mitochondria of a tick.** Note that these mitochondria are large enough to readily accommodate bacteria. (Courtesy of L. Sacchi.)

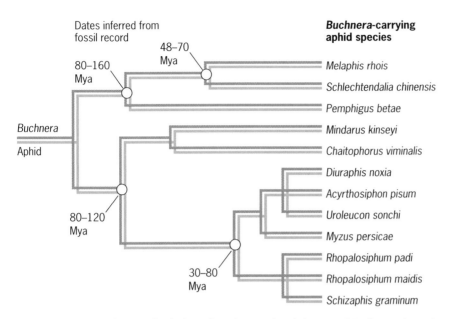

FIGURE 21.8 **Coevolution of aphids and *Buchnera*.** The phylogeny of *Buchnera*, derived from genomic relatedness, is shown in turquoise. The phylogeny of the aphid hosts, derived from the fossil record, is shown in orange. Note the remarkable superimposition of the two phylogenies. Mya, millions of years ago.

amino acids, the ones that the aphids are able to synthesize. The insects also provide precursors that the bacteria can use to make their required amino acids. This is a perfect *quid pro quo*, each partner feeding the other what they need for survival.

Are buchneras on their path to becoming organelles? These bacteria are *transmitted vertically* from progeny host cells and, in this sense, are organelle-like. This contrasts with the typical *horizontal transmission* of many endosymbionts, which are generally passed on to the host's progeny by transovarian transmission, or parasites, which, with some exceptions, must infect each generation of host anew. However, unlike mitochondria and other organelles, buchneras have not transferred genes to the host's nucleus. It is possible that such horizontal gene transfer between endosymbiont and host takes place all the time but only some become stable. Furthermore, because they are loners, strict intracellular symbionts, *Buchnera* cells are also shielded from acquiring genes from other organisms via horizontal transmission. Consequently, their only *obvious source of genetic variation is mutation*. How do these organisms cope with the accumulation of lethal mutations? The answer is not really known, but the aphids seem to insure themselves against a catastrophic loss of endosymbionts by having more than one kind of bacterial endosymbiont, each residing in a separate kind of bacteriocyte. *Clever!*

A symbiosis that changes the insect's sex. Although symbioses between insects and microbes such as *Buchnera* often fulfill nutritional needs, they can do more. In some instances, the partnership modifies the host's reproductive processes. An example is the rickettsia *Wolbachia*, an extremely common bacterium that infects between 25 and 75% of all insect species as well as worms and spiders. *Wolbachia* endosymbionts

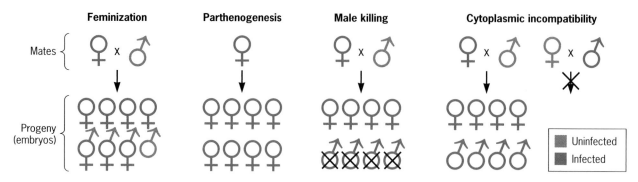

FIGURE 21.9 **Wolbachia and its female sex bias.** (Modified from Werren JH et al., *Nat Rev Microbiol* **6:**741–751, 2008.)

bias the sex of their hosts' offspring, often introducing a female bias (Fig. 21.9). In some invertebrates the bacteria turn the male embryos into females (feminization); in other cases reproduction is controlled so females can make more females (**parthenogenesis**) or the male embryos are killed. And there are *Wolbachia* endosymbionts that allow the male embryos to survive but make their sperm incompatible with the eggs of uninfected females (cytoplasmic incompatibility), so they can have offspring only with infected females. In all cases, the ultimate goal appears to be to increase the chances that the female eggs will carry endosymbionts so they can be passed to the progeny.

You may think that the insect is being harmed in this symbiosis. Not so. *Wolbachia* endosymbionts increase the numbers of females available for mating, thus the males that survive can be picky about who they mate with. This puts considerable evolutionary pressure on the female population and often alters their behavior. In some butterflies, the females congregate in groups where they vie for the attention of the scarce males (please, no snickers!). In this reversal of the often-seen aggregation of males in search of females, it is the males that can be choosy about their mates. And so the best wins.

Nitrogen-Fixing Bacteria and the Legumes

Nitrogen gas (N_2) is the most abundant molecule in our atmosphere but needs to be reduced to ammonia (NH_3), a process called **nitrogen fixation**, to enter biosynthetic pathways and make nitrogen-containing compounds, such as amino acids, purines, and pyrimidines. The process of nitrogen fixation was introduced in chapter 7 and discussed in the context of the nitrogen cycle in chapter 20. Here, we will deal only with the symbiotic associations between nitrogen-fixing bacteria and certain plants such trees (alder), shrubs (bayberry), ferns, and the best studied of them, legumes (peas, alfalfa, and beans).

When a legume is pulled up from the ground, the roots appear to be decorated with small granules, typically 1 mm in diameter, the so-called **root nodules** (Fig. 21.10A). Examine a squashed or sectioned root nodule under a microscope and it will reveal that they are filled with bacterium-like bodies called **bacteroids** (Fig. 21.10B). Bacteroids are the nitrogen-fixing endosymbionts and, thanks to their activities, they turn each root

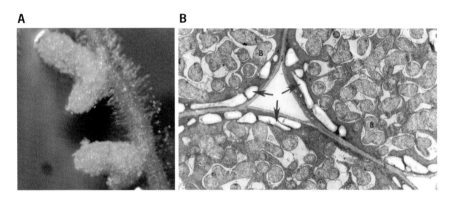

A   **B**

FIGURE 21.10 **Root nodules from legume plants and their nitrogen-fixing endo-symbionts.** (A) Root nodules in an alfalfa plant, showing their pink contents, due to the presence of leghemoglobin, a unique metabolite of this type of symbiosis. (B) Section through a pea root nodule shows the plant cells filled with bacteroids (small B) and starch granules (arrows). (Panel A courtesy of D. Gage; panel B courtesy of D. A. Phillips, from Neidhardt FC et al., *Physiology of the Bacterial Cell*, Sinauer Associates, Inc., Sunderland, MA, 1990.)

nodule into a nitrogen-fixing factory. Root nodules are formed stepwise, starting with the recognition between the plant and the soil-dwelling members of the genus *Rhizobium* and related genera (Fig. 21.11). The plant excretes compounds (flavonoids) through its roots that *attract these bacteria only*. The plant signals also induce the expression of genes (*nod*) in the bacteria that are needed to make **nodulation factors** (fatty acids and chitin-like compounds—polymers containing *N*-acetylglucosamine) that induce nodule formation in the plant.

Once the right bacteria are attracted to the root, they bind to the tip of the root hairs (Fig. 21.11), a process that involves the recognition between a ligand on the bacterial surface and a specific receptor on the root hair. The bacterium then produces enzymes to hydrolyze the tough plant cell wall and invade the root hair, where they reside in intracellular vacuoles. There they secrete nodulation (or Nod) factors that induce a morphological change in the root hair such that it curls up like a shepherd's crook (Fig. 21.11). Nodule formation is not complete yet. It requires that the

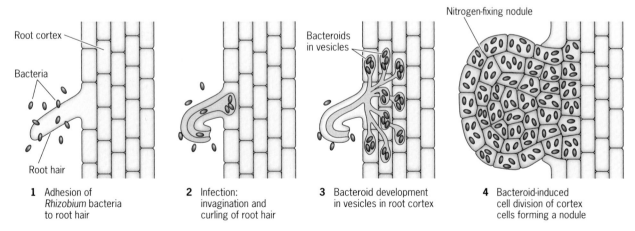

| 1 Adhesion of *Rhizobium* bacteria to root hair | 2 Infection: invagination and curling of root hair | 3 Bacteroid development in vesicles in root cortex | 4 Bacteroid-induced cell division of cortex cells forming a nodule |

FIGURE 21.11 **Morphological changes leading to a nitrogen-fixing nodule.**

bacteria move inward, toward the center of the root. This is a difficult trip because the bacteria have to navigate through tough cellulose-encased plant cells. Travel is facilitated when the plant forms a tube (called an infection thread) that stretches from the root hair to the root's interior. Traveling along this tube, the bacterial cells arrive at a deep location. Here, they exit the infection thread and begin to multiply. The plant cells respond to bacterial stimuli by rapidly proliferating into a tumor-like structure, the nodule. This host response requires a delicate balance because plants have defense mechanisms that could destroy the invading bacteria. Rhizobia and their relatives also do their share; they make capsules and, being Gram negative, possess **lipopolysaccharides** (chapter 2) that help them survive any plant attacks.

Having arrived at their ultimate destination, the bacteria differentiate into nitrogen fixation workshops. They become branched and swollen and are now called the bacteroids (Fig. 21.10B). Before being able to fix nitrogen, the bacteria must address a final problem. Nitrogen fixation is a highly anaerobic process, but rhizobia are aerobes and so are the plant roots. Plants come to the rescue again: they synthesize a form of hemoglobin called **leghemoglobin**, which gives the root nodule a distinctive pink shade (Fig. 21.10A). Leghemoglobin absorbs the oxygen, keeping its concentration at a suitable level for the nitrogen-fixing machinery. Bacteroids are incapable of further growth and are totally dependent on the plant for nutrients. For their side of the bargain, the bacteroids receive from the host organic acids that provide energy and the reducing power needed for nitrogen fixation. In return, they provide to their host ammonia that can be used as the nitrogen source for biosynthesis. This mutualistic relationship requires that both partners undergo profound biochemical and structural adaptations. What prevented nitrogen-fixing bacteria from becoming cell organelles in parallel with mitochondria and chloroplasts? (What would you call them, "nitrochondria?") We simply do not know.

The Rumen and Its Microbes

In contrast to many invertebrates, which rely on symbioses with microbes for survival, vertebrates have developed few essential partnerships with microbes. Perhaps the most outstanding of all is that of ruminants such as cattle, goats, and deer. These animals cannot digest cellulose and many other plant polysaccharides and rely on a complex microbial symbiotic community to process the plant material into digestible products (Fig. 21.12A). Unlike the nodule formation of legumes, this symbiosis does not require major modification of the host but depends on the host providing a large nutrient-filled chamber (an organ called the **rumen**) in which the biochemical transformations can take place.

Ruminants grind and chew the plant material thoroughly and later regurgitate the cud to chew it some more and reduce it to even smaller bits. This allows these animals to first eat rapidly and later process it at their leisure, away from predators. It also makes the plant material easier to hydrolyze and ferment by the rumen microbes. The rumen chamber is filled with liquid (in cattle, its content is about 15 gallons) and is pretty much oxygen free (anaerobic). Here, diverse groups of bacteria, fungi, and

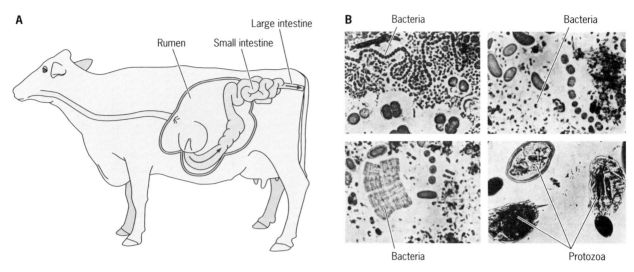

FIGURE 21.12 A cow's digestive system. Cows have a large rumen (A) in their digestive systems teeming with microbes, mainly bacteria and protozoa (B). The "large" intestine of the animal looks relatively small compared to the rumen, in part because the digestion of plant material has already taken place in the rumen. This changes in nonruminant herbivores, such as horses, which do not have a rumen and have instead a very big large intestine. (Panel B courtesy of J. Smiles and M. J. Dobson, from Neidhardt FC et al., *Physiology of the Bacterial Cell*, Sinauer Associates, Inc., Sunderland, MA, 1990.)

protozoa (Fig. 21.12B) live together and form anaerobic consortia that first degrade cellulose and other plant-derived polymers into sugars and then ferment them into volatile fatty acids such as acetic, propionic, and butyric acids. Ruminants absorb the fatty acids through the epithelium and use them for their metabolic needs. To prevent drops of pH as the acids accumulate, ruminants secrete prodigious amounts of well-buffered saliva (25 gallons or more a day per cow). The rumen microbes also provide the animal with essential amino acids and other growth factors. Clearly, the animal and the microbes are symbiotically interdependent; the animal provides the right environment to the microbes and they in turn feed fatty acids, amino acids, and other growth factors to the host. It gets more complicated: the microbes themselves engage in numerous interactions to maintain the productivity of the community and satisfy the needs of their host (Box 21.2).

Once the rumen content is emptied into the next chamber, the stomach, the microbial cells are killed by the acid and their macromolecules degraded by digestive enzymes. Many of the bacteria are broken down by a cell-wall-degrading **lysozyme** that, uniquely in ruminants and befitting its site of action, is *acid resistant*. Ruminants make effective use of their feed via their symbionts, which accounts for the high efficiency of cattle in milk and meat production and their worldwide distribution. It has been said that cattle carry out a fermentation that no industrial microbiologist can match. They use the cheapest substrate (cellulose), gather it themselves, and convert it into valuable products—beef and milk. Match that, chemical engineers!

What about nonruminant herbivores such as horses, rabbits, and elephants? Like their ruminant counterparts they feed on plant material but cannot digest it. Instead of a rumen, they have an extra-large cecum in the large intestine where they house the symbiotic microbes that carry out

BOX 21.2

SYNTROPHY AND PREDATION IN THE RUMEN

The rumen microbiota is made up of highly diverse and specialized populations of bacteria, protozoa, and fungi, which interact to ensure that the community maintains its collective goal of degrading plant material in the rumen and supplying the animal host with essential nutrients. Bacteria are by far the most numerous (as many as 10^{10} per ml of rumen fluid) and include over 200 species. Together with fungi, they cooperate to degrade and ferment the plant-derived polymers to generate essential nutrients for the host. To do this, they establish a food chain, starting with the degradation of cellulose and other plant-derived polymers and ending with the fermentation of sugars. The resulting volatile fatty acids (or VFAs) are absorbed by the animal host through the bloodstream. The fermentative activities also result in the production of large amounts of hydrogen gas (H_2). If this gas were to accumulate, it would inhibit further fermentation. Hence, hydrogen must be removed, lest it compromise the microbe-ruminant symbiosis. For this reason, the consortium includes **methanogens**, a group of *Archaea* that use hydrogen and carbon dioxide (CO_2) to make methane (CH_4). Being a gas, methane can be expelled by belching (formally, "generating an eructation"). If you were to carefully hold a lit match away from a belching cow, you would see a small flame appear. (Do not try this!) The syntrophic removal of hydrogen via *interspecies hydrogen transfer* allows fermentation to efficiently proceed in the rumen. This benefits both the fermentative microbes and the host: the microbes obtain more ATP from fermentation and the animal gets more fatty acids. More ATP for the cells provides more energy for growth and reproduction of microbial cells in the rumen. This, in turn, increases the protein available to the host animal once the rumen contents, including the microbes, are digested in the stomach.

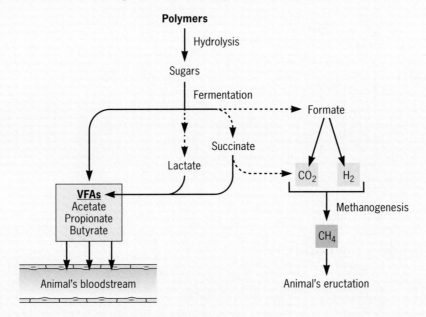

What about the rumen protists? They are less abundant than the bacteria (about 1 million per ml) but, being larger, they make up almost half the total microbial load by weight. They are also involved in cellulose degradation, but their major role in the rumen is to prey on the bacteria. This has been proposed to have a negative influence on the overall fermentation. However, this is a still controversial point. Perhaps predation is necessary here to control the number of bacteria so they can work synergistically rather than competing for the available resources. What do you think?

cellulose digestion. Here, the microbes are not recycled as efficiently as in the rumen and are instead discarded in the feces. This accounts for the fact that some of these animals, e.g., rabbits and rats, are coprophagic, that is, they eat their highly nutrient-rich feces. (Note that the large intestine of humans is of medium size and does not have specialized compartments to house dedicated cellulose-degrading microbial symbionts. This suggests that cellulose degradation is not likely to be a nutritionally essential process.) Microbial cellulose decomposition does take place in the gut of termites, again carried out by their microbial partners in enlarged intestinal compartments. In the case of termites, the host helps by secreting its own cellulases.

HOW DO MICROBES FEED TOGETHER IN SYNTROPHIC ASSOCIATIONS?

LEARNING OUTCOMES:
- Differentiate syntrophy and symbiosis.
- State reasons a syntrophy is beneficial for many fermentation pathways and even necessary for others.
- Use an example of interspecies metabolite transfer and electron transfer to justify the statement: "The interactions of microorganisms among themselves and with their environment are determined by their metabolic abilities."

The staggering number of metabolic activities that microbes carry out on the planet allows them to exploit just about any resource available in their environment. If the reactants needed to carry out a thermodynamically feasible reaction are present, a microbe will almost certainly evolve a way to use it to gain energy. But, by living in communities, microbes can combine their activities to carry out reactions that yield low energy or even consume energy. Their metabolic potential is greatly expanded when they feed together. How is this achieved? The notion is simple: one microbe catalyzes one reaction and another "eats" the reaction product. Both microbes gain from the association: the first partner produces waste products that, if left to accumulate, would eventually inhibit its metabolism; the waste products are rapidly removed by the second partner. This type of interaction, in which the two microbes *feed one another*, is called **syntrophy**. Do you think the bacterial partners described in the case study were feeding each other? It is indeed possible. One of the bacterial partners could metabolize 3-chlorobiphenyl into a product that if not removed rapidly by the second partner would feed back to inhibit its metabolism. The two would grow in close proximity to each other within the biofilm pillar to facilitate the metabolic exchange. To test this, you could examine if each bacterium can grow individually with 3-chlorobiphenyl. Chances are that only one (the first partner) would be able to do it, but its growth would slow down if the reaction product accumulates.

Interspecies Metabolite Transfer

Syntrophic interactions allow microbes to carry out the problematic task of degrading organic matter (polysaccharides, proteins, and lipids) under anaerobic conditions to return the carbon back to the atmosphere as carbon dioxide (CO_2) and methane (CH_4) (Fig. 21.13). The organic material is processed sequentially through the integrated activities of many microbes: (i) organic matter is first hydrolyzed by extracellular enzymes (e.g., cellulases, proteases, lipases) to generate soluble carbon substrates that the same or

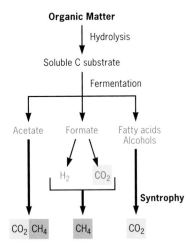

FIGURE 21.13 Syntrophic interactions that drive the anaerobic degradation of organic matter. The anaerobic degradation of organic matter integrates the activities of many microbes, which hydrolyze the complex substrates (polysaccharides, proteins, and lipids) into soluble molecules (sugars, amino acids, and fatty acids) that the cells can take up and ferment. The fermentation by-products (in red) are removed as soon as they are produced via syntrophic interactions (thick arrows) with microbes that return all the carbon (C) back to the atmosphere as CO_2 and CH_4 gases.

other microbes can transport inside their cells; (ii) the carbon substrates are then fermented by the same or different organisms, producing H_2, CO_2, and reduced carbon molecules such as organic acids and alcohols; and (iii) the fermentation by-products are then removed by syntrophic partners. The fate of the fermentation products varies with the type of syntrophic association. Some partners use the reduced molecules produced during fermentation as carbon sources and/or electron donors *for respiration* using alternative electron acceptors such as nitrate, manganese (Mn^{6+}) and iron (Fe^{3+}) minerals, and sulfate. Methanogens also contribute to the removal of the fermentation by-products producing methane (CH_4) from acetate, formate, or H_2 and CO_2. (Have you noticed the similarities between these syntrophic interactions and those that take place in the cow's rumen [Box 21.2]?)

Syntrophic associations provide an efficient strategy for squeezing sustenance out of energetically low-yield reactions. Scavenging waste products of one microbe's metabolism shifts the reaction's equilibrium to yield, rather than consume, energy. In chapter 19, we described for example how "*Methanobacillus omelianskii*," earlier believed to be a single organism that converted ethanol into methane, was in fact a syntrophic consortium of two microbes. The "S organism" ferments the ethanol into acetate and hydrogen, while the methanogen, *Methanobacterium bryantii*, removes the hydrogen to reduce carbon dioxide and make methane (Fig. 21.14). The "S organism" carries out an endergonic reaction, i.e., a reaction that consumes energy, as indicated by the positive change in free energy from the conversion of ethanol to acetate and hydrogen under standard conditions:

$$2CH_3CH_2OH + 2H_2O \rightarrow 2CH_3COOH + 4H_2$$
$$\Delta G^{o\prime} = 19 \text{ kJ (per 2 mol of ethanol)}$$

The methanogen removes the hydrogen to make methane in an exergonic reaction, as indicated by the negative sign of the $\Delta G^{o\prime}$ of the reaction:

$$4H_2 + CO_2 \rightarrow CH_4 + 2H_2O$$
$$\Delta G^{o\prime} = -131 \text{ kJ (per mol of methane)}$$

Hydrogen scavenging by the methanogen keeps partial pressures of the waste gas low enough to result in a combined reaction that yields energy:

$$2CH_3CH_2OH + CO_2 \rightarrow 2CH_3COOH + CH_4$$
$$\Delta G^{o\prime} = -112 \text{ kJ (per mol of methane)}$$

Thus, the syntrophic partners are thermodynamically intertwined and feast on a substrate (ethanol) that neither of them would have been able to metabolize individually.

Interspecies Electron Transfer

The creative ways in which microbes can feed each other take a different spin when *electrons, rather than molecules,* are exchanged. As we learned in chapter 6, microbial respiration harnesses energy from electrons moving down a membrane-bound electron transport chain, which transports the electrons in stepwise favorable reactions that provide the energy to excrete protons. This creates the **proton motive force** that the ATP synthase uses to make ATP from ADP. Some microbes have evolved complex electron transport chains that allow them to transfer electrons *outside the cell*, a process known as extracellular electron transfer. For example, bacteria of the genus *Geobacter* can oxidize acetate anaerobically and transport electrons *across their Gram-negative cell envelope* to reduce insoluble electron acceptors such as Fe(III) oxide minerals (or rust) (chapter 20). These bacteria are well equipped for this respiratory strategy, having evolved conductive **pili** that function as electronic conduits between the cell and the extracellular minerals (Fig. 21.15). They truly "breathe rocks"!

The reduction of iron oxides by *Geobacter* cells transforms the oxide particles into a semiconducting mineral called magnetite. Being conductive, the magnetite becomes a medium for the transport of respiratory electrons from *Geobacter* to an unsuspected syntrophic partner, *Thiobacillus*, which uses them to reduce nitrate (Fig. 21.15). Note how the waste product of one reaction (magnetite) was used to electrically connect two species and allow *Geobacter* cells to respire an electron acceptor (nitrate) with a much higher reduction potential (chapter 6) than its natural electron acceptor, the Fe(III) oxides. *Thiobacillus*, on the other hand, capitalized on the ability of *Geobacter* to oxidize acetate, a common waste product of fermentation of organic matter in anaerobic environments (Fig. 21.13). Hence, in the environment, all of these syntrophic interactions operate coordinately to efficiently decompose organic matter and prevent its accumulation. You may not see them, but they are there, interacting in truly electrifying ways to cycle carbon.

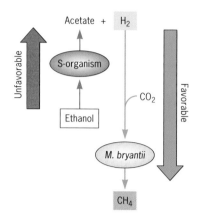

FIGURE 21.14 Interspecies H_2 transfer in "*Methanobacillus omelianskii*." "*Methanobacillus omelianskii*" is a syntrophic consortium between a bacterium (S organism) and the methanogen *Methanobacterium bryantii* based on interspecies H_2 transfer. The syntrophic cooperation between the two keeps low partial pressures of H_2 and makes the otherwise unfavorable conversion of ethanol into acetate and H_2 thermodynamically favorable.

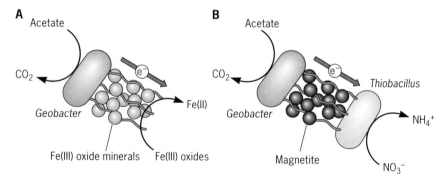

FIGURE 21.15 Extracellular electron transfer alone or with a syntrophic partner. (A) *Geobacter* cells have an electron transport chain spanning the cell envelope and conductive pili that transport electrons generated during the oxidation of acetate between the cell and the insoluble Fe(III) oxide minerals. (B) Once reduced, the Fe(III) oxides are converted into a magnetic mineral called magnetite that conducts electrons to a syntrophic partner such as *Thiobacillus*. This syntrophic interaction allows *Geobacter* to continue to oxidize acetate and to respire an electron acceptor (nitrate) that it cannot use on its own.

WHAT ARE THE PERKS OF BEING A COMMENSAL?

LEARNING OUTCOMES:
- Define commensalism.
- Use examples to help explain why commensal relationships can transition to nonsymbiotic, competitive, or antagonistic relationships when environmental conditions change.

In a commensal relationship one partner benefits (the commensal), while the other is not affected (i.e., it does not benefit or get harmed) (Table 21.1). In the case study, we described two bacteria that grew in intimate association within the biofilm pillars when 3-chlorobiphenyl was added as sole carbon and energy source. *Burkholderia* was able to use this compound for growth, but *Pseudomonas* could not do so on its own. However, they both grew with 3-chlorobiphenyl when cultured together. Is it possible that *Pseudomonas* was feeding on a metabolic product of *Burkholderia* without affecting it? In such case, *Pseudomonas* could be considered a commensal: it benefited from nutrients released by the other bacterial partner without causing harm—an association true to the phrase "Do not bite the hand that feeds you." Note that although the commensal benefits from the association, it can break from its partner when the environmental conditions change that favor its independent growth. In the case study, *Burkholderia* and *Pseudomonas* split up when citrate, a carbon source used by the two, was provided in the medium. In fact, it is even possible that the two became competitors.

Commensalism often involves nutritional benefits. In some cases, the commensals benefit from growth factors such as vitamins and amino acids released by other microbes as part of their metabolism. Often, the commensal benefits from scavenging "waste" products of the other microbe, but without affecting it in any way. This is what happens in our mouth. Even with exquisite dental practices, bacteria colonize our teeth and build multispecies biofilms on the enamel surface (dental plaque) (Fig. 21.16). The first colonizers are commensal bacteria such as streptococci and actinomycetes. These bacteria recognize and bind to specific molecules in the salivary pellicle that coats our teeth. Here they form a thin biofilm that feeds on food debris that sticks to it and on soluble nutrients that permeate through the biofilm matrix. If we are not diligent about brushing, the biofilm grows and recruits other bacteria, getting thicker as the bacteria continue to feed. Eventually, the bacteria form mutualistic interactions designed to stabilize the structure of the biofilm and its inhabitants. If left to grow, potentially pathogenic bacteria join the biofilm, starting out as a commensal. This peaceful interaction can turn antagonistic to the host as the pathogens are allowed to grow and increase in numbers. The troublemakers are usually *Streptococcus mutans* and *Porphyromonas gingivalis*. *S. mutans* starts as a commensal, fermenting sugars and excreting acid. Too much sugar or poor dental hygiene in the oral cavity allows the streptococci to continue to grow, ferment, and produce acid that eventually destroys the tooth's enamel layer and causes cavities. *P. gingivalis* is more adventurous; if allowed to grow, it could invade our gum epithelium, leading to inflammation and forms of periodontal disease. Our saliva also carries

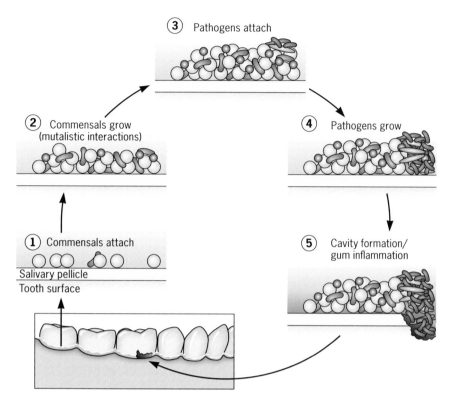

FIGURE 21.16 **Formation of dental plaque biofilms.**

minerals that promote calcification of the biofilms. Now the biofilms are en-cased in a hard, rough surface that some other microbes are apt to attach to. And more microbes join the biofilms, which are now protected by the mineral deposits (tartar), like armor. Brushing will not help you any more. It is time for you to make an appointment with your dental hygienist.

HOW DO PARASITES MANIPULATE THEIR HOST?

LEARNING OUTCOMES:
- Use examples to explain how harming the host with parasitism can be beneficial to the microbe.
- Describe examples of microbe-host parasitic interactions that induce large structural or behavioral changes in the host.

Parasitism is an association in which one microbe benefits at the expense of harming its host (Table 21.1). Parasites exploit their host in various, self-ish ways and without regard for the effect that their lifestyle has on the host. You may be familiar with parasites of humans, such as the tubercle bacillus, an intracellular bacterium in the genus *Mycobacterium* that causes tuberculosis in humans (chapter 25). There are, however, other nuances in the interactions of microbes and other living organisms, some of which are quite startling. For example, the *behavior of animals and plants* may be altered significantly by their interaction with parasitic microbes. A few examples of such manipulations will do better than definitions.

Reckless Rats and Fatal Attraction

The parasitic protist *Toxoplasma gondii* infects rats and, amazingly, causes them to lose their fear of cats. Such suicidal change from normal behavior may make little sense for the rat, but it makes powerful sense for the parasite. To understand why requires knowing about the life cycle of *T. gondii* (Fig. 21.17). The protist is in fact fond of cats: they are its main hosts. Rats, other rodents, and people are incidental hosts. *Toxoplasma* enters the cat digestive system and grows and reproduces in the intestine. Here, it eventually differentiates into tough environmental cellular forms called oocysts, which are eliminated in the feces. Oocysts survive in the soil for long periods, waiting for another cat to eat them and start the cycle again. But sometimes a rodent, rather than a cat, eats them. Once in the rat, the oocysts begin the normal cycle of reproduction but the rat induces a strong immune response to destroy them. Only the more resistant oocysts will survive the immune attack, remaining dormant in the rat's tissue and usually causing no further damage. Similar events take place when people ingest *T. gondii* oocysts, unless people are immune compromised or pregnant, in which case *T. gondii* can cause congenital malformations.

This would seem to be a dead end for the parasite. How can it make it back to a cat? Using outdoor pens, researchers compared the reactions of normal and infected rats to cats' urine. Healthy rats, not surprisingly, were highly averse to the cats' scent, as if knowing what is good for them. Infected rats, on the other hand, appeared to have lost this inhibition. In fact, they even seemed to be attracted to their nemesis' aroma. The significance of this finding has not been confirmed in field studies, but the implications

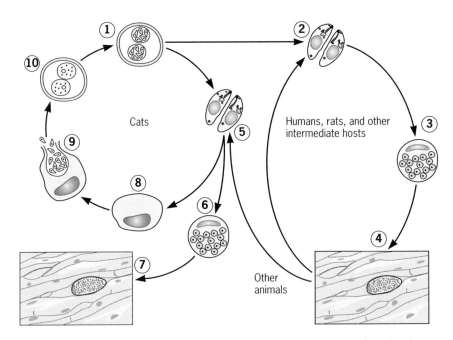

FIGURE 21.17 The life cycle of *Toxoplasma*. Humans and rats become infected with *Toxoplasma* by ingesting cysts (1) in improperly prepared contaminated meat or cat feces. The cysts germinate into active forms (2) that enter the bloodstream and disseminate (3). In most hosts, the immune response eliminates the active form and leaves only dormant cysts in tissues (4). The same happens in cats (5 to 7), but here, some parasites progress to finish their life cycle, which includes a sexual stage (8 to 10).

remain intriguing. Such behavior would hardly be to the rats' advantage, but it would certainly help the parasites complete their life cycle. Chances are that the fearless rats will eventually be eaten by cats. The parasites can now reproduce in the cat's intestine and eventually shed in the feces to start the cycle anew.

Zombie Insects and the Urge To Climb

Spores of the fungus *Ophiocordyceps unilateralis* can infect carpenter ants and other insects and germinate inside their respiratory tract (Fig. 21.18). The mycelia then grow towards the brain and feed on soft tissues until they reach the brain. There, the fungus secretes bioactive compounds that block the areas of the brain in charge of pheromone reception. The insect is now disoriented and under complete control of the fungus, like a zombie. Because the invading fungus develops slowly before killing its host, the infected insects stay active for some time. But their deportment is altered. They acquire the *urge to climb up* the stalks of vegetation and trees. When they have reached a certain height, they bite the stem or the underside of the leaves, attaching themselves to the plant with their mandibles until they die. It is a slow death. The fungus continues to grow and eventually protrudes from the back of the insect's head, forming fruiting bodies filled with spores (Fig. 21.18). Once the spores are sexually mature, they are released from the fruiting bodies from on high, raining down on the path of healthy insects. There they wait to be picked up by a new individual and start a new cycle of infection. Infections in carpenter ants have been well studied but other insect groups—grasshoppers, locusts, aphids, and flies—also exhibit this so-called "summit disease."

Clearly, climbing up the vegetation increases the chances for aerial spore dispersal. However, getting off the forest floor also exposes the zombie insect to sunlight and to temperatures that *are not optimal for the growth of the fungi*. Is the insect fighting back against the fungus? Some entomologists believe that this is an evolutionary safeguard mechanism used by the fungus to control its own dispersal so it does not infect its hosts so fast that they are all killed at once. After all, the fungus needs the host for successful reproduction. In addition, the zombie insect may climb for altruistic reasons. Infected carpenter ants have been proposed to climb to stay away from the path of other ants and avoid infecting other members of its colony. Indeed, infected larvae of certain butterflies and moths crawl into inaccessible spaces, such as beneath the tree bark, as if to get away from their kin. Such fungi must develop long stalks on their fruiting bodies to spread their spores. Whatever the reason, the interplay of signals between the fungus and the insects is extraordinary. Is there a mechanism that keeps the fungus from growing until the insect reaches a certain distance above the ground? Is it all part of the fungus parasitic plan? Who gains and who loses?

When Is a Flower Not a Flower?

For tweaking the host into making a new and elaborate flower-like structure, the prize for "parasite of the year" goes to a fungus, *Puccinia monoica*. This species infects wild plants of the mustard family and induces them to

FIGURE 21.18 **Fungal parasites and zombie insect hosts.** A weevil is firmly attached to the stem of a plant as the result of a fungal infection (*Cordyceps curculionum*) with fruiting bodies of the fungus emerging on stalks. (Courtesy of J. Beach.)

FIGURE 21.19 **A pseudoflower.** This is caused by the growth of a fungus on a wild mustard plant, transforming leaves into what look like petals. (Courtesy of B. A. Roy.)

develop dense clusters of leaves at the tips of stems. To do so, the parasite evades the host defenses and reprograms the host genes to induce the formation of a pseudoflower, a flower-like structure that mimics the shape, color, nectar, and even the scent of flowers from other plants that grow in the same area (Fig. 21.19). The pseudoflowers have, for example, rosettes that look like the petals of a real flower. The pseudopetals get covered with fungal spores and become sticky and sweet smelling. They are a beautiful yellow color, different from the normal flowers of the plant but similar to those formed by neighboring plants. Insects arrive, with nectar on their agenda, and poke around the pseudoflower, collecting fungal spores instead of the desired pollen. And off they go, spreading spores to other plants. As can be seen in the photograph (Fig. 21.19), the impersonation is nearly faultless, and at a distance has fooled even professional botanists.

ARE MICROBES PREY OR PREDATORS?

LEARNING OUTCOMES:
- Contrast the impact on the environment for predation upon oceanic microbes by protists versus viral lysis.
- Identify characteristics of *Bdellovibrio bacteriovorus* that contribute to its success as a bacterial predator.
- Describe examples of bacterial predators that use specialized structures in order to improve predation success.

Microbes are usually considered predators, but in reality, they are more often the prey. In fact, bacteria are a major source of food for many organisms such as many protists and nematodes. Thus, bacteria (and most likely archaea as well) are important constituents of the food web. Predation can explain why many natural microbial populations do not reach the sizes that would be expected based on available food. As we discussed earlier (Box 21.2), protists are abundant in the rumen and prey on the abundant bacteria, likely exerting some sort of population control. Predation is also a frequent event in oceans, where the major predators are phages and protists. There are about 1 million bacteria per ml of seawater throughout the world's oceans. Even though this is a large number, it is smaller than would be expected based on the amount of food in the form of dissolved organic carbon, available to marine microbes. This discrepancy is due to the fact that microbes are kept in check by their predators. In fact, without predation, the ocean's microbes could well be 10-fold more abundant. And these are just but a few examples of what may be a widespread form of microbial interaction.

While half the predation of oceanic microbes is due to protists, the other half is due to viral infection. As we learned in chapter 19, viruses are common wherever there are microbial cells. Take 1 ml of seawater and you will count 10 million viral particles, 10 times greater than the number of microbial cells. (As microbiologist Forest Rohwer pointed out, seawater contains about 10^7 phages per ml but only 10^{-19} great white sharks per ml!) Environmental surveys also suggest that at least half of the oceanic viruses are lytic (chapter 19). Upon infection, lytic viruses take over cellular functions and redirect them towards the synthesis of viral components needed to make new viral particles. Lytic viruses also encode functions to lyse the cells and release the viral progeny, releasing the cell contents on which other bacteria can feed. Indeed, when a virus kills a microbe, it actually stimulates other microbes to grow. By contrast, when a protist eats a bacterium, the bacterial cell is ingested and its contents assimilated by the predator. Yet the protists are themselves prey to other protists and to animals (the bigger fish eats the small one!). Hence, the bacterial prey enters higher trophic levels of the food chain, including fish. Juvenile cod, for example, eat large quantities of protists.

Predation can also be observed *between bacteria*. In this case, the bacterial predator either grazes upon or actually penetrates its prey. The discovery of such predators was as surprising to the discoverers as it may be to you. The researchers searched for phages by a tried and true method, looking for "plaques," or holes, in lawns of susceptible bacteria on agar plates (see chapter 17). They found plaques, all right, but noticed that, unlike those caused by phages, which stop enlarging as the culture stops growing, these holes grew bigger and bigger even after several days of incubation. When they looked at the material from these plaques under a microscope, they saw very tiny bacteria (about one-fifth the size of an *E. coli* cell). The bacterial predators were given a tongue twister of a name, *Bdellovibrio bacteriovorus* ("bdello" means "leech").

The *Bdellovibrio* life cycle is quite peculiar and consists of four defined steps: encounter of the prey (always a Gram-negative bacterium), entry in the periplasm, multiplication, and cell lysis to release the progeny (Fig. 21.20). Each of these steps requires the expression of specialized genes and the silencing of unneeded ones. An example is the fact that bdellovibrios on the prowl are highly motile, which helps them approach their prey (although apparently not by chemotaxis). However, once they penetrate the outer membrane and reach the periplasm of the host, they lose their flagella. How this invasion occurs is not known. We do know however that *only one cell* makes it into the periplasm of the host cell and, once there, the invading bacterium augments its food supply by making the host's inner membrane leaky and introducing hydrolytic enzymes into the host's cytoplasm. Bdellovibrios thrive on degradation products of the host's proteins and nucleic acids. At least 30 of their proteins are specific to the periplasmic stage. With time, the invading cell grows and elongates, forming a long helical bdellovibrio that divides into three to five motile daughter cells. By then, the host cell has weakened so much that it lyses,

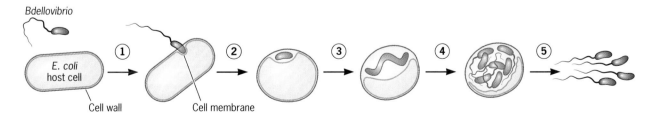

FIGURE 21.20 The life cycle of *Bdellovibrio.* Attachment to the host cell (1) is quickly followed by penetration into the periplasm (2). The parasite grows without dividing, making a helical bdellovibrio (3 and 4). This divides into small rods that make flagella (5). The weakened host cell lyses, releasing progeny bdellovibrios.

releasing the bdellovibrios' progeny. The whole process takes about 3 to 4 hours, which is long in comparison to the growth rate of the host bacteria under nutritionally abundant conditions, but is ample to cause a microepidemic when the host is no longer growing. This explains the gradual increase in size of the plaques on "old" agar plates. In fact, this is the only way to grow bdellovibrios in the lab: within their host. Some mutants can be grown alone, without the host, on artificial media in the laboratory. However, the faster and longer they grow under these conditions, the less invasive they become. Somehow the predator becomes domesticated. This makes their study more challenging.

We know little about their ecological role. Bdellovibrios are often found in ocean water and soils, associated with biofilms, where they prey on just about any bacterial species, as long as they are Gram negative. Does it mean that nothing stops them from preying and killing hosts? What prevents them from killing *all* of their preys? As it is often the case with predators, such as lions and killer whales, their numbers will wane in the absence of adequate prey. Kill all of your prey and you may end up dying of starvation. Some tantalizing ideas have been put forth. There may be nonsusceptible species that act as "decoys" by sopping up on bdellovibrios and preventing them from preying on new hosts. Some preys may form biofilms that the bdellovibrios are unable to invade.

Bdellovibrios are not the only bacteria that prey on other bacteria. Other organisms, such as the myxobacteria, hunt in "wolf packs," swarms of cells that move on surfaces and feed on other kinds of bacteria via the excretion of hydrolytic enzymes (see chapter 13). An analogous kind of predation relies on the excretion of toxic chemicals, which we will discuss in more detail in the next section about antagonism. Other predators display variations on the theme of *Bdellovibrio*. Instead of invading the periplasm, they either penetrate the cytoplasm or establish cell-to-cell "bridges" between themselves and the prey. A unique strategy is used by a marine gliding bacterium called *Saprospira grandis*. This organism catches other bacteria by trapping their flagella, much as flies are trapped on flypaper, or insects in a spider web. Flagella, which are otherwise so useful for the growth and survival of many bacteria, here become their downfall.

These are but a few examples of what is, in fact, a vast repertoire of predation among bacteria. Considering that the majority of microbes have not been cultivated yet, the variety of microbial predatory behaviors may be even vaster. Predation in the microbial world has been a neglected subject, but in partnership with its converse phenomenon, cooperation, it helps explain complex ecological relationships.

HOW DO MICROBES ANTAGONIZE THEIR NEIGHBORS?

LEARNING OUTCOMES:
- Compare and contrast antibiotics and bacteriocins.
- Describe how bacteria that produce bacteriocins are not killed by their own products.
- List examples of mechanisms by which microbes compete with other microbes.

Besides predation, microbial species use subtle (and not so subtle) strategies to gain an advantage over other species in their environment. Often times the best way to antagonize your neighbors is to secrete a compound that is toxic to them but not to you. Perhaps the most notorious example in this category is the secretion of **antibiotics**, which kill or inhibit the growth of pathogenic bacteria. Growth inhibition by antibiotics can be easily tested in the laboratory by growing bacteria as a lawn on the surface of solidified growth media and testing for a halo of inhibition around a paper disk from which the antibiotic diffuses into the medium (Fig. 21.21). Antibiotics have gained their fame because they are used to treat human and animal diseases caused by bacteria. Not surprisingly, we know a great deal about the way they inhibit bacteria both in the lab and in infected hosts (Table 21.2). Antibiotics comprise a large number of different organic molecules, generally with molecular masses in the 300- to 1,000-dalton range. Although many of the antibiotics we use now are synthetic or chemical modifications of natural ones, the first antibiotics to be used were natural products produced by bacteria and fungi. However, we know relatively little about what they actually do in nature. The concentration of antibiotics in soils is low, usually below that required to inhibit microbes in laboratory studies (Fig. 21.21). It is conceivable that the local concentration around individual cells may reach inhibitory levels, permitting the producers to outgrow sensitive organisms present in their vicinities. It is fair to say that we don't know what selective pressures have led to the emergence of antibiotic-producing organisms.

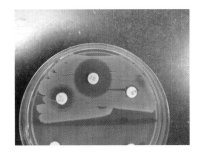

FIGURE 21.21 Antibiotic disk inhibition assay. The diffusion of antibiotics from paper disks into the solidified medium inhibits the growth of susceptible bacteria grown on the agar surface (in this case, *E. coli*). (ASM MicrobeLibrary © Peltier Horn.)

TABLE 21.2 Commonly used antibiotics and how they work

Antibiotic class	Examples	Mechanism of action
Murein synthesis inhibitors		
β-Lactams	Penicillins, cephalosporins, and synthetic derivatives	Interfere with cell wall synthesis, leading to cell lysis
Others	Vancomycin, aztreonam, imipenem	
Protein synthesis inhibitors		
Aminoglycosides	Streptomycin, kanamycin, neomycin, gentamycin, amikacin, tobramycin	Bind small (30S) ribosomal subunit, misread mRNA; some inhibit peptide chain elongation
Tetracyclines	Tetracycline, doxycycline	Bind small (30S) ribosomal subunit, inhibit peptide chain elongation
Chloramphenicol		Bind to large (50S) ribosomal subunit, inhibit peptide chain elongation
Erythromycins	Azithromycin (Zithromax)	Block exit of growing peptide chain from the ribosome
RNA synthesis inhibitors		
Rifampin		Binds bacterial RNA polymerase, blocks transcription initiation
DNA synthesis inhibitors		
Nitrofurans	Nitrofurans and synthetic derivatives (nitrofurantoin and nitrofurazone)	Activated upon reduction by cellular enzymes, cause DNA strand breakage
Metronidazole Nalidixic acid Novobiocin Ciprofloxacin	Cipro and other fluoroquinolones	Inhibit type II DNA topoisomerases (e.g., gyrase and topo IV), interfere with DNA replication
Folate antagonists		
Sulfonamides	Trimethoprim-sulfamethoxazole (Septra)	Block synthesis of tetrahydrofolate and 1-carbon metabolism

TABLE 21.3 Modes of action of some bacteriocins[a]

Forming ion channels in membranes (depolarization)
Breakdown of DNA
Inactivation of ribosomes
Inhibition of cell wall murein synthesis
Hydrolysis of cell wall murein

[a]Some also affect eukaryotic cells

Many bacteria also secrete **bacteriocins,** a diverse type of peptides. Some are active only against relatives of the species that produce them; others have a broad spectrum and affect distantly related species as well. There are even bacteriocins that affect all the Gram-positive bacteria tested. Bacteriocins inhibit sensitive bacteria in a variety of ways (Table 21.3). A common mode of action is to disrupt the cell membrane, making it porous for potassium and phosphate ions. The bacteria then attempt to stock up on these ions using an ATP-dependent transport system and drain their ATP supply, thus stalling growth. Why are bacteriocin-producing strains *not sensitive to the actions of their own compounds*? It turns out that the bacteriocin producers also *make immunity proteins that counteract their activities.* Often, genes for making the bacteriocin and their immunity proteins are on the same operon, so they can be expressed at the same time. Bacteriocins have found practical uses, especially in the dairy industry, where they are used in both a predictable and an unexpected way. The predictable use is to limit the growth of undesirable bacteria, such as *Listeria*, a pathogen that sometimes contaminates dairy products (chapter 24). A handy way to deliver the bacteriocins is to clone their genes into the strains that are used for cheese production. The less obvious use of bacteriocins is to enhance the flavor of cheeses, such as cheddar. When bacteria used in the production of such cheeses lyse, they release enzymes that improve flavor development and speed up the ripening of the cheese. Who said that bacteria are mainly bad guys?

In addition to secreting antibiotics and other antimicrobial products, some microbes use subtle physiological means to compete. One example of a physiological strategy is *competition for iron*, a metal that, although abundant in nature, is quite scarce in a soluble form, which is needed for assimilation. In the environment, as well as in the bodies of animals, competition for iron is crucial for growth (as we will see for Staph infections in chapter 23). A common tactic to acquire the needed iron is to secrete iron-chelating compounds called **siderophores**, small molecules that bind iron with great avidity. The iron-bearing chelates can then be reabsorbed and the iron can be utilized for metabolic purposes, e.g., the synthesis of the cofactors of cytochromes and other oxidative enzymes. As these microbes take up iron, they also reduce its concentration to even lower levels, thus hindering the growth of organisms less adroit at scavenging the mineral.

In other cases, a microbe's waste antagonizes its neighbors. For example, end products of a microbe's metabolism can inhibit the growth of other microbes and, therefore, become antagonists. Ethanol, a common end product of microbial fermentations (chapter 6), inhibits the growth of many microbes, even at relatively low concentrations. Ethanol's primary target is the cell membrane, which loses fluidity and integrity, becoming leaky and triggering a stress response aimed at repairing the membrane and restoring its essential functions. Why is ethanol tolerated by the producing strains? They use several tricks, the most significant being that they evolved lipid membranes with a *fatty acid composition that reduces their fluidity*. By increasing the rigidity of the membrane, they counteract the fluidizing effect of the alcohol. Changes such as this are often observed in ethanol-tolerant yeasts used in wine and beer making. (Coincidentally, the accumulation of ethanol in the fermentation broth inhibits the growth of other microbes and prevents the spoilage of the alcoholic beverage. However, as many beer and wine producers will tell you, that is not always the case.) Microbial fermentations also lead to the accumulation of organic acids and *drops in the pH of the medium*, which can become suboptimal for the growth of other microbes. In this case, the antagonistic effect is indirect and mediated by the low pH rather than directly by secreting chemicals.

THE HUMAN MICROBIOTA: ONE FOR ALL AND ALL FOR ONE?

LEARNING OUTCOMES:
- Describe syntrophy between the human gut and the gut microbes.
- Identify specialized structures and characteristics of gut methanogen *Methanobrevibacter smithii* that allow it to be uniquely adapted to syntrophy in the intestinal environment.
- State how the growth of potentially harmful and unwanted bacteria can be controlled in the human gut.
- Describe an example of how interactions among gut microbiota can impact and respond to human health.

As we learned in chapter 19, many of our body parts carry microbial communities, collectively known as the human microbiota. Notably, each location has selected for a unique microbial combo adapted for specific functions and to the unique environmental conditions that prevail in each location. Thus, there is an oral microbiota (mouth) that is very different from the microbial communities of the nose, skin, or vagina. Microbes are particularly abundant (in the trillions or, if you prefer, a couple of pounds by weight!) in the terminal portion of our intestines (the colon), where they establish complex microbial communities. Collectively, the microbes that live in our gut perform functions that influence many aspects of our physiology and health (chapter 19). To do so, specialized groups of microbes interact with each other and integrate their activities to perform functions as a community that could not be carried by the individuals. Our association with the gut microbes is indeed mutualistic—the benefit is mutual.

We humans may not have a specialized digestive chamber such as the cow's rumen (Box 21.2). Yet the gut microbes that colonize our intestinal mucosa play analogous roles (Fig. 21.22). The anaerobic conditions that prevail in the terminal portion of the intestine (the colon) enrich for strict anaerobes such as *Firmicutes* and *Bacteroidetes*, which become the dominant (>90%) bacterial phyla. Together, bacteria in these two groups break down and ferment complex carbohydrates such as dietary fiber and generate short-chain fatty acids (SCFAs) for the host. *Firmicutes* produce primarily butyrate; *Bacteroidetes* produce acetate and propionate. Just like in cows, most of the fatty acids are adsorbed by the intestinal epithelium and are either consumed by the epithelial cells (butyrate) or distributed

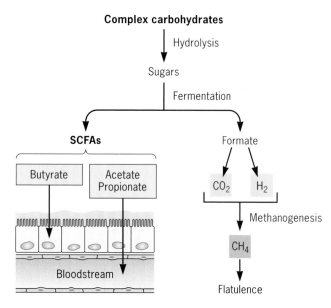

Complex carbohydrates

↓ Hydrolysis

Sugars

↓ Fermentation

SCFAs   Formate

Butyrate | Acetate Propionate CO_2 H_2

Bloodstream   ↓ Methanogenesis

                             CH_4

                             ↓

                             Flatulence

FIGURE 21.22 **Interactions between gut microbes and human host.**

in the bloodstream to peripheral organs (acetate and propionate). About two-thirds of the acetate is delivered to the liver, where it is used as a source of energy and as a biosynthetic precursor of cholesterol, long-chain fatty acids, and amino acids. In addition, SCFAs promote colonic health by acting as inflammation modulators, vasodilators, and contributing to gut motility and wound healing. Note the similarities between syntrophic interactions among microbes and those between the human host and the gut microbiota: we use the end products of their fermentation and, in so doing, we stimulate our microbes and our digestion.

As in the rumen (Box 21.2), hydrogen is also a fermentation waste product in the human intestine and, if not removed, will accumulate and feedback-inhibit the activities of the fermentative bacteria. To prevent this, methanogens (phylum *Euryarchaea*) also live in the gut of most people. These organisms remove the hydrogen and carbon dioxide as well as the formate produced fermentatively and make methane (CH_4) (Fig. 21.22). Passing gas helps expel the methane and any unutilized hydrogen and carbon dioxide gases so fermentation can proceed efficiently. Who are the gut methanogens? Although functionally analogous to the methanogens in the rumen (Box 21.2) or in soils and sediments (Fig. 21.13), the gut methanogens are uniquely adapted to the human intestinal environment. The most prominent one is *Methanobrevibacter smithii*, which carries many more genes for methanogenesis in its genome than non-gut methanogens. Genomic, transcriptomic, and proteomic analyses revealed yet another adaptation to the intestinal environment: *M. smithii* has numerous transporters that allow it to efficiently compete for ammonium. Why is this important? Ammonium is the primary source of nitrogen for our intestinal microbial companions. Hence, competition for ammonium is high in our intestines. *M. smithii* also has many more genes for adhesins and capsule synthesis than non-gut methanogens, which help the archaeal cells attach to the intestinal mucosa and stay in close proximity to their syntrophic partners to promote metabolite exchange. Capsules also create a barrier that prevents the diffusion of toxic dietary substrates (e.g., certain sugars and phenolic compounds derived from plant material). And, just in case this is not enough, *M. smithii* is equipped with many multidrug efflux pups and transporters to excrete any noxious compound that makes it through.

In addition to the dominant bacterial phyla (*Firmicutes* and *Bacteroidetes*) and the archaeal phylum (*Euryarchaeota*), the human gut microbiota includes three low-abundant bacterial phyla (*Proteobacteria*, *Actinobacteria*, and *Verrucomicrobia*). The *Proteobacteria* have a versatile metabolism and can be fine commensal guests in the human gut. However, some of them are opportunistic pathogens (e.g., *E. coli*) and can cause disease if allowed to grow and outcompete the others. How are their numbers kept at bay?

Some of the fatty acids produced during the fermentation of carbohydrates by the gut microbiota do not get adsorbed by the colonic epithelium and help drop the pH in the intestinal lumen to levels low enough to control the growth of potentially harmful bacteria. Should the careful balance of gut microbes be altered, for example after a treatment with antibiotics, opportunistic pathogens can grow without restraint and cause disease.

Predation can also control the growth of unwanted bacteria in the human gut. Microscopic examination of human feces reveals a high abundance of viral-like particles (VPLs) (10^8 to 10^9 per gram of feces). And these are just the particles that are shed with the feces. Their true abundance inside the gut may be higher. Viruses often prey on the "winner," i.e., the microbes that grow and proliferate more rapidly. In doing so, viruses control the relative abundance of bacteria in the microbial community so as to preserve its optimal structure and functionality. And the human host benefits from their dynamic interplay because a "healthy" combo of microbes is present to better satisfy the host's needs. Again, everybody wins.

The synergistic interactions among the gut microbes impact host functions and health in ways that we yet do not fully understand. There is for example a correlation between the structure of the microbial community inhabiting our guts and obesity. On one hand, the relative abundance of methanogenic archaea increases in obese individuals. This suggests that more hydrogen and carbon dioxide (or formate) are available from the fermentation of carbohydrates by *Firmicutes* and *Bacteroidetes* bacteria to feed the methanogens. Consistent with this, obesity has been linked to increases in genes involved in carbohydrate metabolism in metagenomic data. One may think that this is related to calorie intake—more carbohydrates, more fermentative bacteria, and more syntrophic methanogenic partners. However, obese and lean individuals fed the same diet have different gut **microbiomes**; that is, the genetic makeup of their gut microbiota is different. Although the food intake is the same in these individuals, the gut microbiome of an obese host has a higher proportion of *Firmicutes* and lower abundance of *Bacteroidetes*. As a result, the gut microbiome of an obese host absorbs more calories than that of a lean individual. Once an obese person loses excess body weight, the number of *Bacteroidetes* increases again and the community structure stabilizes as in lean individuals. Clearly, the gut microbiome is responding to changes in the host physiology (in this case, body weight). It is well known that body weight has a genetic and environmental component. Some people are genetically programmed to gain more weight from the food they eat than others, perhaps because their intestinal microenvironment has selected for some microbes over others during the early stages of colonization (as a newborn or during the maturation that the gut microbiome undergoes throughout our lives [see chapter 19]). Similarly, environmental cues such as sleep deprivation and stress influence body gain and the composition of the gut microbiome. In some respects, the gut microbes behave like an organ, collectively modulating their activities in response to the host's cues. To do so, the microbial members must integrate their activities in specific ways to behave as one. One for all and all for one!

CASE REVISITED: **Living apart or together?**

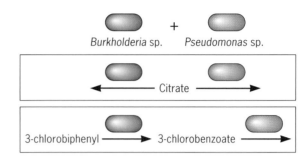

Did you think that the separation or proximity of the two bacteria in the biofilms was due to both competition and cooperation depending on the carbon source? If so, you got it right. Dr. T suspected that the bacteria formed separate pillars because they were competing for the citrate. However, the very same bacteria came together to degrade 3-chlorobiphenyl, suggesting they needed each other to do so (i.e., they cooperated). To test this hypothesis, Dr. T grew each bacterium separately with citrate or 3-chlorophenyl. This is what he observed:

- Citrate supported the growth of biofilms of each bacterium, consistent with the original hypothesis that the two bacteria were competing for this carbon source when grown together. Not surprisingly, they did not grow in close proximity in the biofilms but, rather, they preferred to make their own separate pillars. There is an additional advantage to making a biofilm pillar: they maximized the biofilm surface exposed to the carbon source.

- 3-Chlorobiphenyl only supported the growth of *Burkholderia* sp. (the one labeled with the red FISH probe). Using an HPLC, Dr. T detected 3-chlorobenzoate, a by-product of the metabolism of 3-chlorobiphenyl. When he added 3-chlorobenzoate to the *Pseudomonas* cultures, they grew! Hence, the two grew together in close proximity to team up: one grew with 3-chlorobiphenyl and the other with the metabolic product of the first. As in an assembly line, the two bacteria cooperated to degrade 3-chlorophenyl. Would you classify this interaction as commensalism or as syntrophy? The data shown in this experiment tell us that both grew together, but we do not know if they both benefited from the teamwork or only one. If the accumulation of 3-chlorobenzoate were, for example, toxic to *Burkholderia* sp., its removal by *Pseudomonas* sp. would benefit both greatly (one is not exposed to toxic waste, the other gets its food). But commensalism is also possible: *Burkholderia* sp. is eating its food and is not affected by the accumulation or removal of 3-chlorobenzoate.

This study highlights the transient nature of microbial interactions as a function of the environment (in this case, as a function of the food that is available). The results also help us appreciate that microbes can engage in more than one type of interaction. In this example, they competed to use citrate but cooperated to use 3-chlorobiphenyl. In the end, both grew and reproduced; both survived. That's all that matters.

CONCLUSIONS

The interactions among living organisms are varied and are central to the survival of many species. No biological entity has evolved without being molded by the presence of other organisms. In this chapter, we have considered several mutually beneficial interactions, as well as others that are more one-sided. In the next chapters, we will tackle the weighty topic of microbial pathogenesis, which focuses on interactions that have a deleterious effect on the host.

Representative ASM Fundamental Statements and Learning Outcomes for the Chapter

EVOLUTION

1. Cells, organelles (e.g., mitochondria and chloroplasts), and all major metabolic pathways evolved from early prokaryotic cells.

   a. Use evidence to support the endosymbiotic theory of mitochondria and chloroplasts.

   b. Identify differences between modern-day "ex-microbe" organelles, such as mitochondria and chloroplasts, and their free-living ancestral bacteria.

2. The evolutionary relatedness of organisms is best reflected in phylogenetic trees.

   a. Explain the evolutionary consequences of the unique mode of *Buchnera* symbiont transmission to host cells.

CELL STRUCTURE AND FUNCTION

3. *Bacteria* and *Archaea* have specialized structures (e.g., flagella, endospores, and pili) that often confer critical capabilities.

   a. Use examples to differentiate organelles, endosymbionts, and intracellular parasites.

   b. Describe the function of organelles other than mitochondria and chloroplasts that are suspected to have an endosymbiotic origin.

   c. Speculate an advantage to hosting endosymbionts in specialized bacteriocyte cells.

   d. Identify characteristics of *Bdellovibrio bacteriovorus* that contribute to its success as a bacterial predator.

   e. Describe examples of bacterial predators that use specialized structures in order to improve predation success.

   f. Identify specialized structures and characteristics of the gut methanogen *Methanobrevibacter smithii* that allow it to be uniquely adapted to syntrophy in the intestinal environment.

METABOLIC PATHWAYS

4. The interactions of microorganisms among themselves and with their environment are determined by their metabolic abilities (e.g., quorum sensing, oxygen consumption, nitrogen transformations).

   a. Identify the most common reason for microbes to interact, either in a positive or negative way.

   b. State reasons why many microbial interactions are transient and change over time.

   c. Speculate why neutralism, or living solely by oneself, is rare in the microbial world.

   d. Explain how microbe-microbe interactions make it possible for ruminants, such as cows, to subsist on a cellulose-heavy plant diet.

   e. State reasons why a syntrophy is beneficial for many fermentation pathways and even necessary for others.

f. Use an example of interspecies metabolite transfer and electron transfer to justify the statement: "The interactions of microorganisms among themselves and with their environment are determined by their metabolic abilities."
g. Compare and contrast antibiotics and bacteriocins.
h. Describe how bacteria that produce bacteriocins are not killed by their own products.
i. List examples of mechanisms by which microbes compete with other microbes.

MICROBIAL SYSTEMS

5. Microorganisms, cellular and viral, can interact with both human and nonhuman hosts in beneficial, neutral, or detrimental ways.

a. Define symbiosis.
b. Describe the nutritional mutualism of *Buchnera* and aphids.
c. Identify how *Wolbachia* increases its genetic fitness through its mode of being passed on to the progeny of insects.
d. Describe how interactions between nitrogen-fixing bacteria and legumes lead to the formation of root nodules.
e. Identify how legumes accommodate the physiological needs of their N_2-fixing endosymbionts.
f. Name physiological or structural features of ruminants that make the symbiosis with rumen microbes possible.
g. Differentiate syntrophy and symbiosis.
h. Use examples to help explain why commensal relationships can transition to non-symbiotic, competitive, or antagonistic relationships when environmental conditions change.
i. Use examples to explain how harming the host with parasitism can be beneficial to the microbe.
j. Describe examples of microbe-host parasitic interactions that induce large structural or behavioral changes in the host.
k. Contrast the impact on the environment for predation upon oceanic microbes by protists versus viral lysis.

IMPACT OF MICROORGANISMS

6. Microbes are essential for life as we know it and the processes that support life (e.g., in biogeochemical cycles and plant/animal microbiota).

a. Describe the syntrophy between the human gut and the gut microbes.
b. State how the growth of potentially harmful and unwanted bacteria can be controlled in the human gut.
c. Describe an example of how interactions among gut microbiota can impact and respond to human health.

SUPPLEMENTAL MATERIAL

References

- **Keeling PJ,. McCutcheon JP, Doolittle WF.** 2015. Symbiosis becoming permanent: survival of the luckiest. *PNAS* **112:**10101–10103. doi:10.1073/pnas.1513346112
- **Sieber JR, McInerney MJ, Gunsalus RP.** 2012. Genomic insights into syntrophy: the paradigm for anaerobic metabolic cooperation. *Ann Rev Microbiol* **66:**429–452. doi: 10.1146/annurev-micro-090110-102844
- **Fischbach MA, Sonnenburg JL.** 2011. Eating for two: how metabolism establishes interspecies interactions in the gut. *Cell Host Microbe* **10**(4):336–347. doi: 10.1016/j.chom.2011.10.002

Supplemental Activities

CHECK YOUR UNDERSTANDING

1. A symbiosis can be . . .
   a. mutualistic
   b. parasitic

    c. syntrophic

    d. all of the above

2. **What pieces of evidence support the bacterial origin of mitochondria and chloroplasts?**

3. **What is the function of root nodules in legumes, and how do they arise?**

4. **What are the main microbial activities that take place in the rumen of a cow? How are they similar to those in the human gut?**

5. **Discuss examples of bacteria being used as prey by predators.**

DIG DEEPER

1. Aphids contain specialized bacteriocyte-like cells where they harbor and feed millions of bacterial endosymbionts with the sole purpose of making 10 amino acids that the insect cannot make on its own. This symbiosis allows them to feed on sap, despite its low protein content.

    a. What do you think would happen if you feed the aphids sap that contains antibiotics?

    b. Some aphids also carry a bacterium symbiont (*Hamiltonella defense*) that protects the insects from wasp predators. The wasps inject their eggs inside the aphid, which is eaten alive by the wasp larvae when they hatch. However, aphids that carry strains of *H. defense* infected with the bacteriophage APSE produce a toxin that kills the wasp larvae. How would you demonstrate that the bacteriophage is a secondary endosymbiont essential for the survival of the aphid host against the parasitic wasps?

    Resource: "Virus and bacteria team up to save aphid from parasitic wasp," by Ed Yong, http://blogs.discovermagazine.com/notrocketscience/2009/08/21/virus-and-bacteria-team-up-to-save-aphid-from-parasitic-wasp/#.Vpq4gjaueA8

2. The *Cordyceps* fungus infects and feeds on the brain of carpenter ants and causes them to climb up the vegetation and die while biting the underside of a leaf. However, *Cordyceps* is itself parasitized by a white fungus in the genus *Polycephalomyces*. (This makes the white fungus a parasite of a parasite or a hyperparasite.) Studies show that more than half of the "zombie" ants that congregate to die on the underside of leaves also contain the hyperparasite. In all cases, the hyperparasite prevented the *Cordyceps* fungus from forming a fruiting body and releasing spores.

    a. Could the relationship between the ants and the two fungi be a three-way symbiosis?

    b. Who benefits and who gets harmed?

    Resource: "An evolutionary tale of zombie ants and fungal villains and knights" (http://schaechter.asmblog.org/schaechter/2012/07/an-evolutionary-tale-of-zombie-ants-and-fungal-villains-knights.html)

3. One of the many ways that microbes contribute to human health is by producing antibiotics, which we use to treat bacterial infections. The first antibiotic, penicillin, is produced by the fungus *Penicillium notatum* and was discovered by Sir Alexander Fleming in 1928. Of course not all microbes produce antibiotics, nor are all sensitive to them. Use the questions below to expand your understanding of what antibiotics are and do.

    a. Why would a fungus like *Penicillium notatum* produce antibiotics?

    b. You isolated a new fungal strain from a forest soil and want to test if it produces antibiotics. How would you do it?

    c. After confirming that your fungal isolate produces an antibiotic, you want to test its suitability to treat infections caused by the following bacteria:

        Bacterium A: *Staphylococcus aureus*

        Bacterium B: *Bacillus subtilis*

        Bacterium C: *Escherichia coli*

    How would you test this in the laboratory using petri dishes containing agar-solidified growth medium? Illustrate a result in which bacteria A and B are sensitive but *E. coli* is resistant.

    What control(s) do you need to include in the experiment?

   d. You discover that your antibiotic inhibits the growth of methicillin-resistant *Staphylococcus aureus* (MRSA). Why is this significant?

   Resources:
   - How Alexander Fleming accidentally discovered penicillin. https://www.youtube.com/watch?v=PdWhVwiJWaU
   - MRSA infection. Diseases and conditions. Mayo Clinic. http://www.mayoclinic.org/diseases-conditions/mrsa/basics/definition/con-20024479

4. **Team work:** Working in groups, discuss what may have prevented nitrogen-fixing rhizobia from becoming cell organelles as mitochondria and chloroplasts?

5. **Team work:** Present examples for a lay audience of how the interactions of microbes in the human gut can alter the health of the host.

   Resources:

   "Five ways gut bacteria affect your health," by Rachael Rettner (http://www.livescience.com/39444-gut-bacteria-health.html)

   This Week in Microbiology #71: "Colon Cancer's Little Shop of Horrors," MicrobeWorld podcast (http://www.microbeworld.org/podcasts/this-week-in-microbiology/archives/1614-twim-71-colon-cancer-s-little-shop-of-horrors)

Supplemental Resources

VISUAL

- Paul Andersen discusses endosymbiosis and the origin of eukaryotes in the video "Endosymbiosis" (7 min): https://www.youtube.com/watch?v=-FQmAnmLZtE
- Learn about symbiotic bacteria that help worms get nutrients in the animation "No Guts, No Glory: The Metabolism of Symbiotic Bacteria Inside a Worm" (3:34 min): https://www.youtube.com/watch?v=muOS9QHBvIg
- An animation illustrating how root nodules in legume plants are formed (1:50 min): https://www.youtube.com/watch?v=fl4NhK7AaDQ
- Video showing rumen microbes (0:40 min): https://www.youtube.com/watch?v=FoLJF7cs1Rk
- Animated video illustrating how *Toxoplasma*-infected rats lose their fear of cats to pass the parasite to their preferred host (0:36 min): https://www.youtube.com/watch?v=zCbxdLnYclU

WRITTEN

- Gemma Reguera discusses what is known about dental plaque biofilms and how their composition has changed with diet and dental hygiene in the *Small Things Considered* blog post "Smile (or Not)!" (http://schaechter.asmblog.org/schaechter/2014/03/a-mouthful-of-microbes.html).
- Gemma Reguera reviews how microbes respire via interspecies electron transfer in the *Small Things Considered* blog post "Love at First Zap" (http://schaechter.asmblog.org/schaechter/2012/12/lov.html).
- Moselio Schaechter tells us about the bacterial endosymbionts that live inside the mitochondria of the ticks' ovarian cells in the *Small Things Considered* blog post "All in the Family" (http://schaechter.asmblog.org/schaechter/2007/01/all_in_the_fami.html).
- Lesson on antibiotics by Kahn Academy: "Antibiotics: An Overview" (https://www.khanacademy.org/science/health-and-medicine/current-issues-in-health-and-medicine/Antibiotics-and-antibiotic-resistance/a/antibiotics-an-overview).

Microbial Pathogenesis

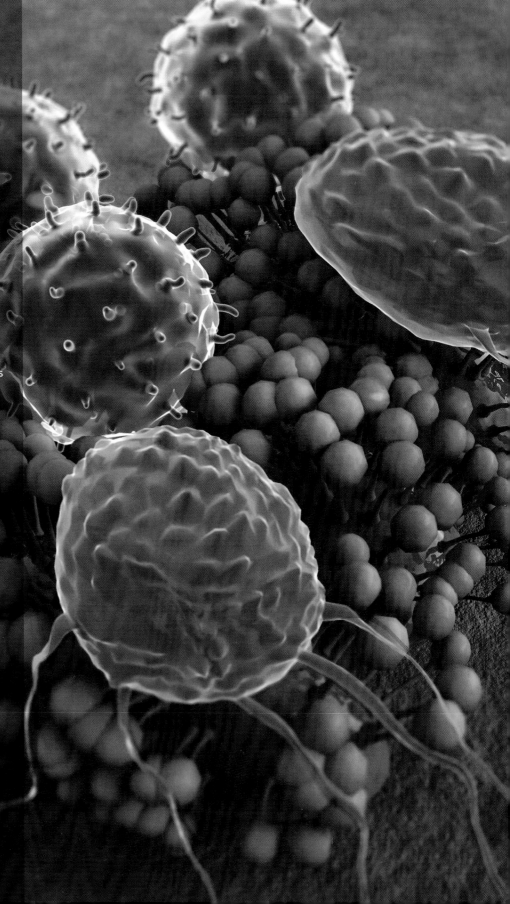

T lymphocytes attacking a colony of fungi

CHAPTER TWENTY-TWO

Infection:
The Vertebrate Host

KEY CONCEPTS

This chapter covers the following topics in the ASM Fundamental Statements.

EVOLUTION

1. Human impact on the environment influences the evolution of microorganisms

CELL STRUCTURE AND FUNCTION

2. Bacteria have unique cell structures that can be targets for antibiotics, immunity, and phage infection

METABOLIC PATHWAYS

3. The growth of microorganisms can be controlled by physical, chemical, mechanical, or biological means

INFORMATION FLOW AND GENETICS

4. Cell genomes can be manipulated to alter cell function

MICROBIAL SYSTEMS

5. Microorganisms and their environment interact and modify each other
6. Microorganisms, cellular and viral, can interact with both human and nonhuman hosts in beneficial, neutral, or detrimental ways.

IMPACT OF MICROORGANISMS

7. Microbes are essential for life as we know it and the processes that support life (e.g., in biogeochemical cycles and plant and/or animal microbioata)

INTRODUCTION

Even without the benefit of a microbiology course, most people have no problem naming a dozen or more infectious diseases. (Go ahead, try it.) Likely to be included in the list will be some of the terrifying scourges of mankind, such as Ebola, AIDS, tuberculosis, and smallpox. Others are a nuisance but not life threatening, for example, athlete's foot and the common cold. Clearly, infectious diseases are common and widespread. Some are deadly, others nearly trivial, and everyone knows something about them from personal experience.

Infectious diseases have caused tremendous hardship in the past, and they still do in many parts of the world. **Pathogens** will continue to shape history, but at least one dreaded disease—smallpox—has been eliminated from the globe. Polio is a candidate for imminent eradication, and, over a longer time frame, so are syphilis and measles. Meanwhile, the struggle between hosts and parasites goes on, making relentless demands on human ingenuity. We possess three main kinds of *external* weapons for controlling infectious diseases: sanitation (including insect control), vaccination, and antimicrobial drugs. A moment's reflection will suggest how these can be applied and how each has potential downsides. However, the most powerful defense by far is *internal*: our own immune system.

The bonds between humans and microbes are forever changing. Some diseases, such as AIDS, appear anew on our horizon; others, like the *Helicobacter* infection of the stomach, have been with us for a long time but have been recognized only relatively recently. In most cases, human behavior increases our exposure to offending microbes by altering the environment, increasing global travel, overusing antimicrobial drugs, or changing our food production and distribution practices (Table 22.1).

There are far too many infectious diseases to learn in any one course. Instead, we begin this chapter on host defenses by describing features all infections have in common. A *conceptual framework* on which to hang particular facts is this: the development of all infectious diseases, whether of humans, animals, or plants, proceeds through the same steps. Before discussing these steps, a word of apology is called for. We shall employ imagery of conflict and war by using terms such as defense, attack, invasion, weapons, and so forth. This usage is handy but misleading because it implies a belligerent in-

TABLE 22.1 Human behavior that contributes to the emergence of infectious diseases

Factor	Circumstances	Example(s)
Poverty in countries and regions	Poor health services	Multidrug-resistant tuberculosis
	Contaminated food and water	Cholera, many others
	Malnutrition	Enhanced severity of measles
Ecological changes	Reforestation (increase in deer and ticks)	Lyme disease
Personal behavior	Sexual behavior and intravenous drug use	AIDS, hepatitis C
Social behavior	Urbanization	AIDS, dengue, tuberculosis
International travel and trade	Shipping	Cholera to South America
	Global food trade	Foodborne outbreaks (many kinds)
Modern production practices	Mass food production	*Escherichia coli* O157:H7 in hamburgers
	Feeding scraps to cattle	"Mad cow" disease
	Superabsorbent tampons	Toxic shock syndrome
Medical and agricultural use of antimicrobials	Overuse of antibacterial and antiviral agents	Microbial drug resistance

tent on the part of the agents. Microbes that cause disease simply go about their business. The goal of every microbe is to become two microbes. Except for a few cases, causing disease is incidental to their quest to multiply.

CASE 1: Why was New Caledonia vulnerable to invasive meningococcal disease?

Compared to mainland France, a group of French islands in the Pacific Ocean reported ~4 times more cases of meningitis caused by *Neisseria meningitidis* during 2005 through 2011. Most patients experienced fever, stiff neck, severe headache, and nausea; many developed small purple "purpuric" skin lesions; and specimens of their cerebrospinal fluid contained **Gram-negative** coccobaccilli and neutrophils. To understand why the quarter million inhabitants of New Caledonia were more vulnerable to invasive meningococcal disease, a team of epidemiologists analyzed a cohort of these patients. They recorded the number of cases per year; the age, mortality, gender, and ethnic group of the patients; and the *N. meningitidis* serogroup the patient had been exposed to. In addition, blood samples obtained from patients were screened for a component of the complement system (Fig. 22.1, Table 22.2). Based on the results of this study, a subset of people in New Caledonia was offered a vaccine comprising surface components specific to the four prevalent **serogroups** of *N. meningitidis*.

Questions arising:

- Where do people encounter *N. meningitidis*?
- How does *N. meningitidis* enter the cerebrospinal fluid?
- How does *N. meningitidis* establish infection?
- How does *N. meningitidis* cause damage?

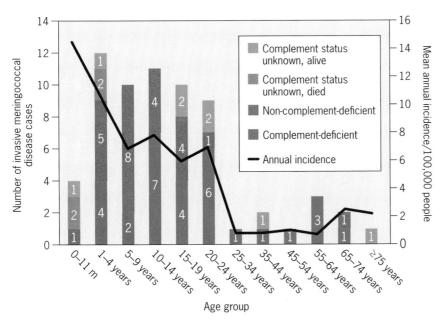

FIGURE 22.1 Age distribution and complement-deficiency status of invasive meningococcal disease cases between 2005 and 2011 in New Caledonia. (Modified from Daures et al., *J Clin Immunol* **35**:47–55, 2015, Fig. 2.)

TABLE 22.2 **Comparison of age, sex, ethnic group, and meningococcal serogroup between patients with and without complement deficiency[a]**

		Complement-deficient patients (n=28)	Non-complement-deficient patients (n=25)	P value
Mean age (standard deviation)		21.8 (17)	11.1 (13.3)	0.01*
Sex ratio M/F		1	0.8	0.7
Origin (%)	Melanesian	26 (92.8%)	14 (58.3%)	0.003*
	European	1 (3.6%)	3 (12.5%)	0.3
	Wallisian	0 (0%)	6 (25.0%)	0.007*
	Mixed race	1 (3.6%)	1 (4.2%)	1
	Unknown	0 (0%)	1 (4.2%)	–
Meningococcal serogroup[a]	A	1 (3.3%)	0 (0%)	0.4
	B	8 (26.7%)	14 (56.0%)	0.02*
	C	1 (3.3%)	5 (20.0%)	0.08
	Y/W135	15 (50.0%)	4 (16.0%)	0.01*
	Unknown	5 (16.7%)	2 (8.0%)	–

[a]Adapted from Daures et al., *J Clin Immunol* **35**:47–55, 2015.
[b]$P<0.05$, $n=30$, including two recurrences.

- Does the complement pathway protect against disease?
- What ethnic group was at higher risk of meningococcal disease? Why?
- Does complement deficiency decrease or increase the benefit of vaccination?

CASE 2: How does the immune system learn from experience?

Following the recommendations of the Centers for Disease Control and Prevention, a college student preparing to study abroad in Vietnam scheduled vaccinations for two pathogens that too frequently contaminate water and food. In June, she received the first dose of the hepatitis A vaccine. Six months later, she received both the second hepatitis A dose and her first vaccine for *Salmonella typhi*. Her immune response to each vaccine is illustrated in Fig. 22.2.

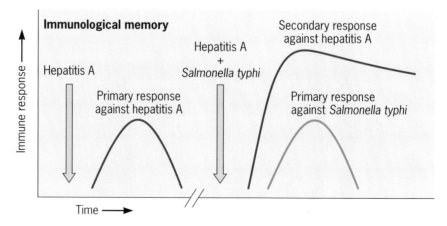

FIGURE 22.2 **Immunologic memory.** The adaptive immune response is equipped to recall a previous encounter with a microbe and then to mount a more rapid and robust specific response. The lag time between exposure to a foreign antigen and the immune response represents the period of activation and proliferation of antigen-specific lymphocytes. Note that the lag time is longer in the primary response of naïve lymphocytes to hepatitis A than in the secondary response when the same antigen is encountered by memory B lymphocytes weeks or years later.

Questions arising:

- Why was there a delay between the time she received her first vaccine dose and her immune response to the hepatitis A antigen?
- Why was her immune response to the second dose of hepatitis A antigen stronger than that to the first dose?
- Why was her response to the second hepatitis A dose stronger than her response to the *S. typhi* vaccine?

WHAT CHALLENGES DO ALL INFECTIOUS AGENTS CONFRONT?

LEARNING OUTCOMES:
- Speculate how a member of ordinary microbiota can cause an opportunistic infection.
- Identify steps in pathogenesis that are common to all infections.
- Predict the mechanisms that might make a pathogen host-specific or able to infect many hosts.
- Define the terms incubation period, inoculum size, and microbiota.
- Explain why the definition of a pathogen is conditional.
- State the main processes that lead to symptoms in infection.
- Infer a mechanism(s) of how a pathogen can cause damage at a site far removed from its point of entry.

Encounter

For an infectious disease to happen, the host and the infecting agent must meet. One may "catch a bug" and get sick within a few days, or an agent may reside in the body for a considerable time before causing disease. This **incubation period** may last days, months, or years (as in leprosy and AIDS, where the symptoms may arise a decade or more after infection). The encounter may even happen before birth. Certain diseases, such as human immunodeficiency virus (HIV) infection, German measles (rubella), and syphilis, can be acquired while in the mother's womb—a dubious bequest.

The source of the organisms may be the inanimate environment (as for *Salmonella* food poisoning and cholera), animals (as in Lyme disease, which is spread by deer ticks), or other humans (as for the common cold, the flu, and the sexually transmitted diseases). Indeed, one need not look far for infectious agents, because many are common inhabitants of our own bodies. All vertebrates and most invertebrates are endowed with large and varied microbial communities, or **microbiota** (chapter 21). (Note that the older term "normal flora" is misleading, because microbes are not plants.) The body's bacterial populations may be very dense, as in the mouth or the large intestine; of moderate size, as on the skin; or virtually absent, as in deep tissues. Although most of these organisms are harmless commensals or even participate in mutualistic relationships, some of them can turn on the host and cause disease. Such diseases are among the **opportunistic infections** (chapter 23). The infectious agent of the first case study, *N. meningitidis*, resides inconspicuously in ~10% of healthy individuals within

TABLE 22.3 Modes of entry

Inhalation (influenza virus, hantavirus, *Bacillus anthracis*)
Ingestion, usually fecal-oral route (*Salmonella, Vibrio cholerae, E. coli*)
Insect bites (malaria, *Yersinia pestis*, Lyme disease agents)
Sexual contact (sexually transmitted disease agents, HIV)
Wound infection (accidental or surgical)
Organ transplants (cornea, blood transfusion)

their oropharynx (the space behind the nose and mouth that connects to the esophagus).

Entry

To cause an infectious disease, the agent must enter its host (Table 22.3). Entry has two meanings. In the more familiar sense, infectious agents penetrate into the host's tissues. On the other hand, serious diseases can take place without penetration. The agent can enter one of the cavities of the animal body, such as the gastrointestinal, respiratory, or genitourinary tract, all of which are essentially tubes that are topologically contiguous with the exterior. Indeed, an agent can pass from the mouth to the anus without crossing any epithelial surface. Consequently, agents of many intestinal diseases, such as cholera (chapter 26), actually reside on the outside of the body; they just adhere to epithelial membranes on the body's surface.

Likewise, *N. meningitidis,* also called meningococcus, typically adheres to the human respiratory epithelium without incident. During a cough or sneeze, this microbe can spread in aerosol droplets from one person to another and thus may colonize a new oropharynx. However, in rare instances some bacteria are translocated across both the respiratory epithelium and the endothelial cells that line blood vessels. Once deposited into the bloodstream, this pathogen gains access to deeper tissues, including the cerebrospinal fluid.

Becoming Established

Invariably, the establishment of every infection requires some *breach in the defense mechanisms of the host.* The breach may be as trivial as a cut through the skin, or it may be a catastrophic event such as the breakdown of the entire immune system in AIDS patients or transplant patients who have received an immunosuppressive regimen. Then begins a series of interactions between invader and host—a choreographed action and reaction or point and counterpoint. The outcome determines whether there will be symptoms of disease and whether the agent will persist, and in which tissues.

Three factors dictate if an infectious agent will become established and cause disease:

- **inoculum size** (the number of invading microbes)
- invasive ability, or **virulence**, of the infectious agent
- strength of the host's defenses

These factors are interrelated (Fig. 22.3). If the invasive ability of the agents is high, fewer of them will be required to establish infection than if the agent were less virulent. The dose of the inoculum refers to not just how many organisms are inhaled or ingested, but how many actually reach the target tissue. Much can happen before getting there. For example, bacteria that are swallowed and cause disease in the bowel must first withstand an acid bath in the stomach. And we know that if a person is deficient in producing stomach acid, fewer cholera bacilli are required to cause disease. The weaker the defensive barriers of the host, the easier it is for a smaller number of invading organisms to tip the balance from health toward disease, and vice versa. As we'll learn from Case 1, people with genetic deficiencies in particular components of the immune system are more prone to infection by microbes that typically colonize the oropharynx without incident.

One may then ask, *what is a pathogen*? A tricky term, that. The same microbes can cause disease in one person but not in another. Again consider Fig. 22.3 to understand why. Patients whose defenses are very low (as in advanced cases of AIDS) may be infected by almost any microbe, including many commensals. Thus, the definition of a pathogen is conditional. Agents that are harmless commensals in a host with strong immune defenses may—given the opportunity—act as pathogens and cause disease in a compromised one. The immune competence of the host is of paramount importance, but it is not the only factor. Location also matters. Anaerobic bacteria that live harmlessly in the appendix can, if the organ becomes inflamed and bursts, cause fatal peritonitis if untreated.

Some microbes are host specific, whereas others can infect a wide variety of hosts. The meningococcus, for example, affects only humans, whereas *Pseudomonas aeruginosa*, a bacterium associated with infections of patients with cystic fibrosis and burns, is one of the least choosy of all pathogens known, causing a variety of diseases in such disparate hosts as plants, insects, worms, and vertebrates.

Causing Damage

The establishment of an agent in the host will not result in disease unless the pathogen causes tissue damage. The impact on the host depends greatly on which tissues are involved. Damage can be life threatening when it affects vital organs—such as the lung—or relatively mild when less essential sites are involved—like a portion of the skin.

Damage occurs in a variety of ways. Tissue trauma may result from cell death, either when a microbe lyses the host cells or when the host cell itself induces **programmed cell death** (**apoptosis** or **pyroptosis**). Alternatively, tissues may be affected by the pharmacological action of microbial toxins. As we'll see in chapter 26, the bacterial diarrheal disease cholera results from a derangement of normal cell function, in this case, ion and water exchange in the intestinal cells. The patient suffers greatly, but the intoxicated intestinal cells remain healthy and intact. More rarely, damage results from mechanical action, as occurs when blood or lymph vessels become blocked by infection.

The location of the damage also varies from one pathogen to another. Damage caused by infectious agents may be seen close to the site of entry, for example, when a cut on the skin becomes inflamed with pus. Alternatively, the targeted tissues may be far away from the site of invasion, as happens in tetanus. In this disease, the offending bacteria may be found in

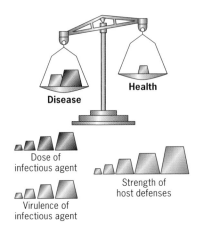

FIGURE 22.3 Whether the outcome of a microbial encounter is health or disease depends on the balance of three factors: the dose of the infectious agent, its virulence, and the strength of the host's defense systems.

TABLE 22.4 **Examples of symptoms caused by host responses**

Disease	Immune response	Symptoms
Causing pus (e.g. gonorrhea, strep throat, acne)	Acute inflammation	Pus, irritation
Brain abscess (e.g., mixed aerobic and anaerobic bacterial infection)	Acute inflammation and effects of having a foreign mass in a confined space (the skull)	Neurological symptoms due to compression of vital sites
Tuberculosis	Chronic inflammation	Cough with bleeding, due to ongoing tissue destruction
Respiratory syncytial virus infection	Allergy-like reaction in bronchioles	Wheezing and difficult breathing (asthma-like)
Malaria	Synchronized release of fever-inducing cytokines	Periodic episodes of shaking chills followed by sweats

a cut on the foot, but their real damage is caused by a toxin that acts on distant neuromuscular junctions. These examples illustrate another of the great themes of infectious diseases. In toxin-mediated diseases, including cholera and tetanus, the damage is caused directly by the agent. In many other infections, such as an infected wound or the meningitis that afflicted New Caledonia in case study 1, most of the damage is caused indirectly by an exuberant response of the host itself, usually **inflammation** (Table 22.4). Most infectious diseases involve both kinds of damage.

HOST DEFENSES

A testimony to the powers of our bodies' defenses is that, despite living in a world teeming with microbes, we are only rarely infected. Our defenses are sophisticated and numerous, and understanding them requires considerable study. But a grasp of the immune response is crucial to understanding infectious diseases. Therefore, we'll review the fundamentals of immunology next.

For convenience, we discuss in turn two systems of defense: **innate** are those that are always available, whereas **adaptive** processes are set in motion by contact with a particular invading agent (Table 22.5). We inherit

TABLE 22.5 **Defenses of the body against microbes**

Constitutive or innate

Mechanical
    Skin, mucous membranes
Chemical (examples only)
    Fatty acids on the skin (due to normal skin bacteria)
    Hydrochloric acid in stomach
    Antimicrobial peptides in tissues and white blood cells
    Lysozymes in cells, tissues, and body fluids
    Bile salts in intestine
    Complement in circulation and tissues (major factor)
Cellular
    White blood cells (neutrophils, macrophages, etc.)

Adaptive or induced

Humoral immunity (antibodies)
Cell-mediated immunity

the innate defenses from our parents, and with few exceptions all the individuals of our species are identical in this regard. The adaptive defenses, on the other hand, are not inherited but are acquired through life experience. Therefore, we each build a unique repertoire of adaptive responses.

WHAT DEFENSES WAS I BORN WITH?

LEARNING OUTCOMES:
- Describe three physical barriers to entry that pathogens encounter.
- Give examples of microbe-associated molecular patterns.
- Explain how the inflammatory response is switched on by microbial pathogens.
- Define the complement system.
- Compare and contrast the alternate pathway and the classical pathway.
- Give evidence for coevolution between humans and microbes.
- Describe how phagocytes sense and move toward bacterial chemotaxins.
- List the steps in phagocytosis leading to bacterial destruction, starting with opsonization.
- Differentiate the action of two types of phagocytic white blood cells, neutrophils and macrophages.
- Compare and contrast toll-like receptors (TLRs) and NOD-like receptors (NLRs).
- State consequences of the activation of complement.
- Draw a concept map about innate defenses, using as many boldfaced vocabulary terms as possible.

Barriers to Entry

Thin as it is, our skin acts like a massive wall surrounding a medieval city, only to be breached upon significant coercion. In addition to providing a physical barrier, the skin is coated with antimicrobial substances, such as salt in high concentrations, certain peptides, and fatty acids. The mucous membranes of the respiratory, digestive, and genitourinary tracts are also bathed in antimicrobial substances; these include **antibodies** and lysozyme (an enzyme that hydrolyzes the cell wall murein of bacteria). In addition, the long tubes in our bodies, namely the respiratory, digestive, and urinary tracts are protected by mechanisms that create currents to sweep bacteria away. Bacteria not firmly attached to the tissue walls are no match for the beating of respiratory cilia, the vehemence of peristalsis, or the torrential flow of urine. Ingested microbes encounter another formidable barrier in the stomach, whose strong acid kills most organisms. Indeed, persons unable to make sufficient hydrochloric acid are at risk of serious intestinal infections.

Inflammation

The host's most powerful reaction to an invading agent is the inflammatory response, an event common to most infections. After suffering a cut or splinter, you've no doubt witnessed inflammation by its classic hallmarks: local reddening, swelling, pain, and pus. Reddening is due to increased blood flow at the site, swelling to an outpouring of fluids in the tissues, and pain to the release of chemical mediators and compression of the nerves. The strongest antimicrobial elements in the inflammatory response are the **phagocytes**, white blood cells that ingest foreign particles and work to digest them. When meningococcus infects the cerebrospinal fluid, the host inflammatory response generates pus in the confined space of the brain cavity, raising the intracranial pressure and causing, among other symptoms, nausea and severe headache.

A central property of the inflammatory response is that it is normally switched off but is switched on by microbial challenge. Were this not so, the body would be in a constant state of inflammation, which would lead to serious tissue damage and death. So, what is the master switch? How does the body recognize the presence of microbial invaders? The answer is that invading microbes are recognizable by some of their uniquely microbial chemical constituents. You may guess what such bacterial visiting cards may be: cell wall **murein**, the **lipopolysaccharide** (endotoxin) of gram-negative bacteria such as the meningococcus, **flagellin** (the major protein of bacterial **flagella**), or **pilin** (the protein of **pili**). Indeed, thanks to the long course of coevolution with microbes, our bodies are endowed with an innate surveillance system that recognizes molecular signatures that are exclusively microbial. Such **microbe-associated molecular patterns (MAMPs)** include surface constituents from bacteria, fungi, and animal parasites and some chemically distinct viral and bacterial nucleic acids. As a general rule, MAMPs are microbial components that are not only common to a wide variety of microbes but also essential for their fitness. With its capacity to recognize about 1,000 distinct MAMPs, the human innate immune system ought to detect most pathogens. A pathogen lacking all MAMPs would, in theory, be able to escape the body's radar screen. But how would such a stripped-down microbe compete in the microbial world?

You may recognize a conundrum. We cohabitate with an impressive microbiota; how do we avoid a constant state of inflammation? Since MAMPs are shared by pathogens and nonpathogens alike, how does the body distinguish, say, beneficial commensals in our gut or on our skin from invasive microbes? This very puzzle motivates a great deal of immunology research today. One pleasing idea is that host cells are designed to sense not just microbial components, but also their context. For example, the cytosolic **pattern recognition receptor** (PRR) NOD2 responds to **peptidoglycan** fragments generated by bacterial lysis but not to peptidoglycan segments shed by growing bacteria. Location also matters. Live pathogens are competent in activating specialized secretion systems to translocate MAMPs into the cytoplasm of host cells where PRRs await; dead microbes tell no such tales.

When MAMPs are recognized at a site of tissue damage, the alarm that invaders are present is sounded, and the defensive responses are set in motion. Two tracks cooperate to generate a robust innate immune response, one involving soluble components, the other mediated by cells. We deal first with a fundamental response mediated by soluble components, the complement system.

The complement system. **Complement** is the name given to a multifunctional defense system in the blood consisting of some 30 different proteins. The name complement reflects the early observation that these proteins "complement" the action of antibodies, although we'll see they also have some unique powers. Working together, these factors elicit the inflammatory response that is essential for health and well-being. An important hint about the role of complement comes from observing people born with hereditary defects in its components. Some individuals with rare **mutations** in one of the components can live healthy lives, but others are highly susceptible to bacterial infections, as well as to noninfectious diseases, such as the autoimmune disease lupus erythematosus. Did you deduce by studying Table 22.2 that the Melanesian people of New Caledonia were more vulnerable to invasive meningococcal disease due to a genetic deficiency in complement components?

Normally dormant, complement must be activated to mount a defense. Various factors activate complement (Fig. 22.4). The most common are MAMPs, such as microbial surface polysaccharides. This "innate" arm of complement activation is termed the **alternative pathway** (for having discovered second). Or, sugars that decorate the surfaces of a variety of microbes are first bound by a host lectin, such as mannose-binding protein. Antibodies can also trigger complement activation by the so-called **classical pathway**. Because it takes a week or more to make sufficient levels of antibodies after antigenic stimulation, the classical pathway does not come into play early in infection.

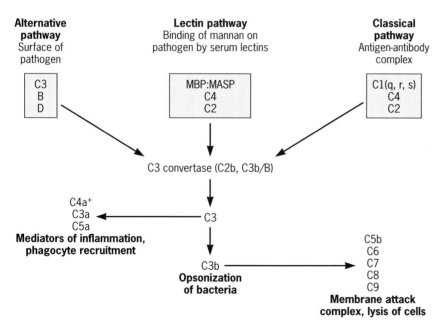

FIGURE 22.4 **Complement activation pathways and their components.** The complement system can be activated in three different ways. In each case, the end result is the formation of proteins involved in mediation of inflammation, phagocyte recruitment, opsonization, and lysis of foreign cells. The alternative pathway is elicited by recognition of surface components of invading organisms. The lectin pathway depends on the recognition of mannan residues on bacterial surfaces by serum lectins. The classical pathway results from the presence of antigen-antibody complexes. The names of the complement proteins often, but not always, start with C (for complement). MBP, mannan-binding protein; MASP, MBP-associated serum protease.

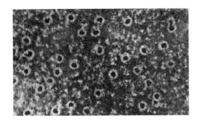

FIGURE 22.5 **The membrane attack complex seen under the electron microscope.** These doughnut-shaped pores, or MACs, are inserted into the membrane of a red blood cell. (Courtesy of J. Tranum-Jensen.)

Activation then sets off a chain reaction of sequential proteolytic cleavages of complement proteins. Normally, when the complement proteins are circulating freely in the blood, they are not susceptible to proteolysis, but they become so once bound to particles, including microbes. The resulting peptides stimulate the cleavage of other proteins, and so on down the line. The products of these proteolytic events are pharmacologically active peptides that coordinate the inflammatory response.

One subset of complement components cooperates to assemble pores in membranes that appear foreign. This **membrane attack complex** (Fig. 22.5) can lyse bacteria, animal cells, and enveloped viruses. The host avoids setting off the complement cascade by adding the amino sugar sialic acid to the surfaces of its cells. The presence of sialic acid marks the cellular membranes as human.

Complement activation also generates **chemotaxins**—proteins that attract white blood cells. To gather at the site of the infection, white blood cells that are normally tumbling through the circulatory system must squeeze through the blood vessel walls, a feat called **diapedesis** from the Greek for "leaping through" (Fig. 22.6). What sets off the alarm? When invading microbes activate complement, a local pool of chemotaxins is generated; white blood cells sense and move up a concentration gradient of these substances. In a similar manner, phagocytes are also equipped to sense and move toward bacterial products. How? Recall that bacterial protein synthesis is unique in that several extra amino acids are incorporated at the start; these are subsequently removed to generate a "mature" protein (see chapter 8). White blood cells are exquisitely sensitive to these short, discarded bacterial peptides. Our immune cells can sense one of these bacterial chemotaxins (N-formylmethionyl-leucyl-phenylalanine) at concentrations of 10^{-11} M! Thus, live bacteria inadvertently—but loudly!—advertise their presence, and the body heeds the call.

Because the complement system is such a potent inducer of inflammation, it must be held in check. Several regulatory molecules monitor the

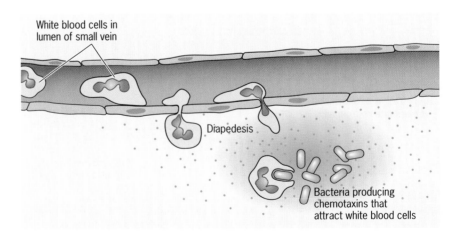

FIGURE 22.6 **Recruitment of white blood cells to a site where microbes are present.** Bacteria produce diffusible substances called chemotaxins that attract white blood cells from the circulation into the sites in tissues where microbes are present. The passage of white blood cells through the walls of the blood vessels is called diapedesis.

extent of complement activation and switch the system off when the stimulus is gone. If unchecked, inflammation can cause damage to tissues and symptoms of disease (Table 22.4). Thus, the complement cascade is a two-edged sword.

Phagocytes. Once recruited to the site of infection, what action do phagocytes take? When at rest, phagocytes are sluggish eaters. However, the complement cascade generates **opsonins**—peptides that "make tasty" (Greek) any particles they bind. Antibodies also turn phagocytic cells into voracious eaters. Once opsonins coat the surface of microbes, opsonin receptors on the phagocyte membrane interact in zipper-like fashion until the prey becomes completely engulfed (Fig. 22.7). Next, a series of membrane fusion events delivers the phagosome to the acidic, digestive lysosomes that efficiently degrade most prey (Fig. 22.8). This process of **phagocytosis** is the most effective of all innate host defenses against bacteria. *The majority of bacteria that invade humans are killed or removed by phagocytosis.* The ability of cells to act as phagocytes—to take up particles—is shared by many cells of the body, but only a few do it with gusto. White blood cells are chief among the cells that earn the title professional phagocytes. Phagocytic white blood cells fall into several classes, each with different functions.

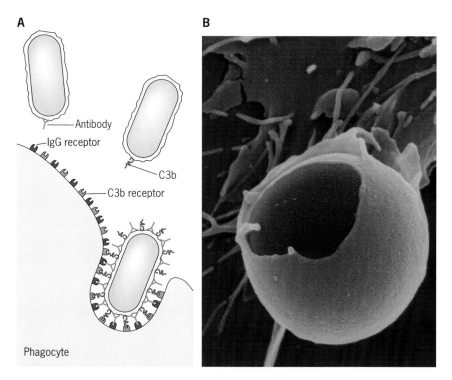

FIGURE 22.7 Opsonization enhances phagocytosis. Microbes and other particles are not readily phagocytized by white blood cells unless they are coated with proteins called opsonins. (A) Several kinds of proteins can act as opsonins. Microbes coated with the complement protein C3b bind to C3b receptors on the surfaces of phagocytic cells by a mechanism that resembles a zipper. An analogous process takes place when microbes are coated with antibodies. (B) A scanning electron micrograph of a macrophage (orange) engaged in phagocytosis of a 3-μm-diameter bead (blue). (Panel B, MicrobeWorld © Darren Brown.)

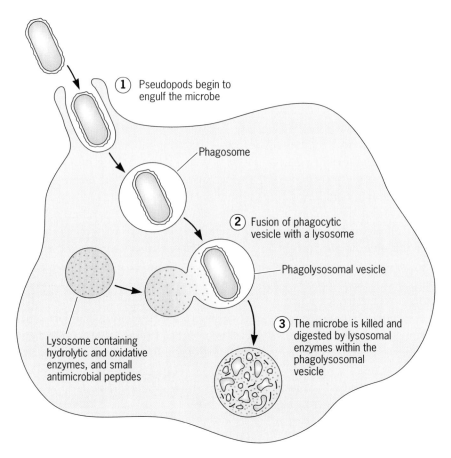

FIGURE 22.8 **Steps in phagocytosis.** A microbe attaches to a phagocyte. The phagocyte forms pseudopods that surround the (opsonized) microbe (1) leading to the formation of a vesicle called a phagosome. Lysosomes filled with antimicrobial compounds fuse with the phagosome and release their contents within the vesicle (2), which is now called a phago-lysosomal vesicle. Microbes within a phagolysosomal vesicle are killed and digested by hydrolytic enzymes (3). Note that this is not an inevitable outcome, because many pathogenic microbes have mechanisms to avoid or withstand phagocytosis.

The most numerous phagocytes are the **neutrophils** (Table 22.6). Because these cells are terminally differentiated, they cannot proliferate, and they typically live for only a few days. Their assignment is to seek out foreign particles, ingest them, and, if alive, try to kill them. Neutrophils are expert first responders. They move about very rapidly, like small amoebas, and are adept at sensing chemotactic gradients that point to their prey. Thus, neutrophils can reach bacterial targets quickly and in large numbers, which leads to inflammation. In case study 1, neutrophils were

TABLE 22.6 **Properties of neutrophils, or polymorphonuclear leukocytes**

Short-lived cells (live a few weeks)
Loaded with large lysosomal granules that contain hydrolytic enzymes that can destroy many
    bacteria and oxidative enzymes that make toxic products, especially hypochlorite (bleach)
Recruited to the site of the microbes by several chemotaxins, some derived from
    complement and others products of bacterial metabolism
Require opsonins to be effective in killing microbes

rapidly recruited when microbial contaminants shed by *N. meningitidis* were detected in the cerebrospinal fluid.

How do neutrophils carry out their activities? They possess large saclike **cytoplasmic granules** (actually gigantic **lysosomes**) that act as veritable bombs. These granules contain antimicrobial peptides called **defensins** plus enzymes that damage bacteria and, if released, even the host cells themselves. Usually, the granules fuse to the phagosomes, assembling a **phagolysosomal vesicle**. When this happens, the enzymes of the granules come in direct contact with the ingested bacteria. These defensive enzymes fall into two classes: (1) hydrolases and other antibacterial proteins capable of directly perturbing the cellular integrity of microbes and (2) oxidative enzymes that act to form hypochlorite, the same powerful chemical found in laundry bleach. Fusion of the granules to the phagosome is so efficient that it may take place before the ingestion of the invaders is complete. Thus, microbial prey may be killed even before being fully internalized—much like poisonous snakes that immobilize their prey before swallowing it.

Nevertheless, the arrival of the neutrophils at the scene does not necessarily mean that the microbes will be killed. Rather, the stage is now set for an epic struggle between microbes and phagocytes, the outcome of which is by no means foreseeable. Each combatant has many weapons to overwhelm the other. Even when the neutrophils prevail, there is a cost to the host. Recruitment of white cells to the site where invading microbes are present must be carefully regulated: If excessive, it causes collateral damage to host tissue. The result is **pus**, the accumulation of live and dead white blood cells, debris, and tissue fluids. As we have said, pus is a characteristic feature of inflammation (the others being reddening, swelling, heat, and pain at a local site). When pus accumulates in a confined space, such as the cranial cavity, the increased pressure can cause dizziness, severe headache, and worse, as we saw for invasive meningococcal disease in case study 1.

The other major class of white blood cells is the **monocytes**. Unlike the neutrophils, these cells are long-lived and capable of differentiating further. Monocytes can settle in tissues, in which case they are known as **macrophages** (Table 22.7). Monocytes arrive at sites of infection after the neutrophils; in fact, monocytes can be recruited to these sites by substances secreted by the neutrophils. Monocytes and macrophages serve various functions. One is to clean up the debris left at the scene of the battle between neutrophils and microbes. Besides being garbage collectors, they *communicate with the other arms of the immune system*. To do so, these professional phagocytes secrete a large assortment of signaling

TABLE 22.7 **Properties of macrophages and monocytes**

Long-lived cells
Arrive at site of microbes more slowly than neutrophils
Act as "garbage collectors;" clear up microbial debris and remaining microbes
Become activated by cytokines, proteins made in response to microbial invasion
Make cytokines that attract and activate neutrophils, thus contributing to acute
   inflammation
Help initiate specific innate immune responses, i.e., cell-mediated immunity

proteins called **cytokines**, some of which activate complement, while others promote inflammation, and yet others set off adaptive immunity. Thus, this class of white blood cells orchestrates multiple aspects of the immune response.

MAMPs, the signatures of invading microbes, and PRRs reenter the discussion here. Macrophages and other cells of the immune system have receptors that specifically bind certain classes of MAMPs. One class of these PRRs are the Toll-like receptors, better known as **TLR**s (Fig. 22.9). (Their formal name reflects an interesting family history. The mammalian proteins resemble so-called Toll receptors on *Drosophila*. It happens that fruit flies with mutations in these proteins behave in odd ways, causing their German discoverers to call them by the German word *toll*, which means, among other things, weird). TLRs are found in many organisms, from fruit flies to humans and even plants, and are likely the product of the ancient coevolution of hosts and parasites. Their reputation as receptors dedicated to microbial recognition was secured by the finding that mice lacking TLRs are normal in all respects other than their ability to fight disease.

TLRs are specific for subsets of MAMPs. As stated already, some recognize the murein of Gram-positive bacteria, others recognize the lipopolysaccharide of Gram-negative bacteria, additional receptors recognize the single-stranded RNA of viruses such as influenza virus or mumps virus,

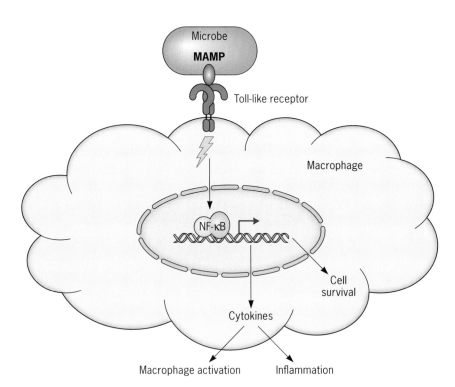

FIGURE 22.9 **Toll-like receptors.** When Toll-like receptors recognize microbe-associated molecular patterns (MAMPs), they trigger a signaling cascade that activates NF-κB, a transcription factor that induces expression of cytokine genes and genes that promote cell survival. Some cytokines promote inflammation, and others activate macrophages to enhance their antimicrobial properties.

TABLE 22.8 **Properties of bacterial endotoxin**

Consists of lipopolysaccharide
Found only in the outer leaflet of the outer membrane of Gram-negative bacteria
  (thus, not found in Gram-positive bacteria)
Acts on:
  Macrophages to produce cytokines and induce fever
  Neutrophils to produce compounds that dilate blood vessels, causing edema and shock
  Complement system to induce its activation, causing inflammation

and so forth. Some TLRs are located on the cytoplasmic membranes of many cells, others on the membranes of the phagocytic vacuoles. Binding of microbial constituents to TLRs sets off a signaling cascade that leads to the activation of a key host cell transcription factor called **nuclear factor κB (NF-κB)** which, in turn, induces expression of cytokine genes and other genes that promote cell survival (Fig. 22.9).

Another consequence of TLRs interacting with MAMPs is the activation of macrophages. Macrophages, like other members of the immune response system, are normally quiescent. To reach their greatest capacity for killing microbes, they must be activated (become "angry"). Angry macrophages are more effective phagocytes and are also better cytokine producers. Consequently, when macrophages recognize the presence of microbes in tissues, they sound alarms that affect all aspects of the immune response.

Prominent among macrophage-activating stimuli is a paradigm of MAMPs, the **lipopolysaccharide (endotoxin)** of the outer membrane of Gram-negative bacteria (Table 22.8). When lipopolysaccharide is present at a high level, the term "endotoxin" becomes fully justified, because the patient may go into severe shock caused by the widening of blood vessels and a drop in blood pressure. This sometimes lethal condition is seen in patients with infections of the blood caused by Gram-negative bacteria (**sepsis**). Release of endotoxin into the cerebrospinal fluid by meningococci triggers the inflammation that is a hallmark in meningitis. The term "endotoxin" denotes that it is also a functional component of the body of the producing bacteria, unlike soluble toxins, which are known as **exotoxins** (chapter 26).

Another class of pattern-recognition receptors is the **NOD-like receptors** or **NLR**s. NLRs are positioned in the cytoplasm to detect microbial products that breach the phagosomal membrane, either as cargo of specialized secretion systems, or after lysis of the vacuole itself (Fig. 22.10). Like TLRs, many of the NLRs are specific for particular MAMPs, like peptidoglycan or flagellin. Others can sense perturbations of the host cell physiology, such as a loss of potassium ions due to damage of the plasma membrane. Depending on the nature and magnitude of the trigger, NLRs can coordinate three host responses (Fig. 22.10). Some NLRs stimulate NF-κB to turn on cytokine genes. Another cohort of NLRs assemble **inflammasomes**, protein complexes that activate caspase-1. This protease not only clips and activates the proinflammatory cytokines IL-1β and IL-18 but also coordinates the cell death program **pyroptosis.** The "fire" in its name recalls the fever of inflammation. As the infected cells lyse, mature IL-1β and IL-18 are released, summoning neutrophils to the site of

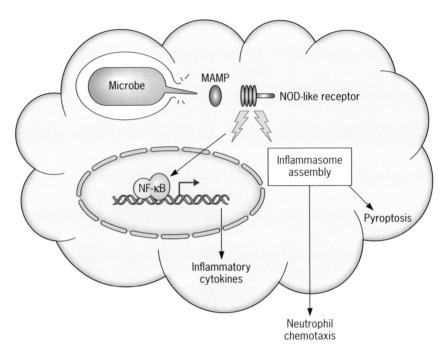

FIGURE 22.10 **NOD-like receptors.** When some NOD-like receptors (NLRs) recognize cytosolic microbe-associated molecular patterns (MAMPs), they activate the transcription factor NF-κB. Other NLRs assemble protein complexes with pro-caspase-1 known as inflammasomes. After activation within this complex, the protease caspase-1 cleaves and activates inflammatory cytokines and also initiates the programmed cell death pathway, pyroptosis. Cytokines released from the porous cell membrane recruit neutrophils to the site of infection.

infection. Thus, PRRs give cells two shots at recognizing MAMPs, first, when the microbes become bound to the host cell membrane and, second, after they are phagocytized and their constituents are released into vacuoles or the cytoplasm.

WHAT DEFENSES DO WE GAIN FROM EXPERIENCE?

LEARNING OUTCOMES:
- Give evidence to support the statement "the adaptive immune response is highly specific and varies from individual to individual."
- Differentiate the adaptive defenses of antibodies and cell-mediated immunity.
- Discuss how antibodies help fend off pathogens.
- Describe how B lymphocytes can generate millions of different kinds of antibodies.
- Discuss how the innate immune system is inherently linked to the adaptive immune system.
- Compare and contrast B lymphocytes with cytotoxic T lymphocytes.
- Identify a type of lymphocyte that is involved with both the innate and adaptive immune response.
- Draw a concept map about the adaptive immune system using as many boldfaced vocabulary terms as possible.

The main adaptive defenses are antibodies and cell-mediated immunity. Both require exposure to their target and, unlike the innate defenses, both are *highly specific* and *vary from individual to individual.* They work extremely well but for one thing; they are not there to protect the host upon first exposure to an agent but require a week or more to develop. On the other hand, protection may have been acquired by previous contact with an agent (whether or not this led to disease symptoms), by immunization, or, for the first few months of life, by the transfer of antibodies from mother to fetus across the placenta and via the colostrum (the high-protein mother's milk secreted during the first few days after giving birth).

How Does Exposure to Microbial Products Protect Us from Disease?
Most children in developed countries receive several vaccines in their early months of life. For example, one of these vaccines, originally called **DTP**, is a multiple-component vaccine that induces the formation of antibodies against diphtheria toxin (the D), tetanus toxin (the T), and the agent of whooping cough, *Bordetella pertussis* (the P). Such vaccination stimulates the formation of antibodies that protect against each of these diseases.

Here is a historical example of how antibodies can clear an infection in the absence of antibiotics. Patients with pneumonia caused by pneumococci (*Streptococcus pneumoniae*) are typically quite ill for a period of a week or two, with symptoms increasing in severity. At this point, the "crisis" occurs, and patients either die or get better rather suddenly. Some patients rise from their deathbed and, within hours, demand a good meal. This miraculous cure occurs once antibodies against the microorganisms reach an effective threshold level. Pneumococci are unusually resistant to phagocytosis because these bacteria are surrounded by a thick, slimy capsule (see Fig. 2.15). Once antibodies are bound to the capsule, they serve as opsonins, allowing phagocytes to bind and ingest the bacteria.

Antibodies are proteins called **globulins** made in response to foreign substances called **antigens**. The emphasis here is on *foreign*. The vertebrate host has an intricate mechanism to differentiate between "self" and "nonself," thus preventing the production of antibodies against its own body components. When this differentiation fails, as happens on occasion, it leads to so-called autoimmune diseases. Many antibodies are amazingly specific and can distinguish between proteins that differ by a single amino acid or polysaccharides that differ only by an α- or a β-linkage, reacting with one but not the other. Normally, antiserum made against a certain antigen is *polyvalent*, that is, it is a mixture of antibodies that recognize different antigenic sites, or **epitopes,** on the same antigen. (An ingenious laboratory trick permits the production of so-called **monoclonal antibodies**, which are specific to one epitope only. Monoclonal antibodies have many uses in research and, increasingly, in diagnostics and therapy.)

Antibodies come in a huge number of varieties, not because there is one gene encoding each antibody, but because immunoglobulin genes can undergo a large number of DNA rearrangements, one of the neatest tricks in genetics. These rearrangements (**site-specific DNA recombination,** chapter 10) take place all the time and engender a large variety of antibody-producing cells, the **B lymphocytes** (**B cells**), each of which produces a different antibody (Fig. 22.11). It has been calculated that recombination can generate in one person upward of 6 *million* different kinds of antibody molecules. Additionally, somatic mutations (not occurring in the germ line) contribute to antibody diversity by another factor of 10 to 100, for a total of at least 108 possibilities. Such a huge number could not possibly result from having that many genes in the genome. (According to some, the human genome contains some 30,000 genes.)

Normally, B lymphocytes lie dormant, but when their cognate antigen is present, each proliferates to make a clone of cells. *Each B-cell clone produces a large amount of a given specific antibody.* The actual amount is about 2,000 molecules per second, a remarkable feat! Clones are set off on their path to become antibody-producing factories by specific cytokines made by macrophages and other cells. In this manner, the innate immune system

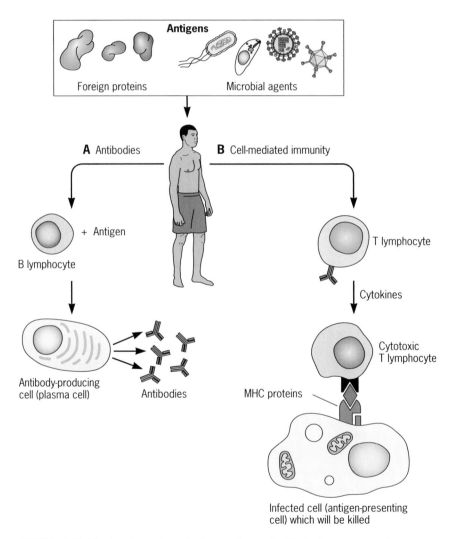

FIGURE 22.11 The two branches of adaptive immunity. (A) In the presence of an antigen, specific B lymphocytes become antibody-forming cells (plasma cells) that proliferate to produce specific antibody-forming clones. (B) T cells differentiate into cytotoxic T lymphocytes (CTLs, or killer T cells) that recognize MHC-associated antigens on the surfaces of antigen-presenting cells (such as those infected by microbes). As a result, the infected cells are killed by the CTL.

is responsible for signaling the activation of antibody formation. We will see that the innate immune system also triggers the other arm of this response, cell-mediated immunity. Therefore, the innate and acquired systems of immunity are intimately linked with each other.

How Do Antibodies Help Combat Infectious Diseases?

1. They neutralize microbial toxins and render them ineffective.
2. They facilitate the removal of infectious agents by:
   - acting as opsonins—when bacteria are coated with antibodies, they are recognized by receptors on the phagocyte surface, much like complement-derived opsonins.
   - clumping bacteria into larger particles, which facilitates their removal.
   - utilizing the filtering mechanisms of the body.
3. They interact with complement to lyse certain bacteria.

How Are Intracellular Pathogens Recognized and Cleared?

Intracellular organisms are shielded from antibodies, complement, and phagocytes. We need a special mechanism to deal with microbes that reside inside host cells: **cell-mediated immunity** (Fig. 22.11). Infected cells are killed by specialized cells, the killer or **cytotoxic T lymphocytes**. Briefly, this is how it works. Infected cells display on their surfaces some of the foreign components—antigens—made by the microbes that reside within them and become transported to the host cell surface. Most cells of the body possess surface glycoproteins called **MHC** (for major histocompatibility complex, a misleading, historical term) that bind and hold foreign antigens in a configuration that can be recognizable. Therefore, these infected cells can serve as antigen-presenting cells. Specialized cells called cytotoxic T lymphocytes recognize the MHC-associated surface antigens and bind to this complex. Just as with antibody-forming cells, such detection triggers T lymphocytes to proliferate into *numerous and specific clones* that are primed to recognize and kill other cells infected with the same agent. *Cell-mediated immunity is therefore as specific as the antibody response.* Indeed, this system must be highly regulated, lest the killer cells attack normal cells of the body. In conclusion, powerful defense mechanisms operate in infectious diseases caused by intracellular parasites, including all viruses, the tubercle bacilli (chapter 24), and the salmonellas.

Cell-mediated immunity requires intense communication among various kinds of cells (Table 22.9). The language used is chemical, and

TABLE 22.9 Some cells involved in adaptive immunity

Cell type	Important function
Macrophages	Antigen presentation; kill microbes directly; kill antigen-presenting cells
Dendritic cells	Antigen presentation
B lymphocytes	Recognize antigens directly; differentiate into antibody-forming cells (plasma cells)
T lymphocytes	Involved in cell-mediated immunity
T helper lymphocytes	Promote differentiation of B lymphocytes; activate macrophages
Cytotoxic T lymphocytes	Kill antigen-presenting cells

TABLE 22.10 **Some important cytokines involved in inflammation and immunity**

Name	Major activities
Interleukin-1 (IL-1)	Induces fever; activation of T and B lymphocytes
Interleukin-2 (IL-2)	Induces proliferation of T and B lymphocytes
Interleukin-4 (IL-4)	Activates B lymphocytes to make antibodies
Interleukin-10 (IL-10)	Modulates functions of macrophages and some lymphocytes
Gamma interferon (IFN-γ)	Activates macrophages, other immune cells
Tumor necrosis factor alpha (TNF-α)	Activates macrophages, neutrophils; induces fever

cytokines deliver the messages (Table 22.10). Some cytokines are made by macrophages, which, to repeat, serve as the sentinels to sound the alarm. One set of lymphocytes, called **T helper cells**, are master communicators that stimulate both branches of adaptive immunity. T helper cells make cytokines that stimulate both B and T lymphocytes, thus serving as middlemen between antibody formation and cell-mediated immunity development. In addition, T helper cells stimulate the innate response by producing cytokines that result in inflammation. The key role of T helper cells is seen in HIV infection, where the virus specifically targets those cells that carry a surface protein called CD4, the CD4 T lymphocytes. Destruction of these cells is the primary mechanism by which HIV causes AIDS.

How Do Vaccines Protect Us from Future Infections?

After the first exposure to an agent or its antigens in a vaccine, we mount a *primary response* manifested by the proliferation of specific B and T lymphocytes. The first dose of hepatitis A and *Salmonella typhi* vaccine triggered an immune response in the college student in Case 2 (Fig. 22.2). These cells are short-lived, and most of them disappear after the antigen is eliminated. But notice that if all of them disappeared, a new encounter with the same agent would require that the immune system start anew. To avoid this, *the immune system retains a memory of its past*. Some of the B lymphocytes differentiate into **memory cells**, which are not destroyed but become quiescent. Should the body encounter the same or a similar antigenic agent, *memory cells proliferate to mount a rapid and efficient immune response*, **the secondary response**. Memory cells in the Case 2 college student account for her more vigorous response to the second dose of hepatitis A vaccine (Fig. 22.2).

Thankfully, memory cells are *long-lived*; they may last for decades. They are poised to proliferate rapidly every time the same antigen is encountered. **Immunological memory** is a key benefit of being naturally exposed to antigens of pathogens as a child, or by vaccination. Without it, vaccination would not make sense. Prior experience with the antigen reduces the risk of the same pathogen causing illness later in life. Thus, the immune response is indeed a gift that keeps on giving.

CASE 1 REVISITED: Why was New Caledonia vulnerable to invasive meningococcal disease?

People encounter *N. meningitidis* when they come into contact with aerosol droplets generated by the cough or sneeze of a person whose oropharynx is already colonized with meningococcus. *N. meningitidis* enters the cerebrospinal fluid in rare cases when the organism crosses the respiratory epithelium and endothelium. *N. meningitidis* is more likely to establish infection when host defenses are impaired. In the New Caledonia Melanesian population, genetic deficiencies in the complement pathway reduced leukocyte chemotaxis and phagocytosis as well as direct lysis of bacteria. Since antibodies are also potent opsonins that promote ingestion and killing of microbes by leukocytes, vaccination is especially beneficial to people with complement deficiencies.

> **CASE 2 REVISITED:** **How does the immune system learn from experience?**

After exposure to a new antigen, the immune system takes a week or more to generate specific antibodies. That delay is evident after the student received hepatitis A and *S. typhi* vaccines (Fig. 22.2). It also explains why, historically, some pneumococcal pneumonia patients who were gravely ill for more than a week suffered a "crisis" and then miraculously recovered. After the student's second dose of hepatitis vaccine, memory cells generated during the primary response to her first hepatitis A vaccine coordinated a secondary response that was more rapid and robust than the response to the first dose of either vaccine.

CONCLUSIONS

Humans and other vertebrates have evolved a large number of defense mechanisms that ensure the maintenance of health and integrity. The advantage of having so many mechanisms is that should one fail, others can fill the breach. Although these mechanisms differ in specificity, strength, and the time when they come into play, they are interactive and work in a cooperative manner. Thus, *neither the adaptive responses nor the innate defenses work independently.* We have seen examples of such interactions, e.g., antibodies help activate complement (the "classical" pathway of complement activation), complement greatly enhances the action of phagocytic cells, and macrophages produce cytokines that activate B and T lymphocytes. Once the adaptive responses—antibodies and cell-mediated immunity—reach an effective level, the body can respond effectively to severe challenges. To appropriate what Shakespeare said (in a different context), the body becomes "a fortress built by Nature for herself against infection. . . ."

Considering our impressive array of host defense mechanisms, you may wonder what strategies microbes have developed to escape detection and elimination by the host. In the next five chapters, we will discuss specific examples of these complex yet fascinating interactions.

Representative ASM Fundamental Statements and Learning Outcomes for the Chapter

EVOLUTION

1. Human impact on the environment influences the evolution of microorganisms.

   a. Give evidence for coevolution between humans and microbes.

CELL STRUCTURE AND FUNCTION

2. Bacteria have unique cell structures that can be targets for antibiotics, immunity, and phage infection.

   a. Predict the mechanisms that might make a pathogen host-specific or able to infect many hosts.
   b. Give examples of microbe-associated molecular patterns.
   c. Explain how the inflammatory response is switched on by microbial pathogens.
   d. Compare and contrast Toll-like receptors (TLRs) and NOD-like receptors (NLRs).

METABOLIC PATHWAYS

3. The growth of microorganisms can be controlled by physical, chemical, mechanical, or biological means.

   a. Describe three physical barriers to entry that pathogens encounter.
   b. Define the complement system.
   c. Compare and contrast the alternate pathway and the classical pathway.
   d. Differentiate the action of following phagocytic white blood cells: neutrophils, monocytes, macrophages.
   e. List the steps in phagocytosis leading to bacterial destruction, starting with opsonization.
   f. Differentiate the adaptive defenses of antibodies and cell-mediated immunity.
   g. Discuss how antibodies help fend off pathogens.
   h. Compare and contrast B lymphocytes with cytotoxic T lymphocytes.
   i. Identify a type of lymphocyte that is involved with both the innate and adaptive immune response.

INFORMATION FLOW AND GENETICS

4. Cell genomes can be manipulated to alter cell function.

   a. Describe how B lymphocytes can generate millions of different kinds of antibodies.

MICROBIAL SYSTEMS

5. Microorganisms and their environment interact and modify each other.

   a. State the main processes that lead to symptoms in infection.
   b. State consequences of the activation of complement.

6. Microorganisms, cellular and viral, can interact with both human and nonhuman hosts in beneficial, neutral, or detrimental ways.

   a. Identify steps in pathogenesis that are common to all infections.
   b. Define the terms incubation period, inoculum size, and microbiota.
   c. Explain why the definition of a pathogen is conditional.
   d. Infer a mechanism of how the location of damage from a pathogen can be very different from the site of entry.
   e. Describe how phagocytes sense and move toward bacterial chemotaxins.
   f. Draw a concept map about innate defenses, using as many boldfaced vocabulary terms as possible.
   g. Give evidence to support the statement "The adaptive immune response is highly specific and varies from individual to individual."
   h. Discuss how the innate immune system is inherently linked to the adaptive immune system.
   i. Draw a concept map of the adaptive immune system, using as many bold-faced vocabulary terms as possible.

IMPACT OF MICROORGANISMS

7. Microbes are essential for life as we know it and the processes that support life (e.g., in biogeochemical cycles and plant and/or animal microbiota).

   a. Speculate how a member of ordinary microbiota can cause an opportunistic infection.

SUPPLEMENTAL MATERIAL

References

- **Daures M et al.** 2014. Relationships between clinico-epidemiological patterns of invasive meningococcal infections and complement deficiencies in French South Pacific Islands (New Caledonia). *J Clin Immunol* **35:**47–55. http://link.springer.com/article/10.1007%2Fs10875-014-0104-6

- **McNeil LK et al.** 2013. Role of factor H binding protein in *Neisseria meningitidis* virulence and its potential as a vaccine candidate to broadly protect against meningococcal disease. *Microbiol Mol Biol Rev.* **77:**234–252. http://mmbr.asm.org/content/77/2/234.long

- **Zipfel PF, Skerka C**. 2009. Complement regulators and inhibitory proteins. *Nat Rev Immunol* **9:**729–740. http://www.nature.com/nri/journal/v9/n10/full/nri2620.html

- **O'Neill LA et al.** 2013. The history of Toll-like receptors—redefining innate immunity. *Nat Rev Immunol* **13:**453–460. http://www.nature.com/nri/journal/v13/n6/full/nri3446.html

- **Blander JM.** 2014. A long-awaited merger of the pathways mediating host defence and programmed cell death. *Nat Rev Immunol* **14:**601–618. http://www.nature.com/nri/journal/v14/n9/full/nri3720.html

Supplemental Activities

CHECK YOUR UNDERSTANDING

1. **Which innate host components initiate an inflammatory response to microbes?**
   a. Toll-like receptors (TLRs)
   b. NOD-like receptors (NLRs)
   c. MHC molecules
   d. Antibodies
   e. Defensins

2. **The complement system does NOT include**
   a. Opsonins, proteins that facilitate phagocytosis
   b. Antibodies, proteins that bind specific antigens
   c. Chemotaxins, proteins that recruit phagocytes
   d. Membrane attack complex, proteins that make pores in membranes
   e. Compounds that enlarge blood vessels and lower blood pressure

3. **Which of the following is an adaptive immune response?**
   a. Opsonin-mediated phagocytosis of *E. coli*
   b. Secretion of cytokines IL-1β and IL-18 in response to microbe-associated molecular patterns (MAMPs)
   c. Generation of chemotaxins to recruit white blood cells to a wound
   d. Microbe-induced programmed cell death of host cells
   e. Rapid proliferation of memory B cells in response to infection

4. **Patients who receive cancer chemotherapy tend to become neutropenic and highly susceptible to infections.**
   a. Why?
   b. Would prior vaccination benefit patients scheduled to receive chemotherapy?

DIG DEEPER

1. After infecting human epithelial cells with *Chlamydia trachomatis* for 24 hours, you detect the cytokines IL-1β and IL-18 in the tissue culture medium.
   a. What control experiments would you perform to determine whether *C. trachomatis* stimulated production of IL-1β and IL-18?
   b. What innate immune components trigger production of IL-1β and IL-18?
      i. How can you determine whether the epithelial cells express any of these components?
      ii. How can you determine which of these components contribute to this cellular response?
   c. How do the cytokines IL-1β and IL-18 contribute to host defense to *C. trachomatis*?

2. Read the article by Kurtz JR et al. (Vaccination with a single CD4 T-cell peptide epitope from a *Salmonella* type III-secreted effector protein provides protection against lethal infection, *Infect Immun* 82:2424-2433, 2014, http://iai.asm.org/content/82/6/2424 .full.pdf+html). Study Fig. 1 to answer the following questions:

Figure 1:
   a. What is the SseI protein? Why was it a good candidate for their vaccine research?
   b. What did the investigators learn by including *S. typhimurium* strain SL3261 in this experiment? OVA peptides?

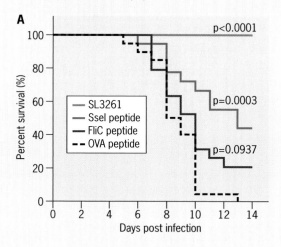

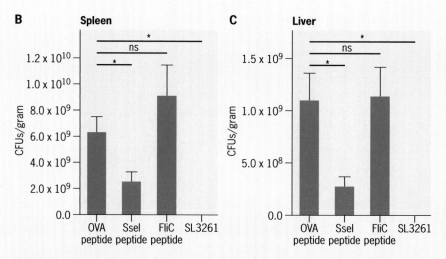

FIGURE 1 (A) Survival of mice first vaccinated with live *S. typhimurium* strain SL3261 or with peptides derived from the *S. typhimurium* SseI or FliC protein, or from the human serum protein ovalbumin (OVA) before oral challenge with live virulent *S. typhimurium*. Colony-forming units (CFUs) of *S. typhimurium* in the spleens (B) and livers (C) of mice vaccinated as shown and then infected orally with live virulent *S. typhimurium*.

   c. What is the FliC protein? What did the investigators learn by including the FliC peptide? Why was it a better choice than, for example, a component of the protein synthesis machinery?

   d. Why did it take a few days after oral infection before mice died?

   e. What conclusion can be drawn from the results?

   f. What additional information was gained by performing the assays in panels B and C?

3. Many infections can now be prevented by vaccination. However, people who are immunocompromised cannot safely receive "live-attenuated" vaccines, but they do benefit from "component vaccines" and from "herd immunity."

   a. How are live-attenuated vaccines different from component vaccines?

   b. In what circumstances are live-attenuated superior to component vaccines? Inferior?

   c. Why are component vaccines safe for immunocompromised people, but live-attenuated vaccines are not?

   d. What is herd immunity? How does it protect immunocompromised people?

4. Multicellular grazing organisms such as *Caenorhabditis elegans* and the slime mold *Dictyostelium discoideum* have been used as model systems to investigate mechanisms of bacterial pathogenesis. Working in groups, use the PubMed search engine to identify an example of this approach. What are the advantages and disadvantages of using these model systems to better understand human diseases? Share your findings with the class using visual aids to teach the key host-pathogen interaction being investigated and the conclusions drawn.

5. **Class debate:** In the United States, the 2015 outbreak of measles at Disneyland stimulated public debate about vaccination, which prevents this viral disease. Use the CDC website to learn about the disease, virus, and vaccine (http://www.cdc.gov/vaccines/vpd-vac/measles/default.htm) and also the unfounded fear of autism (http://www.cdc.gov/vaccinesafety/concerns/autism.html).

Should exemptions to vaccination be legal? Using what criteria? Hold a mock public forum organized to determine whether the measles vaccine should be required to attend the local public schools. What constituencies should be represented? Prepare science- and evidence-based support prior to the hearing.

6. **Class debate:** To improve patient care by saving time and money needed to identify the cause of infectious diseases, clinical microbiology labs are incorporating culture-independent diagnostic tests. In a December 2012 interview, CDC Director Dr. Thomas Frieden discusses the benefits and risks of these new technologies: http://www.cbsnews.com/videos/food-poisoning-fix/. Using the CDC website (http://www.cdc.gov/foodnet/reports/cidt-questions-and-answers-2015.html), an article in *Scientific American* (http://www.scientificamerican.com/article/food-poisoning-outbreaks-become-harder-to-detect/), and other resources, identify the advantages and disadvantages of culture-independent techniques. When should they be applied? When should pure cultures be isolated and analyzed? Who should bear the added expense: the patient, health care facility, insurance companies, or the government? Why?

Supplemental Resources

VISUAL
- *iBiology* lecture "How Can We Sense Infection? Helping to Treat Sepsis" by Jianjin Shi (24 min): http://www.ibiology.org/ibioseminars/jianjin-shi.html
- *iBiology* lecture "Molecular Arms Races between Primate and Viral Genomes" by Harmit Malik (33 min): http://www.ibiology.org/ibioseminars/microbiology/harmit-malik-part-1.html
- *iBiology* lecture "Infectious Disease Overview" by Lucy Shapiro (8 min): http://www.ibiology.org/ibioeducation/exploring-biology/microbiology-ed/bacteria/infectious-diseases-overview.html

- *iBiology* lecture "Host-Pathogen Interaction and Human Disease" by Stanley Falkow (38 min): http://www.ibiology.org/ibioseminars/microbiology/stanley-falkow-part-1.html
- *iBiology* lecture "Pathogenic Bacteria" by Ralph Isberg: http://www.ibiology.org /ibioseminars/ralph-isberg-part-1.html

SPOKEN

- *This Week in Microbiology* podcast episode #45, "Secreted Nucleic Acids RIG a STING": http://www.microbeworld.org/podcasts/this-week-in-microbiology/archives/1307-twim -45-secreted-nucleic-acids-rig-a-sting. (On innate immune sensing of *Listeria*, see also Abdullah Z et al., *EMBO J* **31:**4153–4164, 2012.)
- *This Week in Microbiology* podcast episode #110, "Exploring Unseen Life with Unpronounceable Words," discusses how the composition of the microbial communities in the gut reduce the risk of *C. difficile* disease: http://www.microbeworld.org/podcasts /this-week-in-microbiology/archives/1962-twim-110-exploring-unseen-life-with -unpronounceable-words. (See also Schubert AM et al., *mBio* **6:**4e00974, 2015, doi: 10.1128 /mBio.00974-15.)
- *This Week in Virology* podcast episode #371, "Sympathy for the Devil," discusses the development of new poliovirus strains for the production of inactivated vaccine in the post-eradication era. http://www.microbeworld.org/podcasts/this-week-in-virology/twiv -archives/2052-twiv-371

WRITTEN

- FAQ *Adult Vaccines: A Grown-up Thing to Do*, a colloquium report from the American Academy of Microbiology to educate the general public. http://academy.asm.org/index .php/faq-series/433-adultvaccinesfaq
- Moselio Schaechter tells a story of microbiological heroism during World War II in the *Small Things Considered* blog post "The Louse And The Vaccine" (http://schaechter .asmblog.org/schaechter/2015/08/the-louse-and-the-vaccine.html).
- National Center for Case Study Teaching in Science: Debra Meuler uses the historical account of Warren and Marshall's discovery to expose students to the somewhat messy process of scientific research in "*Helicobacter pylori* and the Bacterial Theory of Ulcers" (http://sciencecases.lib.buffalo.edu/cs/collection/detail.asp?case_id=571&id=571).
- National Center for Case Study Teaching in Science: Caren Shapiro explores the public debate on vaccination requirements of public schools in "To Vaccinate, or Not to Vaccinate" (http://sciencecases.lib.buffalo.edu/cs/collection/detail.asp?case_id=324&id =324).
- National Center for Case Study Teaching in Science: Kate Rittenhouse-Olson focuses on the public health question of vaccination and incorporates epidemiology, etiology and pathology, prevention and treatment, and laboratory diagnosis in "Is it a Lemon or a Lyme" (http://sciencecases.lib.buffalo.edu/cs/collection/detail.asp?case_id=314&id=314).

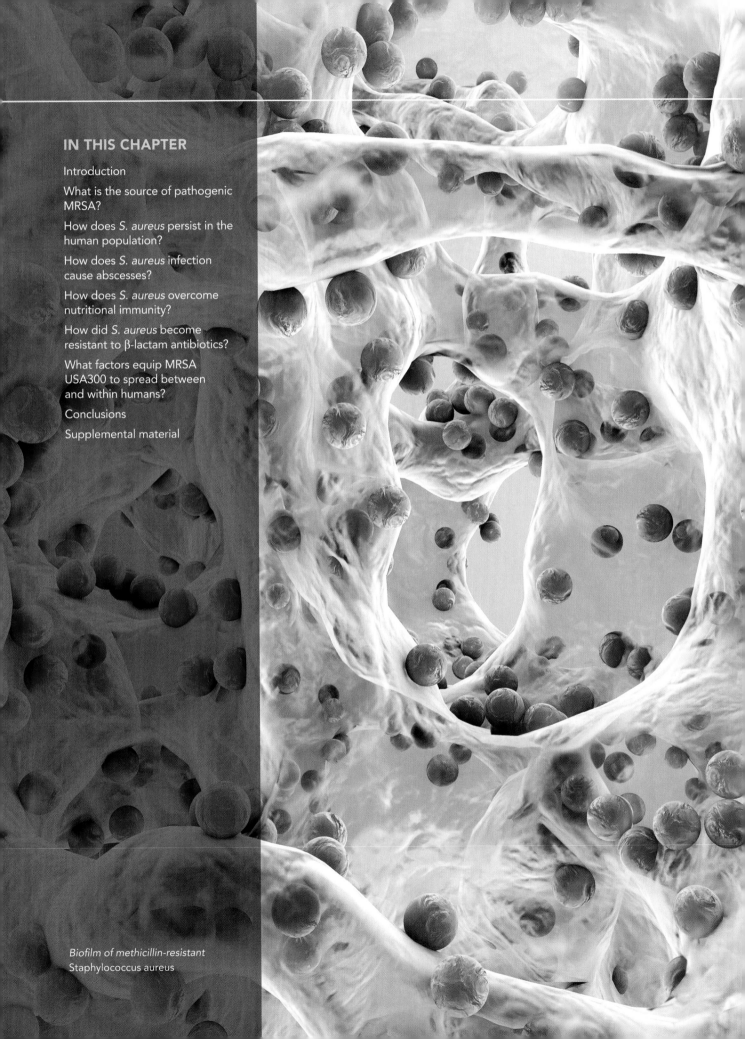

Biofilm of methicillin-resistant Staphylococcus aureus

CHAPTER TWENTY-THREE

Opportunistic Infections by Microbiota: *MRSA*

KEY CONCEPTS

This chapter covers the following topics in the ASM Fundamental Statements.

EVOLUTION

1. Mutations and horizontal gene transfer, with the immense variety of micro-environments, have selected for a huge diversity of microorganisms.
2. Human impact on the environment influences the evolution of microorganisms (e.g., emerging diseases and the selection of antibiotic resistance).

METABOLIC PATHWAYS

3. The survival and growth of any microorganism in a given environment depends on its metabolic characteristics.

INFORMATION FLOW AND GENETICS

4. Genetic variations can impact microbial functions (e.g., in biofilm formation, pathogenicity, and drug resistance).

MICROBIAL SYSTEMS

5. Microorganisms and their environment interact with and modify each other.
6. Microorganisms, cellular and viral, can interact with both human and non-human hosts in beneficial, neutral, or detrimental ways.

INTRODUCTION

As we learned in chapter 22, entry is the first challenge for pathogens. Physical boundaries formed by tight layers of epithelial cells and patrols of professional phagocytes cooperate to protect our tissues from the myriad microbes that colonize our bodies and stow away on the food we eat and the water we drink. At the same time, these same host barriers exert selective pressure on microbes, whose short generation time is a big advantage in evolution's cat-and-mouse game.

As an example of this interplay, many of us are colonized by *Staphylococcus aureus*, a common Gram-positive bacterium with a well-earned reputation for causing a range of serious infections. Alas, the microbial world cannot be divided neatly between friends and foe. In this chapter, *S. aureus* will vividly illustrate that *location matters greatly*. We'll also see that the genome of this **opportunistic pathogen** (a microbe that can cause infection if local or systemic host defenses are weakened) contains a living record of its long history of coevolution with humans. By studying methicillin-resistant *S. aureus* (MRSA) pathogenesis, we'll reinforce fundamentals of innate immunity, plus learn about nutritional immunity, another host strategy to inhibit microbial growth.

CASE: The NFL battles outbreak of skin wound abscesses

As the 2003 professional football season got under way, the St. Louis Rams confronted a formidable opponent that wasn't on their schedule: methicillin-resistant *S. aureus* (MRSA). Trouble started when some players' turf burns and forearm abrasions—long a part of this heavy-hitting sport—were aggravated by swelling, redness, and pus. Soon their skin wounds morphed into abscesses that team doctors had to lance and drain (Fig. 23.1). The five players—all linemen or linebackers (Fig. 23.2)—were given three oral antibiotics: cephalexin, trimethoprim-sulfamethoxazole, and rifampin; two also required intravenous vancomycin and ceftriaxone. When a visiting team later reported that similar wounds developed on their linemen's scraped shins and elbows, investigators from the Centers for Disease Control and Prevention (CDC) were recruited to determine the source and route of transmission of the causative agent, MRSA. The epidemiologists inspected locker rooms, training equipment, whirlpools, and saunas; they also studied antimicrobial use and hygiene practices and collected samples by swabbing the turf field and training facility surfaces, as well as the nasal passages and skin abrasions of affected and unaffected players. Isolates of *S. aureus* were tested for antimicrobial susceptibility (Fig. 23.3), and their relationship to each other and to reference strains in the CDC database was analyzed by pulsed-field gel electrophoresis (Fig. 23.4). The presence of two telling markers, the gene for Panton-Valentine leukoci-

FIGURE 23.1 Cutaneous abscesses caused by methicillin-resistant *S. aureus. (From http://www.cdc.gov /mrsa/community/photos/index.html, credit Gregory Moran.)

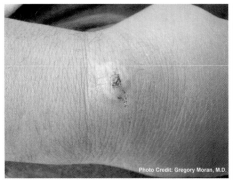
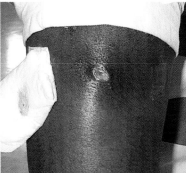

din and the staphylococcal cassette chromosome *mec* element (SCC*mec*), was probed by PCR. In October, the Rams put in place several infection control measures, including increased hand washing, showering before whirlpool use, heightened screening for skin infections, and higher standards for wound care. By November, only one new MRSA skin infection had occurred (Fig. 23.2). Today, educational campaigns run by the National Collegiate Athletic Association (Fig. 23.5) and other organizations work to reduce the risk of MRSA infections among athletes and the general public.

- Where did the football players encounter MRSA?
- What is the mode of entry of MRSA?
- What factors contributed to the spread of MRSA among healthy people?
- How does *S. aureus* cause abscesses?
- Why were the players given multiple antibiotics?
- Why did investigators probe whether the *S. aureus* strains encoded the Panton-Valentine leukocidin and SCC*mec*?
- How does pulsed-field gel electrophoresis analysis contribute to outbreak investigations?

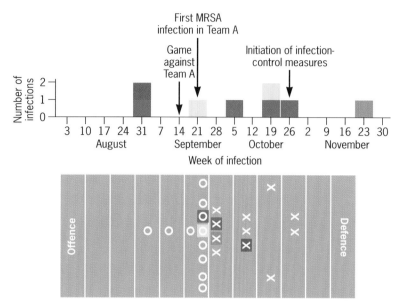

FIGURE 23.2 MRSA infection among St. Louis Rams football players in 2003. Top: Epidemic curve indicates the number of cases diagnosed on each date shown. Each box represents one case, and each color represents a different player; recurrent infections are evident. Bottom: Football field diagram indicates the defensive (X) and offensive (O) position played by each player diagnosed with a MRSA infection (colored squares). (From Kazakova SV et al., *N Engl J Med* **325**:468–475, 2005.)

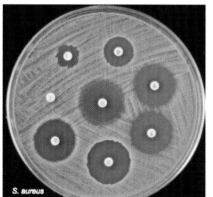
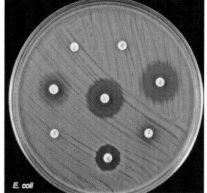
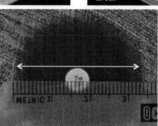

FIGURE 23.3 Kirby-Bauer disk diffusion susceptibility test. Coagulase-negative *S. aureus* and *E. coli* were cultured on Mueller-Hinton agar with disks infused with different antibiotics (tetracycline, 30 µg; cephalothin, 30 µg; erythromycin, 15 µg; chloramphenicol, 30 µg; vancomycin, 30 µg; penicillin, 10 µg; streptomycin, 10 µg; and novobiocin, 30 µg). Zones are measured in millimeters of either the radius or diameter. http://www.microbeworld.org /component/jlibrary/?view=article&id =12255

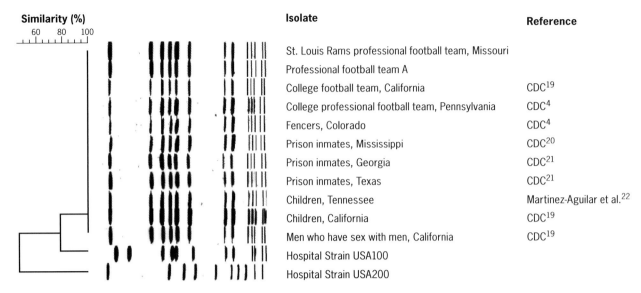

Similarity (%)

Isolate	Reference
St. Louis Rams professional football team, Missouri	
Professional football team A	
College football team, California	CDC[19]
College professional football team, Pennsylvania	CDC[4]
Fencers, Colorado	CDC[4]
Prison inmates, Mississippi	CDC[20]
Prison inmates, Georgia	CDC[21]
Prison inmates, Texas	CDC[21]
Children, Tennessee	Martinez-Aguilar et al.[22]
Children, California	CDC[19]
Men who have sex with men, California	CDC[19]
Hospital Strain USA100	
Hospital Strain USA200	

FIGURE 23.4 Pulsed-field gel electrophoresis analysis of MRSA isolates from the St. Louis Rams, their opponents on Team A, or outbreaks from other community settings and geographic locations, and of the hospital strains USA100 and USA200. (Left) The dendrogram indicates relatedness of one representative MRSA isolate for each setting. (From Kazakova SV et al., *N Engl J Med* **352**:468–475, 2005.)

WHAT IS THE SOURCE OF PATHOGENIC MRSA?

LEARNING OUTCOMES:
- Infer reasons why the both microbial density and microbial diversity are so much higher in the human gastrointestinal tract over skin.
- Use microbiome examples to justify the statement "Microbes are essential for life as we know it and the processes that support life."
- Explain the term "opportunistic pathogen."

FIGURE 23.5 To increase awareness among athletes of the risks of MRSA skin infections, the CDC and NCAA launched an educational campaign. (From NCAA, http://www.ncaa.org/health-and-safety/medical-conditions/skin-issues.)

TABLE 23.1 **Microbiota of human skin and gastrointestinal tract**

	Skin	Gastrointestinal tract
Microbial density	$10^6/cm^2$	$10^{12}/g$ of contents
Microbial diversity	Bacteria dominate, ~40 species/person; up to 10% fungi and 40% viral colonization; shifts at puberty	Bacteria dominate, ~100 species/person; fungi and viruses are rare
Conditions	Aerobic Cool Acidic Ranges from dry to moist, oily, and salty	Anaerobic Warm Moist Ranges from acidic stomach to neutral colon
Nutrients	Poor	Rich: dietary, bacterial metabolites, mucus
Host defenses	Antimicrobial peptides	Antimicrobial, peptides, IgA, leukocytes

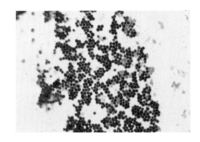

FIGURE 23.6 **Gram-stained *S. aureus* in clusters (magnified ~×1,000).** (From MicrobeWorld, Wistreich Collection, http://www.microbeworld.org/component/jlibrary/?view=article&id=7611)

Humans are home to myriad microbes. Collectively known as the *human microbiome*, these microscopic partners easily outnumber our human cells. Thanks to impressive advances in the speed and efficiency of DNA sequence analysis, it is now feasible to choose a body site and ask: Who's there? The answers are astonishing. Trillions of *Bacteria, Archaea*, viruses, fungi, and protists inhabit our skin, mouth, gut, and urogenital tract (Table 23.1, chapter 19). In fact, scientists estimate that we carry as many as 300 microbial genes for each of our human genes!

What our vast, diverse collections of microbes do to us—and *for* us—is one of the exciting new frontiers in microbiology. The microbiota in our gut can synthesize the vitamins niacin, K, B_{12}, and others. And we rely on one of our residents, *Bacteroides thetaiotaomicron,* to digest the complex carbohydrates that we consume in our diet. It may surprise you to learn that our microbiome also *protects* us from **pathogens**. For example, by forming food chains within our gut, these stable microbial communities make it difficult for interlopers to establish residence. In addition, cell components shed by our **commensal** microbes stimulate the epithelial cells that line our intestinal tracts to produce protective IgA and antimicrobial peptides. Can you think of a circumstance that demonstrates the beneficial effect of the human gut microbiome? (*Quiz spoiler:* Use of oral antibiotics puts people at risk for diarrheal disease caused by another opportunistic pathogen, *Clostridium difficile.*) You can learn more about the fascinating partnerships that evolution forges between us and our microbiota from the American Academy of Microbiology's booklet *FAQ: Human Microbiome* (Supplemental Resources).

Like every other body surface that comes into contact with the outside world, our nasal passages are colonized by microbes. Residing in the nose of approximately one in three people is *S. aureus*, a Gram-positive coccoid bacterium that readily clusters, resembling a bunch of grapes, or "staphyle" in Greek (Fig. 23.6). When streaked on agar, these staphylococci form colonies with a golden hue, hence "aureus." Although harmless in the nostrils, *S. aureus* is a versatile microbe that can wreak havoc if it gains entry to the blood, lungs, bones, or open wounds, as we saw for the athletes in the case study. It can cause not only skin **abscesses** (inflammatory lesions that contain pus) but also life-threatening pneumonia, **sepsis** ("blood poisoning"), and a range of infections in deeper tissue, including bones (osteomyelitis) and heart (endocarditis). Its capacity to switch from harmless commensal merits *S. aureus* the dubious distinction of being an opportunistic pathogen.

HOW DOES *S. AUREUS* PERSIST IN THE HUMAN POPULATION?

LEARNING OUTCOMES:
- Give examples of how MRSA's genome has been influenced by humans.
- Explain why *S. aureus* isolated from nasal passages of healthy people rarely encodes PVL toxin and why not all *S. aureus* (MRSA) strains carry the PVL toxin.

S. aureus causes trouble when it moves from its natural habitat—the nasal passages or skin—into the body proper. As the outbreak of MRSA among football players illustrates, damage to the protective barrier of our skin by an opposing lineman, a splinter, a needle, a burn, or a bug bite creates the opportunity for *S. aureus* to enter deeper tissues. This misplaced microbe immediately encounters innate immune defenses, including the **complement** system and **neutrophils** (chapter 22), which cooperate to sense, mark, and clear most microbial invaders. However, these very defenses also exert strong selective pressure on commensal microbes. Rare genetic variants that can evade or disable host barriers to colonization gain a competitive advantage. Thus, over its long course of coevolution with humans, *S. aureus* has acquired an impressive toolkit to deactivate or surmount innate and acquired immunity processes. To set the stage for this story of host-microbe competition, watch as a neutrophil works valiantly to catch and ingest *S. aureus* in Fig. 23.7.

Indeed, neutrophils are motile cells that are equipped to sense and move quickly toward microbial products or host-produced **chemotaxins**, small signaling molecules that attract white blood cells. As we learned in chapter 22, they're also professional eaters, especially when their prey has been made tasty by **opsonins**, such as C3b or IgG. Is the small, coccoid *S. aureus* defenseless? Far from it! It has an amazing repertoire of factors that counteract host defenses. *S. aureus* secretes **virulence factors** that inhibit opsonization, as well as neutrophil chemotaxis and **phagocytosis** (Table 23.2). For example, *S. aureus* inhibits phagocytosis when one of its surface components, protein A, binds to the conserved Fc region of **immunoglobulins**. Yes, that's backwards from the way **antibodies** usually bind to **antigens**! Still other virulence factors (staphyloxanthin, catalase, and alkyl peroxidase) protect this pathogen from being killed by reactive oxygen species within the phagosome. And remem-

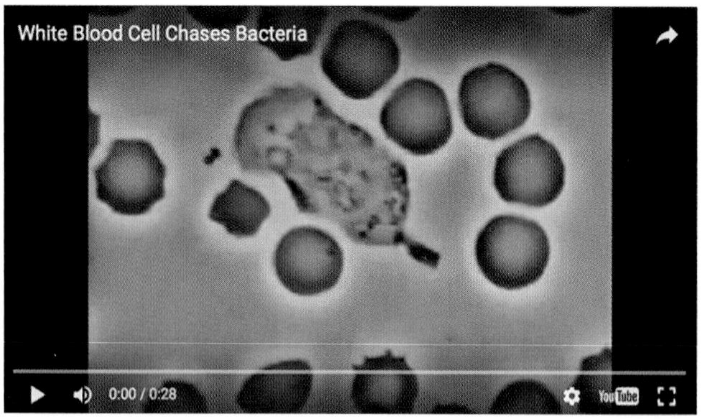

FIGURE 23.7 Still image from a movie of *S. aureus* being pursued by a neutrophil in a blood smear that also contains platelets and red blood cells. The 16-mm film was created by Dr. David Rogers in the 1950s. (From MicrobeWorld, credit David Rogers, http://www.microbeworld.org/component/jlibrary/?view=article&id=2792)

White Blood Cell Chases Bacteria

0:00 / 0:28

ber the aggressive neutrophils (Fig. 23.7). How many factors listed in Table 23.2 inhibit their chemotaxis, migration, and phagocytosis? Even this abbreviated list of *S. aureus* antagonists of the human immune system beautifully demonstrates that a pathogen's genome is molded by the never-ending competition with its host.

An even more aggressive virulence strategy is to kill neutrophils outright, and the staphylococcal **Panton-Valentine leukocidin (PVL)** does just that. It does so in two steps (Fig. 23.8). First, *S. aureus* secretes two proteins, LukS and LukF, that associate with the neutrophil's plasma membrane. Next, eight subunits of LukS and LukF cooperate to self-assemble into a ring-like structure that forms pores in the host membrane. When several pores form on the same cell, the neutrophil lyses, triggering **inflammation** and tissue damage.

In the case of this leukocidin, the real culprit is a **lysogenic phage**. At some point, a phage carrying the *lukS-PV* and *lukF-PV* genes infected an

TABLE 23.2 **Anti-immunity factors of *S. aureus***

Immune defenses	Host component	*S. aureus* factor	*S. aureus* tactic
Innate defenses			
Barriers	Skin	None	Enters through break in skin barrier
Antimicrobial peptides	LL-37	Protease aureolysin	By cleaving antimicrobial peptides, inhibits their incorporation into bacterial cell wall, a lethal event
Complement	Factor H	Positively charged surface	By recruiting this negative regulator to its surface, inhibits activation of the alternative complement pathway
Complement	C3	Protease aureolysin	By cleaving C3, inhibits production of opsonin C3b
Complement	C5a	SCIN	By binding to C3b-Bb complex of complement cascade, inhibits opsonization
Complement	C3b	Staphylokinase	By recruiting and activating plasminogen to form the protease plasmin, cleaves opsonin C3b from bacterial surface
Neutrophils		Phenol-soluble modulin peptides	Lyse neutrophils by disrupting their plasma membrane
Adherence to wall of blood vessel	Integrins	EAP	By blocking adhesion, inhibits neutrophils from migrating from blood to site of infected tissue
Chemotaxis	Chemokine CXCR2	Protease staphopain A	By cleaving chemokine, inhibits neutrophil recruitment
Chemotaxis	Formyl protein receptor	CHIPS	By interfering with binding of host receptor to discarded bacterial *N*-formylmethionyl-leucyl-phenylalanine peptides, inhibits recruitment of neutrophils
Phagocytosis	Complement receptor CR3 and immunoglobulin receptor FcR	Multiple factors described in Complement and IgG section	Blocks opsonization and inhibits ingestion by phagocytes
Oxidative killing	NADPH oxidase	Pigment staphyloxanthin	This antioxidant protects the pathogen from damaging reactive oxygen species (ROS)
Oxidative killing	NADPH oxidase	Catalase	Converts toxic hydrogen peroxide to harmless oxygen and water
Oxidative killing	NADPH oxidase	Alkyl hydroperoxidase	harmless oxygen and water
Adaptive Defenses Antibodies	Fc receptor recognizes conserved Fc region of immunoglobulins	Protein A	By binding to the region of IgG that is recognized by Fc receptor of phagocytes, inhibits phagocytosis and complement activation

FIGURE 23.8 **Multistep mechanism of two-component toxin assembly in host cell membranes.** Some strains of *S. aureus* encode the two-component pore-forming toxin Panton-Valentine leukocidin (PVL). Pink bar represents host receptor.

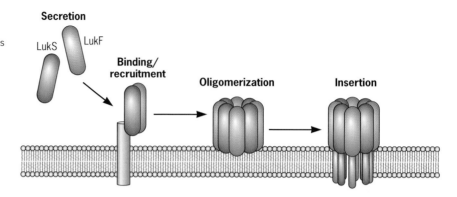

S. aureus strain (Fig. 23.9). But not all *S. aureus* strains carry the prophage nor produce the PVL toxin. Today, *S. aureus* strains isolated from the nasal passages of healthy people rarely encode the pore-forming toxin. On the other hand, *S. aureus* from severe skin and soft tissue infections more frequently produces PVL. Therefore, the PVL toxin genes increase the risk of severe infections. Even more worrisome is that ~85% of MRSA carry the toxin-encoding phage. It's no wonder that, during the *S. aureus* outbreak among the St. Louis Rams, the CDC experts were on high alert for PVL toxin genes!

HOW DOES *S. AUREUS* INFECTION CAUSE ABSCESSES?

LEARNING OUTCOMES:
- State how interactions between *S. aureus* and the host help to create abscesses during *S. aureus* (MRSA) infections.

The struggle between *S. aureus* and the host is also evident to the naked eye (Fig. 23.1). The abscesses that formed on the scraped shins and elbows of the Rams players are a trademark of *S. aureus* infections. Recall that the redness, tenderness, and swelling that first appeared are classical signs of inflammation (chapter 22). To restrict collateral tissue damage caused by

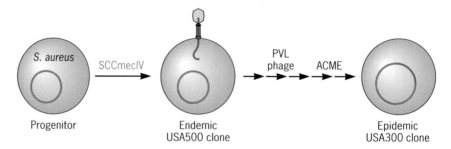

FIGURE 23.9 **Evolution of MRSA clone USA300 by multiple horizontal gene transmission events.** An *S. aureus* progenitor strain acquired by horizontal transfer a methicillin resistance cassette (SCC*mecIV*) carrying the *mecA* gene. Subsequently, this endemic strain (USA500) was infected and lysogenized by a phage that encodes the PVL toxin and acquired by horizontal transmission the arginine catabolic mobile element (ACME), among other genetic changes, generating the epidemic clone USA300.

the array of enzymes and pore-forming proteins that both warring parties release, the host deposits fibrin, a tough fiber-like protein that self-assembles into a protective web. *S. aureus* also contributes; it secretes a **coagulase**, an enzyme that activates host thrombin to generate more fibrin. Now, the clot thickens. But just who is protected by this tough shell of fibrin and leukocytes? True, *S. aureus* cells are trapped within the abscess. But they can survive and replicate there, shielded from complement, antibody, neutrophils, and antibiotics. Eventually, some progeny escape from the abscess, creating an opportunity to seed new focal infections elsewhere. We don't yet fully understand the mechanisms that govern spread of *S. aureus* from abscesses, but what is abundantly clear is the clinical challenge of eliminating *S. aureus* that is walled off within abscesses that form on bone, heart, and other deep tissues. Typically, a combination of surgery and an extended course of antibiotics is required.

HOW DOES *S. AUREUS* OVERCOME NUTRITIONAL IMMUNITY?

LEARNING OUTCOMES:
- Describe how hosts can restrict microbial growth by using nutritional immunity.
- Outline how some *S. aureus* variants are capable of overcoming nutritional immunity.

Surviving the onslaught of complement and neutrophils in inflamed tissue is an impressive feat, but *S. aureus* faces a whole other challenge, that of *acquiring the nutrients and metals* it needs to make new cells (Part I). Consider iron, an essential cofactor of many biosynthetic enzymes (chapter 8). Vertebrates selfishly sequester this precious metal from their microbial neighbors (Fig. 23.10). **Macrophages** pump ferrous iron (Fe^{2+}) away from any microbial cargo in their phagosomes by using the so-called natural resistance-associated macrophage protein 1 (NRAMP1). Within red blood cells, Fe^{2+} is bound by heme, a cofactor that is complexed with hemoglobin. If red blood cells lyse, both hemoglobin and free heme are captured by dedicated host serum proteins. Free ferric iron (Fe^{3+}) is also a rare commodity: this metal is sequestered by the protein lactoferrin on mucosal surfaces, by transferrin in blood, and by ferritin inside cells. Although bacteria are equipped with compounds that avidly bind Fe^{3+}, neutrophils also secrete a protein that captures iron. Likewise, the host jealously protects its supply of zinc and manganese. Together, iron, zinc, and manganese are estimated to be essential cofactors for as many as 30% of bacterial proteins. Thus, by actively sequestering these metals, the host impedes bacterial respiration, DNA synthesis, and other key processes. This broad host strategy for restricting microbial growth is known as **nutritional immunity**.

But once again, evolution writes another side to this story. The very host pathways that sequester critical nutrients from microbes also exert strong selective pressure for genetic variants that overcome this defensive barrier. Once more, the *S. aureus* genome provides clear evidence of how the

FIGURE 23.10 Multiple host mechanisms sequester iron from microbes. At the mucosal surface, Fe^{3+} is bound by lactoferrin. Intracellular Fe^{3+} is complexed with ferritin (F) or bound by serum transferrin. In blood, Fe^{2+} is complexed with heme, which is bound by hemoglobin within red blood cells. When released from lysed red cells, hemoglobin is bound by haptoglobin (Hp) and free heme by hemopexin (Hpx). Neutrophils produce gelatinase-associated lipocalin (NGAL) that binds and sequesters bacterial siderophores. In macrophages, the NRAMP1 transporter pumps Fe^{2+} out of phagosomes. (Modified from Hood MI, Skaar EP, *Nat Rev Microbiol* **10**:525–537, 2012.)

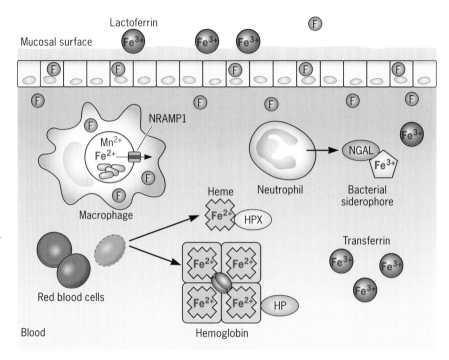

battle for iron is waged (Fig. 23.11). Like many other bacterial species, *S. aureus* senses iron-poor conditions using a transcriptional repressor protein known as Fur, for f̲erric u̲ptake r̲egulator. When iron is limiting, Fur repression is relieved, and multiple genes dedicated to iron acquisition are expressed (chapter 11). In response, *S. aureus* in need of iron secretes **siderophores** that bind iron even more tightly than host transferrin or lactoferrin do. The siderophore-Fe^{3+} complexes are then brought into the cell by dedicated transporters. Many *S. aureus* strains also lyse red blood cells by secreting toxins, aptly named **hemolysins,** and thus release iron-containing hemoglobin and heme. Then, another set of surface proteins capture the iron bound within hemoglobin. First, receptors embedded in the cell wall bind iron-loaded heme and hemoglobin. Next, Fe^{2+}-heme complexes are transported across the cell membrane by a multicomponent protein complex. Within the cytoplasm, heme is degraded by other proteins. At last! *S. aureus* now collects the liberated Fe^{2+} for its many enzymes. Indeed, embedded in nearly every metabolic pathway is a component that requires iron.

HOW DID *S. AUREUS* BECOME RESISTANT TO β-LACTAM ANTIBIOTICS?

LEARNING OUTCOMES:
- State how methicillin works as an antibiotic against *S. aureus*.
- Describe a reason why resistance to antibiotics spreads so quickly in *S. aureus*.
- Use MRSA examples to support the statement "Genetic variations can impact microbial functions."

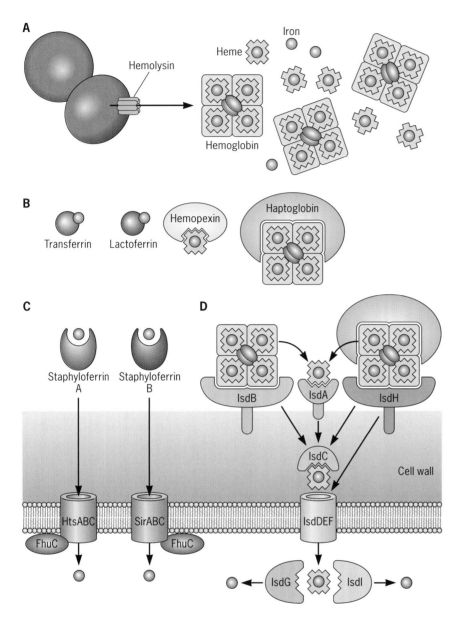

FIGURE 23.11 *S. aureus* **uses multiple mechanisms to obtain iron from the host.** (A) *S. aureus* hemolysins lyse red blood cells to release hemoglobin, which can be degraded to heme and free iron. (B) Host proteins transferrin and lactoferrin sequester free iron, while hemopexin binds heme and haptoglobulin traps hemoglobin. (C) *S. aureus* siderophores staphyloferrin A and staphyloferrin B bind iron with sufficiently high affinity to compete with host iron sequestration proteins lactoferrin and transferrin. The bacterial transporters HtsABC and SirABC import iron bound to staphyloferrin A and staphyloferrin B, respectively, using energy generated by the FhuC ATPase. To acquire heme, the cell surface receptors IsdB and IsdH bind hemoglobin and hemoglobin-haptoglobin, respectively. Next, IsdB and IsdH pass heme either to IsdA or to IsdC or IsdE for transport across the cell membrane by IsdDEF. In the cytoplasm, IsdG or IsdI degrade heme to release iron. (Modified from Cassat JE, Skaar EP, *Semin Immunopathol* **34:**215–235, 2012.)

As a common inhabitant of humans, *S. aureus* has an impressive track record as an opportunistic pathogen. Clinicians, together with the pharmaceutical industry, have long worked to protect patients from infectious diseases. Unfortunately, antibiotic resistance is developing—and spreading—at an alarming rate. Penicillin was first prescribed in 1940; by 1942, penicillin-resistant *S. aureus* were reported. Resistance is conferred by a plasmid-born penicillinase, an enzyme that cleaves the four-membered β-lactam ring of penicillin and many other β-lactam antibiotics. Since **plasmids** are readily transmitted within bacterial populations (chapter 10), penicillin-resistant *S. aureus* strains soon became widespread and remain so today. By the late 1950s, chemists responded with methicillin, a related β-lactam antibiotic that also interferes with assembly of the peptidoglycan cell wall but that is resistant to penicillinases. Just 1 year later, methicillin-resistant *S. aureus* (MRSA) had emerged! To make matters worse, these strains had simultaneously acquired resistance to multiple β-lactam antibiotics, namely the penicillins, cephalosporins, and carbapenems. *Considering what you've learned about bacterial cell structure, physiology, and genetics, how many different mechanisms could confer such multidrug resistance?* In this case, the perpetrator is a novel protein that keeps the cell wall assembly machinery humming, despite the antibiotics. A novel penicillin-binding protein (PBP 2′), encoded by *mecA*, is an alternative transpeptidase that simply does not interact with β-lactam antibiotics (chapter 9). More problematic for us is that the gene *mecA* readily spreads horizontally through bacterial populations. A family of large mobile elements named staphylococcal chromosome cassette *mec* (SCC*mec*) encodes the novel penicillin-binding protein and transfers methicillin resistance between *S. aureus* strains (Fig. 23.9).

As a tool to differentiate MRSA isolates, identify their origin, and track their spread—such as during the 2003 outbreak among football players—clinical microbiologists and epidemiologists apply pulsed-field gel electrophoresis (PFGE; Fig. 23.4 and 23.12). Thanks to the development of an electronic database of the PFGE profiles of prototypical strains, health departments worldwide can now readily share epidemiological and clinical information on MRSA infections. For example, after analyzing hundreds of MRSA strains, CDC scientists identified eight distinct lineages in the United States, named USA100 to USA800. Once knowledge of these genomic profiles was combined with clinical data, a striking pattern emerged. Multiple MRSA lineages are typically associated with health care settings, but the one most often acquired in the community is USA300. Indeed, of the adults whose skin and soft-tissue infections prompted a trip to the emergency ward of 11 different hospitals in 1 month, 97% of their MRSA infections were caused by USA300 strains, and 98% of these *S. aureus* isolates encoded both the PVL toxin and SCC*mec* type IV (Fig. 23.9). In addition, USA300 strains encode *msrA*, a gene that confers resistance to the macrolide class of antibiotics such as azithromycin. These striking epidemiological patterns motivate scientists to try to find out *why* USA300 MRSA strains have become so prevalent in our communities. And why does USA300 now affect healthy people previously considered to be at low risk, such as children in day care, athletes, military personnel, and prison inmates?

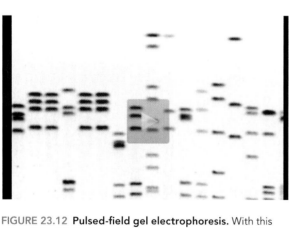

FIGURE 23.12 Pulsed-field gel electrophoresis. With this video (1.3 min), Dr. Brent Gilpin explains how a genomic bar code can be generated by separating very large fragments of DNA generated by treating genomic DNA with restriction enzymes. This molecular typing method is broadly used in epidemiological studies. [From Science Learning Hub, New Zealand, http://sciencelearn.org.nz/Contexts/Enviro-imprints /Sci-Media/Video/Pulsed-field-gel-electrophoresis/(quality) /mp4Video]

WHAT FACTORS EQUIP MRSA USA300 TO SPREAD BETWEEN AND WITHIN HUMANS?

LEARNING OUTCOMES:
- Identify the event that contributed to the emergence of community-acquired MRSA.
- Use evidence to support the role of *speG* in promoting the survival and growth of *S. aureus* on skin. State practices that can decrease your chance of acquiring MRSA.

Historically, *S. aureus* resided in the warm, moist nasal passages of humans, while its cousin *Staphylococcus epidermidis* colonized the skin. But *S. aureus* broke those boundaries when it acquired the large genetic element that had previously equipped *S. epidermidis* to thrive in its dry, acidic, and nutrient-poor habitat. Known (somewhat awkwardly) as the arginine catabolic mobile element, or ACME, this ~30-gene locus has inserted at the same chromosomal site as SCC*mec* islands. Staphylococci that carry ACME can counteract the acid in human sweat by generating ammonia via this element's arginine deaminase system (Arc). Plus, ACME delivers another prize. At sites of damage, inflammation, or wound healing, skin cells (keratinocytes) generate polyamines, a family of compounds derived from arginine with curious toxicity toward *S. aureus*. To thwart this assault, *S. aureus* produces a spermidine acetyltransferase, SpeG, an ACME-encoded enzyme that not only inactivates polyamines (Fig. 23.13) but also triggers other microbial pathways that promote survival and growth on skin. Moreover, since host cells use polyamines to stimulate wound healing SpeG likely prolongs the tissue damage caused by *S. aureus* infection of skin and soft tissues.

By increasing the fitness of *S. aureus* on the skin, the horizontal transmission of the *S. epidermidis* ACME locus contributed to the emergence of community-acquired MRSA. Endowed with enzymes that counteract low pH and polyamines, USA300 strains can now spread easily among healthy people, especially those whose skin comes into direct contact, like athletes, or who share exercise or recreational equipment or towels. Of course, treating MRSA USA300 infections is complicated by their resistance to multiple β-lactam and macrolide antibiotics. That *S. aureus* has become quite a foe, hasn't it?

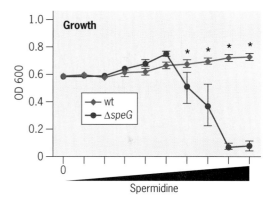

FIGURE 23.13 *speG* **equips *S. aureus* to tolerate spermidine.** Wild-type and *speG* mutant USA300 *S. aureus* strains were cultured for 18 hours at 37°C in nutrient broth that was supplemented with the polyamine spermidine at concentrations ranging from 0 to 11.5 mM. Growth was measured by quantifying optical density at 600 nm. Shown are means of quintuplicate samples; * indicates *P*<0.01. [Modified from Fig. 3D in Planet PJ et al., *MBio* **4(6)**:300889-13, 2013, doi: 10.1128/mBio.00889-13.]

CASE REVISITED: The NFL battles outbreak of skin wound abscesses

The football players may have encountered MRSA during physical contact on the playing field, since a member of the opposing team contracted an infection a week after their game with the Rams (Fig. 23.2). However, since MRSA strains of the same PFGE type are widespread geographically (Fig. 23.4), it is possible that the player from the visiting team was infected from another source. Since the infection control measures reduced the incidence of MRSA infections among the Rams, we can deduce that MRSA was also likely spread by contact with shared training equipment, the whirlpool, or towels—objects that each come into contact with skin. Skin abrasions sustained during play, especially on the offensive and defensive lines, gave MRSA the opportunity to enter the players' soft tissues. Because *S. aureus* had acquired ACME, the horizontally acquired locus that confers resistance to acidic pH and to antimicrobial polyamines, its increased fitness on skin increased the opportunity for MRSA to spread among healthy people. The abscesses that are a classic clinical sign of *S. aureus* infection of skin and soft tissue are triggered by both host defense and bacterial virulence factors. These fibrin webs restrict local spread of *S. aureus*, but the pathogen can escape and seed other tissue sites. The players were treated with multiple antibiotics because MRSA strains carry the SCC*mec* island; its *mecA* gene confers resistance to β-lactam antibiotics, and its *msrA* increases resistance to macrolides. Therefore, physicians prescribed antibiotics of other chemical classes. The MRSA USA300 strain has become a predominant cause of community-acquired skin infections, and various factors contribute to its virulence and spread. Knowledge of the strain genotype can not only guide therapy for these football players but also facilitate epidemiological studies of the outbreak. PFGE patterns indicate relatedness of strains, so comparing the profile of the MRSA isolates provides useful information to identify the source and analyze the spread of epidemic pathogens. Because the MRSA USA300 clone is now widespread geographically, more detailed molecular analysis would be required to definitively determine the chain of transmission within the Rams locker room and to determine if they, in fact, spread MRSA to the visiting team.

CONCLUSIONS

Skin abrasions are just one of many ways microbes can break through a host barrier to infection. Physical trauma, indwelling medical devices, and immunosuppression each create opportunities for microbes to enter and colonize deeper tissues (Table 23.3). When untreated, HIV infection provides a stark case study of the vital capacity of cell-mediated immunity to ward off infections by a wide range of microbes. As the replicating virus depletes CD4+ T cells, patients become vulnerable to a long list of fungi, bacteria, and viruses. In fact, the appearance of each of these opportunistic pathogens can predict the patient's CD4+ T cell count (Fig. 23.14). The strat-

TABLE 23.3 **Opportunistic pathogens**

Opportunity	Pathogen	Disease
Burn	*Pseudomonas aeruginosa*	Soft tissue infection, sepsis
Catheter	*Proteus mirabilis*	Urinary tract infection
Oral antibiotics	*Clostridium difficile*	Diarrhea, colitis
Ruptured appendix	*Escherichia coli*	Peritonitis
Puncture wound	*Clostridium tetani*	Tetanus (spastic paralysis)
Complement deficiency	*Neisseria meningitidis*	Meningitis
T cell deficiency	*Pneumocystis carinii*	Pneumonia
Neutrophil deficiency	*Staphylococcus aureus*	Skin infection, pneumonia

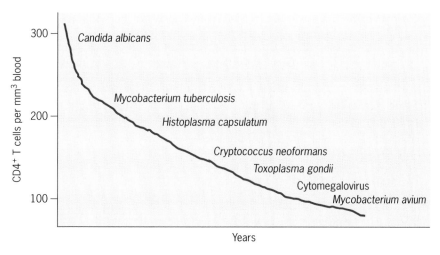

FIGURE 23.14 The risk of infection by the opportunistic pathogens shown increases as the population of CD4$^+$ T cells in the blood declines due to HIV-1 infection that is untreated.

egies each microbe uses to establish infection vary widely. However, each will confront not only innate and acquired immune defenses but also the challenge of acquiring the nutrients essential to fulfill every microbe's goal: to make two microbes.

Representative ASM Fundamental Statements and Learning Outcomes for the Chapter

EVOLUTION

1. Mutations and horizontal gene transfer, with the immense variety of microenvironments, have selected for a huge diversity of microorganisms.

   a. Describe a reason why resistance to antibiotics spreads so quickly in *S. aureus*.
   b. Identify the event that contributed to the emergence of community-acquired MRSA.

2. Human impact on the environment influences the evolution of microorganisms (e.g., emerging diseases and the selection of antibiotic resistance).

   a. Give examples of how MRSA's genome has been influenced by humans.
   b. State practices that can decrease your chance of acquiring MRSA.

METABOLIC PATHWAYS

3. The survival and growth of any microorganism in a given environment depend on its metabolic characteristics.

   a. Infer reasons why both microbial density and microbial diversity are so much higher in the human gastrointestinal tract than on skin.
   b. Use evidence to support the role of *speG* in promoting the survival and growth of *S. aureus* on skin.

4. The growth of microorganisms can be controlled by physical, chemical, mechanical, or biological means.

   a. Describe how hosts can restrict microbial growth by using nutritional immunity.
   b. State how methicillin works as an antibiotic against *S. aureus*.

INFORMATION FLOW AND GENETICS

5. Genetic variations can impact microbial functions (e.g., in biofilm formation, pathogenicity, and drug resistance).

a. Explain why *S. aureus* isolated from nasal passages of healthy people rarely encodes PVL toxin but not all *S. aureus* (MRSA) strains carry the PVL toxin.

b. Outline how some *S. aureus* variants are capable of overcoming nutritional immunity.

c. Use MRSA examples to support the statement "Genetic variations can impact microbial functions."

MICROBIAL SYSTEMS

6. Microorganisms, cellular and viral, can interact with both human and nonhuman hosts in beneficial, neutral, or detrimental ways.

a. Explain the term "opportunistic pathogen."

b. State how interactions between *S. aureus* and the host help to create abscesses during *S. aureus* (MRSA) infections.

IMPACT OF MICROORGANISMS

7. Microbes are essential for life as we know it and the processes that support life (e.g., in biogeochemical cycles and plant and/or animal microbiota).

a. Use microbiome examples to justify the statement "Microbes are essential for life as we know it and the processes that support life."

SUPPLEMENTAL MATERIAL

References

- **Kazakova SV et al.** 2005. A clone of methicillin-resistant *Staphylococcus aureus* among professional football players. *N Engl J Med* **325:**468–475. http://www.nejm.org/doi/full/10.1056/NEJMoa042859

- **Cheng AG et al**. 2011. A play in four acts: *Staphylococcus aureus* abscess formation. *Trends Microbiol* **19:**225–232. http://www.ncbi.nlm.nih.gov/pmc/articles/PMC3087859/

- **Spaan AN et al.** 2013. Neutrophils versus *Staphylococcus aureus*: a biological tug of war. *Ann Rev Microbiol* **67:**629–650. http://www.annualreviews.org/doi/full/10.1146/annurev-micro-092412-155746?url_ver=Z39.88-2003&rfr_id=ori%3Arid%3Acrossref.org&rfr_dat=cr_pub=pubmed&

- **Thammavongsa V et al.** 2015. Staphylococcal manipulation of host immune responses. *Nat Rev Microbiol* **13:**529–543. http://www.ncbi.nlm.nih.gov/pmc/articles/PMC4625792/

- **DuMont AL, Torres VL.** 2014. Cell targeting by the *Staphylococcus aureus* pore-forming toxins: it's not just about lipids. *Trends Microbiol* **22:**21–27. http://www.ncbi.nlm.nih.gov/pmc/articles/PMC3929396/

- **Hood MI, Skaar EP.** 2012. Nutritional immunity: transition metals at the pathogen-host interface. *Nat Rev Microbiol* **10:**525-537. http://www.nature.com/nrmicro/journal/v10/n8/full/nrmicro2836.html

Supplemental Activities

CHECK YOUR UNDERSTANDING

1. **Based on the myriad anti-immunity factors expressed by *S. aureus* (Table 23.2), which host factors provide the first line of defense during *S. aureus* infection?**

a. complement

b. antibodies

c. neutrophils

d. macrophages

2. **Competitive binding for host nutrients is a mechanism employed by which of the following *S. aureus* factors?**

a. Fur regulator

b. siderophores

c. catalase

d. coagulase

3. **Which of the following is NOT an example of nutritional immunity?**
   a. macrophages pumping ferrous iron (Fe^{2+}) out of their phagosomes
   b. human serum proteins binding hemoglobins and free heme released from lysed red blood cells
   c. *S. aureus* transcriptional repressor protein Fur sensing iron-poor conditions
   d. lactoferrin sequestering free ferric iron (Fe^{3+}) on mucosal surfaces

DIG DEEPER

1. Study Fig. 23.4. Outline the basic steps in pulsed-field gel electrophoresis (PFGE) and PCR assays. Is USA100 more closely similar to USA200 or USA300 (the strain isolated from the St. Louis Rams)? Design an experiment to determine which strains have the *mec* cassette. What positive and negative controls do you need?

2. *S. aureus* frequently colonizes humans, but not mice or rabbits. Use the PubMed search engine (http://www.ncbi.nlm.nih.gov/pubmed) to identify a research report that details how a specific virulence factor contributes to this tropism. Discuss each of the virulence factors identified, including their molecular mechanisms. How does this species specificity impact experimental research? What experimental strategies could overcome this barrier to research?

3. Read Vitko NB et al. ("Glycolytic dependency of high-level nitric oxide resistance and virulence in *Staphylococcus aureus*," *mBio* 6(2)e00045-15, 2015, http://mbio.asm.org /content/6/2/e00045-15). Activated phagocytes produce nitric oxide (NO), a broad-spectrum antimicrobial molecule that inactivates enzymes that require metal cofactors, including those of the respiratory chain. Unlike most microbes, *S. aureus* can tolerate high levels of NO. To do so, this opportunistic pathogen requires lactate dehydrogenase enzymes. Study Fig. 4 of the article (below) to answer the following questions:

   - What is L-NIL and why was it included in this experiment?
   - What was the benefit of also analyzing *S. haemolyticus* and *S. saprophyticus*?
   - How do lactate dehydrogenase enzymes confer NO resistance to *S. aureus*?
   - How do these enzymes increase the versatility of *S. aureus*?
   - Draw a schematic to illustrate how *ldh1*, *ldh2*, and *ddh* contribute to the fitness of *S. aureus*.

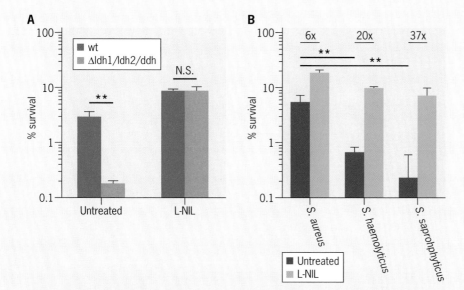

SUPPLEMENTAL MATERIAL QUESTION 3 Lactate fermentation enables *S. aureus* NO⁻· resistance during phagocytosis. (A) Percent survival of wild-type *S. aureus* COL and *Δldh1Δldh2Δddh* mutant in activated RAW 264.7 macrophages with L-NIL (100 nM) or without L-NIL at 20 h postinoculation (MOI = 10:1) (*n*=3; error bars show pooled SD). (B) Percent survival of wild-type *S. aureus* COL (SA), *S. haemolyticus* (SH), and *S. saprophyticus* (SS) in activated RAW 264.7 macrophages with or without L-NIL (100 nM) at 20 h postinoculation (MOI = 10:1) ($3 \leq n \leq 4$). The fold effect of L-NIL treatment on bacterial survival is indicated above the bars. Statistical significance was assessed using a Student's two-sided *t* test (**, $P \leq 0.01$; N.S., not significant).

4. **Team work:** Iron plays a central role in bacterial physiology and pathogenesis. How might this knowledge be harnessed to treat certain bacterial infections? Begin by outlining key features of this host-pathogen competition. Which component(s) are good candidates to manipulate therapeutically? Why?

5. **Team work:** Select an opportunistic pathogen from Table 23.3. Prepare a 10-minute presentation to teach classmates about this microbe, including the following:

    - Encounter
    - Entry
    - Becoming established
    - Causing damage
    - Prevention and treatment

6. "This Week in Microbiology" podcast episode #101 discusses the *mBio* 2014 paper by Alam et al. "Transmission and microevolution of USA300 MRSA in U.S. households: evidence from whole-genome sequencing." Listen to the episode (http://www.microbeworld .org/podcasts/this-week-in-microbiology/archives/1877-twim-101-the-mrsa-in-your -home), read the paper (http://mbio.asm.org/content/6/2/e00054-15), and write a newspaper-style article to share this research with the general public. What is the major finding? Why is it important? What questions should be investigated next? What advice can you offer?

7. **Team work:** Swine workers are ~six times more likely to carry multidrug-resistant *S. aureus* than those without swine exposure (Wardyn et al., *J Clin Infect Dis* **61**:59, 2015; Smith TC, *PLoS Pathog* 11(2): e1004564, 2015). These and other livestock-associated MRSA (LA-MRSA) zoonotic infections have been reported worldwide, with the majority being skin and soft tissue infections. It is currently not known what risk LA-MRSA poses to consumers.

    - What farming, processing, and kitchen practices may increase risk to consumers?
    - How would you design a study to test whether one of these practices increases the risk of LA-MRSA infection? Would you focus on the population at large or a specific sub-population of people?
    - What methods would you use to determine if an *S. aureus* strain identified on a commercial farm was the same or different from one isolated from packaged meat?
    - What, if any, surveillance mechanisms would you recommend that public health departments put in place?
    - What group(s) should bear the added costs?

Supplemental Resources

SPOKEN

- *Small Things Considered* blog post "A Mouthful of Microbes" by Gemma Reguera: http://schaechter.asmblog.org/schaechter/2014/03/a-mouthful-of-microbes.html
- *This Week in Microbiology* podcast episode #107 "The Battle in Your Bladder" considers the paper by Shields-Cutler RR et al. ("Human urinary composition controls antibacterial activity of siderocalin," *J Biol Chem.* **290**:15949-15960, 2015) and discusses a host factor that sequesters iron from urinary tract pathogens: http://www.microbeworld.org /podcasts/this-week-in-microbiology/archives/1942-twim-107-the-battle-in-your -bladder
- *This Week in Microbiology* podcast episode #113 "Waves of Change: Emerging Technoloties for Antimicrobial Susceptibility Testing, Sequencing and Advanced Molecular Diagnostics" discusses innovative molecular techniques for diagnosing infectious diseases: http://www.microbeworld.org/podcasts/this-week-in-microbiology/archives/2012 -twim-113-waves-of-change

WRITTEN

- CDC Fact Sheet "Methicillin-Resistant *Staphylococcus aureus*" http://www.cdc.gov/mrsa /community/
- American Academy of Microbiology FAQ "The Threat of MRSA": Experts educate the general public about MRSA infections with this booklet. http://academy.asm.org/index .php/faq-series/5292-faqmrsa

- National Center for Case Study Teaching in Science "Antibiotic Resistance: Can We Ever Win" by Maureen Leonard builds from a published report on restoring antibiotic sensitivity to MRSA and incorporates quantitative reasoning. http://sciencecases.lib.buffalo.edu/cs/collection/detail.asp?case_id=666&id=666

IN THIS CHAPTER

Listeria *colonies on a petri plate*

CHAPTER TWENTY-FOUR

Intracellular Pathogens:
Listeria and *Mycobacterium*

KEY CONCEPTS

This chapter covers the following topics in the ASM Fundamental Statements.

EVOLUTION

1. Human impact on the environment influences the evolution of microorganisms (e.g., emerging diseases and the selection of antibiotic resistance).

CELL STRUCTURE AND FUNCTION

2. *Bacteria* and *Archaea* have specialized structures (e.g., flagella, endospores, and pili) that often confer critical capabilities.

INFORMATION FLOW AND GENETICS

3. The regulation of gene expression is influenced by external and internal molecular cues and/or signals.

MICROBIAL SYSTEMS

4. Microorganisms, cellular and viral, can interact with both human and non-human hosts in beneficial, neutral, or detrimental ways.

IMPACT OF MICROORGANISMS

5. Microorganisms provide essential models that give us fundamental knowledge about life processes.

INTRODUCTION

To protect our tissues from the multitude of microbes that colonize our bodies, inhabit the soil, water, and air, and hitch a ride on the food we eat, physical boundaries are formed by tight layers of epithelium. Not only that, these borders are patrolled by professional phagocytes that efficiently eat and kill microbes. On the other hand, as chapter 23 illustrated, host barriers also exert selective pressure on microbes, favoring the emergence of variants that can evade host defenses. In this chapter, two bacterial pathogens that thrive inside host cells—*Listeria monocytogenes* and *Mycobacterium tuberculosis*—will illustrate key **cell-mediated defenses** that intracellular microbes must overcome to establish infection.

CASE 1: Listeriosis outbreak triggers international recall of frozen dairy desserts

While conducting routine screens of ice cream products received at a distribution center, South Carolina's Department of Health and Environmental Control detected *L. monocytogenes*, a known cause of foodborne illness. Alerted to this threat, health departments in Texas inspected the ice cream maker's local production facility, where tests again yielded *L. monocytogenes*. After isolating bacterial strains from contaminated desserts found in both states and then analyzing their DNA, microbiologists submitted their data to PulseNet, a national network of government laboratories that manage a database of microbial pulsed-field gel electrophoresis (PFGE; chapter 23) profiles and whole-genome sequences. Meanwhile, extensive recalls were announced by the ice cream manufacturer because the same production lines made the ice cream, frozen yogurt, sherbet, and frozen snacks sold in many countries and 23 states. Working with the PulseNet database, epidemiologists ultimately identified 10 patients hospitalized over a 5-year period in Arizona, Kansas, Oklahoma, and Texas who had severe systemic *L. monocytogenes* infections (Fig. 24.1), three of whom died. Five of these patients had been hospitalized before contracting **systemic listeriosis** infections. In each case, the genome profiles of the disease isolates matched those identified at the frozen dessert production facilities in Texas and Oklahoma.

- Where did the individuals encounter *L. monocytogenes*?
- Is it significant that half of the cases occurred in a hospital?
- How did investigators identify the source of this multistate outbreak of listeriosis?
- How did ingested *L. monocytogenes* spread beyond the gastrointestinal tract?

WHERE DO PEOPLE ENCOUNTER *L. MONOCYTOGENES*?

LEARNING OUTCOMES:
- Describe where humans encounter *Listeria monocytogenes*.
- Name populations that are more susceptible to *L. monocytogenes* infection.

Despite its well-earned reputation as a human pathogen, *L. monocytogenes* has humble roots. This Gram positive bacterium lives in the soil, where it feeds on dead and decaying matter—the diet of **saprophytes.** In the environment, *L. monocytogenes* can infect a wide variety of crea-

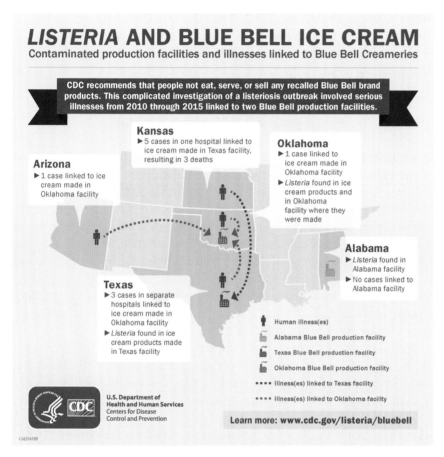

FIGURE 24.1 **A multi-year, multistate outbreak of listeriosis occurred when commercial facilities that produce frozen desserts were contaminated by the pathogen** *Listeria monocytogenes.* Learn more at www.cdc.gov/listeria/bluebell

tures, including protists, arthropods, and cold- and warm-blooded animals. Indeed, "monocytogenes" emphasizes the dramatic accumulation of monocytes in the blood of infected rabbits. ("Listeria" honors Joseph Lister, the microbiologist whose pioneering work to sterilize open wounds and medical instruments revolutionized surgery in the late 1800s.) Humans are exposed by ingesting contaminated food products. The most common culprits are processed meat and milk products. Unfortunately, this versatile microbe tolerates refrigeration and freezing, high salt, and low moisture—conditions typically used to preserve many commercial foods. Although healthy children and adults who consume *L. monocytogenes* rarely experience more than diarrhea, other populations are at high risk. Fatal sepsis, meningitis, or encephalitis can afflict adults with weakened cell-mediated immune systems, such as the elderly or people with HIV, cancer, or organ transplants requiring immunosuppressive therapy. Pregnant women may develop mild, flu-like disease, but they are in danger of a miscarriage, premature or stillbirth, or severe neonatal infection. To protect the public from rare, but life-threatening, foodborne listeriosis, the U.S. Department of Agriculture produced a video to teach what types of foods should be avoided during pregnancy (see Supplemental Resources, Visual).

HOW CAN *L. MONOCYTOGENES* BE TRACED IN THE FOOD SUPPLY?

LEARNING OUTCOMES:
- Explain how the Centers for Disease Control and Prevention traces *Listeria* outbreaks in the complex and global food supply.
- List reasons why it is important to track *Listeria* outbreaks.

In a typical year in the United States, the Centers for Disease Control and Prevention (CDC) estimates that approximately one of every six people becomes ill from foodborne pathogens. When family and friends share a homemade meal that goes awry, even amateurs can play at being epidemiologists by quickly comparing notes to zero in on the offender. However, as the 5-year, multistate outbreak of listeriosis illustrates (Fig. 24.1), the increasingly complex and global food supply poses substantial challenges to public health agencies. Industrial food production and distribution chains broadly disperse foods prepared from multiple ingredients obtained from a range of sources. To complicate matters further, some processed foods are stored for weeks or months before sale or consumption.

Technology and public education must save the day. In the United States, the CDC is one source for current information on infectious diseases, including outbreaks. For example, trends in foodborne illnesses are tracked through their Foodborne Diseases Active Surveillance Network, FoodNet (http://www.cdc.gov/foodnet/index.html). You can also find out what pathogens have caused multistate foodborne illness outbreaks each year (http://www.cdc.gov/foodsafety/outbreaks/multistate-outbreaks/outbreaks-list.html), or learn more about the offending microbes (e.g., http://www.cdc.gov/listeria/index.html). Check it out!

Fortunately, the major foodborne pathogens can now be rapidly "fingerprinted," even to the level of single nucleotides. By coupling classical microbiological techniques with modern molecular biology, image analysis software, electronic databases, and networks that connect states and countries, genomic profiles of disease isolates are efficiently shared and compared (http://www.cdc.gov/pulsenet/about/works.html). In the United States, each state has at least one laboratory that forwards molecular epidemiological data to PulseNet, another network managed by the CDC. In addition, infectious disease experts in more than 80 countries are connected through PulseNet International. Since establishment of these networks in 1996, more than 500,000 bacterial isolates have been catalogued—a rich resource to track ongoing outbreaks and conduct retrospective studies of infectious diseases. These public surveillance programs have four goals: to investigate and control local disease; to assess disease trends and control measures across populations; to detect outbreaks; and to increase basic knowledge of pathogens and their spread (the subject we turn to next).

HOW DOES *L. MONOCYTOGENES* ALTERNATE BETWEEN THE ENVIRONMENT AND HOST CELLS?

LEARNING OUTCOMES:
- Use examples from the lifestyle of *L. monocytogenes* to justify the statement "Microbes are adept at tailoring their protein profile to suit their surroundings."

When ingested via contaminated food, *L. monocytogenes* can colonize the digestive track. In most healthy people, the microbe will be shed without incident. However, as we saw in Case 1, *L. monocytogenes* can threaten the lives of the elderly, immunocompromised, or neonates. How does this pathogen make its way from the gastrointestinal tract into the blood, brain, or fetus? First, the pathogen must cross the intestinal barrier. After reaching nearby lymph nodes, the microbe can travel through the lymphatic circulation system and gain access to other tissues. Nevertheless, 3 to 4 weeks may pass before symptoms develop. (You can imagine how such a long incubation period complicates epidemiological tracking of the source of contamination!) If *L. monocytogenes* enters the bloodstream, septicemia develops. When pathogens breach the blood-brain barrier, they can bring about meningitis or encephalitis. If *L. monocytogenes* is ingested during pregnancy, the bacteria may cross the placental barrier and cause miscarriage or neonatal infection. As we'll soon see, this pathogen is armed to invade nonphagocytic cells and establish a replication niche in their cytoplasm.

But first, are you impressed by the versatility of *L. monocytogenes*? This Gram-positive bacillus can thrive in soil and infect protists, arthropods, and cold- and warm-blooded mammals and can breach intestinal, placental, and blood-brain barriers. Clearly, the pathogen carries an impressive toolkit. But how does it know *when* to deploy virulence factors that promote entry, replication, and spread in the host? As we learned in chapter 11, microbes are adept at tailoring their protein profile to suit their surroundings.

One strategy that *L. monocytogenes* deploys is to choose which genes are expressed. After entering host cells, the transcription factor PrfA increases the expression of a panel of virulence factors that equip *L. monocytogenes* to take up residence in the cytosol. In the next section we'll learn about a few of them—the **toxin** listeriolysin O, phospholipases, and an actin-recruiting protein ActA. While growing freely in the soil, *L. monocytogenes* doesn't need to make these proteins, and doesn't. Economical!

How the pathogen can tell that it is within a host cell is a topic of much research. One exciting discovery is that *L. monocytogenes* not only *secretes* but also *imports* a small lipoprotein (pPplA, for pheromone-encoding lipoprotein A). Does this pattern ring a bell? In fact, the secreted factor resembles peptide **pheromones**, those signaling molecules that coordinate activities of other Gram-positive bacteria (chapter 19). *L. monocytogenes* doesn't need this peptide to grow extracellularly. However, when mutants that lack this lipoprotein are internalized by host cells, the bacteria are slow to establish a replication niche. On the other hand, if the bacteria also encode a PrfA transcription factor that is engineered to be active constantly, they don't need the peptide to thrive in host cells. From this experimental data, can you develop a hypothesis to explain how *L. monocytogenes* might exploit this peptide pheromone to ensure its virulence factors are abundant in host cells but not in food? (*Quiz spoiler:* By some mechanism, the peptide triggers PrfA to express the virulence factors that equip this pathogen to establish a replication niche inside host cells.)

HOW DOES *L. MONOCYTOGENES* CAUSE INVASIVE DISEASE?

LEARNING OUTCOMES:
- Discuss how specialized structures and toxins allow *L. monocytogenes* to enter and be maintained in a cell.
- State the benefit of actin-based motility for *L. monocytogenes*.
- Identify key eukaryotic cellular processes that have been elucidated by studying intracellular pathogens.

How does *L. monocytogenes* cross the epithelial barriers and evade the phagocytes that guard vital tissues like blood, the brain, or a developing fetus? Entry is the first challenge. *L. monocytogenes* displays on its surface two proteins that coerce epithelial cells to take them in. Aptly named **internalins A and B**, these two virulence factors bind to particular proteins on the surface of epithelial cells, triggering a signal transduction cascade that rearranges the host's actin cytoskeleton. A **phagosome** forms, pulling the microbe inside (Fig. 24.2, step 1). At the intestinal boundary, internalin A initiates invasion by the bacterium; internalin B induces its transport across the placental barrier. These are impressive feats, since epithelial cells normally are not phagocytic. Once internalized within a phagosome, most microbes are efficiently delivered to acidic **lysosomes** for digestion (chapter 22, Fig. 22.8), but not this pathogen!

L. monocytogenes can escape into the more hospitable cytoplasm by poking holes in its phagosome (Fig. 24.2, step 2). To do so, the bacteria secrete a pore-forming toxin, plus enzymes that digest host phospholipids (phospholipases). The toxin, **listeriolysin O**, is actually assembled from dozens of monomers in vacuoles with an acidic pH (Fig. 24.3). First, each monomer binds to the cholesterol-rich membranes of its host vacuole. There the subunits gradually assemble into linear oligomers that undergo coordinated conformational changes sufficient to drive the protein complex into the hydrophobic membrane. Initially, ions pass freely; ultimately, the vacuole ruptures. Structurally similar pore-forming toxins, collectively known as **cholesterol-dependent cytolysins**, are deployed by a variety of other pathogens, including *Clostridium perfringens*, *Streptococcus pneumoniae*, and *Streptococcus pyogenes* (Table 24.1). Each toxin is capable of forming pores in membranes that contain cholesterol, but the outcome depends on the host cell type, the toxin concentration, and the presence of other microbial components.

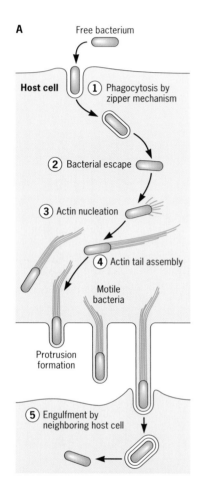

A

Free bacterium

Host cell

(1) Phagocytosis by zipper mechanism

(2) Bacterial escape

(3) Actin nucleation

(4) Actin tail assembly

Motile bacteria

Protrusion formation

(5) Engulfment by neighboring host cell

B

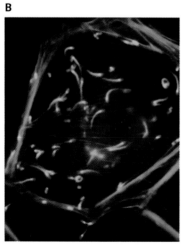

FIGURE 24.2 (A) **The life cycle of intracellular *L. monocytogenes*, as described in the text.** (B) Actin tails assembled at one pole of *L. monocytogenes* propel the pathogen through the cytoplasm and into adjacent cells. The fluorescence micrograph shows actin filaments (green) behind each bacterium (red); yellow appears where the green actin and the red bacteria overlap. (http://www.microbeworld.org/component/jlibrary/?view=article&id=7627)

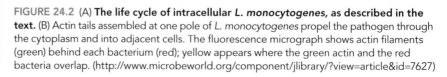

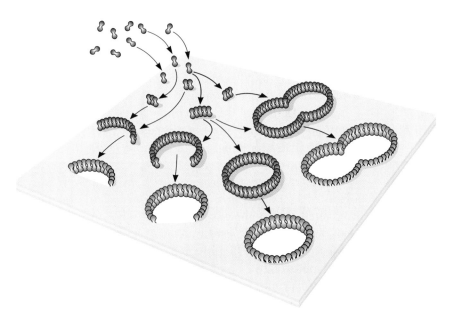

FIGURE 24.3 Assembly of the pore-forming toxin listeriolysin O. Monomers of the protein first bind to host membranes and form linear oligomers (blue). Next, conformational changes drive insertion of the oligomer into cholesterol-rich membranes, forming pores that eventually trigger lysis of the phagosomal vacuole (vacuole (green). (Adapted from Leung et al., *eLife* **3**:e04247, 2014, http://elifesciences.org/content/elife/3/e04247.full.pdf.)

Once liberated into the cytoplasm of an epithelial cell or **macrophage**, *L. monocytogenes* begins to multiply. The cytoplasmic bacteria are free to scavenge host nutrients, while being shielded from professional phagocytes, **complement**, and **antibodies** (chapter 22). However, the progeny face another problem: One host cell can supply only so much in the way of nutrients. The pathogen's solution is ingenious, as you can see for yourself with Fig. 24.4! *L. monocytogenes* spreads into neighboring cells, once again by manipulating the host's actin cytoskeleton. The pathogen displays on its surface the virulence factor ActA, which stimulates actin polymerization (Fig. 24.2, step 3). Soon, short actin filaments accumulate adjacent to one pole of the bacterial cell. As the bacillus jiggles in perpetual Brown-

TABLE 24.1 A variety of bacteria produce cholesterol-dependent cytolysin toxins

Species	Toxin	Effects[a]
Bacillus anthracis	Anthrolysin O	Disrupts epithelial barriers
Clostridium botulinum	Botulinolysin	Inhibits relaxation of blood vessel endothelium
Clostridium perfringens	Perfringolysin O	Inhibits migration of leukocytes
Clostridium tetani	Tetanolysin	Damages local tissue
Listeria monocytogenes	Listeriolysin O	Lyses phagosome
Streptococcus pneumolysin	Pneumolysin	Induces inflammation and apoptosis
Streptococcus pyogenes	Streptolysin O	Induces apoptosis of phagocytes and epithelial cells

[a]Some major effects are listed, but the impact of pore-forming cytotoxins can vary depending on the target cell type, toxin concentration, and the presence of other microbial products.

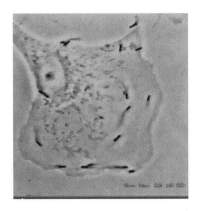

FIGURE 24.4 Still image from a movie of *L. monocytogenes* moving by actin-based motility in an epithelial cell. The movie speed is 150 × real time and was made by Julie Theriot and Dan Portnoy. (http://cmgm.stanford.edu/theriot /movies/seq1LmPtK2.mov)

ian motion, any gaps are immediately stuffed with new actin filaments, ensuring the microbe's forward progress (Fig. 24.2, step 4). Although *L. monocytogenes* appears to be riding on a rocket tail of actin, looks can be deceiving. Its actin tail is actually composed of stationary short filaments that continually assemble at the bacterial pole, while disassembling at the distal tip. That means the length of the actin tail is actually dictated by the half-life of the actin filaments in the tail. Actin-based motility does more than delight microbiologists. Seemingly at random, some bacilli are propelled out of one cell and into another (Fig. 24.2, step 5). After entering a neighboring cell, *L. monocytogenes* is wrapped in a *double layer of plasma membranes*, courtesy of its old and new host cells. Once again, the pathogen relies on its secreted phospholipases and listeriolysin O toxin to escape into the cytoplasm, where the cycle repeats. After *L. monocytogenes* breaches the epithelial barrier of a person with impaired cell-mediated immunity, it can spread via the bloodstream or lymphatic system to other organs, including the brain and placenta, and cause life-threatening disease, as we saw in Case 1.

Remarkably, although a variety of intracellular microbes are propelled by actin tails, including *Shigella flexnerii*, *Rickettsia riskettsii*, *Burkholderia pseudomallei*, and *Mycobacterium marinum*, each species deploys a distinct mechanism to recruit host actin, indicating **convergent evolution**. Nevertheless, the structure and dynamics of the actin filaments stimulated by each of these pathogens are quite similar to the cytoskeletal mechanics that propel migrating leukocytes and tumor cells. As a result, *L. monocytogenes* and other genetically tractable microbes are excellent experimental systems to study fundamental questions in eukaryotic cell biology. Indeed, the microbiologist Stanley Falkow once observed, "The examination of the cell biology of microbial invasion reveals not so much a belligerent event, but rather a view that many of the bacteria we associate with disease are actually quite remarkable cell biologists who possess admirable insight into the fine points of eukaryotic life."

Inflammatory cells

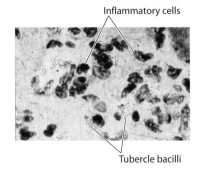

Tubercle bacilli

FIGURE 24.5 Acid-fast stain of a sputum sample reveals leukocytes (purple) and acid-fast *M. tuberculosis* bacilli (red). (ASM Microbe Library, © G. Delisle and L. Tomalty, http://www.microbelibrary.org.)

CASE 2: *Mycobacterium tuberculosis* **infection, an outcome determined by the host response**

On a snowy winter evening, a 32-year-old man living on the streets came to a walk-in clinic in Boston complaining of a cough that had bothered him for several months, plus fever and night sweats. He appeared chronically malnourished and had a temperature of 102°F. Examination of his chest as he breathed deeply revealed crackly noises, indicating fluid in his air sacs, a sign of pneumonia. After getting a chest X-ray, a blood test for human immunodeficiency virus (HIV), and depositing a sputum sample, he left abruptly and spent the night in an abandoned building he shared with several friends. Laboratory examination of the sputum revealed acid-fast bacilli, some within leukocytes, a profile consistent with *M. tuberculosis* (Fig. 24.5). The X-rays added further credence to the diagnosis of tuberculosis (TB). The patient was HIV negative. When he returned to the clinic 4 months later with similar complaints, he was given a cocktail of several antibiotics with the stern and explicit advice that it was essential that he not skip a single dose. This treatment was to last for 9 months. The personnel at the clinic were concerned that the patient would have difficulty complying with this regimen.

- Where do people encounter *M. tuberculosis*?
- How is TB diagnosed?
- Why do TB patients experience fever, night sweats, and weight loss?
- Why was this patient tested for HIV?
- Why are multiple antibiotics prescribed at once and for such a long period?
- Why are health care workers concerned that patients complete their full antibiotic treatment?

WHAT IS THE HISTORY OF TUBERCULOSIS IN HUMAN POPULATIONS?

LEARNING OUTCOMES:
- Describe why declines in tuberculosis cases in the United States can be credited to increases in successful use of HIV drugs.
- Discuss how humans have influenced the emergence and evolution of drug-resistant mutant strains of *Mycobacterium tuberculosis*.

TB conjures up the image of a severe lung disease that can be fatal unless treated. Movie-goers may remember Nicole Kidman's Satine in *Moulin Rouge*, some book lovers will be reminded of Thomas Mann's *Magic Mountain* and Emily Brontë's *Wuthering Heights,* and opera fans will recall the dying heroines in *La Traviata* and *La Bohème* ("Che gelida manina . . ." ["What an icy little hand . . ."]). Fossil records indicate early humans carried *M. tuberculosis* when they migrated out of Africa ~70,000 years ago. In the 17th and 18th centuries in Europe, *M. tuberculosis* caused the "White Plague" (in contrast to the Black Plague, spread by the plague bacillus *Yersinia pestis*; chapter 27). During that period, virtually everyone in Europe was infected, and an astonishing 25% of all adult deaths were due to TB. It is no wonder this ancient disease has left its mark on literature, theater, and the arts!

Although now relatively infrequent in developed countries, TB remains the second leading cause of death from an infectious disease worldwide. In the United States, the number of TB cases dropped almost every year since records were kept, but in the late 1980s, the number of cases rose somewhat, to fall again beginning around 1992 (Fig. 24.6). The rise has been largely attributed to the increase in the number of homeless persons (as in Case 2). Globally, greater severity and incidence of TB cases have coincided with the emergence of HIV infections, which impair cell-mediated immune defenses. The impact of HIV on TB incidence is captured by Worldmapper, which resizes geographical territory according to data of interest, in this case the number of people with TB and HIV (Fig. 24.7). The recent decline in TB in the United States may be credited in part to the more successful use of anti-HIV drugs. In this country today, foreign-born people account for about two-thirds of the TB cases.

In recent times, multidrug resistance has emerged among strains of *M. tuberculosis*, making them particularly threatening. A big reason for

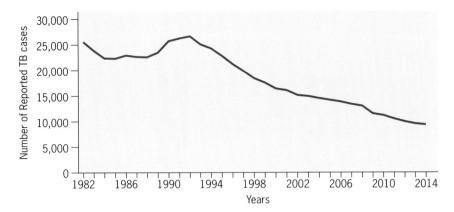

FIGURE 24.6 The number of cases of tuberculosis in the United States is on the decline after a rise during the late 1980s attributed to an increase in HIV infections and homelessness. (http://www.cdc.gov/tb/publications/factsheets/statistics/TBTrends.htm)

drug resistance is noncompliance with long-term antibiotic treatment, as in the case described, which leads to the selection of drug-resistant mutant strains. Taking a combination of multiple antibiotics with different mechanisms of action increases the likelihood that all mycobacterial cells will

Tuberculosis Cases 2003

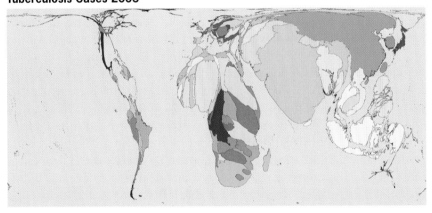

HIV Cases 2003

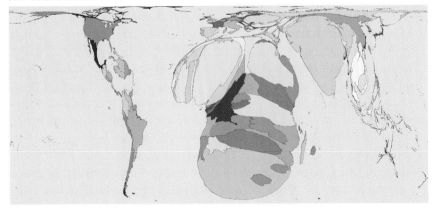

FIGURE 24.7 Worldmapper resizes geographic territory according to the subject of interest. In the upper panel, territory size shows the proportion of worldwide tuberculosis cases found there. In the lower panel, territory size shows the proportion of all people aged 15 to 49 with HIV worldwide, living there. (© Sasi Group, University of Sheffield, and Mark Newman, University of Michigan, http://www.worldmapper.org/index.html.)

be killed. In the United States in 2014, 88% of cases of multidrug-resistant TB occurred among people who acquired it abroad.

WHAT HAPPENS WHEN PEOPLE ENCOUNTER *M. TUBERCULOSIS?*

LEARNING OUTCOMES:
- Identify the factors that contribute to primary and secondary TB infection.

When people are crowded together, as in poor housing, jails, or hospitals, the likelihood of contracting TB increases. It is easy to understand why; the bacillus spreads in airborne respiratory secretions, although the number of tubercle bacilli required to cause disease is actually quite high. It seems plausible that the homeless man in Case 2 inhaled a sufficient inoculum from the persons with whom he shared shelter. Once he developed severe pneumonia and coughed sputum that contained bacilli, he could spread the infectious microbe to others. Note that the ability of the tubercle bacillus to induce cough gives it a selective evolutionary advantage by promoting its transmission between hosts.

Pulmonary TB typically has two stages, which is a crucial point in understanding the disease. In the first stage (**primary TB**), exposure to the tubercle bacilli typically leads to a mild, self-limiting disease that may be totally imperceptible or as mild as a cold. Most healthy people won't have symptoms and are not infectious. However, if their immune defenses are impaired, noticeably or imperceptibly, a much more serious disease emerges: **secondary TB**. In this infectious stage, people experience the symptoms that are classically associated with the deadly image of TB—persistent cough, fever, and emaciation. Between the two stages, the tubercle bacilli were successfully restrained by the immune system and were little noticed by the host. The time span between the primary and the secondary stages ranges from months to many years, depending on the infectious dose and the strength of the host's cell-mediated immune response (chapter 22, Fig. 22.3). In fact, TB is not a single disease but has many manifestations, depending on previous exposure, nutrition, and other health factors.

WHAT FEATURES OF *M. TUBERCULOSIS* CONTRIBUTE TO ITS SPREAD?

LEARNING OUTCOMES:
- List characteristics of *Mycobacterium tuberculosis* that contribute to high infectivity.

Tubercle bacilli have several distinctive characteristics that help explain the disease process. *M. tuberculosis* cells grow very slowly, doubling once about every 14 hours. Since colonies take several weeks to form, laboratory diagnosis by traditional culture methods is not ideal. Furthermore, the treatment course must be prolonged because most antibiotics are active against replicating bacteria. The tubercle bacilli are also **acid-fast** (see chapter 2), meaning that once they are stained, they retain the dyes even

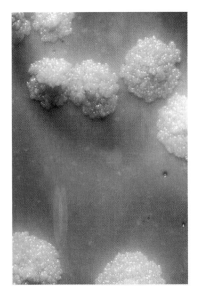

FIGURE 24.8 Production of long-chain mycolic acids by *M. tuberculosis* gives their colonies a waxy appearance. (George Kubica, CDC, http://www.microbeworld.org /component/jlibrary/?view=article &id=6641)

after treatment with acids. Acid-fastness, an uncommon property among bacteria, is related to their unusual resistance to harsh chemicals. Acid-fast organisms are surrounded by a waxy outer membrane (see "The acid-fast solution" in chapter 2) that makes them impermeable to many polar molecules, including the germicides typically used to sanitize hospital surfaces.

To be sure, mycobacteria are singularly preoccupied with the metabolism of lipids, both their synthesis and utilization. A disproportionately large number of their genes are dedicated to this metabolism. To drive this point home, take a look at the distinctive colonies *M. tuberculosis* forms on agar, which look like yellowish lumps of wax (Fig. 24.8). Linked to the mycobacterial cell wall are long-chain waxes about 80 carbons long, called **mycolic acids** (Fig. 24.9; chapter 2). This unusual outer membrane makes tubercle bacilli resistant to drying. In advanced stages of pulmonary TB, patients cough up huge numbers of the organisms, which can persist in the air in aerosols or as "dust" particles. These two properties, causing a pulmonary disease and being resistant to drying, conspire to increase the chances of the microbes being transmitted to other persons. In fact, humans are the only known reservoir of *M. tuberculosis*. Other pathogenic mycobacteria have animal or environmental reservoirs.

WHAT HAPPENS WHEN MACROPHAGES INGEST *M. TUBERCULOSIS*?

LEARNING OUTCOMES:
- Explain the mechanism by which individuals with primary TB are not infectious and TB is readily controlled by healthy individuals.
- Contrast the outcome of macrophage ingestion for *Listeria monocytogenes* versus *Mycobacterium tuberculosis*.

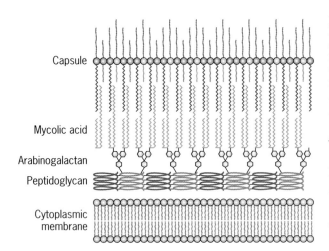

FIGURE 24.9 Structure of the mycobacterial cell wall. The cell wall is composed of covalently linked peptidoglycan (brown), arabinogalactan (green and yellow), and mycolic acids (orange), which together form a fluid membrane-like structure. Intercalated with the mycolic acids are other lipids, many specific to the mycobacteria. The outermost capsule-like layer contains polysaccharides such as glucan and arabinomannan.

Primary TB is readily controlled by most people. Once inhaled, microbes are ingested by macrophages that patrol the lung. However, this pathogen is not killed. Instead *M. tuberculosis* survives and replicates within macrophages by actively preventing fusion of its phagosome with the acidic, digestive lysosomes (Fig. 24.10). Their strategy is different from that of *L. monocytogenes*; the tubercle bacilli *remain in phagosomal vacuoles that fail to acidify or mature.* How *M. tuberculosis* blocks phagosome-lysosome fusion is being actively investigated, and a few themes have emerged. First, over its long course of coevolution with humans, *M. tuberculosis* has acquired multiple factors that interfere with specific components of cell-mediated immune pathways. Second, the pathogen's collection of unusual fatty acids contribute to its survival in humans, both short and long term. For example, in order to establish a persistent long-term infection, the microbe requires a gene that directs the addition of cyclopropane residues to mycolic acids. Furthermore, some

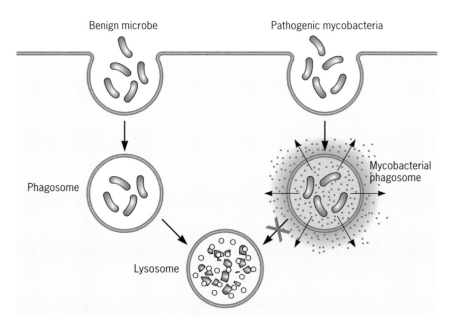

FIGURE 24.10 Unlike benign microbes, *M. tuberculosis* prevents fusion of its phagosome with lysosomes by releasing cell wall lipids and other factors (green).

cell wall lipids released from the *M. tuberculosis* phagosome inhibit the vacuole's fusion with lysosomes. One of these, the glycolipid mannose-capped lipoarabinomannan, is also recognized by the innate immunity's surveillance system. In response to this **microbe-associated molecular pattern (MAMP)**, leukocytes secrete **cytokines** (chapter 22, Fig. 22.9) that coordinate formation of **granulomas** (Fig. 24.11). These structures have walls of fibroblasts and leukocytes that surround and contain—but do not kill—*M. tuberculosis*, which can *persist within lung granulomas for decades*.

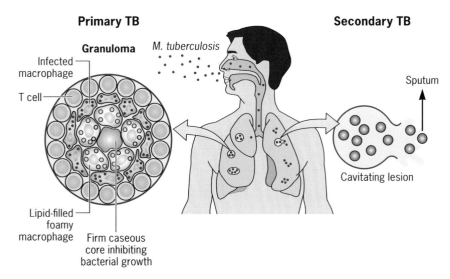

FIGURE 24.11 *M. tuberculosis* **pathogenesis.** After inhaling the pathogen in aerosols, most healthy humans contain the microbe within granulomas for decades. If cell-mediated immunity weakens, the granulomas weaken and rupture, releasing infectious microbes in sputum and respiratory droplets.

Previous exposure of a person to tubercle bacilli can be readily demonstrated by the *tuberculin test.* When a mixture of bacterial antigens, collectively called purified protein derivative or PPD, is introduced into the skin, a cell-mediated immunity reaction is manifested only in persons who harbor or have harbored the organisms. A positive test is recognized as local reddening and hardening of the skin at the site of inoculation. Additional tests are needed to distinguish between a latent TB infection and TB disease. In fact, many people who have had contact with tubercle bacilli in their youth remain tuberculin positive for years. This attests to the long-range survival of the organisms in the body.

WHY DOES *M. TUBERCULOSIS* MAKE SOME PEOPLE SICK, BUT NOT OTHERS?

LEARNING OUTCOMES:
- Explain the mechanism by which persons with defects in the immune system are more susceptible to secondary TB infection.
- State how interactions between *Mycobacterium tuberculosis* and the host help to create granulomas that contain infections.
- Identify the source that accounts for most secondary TB symptoms.

Secondary TB is usually seen in persons with defects in the immune system, even though the defects may be mild or even unrecognized. In the elderly, transplant patients who receive immunosuppressive therapy, or HIV-positive individuals who develop AIDS, the balance between microbe and host tilts in favor of the microbe. Because *cytokines are required to maintain granulomas,* a weakened immune system relaxes the constraints on bacterial replication. But, unlike *L. monocytogenes,* tubercle bacilli *do not make **toxins** or other products* that damage cells. Their slow growth signals no intent to rapidly overwhelm the host—they are merely innocent passengers. However, their very presence is noticed by the immune surveillance system of the host.

In response to a higher microbial load, the body answers with a vigorous cell-mediated response (chapter 22) that ultimately damages tissues. Some of the bacterial products that trigger the response are components of the envelope, such as mycolic acids from the cell wall (Fig. 24.9). Both are MAMPs that bind to receptors on macrophages and stimulate cytokine production (chapter 22, Fig. 22.9). And it is the cytokines stimulated in response to these organisms that account for most of the disease symptoms. For example, *high levels of the cytokine TNFα* cause wasting, a weight loss known as cachexia. (When first discovered, TNFα was named cachectin.) Other cytokines cause fever and night sweats. Damage is also caused by the release of toxic lysosomal components from the macrophages that are trying to kill *M. tuberculosis.* The result is necrosis, or death of host cells. When a granuloma breaks down, it turns into a cheesy-looking material that contains few host cells but many bacteria. In the lung, such a lesion may break through into the airways (Fig. 24.12). When the contents are coughed up, a cavity is left behind (Fig. 24.12). At this stage, the disease is

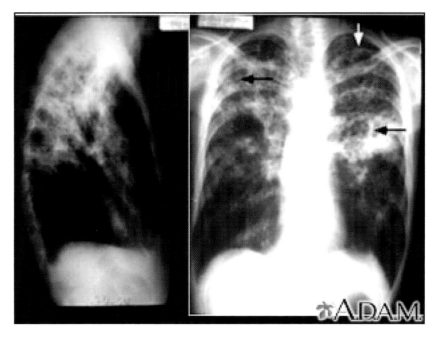

FIGURE 24.12 Chest X-rays of a patient with advanced pulmonary tuberculosis. Regions of inflammation appear as light areas (white arrow on right). Within these areas, cavities can form (black arrows). (https://www.nlm.nih.gov/medlineplus/ency/imagepages /1607.htm)

highly infectious and progresses rapidly. This final phase of disease, marked by severe weight loss, earned TB the name "galloping consumption."

Because the immune response is responsible for so much damage, one may ask if it does more harm than good and whether we would be better off without it. Obviously, *the immune system is capable of both damaging and healing*. A healthy immune system does control TB within granulomas, and people with TB can live with it for many years, even without treatment. Contrast this with TB in an AIDS patient whose cell-mediated immune system is damaged. Here, the latent tubercle bacilli rapidly cause a severe, disseminating, fast-progressing, and life-threatening disease. The choice between a functioning immune system and an impaired one is clear.

WHAT CHALLENGES ARE UNIQUE TO INFECTIONS WITH INTRACELLULAR PATHOGENS?

LEARNING OUTCOMES:
- State reasons why infections caused by intracellular pathogens can be difficult to treat and prevent.

L. monocytogenes and *M. tuberculosis* are just two of many intracellular pathogens that cause disease worldwide. To do so, the pathogens must either avoid or tolerate lysosomes. At one extreme is *Plasmodium falciparum*, a malaria parasite that invades red blood cells, which lack lysosomes and other organelles. At the other extreme is the agent of Q fever, *Coxiella burnetti,* a Gram-negative bacterium that thrives in the acidic lysosomes of

TABLE 24.2 **Some of the microbes that establish a replication niche within host cells**

Replication site	Species	Type
Cytoplasm	Listeria monocytogenes	Gram-positive bacterium
	Mycobacterium marinum	Acid-fast bacterium
	Rickettsia rickettsii	Gram-negative bacterium
	Burkholderia pseudomallei	Gram-negative bacterium
	Shigella flexnerii	Gram-negative bacterium
	Francisella tularensis	Gram-negative bacterium
Vacuole[a]		
Recycling endosome	Mycobacterium tuberculosis	Acid-fast bacterium
Late endosome	Salmonella typhimurium	Gram-negative bacterium
Endoplasmic reticulum	Brucella abortus	Gram-negative bacterium
Endoplasmic reticulum	Legionella pneumophila	Gram-negative bacterium
Golgi	Toxoplasma gondii	Protist
Lysosome	Chlamydia trachomatis	Gram-negative bacterium
	Coxiella burnetti	Gram-negative bacterium
	Histoplasma capsulatum	Fungus
	Leishmania donovani	Protist

[a]Host organelles that interact with the pathogen vacuole.

mammalian cells. Others are escape artists that, like *L. monocytogenes*, force their way into the nutrient-rich cytoplasm. In between is a full spectrum of vacuolar habitats that harbor replicating microbes (Table 24.2).

Pathogens that establish their replication niche within host cells provide distinct challenges to both the host and the biomedical community. First, they are shielded from many antimicrobial factors. Antibiotics must not only cross the bacterial membrane(s), but also the host cell's plasma membrane and, in some cases, a phagosomal membrane. Not all can do that. Complement and antibody are also ineffective, so the host must mount a cell-mediated immune response (chapter 22). Once activated, macrophages can restrict replication of many intracellular microbes, including *L. monocytogenes* and *M. tuberculosis*. Infected cells can also be *detected and killed by cytotoxic T cells* that have been primed to recognize the pathogen's antigens.

What about vaccines? To stimulate the cell-mediated arm of the immune system, the MHCI pathway (chapter 22, Fig. 22.11) first scavenges microbial antigens that contaminate the host cell cytoplasm. But vaccines composed of purified microbial products, or whole killed bacterial cells, are efficiently delivered to lysosomes, a compartment patrolled by the MHCII pathway, which stimulates antibody production (chapter 22, Fig. 22.11). One strategy to recruit the cell-mediated immune pathway is to exploit *live but avirulent microbes* that can survive in macrophage vacuoles, but not establish persistent infection. That's the basis for the TB vaccine named **BCG**, or Bacille de Calmette et Guérin, after the pair of French scientists who developed the therapeutic in the early 1900s. However, the efficacy of the BCG vaccine varies widely, and live attenuated vaccines are not safe for immunocompromised people. BCG also stimulates a positive *M. tuberculosis* PPD/tuberculin skin test, interfering with its use as a diagnostic tool. Therefore, the World Health Organization recommends that BCG vaccine be used only in regions where TB is endemic, because young children are at risk of developing severe, disseminated disease.

CASE 1 REVISITED: **Listeriosis outbreak triggers international recall of frozen dairy desserts**

Listeria monocytogenes is common in the soil. People encounter this opportunistic pathogen when they eat produce or processed foods that have been contaminated during cultivation or processing. Public health departments conduct surviellance on points of food distribution using culturing, genomic profiling, and computer networks that connect states and countries. *L. monocytogenes* causes invasive disease by inducing its own uptake by nonphagocytic epithelial cells. Internalins A and B equip the pathogen to cross gastrointestinal, placental, and blood-brain barriers; listeriolysin O and phospholipases enable the pathogen to lyse its phagosomal vacule to escape lysosomal degradation and replicate in the cytoplasm. By recruiting host actin filaments, *L. monocytogenes* can move within and between cells. Virulence factors that promote replication within cells and spread between cells are expressed when the bacterial transcriptional regulator PrfA senses the phagosomal environment.

CASE 2 REVISITED: ***Mycobacterium tuberculosis* infection, an outcome determined by the host response**

Humans and *M. tuberculosis* share a long and storied history. After primary TB infection, most healthy people silently retain this pathogen within lung granulomas for decades. But microbial products released from their infected macrophages continue to stimulate cell-mediated immunity, which can be detected as a local and mild inflammatory response to the mycobacterial antigens injected under the skin in the diagnostic PPD test. If the cell-mediated immune system is weakened by advanced age, immunosuppressive therapy, or HIV infection, granulomas break down and robust *M. tuberculosis* replication resumes. The vigorous inflammatory response to mycobacterial MAMPs that ensues triggers local fluid accumulation that can be visualized on chest X-rays. Inflammatory cytokines also trigger weight loss and fever, two classic signs of TB disease. Because cell-mediated immunity is required to contain *M. tuberculosis*, untreated HIV infection is a major risk factor for TB disease. To eliminate this slow-growing intracellular pathogen and reduce the odds that antibiotic-resistant variants will emerge, a long course of multiple antibiotics is required. People who lack access to health care and a stable residence are at high risk of developing and spreading antibiotic-resistant *M. tuberculosis*.

CONCLUSIONS

We conclude these descriptions of two intracellular pathogens with the awareness that they are merely examples of the vast number of ways in which humans interact with bacteria, fungi, and other parasites, including protozoa and worms. To consider such vast amounts of information, it helps to keep in mind that all infectious diseases share the same steps in pathogenesis and vary only in detail. In the chapter that follows on herpesviruses, you'll recognize a number of parallels with *M. tuberculosis* infections, as both of these intracellular pathogens establish silent, chronic infections.

Representative ASM Fundamental Statements and Learning Outcomes for the Chapter

EVOLUTION

1. Human impact on the environment influences the evolution of microorganisms (e.g., emerging diseases and the selection of antibiotic resistance).

   a. Discuss how humans have influenced the emergence and evolution of drug-resistant mutant strains of *M. tuberculosis*.

CELL STRUCTURE AND FUNCTION

2. *Bacteria* and *Archaea* have specialized structures (e.g., flagella, endospores, and pili) that often confer critical capabilities.

   a. Discuss how specialized structures and toxins allow *Listeria monocytogenes* to enter and be maintained in a cell.
   b. State the benefit of actin-based motility for *Listeria monocytogenes*.
   c. List characteristics of *Mycobacterium tuberculosis* that contribute to high infectivity.

INFORMATION FLOW AND GENETICS

3. The regulation of gene expression is influenced by external and internal molecular cues and/or signals.

   a. Use examples from the lifestyle of *Listeria monocytogenes* to justify the statement "Microbes are adept at tailoring their protein profile to suit their surroundings."

MICROBIAL SYSTEMS

4. Microorganisms, cellular and viral, can interact with both human and nonhuman hosts in beneficial, neutral, or detrimental ways.

   a. Describe where humans encounter *Listeria monocytogenes*.
   b. Name populations that are more susceptible to *Listeria monocytogenes* infection.
   c. Explain how the Centers for Disease Control and Prevention traces *Listeria* outbreaks in the complex and global food supply.
   d. List reasons it is important to track *Listeria* outbreaks.
   e. Describe why declines in tuberculosis cases in the United States can be credited to increases in successful use of HIV drugs.
   f. Identify the factors that contribute to primary and secondary TB infection.
   g. Explain the mechanism by which individuals with primary TB are not infectious and TB is readily controlled by healthy individuals.
   h. State how interactions between *Mycobacterium tuberculosis* and the host help to create granulomas that contain infections.
   i. Identify the source that accounts for most secondary TB symptoms.
   j. State reasons infections caused by intracellular pathogens can be difficult to treat and prevent.

IMPACT OF MICROORGANISMS

5. Microorganisms provide essential models that give us fundamental knowledge about life processes.

   a. Identify key eukaryotic cellular processes that have been elucidated by studying intracellular pathogens.

SUPPLEMENTAL MATERIAL

References

- **Cossart P.** 2011. Illuminating the landscape of host-pathogen interactions with the bacterium *Listeria monocytogenes*. *Proc Natl Acad Sci USA* **108:**19484–19491. http://www.ncbi.nlm.nih.gov/pmc/articles/PMC3241796/

- **Goldberg MB.** 2001. Actin-based motility of intracellular microbial pathogens. *Microbiol Mol Biol Rev* **65(4):**595–626. http://mmbr.asm.org/content/65/4/595.long
- **Stewart GR.** 2003. Tuberculosis: a problem with persistence. *Nat Rev Microbiol* **1:**97–105 http://www.nature.com/nrmicro/journal/v1/n2/full/nrmicro749.html
- **Russell DG.** 2010. *Mycobacterium tuberculosis* wears what it eats. *Cell Host Microbe* **8:**68–76. http://www.sciencedirect.com/science/article/pii/S1931312810001800

Supplemental Activities

CHECK YOUR UNDERSTANDING

1. **Acid-fast organisms are NOT**
   a. resistant to harsh chemicals and germicides
   b. the most common type of bacteria
   c. washed free of stain by acid treatment
   d. enclosed within a waxy outer membrane

2. **Tuberculin skin tests are positive after**
   a. active but cured disease
   b. primary infection but no disease
   c. BCG vaccination
   d. all the above

3. ***Listeria monocytogenes* infection does NOT cause**
   a. abortion
   b. sepsis
   c. meningoencephalitis
   d. pneumonia
   e. gastroenteritis

DIG DEEPER

1. Rifamycin is one member of a family of antibiotics that are commonly included in antituberculosis therapy. Research its mechanism of action to answer the following questions:

   a. What is rifamycin's mechanism of action?

   b. List three characteristics of rifamycin that are important for selective killing of *M. tuberculosis* in host cells.

   c. Drug resistance to rifamycin is commonly observed in patients infected with *M. tuberculosis*. Postulate how *M. tuberculosis* develops drug resistance to rifamycin, and design experiments to test your hypothesis.

2. The regulatory protein PrfA coordinately activates expression of multiple virulence factors that manipulate host cell pathways to promote intracellular replication and spread of *L. monocytogenes*. Does the microbe also benefit from *not* expressing this panel of virulence factors while growing on decaying plants? How would you address this question experimentally? What bacterial genotypes would you compare? What growth conditions?

3. *L. monocytogenes* synthesizes the protein ActA to induce actin-based motility. Experimentalists have also identified and purified the four host components that *L. monocytogenes* requires for actin-based motility: actin monomers, Arp2/3 complex (a weak actin nucleator), ADF/cofilin (which severs filaments to promote recycling), and capping protein (whose role is not yet clear). Design an experiment to test if *L. monocytogenes* requires bacterial proteins other than ActA for actin-based motility. How would you measure motility? What particle(s) would you prepare for your analysis? What would you include as a positive control? Negative controls?

4. In their study "Identification of a small molecule with activity against drug-resistant and persistent tuberculosis" (*Proc Natl Acad Sci USA* **110:**e2510–e2517, 2013), Wang et al. identified a compound (TCA1) that inhibits *M. tuberculosis* biofilm formation and is also bactericidal to the pathogen. As one strategy to identify its microbial target, they first isolated mutants that had acquired resistance to the drug. Spontaneous TCA1-resistant

mutants of *M. tuberculosis* arose at a rate of 10^{-8} to 10^{-9}. How many bacteria would the scientists need to screen to isolate 10 TCA1-resistant mutants?

5. **Team work:** In the April 15, 2015 podcast (8 min) "Precision Immunology: The Promise of Tumor Immunotherapy for the Treatment of Cancer," *Journal of Clinical Oncology* (http://jco.ascopubs.org/site/podcasts/archive/2015/april-june2015.xhtml) Editor Howard L. Kaufman discusses the research article by Le et al. "Safety and Survival with GVAX Pancreas Prime and *Listeria monocytogenes*—Expressing Mesothelin (CRS-207) Boost Vaccines for Metastatic Pancreatic Cancer" in *J Clin Oncol* **33:**1325–1333, 2015, http://jco.ascopubs.org/content/33/12/1325.long. Draw a schematic to illustrate how strain CRS-207 of the intracellular pathogen *L. monocytogenes* was engineered to recruit the immune system to fight cancer. Write a newspaper-style article to share this exciting science with the general public.

6. **Class debate.** In 2011, cantaloupe fruit grown on Jensen Farm in Colorado transmitted listeriosis to 144 people in 28 states, resulting in 143 hospitalizations and 33 deaths, including one miscarriage. In response to this outbreak, the deadliest outbreak of foodborne illness in the United States recorded by the CDC, you organize a mock congressional hearing to investigate risk and responsibility. Constituencies represented may include:

   - the Jensen Farm owners
   - a national grocery chain that sold contaminated cantaloupe
   - a food producer whose packaged cut fruit included contaminated cut cantaloupe
   - FDA officials who participated in the outbreak investigation
   - CDC officials who oversee the PulseNet database used to track the outbreak
   - microbiologists who study *Listeria monocytogenes* pathogenesis
   - patients who were sickened by consuming cantaloupe
   - adult children of patients who died from listeriosis

   Read the CDC's report on this *Listeria* outbreak (http://www.cdc.gov/listeria/outbreaks/cantaloupes-jensen-farms/082712/index.html) and step-by-step guide on food outbreak investigations (http://www.cdc.gov/foodsafety/outbreaks/investigating-outbreaks/investigations/index.html). Work in groups that assume each role, including the congressional panel. Prepare talking points before the public hearing is held in class. The congressional panel's mission is to evaluate what measures should be taken to reduce public risk of foodborne listeriosis and to issue recommendations.

7. **Class debate:** Pasteurization of dairy products extends their shelf life and makes it possible to transport the products longer distances, leading to commercialization of these processed foods. Currently it is illegal to sell unpasteurized or "raw" milk to people because of the risk of contamination. On the other hand, recent studies have identified multiple health benefits associated with diverse microbial communities in the human gastrointestinal tract, expanding the market for probiotics. Do the benefits of unpasteurized milk outweigh the risks? How is the probiotic industry regulated? Use data and statistics from primary literature sources to argue for or against the sale of unpasteurized milk.

8. **Class discussion:** In this 6-min video, Carl Nathan, M.D., Professor and Chairman of the Department of Microbiology and Immunology at Weill Cornell Medical College, discusses the urgent worldwide problem of antibiotic resistance, and their drug discovery program that sought known and approved drugs that could be repurposed to fight tuberculosis: http://www.microbeworld.org/component/jlibrary/?view=article&id=8093

   Who bears responsibility for developing drugs to cure global diseases? For establishing safe antibiotic usage policies? Pharmaceutical companies? Governments of the countries most affected? International bodies? Philanthropists?

Supplemental Resources

VISUAL

- *iBiology* lecture "The Bacterial Pathogen *Listeria monocytogenes*" by Pascale Cossart (24 min): http://www.ibiology.org/ibioseminars/pascale-cossart-part-1.html
- *iBiology* lecture "Intracellular Parasitism by *Trypanosoma cruzi* and *Leishmania*" by Norma Andrews (20 min): http://www.ibiology.org/ibioseminars/microbiology/norma -andrews-part-1.html
- *iBiology* lecture "Tuberculosis: A Persistent Threat to Global Health" by John McKinney: http://www.ibiology.org/ibioseminars/microbiology/john-mckinney-part-1.html
- U.S. Department of Agriculture public service video teaches how pregnant women can reduce their risk of listeriosis from contaminated foods: https://www.youtube.com /watch?v=BW6wNYyQe7w
- "Cheese and Microbes" video (7 min) on food safety and artisanal cheese practices: http://www.microbeworld.org/component/content/article?id=429

SPOKEN

- *Meet the Scientist* podcast episode #15 "The Science of Foodborne Pathogens" (13 min): Merry Buckley interviews Kathryn Boor, Professor and Chair of the Food Science Department at Cornell University, who discusses food safety, pasteurization, and the contribution of stress response pathways to *Listeria monocytogenes* virulence: http://www.microbeworld.org/component/content/article?id=271
- *This Week in Microbiology* podcast episode #44 discusses "Phage Contribution to Listeria Escape from Phagosome" (Rabinovich L et al., Cell **150**:792–802, 2012): http://www .microbeworld.org/podcasts/this-week-in-microbiology/archives/1300-twim-44
- *This Week in Microbiology* podcast episode #83 "Illuminating tuberculosis and *Cryptococcus*" discusses a new fluorogenic diagnostic test for tuberculosis bacteria (Cheng Y et al., *Angew Chem Int Ed Engl* **53**:9360–9364, 2014): http://www.microbeworld .org/podcasts/this-week-in-microbiology/archives/1731-twim-83

WRITTEN

- CDC fact sheet on *Listeria*: http://www.cdc.gov/listeria/index.html
- CDC report on the listeriosis outbreak discussed in case study 1: http://www.cdc.gov /listeria/outbreaks/ice-cream-03-15/
- CDC list of multistate foodborne illness outbreak investigations, by year and by pathogen: http://www.cdc.gov/foodsafety/outbreaks/multistate-outbreaks/outbreaks-list .html
- US Public Health Service "Fight BAC!" campaign: http://www.fightbac.org/food-poisoning /foodborne-pathogens/
- In the *Small Things Considered* blog post "Why *Listeria* Is Competent to Be Virulent," Marvin Friedman discusses how this intracellular parasite copes within the host cells: http://schaechter.asmblog.org/schaechter/2012/11/why-listeria-is-competent-to-be -virulent-.html
- CDC Basic TB Facts: http://www.cdc.gov/tb/topic/basics/default.htm
- CDC Reported TB in USA 2011: http://www.cdc.gov/tb/statistics/reports/2011/default.htm
- WHO Global Tuberculosis Report 2015: http://www.who.int/tb/publications/global _report/en/
- In the *Small Things Considered* blog post "TB or Not TB?" Jamie Zlameal, Andy Cutting, and Steven Quistad discuss the finding that tubercle bacilli do not divide symmetrically, and why does this matter?: http://schaechter.asmblog.org/schaechter/2012/03/tb-or-not -tb.html
- *Global Food Safety: Keeping Foods Safe from Farm to Table*, a 2010 colloquium report from the American Academy of Microbiology: http://academy.asm.org/index.php /browse-all-reports/439-global-food-safety-keeping-food-safe-from-farm
- *Moving Targets: Fighting the Evolution of Resistance in Infections, Pests, and Cancer*, a 2013 colloquium report from the American Academy of Microbiology: http://academy .asm.org/index.php/browse-all-reports/666-moving-targets-fighting-the-evolution-of -resistance-in-infections-pests-and-cancer

- National Center for Case Study Teaching in Science: "The Benign Hamburger" by Graham Peaslee, Juliette Lantz, and Mary Walczak highlights food safety issues: http://sciencecases.lib.buffalo.edu/cs/collection/detail.asp?case_id=421&id=421
- National Center for Case Study Teaching in Science: "African Ilness: A Case of Parasites?" by Kevin Bonney incorporates parasite biology, host response, clinical treatment options, and public health measures: http://sciencecases.lib.buffalo.edu/cs/collection/detail.asp?case_id=648&id=648
- National Center for Case Study Teaching in Science: "MDR Tuberculosis" by Katayoun Chamany focuses on social justice issues and incorporates bacterial genetics, evolution, immune defenses, and vaccines: http://sciencecases.lib.buffalo.edu/cs/collection/detail.asp?case_id=575&id=575

Herpes simplex virus 1

CHAPTER TWENTY-FIVE

Herpes:
An Ancient Human Virus Prevails

KEY CONCEPTS

EVOLUTION

1. Mutations and horizontal gene transfer, with the immense variety of micro-environments, have selected for a huge diversity of microorganisms.

CELL STRUCTURE AND FUNCTION

2. The replication cycles of viruses (lytic and lysogenic) differ among viruses and are determined by their unique structures and genomes.

INFORMATION FLOW AND GENETICS

3. The synthesis of viral genetic material and proteins is dependent on host cells.

MICROBIAL SYSTEMS

4. Microorganisms, cellular and viral, can interact with both human and non-human hosts in beneficial, neutral, or detrimental ways.

INTRODUCTION

Viruses are the ultimate intracellular pathogen since they are utterly dependent on a host cell to provide the energy and machinery they need to generate new viral particles. We'll revisit herpesvirus—CJ's unwelcome souvenir from his spring break vacation (chapter 17)—to build on fundamental concepts of virology (chapters 17, 18), intracellular pathogens (chapter 24), and host defenses (chapter 22). Using a case of neonatal herpes infection as a lens, we'll examine how this insidious virus exploits a host innate defense against foreign DNA to persist in a latent state under the radar of the innate immune system, only to reactivate, catch a ride on microtubules, trigger new blisters, and spread from one person to another.

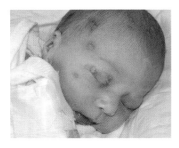

FIGURE 25.1 Neonatal HSV infection often causes blisters on the skin, eyes, and mouth of newborn babies. (From Brady MT, https://www2.aap.org/sections /perinatal/pdf/2014%20NCE%20 Presentations/BradyNeonatal _HSV-brady.pdf)

CASE: An unexpected neonatal herpesvirus infection

Ten days after bringing their first baby home, a couple was shocked to see blisters erupting on the face and scalp of their conspicuously irritable son (Fig. 25.1). Acting on their pediatrician's advice, the worried parents quickly brought the infant to clinic. After taking the infant's medical history, health care workers asked if either parent had a history of genital herpes. The parents nodded, but explained that it had been years since either had suffered a bout of painful sores; physical examination confirmed that both parents were free of genital lesions. Immediately the infant was started on intravenous acyclovir, and scrapings from blisters on his scalp were collected. In the clinical lab, his specimens were inoculated into cell cultures that were then monitored for cytotoxicity (Fig. 25.2). Two days later, the lab confirmed that the cultured cells had died and herpes simplex virus 2 (HSV-2) was identified by PCR. Meanwhile, after several days of lethargy, feeding poorly, and intermittent seizures, the newborn gradually recovered and was discharged. At the baby's 1-year pediatrician appointment, a comprehensive physical exam verified that the child was developing normally, much to his parents' relief.

- Why were the infant's blister specimens inoculated into cell culture, not growth medium?
- Where did the infant encounter HSV-2 virus?
- What caused the infant's lethargy and seizures?
- Why was it urgent that the infant receive intravenous acyclovir?
- Why were the parents and physician concerned about the baby's development?
- Why did the infant's disease follow a dramatically different course than his parents' infections?
- Could the infant's infection have been prevented?

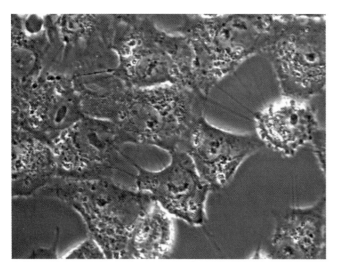

FIGURE 25.2 Time-lapse movie of a permissive cell line infected with HSV-1, a standard clinical diagnostic test. Several hours after infection, images were taken every 3 minutes over a period of several hours. As the virus replicates, the host cell's cytoskeleton collapses, causing the cell to round up, and the cell eventually dies. (Credit: The Virologist's Channel, https://www.youtube.com /watch?v=lwRX53Y6pAg)

WHAT ARE HERPESVIRUSES?

LEARNING OUTCOMES:
- List traits shared by all viruses in the *Herpesviridae* family.
- Differentiate HSV-1 and HSV-2.

Herpesviruses are extraordinarily successful, causing recurrent infections in a third of the world's population today. The written record corroborates that these viruses also plagued the ancient Greeks, who named the malady *herpein,* meaning "to creep," or move slowly. These insidious viruses persist for the entire life of humans, their sole reservoir. Indeed, scientists estimate that the human lineage has carried some herpesviruses for *8 million years*! Before learning what tricks this pathogen has acquired during its prolonged coevolution with humans, let's review the herpesvirus family tree and get some terminology straight.

The herpesviruses that aggravated CJ (chapter 17) and distressed the newborn's parents are members of the *Herpesviridae* family. In addition to herpes simplex viruses 1 and 2 (HSV-1, HSV-2), the eight members include other pathogens you may recognize: varicella-zoster virus (the agent of chickenpox and shingles; case study, chapter 18) and Epstein-Barr virus (the cause of mononucleosis; chapter 18). One trait every member of the *Herpesviridae* family shares is the capacity to establish *latent infections* in humans. Also, all form large particles wrapped in an envelope of phospholipids. Their linear double-stranded DNA genome encodes approximately 70 genes packed in an icosahedral **capsid**, or protein coat. Because the large enveloped herpesviruses are relatively fragile (check out the delicate appearance of their membranes in Fig. 25.3), they depend on intimate contact between mucous membranes to spread from one human to another. Although HSV-1 and HSV-2 are close cousins whose genomes share ~70% identity, their lifestyles have diverged somewhat. HSV-1 is typically acquired through oral contact during childhood, whereas HSV-2 is primarily spread by intimate genital interactions. (Alas, this distinction is not absolute: genital infections by HSV-1 are on the rise.) For simplicity, we'll use the term "HSV" when describing steps these two viruses have in common.

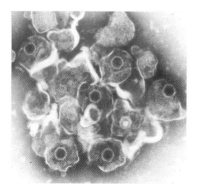

FIGURE 25.3 Electron micrograph of herpes simplex virus. Note its uniform icosahedral nucleocapsid surrounded by a phospholipid membrane. (MicrobeWorld, http://www.microbeworld.org/component/jlibrary/?view=article&id=1691)

HOW DOES HSV INVADE MUCOUS MEMBRANES?

LEARNING OUTCOMES:
- Describe the mechanism by which HSV invades mucous membranes.
- Determine a viral protein(s) that may be a good target for HSV drugs or vaccines.

When HSV comes into contact with mucous membranes, the virus can bind and enter the epithelial cells to establish a replication niche. Embedded in the viral envelope are multiple glycoproteins that cooperate to breach the host barrier (chapter 17, Fig. 17.8). These viral surface proteins first bind to receptors on the epithelial cell, tethering the infectious particle to the host plasma membrane. Upon binding to receptors, these viral proteins cluster and then change their conformation, releasing sufficient energy to drive fusion of the viral and host phospholipid membranes. This nifty trick generates a membrane pore, which the viral **nucleocapsid** slips through to enter the host cell cytoplasm.

Because viral entry is essential to initiate infection, detailed knowledge of these fusion machines may enable scientists to design tactics to treat or prevent infections. One way to reveal the workings of HSV's invasion equipment is to determine the three-dimensional structures of its fusion proteins. A remarkable discovery is that the architecture of a key HSV-1 fusion protein (gB) resembles that of the fusion proteins of two unrelated viruses, vesicular stomatitis virus and baculovirus. (How could you determine whether this striking similarity is a product of ancient ancestry or **convergent evolution**?) Another fruitful strategy to decipher fusion pore formation is cryoelectron microscopy. This technique preserves phospholipid membranes by rapidly freezing the infected cells at −180°C with liquid ethane cooled with liquid nitrogen, which prevents disruptive ice crystals

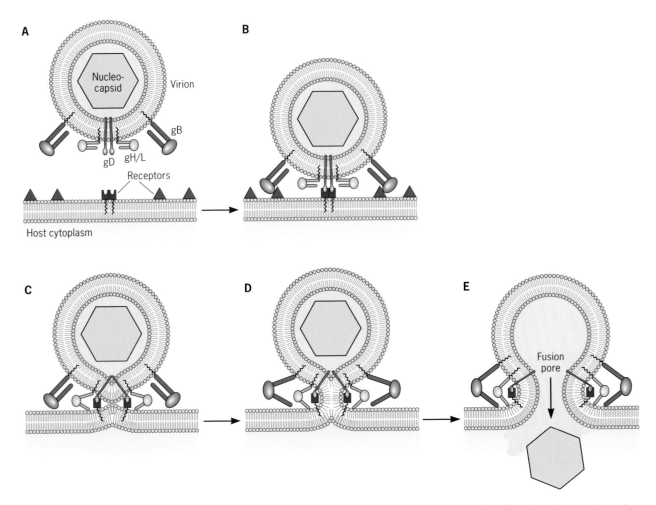

FIGURE 25.4 Model for the HSV-1 fusion mechanism deduced from cryoelectron microscopy. (Modified from Maurer UE et al., *Proc Natl Acad Sci USA* **105**:10559–10564, 2008.)

from forming (see chapter 17, Box 17.1). By studying images of cells trapped at different stages of infection, scientists develop working models for how the viral tethering proteins (gB, gC, and gD) cooperate with the fusion machinery (gB and a gH-gL complex) to trigger formation of a portal (Fig. 25.4). Which viral protein(s) would you target in a quest for an HSV drug or a vaccine?

HOW DOES HSV GENERATE NEW VIRIONS IN EPITHELIAL CELLS?

LEARNING OUTCOMES:
- Describe the replication cycle of HSV in epithelial cells.
- Identify adaptations of HSV that allow it to evade the host's defenses.

Once an HSV nucleocapsid has made its way out of its envelope and into the cytoplasm of an epithelial cell, it must coerce host RNA polymerase to transcribe the viral DNA. First, the nucleocapsid docks on a nuclear pore complex and delivers a few potent viral factors along with its DNA, which quickly circularizes. But the virion is not yet home free: innate immune

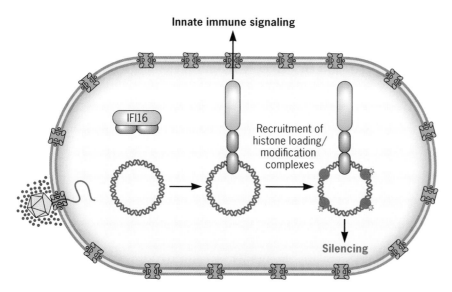

FIGURE 25.5 An innate immune surveillance system detects and silences foreign DNA. DNA in the nucleus that lacks host nucleosomes is detected by the pattern recognition receptor IFI16. Binding to DNA triggers a conformational change in IFI16. So activated, IFI16 then recruits chromatin modification complexes that silence viral genomes and also trigger innate immune signaling pathways that generate proinflammatory cytokines.

sensors patrol the nucleus for foreign DNA. To avoid any monkey business, host cells quickly package naked DNA within tight bundles of chromatin, limiting its access to the transcription machinery, while also triggering cytokine production (Fig. 25.5). But HSV is not easily silenced. One viral protein degrades a key component of this host surveillance system, while others keep viral promoters clear of repressive chromatin. Still other HSV factors hijack the host transcription machinery by modifying its RNA polymerase. What follows is a choreographed cascade of viral gene expression, beginning with the "immediate early" genes and progressing to "late" genes (chapter 17). Together these factors orchestrate the virus's **lytic cycle**: HSV proteins are synthesized, its DNA is replicated, and progeny nucleocapsids are assembled.

But where do these new particles get their envelope? In a brilliant display of viral resourcefulness, the progeny nucleocapsids robe, disrobe, and robe once more as they transit first across the nuclear membrane, then into the secretory system, and finally exit from the cell by exocytosis. Wrapped within membranes pilfered from its first host cell and studded with viral glycoproteins, the new virions are equipped to infect adjacent epithelial cells. Eventually, such lytic viral cycles generate the hallmark blisters on the mouth (HSV-1), the genitals of CJ (HSV-2; chapter 17), or the face and scalp of the infant in the case study (Fig. 25.1).

WHY DO HSV LESIONS REAPPEAR, AGAIN AND AGAIN?

LEARNING OUTCOMES:
- Explain how HSV nucleocapsids travel from their site of entry on the surface of nerve cells to the cell nucleus.
- State the role and impact of HSV latency-associated transcript.
- Identify the mechanism by which HSV lesions disappear and reappear.

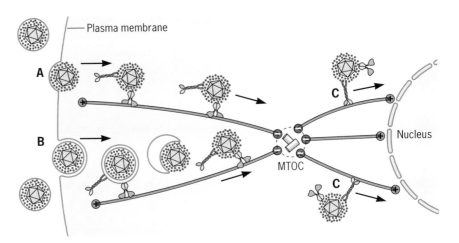

— Plasma membrane

A

B

C

C

MTOC

Nucleus

FIGURE 25.6 **Viruses exploit microtubule motor proteins to move between the plasma membrane and the nucleus.** Illustrated here is a nonpolarized cell; a similar process transports viruses between the cell body and axon terminus in polarized neuronal cells. Host dynein protein motors (blue) move cargo toward the minus ends of microtubules, whereas kinesin protein motors (beige) move toward their plus ends. MTOC: microtubule organizing center. (Modified from Dodding MP, Way M, *EMBO J* **30:**3527–3539, 2011.)

Quite a different story unfolds when HSV encounters the sensory neurons that innervate mucosal tissues of the mouth or genitals. The viral fusion machinery can also work its magic at the tips of peripheral nerve cells. In fact, entry into these cells is so efficient that HSV has earned the designation *neuroinvasive*. But how do HSV nucleocapsids make the long journey through the nerve axon to the remote nucleus?

These shrewd pathogens hitch a ride on a transportation system that carries cargo between the neuronal cell body and its distant branched tips. The tracks are composed of long cylinders of assembled tubulin that radiate from an anchor adjacent to the nucleus (aptly named the *microtubule organizing center*) and extend toward the cell periphery. Pushing cargo along these tracks in neuronal and non-neuronal cells alike are two motor proteins: **dynein** drives organelles toward the nucleus, while **kinesin** moves particles toward the cell periphery (Fig. 25.6). By latching onto these motor complexes, HSV-1 nucleocapsids hitchhike along microtubules to and from the host nucleus—a splendid sight that virologists have captured on video (Fig. 25.7). Indeed, during their coevolution with eukaryotic

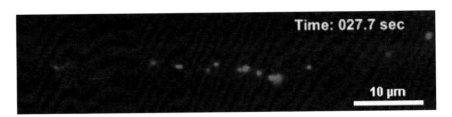

Time: 027.7 sec

10 μm

FIGURE 25.7 **Time-lapse movie of individual HSV-1 nucleocapsids moving in axons.** The viral particles are visible because they incorporate many copies of an mRFP-capsid fusion protein expressed by this recombinant HSV-1 virus. Accelerated 7×. Movie made by Esteban Engel and L.W. Enquist, Department of Molecular Biology and the Princeton Neuroscience Institute, Princeton University.

TABLE 25.1 **Some viruses that exploit host motor proteins**

Driven by dynein motor	Egress driven by a kinesin motor
HSV-1	Kaposi's sarcoma-associated herpesvirus
CMV	Vesicular stomatitis virus
Kaposi's sarcoma-associated herpesvirus	HIV-1
HIV-1	Sendai virus
HPV	African swine fever virus
Adenovirus	Vaccinia virus
Parvovirus	
Influenza A virus	

hosts, many viruses have become adept at binding to cytoskeletal motor proteins (Table 25.1). Moreover, like *Listeria monocytogenes* and other intracellular bacterial pathogens (chapter 24), a number of viruses actively manipulate tubulin or actin cytoskeletal networks to spread between or within host cells. This virulence strategy is especially critical for the neurotropic viruses, whose journey is long.

Once delivered to the nucleus, rather than dodge the host innate defenses that detect and silence foreign DNA, in nerve cells HSV actively *recruits* chromatin assembly (Fig. 25.8)! Consequently, only one HSV gene is highly expressed, that producing the **latency-associated transcript** (LAT). As its name implies, this viral RNA not only stimulates tight packaging of the now circularized viral genome, but also *prevents the infected cell from committing*

A. Epithelial cell

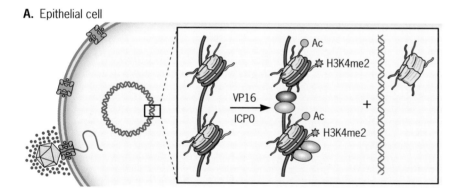

B. Neuron

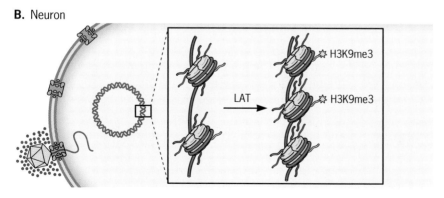

FIGURE 25.8 **HSV avoids silencing within heterochromatin in epithelial cells (A), but promotes silencing in neuronal cells (B).**

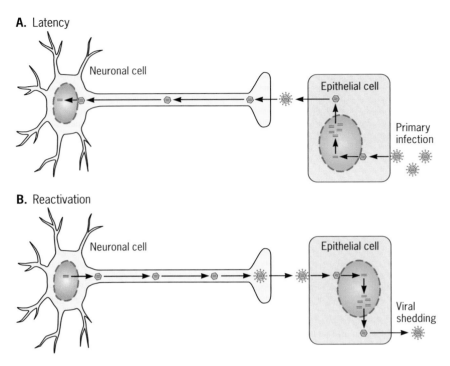

A. Latency

Neuronal cell

Epithelial cell

Primary infection

B. Reactivation

Neuronal cell

Epithelial cell

Viral shedding

FIGURE 25.9 **HSV life cycle in epithelial and neuronal cells.**

apoptosis. Although the exact mechanisms are still being discovered, the impact of viral LAT is clear: for the duration of its long life, the neuronal cell unwittingly harbors a silent HSV genome. The latent virus is hidden from the innate immune system. HSV is not the only virus that acquired, through coevolution with humans, tools to manipulate this innate defense pathway to mute telltale signaling by foreign DNA (see chapter 18). In fact, all of the herpesviruses family members depend on chromatin formation to silence their genomes and establish latent infections.

Is a chromatin cage in neurons the end of the line for HSV? Far from it! The latent virus can *reactivate* when its host cell senses stress, be it emotional, physical (like fever, tissue damage, or the UV rays in strong sunlight), clinical (such as chemotherapy for cancer or transplantation, certain genetic disorders, or HIV-1 infection), or simply aging. In response to stress, some as yet unknown signal prompts nucleocapsids to de-repress chromatin silencing, reinitiate viral gene expression, and restart the replication cycle. The progeny viral particles then retrace their transport steps back to the epithelial surface and begin the long journey along microtubules extending from the cell body to the tips of the sensory neuron, where the new virions are shed (Fig. 25.9). Finally, the released HSV particles can once again infect the adjacent epithelium, initiating a new round of lytic infection. So begins another episode of "cold sores" around the mouth (HSV-1) or lesions on the genitalia (HSV-2).

HOW DOES HSV REGULATE ITS DISAPPEARING ACT?

LEARNING OUTCOMES:
- Compare and contrast epigenetic effects in host neuronal cells with that of epithelial cells infected by HSV.
- Identify the consequence of incomplete silencing of the HSV-2 genome within sensory neurons.
- Describe how the antiviral drug acyclovir treats HSV infections.

In chapters 11, 12, 13, and 19 we learned about a number of mechanisms that equip microbes to alter their pattern of gene expression in response to external cues. (Remember the cascade of sigma factors that control *Bacillus subtilis* spore formation, the cyclic-di-GMP that coordinates biofilm formation, the transcriptional repressor that regulates *Escherichia coli* iron acquisition, or the two-component regulatory system that regulates phosphate uptake by *E. coli*?) The HSV strategy is something entirely different: **epigenetic** regulation. This term refers to any process that not only turns genes on or off in one cell, but also maintains the pattern of gene expression in its daughter cells, *without* changing the DNA code (hence *epi*, meaning "near to" or "on top of"—genetics). One way to establish such heritable patterns of gene expression is by altering the 3D structure of the DNA strands. Recall that eukaryotic cells wrap their DNA fibers around clusters of histone proteins, forming complexes called **nucleosomes**. By altering the density of nucleosomes, or by adding acetyl or methyl groups to particular sites on the proteins themselves, cells dictate the architecture of stretches of DNA, which, in turn, determines accessibility to transcription factors. Tightly packed **heterochromatin** represses gene expression, whereas the more flexible **euchromatin** is accessible to the transcription machinery.

As you know by now, in neuronal cells HSV actively *promotes* epigenetic silencing to establish a latent infection. But in epithelial cells, the virus deliberately *blocks* heterochromatin formation to launch a lytic infection (Fig. 25.8)—an ingenious maneuver, especially for a virus that must coordinate expression of a large cohort of ~70 genes. But here's a wrinkle: human cells don't strictly abide by our tidy textbook drawings. Using sensitive techniques, researchers have discovered that some people shed HSV-2 particles from their genital tissue even when they lack visible signs of infection. As you have likely figured out, the mother in our case study was carrying HSV asymptomatically, and so was CJ's spring break contact (chapter 17). And they are not alone. In the United States, approximately two-thirds of the women who pass HSV to their infants have no signs of genital infection at the time of childbirth, nor do the mothers or fathers of these infants report a history of genital herpes.

We deduce, then, that silencing of the HSV-2 genome within the population of sensory neurons is not complete ("stochastic" is a term scientists use to describe such regulatory switches that fluctuate at a low rate, seemingly at random). The study case of the newborn infant—and likely too CJ's spring break infection (chapter 17)—make vividly clear the dreaded consequences to humans of this leaky epigenetic regulation of HSV latency within neuronal cells. But stop for a moment to weigh the impact of such asymptomatic shedding on this sexually transmitted virus. *Topic for*

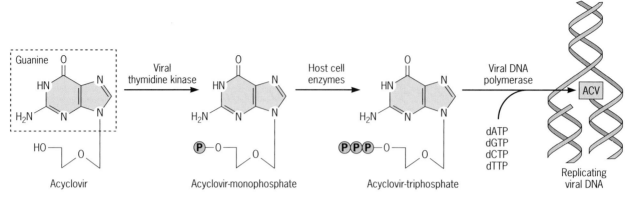

FIGURE 25.10 **Acyclovir mechanism of action**. A derivative of guanosine, acyclovir is membrane permeable until it is phosphory-lated by an HSV thymidine kinase enzyme. Therefore, the drug accumulates specifically in infected cells. Once incorporated into a newly replicated strand of DNA, this phosphorylated guanosine analog blocks elongation by viral polymerase.

debate: The leaky epigenetic silencing of virions in neuronal cells is evidence that the potency of the HSV LAT virulence factor is not yet optimal, despite millions of years of coevolution with humans. Affirmative or negative?

What do we know of the mechanism of epigenetic silencing of HSV? The question is more than academic. Scientists have devised an array of drugs that block specific viruses from entering host cells, inhibit replication of their genomes, or prevent assembly of new virions. For example, the acy-clovir that successfully treated both CJ and the newborn baby is a nucleo-side derivative that, once activated by an HSV enzyme, blocks elongation of DNA chains during viral replication (Fig. 25.10; chapter 17). As the infant's parents would surely attest, acyclovir is a wonder drug for acute HSV infec-tions. In stark contrast, we have *no therapeutic agents that target latent viruses*.

Scientists have discovered that HSV is proficient at manipulating the innate defense pathway that recognizes and silences foreign DNA, and a number of the key viral and host components that dictate the outcome of this host-pathogen interaction have been identified. However, the molecu-lar details of HSV latency and reactivation remain to be elucidated. A ma-jor challenge is that we don't yet have a tissue culture system that accurately recapitulates the latent HSV infection of neuronal cells. Instead, virologists must study this crucial phase of the HSV life cycle using more onerous small animal models of infection. As a consequence, scientists are still grappling with some fundamental questions, such as how HSV virions assemble within neuronal cells and spread from one neuron to another.

WHY DOES HSV CAUSE SEVERE INFECTIONS IN NEWBORNS?

LEARNING OUTCOMES:
- Explain why newborns and immunocompromised adults are more likely to have severe HSV infections.

Because epigenetic silencing of HSV-2 within neuronal cells is not com-plete, the newborn in our case study was inadvertently exposed to virus shed asymptomatically into his mother's birth canal. The consequences can be devastating. Lesions on the skin, eye, and mouth are typical, but in one out of three cases this neurotropic virus also spreads to the central nervous system. Inflammation in the membranes that enclose the brain or spinal cord (meningitis) or in the brain itself (encephalitis) is life threatening. When untreated, >80% of neonatal HSV infections are fatal. Likewise, immunocompromised adults are also vulnerable to infections of the central nervous system or other tissues. By infecting the eye cornea, HSV-1 causes blindness in 50,000 people in the U.S. each year. Presumably, in infants the cell-mediated immune system is not fully developed. Without mature **cytotoxic T cell**s and natural killer cells, which recognize and lyse infected cells (chapter 22), infants are poorly equipped to restrict HSV replication and spread.

CAN NEONATAL HSV INFECTIONS BE PREVENTED?

LEARNING OUTCOMES:
- Explain why HSV latency in neuronal cells is a major obstacle to ridding HSV.
- State a reason why an HSV vaccine has yet to be developed.

Because infants are typically exposed to HSV during birth, not in the womb, obstetricians do have a window of opportunity to reduce the risk of this devastating disease. When detected early, either by viral culture within cells or by PCR, a prolonged course of acyclovir, delivered in high doses intravenously, protects most infants. Since this drug became available, mortality has fallen from approximately 50% to 4% of newborns who suffer HSV infections of their central nervous system. Unfortunately, even with treatment, developmental problems persist in ~15% of these babies.

Microbiologists remain highly motivated to improve methods to detect, treat, and prevent HSV infections. For pregnant women who experience an outbreak of genital HSV, surgical delivery by cesarean section coupled with oral acyclovir reduces transmission to the newborn. But a major clinical challenge is the extraordinary capacity of this human pathogen to remain concealed within neuronal cells, which then seed the inconspicuous viral shedding from mucosal epithelium. One therapeutic strategy that has been considered is to develop drugs that relax the host cells' heterochromatin to reactivate the latent viral genomes, while simultaneously treating with antiviral compounds that interrupt the lytic cycle. What do you suppose are the advantages and risks of this clinical tactic?

Since prevention is always better than treatment, the holy grail is an HSV vaccine. Ideally, the vaccine would block intermittent release of virions from neuronal cells and their subsequent rapid amplification within epithelial cells. But again the virus has erected potent barriers. Remember that, once infected, humans remain so for life. What does this tell you about HSV? (*Quiz spoiler:* HSV encodes potent factors that actively block the development of a specific immune response.) Subunit vaccines designed to stimulate production of antibodies that inhibit the viral fusion machinery (gB, gD) have been tested in human trials but to no avail. A clue to this failure may be that, in mouse models of HSV infection, antibodies alone are not sufficient to block disease; however, injection of antigen-primed T cells provides some protection. Therefore, one alternative vaccine design strategy is to engineer an attenuated virus that also lacks its immune evasion factors. A genetic approach that scientists are also pursuing is to study atypical couples in which one partner is infected with HSV, but the other remains virus-free (formally, "seronegative"). Hopefully, detailed knowledge of the immune responses and genetic profiles of these rare lucky partners will reveal the host and viral factors that can prime all humans to develop resistance to HSV infection.

CASE REVISITED: An unexpected neonatal herpesvirus infection

Clinical labs use cultured mammalian cells to test patient specimens for virus since these infectious agents cannot replicate on their own. Because epigenetic silencing of the latent HSV-2 genome within neuronal cells is incomplete, the infected but asymptomatic mother was shedding virus from her genital mucosal epithelium during delivery. As a neurotropic virus, HSV efficiently spreads between neuronal tissues in individuals who lack a mature or robust immune system. If spinal cord or brain tissues become inflamed, patients may suffer lethargy or seizures, and development of infected infants may be impaired. To limit the damaging inflammatory response to HSV-2 infection in the central nervous system, patients are treated immediately with acyclovir, a nucleoside derivative that inhibits DNA replication in infected cells. Based on both epidemiological data and small model laboratory experiments, it is evident that the cell-mediated immune system of infants is not sufficiently developed to restrict systemic spread of HSV. Although no vaccine is available to prevent HSV infection, by increasing awareness that this virus can be shed from genital tissue whether or not lesions are present, more parents and their obstetricians can reduce the risk of neonatal infection by treating the expectant mother with oral acyclovir when delivery is imminent. One reliable source of information is the CDC's Fact Sheet on Genital Herpes (see Supplemental Resources).

CONCLUSIONS

HSV's ongoing relationship with the human race illustrates a number of fundamental principles of microbial pathogenesis. As we learned for MRSA (chapter 23) and *Mycobacterium tuberculosis* (chapter 24), any microbe that relies on humans as its sole reservoir must be equipped to evade or disarm the innate and acquired immune systems. A more subtle but no less significant truth is this: rather than strike their host dead, the most successful pathogens are in *balance* with their host. Remember, *M. tuberculosis,* HSV, and *Staphylococcus aureus* inhabit approximately one-third of the people on Earth, mostly without symptoms. Their strategies differ, but their goals are shared: to persist under the radar, causing only the inflammation or damage necessary to acquire the nutrients the microbe needs to survive, replicate, and spread to a new host. By investigating the intimate interactions between pathogens and their hosts, microbiologists not only expand knowledge of the microbial world, but also uncover the inner workings of host cells and immune defenses. From this strong foundation, scientists can design new schemes to prevent and treat infectious diseases.

Representative ASM Fundamental Statements and Learning Outcomes for the Chapter

EVOLUTION

1. Mutations and horizontal gene transfer, with the immense variety of microenvironments, have selected for a huge diversity of microorganisms

   a. List traits shared by all viruses in the *Herpesviridae* family.
   b. Differentiate HSV-1 and HSV-2.
   c. Identify adaptations of HSV that allow it to evade the host's defenses.

CELL STRUCTURE AND FUNCTION

2. The replication cycles of viruses (lytic and lysogenic) differ among viruses and are determined by their unique structures and genomes.

   a. Describe the mechanism by which HSV invades mucous membranes.
   b. Describe the replication cycle of HSV in epithelial cells.
   c. Explain how HSV nucleocapsids travel from the site of entry in nerve cells, the cell periphery, to the nucleus.
   d. State the role and impact of HSV latency-associated transcript.
   e. Identify the mechanism by which HSV lesions disappear and reappear.
   f. Identify the consequence of incomplete silencing of the HSV-2 genome within sensory neurons.
   g. Explain why HSV latency in neuronal cells is a major obstacle to ridding the patient of HSV.

INFORMATION FLOW AND GENETICS

3. The synthesis of viral genetic material and proteins is dependent on host cells.

   a. Describe how the antiviral drug acyclovir treats HSV infections.

MICROBIAL SYSTEMS

4. Microorganisms, cellular and viral, can interact with both human and nonhuman hosts in beneficial, neutral, or detrimental ways.

   a. Determine a viral protein(s) that may be a good target for HSV drugs or vaccines.
   b. Compare and contrast epigenetic effects in host neuronal cells with those in epithelial cells infected by HSV.
   c. Explain why newborns and immunocompromised adults are more likely to have severe HSV infections.
   d. State a reason why an HSV vaccine has yet to be developed.

SUPPLEMENTAL MATERIAL

References

- **Kimberlin DW.** 2004. Neonatal herpes simplex infection. *Clin Microbiol Rev* **17:**1–13. http://cmr.asm.org/cgi/pmidlookup?view=long&pmid=14726453
- **Connolly SA, Jackson JO, Jardetzky T, Longnecker R.** 2001. Fusing structure and function: a structural view of the herpesvirus entry machinery. *Nat Rev Microbiol* **9:**369–381. http://www.ncbi.nlm.nih.gov/pmc/articles/PMC3242325/
- **Kramer T, Enquist LW.** 2013. Directional spread of alphaherpesviruses in the nervous system. *Viruses* **5:**678–707. http://www.mdpi.com/1999-4915/5/2/678
- **Knipe DM.** 2015. Nuclear sensing of viral DNA, epigenetic regulation of herpes simplex virus infection, and innate immunity. *Virology* **479-480:**153–159. http://www.sciencedirect.com/science/article/pii/S0042682215000525

Supplemental Activities

CHECK YOUR UNDERSTANDING

1. **From which host cell membrane do enveloped viruses obtain their envelope?**
   a. Endoplasmic reticulum
   b. Nuclear membrane
   c. Mitochondrial membrane
   d. Plasma membrane

2. **Which statements accurately describe herpes virus entry?**
   a. After binding, a cluster of folded viral proteins change conformation.
   b. HSV-1, vesicular stomatitis virus, and baculovirus all use a similar fusion protein.
   c. To form a portal, HSV-1 needs both viral tethering proteins and a fusion machinery.
   d. All of the above.
   e. None of the above.

3. **What is a potential strategy to prevent HSV from invading mucous membranes?**
   a. Enhance innate immune sensing of HSV DNA by epithelial cells.
   b. Inhibit viral protein gB activity.
   c. Inactivate the latency-associated transcript (LAT).
   d. Inhibit microtubule polymerization.

DIG DEEPER

1. Read the research article by Linehan et al., *"In vivo* role of nectin-1 in entry of herpes simplex virus type 1 (HSV-1) and HSV-2 through the vaginal mucosa" in *J Virol* **78:**2530–2536, 2004 (http://jvi.asm.org/content/78/5/2530.full) to answer the following questions:
   - What epidemiological data motivated the investigators to analyze HSV-1 entry using vaginal epithelial cells?
   - What is nectin-1?
   - In Fig. 3 and 4 of the article, what approaches were used to analyze the contribution of nectin-1 to HSV-1 entry? Draw a schematic to illustrate the experimental design.
   - In Fig. 5, what approach was used to analyze the contribution of nectin-1 to HSV-1 entry? Draw a schematic to illustrate the experimental design.
   - The authors used a mouse model to analyze HSV-1 infection. Is this an ethical use of animals? Why or why not?
   - Draw a schematic to illustrate their key finding.
   - What key questions remain to be addressed?

2. In their study "CD200R1 supports HSV-1 viral replication and licenses pro-inflammatory signaling functions of TLR2" [*PLoS One* **7(10):**e47740, 2012], Soberman et al. analyzed HSV infection in macrophages that encoded or lacked the host protein CD200R1.
   - What is CD200R1? Draw a schematic to illustrate the contribution of the CD200R1 pathway to the immune response.
   - For Fig. 3, what approach was used to analyze the contribution of CD200R1 to HSV-1 pathogenesis? What conclusion can be drawn?
   - For Fig. 5C, what approach was used to analyze the contribution of CD200R1 to HSV-1 pathogenesis?
   - What additional value is gained by using both of these approaches?
   - What key questions remain to be addressed?

3. Listen to "This Week in Virology" podcast episode #339 (http://www.microbe.tv/twiv/twiv-339/; 20:30–54:00) and discuss the research article by Rutkowski et al., "Widespread disruption of host transcription termination in HSV-1 infection," in the open access journal *Nature Commun* **6:**7126, 2015, http://www.nature.com/ncomms/2015/150520/ncomms8126/full/ncomms8126.html
   - What technology enabled the authors to study how HSV-1 lytic infection alters transcription termination by host cells?
   - Draw a schematic to illustrate what the authors called "read-out" and "read-in."
   - What fitness advantage might the HSV-1 virulence strategy they discovered confer to this pathogen?
   - How do their data differ from the conclusion of the 1989 report by McLauchlan et al.? What accounts for this difference?

- Describe their key findings for the general public in a newspaper-style article, and include one schematic to illustrate your main point.

4. Read the 2013 NIH news release "Phase 1 trial for the HSV-2 candidate vaccine HSV529" (http://www.niaid.nih.gov/news/newsreleases/2013/Pages/HSV11-8-13.aspx) and learn at the National Institutes of Health website ClinicalTrials.gov (https://clinicaltrials.gov/ct2/show/NCT02571166) about the subsequent clinical trial of a vaccine developed during a 10-year collaboration by David Knipe at Harvard Medical School and Jeffrey Cohen of NIAID and manufactured by Sanofi Pasteur.

- What is HSV529? What is the rationale for its design?
- What is valacyclovir? How is it used in this study?
- Draw a schematic to illustrate the experimental design of this clinical trial. Why are biopsies also performed on the upper arm? A non-lesion genital site?
- Why is this clinical study enrolling people who are HIV-1 seronegative? HSV-2 seropositive?
- Why are only those women who lack the potential to bear a child being enrolled?
- Why are only people willing to avoid contact with infants and immunocompromised adults for 3 days after each HSV529 injection being enrolled?
- Is this clinical trial ethical? Why or why not?

Supplemental Resources

VISUAL

- *iBiology* lecture "Virus Infection and the Biology of Kaposi's Sarcoma" by Don Ganem (23 min): file://localhost/http/::www.ibiology.org:ibioseminars:microbiology:don-ganem-part-1.html

SPOKEN

- *Meet the Scientist* audio (15 min): Mary Buckley interviews David Knipe, a herpes simplex virus 2 researcher whose research group developed a candidate vaccine. http://www.microbeworld.org/careers/audio-interviews/410-mts22-david-knipe-herpes-simplex-virus-2
- *This Week in Virology* podcast episode #84: Hosts Vincent Racaniello and Rich Condit interview Dave Bloom and discuss his lab's studies of HSV latency (min 18:08–32:45). http://www.microbeworld.org/podcasts/this-week-in-virology/twiv-archives/664-twiv-84-gators-go-viral
- *This Week in Virology* podcast episode #369 discusses a neuronal stress pathway that reactivates HSV from the research article by Cliffe et al., "Neuronal stress pathway mediating a histone methyl/phospho switch is required for herpes simplex virus reactivation," *Cell Host Microbe* **18:**649–658, 2015, http://www.microbe.tv/twiv/twiv-369/
- *This Week in Virology* podcast episode #269 discusses a host protein that senses HSV DNA in the nucleus from the research articles by Kerur et al. ("IFI16 acts as a nuclear pathogen sensor to induce the inflammasome in response to Kaposi Sarcoma-associated herpesvirus infection," *Cell Host Microbe* **19:**363–375, 2011) and Orzalli et al. ("Nuclear IFI16 induction of IRF-3 signaling during herpesviral infection and degradation of IFI16 by the viral ICP0 protein," *Proc Natl Acad Sci USA* **109:**e3008–3017, 2012). http://www.microbe.tv/twiv/twiv-269-herpesvirus-stops-a-nuclear-attack/
- *This Week in Virology* podcast episode #339 discusses the article by Rutkowski et al. "Widespread disruption of host transcription termination in HSV-1 infection," *Nat Commun* **6:**7126, 2015. http://www.microbeworld.org/component/content/article?id=1913 http://www.nature.com/ncomms/2015/150520/ncomms8126/full/ncomms8126.html

WRITTEN

- CDC Genital Herpes website: Includes fact sheet, statistics, and treatment information. http://www.cdc.gov/std/herpes/
- NIH news release: Phase 1 trial for HSV-2 candidate vaccine, HSV529, developed during a 10-year collaboration by David Knipe at Harvard Medical School and Jeffrey Cohen of NIAID and manufactured by Sanofi Pasteur. The small safety study is expected to be completed by October 2016. http://www.niaid.nih.gov/news/newsreleases/2013/Pages/HSV11-8-13.aspx

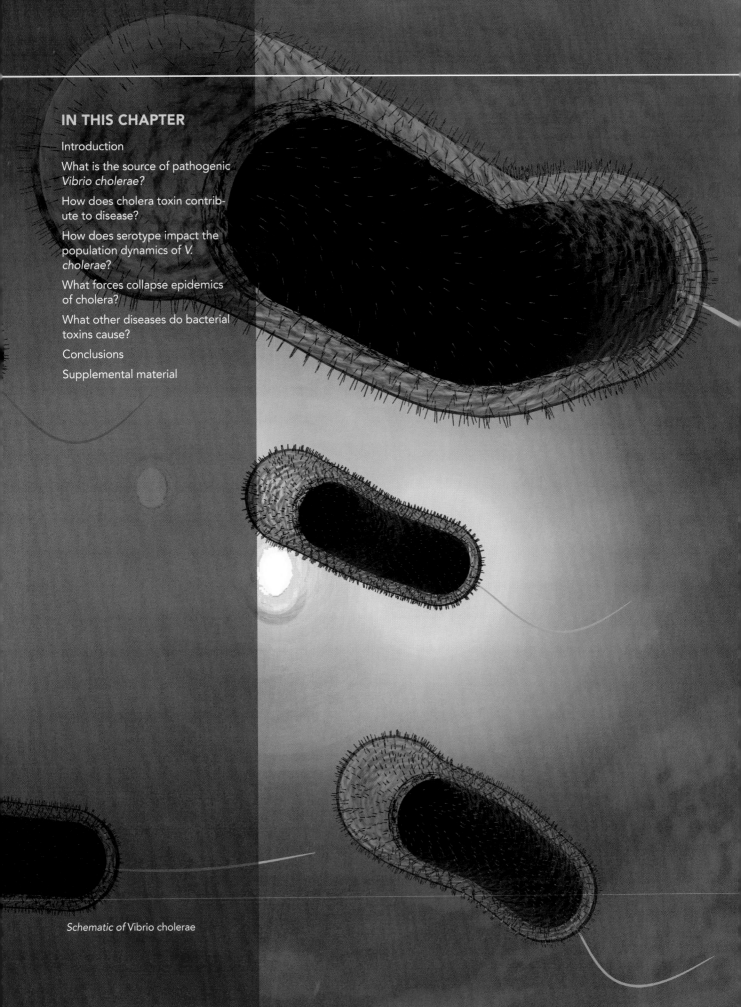

Schematic of Vibrio cholerae

CHAPTER TWENTY-SIX

Toxins and Epidemic Cholera:
Phage Giveth, and
Phage Taketh Away

KEY CONCEPTS

This chapter covers the following topics in the ASM Fundamental Statements.

EVOLUTION

1. Mutations and horizontal gene transfer, with the immense variety of microenvironments, have selected for a huge diversity of microorganisms.

CELL STRUCTURE AND FUNCTION

2. Bacteria have unique cell structures that can be targets for antibiotics, immunity, and phage infection.

INFORMATION FLOW AND GENETICS

3. The synthesis of viral genetic material and proteins is dependent upon host cells.
4. Cell genomes can be manipulated to alter cell function.

MICROBIAL SYSTEMS

5. Microorganisms, cellular and viral, can interact with both human and nonhuman hosts in beneficial, neutral, or detrimental ways.

INTRODUCTION

One way that microbes cause disease is by secreting **toxins**. If, only hours after dining out, you've endured an acute episode of nausea and vomiting, the culprit may well have been a heat-stable toxin deposited in your food by *Staphylococcus aureus*. In fact, microbes produce an impressive variety of toxins. Some are generalists, such as the **cytotoxins** that kill host cells by either degrading or forming pores in

FIGURE 26.1 Rice water stool specimen from a cholera patient. The watery appearance is characteristic of a secretory diarrhea. http://sackler.tufts.edu/Faculty-and -Research/Faculty-Research-Pages /Andrew-Camilli

their membranes. Others act with exquisite specificity, as the paralysis caused by *Clostridium tetani*'s neurotoxin dramatically illustrates. Certain toxins are encoded on mobile genetic elements acquired by some but not all strains of the species; the genes for other toxins are embedded in the species' core genome. To illustrate the power of a toxin and horizontal gene transmission, here we will focus on one pathogen, *Vibrio cholerae*. By analyzing cholera epidemics, we'll also see how lysogenic and lytic phage can drive the evolution of their bacterial hosts. In addition, we'll appreciate the impact of LPS recognition by host antibodies and by predator phage.

CASE 1: Why was Haiti vulnerable to epidemic cholera?

On January 12, 2010, a catastrophic earthquake of magnitude 7.0 devastated Haiti, killing a quarter million of the island's 8.7 million inhabitants and displacing hundreds of thousands more. Amid the chaos of damaged homes and schools, hospitals and government buildings, airport and roads, power and telecommunications, an urgent challenge was to provide shelter, food, clean water, and medical care to the traumatized population of this developing country. Aid workers and supply flights arrived from around the world to address the survivors' basic needs. Just 10 months later, a second disaster struck: cholera, a severe diarrheal disease caused by the waterborne bacterial pathogen *V. cholerae*. Haiti had not seen cholera in the past century, yet 350,000 cases of the disease were reported over the next year. Nearly 80% of cholera patients who do not receive medical care die from dehydration after discharging voluminous watery diarrhea flecked with mucus—"rice water stools" (Fig. 26.1). The death rate falls to <5% for patients who drink solutions of clean water, salts, and glucose. By 2013, the number of cholera cases reported in Haiti had declined to 50,000, but the epidemic had spread to the Dominican Republic, Cuba, and Mexico.

- How did the earthquake increase the Haitians' risk of cholera?
- Where did Haitians encounter *V. cholerae*?
- What is the mode of entry of *V. cholerae*?
- What accounts for the explosive spread of the disease?
- How does *V. cholerae* trigger severe diarrhea?
- Are antibiotics useful for treatment?
- Does *V. cholerae* cause damage?
- How did cholera reach Haiti?
- How did the disease spread to neighboring countries?
- What brought the Haitian epidemic under control?
- What could have been done to minimize the spread of cholera in this situation?

WHAT IS THE SOURCE OF PATHOGENIC *VIBRIO CHOLERAE*?

LEARNING OUTCOMES:
- List the contributions of the *V. cholerae* TCP to its pathogenesis and evolution.
- Propose an experiment or assay to distinguish whether vibrios acquired the 7-kb *ctxAB* element by transformation or by phage transduction.
- Give evidence for the horizontal gene transmission that equipped a pathogen to evolve from benign environmental *V. cholerae*.

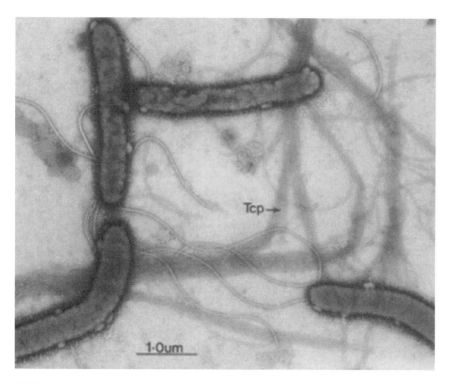

FIGURE 26.2 Toxin-coregulated pilus (TCP) of *V. cholerae*, visualized by electron microscopy. (From Herrington D et al., *J Exp Med*, 1988, http://jem.rupress.org/content /168/4/1487.long.)

V. cholerae is a motile microbe that inhabits lakes, rivers, and estuaries. Like all Gram-negative bacteria, the *V. cholerae* outer membrane is rich in **lipopolysaccharide (LPS)** (chapter 2). In natural waterways, a wide variety of strains persist yet rarely cause severe disease. More than 200 environmental strains are distinguishable by the structure of their LPS, the dominant surface antigen that defines their **serotype**. These vibrios alternately swim freely or reside within **biofilms** attached to crustaceans, large and small. To aggregate on chitin-containing surfaces, vibrios rely on a filamentous appendage, the **toxin-coregulated pilus**, or **TCP** (Fig. 26.2).

But the vibrios' aquatic life is not as idyllic as you may think! Recall from chapter 17 that waters are also teeming with phage, viruses whose ability to multiply depends utterly on infecting susceptible bacteria. As we will see in this chapter, a ceaseless battle between phage and *V. cholerae* plays out not only in the microscopic underwater world, but also in human civilizations.

With some hard-nosed sleuthing, bacterial geneticists tracked down enticing clues to the evolution of pathogenic *V. cholerae*. Unlike benign environmental strains, the genomes of *V. cholerae* isolated during disease outbreaks harbor a ~7-kb element whose content of G+C nucleotides is significantly below the genome average (Fig. 26.3). This foreign "pathogenicity island" carries not only the dreaded **cholera toxin**, encoded by the *ctxAB* locus, but also an enzyme called a **site-specific DNA recombinase**. After inserting an antibiotic resistance gene into the 7-kb pathogenicity island and using a growth medium that contained that antibiotic, microbiologists detected transfer of the genetically marked *ctxAB* island from one *V. cholerae* strain to another. By designing a few more clever

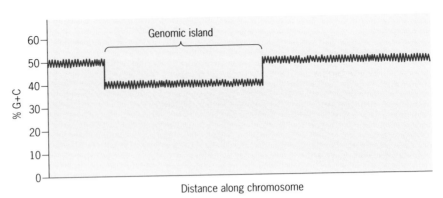

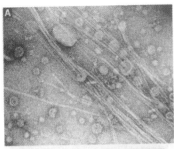

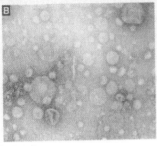

FIGURE 26.3 A hallmark of horizontally acquired DNA is an average-G+C content that deviates from that of the adjacent chromosomal DNA.

FIGURE 26.4 **Supernatants prepared from a culture of V. cholerae carrying the CTX pathogenicity island contain filamentous particles (top), but those from a V. cholerae strain that lacks the genetic element do not (bottom).** Electron microscopy images at 100,000× magnification. (From Waldor MK, Mekalanos JJ, *Science* 272:1910–1914, 1996, with permission.)

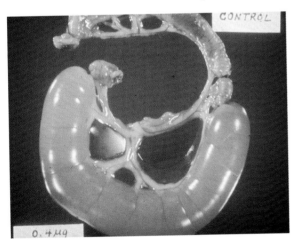

FIGURE 26.5 **Inoculating 0.4 µg of cholera toxin into a section of rabbit small intestine triggers a massive efflux of water into the lumen of the intestine; injection of a control protein does not.** (From MicrobeWorld, http://www.microbeworld.org/component/jlibrary/?view=article&id=1643)

experiments, the scientists were able to distinguish whether vibrios acquired the 7-kb *ctxAB* element by **transformation** or by phage **transduction**. Can you?

Our first clue is that this **horizontal gene transfer**—passage of DNA between two bacterial cells—occurs even when DNase is added to the cell suspension. Second, vibrios readily transfer the 7-kb locus to *V. cholerae* endowed with TCP, but not to cells without these appendages. And, transfer of the *ctxAB* island to TCP-expressing vibrios is blocked if the bacteria are incubated with antibody that binds specifically to TCP filaments. Moreover, cell-free supernatants collected from cultures of the pathogenic *V. cholerae* strain are capable of transferring antibiotic resistance to other vibrios that have TCP. Even more striking, when mice are fed benign vibrios mixed with virulent *V. cholerae* that carry the genetically marked pathogenicity island, the animals shed large numbers of *V. cholerae* that had acquired the *ctxAB* pathogenicity island during passage through the gastrointestinal tract! The astute bacteriology student will not be surprised to learn that cell-free supernatants obtained from virulent *V. cholerae* cultures contain filamentous infectious particles, now known as **CTX phage** (Fig. 26.4).

Based on these experiments and others, it is possible to reconstruct the genetic steps that led to the emergence of pathogenic *V. cholerae* strains from their more benign environmental cousins. (Remember, knowledge is the first step toward prevention.) Using TCP pili as its receptor, CTX phage can bind and enter *V. cholerae*, and then integrate its viral DNA into the host genome to be replicated by the vibrio machinery. TCP serves not only as the phage receptor and to mediate vibrio aggregation on plankton; it also promotes bacterial colonization on our intestinal lining. To compound matters for humans, the same external cues that induce expression of TCP also trigger production of cholera toxin, which readily latches on to the underlying epithelial cells. That's efficiency! This potent toxin is sufficient to induce profuse watery diarrhea in humans and in rabbit models of disease (Fig. 26.5). How?

HOW DOES CHOLERA TOXIN CONTRIBUTE TO DISEASE?

LEARNING OUTCOMES:
- Describe the biochemical mechanism for the watery diarrhea induced by cholera toxin.
- Speculate how to prevent epidemics from becoming pandemics.

Even knowing the havoc it wreaks, it is difficult not to be impressed by cholera toxin's mechanism of action. This key virulence factor is made of two subunits, "A" and "B." The B subunits (five per toxin) *bind* to particular oligosaccharides abundantly expressed on cells that line our gastrointestinal tract. The A subunit is an *active* enzyme that attaches a ribose sugar onto G_s, a host regulatory protein that activates the enzyme adenylate cyclase. Once G_s is modified by this new ribosyl group, it continuously activates adenylate cyclase, which soon fills the intoxicated epithelial cells with excess cAMP. This small nucleotide is a key regulator of the ion channels that control water flux across our epithelial cells. In the presence of excess cAMP, the channels constantly pump chloride ions out of the cell and into the lumen of the gastrointestinal tract (Fig. 26.6). The problem is that water follows this steep chloride concentration gradient. Fluid from tissues pours in abundance into the lumen of the intestinal tract, causing profuse watery diarrhea.

Just how potent is cholera toxin? See for yourself: Compare the volume of stools collected from people who voluntarily drank a dose of *V. cholerae* that either encoded or lacked the phage-borne *ctxAB* genes (Table 26.1). Realize that patients can discharge 5, 10, or more liters of stool, and each milliliter of rice water stool can contain ~10^8 CFU of the pathogen. Some basic math goes a long way toward explaining cholera **epidemics**—sudden, widespread increases in disease—especially when natural disaster, or poverty, increases crowding and reduces barriers between drinking and wastewater.

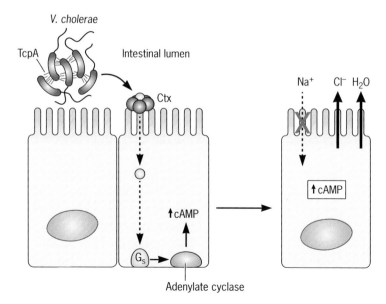

FIGURE 26.6 Mechanism of action of cholera toxin. The five B subunits bind to host cells and trigger translocation into the cytoplasm, where its catalytic A subunit adds an ADP-ribose unit to the activating factor Gs. As a consequence, adenylate cyclase remains "ON," the cell fills with cAMP, and chloride is pumped into the intestinal lumen. Water then follows the chloride gradient.

TABLE 26.1 Volunteer study demonstrates the contribution of cholera toxin to watery diarrhea (dose in CFU is shown)[a]

Strain and dose ingested	No. of volunteers	No. (%) with diarrhea	Mean diarrheal stool vol/ill volunteer (ml)	No. (%) with stool vol ≥ 5 liters
El Tor O1 *ctxAB* mutant				
10^6	4	1 (25)	543	0
10^8	5	2 (40)	1,180	0
10^{10}	5	4 (80)	802	0
El Tor O1 (10^6)	38	35 (92)	4,227	10 (26)

[a]Levine et al. *Infect Immunity* 1988 http://iai.asm.org/content/56/1/161.long.

As we learned from the recent Haiti epidemic, *V. cholerae* has the power to sweep across countries and continents. Indeed, the past 200 years have been marked by seven **pandemics**—disease outbreaks that cross national boundaries. To appreciate how quickly and far *V. cholerae* can spread through natural waterways, examine the transmission map deduced for the seventh pandemic (Fig. 26.7).

HOW DOES SEROTYPE IMPACT THE POPULATION DYNAMICS OF *V. CHOLERAE*?

LEARNING OUTCOMES:
- Distinguish among epidemic, pandemic, and endemic disease.
- Discuss how two mechanisms of horizontal gene transfer have contributed to epidemic cholera.
- Explain why endemic cholera primarily affects children, whereas both children and adults were vulnerable in Haiti and to infection by O139 *V. cholerae*.
- Generate a hypothesis to account for the seasonal cycle of cholera in the Bay of Bengal.

Although >200 serotypes of *V. cholerae* exist in the environment, the culprit in all seven recorded pandemics has been a *V. cholerae* strain with the LPS O1 serotype and an integrated CTX phage. However, beginning in 1992, a new virulent strain emerged in Southeast Asia. In addition to a sharp rise in cases of severe watery diarrhea, health care workers recognized that even adults who had previously been sickened by cholera and had recovered were vulnerable: That is, they lacked **protective immunity,** which is conferred by antibodies specific to surface structures, especially LPS. For example, in the Madras region of India, a new *V. cholerae* serotype, named O139, was isolated 10 times more frequently than the familiar O1 strain. How did this new epidemic strain emerge?

Once again, horizontal gene transfer is the perpetrator. The genome of the new pathogenic O139 strain of *V. cholerae* closely resembles the O1 strain that is **endemic**—or continuously present—in Southeast Asia. As you would expect, the O139 strains isolated from the Madras patients all produced cholera toxin. But there is a striking difference: O139 serotype vibrios have lost the 22-kb region that encodes enzymes that generate the LPS O1 antigen. Inserted in its place is a 35-kb fragment that codes for the novel

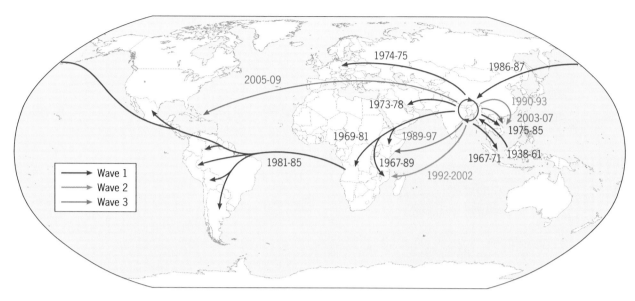

FIGURE 26.7 Transmission pattern deduced for pathogenic *V. cholerae* during the seventh pandemic of cholera. (From Mutreja A, *Nature* **477**:462–465, 2011, doi: 10.1038/nature10392.)

O139 antigen on LPS and a polysaccharide capsule. Are phage to blame, or is another mechanism of horizontal DNA transfer at work?

When growing within biofilms attached to crustaceans, *V. cholerae* are primed to take up DNA in their vicinity. Fragments 20 to 40 kb in size can be internalized and integrated into the genome by homologous recombination. If this process of **natural transformation** (chapter 10) swaps the LPS biosynthetic loci, vibrios that have acquired the O139 antigen escape recognition by antibodies that bind specifically to O1 LPS. Therefore, the O139 strain afflicts not only the young, but also adults who were previously infected with the O1 strain and had acquired O1 type-specific immunity. Moreover, vibrios transformed with the O139 locus also become resistant to phage that prey on serotype O1 *V. cholerae*. Together these observations build the provocative case that a combination of transformation and phage predation likely contributes to the evolution of the many diverse serotypes of *V. cholerae* that persist in the environment (Fig. 26.8).

CASE 2: Why is cholera seasonal in the Bay of Bengal region?

In contrast to Haiti's explosive, and unexpected, epidemic of cholera, seasonal outbreaks are endured each year by people living in the Ganges River Delta region of Bangladesh and India. Peak incidence of cholera disease typically follows autumn monsoon season, with a second wave in the spring; children are primarily affected. Aware that natural waters are rich in microbes, experts at the International Centre of Diarrhoeal Disease Research in Bangladesh and research universities in Boston collaborated to investigate the impact of phage on seasonal endemic cholera. Throughout a 3-year period, they tracked the number of cholera cases caused by either the O1 or the O139 serotype of *V. cholerae*. In parallel, the scientists measured the Bay of Bengal's concentration of phage that can infect and lyse one or the other serotype of this bacterial pathogen (Fig. 26.9).

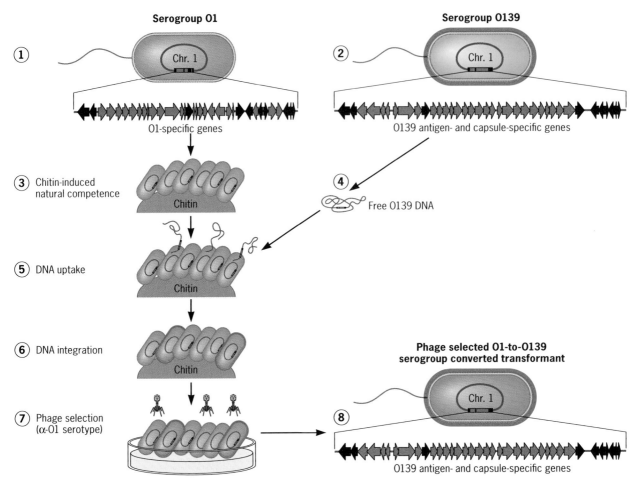

FIGURE 26.8 Model for how DNA transformation and phage predation could promote emergence of a new dominant serotype of *V. cholerae.* Green indicates serotype O1 LPS and genes; red indicates serotype O139. (Modified from Blokesch M, Schoolnik GK, *PLoS Pathog* **3:**e81, 2007, http://journals.plos.org/plospathogensarticle?id=10.1371/journal.ppat.0030081.)

- During this 3-year period, was cholera seasonal?
- Was phage density seasonal?
- Is there a relationship between burden of disease in humans and phage in the Bay?
- Can you generate a hypothesis that accounts for seasonal cholera in this region?

WHAT FORCES COLLAPSE EPIDEMICS OF CHOLERA?

LEARNING OUTCOMES:
- Explain why LPS structure dictates whether vibrios are sensitive or resistant to this particular family of phage.
- Describe the impact of lytic phage on the population dynamics of *V. cholerae* in natural waters.
- Discuss some scientific, economic, and political factors that contribute to the gap between our knowledge of the mechanisms of *V. cholera* pathogenesis and the public health measures enacted to prevent and treat cholera disease.

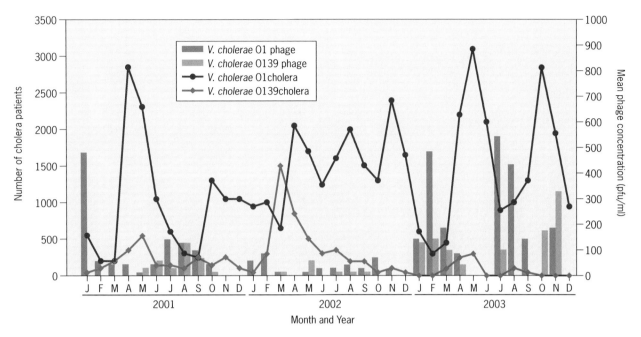

FIGURE 26.9 Mean concentration of lytic phage specific to O1 or O139 *V. cholerae* and estimated number of cases of cholera due to each serotype reported during a 3-year period in Dhaka, India. PFU, plaque-forming units of phage. (Modified from Faruque SM et al., *Proc Natl Acad Sci USA* **102:**1702–1707, 2005, http://www.pnas.org/content/102/5/1702.long.)

Subsequent analysis of stool samples revealed that cholera patients excrete not only copious numbers of vibrios, but also phage (Fig. 26.10). One type of these phage forms plaques on *V. cholerae* by utilizing as its receptor LPS of serotype O1; a different lytic phage attacks *V. cholerae* of the LPS O139 serotype. Can you understand how the abundance and specificity of the phage population can affect the burden of cholera disease (Fig. 26.9) and also drive the evolution of *V. cholerae* (Fig. 26.11)?

Today an estimated 1.4 billion people remain at risk of contracting cholera. Each year, 100,000 to 200,000 die of cholera; the majority are children. Although a few vaccines comprising killed whole vibrio cells have been developed and tested, their stability in the absence of refrigeration is poor, multiple doses are required, and the immune response generated by young children is weak. As a pragmatic, low-cost alternative, in a region of the developing world where cholera is endemic, a program encouraging villagers to filter pond and river water through layers of sari cloth before use reduced the incidence of cholera by nearly half.

WHAT OTHER DISEASES DO BACTERIAL TOXINS CAUSE?

LEARNING OUTCOMES:
- Describe how vaccines prevent toxin-mediated diseases in humans.
- Speculate on how phage benefit from encoding bacterial toxins.

Unfortunately for us, phage spread a number of other toxins among bacterial populations; Table 26.2 lists a few examples. Some toxins are sufficient to convert a benign commensal microbe of humans into a dreaded pathogen, as illustrated by the emergence of **toxigenic strains** of the

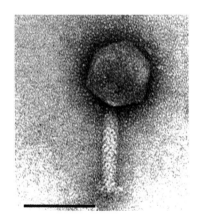

FIGURE 26.10 *V. cholerae* phage isolated from stool sample of cholera patient residing in the Bay of Bengal region. Bar=100 nm. [Modified from Seed KD et al., *mBio* **2(1):**00334-10, 2011, http://mbio.asm.org/content/2/1/e00334-10?ijkey=21db271b5a66ffb76292676be6f1d728a8719426&keytype2=tf_ipsecsha.]

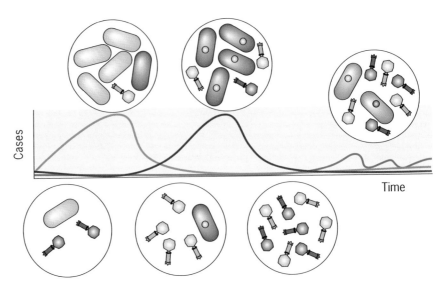

FIGURE 26.11 **Model for how lytic phage drive the emergence and decline of distinct serotypes of *V. cholerae.*** Each color indicates a compatible phage-LPS serotype pair. Above the line are microbial profiles during epidemics and below the line are environmental profiles. (Modified from Faruque SM et al., *Proc Natl Acad Sci USA* **102:**1702–1707, 2005, http://www.pnas.org/content/102/5/1702.long.)

respiratory bacterium *Corynebacterium diphtheriae*. In colonial America, epidemics of diphtheria killed one in three children. On the bright side, people are fully protected from **toxin-mediated diseases** by antibodies that bind and inactivate the toxin. Accordingly, diseases whose symptoms can be mimicked by pure toxin can theoretically be eliminated by widespread vaccination, particularly when humans are the microbe's only reservoir. Indeed, diphtheria was nearly eliminated by the mid 1900s thanks to broad use of "DTP" vaccines (diphtheria toxoid is combined with tetanus toxoid and *Bordetella pertussis* to protect from three different diseases). However, disease resurged when geopolitical strife during the dissolution of the Soviet Union increased crowding and decreased access to health care. The >150,000 cases and >4,000 deaths that struck eastern Europe in the 1990s are a stark reminder that we share the world with microbes that can evolve and spread at an alarming pace.

Numerous bacteria encode toxins, but their effects on host cells differ (Tables 26.2 and 26.3). Similar to cholera toxin, the **ADP-ribosylase** activity of other toxins modifies host factors to commandeer cellular processes. Distinct mechanisms of action have also evolved. **Pore-forming toxins** puncture host cell membranes, which at high doses kills the cell. **Superantigens** bridge MHCII molecules and T cell receptors, irrespective of their antigen specificity. By doing so, this class of toxins stimulates a nonspecific inflammatory immune response and, at high doses, can cause systemic, and lethal, toxic shock. Botulinum toxin is a protease that cleaves a component of the protein machinery that delivers vesicles of neurotrans-

TABLE 26.2 **Phage-encoded toxins**

Pathogen	Toxin	Phage	Symptoms
Vibrio cholerae	Cholera toxin	CTX phage	Watery diarrhea
Corynebacterium diphtheriae	Diphtheria toxin	β phage	Tissue necrosis in respiratory tract
Escherichia coli	Shiga toxins	H19B	Bloody diarrhea
Streptococcus pyogenes	SpeA toxin	T12	Superantigen—nonspecific immune response; shock
Clostridium botulinum	Botulinum toxin	Phage C1	Flaccid paralysis
Staphylococcus aureus	Enterotoxin A	Phage 13	Gastroenteritis, shock
Staphylococcus aureus	Exfoliative toxin A	Phage ETA	Scalded-skin syndrome

TABLE 26.3 **Bacterial toxins and their targets in host cells**

Pathogen	Toxin(s)	Mechanism	Outcome
Streptococcus pyogenes	SpyA toxin	ADP-ribosylates vimentin and actin	Perturbs host cell shape
Corynebacterium diphtheriae	Diphtheria toxin	ADP-ribosylates elongation factor 2	Inhibits protein synthesis
Shigella dysenteriae	Shiga toxin	Inactivates 60S ribosomes	Inhibits protein synthesis
Bordetella pertussis	Pertussis toxin and adenylate cyclase toxin	Elevate cAMP	Pertussis, upper respiratory disease
Clostridium tetani	Tetanus toxin	Inhibits release of inhibitory neurotransmitters	Spastic paralysis
Clostridium botulinum	Botulinum toxin	Inhibits release of excitatory neurotransmitters	Flaccid paralysis
Listeria monocytogenes	Listeriolysin	Pore-forming toxin	Escape from degradative phagolysosome
Staphylococcus aureus	Toxic shock syndrome toxin	Superantigen binds nonspecifically to T-cell receptors and MHCII	Stimulates ineffective immune response

mitter to neighboring neurons, disrupting excitatory signaling. Ironically, clinicians have harnessed "botox" to treat certain spastic motor disorders or, for purely cosmetic purposes, to disable the facial muscles that generate smile and frown lines. Despite their well-deserved reputation as villains of infectious disease, microbial toxins can also be harnessed to develop protective vaccines, potent therapeutics, and powerful laboratory tools to analyze fundamental cellular processes.

CASE 1 REVISITED: **Why was Haiti vulnerable to epidemic cholera?**

Did you recognize multiple ways natural disasters increase the risk of infectious disease, especially cholera? Crowded refugee camps that lack clean water and sanitation are especially vulnerable to outbreaks of diseases that spread by the fecal-oral route. Unfortunately, aide workers newly arrived from Nepal, a country suffering its own cholera epidemic, inadvertently shed *V. cholerae* O1 into the Haitian waters; whole genome sequence analysis revealed that the Nepalese and Haitian epidemic strains were nearly identical to each other and distinct from isolates circulating elsewhere in the world. By triggering profuse watery diarrhea, cholera toxin promotes spread of *V. cholerae* in waterways, irrespective of national borders. Because the secretory diarrhea is stimulated by toxin, not bacterial cells, antibiotics are not necessary for treatment. The toxin does not damage tissues, and dehydration therapy alone saves patients: By drinking solutions of glucose and salts, our tissue fluids and electrolytes can be replaced through the action of glucose-driven ion pumps on the epithelial cells that line the small intestine. A number of factors likely reduced the number of cholera cases in Haiti, including rebuilding of the infrastructure to restore sanitation and clean drinking water, a bloom in the population of lytic phage specific to the O1 serotype of *V. cholerae*, and immunity acquired naturally during infection with the O1 strain.

CASE 2 REVISITED: **Why is cholera seasonal in the Bay of Bengal region?**

After learning about lytic phage and their LPS receptors, and studying the pattern of disease and phage burden in Dhaka, India, did you generate a hypothesis to explain why cholera cases rise and fall every year in the Bay of Bengal? The fall monsoons apparently reduce the number of infectious lytic phage in the waters, creating an opportunity for the *V. cholerae* population to expand. When contaminated water is ingested by members of the community who had not been exposed previously, the patients shed large numbers of infectious vibrios (green cells in Fig. 26.11) into the water supply, and the epidemic expands. In time, the population of lytic phage that bind specifically to the LPS of the virulent strain (green phage) increases. By infecting and lysing *V. cholerae*, these predator phage decrease the concentration of the pathogenic serotype in the water supply, eventually collapsing the epidemic. When environmental conditions once again decrease the concentration of phage, the population of *V. cholerae* can expand (red), and the cycle repeats. *V. cholerae* mutants that lack LPS side chains may arise due to the selective pressure exerted by lytic phage. However, because LPS is critical to the integrity of Gram-negative bacteria, such loss-of-function LPS mutants lose fitness in the human intestine, so the variants are not amplified during epidemics.

CONCLUSIONS

Epidemic cholera vividly illustrates the power of phage-bacteria dynamics. The lysogenic CTX bacteriophage endowed a *V. cholerae* O1 serotype strain with a toxin so potent it can kill humans within days. Fortunately for us, viruses—specifically **lytic bacteriophage**—can, and do, attack specific lineages of *V. cholerae*, depending on their LPS structure. Thus, unending conflicts between phage, bacteria, and humans drive bacterial evolution, as vividly illustrated by endemic, epidemic, and pandemic cholera.

Representative ASM Fundamental Statements and Learning Outcomes for the Chapter

EVOLUTION

1. Mutations and horizontal gene transfer, with the immense variety of microenvironments, have selected for a huge diversity of microorganisms.

   a. Discuss two mechanisms of horizontal gene transfer and how they have contributed to epidemic cholera.
   b. Give evidence for the horizontal gene transmission that equipped a pathogen to evolve from benign environmental *V. cholerae*.

CELL STRUCTURE AND FUNCTION

2. Bacteria have unique cell structures that can be targets for antibiotics, immunity, and phage infection.

   a. List the contributions of the *V. cholerae* TCP to its pathogenesis and evolution.
   b. Explain why endemic cholera primarily affects children, whereas both children and adults were vulnerable in Haiti and to infection by O139 *V. cholerae*.
   c. Explain why LPS structure dictates whether vibrios are sensitive or resistant to this particular family of phage.
   d. Describe two consequences of LPS structural changes on epidemic cholera.

INFORMATION FLOW AND GENETICS

3. The synthesis of viral genetic material and proteins is dependent upon host cells.

   a. Describe the impact of lytic phage on the population dynamics of *V. cholerae* in natural waters.

   b. Speculate whether it is possible to harness bacteriophage to eliminate pathogenic *V. cholerae* from infected waters.

4. Cell genomes can be manipulated to alter cell function.

   a. Propose an experiment or assay to distinguish whether vibrios acquired the 7-kb *ctxAB* element by transformation or by phage transduction.

MICROBIAL SYSTEMS

5. Microorganisms, cellular and viral, can interact with both human and nonhuman hosts in beneficial, neutral, or detrimental ways.

   a. Speculate how to prevent epidemics from becoming pandemics.

   b. Distinguish among epidemic, pandemic, and endemic disease.

   c. Generate a hypothesis to account for the seasonal cycle of cholera in the Bay of Bengal.

   d. Discuss some scientific, economic, and political factors that contribute to the gap between our knowledge of the mechanisms of *V. cholerae* pathogenesis and the public health measures enacted to prevent and treat cholera disease.

   e. Describe how vaccines prevent toxin-mediated diseases in humans.

   f. Describe the biochemical mechanism for the watery diarrhea induced by cholera toxin.

SUPPLEMENTAL MATERIAL

References

- **Herrington DA, Hall RH, Losonsky G, Mekalanos JJ, Taylor RK, Levine MM.** 1988. Toxin, toxin-coregulated pili, and the *toxR* regulon are essential for *Vibrio cholerae* pathogenesis in humans. *J Exp Med* **168:**1487–1492. http://jem.rupress.org/content/168/4/1487.long

- **Faruque SM, Islam MJ, Ahmad QS, Faruque AS, Sack DA, Nair GB, Mekalanos JJ.** 2005. Self-limiting nature of seasonal cholera epidemics: role of host-mediated amplification of phage. *Proc Natl Acad Sci USA* **102:**6119–6124. http://www.pnas.org/content/102/17/6119.long

- **Hendrikson RS et al.** 2011. Population genetics of *Vibrio cholerae* from Nepal in 2010: evidence on the origin of the Haitian outbreak. *mBio* **2(4):**e00157–11 http://mbio.asm.org/content/2/4/e00157-11

- **Faruque SM, Mekalanos JJ.** 2012. Phage-bacterial interactions in the evolution of toxigenic *Vibrio cholerae*. *Virulence* **3:**556–565. http://www.tandfonline.com/doi/full/10.4161/viru.22351

Supplemental Activities

CHECK YOUR UNDERSTANDING

1. **What does the lysogenic CTX phage require to persist in *V. cholerae* populations?**
   a. LPS
   b. Toxin coregulated pilus (TCP)
   c. *ctxAB* locus and DNA recombinase
   d. Recognition sequence for site-specific DNA recombinase
   e. TCP and recognition sequence for site-specific DNA recombinase

2. **Infectious diseases are prevalent in many parts of the world. Which of these measures is most likely to help the greatest number of the people affected?**
   a. Widespread administration of antimicrobial drugs
   b. Vaccination against a few specific agents

   c. Sanitation of the water and food supply
   d. Urbanization of the rural population
   e. Mosquito control

3. **During seasonal cholera outbreaks in the Bay of Bengal region, children are primarily affected because they**
   a. are less likely to practice good hygiene.
   b. lack a mature cell-mediated immune response.
   c. are more often exposed to contaminated water.
   d. are more often exposed to people with cholera.
   e. have had less natural exposure to *V. cholerae*

4. **Prepare a schematic or animation to teach the mechanism of action of cholera toxin. Include toxin, toxin receptor, Gs protein, adenylate cyclase, cAMP, chloride, chloride transporter, glucose-driven ion pumps, and water. Use your visual aid to explain how oral rehydration therapy treats this disease.**

DIG DEEPER

1. The *V. cholerae* pathogenicity island (VPI) was identified by its dramatically different GC content compared to the bacterial chromosome. In this *in silico* research exercise, you will use publicly available resources to:
   a. Locate in the bacterial genome the base pair range for the VPI.
   b. Identify the numbers assigned to the genes that encode the toxin-coregulated pilus (TCP).
   c. Determine how the VPI GC content compares to the rest of the genome.
   d. Locate the *ctxA* gene in the *V. cholerae* genome sequence.
   • Download the *V. cholerae* O1 biovar El Tor str. N16961 chromosome I sequence: http://www.ncbi.nlm.nih.gov/nuccore/15640032?report=fasta
   • Analyze the GC content of the chromosome using GC-Profile: http://tubic.tju.edu.cn/GC-Profile/
     • Upload the FASTA file (.txt).
     • Leave standard settings, but check "label the coordinates to cumulative GC profile."
     • Submit analysis.
     • View the plot.
     • Hover over each gene to see a description of the predicted function of the open reading frame.
     • The *ctxA* gene is not annotated in the *V. cholerae* genome. To identify this phage gene, do an NCBI gene search (http://www.ncbi.nlm.nih.gov/gene) for *ctxA*. Next, use BLAST (http://blast.ncbi.nlm.nih.gov/Blast.cgi) to interrogate the *V. cholerae* El Tor genome using the *ctxA* phage gene sequence.

2. Review the data from the volunteer study described in Table 26.1.
   a. Develop a hypothesis to account for the observation that larger doses of *ctxAB* mutant *V. cholerae* caused a larger percentage of volunteers to have diarrhea.
   b. Design an experiment to test your hypothesis.
   c. Under what circumstances, if any, is it ethical to recruit volunteers to ingest live *V. cholerae*? To participate in the experiments you designed to identify why the *ctxAB* mutant caused diarrhea?

3. Read the study by Kato et al., "Enhanced sensitivity to cholera toxin in ADP-ribosylarginine hydrolase-deficient mice" (*Mol Cell Biol* **27**:5534–5543, 2007, http://www.ncbi.nlm.nih.gov/pmc/articles/PMC1952103/), which examines the impact of a host enzyme on the potency of cholera toxin, and study Fig. 4A and B (below).
   a. What motivated the scientists to investigate the host enzyme ADP-ribosylarginine hydrolase?
   b. Draw a schematic to illustrate the experimental design for Fig. 4A and B.
   c. Why was fluid accumulation reported as a ratio of milligrams per centimeter?
   d. Why was PBS (A and B) and 8-bromo-cAMP (B) analyzed?
   e. What are *P* values, and why are they included?
   f. What conclusions can be drawn from these data?

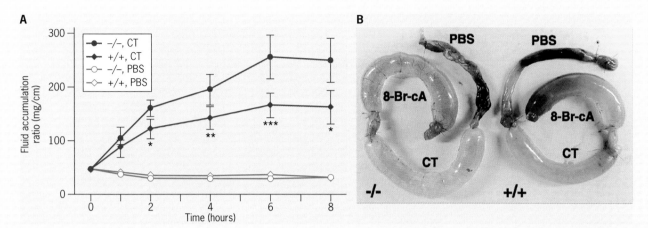

FIGURE 4 Effect of cholera toxin (CT) on accumulation of fluid in intestinal loops of ADPRH wild-type and mutant (KO) mice. (A) In each wild-type (square) or mutant (circle) mouse, intestinal loops were injected with 0.2 ml of PBS without (◇ or ○, respectively) or with (◆ or ●, respectively) CT (0.5 µg) and were separately excised at the indicated time thereafter. The weight and length of each loop reported as milligrams per centimeter were used to measure of fluid accumulation. Data are means±SEM (*n*=6) of values for fluid accumulation in intestinal loops of ADPRH+/+and ADPRH−/− mice. For the difference between two alleles incubated with CT, the *P* values are as follows: *, *P*=0.004; **, *P*=0.007; ***, *P*=0.002. (B) Appearance of intestinal loops 6 h after injection of 0.2 ml of PBS without or with 0.5 µg of CT or 5 mM 8-bromo-cAMP (8-Br-cA). (Modified from Kato et al., *Mol Cell Biol* **27**:5534–5543, 2007.)

g. Draw a schematic to illustrate the main biological finding from the data in Fig. 4A and B.

4. **Team work:** Select a toxin from Table 26.2. Research and then teach your classmates its mechanism of action using visual aids. Alternatively, act out the cellular pathway by assigning roles for the key bacterial and host components.

5. "General Hospital": Using skit-writing and role-playing to teach pathogenesis, by Adrienne Ann Dolberry, *JMBE* 12 (1), 2011. http://jmbe.asm.org/index.php/jmbe/article/view/214/html_101

6. Would it be possible to treat cholera patients with a predator phage? Discuss potential designs and pros and cons of this type of treatment. In a given geographical region, would the treatment become more or less effective over time? Would it be ethical? Use as resources Fig. 25.9 and the American Academy of Microbiology report "Harnessing the Power of Microbes as Therapeutics: Bugs as Drugs" (http://asmscience.org/content/colloquia.52;jsessionid=3egigi2x62mul.x-asm-books-live-01).

Supplemental Resources

VISUAL

- *iBiology* lecture "Escalating Infectious Disease Threat" by Lucy Shapiro (40 min): http://www.ibiology.org/ibioseminars/microbiology/lucy-shapiro-part-2.html
- TEDSalon video (10 min): Author Steven Johnson discusses his book *The Ghost Map* which describes how John Snow discovered fundamental *V. cholerae* biology as he investigated the 1854 cholera outbreak in London. http://www.ted.com/talks/steven_johnson_tours_the_ghost_map?language=en
- *Harvard University Water Lecture Series* video (52 min): In "River Monster: The Epidemiology, Ecology, and Pathobiology of Cholera || Radcliffe Institute," John Mekalanos discusses the biology of cholera. With his many colleagues in Bangladesh, Haiti, and elsewhere, he has provided strong evidence for how this organism emerged as a human pathogen and has recently become more pathogenic, as well as for why epidemics begin and end so abruptly. He applied this knowledge to the construction of genetically stable cholera vaccines that have been successfully tested in the United States and Bangladesh. http://www.microbeworld.org/component/jlibrary/?view=article&id=9605

SPOKEN

- McMaster University interview (1.5 min) with a scientist who sequenced ancient *V. cholerae* strain from a 200-year-old intestinal specimen archived by the College of Physicians of Philadelphia, a study published by Devault AM et al. (N *Engl J Med* **370:**334–340, 2014). http://dailynews.mcmaster.ca/article/scientists-unlock-evolution-of-cholera/
- *This Week in Microbiology* podcast episode #108 "Vaccine in the Time of Cholera" discusses results of a 2015 oral cholera vaccine trial in urban Bangladesh by Qadri et al. (*Lancet* **386:**1362–1371, 2015). http://www.microbeworld.org/podcasts/this-week-in-microbiology /archives/1948-twim-108-vaccine-in-the-time-of-cholera
- *This Week in Microbiology* podcast episode #15 "Microbial Long Distance Relationships" discusses the epidemiological link between the 2010 cholera outbreak Haiti and peacekeepers from Nepal presented in a study published by Hendrikson RS et al. in *mBio* 2(4):e00157 11, 2011). http://www.microbeworld.org/component/content/article?id=1014.

WRITTEN

- CDC public resources on cholera: http://www.cdc.gov/cholera/index.html
- CDC's Cholera and Other Vibrio Illness Surveillance System (COVIS): Learn about the incidence of cholera in the United States (http://www.cdc.gov/cholera/usa/surveillance .html), Africa (http://www.cdc.gov/cholera/africa/index.html), Southeast Asia (http:// www.cdc.gov/cholera/asia/index.html), and Haiti (http://www.cdc.gov/cholera/haiti/index .html).
- *From Outside to Inside: Environmental Microorganisms as Human Pathogens*, a 2005 colloquium report from the American Academy of Microbiology. http://academy.asm .org/index.php/browse-all-reports/178-from-outside-to-inside-environmental -microorganisms-as-human-pathogens
- *Clean Water: What Is Acceptable Microbial Risk?* A 2007 colloquium report from the American Academy of Microbiology. *http://academy.asm.org/index.php/browse-all -reports/521-clean-water-what-is-acceptable-microbial-risk*
- *Microbes in Pipes: The Microbiology of the Water Distribution System*, a 2013 colloquium report from the American Academy of Microbiology. http://academy.asm.org/index .php/browse-all-reports/520-water-distribution-system
- National Center for Case Study Teaching in Science: "Danielle's Difficulty: Risks, Treatments, and Prevention of *Clostridium difficile*" by Dorothy Debbie incorporates the epidemiology, pathogenesis, treatment, and prevention of this toxin-mediated disease. http://sciencecases.lib.buffalo.edu/cs/collection/detail.asp?case_id=699&id=699
- National Center for Case Study Teaching in Science: "Disease Along the River: A Case Study and Cholera Outbreak Game" by Andrea Nicholas simulates a cholera outbreak and incorporates how geographic, farming, and sanitation practices impact spread of infectious disease. http://sciencecases.lib.buffalo.edu/cs/collection/detail.asp?case_id =766&id=766

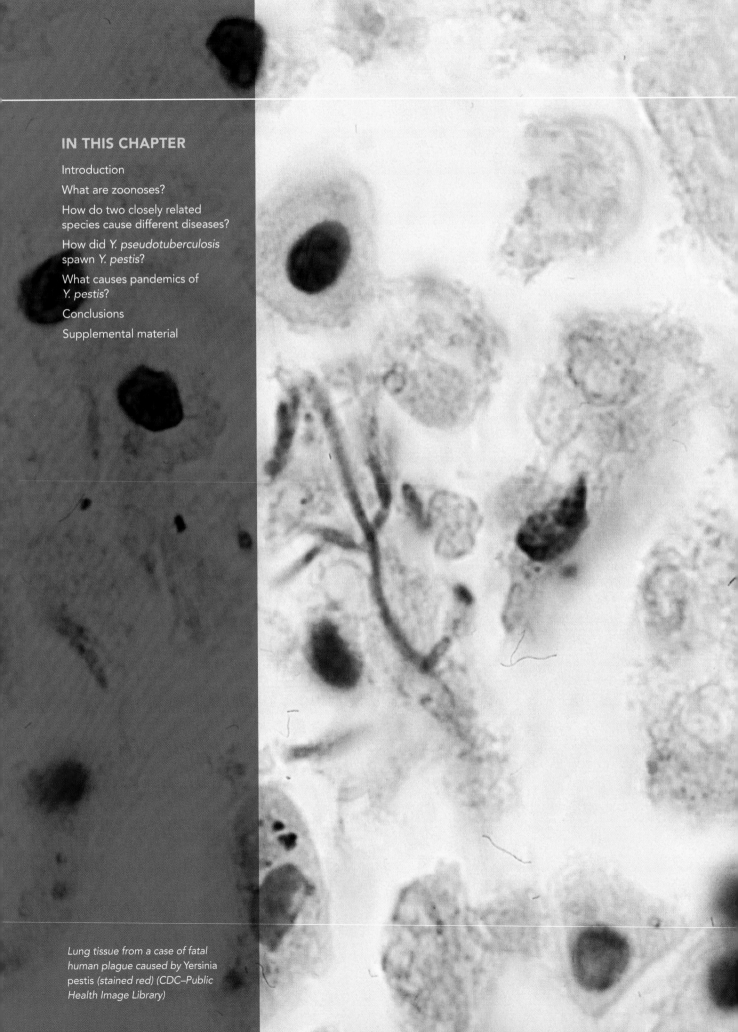

Lung tissue from a case of fatal human plague caused by Yersinia pestis (stained red) (CDC–Public Health Image Library)

CHAPTER TWENTY-SEVEN

Zoonoses:
How Plague Emerged from a Foodborne Illness

KEY CONCEPTS

This chapter covers the following topics in the ASM Fundamental Statements.

EVOLUTION

1. Mutations and horizontal gene transfer, with the immense variety of micro-environments, have selected for a huge diversity of microorganisms.

CELL STRUCTURE AND FUNCTION

2. *Bacteria* and *Archaea* have specialized structures that often confer critical capabilities.

METABOLIC PATHWAYS

3. The interactions of microorganisms among themselves and with their environment are determined by their metabolic abilities.

INFORMATION FLOW AND GENETICS

4. Genetic variations can impact microbial functions.
5. Cell genomes can be manipulated to alter cell function.

MICROBIAL SYSTEMS

6. Most bacteria in nature live in biofilm communities.
7. Microorganisms and their environment interact with and modify each other.
8. Microorganisms, cellular and viral, can interact with both human and non-human hosts in beneficial, neutral, or detrimental ways.

INTRODUCTION

Although animals provide sustenance and joy, they can also make us sick. Whether in the wild, on farms, or in our homes, animals are a source of an impressive variety of infectious diseases, collectively referred to as "zoonoses." In addition to familiar illnesses, like diarrhea caused by foodborne *Salmonella typhimurium* or rabies acquired from the bite of an infected bat, alarming new diseases emerge from animal reservoirs, such as the 2012 outbreak of the severe respiratory illness Middle East respiratory syndrome (MERS). Because they typically circulate in free-roaming animals, **zoonotic** pathogens can be challenging to track down and control. In this chapter two highly related zoonotic bacterial pathogens that cause strikingly different illnesses—*Yersinia pseudotuberculosis*, the cause of foodborne gastrointestinal illness, and *Yersinia pestis*, the flea-borne agent of the plague—will vividly illustrate the power of such microbial attributes as type III secretion systems, horizontal gene transmission, the second messenger cyclic di-GMP, and biofilms to transform pesky fleas into agents of the Black Death. This highly contagious disease devastated Europe in the 14th century, killing tens of millions of people and leaving indelible marks on economies, societies, art, and religion.

CASE 1: A foodborne infection masquerading as appendicitis

Clinical microbiologists in multiple laboratories across the southeastern United States were surprised when, during a 2-week period, there was an outbreak of a relatively rare intestinal disease. The 17 patients ranged in age from 12 to 72 years, and each reported fever and abdominal pain; three also had diarrhea. All recovered, whether treated with antibiotics or not. Extensive interviews revealed that 11 had consumed iceberg lettuce at their school or workplace cafeteria; the 12th was kitchen staff in one of these establishments. Investigation by each state's Department of Public Health ascertained that earlier that month, each of the cafeterias had received a lettuce shipment from the same produce distributor. Inspection of the distributor's farms found fields irrigated with water from reservoirs supplied by rainwater and runoff; the reservoirs were not fenced off, and the local deer population was robust. *Yersinia pseudotuberculosis* was isolated from the patients, cafeteria lettuce, and deer feces collected from the distributor's farmland. Genomic DNA of bacteria isolated from patients, cafeteria lettuce, and deer feces was analyzed by pulsed-field gel electrophoresis to determine if the strains were related (see chapter 23, Fig. 23.12).

- Where did the individuals encounter *Y. pseudotuberculosis*?
- How did *Y. pseudotuberculosis* enter these individuals?
- How did the individuals recover without antibiotic treatment?
- How did pulsed-field gel electrophoresis analysis contribute to the outbreak investigation?

CASE 2: A hunter's narrow escape

A 57-year-old man living in rural eastern California was brought to a hospital after 3 days of high fever, a painful swelling in his groin, and necrotic tissue on his toes (Fig. 27.1). Blood was drawn, the groin lesion was aspirated, and intravenous antibiotics were immediately administered. After 2

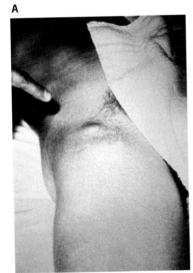

A

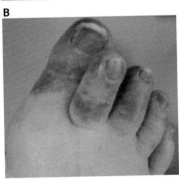

B

FIGURE 27.1 **Clinical presentation of** **Y. pestis infection.** (A) Characteristic swelling of the regional lymph node, forming a "bubo" as a result of a robust inflammatory response to the bacteria. (B) Tissue necrosis characteristic of *Y. pestis* bloodstream infection. (Panel A, MicrobeLibrary, http://www.microbe library.org/images/cdcphil/yersinia %20pestis%20fig4.jpg; panel B, from Margolis DA et al., *Am J Trop Med Hyg* **78:**868–871, 2008.)

days of culture, the groin aspirate and blood culture were Gram stained (Fig. 27.2); subsequent immunological and molecular tests confirmed infection by *Yersinia pestis*. After several days in the intensive care unit and amputation of his necrotic toes, the man recovered fully. Interviews determined that, during the previous week, the patient had hunted, cleaned, and skinned wild rabbits. After taking appropriate safety precautions, the county's Department of Public Health staff inspected the hunting site, where they identified rotting rabbit carcasses that contained the bacteria with the same pulsed-field gel electrophoresis profile as the patient. Health care workers and close contacts of the case were screened and offered antibiotic prophylaxis.

- Where did the individual encounter *Yersinia pestis*?
- How did *Y. pestis* enter the individual?
- How did *Y. pestis* cause damage?
- Why was antibiotic treatment urgent for the patient and his close contacts?
- How did pulsed-field gel electrophoresis analysis contribute to the epidemiological investigation?

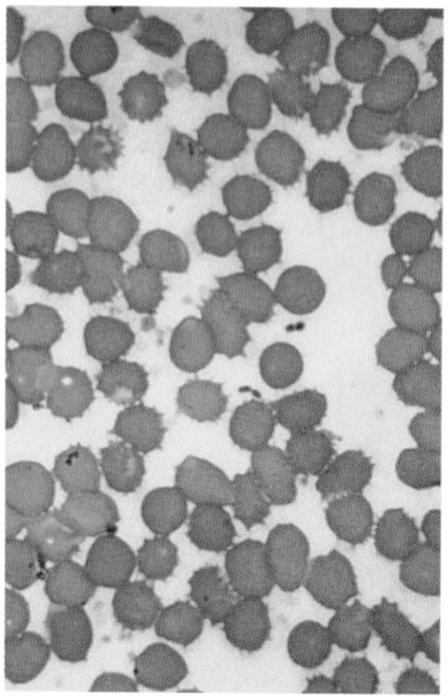

FIGURE 27.2 A patient's blood specimen is Gram stained as a first step to identify the causative agent of the infection.

WHAT ARE ZOONOSES?

LEARNING OUTCOMES:
- List two direct and two indirect ways that humans can acquire zoonotic pathogens.
- Speculate about zoonoses for which you might be at risk (see Table 27.1).

While the cafeteria patrons and hunter suffered very different diseases, the microbes that caused their infections share a number of key traits. The gastrointestinal illness and life-threatening blood infection were caused by two very closely related bacteria, *Yersinia pseudotuberculosis* and *Yersinia pestis,* infectious agents that are each transmitted from animals to humans.

Exposure to wild and domesticated animals increases our risk of acquiring pathogenic viruses, bacteria, fungi, and parasites that circulate in vertebrate populations (Fig.27.3). Humans may acquire zoonotic pathogens

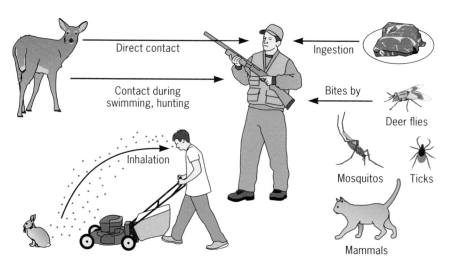

FIGURE 27.3 Humans can acquire zoonotic infections by direct contact, inhalation, or ingestion, or from bites by infected insects or animals.

by direct contact with an infected animal, such as while cleaning wild game (the hunter of Case 2), or by being bitten or scratched. Others may be exposed indirectly, through aerosols of animal urine or fur, or by ingesting food or water that had been contaminated by an infected animal. Deer that carry *Y. pseudotuberculosis* or other pathogens in their gastrointestinal tract are an indirect source of foodborne human disease, as we saw in Case 1. Alternatively, a biting insect that previously fed on an infected animal may transmit the pathogen to vertebrates of either the four- or two-legged variety. Indeed, fleas are a critical agent, or "vector," that spreads *Y. pestis* among its animal **reservoir** and, more rarely, to humans. In some cases, such as the rabies virus, the pathogen causes no symptoms in its natural host, or reservoir (bats). In other instances, including *Y. pestis*, one vertebrate host (rats) may tolerate infection whereas another animal species (the wild rabbits of Case 2) is vulnerable. Considering the various ways we interact with many different animals, it is easy to understand why there are so many zoonotic diseases! Table 27.1 lists some examples.

TABLE 27.1 Zoonoses: some of the many microbial infections humans acquire from animals

Encounter	Disease	Microbe	Group	Common reservoirs
Ingestion	Gastroenteritis	*Yersinia pseudotuberculosis*	Gram-negative bacterium	Deer, sheep, pigs, birds
	Gastroenteritis	*Salmonella* species (except *S. typhi*)	Gram-negative bacterium	Fowl, domestic mammals, reptiles
	Gastroenteritis	*Campylobacter jejuni*	Gram-negative bacterium	Fowl, domestic mammals
	Gastroenteritis	*Listeria monocytogenes*	Gram-positive bacterium	Domestic mammals, rodents
	Toxoplasmosis	*Toxoplasma gondii*	Protist	Cats
	Trichinosis	*Trichinella spiralis*	Nematode	Pigs, domestic and wild mammals, rodents
	Variant Creutzfeldt-Jakob disease	Not applicable	Prion	Cattle
Arthrodpod bite	Plague	*Yersinia pestis*	Gram-negative bacterium	Rodents, rabbits; flea vector
	Lyme disease	*Borrelia burgdorferi*	Spirochete	Rodents, deer; tick vector
	Rift Valley fever	Bunyavirus	Virus	Sheep, goats, cattle; mosquito vector
	African sleeping sickness	*Trypanosoma cruzi*	Protist	Cattle, wild animals; tsetse fly vector
Animal bite or scratch	Rabies	Rabies virus	Virus	Bats, skunks, foxes, opossums, domestic mammals
	Cat scratch disease	*Bartonella henselae*	Gram-negative bacterium	Cats, dogs
	Pasteurellosis	*Pasteurella multocida*	Gram-negative bacterium	Cats, dogs, wild mammals
Inhalation	Anthrax	*Bacillus anthracis*	Gram-positive bacterium	Goat, sheep, and cattle wool or skin
	Histoplasmosis	*Histoplasma capsulatum*	Fungus	Bird and bat excrement
	Hantavirus pulmonary syndrome	Sin Nombre virus and others	Virus	Mice and rodent excrement

HOW DO TWO CLOSELY RELATED SPECIES CAUSE DIFFERENT DISEASES?

LEARNING OUTCOMES:

- Describe the reservoir, vector, mode of transmission, and symptoms of infection for *Y. pseudotuberculosis* and *Y. pestis*.
- Explain how the large plasmid carried by both bacteria plays a role in disease in both cases.
- Explain why adhesins are an important virulence factor for *Y. pseudotuberculosis*.
- Trace the steps from the entry of *Y. pseudotuberculosis* into a cell to its spread within a patient's body.
- Describe why antibiotics are not necessary to treat patients with *Y. pseudotuberculosis* gastroenteritis.

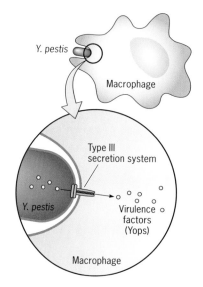

FIGURE 27.4 The steps of phagocytosis by macrophages of benign bacteria are illustrated in the first animation. Next, see how *Y. pestis* disarms macrophages using type III secretion effectors and how *Mycobacterium tuberculosis* establishes a replication niche inside macrophages. http://www.microbelibrary.org/images/tterry/anim/phago053.html (credit Thomas Terry)

Y. pseudotuberculosis and *Y. pestis* are genetically quite similar, yet their lifestyles and impact on humans differ drastically. *Y. pseudotuberculosis* resides in the gastrointestinal tract of a wide range of animals, including deer, sheep, pigs, and birds, but this Gram-negative member of the *Enterobacteriaceae* is rarely identified as a cause of human illness. If swallowed in contaminated food or water, it can cause an abdominal infection that, though painful, typically resolves on its own, as we saw in the foodborne outbreak of Case 1. In contrast, the dreaded *Y. pestis* is carried from the bloodstream of one rodent to another by the bite of a flea. *Y. pestis* can also infect humans; indeed, historically it caused deadly epidemics that terrorized continents. When teamed up, these two *Yersinia* species become compelling teachers of a number of processes common in the microbial world.

Think of *Y. pestis* as the unruly younger cousin of *Y. pseudotuberculosis*. *Y. pestis* separated from the more ancient *Y. pseudotuberculosis* only within the past 20,000 years; the majority of their genes still retain >97% identity. Both carry a large plasmid that encodes a type III secretion system (chapter 9, Fig. 9.9), a delivery mechanism that equips both pathogens to disarm multiple defense pathways of their hosts. The payload this apparatus injects into the cytoplasm of eukaryotic cells includes virulence factors capable of interfering with ingestion by macrophages (Fig. 27.4) and signaling pathways that modulate innate immune responses.

Only a handful of genetic differences between these cousins accounts for their distinct lifestyles (Table 27.2). *Y. pseudotuberculosis* is endowed

TABLE 27.2 Four genetic changes critical to the evolution of flea-borne *Y. pestis* from its gastrointestinal tract predecessor, *Y. pseudotuberculosis*

Function	Gene	Contribution	Presence in:	
			Y. pseudotuberculosis	*Y. pestis*
Phospholipase D	*ymt*	Promotes survival in flea midgut	−	+
Phosphodiesterase 3	*pde3*	Degrades cyclic di-GMP, an inducer of biofilm formation	+	−
Phosphodiesterase 2	*pde2*	Degrades cyclic di-GMP, an inducer of biofilm formation	+	−
RscA regulatory protein	*rscA*	Represses expression of a cyclic di-GMP synthetase gene	+	−

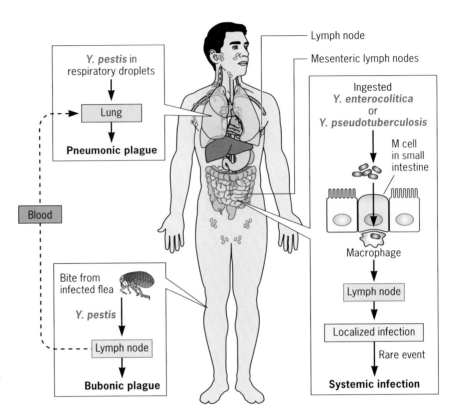

FIGURE 27.5 **The steps of pathogenesis for the zoonotic pathogens** *Y. pseudotuberculosis* and *Y. pestis.* (Modified from Wren BW, *Nat Microbiol Rev* **1**:55–64, 2003.)

with multiple **adhesins** that promote tight binding to tissues and even invasion into mammalian cells. As a consequence, within the intestine, this pathogen traverses M cells, a gateway across this epithelial barrier (Fig. 27.5). From there, the bacteria can travel to abdominal lymph nodes and multiply. Fortunately, our innate immune system's Toll-like receptor 4 easily recognizes the **LPS** coat of *Y. pseudotuberculosis* to alert host defense pathways (chapter 22, Fig. 22.9). Dramatic enough to mimic painful appendicitis, the ensuing robust immune response typically gets the upper hand, and the infection is contained and cleared. Many wild animals are adept at restricting *Y. pseudotuberculosis* to their gastrointestinal tract and shedding the microbe back into the environment, uneventfully. Because of this natural reservoir, sporadic outbreaks of human gastroenteritis due to *Y. pseudotuberculosis* are reported worldwide; in the United States today, typically <10 cases are recorded each year.

HOW DID *Y. PSEUDOTUBERCULOSIS* SPAWN *Y. PESTIS*?

LEARNING OUTCOMES:
- Describe the key events that led to the speciation of *Y. pestis* from *Y. pseudotuberculosis*.
- Explain the genetic mechanism by which bacteria acquire or lose traits.
- Explain how the genetic differences between *Y. pestis* and *Y. pseudotuberculosis* influence biofilm formation.
- Identify how the capacity to form biofilms promotes the transmission of *Y. pestis*.

- List three traits that *Y. pestis* and *Y. pseudotuberculosis* have in common and three that are different.
- Discuss why the risk to humans is higher for *Y. pestis* than for *Y. pseudotuberculosis*.

So how did the reasonably well-behaved foodborne pathogen *Y. pseudotuberculosis* spawn the dreaded killer *Y. pestis*? Impressive sleuthing by teams of paleobiologists, genomics specialists, and bacterial geneticists—with some help from flea anatomy experts—reconstructed their evolutionary trail. Like humans, rodents that encounter *Y. pseudotuberculosis* occasionally fail to contain the microbe in their intestinal tracts; in these rare instances, the infection can spread to the rodent's bloodstream. When this **septic** rodent's entourage of fleas take their usual blood meal, they also ingest *Y. pseudotuberculosis*. These bacteria pass through the flea's intestinal tract, replicate in the hindgut, and are shed in the feces.

Big trouble began when *Y. pseudotuberculosis* acquired a new plasmid, one encoding an enzyme—phospolipase D—that increases survival of the bacteria in the flea's midgut. Because this new niche is relatively free of microbial competitors, the newly endowed pathogen could replicate to high densities. Over time, in a series of evolutionary steps, these midgut residents accumulated mutations in three genes—*PDE3*, *PDE2*, *rcsA*—whose products each function to *decrease* the bacterial supply of a small signaling molecule, the second messenger cyclic di-GMP. Because this molecule induces the elaborate pathway of biofilm formation (chapter 13), *Y. pseudotuberculosis* variants that lack functional PDE3, PDE2, and RscA have *increased* cyclic di-GMP and therefore make *thicker* biofilms. Consequently, when ingested in a blood meal, this new *Yersinia* lineage is more apt to get stuck in the flea's foregut, forming a luxuriant biofilm on the valve that connects the flea's esophagus to its midgut (Fig. 27.6). As you might imagine, these massive globs of sticky bacteria don't go unnoticed. As the insect feeds, its meal backs up in its constricted esophagus, increasing the likelihood that some blood—now laced with *Yersinia*—is regurgitated into the unsuspecting rodent, where the cycle continues. Thus, by sequential gene gain and gene loss, the flea-borne species *Y. pestis* emerged

A

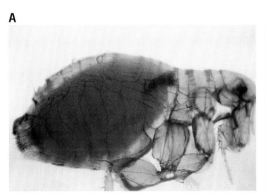

B

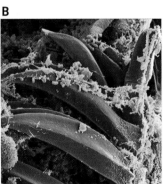

FIGURE 27.6 *Y. pestis* **forms biofilms in fleas.** (A) *Y. pestis* forms biofilms (dark mass) on a valve in the foregut of the flea, *Xenopsylla cheopis*. (B) A mat of bacteria (yellow) forms on the proventricular spines of the flea (purple), visualized by scanning EM. (Panel A, from Public Health Image Library; panel B, from MicrobeWorld.)

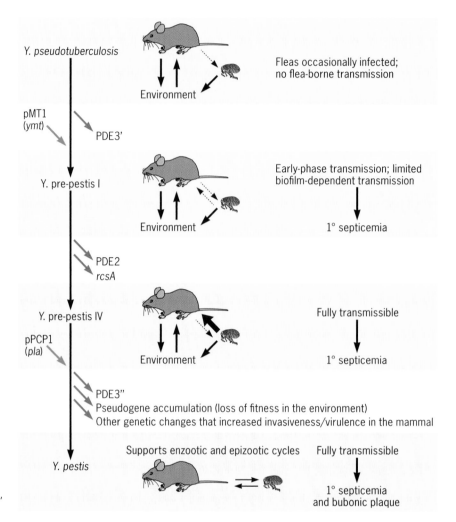

FIGURE 27.7 Over the past ~20,000 years, flea-borne *Y. pestis* evolved from gastrointestinal *Y. pseudotuberculosis* by gene gain and loss. (Modified from Sun YC et al., *Cell Host Microbe* **15**:578–586, 2014.)

from the gastrointestinal species *Y. pseudotuberculosis* (Fig. 27.7). Both species retain several features that equip the pathogens to evade host defenses. Key traits include a large plasmid encoding a type III secretion system with a cargo of virulence factors, a regulatory system responsive to mammalian body temperatures, mechanisms to kill macrophages and to suppress immediate inflammatory responses, and the ability to replicate in the lymph nodes.

WHAT CAUSES PANDEMICS OF *Y. PESTIS*?

LEARNING OUTCOMES:
- Explain how conditions in urban areas during the 1300s contributed to the Black Plague pandemic.
- Speculate whether zoonotic pathogens or pathogens that persist only in human populations are more difficult to eradicate.

The rest of this story is history (literally). Once *Y. pestis* became established, or **endemic**, among rats and other small rodents, humans were at greater risk of exposure (Fig. 27.8). People can become infected with *Y. pestis* di-

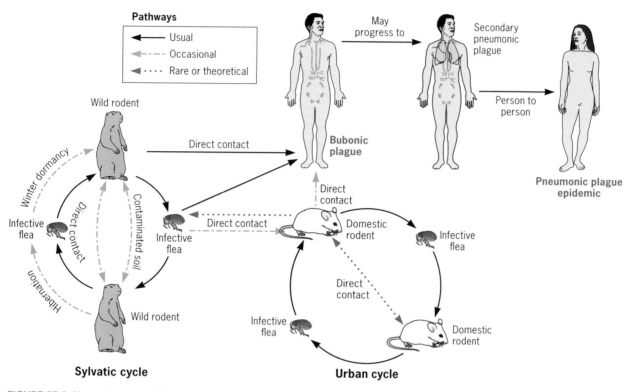

FIGURE 27.8 *Y. pestis* routes of transmission among rodents and humans. The sylvatic cycle occurs between rodents and fleas in natural environments. In areas with a high density of rodents, fleas, and people, the urban cycle can cause outbreaks of bubonic plague. If *Y. pestis* spreads from the bloodstream to the human lung, the pathogen can spread among the human population by the respiratory route, causing epidemic pneumonic plague. (Modified from ASM MicrobeLibrary.)

rectly from carrier fleas or lice, from bites, scratches, or aerosols from infected hosts, or by ingesting contaminated food. When the pathogen invades the bloodstream and spreads to the lung, the situation is especially dire. A cough that transmits as few as 10 bacteria is sufficient to spread this respiratory disease, known as pneumonic plague.

Indeed, crowded areas with poor sanitation become ground zero for epidemic disease. The first **pandemic** (an epidemic that crosses country borders) started ~500 AD, as verified by DNA analysis of the contents of teeth from corpses dated to that era. In the 1300s, a second pandemic began in Europe, killing nearly a third of the population. The rampant tissue necrosis clearly recorded in medieval records earned *Y. pestis* its reputation as the agent of the Black Death or bubonic plague, so named for the dramatically swollen lymph nodes called "buboes" (Fig. 27.1). In the early 1900s, the third pandemic arrived in the United States, likely carried by rodent stowaways on ships that sailed into San Francisco harbor. Public health measures to improve sanitation and reduce the rat population shifted plague from a dreaded urban disease to primarily a zoonotic disease acquired from small animals in the wild, such as prairie dogs in the southwestern United States or rabbits in eastern California (Case 2). The pathogen can also persist in soil. When promptly diagnosed and treated with antibiotics, the disease can be cured. Between 2000 and 2009, ~22,000 cases were reported worldwide, and ~7% were fatal; the United States recorded 57 cases and 7 deaths. Since exposure to wild animals is the major risk factor, a clinician's skills at obtaining a complete patient history can save a limb or a life.

CASE 1 REVISITED: A foodborne infection masquerading as appendicitis

In natural environments, *Y. pseudotuberculosis* colonizes the gastrointestinal tract of a variety of animals, including the deer that roamed on the commercial farm that supplied the lettuce to cafeterias in multiple states. When not properly washed, fresh produce contaminated with animal feces can spread the pathogen to humans. In most individuals, the innate immune system can recognize and mount an inflammatory response that is sufficient to clear *Y. pseudotuberculosis* infection. To prevent additional cases of foodborne disease, epidemiologists used pulsed-field gel electrophoresis (chapter 23, Fig. 23.12) to type the disease isolates and deduce the likely chain of transmission from the deer to the lettuce that was distributed broadly to cafeterias.

CASE 2 REVISITED: A hunter's narrow escape

In natural environments, *Y. pestis* can be transmitted by fleas among rodents (sylvatic cycle, Fig. 27.8), including the wild rabbits the hunter had handled. The pathogen likely entered the individual when rabbit blood came into contact with a break in his skin. By depositing virulence factors (Yops; Fig. 27.4 animation) into the cytoplasm of macrophages using its type III secretion system, *Y. pestis* inhibits phagocytosis and cytokine signaling, disarming the innate immune response. As the bacteria replicate, their LPS stimulates inflammation, producing swollen lymph nodes, or buboes. If bacteria enter the bloodstream and spread throughout the body, a robust inflammatory response causes necrosis of small blood vessels and tissue death, which may require amputation. Therefore, it is urgent to halt the infection by treating patients with antibiotics. Close contacts are also treated upon suspicion of exposure to prevent tissue damage that can be fatal. To reduce the risk of additional cases of zoonotic disease, epidemiologists used pulsed-field gel electrophoresis (chapter 23, Fig. 23.12) to type the disease isolates and deduce the likely chain of transmission from the wild rabbits to the hunter. Public health officials could then plan a targeted education and flea eradication campaign in the geographical region affected.

By comparing Cases 1 and 2, the power of a few extra genes is clear. Despite their genomic DNA sequence being nearly 97% identical, these two *Yersinia* species have distinct modes of transmission. Although one species causes mild intestinal distress, the other is deadly.

CONCLUSIONS

Both humans and animals are hosts to an enormous number and diversity of microbes, as we learned in chapter 21. If members of the human **microbiota** breach defensive barriers within their own host, *opportunistic* **infections** can develop, as we learned for MRSA (chapter 23). When humans come into contact with the microbiota of animals, the risk of *zoonotic* infections rises. Numerous types of bacteria, viruses, prions, fungi, and protists are all known to make the leap from their natural animal reservoir to humans, where they cause a wide variety of diseases (Table 27.1). Globalization of our food supply, international travel, and living in previously undeveloped natural habitats all increase the burden of zoonotic disease and heighten future risks. Currently it is estimated that >70% of newly emerging or re-emerging infectious diseases worldwide originate in animals. Fortunately, humans are often dead-end hosts for microbes that are adapted to other animals; typically the microorganisms lack mechanisms to spread efficiently from one human to another. But not always: witness the tragic 2014 Ebola pandemic. Local, national, and international surveillance networks are key to recognizing and containing outbreaks of zoonotic disease.

Representative ASM Fundamental Statements and Learning Outcomes for the Chapter

EVOLUTION

1. Mutations and horizontal gene tranfer, with the immense variety of microenvironments, have selected for a huge diversity of microorganisms.

    a. Describe the key events that led to the speciation of *Y. pestis* from *Y. pseudotuberculosis*.
    b. Explain the genetic mechanism by which bacteria acquire or lose traits.

CELL STRUCTURE AND FUNCTION

2. *Bacteria* and *Archaea* have specialized structures that often confer critical capabilities.

    a. Explain why adhesions are an important virulence factor for *Y. pseudotuberculosis*.

METABOLIC PATHWAYS

3. The interactions of microorganisms among themselves and with their environment are determined by their metabolic abilities.

    a. List three traits that *Y. pestis* and *Y. pseudotuberculosis* have in common and three that are different.

INFORMATION FLOW AND GENETICS

4. Genetic variations can impact microbial functions.

    a. Explain how the genetic differences between *Y. pestis* and *Y. pseudotuberculosis* influence biofilm formation.

5. Cell genomes can be manipulated to alter cell function.

    a. Explain how the large plasmid carried by both bacteria plays a role in disease in both cases.

MICROBIAL SYSTEMS

6. Most bacteria in nature live in biofilm communities.

    a. Identify how the capacity to form biofilms promotes the transmission of *Y. pestis*.

7. Microorganisms and their environment interact with and modify each other.

    a. List two direct and two indirect ways that humans can acquire zoonotic pathogens.
    b. Speculate whether zoonotic pathogens or pathogens that persist only in human populations are more difficult to eradicate.
    c. Explain how conditions in urban areas during the 1300s contributed to the Black Plague pandemic.

8. Microrganisms, cellular and viral, can interact with both human and nonhuman hosts in beneficial, neutral, or detrimental ways.

    a. Speculate which zoonoses (see Fig. 27.1) for which you might be at risk.
    b. Describe the reservoir, vector, mode of transmission, and symptoms of infection for *Y. pseudotuberculosis* and *Y. pestis*.

c. Trace the entry of Y. *pseudotuberculosis* into a cell and its spread within a patient's body.

d. Describe why antibiotics are not necessary to treat patients with Y. *pseudotuberculosis* gastroenteritis.

e. Discuss why the risk to humans is higher for Y. *pestis* over Y. *pseudotuberculosis*.

SUPPLEMENTAL MATERIAL

References

- **Long C et al.** 2010. *Yersinia pseudotuberculosis* and *Y. enterocolitica* infections, Food-Net, 1996–2007. *Emerg Infect Dis* **16:**566–567. http://www.ncbi.nlm.nih.gov/pmc/articles/PMC3322025/

- **Butler T.** 2013. Plague gives surprises in the first decade of the 21st century in the United States and worldwide. *Am J Trop Hyg* **89:**788–793. http://www.ncbi.nlm.nih.gov/pmc/articles/PMC3795114/

- **Raoult D et al.** 2013. Plague: history and contemporary analysis. *J Infect* **66:**18–26. http://www.sciencedirect.com/science/article/pii/S0163445312002770

- **Chouikha I, Hinnebusch BJ.** 2012. Yersinia–flea interactions and the evolution of the arthropodborne transmission route of plague. *Curr Opin Microbiol* **15:**239–246. http://www.ncbi.nlm.nih.gov/pmc/articles/PMC3386424/

- **Sun Y-C et al.** 2014. Retracing the evolutionary path that led to flea-borne transmission of *Yersinia pestis*. *Cell Host Microbe* **15:**578–586. http://www.sciencedirect.com/science/article/pii/S1931312814001371

- **Bliska JB et al.** 2013. Modulation of innate immune responses by *Yersinia* type III secretion system translocators and effectors. *Cell Microbiol* **15:**1622–1631. http://onlinelibrary.wiley.com/doi/10.1111/cmi.12164/abstract;jsessionid=FE5BD5F6A40DABCC05CA6C8BCAE02620.f01t03http://onlinelibrary.wiley.com/doi/10.1111/cmi.12164/abstract;jsessionid=FE5BD5F6A40DABCC05CA6C8BCAE02620.f01t03

Supplemental Activities

CHECK YOUR UNDERSTANDING

1. **You are a public health official in the United States worried about a recent bubonic plague outbreak in Canada. What strategies would best protect U.S. residents from plague?**

   a. Restrict border crossings between the United States and Canada.

   b. Invest in the research and development of a vaccine.

   c. Initiate a flea eradication campaign in the affected region.

   d. Cooperate with Canada to enact a quarantine protocol for all those infected.

   e. Launch a public health education campaign.

2. **Why is Y. *pestis* infection relatively rare in high-resourced countries?**

   a. There is no natural reservoir.

   b. Contact between fleas and humans is less frequent.

   c. Fleas do not transmit a high enough density of bacteria to cause disease in humans.

   d. Past epidemics have increased the frequency of resistance alleles in the human population.

3. **What trait did NOT contribute to the evolution of Y. *pestis* from Y. *pseudotuberculosis*?**

   a. Survival in the flea midgut.

   b. Increased production of cyclic di-GMP.

   c. Biofilm-mediated blockage of the flea esophagus.

   d. Loss of adhesins that promote invasion of human epithelial cells.

DIG DEEPER

1. *Y. pestis* produces a flea enterotoxin, a protein toxin that targets the intestines, but *Y. pseudotuberculosis* does not. You recently developed efficient assays to detect and purify the toxin. Which experimental strategy would you use to identify the gene that encodes the toxin? Explain your reasoning.

   a. Delete each *Y. pestis* gene of unknown function, and then determine which mutation disrupts enterotoxin production.

   b. Use mass spectrometry to determine the peptide sequence of the toxin, and then deduce from this peptide sequence the DNA sequence that encodes the toxic peptide. Finally, use publicly available DNA analysis software to interrogate the *Y. pestis* genome sequence *in silicio* with the toxin DNA sequence.

   c. Perform B, and then confirm its identity by deleting the gene(s) as described in A.

   d. Perform A first, and then confirm its identity as described in B.

   e. None of the above. Design a better alternative.

2. **Team work:** *Y. pestis* is best known as the cause of the bubonic plague, a disease that between 1347 and 1353 killed approximately one-third of the population of Europe. However, *Y. pestis* has an even more sinister history. In one of the first documented cases of biological warfare, Mongols are accused of spreading the plague in 1347 by catapulting the corpses of plague victims over the walls of a city in Crimea (http://wwwnc.cdc.gov/eid/article/8/9/01-0536_article). During World War II, rice and wheat infested with rat fleas carrying *Y. pestis* were dropped by Japanese planes onto a city in China, causing a local outbreak that killed 121 people (Harris SH, *Factories of Death*, Routledge, New York NY, 1994, p 78, 96).

   You are microbiologists in the Department of Defense bioterrorism scientific research unit. Recently, vague but reliable intelligence indicates that a terrorist group is planning to expose civilian populations to *Y. pestis*. You receive an urgent request for knowledge of most likely exposure scenarios and measures to counteract a *Y. pestis* outbreak. After consulting the CDC resource *Recommendations from the Centers for Disease Control Bioterrorism Unit on Preparation and Response to a Bioterrorist Attack with an Aerosolized Form of Y. pestis* (http://www.bt.cdc.gov/agent/plague/consensus.pdf), your team immediately begins brainstorming.

   - What are potential routes and vectors of *Y. pestis* dispersal? What is the relative risk of each?
   - After an attack, what are the best means to limit transmission? Should you target the bacteria, flea, or rodent population, or a combination of the three?
   - Other than zoonoses, are there additional ways *Y. pestis* can be spread?

   Develop a strategy to present to your Department of Defense colleagues. Time is of the essence, so keep it short. Emphasize the relevant microbiology that explains why and how your strategy to contain an attack would break the *Y. pestis* life cycle and its spread to humans.

3. In 2009, an accomplished 60-year-old male researcher (http://jb.asm.org/content/192/17/4261.long) studying *Y. pestis* passed away within hours of presenting at the hospital. During his intake interview, it was not noted that he worked with *Y. pestis*, as the lab strain he handled had previously been attenuated by inactivating the iron acquisition machinery the pathogen requires for full virulence. This lab strain was in routine use in the research community without prior incidence. Nonetheless, the attenuated strain of *Y. pestis* was detected in the scientist's blood and determined to be the cause of his death. On autopsy, it was determined that the scientist also had hemochromatosis, a genetic disorder. Learn more from the following news article, http://www.bloomberg.com/news/articles/2011-02-25/plague-kills-u-s-scientist-in-first-laboratory-case-in-50-years-cdc-says, and other sources to answer these questions:

   - Draw schematics to teach the general public the differences between infection of wild-type *Y. pestis* in healthy hosts, attenuated *Y. pestis* in healthy hosts, and attenuated *Y. pestis* in a host with the genetic disorder of this research scientist.
   - Hemochromatosis is common in people of Western European descent. What selective pressures are thought to account for the prevalence of this mutation in this population?

- What safety measures to protect *Y. pestis* researchers do you recommend?

4. **Team work:** Biotechnologists are developing antimicrobial drugs that target type III secretion systems. For examples, read Gong et al. (*Infect Immun* **83:**1276–1285, 2015, http://iai.asm.org/content/83/4/1276.long) and Zetterstrom et al. (*PloS One* 8(12):e81969, http://journals.plos.org/plosone/article?id=10.1371/journal.pone.0081969). Your investment club has an opportunity to invest in this technology and wants your expert advice. Discuss the questions below before preparing your recommendation and its scientific rationale.

   - How antibiotic resistance develops, spreads, and increases the need for alternative therapeutic strategies
   - Pros and cons of targeting type III secretion systems
   - Pros and cons of antibiotics that kill the bacteria versus disarming the bacteria

5. **Team work:** National Center for Case Study Teaching in Science "Farmville Future?" by Stephanie Luster-Teasley and Rebecca Ives uses concentrated animal feeding operations to teach the impact of environmental contamination on animal and human health and the intersection of microbiology and public policy. http://sciencecases.lib.buffalo.edu/cs/collection/detail.asp?case_id=687&id=687

6. Research a zoonotic disease caused by a bacterium, virus, eukaryote, or protein (prion) from Table 27.1 or elsewhere using vetted resources, such as those prepared by the CDC, American Society for Microbiology, or peer-reviewed journals. Argue why your zoonotic disease is in most need of a $20 million research grant. In your oral or written presentation, describe mechanisms of encounter, entry, spread and damage; epidemiology; public health preventive measures and treatment options; and your point of intervention.

Supplemental Resources

VISUAL

- American Society for Microbiology General Meeting interview (31 min): Listen to four experts discuss "One Health," the concept that the health of humans, animals, and the environment is interconnected. What measures can be taken locally and internationally to reduce and manage the risks of zoonotic disease? http://gm.asm.org/index.php/scientific-activities/asm-live/archives/371-one-health-humans-animals-and-the-environment
- *Microbes After Hours* lecture "West Nile Virus" from the American Society for Microbiology to educate the general public (1 h 16 min): https://www.youtube.com/watch?v=QuDYrenM-0c&feature=c4-overview-vl&list=PLD2Sv5NskYtwPnpiWnu63hANr0Bo0Xzu0
- TedEd video (5.13 min) "How parasites change their host's behavior": Jaap de Roode illustrates four amazing examples of microbes that promote their own replication and spread by altering the behavior of their hosts. https://www.youtube.com/watch?v=g09BQes-B7E
- "Under Our Skin" (2008, 100 min) and its sequel "Under Our Skin 2: Emergence" (2014, 60 min): These acclaimed documentaries examine the microbiology and the medical and public response to epidemic Lyme disease, another zoonotic disease. http://underourskin.vhx.tv/

SPOKEN

- *This Week in Microbiology* podcast episode #80 "Hurling fleas and designer chromosomes": Scientists discuss the article by Sun Y-C et al. (*Cell Host Microbe* **15:**578–586, 2014) that recapitulates in the laboratory four discrete steps in the evolution of *Y. pestis* from *Y. pseudotuberculosis*.

WRITTEN

- CDC facts on *Yersinia:* http://www.cdc.gov/nczved/divisions/dfbmd/diseases/yersinia/
- CDC facts on plague: http://www.cdc.gov/plague/

- CDC Morbidity and Mortality Monthly Report: Timely and authoritative public health information and recommendations, including outbreaks of zoonotic disease. http://www .cdc.gov/mmwr/
- In the *Small Things Considered* blog post "A Pestis from the Past," S. Marvin Friedman discusses not only what paleobiologists have learned about plague, but also what they haven't. Throughout history, in addition to a pathogen's genetic composition, a variety of human factors impact the severity and spread of infection. http://schaechter.asmblog.org/schaechter/2011/11/a-pestis-from-the-past.html
- FAQ *West Nile Virus*, a colloquium report from the American Academy of Microbiology to educate the general public. http://academy.asm.org/index.php/faq-series/793-faq -west-nile-virus-july-2013
- *Bioinformatics and Biodefense: Keys to Understanding Natural and Altered Pathogens, 2009,* a colloquium report from the American Academy of Microbiology. https://eh.uc .edu/assets/news/Yadav_11_09.pdf

Putting Microbes to Work

IN THIS CHAPTER

Scientists examining viral plaques

CHAPTER TWENTY-EIGHT

A Past, a Present, and a Future with Microbes

KEY CONCEPTS

This chapter covers the following topics from the ASM Fundamental Statements.

METABOLIC PATHWAYS

1. *Bacteria* and *Archaea* exhibit extensive, and often unique, metabolic diversity.

INFORMATION FLOW AND GENETICS

2. Cell genomes can be manipulated to alter cell function.

MICROBIAL SYSTEMS

3. Microorganisms are ubiquitous and live in diverse and dynamic ecosystems.
4. Microorganisms and their environment interact with and modify each other.

IMPACT OF MICROORGANISMS

5. Humans utilize and harness microorganisms and their products.
6. Because the true diversity of microbial life is largely unknown, its effects and potential benefits have not been fully explored.

INTRODUCTION

Most of our interactions with microbes are involuntary. They do their thing; we do ours. This usually works to our advantage because microbes are essential participants in this planet's metabolism (see chapter 1). Other times, the microbes' activities are noxious: they spoil our food, destroy our crops, or make us sick. We reciprocate: certain human pursuits, such as habitat modification, greatly affect the microbial biota, but these are usually unintended consequences. On the other hand, we intentionally put microbes to work in a variety of ways and have done so since the beginning of human culture.

Our past and our present are intertwined with microbes. Indeed, the quality of our future depends on microbes and how our activities affect theirs. Our future is a microbial future. Let us use this last chapter to think critically about our past and present with microbes and to consider how our activities on planet Earth will impact theirs. Why not be their allies?

WHAT THINGS DO MICROBES MAKE FOR US?

LEARNING OUTCOMES:
- Name several microbial products that you use or are affected by in your everyday life.
- List reasons researchers use cloning in *Escherichia coli* to make significant quantities of human insulin.
- Outline a cloning method to engineer microbes that synthesize human proteins.
- Identify the range of real-world purposes of synthetic biology.
- Describe how genetic engineering of the Bt toxin has benefited agriculture.
- Discuss the mechanism by which Bt toxins are poisonous for insects but safe for human consumption.
- Identify several differences in the process of making beer versus making wine.
- Discuss how the succession of yeasts and lactobacilli helps make and refine wine

The list of microbial domestications by people is long and varied (Table 28.1). For fun, name a half dozen common food or drink items that are the product of microbial activity. We bet you can. Some of these transactions require deliberate biological engineering (e.g., the production of antibiotics); others happen without sophisticated human intervention (e.g., the production of bread, cheese, pickles, vinegar, sauerkraut, soy sauce, wine, or beer in traditional ways). As the microbial sciences have advanced, we have increased our understanding of how to harness the power of microbes. Usually, progress has come from recognizing which microbe is involved, whether acting alone or in a consortium; such identification is frequently followed by selecting a better strain or modifying it genetically. Of course, this approach is not unique to microbes and has been used since time immemorial to select for better strains of domestic animals, such as cattle and chickens, and plants, including food crops and decorative flora. In some cases, the enzymes that the microbe produces are all that is needed to make a product (Table 28.2). For example, this is how cornstarch is converted into high-fructose corn syrup (Table 28.3).

We will begin with a story of selecting better strains and go on to discuss a number of other instances where humans have put microbes (or their enzymes) to work.

TABLE 28.1 **Some uses of domesticated microbes**

Purpose	Comments
Making, preserving, or enriching food	
Bread	The carbon dioxide produced when baker's yeast, *Saccharomyces cerevisiae*, ferments sugar causes bread to rise.
Cheese	Lactic acid bacteria cause milk to curdle, the first step in making certain cheeses; many different bacteria, through their proteolytic and lipolytic activities, contribute to the ripening of cheese.
Yogurt	Yogurt is just one of many dairy products made by the fermentation of the lactose in milk by various lactic acid bacteria.
Pickles	Pickles are made by lactic acid bacteria fermenting the sugars that cucumbers contain. Many other vegetables, including olives, are similarly preserved.
Vinegar	Vinegar is made when acetic acid bacteria oxidize ethanol to acetic acid. Starting materials are usually made by yeast fermentation of various fruits, principally grapes and apples.
Sauerkraut	Sauerkraut is the product of the action of lactic acid bacteria on cabbage (see chapter 19).
Silage	Silage is preserved animal forage made by the action of lactic acid bacteria.
Beer	Beer is the product of a yeast fermentation of grain that has been saccharified.
Wine	Wine is fermented fruit, principally grapes.
Vitamins	Many vitamins, including riboflavin, vitamin C, and vitamin B_{12}, are made by microbial fermentations.
Amino acids	Several amino acids, including lysine, methionine, and monosodium glutamate, are made by microbial fermentations; they are added to human and animal food to increase the nutritional value or to enhance flavor.
Enzymes	Many microbial enzymes are used industrially or therapeutically (Table 28.2).
Medicinal drugs	
Antibiotics	With very few exceptions, antibiotics in use today are made by microbial fermentations.
Probiotics	Various live microbes or mixtures of them are used to treat animal diseases or enhance the growth of crops. For example, mixtures of bacteria approximating a chicken's normal intestinal flora are given to chicks to prevent infection by *Salmonella* spp., and legume seeds are coated with nitrogen-fixing bacteria to ensure the developing plant will be able to fix nitrogen.
Therapeutic proteins	Bacteria into which human genes have been inserted are used to produce certain therapeutic proteins, including insulin and human growth hormone.
Steroids	Microbes are used to mediate certain steps in the chemical synthesis of steroids.
Waste treatment	
Sewage disposal	Sewage treatment involves the oxidation of the organic components of sewage by microbes. Subsequent microbial processes are also employed to reduce the nitrogen and phosphorus content of the effluent.
Composting	Microbial action converts plant waste into enriched material to augment soil.
Bioremediation	Various microbes are used to eliminate toxic waste.
Mining	In certain ores, minerals such as gold are trapped in insoluble sulfides; bacteria are used to oxidize the sulfides, thereby releasing the mineral.
Making fuel	Sewage and animal waste are used as substrates for archaea to make methane, which is used domestically and municipally as fuel. Starch from corn is converted to sugar and fermented to form alcohol, to be used in gasohol.

Using Microbes To Make Drugs: Insulin and Human Growth Hormone

Beginning in the mid-20th century, microbes have contributed mightily to the manufacture of medicinal drugs. Antibiotics, with some exceptions, are made by microbes; furthermore, microbes are used to carry out difficult steps in the chemical synthesis of some drugs such as certain steroids. Each of these uses of microbes exploits their natural capacities. Thus, when antibiotic-producing microbes are first isolated from nature, they synthesize antimicrobial compounds in small amounts. Subsequent genetic and

TABLE 28.2 **A sampling of microbial enzymes that are used commercially**

Enzyme	Microbial source	Activity	Use
Invertase	*Saccharomyces cerevisiae*	Hydrolyzes sucrose into glucose and fructose	Baking, stabilizing syrup, candy making
β-Glucanase	*Bacillus subtilis, Aspergillus niger, Penicillium emersonii*	Hydrolyzes β-glucans	Clarifying beer
Lactase	*Saccharomyces lactis, Aspergillus niger, Aspergillus oryzae, Rhizopus oryzae*	Hydrolyzes lactose into glucose and galactose	Digestive aid to those with lactose intolerance; dairy industry
Pectinase	*Aspergillus niger, Aspergillus oryzae, Rhizopus oryzae*	Hydrolyzes pectin	Clarification of fruit juices and wines
Rennin	*Escherichia coli* with cloned gene from cattle	Hydrolyzes a bond in the milk protein, casein, causing it to coagulate	Cheese making
Neutral protease	*Bacillus subtilis, Aspergillus niger*	Hydrolyzes proteins at neutral pH	Enhancing flavor of meats and cheese
Alkaline protease	*Bacillus licheniformis*	Hydrolyzes proteins at an alkaline pH	An additive to detergents that removes protein-based stains
Lipase	*Aspergillus niger, Aspergillus oryzae, Rhizopus oryzae*	Hydrolysis of ester bonds in fats and oils	Dairy industry; an additive to detergents that removes fat-based stains
Cellulase	*Trichoderma konigi*	Hydrolyzes cellulose	A digestive aid
α-Galactosidase	Various lactobacilli	Hydrolyzes α-galactosides	Treatment of legumes to decrease their ability to cause flatulence

cultural manipulations will induce the microbes to synthesize the larger quantities of the drugs necessary for commercialization. With the advent of recombinant DNA technology in the 1970s, the uses of microbes in drug manufacturing changed drastically: the therapeutic agents that could be produced were no longer limited by the microbes' natural abilities. Rather than searching for a microbe that could produce a useful agent, a microbe could be genetically altered to produce the desired product. The hope is that a gene from any organism can be introduced into a microbe to exploit its remarkable biosynthetic capacity to churn out great quantities of that gene's product. However, there are hurdles in the track.

TABLE 28.3 **Enzymes used to convert cornstarch to high-fructose corn syrup**

Enzyme	Microbial sources	Activity
Amylase	*Bacillus licheniformis, Bacillus subtilis*	Cleaves starch (a mixture of amylose, a straight-chain polymer of α-D-glucopyranose, and amylopectin, a branched form that contains some α-1,6 bonds) to maltodextrins, which are short chains
Glucoamylase	*Aspergillus niger, Aspergillus oryzae*	Splits glucose from maltodextrins by cleaving α-1,4 bonds
Pullulanase	*Klebsiella aerogenes, Bacillus* spp.	Cleaves α-1,6 bonds in maltodextrins
Glucose isomerase	*Bacillus coagulans, Actinoplanes missouriensis, Streptomyces* spp.	Converts D-glucose into D-fructose

Using bacteria to make human insulin. Even before launching a commercial process, there are some obvious restrictions about which compound to make. The compound has to be sufficiently valuable to justify the cost of development and manufacture, and it should not be readily available from another source. Many therapeutically desirable human proteins are obvious choices. Insulin, used for treating patients with diabetes, was the first therapeutic agent made by a genetically engineered microbe. Insulin has been available for therapeutic use since shortly after its discovery in the 1920s. Initially, it was purified from the pancreas of animals—largely from cows in the United States and pigs in Europe. Although most human patients who received repeated doses of these foreign proteins tolerated them well, there were complications. Some patients developed insulin resistance because animal insulins, although very similar, are not antigenically identical to the human one. Also, the supply of insulin was limited because feedlot cattle, the major source of bovine pancreases for animal insulin, produce smaller amounts of insulin than range cattle do. The solution to the problem was to clone the insulin-encoding human genes in *Escherichia coli* and scale up the cultures to produce practically unlimited amounts of insulin. Because the bacterium makes human insulin, there is no immune response against it.

Using bacteria to make human growth hormone. Human growth hormone (hGH) also presented a compelling need. This hormone is essential for children whose pituitary glands make it in insufficient amounts (and who if untreated are destined to become pituitary dwarfs). Before microbes came to the rescue, the hormone was extracted from the pituitary glands of human cadavers. This source presented clear dangers, as there was evidence that a few children had contracted the prion Creutzfeldt-Jakob disease (chapter 18) from such treatments. Thus, scientists used recombinant techniques to produce the hGH in bacteria (Fig. 28.1).

The methodology used to clone and express the hGH gene in bacteria is typical of that used for other human proteins and illustrates how important it is to understand the fundamental differences between prokaryotic and eukaryotic protein synthesis. All human cells carry the gene that encodes hGH, but how do you isolate that gene from all others in order to clone it? One handy way is to start with messenger RNA (mRNA) instead of DNA. Human pituitary gland tissue makes the body's supply of hGH, so its cells contain large amounts of the corresponding mRNA. These mRNA molecules can be identified with an appropriate probe (a short DNA molecule homologous to part of the hGH gene). The probe will hybridize with the target mRNAs, and the targeted transcripts (those encoding hGH) can be retrieved. There is a second and more compelling reason for isolating mRNA rather than isolating the gene directly. Genes from eukaryotes, including humans, contain introns (stretches of DNA that do not encode the protein but are spliced out of mRNA as it matures). Because prokaryotes lack the enzymes to eliminate introns, they would synthesize an incorrect protein from intron-carrying eukaryotic DNA.

How do you make the required DNA probe? First, you have to know the sequence of amino acids in the hGH protein. Then, the sequence of bases for the probe can be deduced by referring to the genetic code (chap-

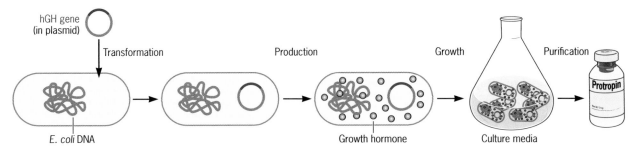

FIGURE 28.1 Recombinant DNA technology at work to make human growth hormone (hGH). (Modified from source: https://www.ied.edu.hk/biotech/eng/classrm/explain/health5.jpg.)

ter 8). Recall that the genetic code is redundant—most amino acids are encoded by several different codons. The genetic code states precisely which amino acid sequence is encoded by a particular sequence of bases in DNA, but the reverse is not the case. Because of the redundancy of the code, *all the DNA sequences* that designate the desired amino acid sequence must be represented. Therefore, a mixture is used as a probe to fish out the correct mRNAs. Once these molecules have been isolated, they are used to make DNA via **reverse transcriptase** (the enzyme from retroviruses that uses RNA as a template to make DNA [see chapter 17]). The product is called **complementary DNA** (cDNA) to indicate that it is a copy of mRNA, not the DNA in the gene itself. Then, the cDNA is cut with restriction endonucleases, ligated into a bacterial **plasmid**, and inserted into the bacterial host by transformation (Fig. 28.1). In the microbial host, the recombinant DNA molecule is replicated and expressed. The transformed bacterial cell now becomes an hGH factory.

This cloning method is not the only way microbes can be engineered to synthesize human proteins. Many variations are possible, and new techniques are developed almost daily. The most fundamental changes are in the cloning vector and the host cell. As we learned in chapter 10, a plasmid can be used to clone and express the gene of interest (hGH), but viral DNA can also be used. Almost any kind of cell or intact organism can now be engineered to encode a recombinant DNA molecule. Valuable products are being manufactured in cloned yeasts, plants, and even some animals. However, no matter which host cell is finally selected to make the protein product, bacteria are typically used in the cloning process, because they are relatively easy and inexpensive to manipulate and culture. As industrial microbiologists will tell you, the job of producing the hormone commercially has just begun. The recombinant strain needs to be grown at large scales (sometimes thousands of liters) under conditions that maximize hGH production. Furthermore, the hormone must be purified in active form. Thus, the hGH protein must be properly folded and post-translationally modified (glycosylation). But in the end, the active hormone is made in quantities large enough to commercialize. And in doing so, many children with pituitary dwarfism safely receive hormone replacement therapy.

Synthetic Biology: A Microbiological Science
Imagine that in 10 or so years some of you may be teaching microbiology. (We hope so!) You may recall for your students that some 40 years previously it became possible to clone genes into organisms that didn't carry

them. These techniques created the opportunity to generate new genomic combinations, producing organisms that had not existed before. You may ask your students whether the originators of this technology should have worried about the possibility of creating dangerous forms of life. In fact, in 1975 leading researchers got together with lawyers and clinicians to discuss how to safeguard against this possibility (the so-called Asilomar Conference, named for the site where this meeting took place). The group generated a set of guidelines concerning physical and biological restraints on the use of genetic engineering. With more experience and the lack of documented dangerous outcomes, these restrictions were gradually relaxed. However, similar concerns emerged again in the early 2010s, this time with the construction of expressly new combinations of pathogens. "Gain of function" experiments came under scrutiny once again.

Before long, new technologies will lead to prodigious feats of data gathering on a gigantic scale. Rapid and economical *DNA sequencing* and *synthesis* together will allow the cataloging of almost all prototypical biological parts and the construction of strains with improved modes of gene expression. In fact, the making of new viruses and microbes will become commonplace. In 2016, scientists reported the construction of a synthetic bacterium. The novel bacterium, designated Syn 3.0, grows and reproduces with a synthetic genome containing just 473 genes! Such feats of genetic engineering are now subsumed under the term **synthetic biology**. Note the difference between synthetic and **systems biology**. The former's aim is to build artificial biological systems useful for engineering applications, whereas systems biology is about the integration of the major activities of the cell using tools that include modeling and simulations.

Here is another example of synthetic biology at work: the redirecting of yeast cells to make the antimalarial drug artemisinin. This compound is made by a plant, which makes industrial production erratic. To make the drug by fermentation, scientists engineered into yeast the genes for several pathways leading to the making of nearly 20 intermediates in artemisinin synthesis. These genes were derived from various sources including the plant that makes the drug, and some of them were genetically modified. Using improved fermentation techniques, enough biosynthetic artemisinin was produced by 2015 at a reasonable price for 40 million treatments. Meanwhile, the malaria parasite has become resistant to this drug, which emphasizes the need for further synthetic biological approaches.

Armed with powerful tools, synthetic biologists approach biological problems in several ways. One is to rebuild natural systems from the ground up (as with the construction of the synthetic bacterium Syn 3.0); another approach is to construct toolkits for the ready introduction of new functions. In addition, these approaches make it possible to store vast amounts of information in the DNA itself. With such tools at hand, the only limits seem to be the powers of human imagination. However, with these developments come renewed concerns for biosafety. The debate continues.

Biological Insecticides: Bt

Bacillus thuringiensis is one of a number of spore-forming soil bacteria that are useful in agriculture as pesticide producers. This bacterium makes a set of powerful protein toxins that are used widely by farmers to control a variety of insects. Known as **Bt toxins**, they kill over 150 different spe-

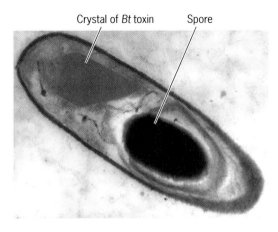

Crystal of *Bt* toxin   Spore

FIGURE 28.2 **A sporulating cell of *Bacillus thuringiensis* with a large crystal of Bt toxin.** (Courtesy D. Lereclus.)

cies of noxious insects. The toxins are proteins made by the bacteria during sporulation and are produced in such large amounts that they make crystals readily seen under the electron microscope (Fig. 28.2).

Bt toxins come in a large variety, each specific for a different range of insects. When the insect ingests Bt toxin-forming bacteria, the toxins bind specifically to receptors in the insect's intestines. These receptors differ with the insect species, which accounts for the specificity of the various toxins. Humans and other vertebrates do not have these receptors; hence, the toxins are inactive in larger animals and are considered safe for food production. The protein, as it is made in the bacterium, should more accurately be called a **protoxin**, because in itself it is not toxic. It is converted into lethal form only when ingested by the insect. The protein crystals become solubilized in the alkaline, reducing conditions of the midguts of susceptible insect larvae. The soluble protein, which is still inactive, is converted into an active form by proteolysis. The now active toxin inserts into the membranes of the cells lining the midgut to form cation-selective channels, which leads to death of the cell and eventually the insect.

We know little about the role of these Bt toxins in nature. What would be the selective advantage for a soil-dwelling bacterium to kill insects? In nature, *B. thuringiensis* rarely causes epidemics among insects. However, Bt toxins can also kill nematodes, little worms that are abundant in soil and feed on bacteria. Could Bt toxin be a mechanism for converting a potential predator into a source of nutrients? If so, its insect-killing activity might be unintended. This speculation is plausible, because the toxins kill nematodes and insects by the same mechanism.

Genetic engineers have inserted the genes for Bt toxin production directly into plants such as corn, making it unnecessary to apply toxins to the field. Also, these genes have been inserted into rhizosphere bacteria (bacteria that are abundant in the soil surrounding a plant's roots). The bacteria then deliver the toxins to the plant, protecting it from attacking insects. Appealing though these alternative methods of delivering Bt may seem, they may lead to increased risk of resistance, because the toxin would persist for longer periods. Also, because most plant parts, including pollen, contain toxin, some desirable insects may be killed.

Making Better Wines: The Malolactic Fermentation

Enology, the science of wine making, has become a sophisticated enterprise. University departments and industrial labs are dedicated to the underlying science that supports this worldwide industry—in spite of the fact that, at its heart, making wine is simply the inevitable microbiological consequence of crushing grapes. The powdery bloom on a grape's surface contains a population of yeasts capable of mediating an alcoholic fermentation; the grape's contents constitute a near-ideal growth medium and a rich source of fermentable sugar. Introduce one to the other by crushing the grape, and wine results. Beer and vodka are different: yeasts cannot ferment the starch in barley or wheat and require an initial human intervention. The process of converting the starch into the fermentable sugars glucose and maltose starts with *malting*, the germination of the grain. Malt is a source of the enzyme amylase that hydrolyzes starch into sugars (Table 28.3). In contrast to grains, however, grape juice is ready to go. Along with all the nutrients that yeast needs to grow, grapes contain an equimolar mixture of glucose and fructose, both of which yeasts can ferment.

Crushing grapes produces wine, but not necessarily good wine. Making a high-quality "grand cru" or the like requires careful attention to myriad details, many of which are microbiological. Traditional practices for making wine, developed by trial and error over centuries, have yielded remarkably good results, but modern microbiology has made them better and much more reliable. Now, the quality of wines still varies from year to year due to changes in growing conditions, not in the wine maker's luck. Microbiology's best-known contributions to enology are controlling the alcoholic fermentation by favoring desirable strains of yeasts and preventing spoilage, principally by acetic acid bacteria and lactic acid bacteria. Louis Pasteur developed a moderate heating procedure called **pasteurization** to eliminate spoilage microorganisms. Pasteurization of milk, which has had major public health benefits, derives from wine making.

Although some can spoil wine, bacteria also carry out a fermentation known as the **malolactic fermentation** that is critical to making high-quality red wines (chapter 6). Traditionally, this fermentation begins spontaneously in midwinter, several months after the grapes are crushed. Grape juice contains two organic acids, tartaric and malic acid. The primary alcoholic fermentation leaves both of them untouched, but the malolactic fermentation converts all the malic acid to lactic acid:

$$\text{HOOC-CH}_2\text{-CHOH-COOH} \rightarrow \text{CH}_3\text{-CHOH-COOH} + \text{CO}_2$$
$$\text{Malic acid} \qquad\qquad\qquad \text{Lactic acid}$$

This conversion yields minuscule amounts of energy, but it is sufficient to support slow growth of the lactic acid bacteria that drive the chemical reaction. This transformation of malic acid to lactic acid decreases the wine's acidity and softens it, an essential process for fine red wines. Other products of the lactic acid bacteria add to the wine's flavor complexity. (Aha! Lactic acid bacteria also contribute to the flavor of sauerkraut [chapter 19].)

If the malolactic fermentation occurs regularly, why should the winemaker want to control it? There are several reasons. If the metabolism of malic acid was delayed too long and occurred after bottling, some volatile components would be trapped, giving the wine a foul taste. The solution would seem straightforward: isolate the malolactic fermenting bacteria from a high-quality wine and use them as an inoculum. However, that must be done carefully. Yeasts are sponges for nutrients: when they stop growing, they take up nutrients from their environment and store them. Lactic acid bacteria, including those that mediate the malolactic fermentation, require a long list of growth factors, such as amino acids and vitamins that are sequestered by the yeasts. When this fact was recognized, it became possible to time the malolactic fermentation by adding the lactobacilli before the yeasts have a chance to deplete the wine of nutrients. This story is yet another good example of how bacterial populations succeed one another in the environment. In chapter 19, we described another example of sequential bacterial activities, during the making of sauerkraut.

HOW ARE MICROBES HELPING US REPLACE FOSSIL FUELS?

LEARNING OUTCOMES:
- Name four ways in which microbes may be used to generate biofuels.
- Identify benefits and potential drawbacks to bioethanol generation by yeast.
- Describe how a microbial fuel cell works.
- Explain how *Geobacter* produces energy.

The finite supply of fossil fuels in our planet and the deleterious impact that fossil fuels have on the environment have stimulated research into alternative modes of energy. As always, we have turned our attention to microbes. Why? Many of the chemicals that are currently made in petrochemical refineries can be made in biological reactions evolved by microbes as part of their versatile metabolism. Fuels are no exception. In fact, the ability of microbes to degrade and ferment plant material shows promise to replace a range of petrochemical fuels such as ethanol, diesel, and even gasoline with significant reductions in greenhouse emissions and low environmental impact. Microbes can also make electricity—perhaps not enough to power your home but enough to power sensors and other devices. They can also bring light and electricity to disadvantaged regions of the planet that lie outside the electrical grid. Let us consider these examples in more detail to better understand how microbes are helping us replace fossil fuels.

Microbes That Make Biofuels

Natural gas (methane). Included in their amazingly varied microbial metabolic repertoire is the ability of some species to produce hydrocarbons that can be used industrially. An example is the production of methane in municipal landfills (Fig. 28.3). In chapter 20 we learned that this metabolic activity, called methanogenesis, is carried out by a group of archaea. **Methanogens** are busy in landfills. Typically, methane production takes place about 1 year after municipal solid wastes are deposited. Over time, the conditions at the site go from aerobic, with little production of methane, to anaerobic. Now, a mixture of gases is produced, typically consisting of about 50% methane. Capturing these vapors not only reduces the use of fossil fuels but also avoids the unwelcome release of large amounts of this particularly strong greenhouse gas. In the United States, landfills produce an estimated 100 million tons of methane per year. Methane generation within landfills takes place spontaneously and does not usually involve modifying the local microbes. Thus, recovery of methane gas is largely an engineering concern, leaving microbiologists as bystanders for a change.

Bioethanol. Ethanol is made by fermentation on a gigantic industrial scale to generate fuel for automobiles. Much of the crops of corn in the United States, sugar cane in Brazil, and sugar beets in France are now directed toward bioethanol production, which brings up questions about the proper use of foodstuffs and land acreage for making ethanol fuel. An overall question arising is whether this endeavor eventually results in greater or lesser emission of carbon dioxide into the atmosphere than burning fossil fuels.

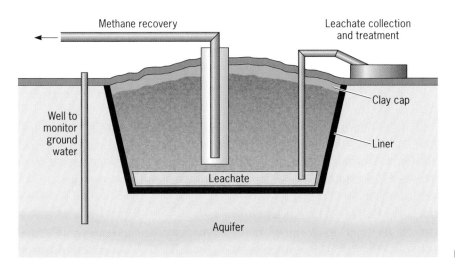

FIGURE 28.3 **Methane from landfill.**

Yeasts carry out the large-scale ethanol fermentation and, as you can imagine, considerable effort has gone into making them more efficient. The relevant metabolic pathways involved can be optimized at almost all stages in macromolecular synthesis: increasing the number of copies of genes, placing the genes under the control of strong promoters, altering the stability of the mRNAs, using strong ribosome-binding sites, and many more.

Industrial microbiologists have succeeded at generating yeast strains that excel at fermenting sugars into ethanol. The goal now is to reduce the cost of the ethanol product by making the yeasts break down plant material, particularly dedicated bioenergy crops and agricultural wastes, prior to fermentation. The structural components of plants—celluloses, hemi-celluloses, and lignins—are made in huge amounts but are only selectively utilized by humans. True, we use wood for construction and making furniture, but we make scant use of the waste material generated in the process. Likewise, we only partially use agricultural waste products. Their biomass produced annually in United States alone amounts to over 1 billion tons. This material is energy rich and can be converted into usable fuels. However, the polymers that make up the structures of plants are generally recalcitrant to breakdown, as can be seen by a stroll through an old forest. It takes dedicated fungi and bacteria to hydrolyze these tough polymers in nature. Much attention is being focused on this topic, including developing innovative technologies that harness the natural activities of the microbes that break down the biomass in the environment. As an example, consider how cellulose, the most abundant polymer in the plant cell wall, is broken down by microbes. The conversion of cellulose into ethanol requires an initial breakdown into sugars and then their fermentation. In nature, these steps are carried out by complex consortia of fungi and bacteria. By studying the natural microbes and their activities, scientists have identified enzymes that can be used for the industrial conversion of cellulose into fermentable sugars. Yeasts can use glucose generated during the enzymatic hydrolysis to make ethanol. As the plant cell wall also includes polymers (e.g., hemicellulose) that contain pentose sugars, scientists are trying to genetically engineer ethanologenic yeasts to ferment cellulose-derived glucose and the hemicellulosic pentose sugars simultaneously.

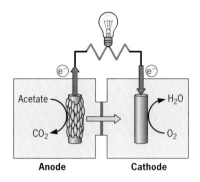

FIGURE 28.4 Schematic of a microbial fuel cell powered by a microbe such as *Geobacter* growing on the anode electrode. The bacterium metabolizes an electron donor (e.g., acetate) to CO_2 in the anode chamber and transfers the respiratory electrons to the anode electrode. The electrons then flow to the electron acceptor in the cathode (in this case, O_2). Protons also travel from the anode to the cathode chamber across a separating proton-permeable membrane to balance the system.

Other biofuels. Microbes make a variety of biofuels, other than ethanol. These include fatty acids that can be converted into biofuels. A favorite group of organisms for making biofuels are the **cyanobacteria**, because they are photosynthetic, grow fast, and can be readily manipulated genetically (they are naturally transformable). By cloning just three genes into these organisms, scientists generated a new strain of bacteria that now makes 2,3-butanediol, a starting material in the synthesis of certain plastics. Likewise, *E. coli* has been engineered to make propane via an enzyme, aldehyde-deformylating oxygenase (ADO), that has been redirected from its usual function of making fatty acids toward synthesizing propane.

Microbes That Make Electricity

Believe it or not, some microbes also make electricity. How? Simply give them an electrode poised to accept electrons that the cells generate during respiration! As the microbes grow on the electrode, they generate an electrical current. This process can be set up in the lab using a device known as a **microbial fuel cell** (Fig. 28.4). Typically, these living fuel cells consist of two chambers separated by a membrane that allows protons to flow freely. Each chamber is equipped with an electrode, the anode and the cathode, wired to each other so electrons produced by microbes growing on the anode electrode can travel to the cathode and react with an electron acceptor. For example, when you bubble the cathode chamber with oxygen, it gets reduced to water (or hydrogen peroxide, H_2O_2) by the electrons flowing from the anode electrode (Fig. 28.4).

The most efficient electricity-generating microbes known to date are those in the genus *Geobacter* (chapters 20 and 21), because of their ability to respire minerals and mediate **interspecies electron transfer**. These microbes have evolved unique strategies to respire solid-phase electron acceptors such as iron and manganese minerals. Their cell envelope is loaded with metal-containing proteins called *c*-cytochromes, which accept respiratory electrons and store them in the cell envelope until ready to be discharged. They also produce protein filaments similar to the type IV pili other bacteria use to attach to and move on surfaces (chapter 2). However, in *Geobacter* the pili are *conductive* and transfer electrons from the cell envelope to the extracellular electron acceptors (chapter 20). Using these remarkable conductive appendages, the electrifying bacteria attach to the electrode and form a thick **biofilm** whose exopolysaccharide matrix is loaded with *c*-cytochromes and conductive pili. The matrix is so conductive that the cells can continue to grow while layers of cells accumulate on the electrode (Fig. 28.4). The thicker the biofilm, the more electricity is harnessed!

Geobacter bacteria and other electrode-reducing bacteria really could not care less whether a mineral or an electrode is provided, as long as they can dump respiratory electrons onto them to gain energy. In fact, scientists realized that they could bury the anode electrode in the ocean sediments and wire it to a cathode maintained in the top oxygen-rich surface waters with a floater. In a few days, *Geobacter* bacteria attached to the buried electrode, built a biofilm, and generated electricity. The electrical energy was used to power a sensor and monitor the presence of hard-to-track sea animals, such as whales. Can you imagine other applications, particularly for remote locations where changing batteries or depending on sunlight to power solar cells is not an option? Scientists are also beginning to use these bioelectrodes in fermentations or to treat industrial wastewater, to remove unwanted waste products and produce electricity. New microbes that make electricity continue to be discovered. And some species have been isolated that do the reverse process: they use the cathode electrode as an electron donor to support their metabolism, pretty much the way some microbes oxidize Fe(II) in nature. These microbes literally "eat" electricity. With the development of diverse processes that exploit microbial electricity, a new field in microbiology called electromicrobiology has emerged. And the best is yet to come. Will you be one of the future electromicrobiologists?

CAN MICROBES HELP US CLEAN UP POLLUTION?

LEARNING OUTCOMES:
- Define bioremediation.
- Summarize reasons why genetically modified microbes are a good solution to cleaning up toxic materials in the environment.
- Describe how methanotrophs have been used to clean up toxic solvents.

The introduction, unintentional or intentional, of toxic and radioactive chemicals into the environment causes biological, physical, and financial damage of massive proportions. Such pollution occurs in many ways, including dumping of contaminants into rivers, lakes, or soils; maritime disasters; and leakage of millions of gallons of highly radioactive waste from buried tanks also containing other toxic chemicals. Cleaning up toxic-waste sites is a formidable task that challenges the limits of our societal capabilities and resolve. It is difficult and expensive.

Cleanup operations that purposefully involve microbes or plants are called **bioremediation**. This process is being used at some of the 50,000 or so hazardous-waste sites in the United States, with the focus on the 1,200 designated Superfund sites. These have been selected for special attention by the Environmental Protection Agency because of the grave threat they represent. There are several routes to bioremediation. In some cases, the aim is to lower the level of pollutants at the site of contamination; in others, the polluted material is extracted and removed to special locations, such as containment tanks or lagoons, there to be treated biologically.

The vast bulk of the radioactive waste is in the form of spent fuel from nuclear power plants, with U.S. inventory exceeding 80,000 metric tons. Almost all of this material is encapsulated and stored in containment sites. To give an idea of the cost, 25 years of cleanup at a single site in Washington State has cost more than $50 billion, a figure that is expected to triple. Japan's Fukushima-Daiichi disaster is predicted to cost at least as much. Storage of contaminated material, for which the United States has not yet found a permanent solution, is costly and carries with it the danger of leakage. Incineration is not an option as it causes air pollution.

Fortunately, microbes, with their enormous repertoire of chemical activities, are well suited to rid the environment of toxic material. A tremendous amount of microbial activity takes place spontaneously, although usually at a pace too slow to provide adequate relief. However, we can improve the microbes' effectiveness by studying natural ways in which they help clean the environment. The ecological principle that the *chemical and physical properties of the environment dictate the physiology of the organisms* applies here. For example, microbes that may help clean up sites contaminated with some of the radioactive materials are present in effluents from nuclear power plants. In such waters, one indeed finds bacteria that are not only resistant to high levels of radiation (like the world champion, *Deinococcus radiodurans*) but that can also render radioactive compounds insoluble, thus easier to dispose of. Such organisms can be cultured from the environment and then genetically modified to become even more effective. Air emissions, however, remain an unresolved problem.

Bioremediation has been moderately successful at the U.S. government's 310-square-mile Savannah River site in South Carolina, where radioactive materials for nuclear weapons were made for over 4 decades (Fig. 28.5). This site and the Hanford Nuclear Reservation in Washington State are among the most polluted tracts of land in the United States. At both sites, solvents used in the process were conveyed through buried pipes and stored in underground tanks. The pipes eventually leaked, and the solvents seeped underground. Among the most abundant of these solvents is trichloroethylene (TCE), which is quite toxic and, like other chlorinated solvents, difficult to clean up. TCE is one of the most abundant of toxic solvents because it is widely used for dry cleaning and other industrial applications. Microbiologists have devised a way to increase the number of bacteria that can oxidize TCE. They enlisted the help of soil bacteria called methanotrophs, which can use methane and other hydrocarbons, including TCE, as their sole source of carbon and energy.

Methanotrophs produce an enzyme called methane mono-oxygenase that is not highly specific and thus can degrade TCE and other compounds besides methane. Normally, these indigenous organisms are present in

FIGURE 28.5 Diagram of the arrangement for bioremediation of the TCE-polluted Savannah River site in South Carolina.

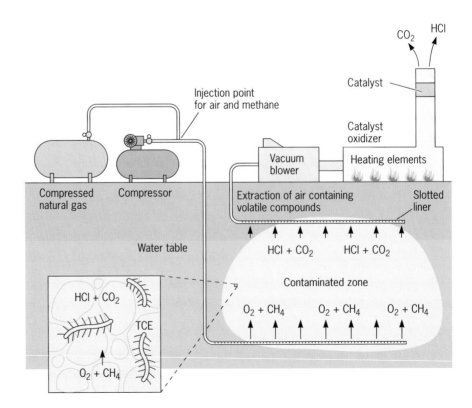

relatively small amounts. How to enrich for them? The contaminated soils were pumped with methane and also air, because hydrocarbon oxidation requires oxygen. As a result, the number of methanotrophs at some sites increased by 7 orders of magnitude! These bacteria then proceeded to reduce the concentration of TCE. The process is not quite as simple as it sounds, because considerable engineering is required to effectively deliver the gases into the soil (Fig. 28.5). One advantage of this process is that its biology is reversible: once the gases are no longer pumped into the soil, the methanotrophs return to their natural level.

This experience illustrates that enrichment for desired microbes requires an understanding of their nutritional requirements. In fact, in the case just described, methane and oxygen were not sufficient to maximally degrade the TCE. Controlled addition of sources of phosphorus and nitrogen further enhanced the growth of the methanotrophs. Another example of the effect of such nutritional supplementation was the 1989 oil spill caused by the tanker *Exxon Valdez* in Alaska (Fig. 28.6). There, the addition of commercial fertilizer supplied needed phosphorus and nitrogen and enhanced the degradation of the petroleum pollutants by resident microbes.

Of course, these approaches reduce the levels of only certain pollutants. The Savannah River site (Fig. 28.5) remains contaminated with heavy metals (lead, chromium, mercury, and cadmium), radioactive compounds (tritium, uranium, fission products, and plutonium), and others. Bioremediation techniques have been developed for the removal of some of these pollutants, including the radioactive nuclides. The same can be said for pesticides, PCBs (polychlorinated biphenyls), and other environmentally damaging substances. Consequently, one can be optimistic for the future of bioremediation. With further research, it seems likely that this approach could become helpful in solving one of modern humanity's most daunting problems.

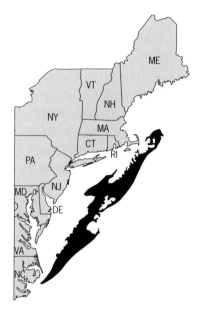

FIGURE 28.6 **Extent of the 1989** *Exxon Valdez* **oil spill drawn to scale on a map of the Eastern United States, for comparison.** (Modified from an illustration provided by the Alaska Wilderness League.)

WHAT DO MICROBES HAVE TO DO WITH CLIMATE CHANGE?

LEARNING OUTCOMES:
- Support the following statement in light of climate change studies: "Because the true diversity of microbial life is largely unknown, its effects and potential benefits have not been fully explored."
- Justify the statement "Climate change cannot be understood without consideration of the effects of microbes."
- Identify reasons why it is difficult to determine the roles of microbes in climate change.
- Describe three ways in which microbes in different environments affect climate change.
- List ways in which climate change has affected human-associated microbes and zoonotics.

By the time you have reached this chapter, if we have convinced you of anything, it should be that microbes play a pervasive role in all the key transactions of this planet. In chapter 20, we discussed how important microbes are in the cycle of elements in the planet, and we learned how their activities influence climate. This knowledge should enable us to consider how microbes participate in the phenomenon of climate change. But

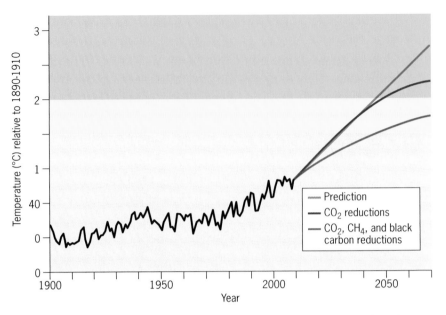

FIGURE 28.7 Microbes and global warming. Soil microbial communities cycle carbon and generate carbon dioxide (CO_2) and methane (CH_4), greenhouse gases that contribute to global warming and influence precipitation and extreme climatic events. Increases in atmospheric CO_2 promote plant productivity, which provides more organic carbon for the soil communities and leads to more carbon losses to the atmosphere. Human activities also contribute greatly to CO_2 and CH_4 releases. The incomplete combustion of fossil fuels, biofuels, and biomass also releases black carbon (or soot), a dark particulate matter that is lost to the atmosphere as an aerosol and contributes greatly to global warming. But control of emissions, particularly CO_2, could have significant long-term effects in mitigating global warming. (Modified from http://climate.nasa.gov/system/image_uploads/main /Shindell_Graph_1966x1390.jpg.)

the task is challenging because we still don't know that much about all that microbes do in the environment. What we do know is that the Earth is warming fast and temperatures are predicted to increase linearly due to the carbon emissions in the form of carbon dioxide, methane, and the carbon-containing particles released from incomplete combustions (as when burning coal, running a diesel engine, or in forest fires) (Fig. 28.7). It is not easy to estimate if higher global temperatures will alter the overall size of microbial biomass, thus changes in its composition cannot be foretold with certainty. Some things are clear in principle:

- The amount of carbon dioxide released into the atmosphere depends ultimately on the balance between its *consumption by photosynthesis and chemosynthesis* on the one hand and its *production by respiration and fermentation* on the other (chapter 20). Temperature changes affect these two sets of processes distinctly.
- Vast quantifies of carbon are stored in living matter, and changes in temperature will affect the size of these carbon sinks. Increased microbial activity will lead to heightened breakdown of organic matter in soils and water, thus increasing the amount of atmospheric carbon dioxide.
- Nitrogen, the main component of our atmosphere, is constantly recycled, to a large extent due to microbial activities (chapter 20). Some bacteria fix atmospheric dinitrogen (N_2), making their nitrogenous organic material; others decompose these compounds and release nitrogen back into

the atmosphere. Particularly relevant to climate change is the microbial conversion of dinitrogen into the greenhouse gas nitrous oxide (N_2O).

We cannot claim to have enough knowledge about all these phenomena to predict their effects, but there is sufficient basis for believing that climate change cannot be understood without considering the effects of microbes. Because there will be an increased demand for studies in these fields, microbial ecology may well be a "growth industry." But it won't be easy.

Why is this area of investigation difficult? The number of variables that affect microbes in the environment is huge. In this spirit, we can try to find out how the world's microbiota is affected by changes in temperature, pH of the oceans, or the amount of precipitation on the ground. We can further ask what will result from changes in the amount of plant growth, the structure of soils, or the layering in the oceanic water columns.

Microbes play numerous roles in the matter by fostering the growth of plants, capturing atmospheric carbon dioxide in the process. Consider a few examples. Field studies done in Colombia showed that introducing mycorrhizal fungi in agricultural soils (chapter 15) stimulated by 20% the growth of the widely consumed tuber staple, cassava. The key is a microbial enzyme, aminocyclopropane-carboxylate deaminase (ACC). Plants make *ethylene* in response to stress, and it stops roots from growing and induces leaves to fall. ACC *inactivates an ethylene precursor*; thus, in the absence of this downregulating signaling molecule, the plants can continue to grow. Similarly, some bacteria make the disaccharide *trehalose* that allows certain plants to withstand drought. Engineering plant-adapted strains to produce trehalose has allowed a 50% increase in the productivity of bean plants and enhanced their resistance to drought. And beyond this, some bacteria contribute to greater plant growth by protecting them from insects and other pathogens. Such beneficial activities not only greatly influence the food supply, but they also contribute to ameliorating climate change because more plants means less carbon dioxide in the atmosphere.

We have already noted that methane, a very powerful greenhouse gas, is made in landfills (Fig. 28.3). The culprits are methanogenic archaea, which are unique in biology in their ability to make this gas. But methane from these sources doesn't have to go into the atmosphere because, with some suitable engineering, this greenhouse gas can be recovered from landfills and used as fuel. However, methane production elsewhere is not so easy to control. A lot of this volatile compound is made by methanogenic archaea in the rumen of cattle and other ruminants, from where it is released when the livestock "belch" from both ends (chapter 21). About 20% of all methane production takes place in these farm animals. Consequently, efforts are directed toward reducing the greenhouse gases generated naturally by livestock.

Rising global temperatures can also lead to further increases in carbon dioxide emissions. Consider the permafrost of the northern polar regions. As the ice in these habitats melts, microbes have a new opportunity to break down previously frozen organic material. By their metabolism, microbes produce heat directly, further melting the ice. And in the process, they make both carbon dioxide and methane. However, some of the methane, the more important greenhouse gas of the two, will be utilized by bacteria, which may help keep it in a constant flux. The magnitude of these effects is being studied, and there are predictions that the thawing of the permafrost and subsequent increase in microbial metabolism will have enormous potential to release vast amounts of carbon dioxide into the atmosphere.

Results from environmental measurements are not always intuitive. For example, studying a forest in Colorado, researchers found that the rate of respiration in soils was higher when the snow cover was thick and lower when it was reduced. It turned out that this surprising result was due to a soil microbial community at this site that was more metabolically active in the cold and thus was influenced by the depth of the snow layer. How do you imagine that some bacteria prefer cold to warmth? (*Quiz spoiler*: they are likely to outcompete species that do not have this ability.) At a glance, this interaction between climate and these microbial communities should function as a feedback loop—higher temperatures result in less carbon dioxide being released from such sites, which would result in lower temperatures. This phenomenon is not likely to work on a large enough scale to maintain global temperatures and is likely to be nearly insignificant in magnitude when compared to the consequences of thawing the permafrost. However, it does illustrate the complexities of the relationship between microbes and climate change.

Consider now how climate change affects the microbes themselves. Particularly important to us are effects on pathogens of humans, animals, and plants. Even in the early stages of climate change, notable consequences have been uncovered. For example, some zoonotic infectious diseases have now been noted in places where they were not known to occur, as their vectors migrate into new territory. An effect seen in many oceans of the world is the bleaching of coral reefs due to increased growth of bacterial pathogens. Many disease-causing insects now live in areas where they had previously been barred by cold temperatures. For example, mosquitoes in Hawaii are now found at higher mountain elevations, carrying a malaria parasite that infects native birds. Marine vibrios that affect fish, shellfish, and corals are now seen in new places, such as parts of the North Sea. The list of human diseases that are emerging in new areas is long and includes such deadly diseases as cholera, dengue fever, West Nile virus, hantavirus, and many others. Unless the current trend toward climate change is mitigated, the future demand for microbiologists and infectious disease specialists looks favorable.

IS THERE LIFE BEYOND OUR PLANET?

LEARNING OUTCOMES:
- Identify specific microbes that may have properties relevant to potential life on other worlds.
- Speculate on why it has been so difficult to eliminate all microbes from spacecraft.

Earth is the only place in the universe that we know to harbor life, but that has not kept people from speculating about its occurrence elsewhere. The discovery of many planets outside our solar system fosters such thoughts. These speculations are based on several assumptions. One is that life elsewhere is also based on carbon. This is not a bad idea because carbon is the fourth most abundant element in the universe and has the right properties for building biological molecules (and plausible alternatives are few, if any). Also assumed is that living things require liquid water. It is further thought that life is most likely on planets orbiting suns, which themselves must fall within a narrow range of size and longevity. But given the vastness of the universe, finding a place with just the right conditions seems plausible.

Of course, the great problem is being able to communicate with faraway sites. Distances across the universe are so large that they cannot be spanned except over extremely long periods of time when compared with a human lifetime. But intelligent living things may have sent out messages in the remote past, which has prompted attempts to detect their signals via radio telescopes. Such efforts, known as "Search for Extra-Terrestrial Intelligence" or SETI, have been under way for some time. To date, none of them has yielded auspicious results.

In parallel with such efforts, scientists have asked whether organisms living on Earth may have properties relevant to life elsewhere. The favored organisms to consider are the extremophilic *Bacteria* and *Archaea*, which thrive in environments that may be found on other planets in our or other solar systems. For one thing, many bacteria, archaea, fungi, plant seeds, and even animals such as tardigrades can survive radiation and vacuum typical of outer space.

Thinking about the possible existence of life beyond this planet is intimately related to thoughts about the origin of life. Knowing how life originated, and assuming that life elsewhere emerged the same way, could give hints about where and even how to look.

Alongside considerations of life in distant planets, there is the perhaps more tractable question of life elsewhere in our solar system. Our planets and some of their moons can be visited with current space technology. As you probably know, efforts have been under way to look for life on Mars. But a problem looms large: sending probes to other planets means that we will almost surely contaminate them with earthly microbes. This is a well-recognized problem that has caused the promulgation of a 1967 Space Treaty that forbids the "harmful contamination" of other worlds with Earth's biology. This is easier said than done. Space probes and their parts are routinely disinfected using radiation and strong chemicals such as hydrogen peroxide. Despite all efforts, these machines remain contaminated with bacteria before their departure. Moreover, such unwanted travelers survive surprisingly well. Thus, Apollo astronauts found bacteria on the moon inside the Surveyor 3 lander that had touched down more than 2½ years previously. Keep in mind that sites on Mars, such as those with briny water, may be much more hospitable.

Dozens of microbial species accompanied the 2012 Curiosity rover to Mars (Fig. 28.8). These stowaways survived the extensive cleaning of the spacecraft before it was launched. Some 65 bacterial species and over 300 strains were found by culturing material swabbed from various surfaces. Many of them were indeed resistant to the treatment used for cleaning the spacecraft. Most of the isolates were of the spore-forming genus *Bacillus*

FIGURE 28.8 **The Mars Curiosity Rover, self-portrait at the "Big Sky" drilling site.**
(Source: NASA, http://photojournal.jpl.nasa.gov/catalog/PIA19920.)

but about a third were not, and the reason for their unusual resistance to radiation and harsh chemical remains unknown. The problems may be compounded when sending astronauts aloft—they obviously cannot be sterilized. These concerns add huge costs and efforts to the exploration for life elsewhere. We can assume that Mars has already experienced several waves of Earthly invaders (Fig. 28.8).

CONCLUSIONS

In this chapter, we have barely sampled the long history of the many ways humans have enlisted the cooperation of microbes for useful purposes. Others abound, and new uses are constantly being discovered. This field of microbiology will undoubtedly expand as new microbes are discovered and our ability to genetically engineer known microbes improves. Our future is as intertwined with microbes as our past was and as our present is. Perhaps even more so, because we face challenges too great to overcome without the help of microbes. This is an exciting time to be studying microbiology.

Representative ASM Fundamental Statements and Learning Outcomes for the Chapter

METABOLIC PATHWAYS

1. *Bacteria* and *Archaea* exhibit extensive, and often unique, metabolic diversity.

    a. Describe how a microbial fuel cell works.

    b. Explain how *Geobacter* produces energy.

    c. Describe how methanotrophs have been used to clean up toxic solvents.

INFORMATION FLOW AND GENETICS

2. Cell genomes can be manipulated to alter cell function.

   a. Outline a cloning method to engineer microbes that synthesize human proteins.

MICROBIAL SYSTEMS

3. Microorganisms are ubiquitous and live in diverse and dynamic ecosystems.

   a. Speculate on why it has been so difficult to eliminate all microbes from spacecraft.

4. Microorganisms and their environment interact with and modify each other.

   a. Justify the statement "Climate change cannot be understood without consideration of the effects of microbes."
   b. Identify reasons why it is difficult to determine the roles of microbes in climate change.
   c. Describe three ways in which microbes in different environments affect climate change.
   d. List ways in which climate change has affected human-associated microbes and zoonotics.

IMPACT OF MICROORGANISMS

5. Humans utilize and harness microorganisms and their products.

   a. Name several microbial products that you use or are impacted by in your everyday life.
   b. List reasons why researchers use cloning in *E. coli* to make significant quantities of human insulin.
   c. Identify the real-world purposes of synthetic biology.
   d. Describe how genetic engineering of the Bt toxin has benefited agriculture.
   e. Discuss the mechanism by which Bt toxins are poisonous for insects but safe for human consumption.
   f. Identify several differences in the process of making beer versus making wine.
   g. Discuss how the succession of yeasts and lactobacilli helps make and refine wine.
   h. Name four ways in which microbes may be used to generate biofuels.
   i. Identify benefits and potential drawbacks to bioethanol generation by yeast.
   j. Define bioremediation.
   k. Summarize reasons why genetically modified microbes are a good solution to cleaning up toxic materials in the environment.

6. Because the true diversity of microbial life is largely unknown, its effects and potential benefits have not been fully explored.

   a. Support the following statement in light of climate change studies: "Because the true diversity of microbial life is largely unknown, its effects and potential benefits have not been fully explored."
   b. Identify specific microbes that may have properties relevant to potential life on other worlds.

SUPPLEMENTAL MATERIAL

References

- **Sumby KM, Grbin PR, Jiranek V.** 2014. Implications of new research and technologies for malolactic fermentation in wine. *Appl Microbiol Biotechnol* **98**:8111–8132.
- **Schmidt FR.** 2004. Recombinant expression systems in the pharmaceutical industry. *Appl Microbiol Biotechnol* **65**:363–372.
- **Melo AL, Soccol VT, Soccol CR.** 2016. *Bacillus thuringiensis*: mechanism of action, resistance, and new applications: a review. *Crit Rev Biotechnol* **36**:317–326.
- **Moissl-Eichinger C, Pukall R, Probst AJ, Stieglmeier M, Schwendner P, Mora M, Barczyk S, Bohmeier M, Rettberg P.** 2013. Lessons learned from the microbial analysis of the Herschel spacecraft during assembly, integration, and test operations. *Astrobiology* **13**:1125–1139.

Supplemental Activities

CHECK YOUR UNDERSTANDING

1. How do bacteria contribute to improving the quality of wines? What do they get in return?

2. How do bacterial enzymes contribute to the production of high-fructose sweeteners used in industry?

3. What are the steps in engineering a bacterium to produce a human protein for industrial purposes?

DIG DEEPER

1. *B. thuringiensis* produces toxins that kill insects. However, the bacterium inhabits the soil, where there are not many insects. How do you think the ability to make this toxin evolved?

2. Search the Web to find information about the contaminants listed below and how microbes interact with them. Discuss what physiological attributes of bacteria may suggest a strategy for their bioremediation.

   a. Uranium
   b. Arsenic
   c. PCE
   d. Mercury

3. **Team work:** Genetically modified organisms, friend or foe?

   Consider the following two examples of genetically modified organisms (or GMOs) used in real-life applications and examine the pros and cons of their use:

   Example 1: *Pseudomonas* species used to melt or make ice

   - What do you think of the public concern over the use in the field of genetically engineered *P. syringae* for the purpose of delaying ice formation on cultivated plants?
   - Why is this not a concern in the case of using bacteria for the opposite effect, making snow?

   Example 2: Toxin-producing *B. thuringiensis* as an insecticide

   - How have Bt toxins contributed to the agriculture industry?
   - What are the risks?

   Can these and other GMOs be used safely, or not?

   Resource: American Academy of Microbiology Report "100 Years of *Bacillus thuringiensis*: A Critical Scientific Assessment, 2002" http://academy.asm.org/index.php/browse -all-reports/467-100-years-of-bacillus-thuringiensis-a-critical-scientific-assessment

4. **Team work:** In your lifetime, the search for life beyond Earth will continue to hold much attention and concerted efforts will be made to seek answers. Consider possible outcomes of being successful in finding extraterrestrial life:

   - What do you expect life elsewhere will teach us about our own biology?
   - Would such findings unite or further disrupt human society?
   - What could extraterrestrial societies teach us about technology and ethics?
   - Could such discoveries help us deal with global phenomena such as climate change?

Supplemental Resources

VISUAL

- Bacterial production of human insulin and growth hormone is highlighted in the FASEB 2013 "Stand Up for Science" award-winning video: https://www.youtube.com/watch?v =GmhD-RWNL6c
- American Society for Microbiology *Microbes After Hours* public lecture "Microbiology of the Bioeconomy" by Sonny Ramaswamy of the National Institute of Food and Agriculture and Joy Doran Peterson of the University of Georgia. http://www.microbeworld.org

/podcasts/asm-after-hours/1403-the-microbiology-of-the-bioeconomy-an-asm-microbes
-after-hours-event-monday-june-3-2013-6-8-p-m-et

- American Society for Microbiology *Microbes After Hours* public lecture "The Microbiology of Cheese" by Rachel Dutton of Harvard University and Mateo Kehler of Jasper Hill Farms. http://www.microbeworld.org/podcasts/asm-after-hours/1698-the-microbiology -of-beer-the-microbes-after-hours-series-6-8-pm-thursday-october-10-2014the -microbiology-of-cheese-live-june-10-at-asm-headquarters

- American Society for Microbiology *Microbes After Hours* public lecture "The Microbiology of Beer" by Charles Bamforth of University of California-Davis and Rebecca Newman, a craft brewery specialist. http://www.microbeworld.org/podcasts/asm-after-hours/1472 -the-microbiology-of-beer-the-microbes-after-hours-series-6-8-pm-thursday-october-10 -2013

- *Microbe World* Meet the Scientist video (39 min): "James Liao–Turning Microbes into Fuel Refineries" http://www.microbeworld.org/component/content/article?id=668

- Microbe World video #30 (7 min) "Biofuels in Puerto Rico": Nadathur S. Govind and William Rosado of the University of Puerto Rico describe the sustainable biofuel program they are launching in southwestern Puerto Rico. http://www.podcastdirectory.com /episodes/mwv-episode-30-biofuels-in-puerto-rico-26696325.html

- In 2010, an explosion on the Deepwater Horizon oil drilling rig spilled 210 million gallons of oil into the Gulf of Mexico. On the ASM LIVE episode (32 min) recorded at the 2010 American Society for Microbiology General Meeting, Jeff Fox discusses the impact of marine microbes on remediation efforts with Jay Grimes of the University of Southern Mississippi and Ronald Atlas of the University of Louisville. http://gm.asm.org/index.php /scientific-activities/asm-live/archives/390-the-gulf-oil-spill-microbes-to-the-rescue

SPOKEN

- *This Week in Microbiology* #6: Bacteriophage therapy discussion. http://www.microbe world.org/podcasts/this-week-in-microbiology/930-twim-6-the-good-and-the-bad-of -bacteriophage

WRITTEN

- American Academy of Microbiology Report "Harnessing Microbes as Therapeutics: Bugs as Drugs": http://academy.asm.org/index.php/browse-all-reports/5296-harnessing -the-power-of-microbes-as-therapeutics-bugs-as-drugs

- American Academy of Microbiology FAQ booklet "Microbes Make the Cheese": http:// academy.asm.org/index.php/faq-series/5282-faq-microbes-make-the-cheese

- American Academy of Microbiology FAQ booklet "If the Yeast Ain't Happy, Ain't Nobody Happy: The Microbiology of Beer": http://academy.asm.org/index.php/faq-series /435-beer

- American Academy of Microbiology Report "Incorporating Microbial Processes into Climate Models, 2012." http://academy.asm.org/index.php/browse-all-reports/167-incor porating-microbial-processes-into-climate-models

- American Academy of Microbiology Report "Microbial Energy Conversion, 2006": http:// academy.asm.org/index.php/browse-all-reports/176-microbial-energy-conversion

- American Academy of Microbiology Report "Microbe-Powered Jobs: How Microbiologists Can Build the Bioeconomy." http://academy.asm.org/index.php/browse-all-reports /5128-microbe-powered-jobs-how-microbiologists-can-help-build-the-bioeconomy

Coda

We are becoming increasingly aware that microbes are essential for all forms of life on Earth, including humans. Microbes are our heritage because they are the ancestors of all other living things. And right now, as we live and breathe, our lives are still critically dependent on microbial activities. As we have seen, microbes process the nutrients and elements necessary for life, they influence our climate and weather, and they sculpt our planet's rocks and bodies of water. Not all of our interactions with microbes are benign, and certain microbes pose a threat to human health and to the health of plants and animals. To us, this may seem like yin and yang, but this is an anthropocentric point of view. The microbes are simply trying to make a living, indifferent to whether they help or hinder human beings.

Microbes then are the foundation of the biosphere and major determinants of human health. Hence, the study of microbes—microbiology—is a fundamental subject and is essential for the study and understanding of all life on Earth. We hope that your excursion into the microbial world will serve as the groundwork for further studies and, more broadly, in your sojourn on this planet.

This book ends with a wish that what we have learned so far and will continue to learn about the microbial world will, for the sake of this planet and all its inhabitants, be put to a sane and prudent use. This requires not only good intentions but also a deepening of our understanding of all that microbes do.

Glossary

3′ → 5′ exonuclease A component of DNA polymerase III that proofreads the new strand as it is synthesized. It detects and then excises **mismatches** by removing nucleotides in a 3′ → 5′ direction so that replacement synthesis can proceed 5′ → 3′ as usual.

A site One of the three tRNA binding sites on ribosomes, specifically the one bound by an incoming aminoacyl-tRNA.

ABC transport (ATP-binding cassette transport) A form of **active transport** bringing solutes into the cell facilitated by solute-specific binding proteins that convey the solute to the membrane and there trigger pore opening by ATP hydrolysis.

Abortive transcription Transcription that stops after approximately a dozen nucleotides have been added due to failure of **RNAP** to separate from the **sigma factor** (σ).

Abscess An inflammatory lesion that contains **pus**.

Activator A regulatory protein that binds to a **DNA control region** and increases the rate of initiation of transcription.

Absorbance The attenuation of light at a particular wavelength by a standard thickness of a solution or suspension. Measurement of the transmitted light can be used to quantitate the number of microbial cells in a culture.

Accommodation In the bacterial chemosensory system that regulates flagellar rotation, the decreased sensitivity of the chemoreceptors for a period following a cycle of stimulation that increases the concentration of the attractant/repellent required to elicit the same response.

Acid-fast The property of a stained cell to resist destaining by mild acids that is characteristic of mycobacteria and some actinomycetes.

Acidophile An organism that thrives in highly acidic environments and that typically grows optimally between pH 1 and pH 5.

Actinobacteria A phylum and a class of Gram-positive bacteria with high guanine-cytosine (GC) content that includes the genus *Streptomyces*, soil bacteria that produce **secondary metabolites** of medical and industrial importance.

Active transport The movement of solutes across a membrane against a concentration gradient powered by cellular energy.

Acyl carrier proteins Cytoplasmic proteins that facilitate **phospholipid** synthesis by preventing aggregation of the hydrophobic fatty acids and ferrying them to the waiting polar head group.

Adaptation (1) Physiological adjustments made by an organism in response to a change in environmental conditions; (2) a trait acquired through genetic variation and maintained by natural selection because it increases the fitness of the organism under particular environmental conditions.

Adaptive immunity **Antigen**-specific immune defenses that are acquired by exposure to the antigen.

Adaptive mechanism A strategy used to cope with fluctuating environmental stresses, particularly those used to counter multiple stresses.

Adaptive response A strategy for coping with stresses such as heat shock and phosphate starvation.

Adenylate cyclase The enzyme that converts ATP into cAMP (cyclic adenosine monophosphate).

Adventurous motility Independent movement of individual cells away from the group, as in myxobacteria.

Adhesins A component of the cell surface or a cellular appendage, typically a protein, that promotes the adhesion of a cell to other cells or surfaces.

Adsorption During a viral infection, the attachment of a virus to its specific **receptor** on a potential host cell.

Aerobic respiration A process by which electrons, either alone or as hydrogen atoms, are passed from an electron donor to oxygen (terminal electron acceptor) while simultaneously generating a transmembrane proton gradient.

ADP-ribosylase An enzyme that post-translationally modifies another protein by adding one or more ADP-ribose moieties.

Akinete A thick-walled, dormant bacterial cell derived from a vegetative cell and capable of reversion to its previous form, produced in response to adverse conditions by some cyanobacteria and other bacteria.

Alkaliphile An organism that thrives in alkaline environments (approximately pH 8.5 to pH 11) and that typically grows optimally at about pH 10.

Allosteric protein A protein whose conformation and catalytic function are modulated by the binding of a specific molecule to a binding site distinct from the catalytically active site. Examples: transcriptional repressors, **enzymes.**

Alphaproteobacteria A diverse class of *Proteobacteria* that includes plant symbionts, insect endosymbionts, and intracellular pathogens, among others.

Alternative pathway The activation of **complement** by **MAMP**s as part of the rapid **innate immune** response.

Aminoacyl-tRNA synthetase The family of **enzymes** that link amino acids to their specific tRNAs for subsequent insertion into a growing polypeptide chain, thereby translating the base sequence in an mRNA into the amino acid sequence of a synthesized protein.

Amplicon A segment of DNA or RNA that is amplified by a natural process or a laboratory procedure such as the **polymerase chain reaction** (PCR).

Amyloid A waxy deposit of protein and carbohydrate formed in some diseased tissues, including the **prion**-associated fibrils of **spongiform encephalopathies**.

Anaerobic respiration A process by which electrons, either alone or as hydrogen atoms, are passed from an electron donor to a terminal electron acceptor (an organic or inorganic molecule other than oxygen) while simultaneously generating a trans-membrane proton gradient.

Anammox The microbial anaerobic oxidation of ammonia to dinitrogen.

Annotation The identification of potential **open reading frames (ORFs)** within a sequenced genome and the assignment of putative functions to them based on comparisons to ORFs with experimentally determined functions.

Anoxygenic photosynthesis A process in which light energy is harnessed to generate ATP and **reducing power** while simultaneously oxidizing inorganic molecules (other than water) or organic compounds.

Antibody Highly diverse blood proteins of the **adaptive immune** system that are produced in response to exposure to a specific **antigen** and are highly antigen-specific in their action.

Antibiotic A natural organic molecule produced by a microbe that kills or inhibits the growth of bacteria; synthetic compounds used to treat bacterial infections or kill environmental bacteria.

Anticodon The nucleotide triplet in each tRNA that matches the amino acid carried by the tRNA to a complementary **codon** in an mRNA molecule during protein synthesis.

Antigen A molecule, often a protein or polysaccharide, which triggers the **adaptive immune** system to produce **antibodies** against it.

Antisense (adj.) A nucleotide sequence that is complementary to a coding (**positive sense**) sequence.

Apicoplast An essential, membrane-bounded, DNA-containing organelle of unknown function found in *Plasmodium* species.

Apoptosis A genetically programmed sequence of events leading to cell death that occurs during normal development of multicellular organisms and also in response to some invading pathogens.

Archaellum The archaeal functional equivalent of the bacterial flagellum.

Ascus (pl. asci) In Ascomycete fungi, the structure that houses the four or eight haploid cells produced by the meiosis.

Assimilation The incorporation of an environmental element, such as phosphorus, into precursor metabolites.

ATP synthase An enzyme complex embedded in the cell membrane that uses the **proton motive force** to drive the phosphorylation of ADP to form ATP.

ATPase *See* **ATP synthase.**

ATP-binding cassette transport *See* **ABC transport**.

Attenuation The pausing during transcription typically prompted by formation of a **hairpin loop** in the nascent RNA chain. *See also* **pausing site**.

Autoinducer A signaling molecule used in **quorum sensing** that is recognized in response to changes in bacterial cell population density.

Autoregulation Regulation of gene expression by the cellular level of the protein product of that gene.

Autotroph An organism that can use an inorganic carbon molecule such as carbon dioxide as its sole carbon source.

Auxotroph An organism that, due to a mutation, has lost the ability to synthesize a particular metabolite or nutrient; alternatively, an organism that requires an external source of one or more **building blocks**.

Axial filament During **endospore** differentiation, the elongated form of the **nucleoid**.

B lymphocyte (B cell) An **antibody**-producing white blood cell of the **adaptive immune** system.

Bacille de Calmette et Guérin (BCG) A TB vaccine that contains avirulent (attenuated) live bacteria.

Bacteriochlorophyll A class of photosynthetic pigments found in *Bacteria*, members of which vary in their ability to harvest light of different wavelengths.

Bacteriocin Any of several classes of peptides that are produced and released by *Bacteria* and *Archaea*, which kill or inhibit the growth of members of the same or closely related species by a variety of tactics.

Bacteriocyte Specialized cells of some insects that harbor large numbers of endosymbiotic bacteria.

Bacteriophage (phage) A virus that infects *Bacteria*.

Bacteriorhodopsin A membrane protein in some **haloarchaea** that uses energy from sunlight to pump protons out of the cell, thus creating a **proton motive force** to drive the production of ATP.

Bacteroid A **differentiated** bacterial cell in a **root nodule** that carries out **nitrogen fixation**.

Balanced growth A growth state characterized by (1) exponential growth, (2) a constant mean cell size, and (3) the same proportional increase of all cell constituents over any given interval of time.

Barophile An organism that grows optimally at pressures exceeding 1 atm (*see piezophile*).

Basidium In Basidiomycete fungi, the structure that produces haploid spores (basidiospores) by budding.

BCG vaccine *See **Bacille de Calmette et Guérin** vaccine.*

Benthic microbe A microbe that lives in the sediments on the ocean floor. *See also pelagic microbe*.

Betaproteobacteria A class within the phylum *Proteobacteria* that includes pathogens such as *Bordetella pertussis* and *Neisseria gonorrhoeae*.

Biased random walk Random motion in which the probability of movement is greater in one direction than in the opposite direction.

Binary cell division (binary fission) A common mechanism whereby cells approximately double in size and then divide to produce two equal daughter cells.

Biofilm A community of microbes that adhere to each other and to a surface, are typically embedded in a secreted matrix, practice division-of-labor, and often adopt a characteristic architecture.

Biogeochemical cycles The complex interconversions of bioelements (carbon, oxygen, nitrogen, sulfur, and phosphorus) that balance utilization with replenishment to maintain relatively stable global levels.

Bioluminescence The production and emission of light by some bacteria and other organisms.

Biomineralization The process by which cells transform elements from the environment into minerals.

Bioremediation The use of microbes, or other organisms, to remove or detoxify anthropogenic environmental contaminants.

Box A **DNA control region** that is typically the binding site for a regulatory protein.

Broad-spectrum antibiotic An antimicrobial that affects a cell constituent that is vital to diverse *Bacteria*, including both Gram-positive and Gram-negative cells.

Bt toxin Any member of a group of proteins with insecticidal properties made by *Bacillus thuringiensis* during sporulation.

Bud scar A differentiated region on the surface of a budding yeast left when a daughter cell detaches from the mother cell.

Budding Asymmetrical cell division found in some yeasts in which a small daughter cell forms as a protuberance (a "bud") from the mother cell.

Building blocks Molecules that are synthesized from **precursor metabolites** and then polymerized to build cellular macromolecules. Example: amino acids are the building blocks of proteins.

CAP *See **catabolite activator protein**.*

Capsid The protective proteinaceous shell surrounding the viral chromosome(s) of a **virion.**

Capsomer A structural unit of a **virion** composed of multiple copies of the main **capsid** protein(s) that self-assemble to form the capsid.

Capsule A diffuse layer of **exopolysaccharide** or polypeptide surrounding, and attached to, the cell envelope of some *Bacteria*.

Carboxysome Intracellular compartments in *Bacteria* that are bounded by a protein shell and that contain the **enzymes** required to fix carbon dioxide, including **RuBisCO.**

Catabolic repression A regulatory strategy that enables the cell to utilize premium substrates first, then switch to lesser substrates when the preferred one is depleted.

Catabolite activator protein (CAP) A regulatory protein activated by cAMP that regulates transcription of **operons** within the **catabolite repression modulon.**

Catabolite repression protein (CRP) A regulatory protein that binds to and regulates transcription of **operons** within the **catabolite repression modulon.** Also known as cAMP repression protein (CRP) or catabolite activator protein (CAP).

Cell envelope In *Bacteria* and *Archaea*, all the layers that bound the cell including the cell membrane and surrounding structures such as **cell wall**, **outer membrane**, and **S-layer**.

Cell transformation The alteration of properties and processes of eukaryotic cells caused by viral infection, including **oncogenesis**.

Cell wall A rigid layer made of a polymer of sugars and amino acids that surrounds the cell membrane of most *Bacteria* and some *Archaea* and provides structural support.

Cell-mediated immunity An **adaptive immune** response by white blood cells that is effective against intracellular invading organisms.

Central Dogma The tenet in molecular biology that information flows in one direction only: from DNA to RNA to proteins.

Chaperone A protein that facilitates the correct folding or refolding of other proteins or prevents them from folding prematurely.

Chemiosmosis The movement of ions across a biological membrane down a concentration gradient, such as the movement of protons used by cells to create ATP.

Chemoautotroph An organism that gains energy for growth from the oxidation of chemicals (organic or inorganic molecules) and carbon from an inorganic carbon molecule (such as carbon dioxide).

Chemoheterotroph An organism that gains energy for growth from the oxidation of chemicals (organic or inorganic molecules) and carbon from an organic carbon source.

Chemolithotroph *See* ***chemoautotroph.***

Chemostat A device that maintains a culture at a constant growth rate by the continuous addition of nutrient-containing medium and proportional removal of medium containing cells and accumulated wastes.

Chemotaxin A protein that attracts white blood cells as part of **innate immunity.**

Chemotaxis Directed movement of a microbe in response to a sensed concentration gradient of a specific chemical.

Chemotroph An organism that obtains its energy from oxidation of organic or inorganic molecules.

Chlorophyll A class of photosynthetic pigments found in numerous bacterial phyla as well as in chloroplasts.

Cholera toxin A protein complex encoded by a temperate phage of *Vibrio cholerae* that produces the diarrhea of cholera.

Cholesterol-dependent cytolysin A class of pore-forming **cytotoxins** secreted by a variety of Gram-positive bacterial pathogens, including *Listeria monocytogenes*.

Cistron A sequence in DNA or mRNA that codes for one polypeptide chain.

Classical pathway The activation of **complement** by **antibodies** as part of **adaptive immunity**.

Coagulase An enzyme produced by *Staphylococcus aureus* and other microbes that converts the fibrinogen in blood into fibrin, thus forming blood clots.

Codon Ribonucleotide triplet in an mRNA that is complementary to the **anticodon** sequence in the cognate tRNA.

Cold shock The death of a microbe due to a sudden drop in temperature even if the microbe can normally tolerate the lower temperature.

Comammox Complete ammonia oxidation to nitrate via nitrite by a single microbe.

Commensal An organism that lives in association with and benefits from another organism without affecting (neither harming nor benefiting) that partner.

Commensalism Intimate interaction between two organisms of different species in which one benefits without harming or benefiting the other.

Compatible solute Molecules accumulated intracellularly to balance high-osmotic environments with minimal damage to cellular constituents such as proteins.

Competence The ability of a prokaryotic cell to take up DNA from its surroundings, which is the first step in **transformation**.

Complement A multifunctional group of proteins in the blood that elicits inflammation as part of **innate immunity** and that complements the actions of **antibodies**.

Complementary DNA (cDNA) DNA made synthetically from an RNA template using **reverse transcriptase**.

Composite transposon A **transposon** composed of one or more genes encoding functions other than those needed for transposition and that is flanked by two different **insertion sequences**.

Concatenate (verb) To connect two or more rings like links in a chain; (noun) two or more chains so linked.

Conjugation Transfer of DNA from one bacterial cell to another via direct cell-to-cell contact or a temporary connecting cytoplasmic structure (e.g., a conjugative or **sex pilus**).

Conjugative plasmid A **plasmid** capable of self-transfer to another cell by promoting **conjugation** between its host and another cell.

Contig A DNA or RNA sequence derived by computer assembly of shorter but overlapping sequences.

Convergent evolution The independent evolution by different organisms of similar phenotypic traits to solve similar challenges.

Core genes In a microbiome, the genes that are found in all microbiomes from a particular environment. In a species, the genes that are found in all sequenced strains.

Cortical inheritance The inheritance of cellular surface properties independent of cellular genes.

Crenarchaeota A major archaeal phylum that includes acidophiles and thermoacidophiles, as well as marine mesophiles.

CRISPR (clustered regularly interspersed short repeats) A prokaryote form of **adaptive immunity** that creates a memory of prior infections and uses this to target matching incoming foreign DNA for cleavage.

Crown gall A plant tumor caused by infection by *Agrobacterium tumefaciens*.

CRP *See* **catabolite repression protein.**

Cryo-electron microscopy A transmission electron microscopy (TEM) variant in which samples are prepared by rapid freezing, often at liquid ethane temperature, and imaged without use of a stain or fixative.

Cryo-electron tomography A version of **cryoelectron microscopy** in which a series of images are recorded as the specimen is tilted, and the images are computationally combined to yield a three-dimensional structure.

Crystalline surface layer (S-layer) The outermost layer of the **cell envelope** of some *Bacteria* and *Archaea* composed of a regular lattice of a single protein or glycoprotein.

CTX phage A **temperate phage** that infects *Vibrio cholerae* via the toxin-coregulated **pilus** (TCP), encodes the cholera toxin, and can be transmitted horizontally between bacteria.

Cultivation The growing of microbes under controlled conditions in the lab.

Culture-independent method Or **cultivation-independent method**, any method that assesses the composition and properties (e.g., relative abundance of taxa, genes, or proteins expressed) of microbial communities in an environment without requiring culturing of the microbes.

"Cut and paste" transposition A mode of transposition in which the **transposon** is cut from its original location and pasted into a new location.

Cyanobacteria A bacterial phylum whose members contain chlorophyll *a* and carry out **oxygenic photosynthesis**.

Cyclic di-GMP In **signal transduction** in bacteria, a molecule that interacts with a variety of regulatory proteins to effect the regulation of transcription that is required for many complex cellular processes and responses.

Cyclic photosynthesis Photosynthesis in which the terminal electron acceptor is the same chlorophyll that was light excited and donated the electron at the start of the **electron transport chain**.

Cytokine Diverse signaling proteins released by a variety of host cell types that affect the behavior of other host cells, including those of the both the **innate** and **adaptive immune** systems.

Cytoplasmic granule A type of lysosome found in **phagocytes** that contain both **defensins** and destructive **enzymes** and that, by fusing with a **phagosome**, contacts and destroys the engulfed bacteria.

Cytotoxic T lymphocyte (killer T cell) A type of white blood cell that selectively kills specific infected cells, particularly cells infected by viruses.

Cytotoxin A **toxin** that acts by degrading or perforating the cell membrane.

Decimal reduction time (*D*-value) The time required, under specified treatment conditions, to decrease the viable microbial population by 1 log unit, i.e., to kill 90% of the viable cells that were present at the start of the treatment interval.

Deconcatenation The unlinking of the rings in a chain.

Defensin A small protein found within the **cytoplasmic granules** in **phagocytes** that defend against diverse microbial invaders.

Dehydrogenation A chemical reaction in which an electron and a proton (i.e., a hydrogen atom) are removed from one of the reactants.

Deletion A mutation in which one or more base pairs are lost from the DNA.

Deltaproteobacteria A class within the phylum *Proteobacteria* that includes bacteria that respire sulfur or iron, and also the socially motile myxobacteria.

Denitrification The microbial conversion of nitrate to dinitrogen.

Deoxyribonucleoside Precursor of a **deoxyribonucleotide** that contains a deoxyribose bound to a nitrogenous base.

Deoxyribonucleotide Phosphorylated **deoxyribonucleoside** that serves as precursor of DNA.

Diapedesis The movement of white blood cells from a capillary, through the vessel wall, and into the tissue as part of **innate immunity**.

Diauxic growth Microbial growth characterized by two phases with different growth rates.

Differentiation Coordinated phenotypic changes of a cell typically toward greater specialization or complexity.

Dikaryon A hyphal compartment formed after sexual fusion that contains two **haploid** nuclei, one from each parent, that postpone fusion and instead persist as haploid nuclei for a period of time.

Dimethyl sulfide A volatile sulfur compound, produced abundantly by marine bacteria, that escapes into the atmosphere where it nucleates cloud formation.

Dimorphism The property of some fungal species to grow in either a **hyphal** or "yeast" form.

Dipicolinic acid A low-molecular-weight compound accumulated in the **forespore** during **endospore differentiation**, where it dehydrates the endospore and protects the DNA from heat denaturation.

Dissimilation The microbial use of a molecule, such as nitrate and sulfate, as an energy source.

DNA control region A sequence in DNA that regulates initiation of transcription at a remote site.

DNA-dependent RNA polymerase (RNAP) The enzyme that synthesizes all classes of RNA by transcribing DNA templates.

DNA gyrase A **topoisomerase** that introduces **supercoils** into circular DNA.

DNA ligase The enzyme that joins two DNA strands, such as is needed to join sequential **Okazaki fragments** during **lagging strand** synthesis.

DNA polymerase I (Pol I) The enzyme complex that catalyzes the addition of nucleotides to fill the gaps between **Okazaki fragments** during **lagging strand** synthesis.

DNA polymerase III (Pol III) The enzyme complex that catalyzes the addition of nucleotides to a growing DNA strand during replication of bacterial DNA.

DnaA The DNA replication initiation protein in *Bacteria* that promotes DNA unwinding at the origin of replication and regulates the initiation of DNA replication.

DnaB *See* **helicase**.

Domain *See* **protein domain**.

DTP vaccine A vaccine typically given early in life that immunizes the child against diphtheria, tetanus, and whooping cough (caused by *Bordetella pertussis*). Also, DTaP to indicate acellular pertussis vaccine.

Dynein A eukaryote motor protein that uses ATP to power the transport of cargo along microtubules, usually toward the center of the cell.

E site One of the three tRNA binding sites on ribosomes, specifically the one bound by the uncharged tRNA as it exits the ribosome.

Eclipse period During the viral **lytic cycle**, the interval between adsorption and virion assembly during which there are no intact virions present.

Ectosymbionts Symbiotic *Bacteria* that live attached to a body surface of their host.

Electrochemical gradient A gradient of electrochemical potential across a membrane created by the accumulation of ions, usually protons, outside the membrane. *See also* **proton motive force**.

Electron cryotomography *See* **cryoelectron tomography**.

Electron transfer system (ETS) *See* **electron transport chain**.

Electron transport chain A series of membrane proteins with different redox potentials that move electrons sequentially from an electron donor to a terminal electron acceptor down an energy gradient while simultaneously exporting protons across the membrane to generate the **proton motive force** used to generate ATP.

Electroporation For genetic transformation, a procedure that exposes cells to millisecond jolts of tens of thousands of volts to make transient holes in the cell membrane so DNA molecules can pass through.

End product inhibition *See* **feedback inhibition**.

Endemic disease A disease continuously present in a particular region or population.

Endergonic reaction A nonspontaneous chemical reaction that requires an energy input to proceed (the change in Gibbs free energy is positive).

Endocytosis The uptake by a eukaryotic cell of substances from the environment by membrane invagination and subsequent vacuole formation.

Endogenous retroviral elements Sequences found in the genomes of some vertebrates that were derived from retroviruses that had integrated into a chromosome in a germ line cell and were subsequently inherited over evolutionary time.

Endonuclease A nucleic acid degrading enzyme that cleaves a strand of DNA or RNA. *See also* **exonuclease**.

Endospore A spore produced inside another cell (a **mother cell**).

Endosymbiont A symbiotic bacterium that lives within the cells of its host.

Endosymbiotic theory The well-supported and now accepted theory that some organelles of eukaryotic cells (e.g., mitochondria, chloroplasts) originated from bacteria living inside another cell and their subsequent coevolution.

Endotoxin *See* **lipopolysaccharide**.

Energy charge A ratio representing the relative amounts of ATP, ADP, and AMP in a cell, thus a measure of the cell's energy status.

Energy-coupled transport Transport of specific molecules across the **outer membrane** of Gram-negative bacteria against a concentration gradient.

Enhancer A **DNA control region**, located outside the **operon**, that acts to increase transcription initiation without the binding of a regulatory protein.

Enterochelin A **siderophore** made by *E. coli* to import iron into the cell.

Enterosome Intracellular compartments of enteric bacteria that are bounded by a protein shell and contain the **enzymes** required to metabolize specific compounds such as 1,2-propanediol and ethanolamine.

Envelope The lipid-carbohydrate-protein membrane surrounding the **capsid** of some viruses.

Environmental stress Any significant change in a microbe's environment, e.g., a shift in temperature, pH, nutrient availability.

Enzymes Molecules (proteins or RNAs) that catalyze chemical reactions.

Epibiont An organism that lives on the surface of a host organism, typically without harm to the host.

Epidemic An outbreak of a disease within a particular region that affects more people (or other organisms) than normal.

Epigenetics The inheritance of patterns of gene expression or other phenotypic characters that vary independent of any variation in the cell's DNA sequence.

Episome An extrachromosomal genetic element, typically a circular double-stranded DNA molecule, that can replicate independently or can integrate into the host cell chromosome.

Epitope A part of an **antigen** that is recognized by a component of the **adaptive immune** system such as an **antibody**.

EPS matrix An essential structural component of biofilms that is composed of extracellular polymeric substances (EPS).

Epsilonproteobacteria A class of *Proteobacteria* that includes microaerophilic species and some chemolithotrophs.

Euchromatin A more dispersed structure of eukaryotic chromosomes that facilitates access by the transcription machinery.

Euryarchaeota The largest archaeal phylum, whose members include **methanogens** and extreme **halophiles (haloarchaea)**.

Eutrophication The overabundance of essential inorganic nutrients such as phosphate in an aquatic ecosystem that leads to an overabundance of algae whose decomposition depletes the water of oxygen and thus kills other organisms, most notably animals.

Excisionase A gene in **temperate** phages whose product, in combination with **integrase**, excises an integrated **prophage** from the host chromosome.

Exergonic reaction A spontaneous chemical reaction that releases energy (the change in Gibbs free energy is negative).

Exonuclease A nucleic acid degrading enzyme that attacks exposed ends of linear DNA or RNA molecules. *See also **endonuclease**.*

Exopolysaccharide (EPS) Polysaccharides secreted outside the cell that form a protective **capsule** or **slime layer**.

Exospore A spore produced by some filamentous bacteria by pinching off the tips of the filamentous cells.

Exosporium In endospores, a thin membranous layer sometimes present surrounding the proteinaceous spore coat.

Exotoxin A toxin secreted by a bacterial cell.

Exponential phase A bacterial growth phase characterized by balanced growth in which a constant rate of growth and cell division produces an exponential increase in the number of cells.

Extremophile An organism that thrives in and may require what, from the anthropocentric point of view, is an inhospitable or "extreme" environment.

Facilitated diffusion Selective **passive transport** of molecules across a membrane that is aided by specific interactions with membrane proteins.

Facultative aerobe An organism that can grow in the presence or the absence of oxygen.

Facultative anaerobe An organism that can grow aerobically or anaerobically.

Feedback inhibition Negative regulation of the first enzyme in a biosynthetic pathway by the end product of the pathway acting as a negative **allosteric** effector.

Fermentation A metabolic process that recycles NADH generated in fueling reactions to reduce pyruvate and other compounds derived from it, thereby converting organic substrates into acids, alcohols, etc.

Fimbria (plural, fimbriae) *See* ***pilus***.

Flagellin The peptide subunit that self-assembles to form flagellar filaments.

Flagellum (plural, flagella) Cellular appendage of *Bacteria* composed of a filament anchored by a complex structure in the cell membrane that, fueled by ATP, rotates propeller-like to propel the cell.

Flippase A lipid transporter protein located in the membrane that facilitates the movement of **phospholipid** molecules between the two leaflets of a bilayer membrane.

Flow cytometer An instrument that counts individual cell types flowing past a laser beam and sorts them based on some physical (size, shape, surface topography, internal complexity, etc.) and/or chemical property.

Fluorescence *in situ* hybridization (FISH) An imaging method that uses a sequence-specific fluorescently labeled DNA probe to visualize DNA with the complementary sequence under a fluorescence microscope.

Fluorescent antibody probe An **antibody** carrying a fluorescent tag that can be employed to observe the location of its specific **antigen** using fluorescence microscopy.

Fluorescent stain A fluorescent dye that selectively binds molecules or cell components and can be visualized using a fluorescence microscope, e.g., DAPI that binds DNA.

Fluorescence recovery after photobleaching (FRAP) An imaging method that measures the rate of diffusion of fluorescently labeled molecules such as small proteins inside a cell. A tiny intracellular zone is photobleached and the interval before more fluorescent molecules populate that zone is measured.

Fluorochrome A chemical that absorbs light at a specific wavelength and reemits it at a longer wavelength.

Forespore During **endospore** differentiation, the smaller daughter cell following cell division that will give rise to the endospore.

Frameshift mutation A **mutation**, such as **insertion** or **deletion** of 1 or 2 base pairs, that alters the **reading frame** for all downstream codons.

FRAP *See* ***fluorescence recovery after photobleaching***.

FtsZ A prokaryotic protein, filamentous temperature sensitive protein Z, that assembles during cell division to form the **Z-ring** at the site of the future septum.

Fueling reactions The metabolic reactions that provide the **precursor metabolites**, energy, and **reducing power** required for growth and maintenance.

Gamete In eukaryotes, a **haploid** cell that fuses with another gamete to form a diploid **zygote**.

Gametocytes Cells in the germ line of *Plasmodium* species that give rise to the gametes and that mediate transmission of the parasite from human to human, via a mosquito vector.

Gammaproteobacteria A large class within the phylum *Proteobacteria* that includes the enterobacteria, vibrios, pasteurellas, and pseudomonads.

Gas vesicles Intracellular structures of some aquatic *Bacteria* and *Archaea* that provide buoyancy and allow the cells to maintain a desired position in the water column.

Gene transfer agent (GTA) A particle that resembles a **virion**, produced by some marine bacteria using coopted phage genes, and used to transfer cellular DNA to a closely related bacterium.

Generalized transduction **Transduction** in which any fragment of the host cell DNA can be packaged into the **virion** and thus transferred to a recipient cell.

Generation time (*g*) The time required for a population of cells to double in number.

Genomics The comparative study of whole genomes, including their sequence, functions, and evolution.

Germination *See* **spore germination**.

GFP *See* **green fluorescent protein.**

Gliding motility Several types of nonflagellar bacterial motility produced by groups or individual cells.

Global regulation The coordinated regulation of transcription at the multi**operon** and multigene level.

Globulins A group of proteins characterized by their solubility in dilute salt solutions and that includes many of the proteins in blood serum.

Gluconeogenesis A metabolic pathway that makes hexoses, such as glucose, from one-, two-, three-, or four-carbon substrates.

Glycolipid A lipid to which a carbohydrate group is attached.

Glycolysis A fueling pathway in which glucose is converted into two molecules of the **precursor metabolite** pyruvate, while also generating energy (ATP), **reducing power** (NADH), and **precursor metabolites** needed to synthesize sugars, amino acids, and fatty acids.

Glycoprotein A protein to which a carbohydrate group is attached.

Gram-negative bacteria Bacteria that are stained red by the Gram staining procedure and that typically have a thin **peptidoglycan** cell wall and an **outer membrane**.

Gram-positive bacteria Bacteria that are stained purple by the Gram staining procedure and that typically have a thick **peptidoglycan** cell wall and no **outer membrane**.

Granuloma A mass of tissue formed of fibroblasts and leukocytes during an **inflammation** response that walls off, but does not necessarily kill, the invading pathogen.

Green bacteria A group of anoxygenic phototrophs that contain chlorophyll pigments.

Green fluorescent protein (GFP) A protein that emits bright fluorescent green light when excited by light of shorter wavelengths and is widely used in research to monitor gene expression and to determine the cellular location of macromolecules tagged with GFP.

Group translocation A type of active transport that brings sugars into some bacteria using the phosphate group from phosphoenolpyruvate to phosphorylate the imported sugar and thereby render the sugar unable to penetrate the membrane to escape from the cell.

Growth metabolism Metabolic reactions that directly contribute to the production of new cells.

GTA *See gene transfer agent.*

Guanosine tetraphosphate (ppGpp) The nucleotide that regulates the **stringent control system** and some other proteins.

Hairpin loop (stem-loop) A secondary structure of single-stranded RNAs formed by intramolecular base-pairing between two regions of the RNA.

Haloarchaea Extremely halophilic members of the *Euryarchaeota*.

Halophile An organism that thrives in and may require a high salinity environment.

Haploid (adj.) The condition of having a single copy of every gene in the genome, in contrast to diploid eukaryotic cells that contain two complete sets of chromosomes.

Haploid spore A fungal spore that contains a **haploid** nucleus.

Heat shock proteins Proteins that protect other proteins from heat denaturation.

Helicase (DnaB) The first protein of the **replisome** to associate with DNA during initiation of DNA replication and that assists with local melting of the DNA to allow replisome access.

Hemolysin Toxins produced by some pathogens, including *Staphylococcus aureus*, that lyse red blood cells.

Heterochromatin Eukaryotic chromosome structure characterized by tighter DNA packing and decreased accessibility to transcription machinery.

Heterocyst A terminally **differentiated** cell present in some filamentous cyanobacteria that is specialized for **nitrogen fixation**.

Heterotroph A microbe that requires an organic carbon source.

Hfr See *high frequency recombination cell.*

High frequency recombination cell (Hfr) A bacterium whose chromosome contains a **conjugative plasmid** and that will therefore transfer part of its own chromosome to the recipient cell during **conjugation**.

High frequency transductant During **generalized transduction**, a viable phage whose chromosome contains host DNA acquired from a previous infection and that can transduce all host cell genes at nearly equal frequencies.

Histidine kinase A common component of two-component regulatory systems that senses an environmental stressor and in response transfers a phosphate from ATP to one of its own histidine residues and then uses it to phosphorylate a **response regulator protein**.

Homolactic fermentation An auxiliary fueling fermentation that reduces pyruvate to lactate to recycle NADH.

Homologous recombination Recombination occurring between two homologous DNA molecules.

Homoserine lactone A class of small organic molecules synthesized and used by many bacteria as intercellular signals for **quorum sensing** and other communication.

Horizontal gene transfer The lateral exchange of genetic material between organisms of the same or different species.

Hormogonium (pl. hormogonia) A reversibly **differentiated** motile cell of filamentous cyanobacteria that functions as an agent of local dispersal.

Hydrogenosome A membrane-bounded organelle found in some protists and fungi that was derived from mitochondria and that now generates ATP and produces hydrogen as a by-product.

Hyperthermophile An organism that thrives in extremely hot environments and that typically grows optimally above 80°C.

Hypha (pl. hyphae) Each of the individual vegetative fungal filaments that make up a **mycelium**.

Immunoglobulin *See* **antibody**.

Immunological memory A key feature of **adaptive immunity** that facilitates a rapid, effective response upon subsequent encounters with a particular **antigen**.

Inclusion A general term for internal structures in some prokaryotes that includes **enterosomes**, **carboxysomes**, **magnetosomes**, and **gas vesicles**, among others.

Incubation period During a bacterial infection, the time between initial encounter with the pathogen and the onset of disease symptoms.

Inducer In prokaryote transcription, a small molecule that binds the **repressor** and releases it from the **promoter** or **operator**, thereby allowing initiation of transcription.

Inflammasome A multiprotein cytoplasmic complex that activates components of the **innate immune** system, including maturation of interleukin 1, and thus triggers **inflammation**.

Inflammation A localized tissue response to infection or damage, often characterized by redness, swelling, pain, and heat.

Initiation complex The association of both ribosomal subunits, an mRNA, and the first aminoacyl-tRNA that together initiate protein synthesis.

Innate immunity Inherited, nonspecific defenses that constitute the initial rapid response to infection.

Inoculum size The number of starting cells in a culture. During a bacterial infection, the number of pathogens that enter the host organism and reach their target tissue.

Insertion A **mutation** in which one or more base pairs are inserted in the DNA.

Insertion sequence (IS element) A type of **transposon** that contains only the minimum essential components needed for **transposition** (the **inverted repeats** and the **transposase** gene).

Integrase A viral enzyme that catalyzes the insertion of a viral DNA chromosome into the host chromosome and its subsequent excision.

Internalins A and B Surface proteins of *Listeria monocytogenes* that function as virulence factors by assisting pathogen entry into epithelial cells.

Interspecies electron transfer The process whereby members of one microbial species transfer electrons to the members of another species.

Inversion A **mutation** that reverses the order of a segment of DNA.

Inverted repeat A nucleotide sequence whose reverse complement (i.e., complementary nucleotides in reversed sequence) is found downstream in the same chromosome.

Ion-coupled transport The active transport of solutes across a membrane driven by the preexisting electrochemical gradient across that membrane.

IS element *See **insertion sequence**.*

Isoprenoid A class of long-chain hydrocarbons found in archaeal membranes, where they take the place of the fatty acids found in bacterial membranes.

Isotopes atomic variations of an element having different numbers of neutrons.

Kinesin A eukaryote motor protein that uses ATP to power the transport of cargo along microtubules toward the periphery of the cell.

Kinetoplast A specialized mitochondrion found in trypanosomes whose DNA consists of interlocking circular molecules.

Lag phase When stationary-phase cells are provided adequate nutrients, the period of time before the cells resume growth.

Lagging strand During DNA replication, the DNA strand that is replicated discontinuously as a series of short fragments (**Okazaki fragments**), each synthesized in the 5′ to 3′ direction, that are then ligated together.

Large ribosomal subunit The subunit of the ribosome that catalyzes the addition of amino acids to the growing protein chain.

Latency *See **viral latency**.*

Latency-associated transcript (LAT) During infection of a neuronal cell by herpes simplex virus 1 or 2, the RNA transcript of the single gene that is active while the virus is latent. These transcripts accumulate and block apoptosis by the infected cell.

Leader peptide A short peptide, of typically 5 to 30 amino acids, that results from cleavage of a polypeptide chain near the N-terminus.

Leading strand During DNA replication, the DNA strand that is replicated by the continuous addition of nucleotides starting at the 5′ end and advancing toward the 3′ end.

Leghemoglobin A form of plant hemoglobin found in **root nodules** that binds oxygen to protect the **bacteroid nitrogenase**.

Ligand Any molecule that binds to a specific biomolecule, such as a protein, to carry out a biological function, e.g., the **allosteric** regulation of an enzyme.

Lipid A A highly immunogenic, phosphorylated glycolipid component of all bacterial **lipopolysaccharides**.

Lipopolysaccharide (LPS) A large molecule composed of a lipid (**lipid A**) and a polysaccharide chain, which together are a major component of the outer layer of the **outer membrane** of Gram-negative bacteria. Also known as endotoxin.

Listeriolysin O A **cholesterol-dependent cytolysin** secreted by *Listeria monocytogenes* that forms pores in the **phagosomal** membrane, thus enabling the pathogen to escape into the host cell cytoplasm.

Lithotroph *See **chemoautotroph**.*

Logarithmic growth (log phase) *See **exponential phase**.*

LPS *See **lipopolysaccharide**.*

LPS core The portion of the LPS polysaccharide chain that is attached to **lipid A** and whose composition is similar in all Gram-negative bacteria.

Luciferase An enzyme present in **bioluminescent** cells that produces light using energy from ATP.

Lysogen A prokaryote that contains one or more **prophages**.

Lysogenic phage A phage that, at the start of each infection, has the option to either replicate immediately or choose **lysogeny**.

Lysogeny A **phage** life cycle in which the phage postpones lytic replication and resides in the host as a **prophage**.

Lysosome A host membrane-bounded vesicle containing degradative **enzymes**.

Lysozyme A muramidase (a **murein**-degrading enzyme) found in tears, **macrophages**, etc., where its antibacterial action contributes to **innate immunity**.

Lytic cycle The series of actions of a **lytic phage** beginning with cell entry and culminating with release of progeny **virions** by host lysis.

Lytic phage A **phage** for which infection of a host cell always leads immediately to replication followed by lysis of the host.

Macronucleus In ciliates, a large nucleus that contains multiple copies of a subset of the cell's genes and where very active gene transcription takes place.

Macrophage A **monocyte** that resides in body tissues.

Magnetosome A specialized, membrane-bounded intracellular structure of some bacteria that contains iron crystals that function as magnets used to orient the cell relative to the Earth's magnetic field.

Maintenance metabolism Metabolic reactions that do not directly contribute to the production of new cells, but rather maintain cellular functions.

Major histocompatibility complex (MHC) Glycoproteins on the surface of most host cells that bind and hold foreign **antigens** for presentation to **cytotoxic T lymphocytes** for destruction.

Malolactic fermentation In wine making, the bacterial conversion of the tart-tasting malic acid present in grapes to the milder lactic acid.

MAMP *See microbe-associated molecular pattern*.

Marine snow Particulate matter, mostly organic, that is produced in the upper regions of the oceans and slowly falls to deeper regions, and which at all depths is a source of nutrients for the attached pelagic microbes.

Mating type In fungi, the equivalent of gender; **haploid** cells of one type cannot mate with cells of the same type.

Membrane attack complex (MAC) A complex of complement proteins that recognizes foreign membranes that lack sialic acid and produces destructive pores in them.

Memory cell A long-lived **B lymphocyte** that retains the ability to produce its specific **antibody** and will proliferate in response to the presence of its target **antigen**. *See also secondary response*.

Mesophile An organism that typically grows optimally between 20°C and 45°C.

Metabolome All the metabolites present in an environmental sample.

Metabolomics The analysis of all of the metabolites being synthesized in an environmental sample.

Metagenome All the DNA sequences retrieved from an environmental sample, which in complex communities typically represents only the more abundant community members.

Metagenomics A culture-independent method in which DNA is extracted from an environmental sample and then sequenced without relying on primer-based amplification or cloning.

Metaproteome All the proteins present in an environmental sample.

Metaproteomics The analysis of all of the proteins expressed in an environmental sample, thus the identification of the genes being actively transcribed and translated.

Metatranscriptome All the mRNA retrieved from an environmental sample.

Metatranscriptomics The sequencing of all of the mRNAs present in an environmental sample, thus the identification of the transcriptionally active genes.

Methanogen An archaeon that produces methane as a metabolic by-product.

Methylase An enzyme that catalyzes the addition of methyl groups to molecules such as DNA and RNA.

Methylotroph An organism that uses reduced one-carbon compounds, such as methane, as their carbon source.

Methyl-accepting proteins Chemoreceptors on the bacterial cell membrane that detect the concentration of specific chemicals (attractants or repellents) in the environment and transmit signals to the **flagellum** to control its rotation and the cell's swimming direction.

Methyl-directed MMR A **mismatch repair system** in *Bacteria* that distinguishes the new incorrect strand from the old correct strand by its lack of methylation, there being a lag between synthesis and methylation.

Microaerophile An organism that requires oxygen but cannot tolerate high oxygen concentrations.

Microbe-associated molecular pattern (MAMP) Essential molecular components of a wide variety of microbes that are found only in microbes and that are recognized as foreign by the **innate immune** system.

Microbial biogeography The study of the distribution of microbes in space, time, and across environmental gradients.

Microbial fuel cell An electrochemical device that harnesses electrical power from the oxidation of organic or inorganic substrates coupled to the reduction of an electrode.

Microbiome All the microbes that inhabit a particular environment, such as the human body, or all of the genetic information that they contain.

Microbiota The community of diverse microbes living in or on the body of a multicellular organism.

Micronucleus In ciliates, the small nucleus that contains the entire genome, is inherited intact, and undergoes meiosis during occasional sexual reproduction.

Minicell Small prokaryotic cells that lack a chromosome and that are formed by mutants that divide abnormally near a pole, rather than at the midline.

Mismatch repair system (MMR) A system found in all cells to detect and repair **mismatches** that are not corrected during proofreading by DNA polymerase.

Mismatches In DNA replication, insertion of a nucleotide in the new strand that is not correctly base paired (A with T or G with C).

Mitosome A membrane-bounded organelle of some parasitic protists living under anaerobic or microaerophilic conditions, thought to be derived from mitochondria although they no longer produce ATP.

MMR *See **mismatch repair system**.*

Mobile element (mobile genetic element) A segment of DNA capable of moving from one position to another within the same or different chromosomes or cells.

Modular rearrangement A mechanism of protein evolution which reshuffles or recombines functional modules of proteins, often corresponding to protein structural domains, to create new capabilities.

Modulator A protein or RNA that interacts with an enzyme to regulate its activity.

Modulon A group of **regulons** whose expression is controlled by the same regulator.

Monoclonal antibodies Antibodies that recognize only one **epitope** on one specific **antigen** and that are produced by a single clone of identical B cells.

Monocyte A long-lived, phagocytic white blood cell that is recruited to the site of **inflammation** by the neutrophils and both cleans up debris and secretes **cytokines**.

Mother cell A bacterial cell that differentiates irreversibly into an **endospore**.

Motif In DNA, a sequence presumed to have a particular function. In proteins, an amino acid sequence presumed to have a particular function (*see also **protein domain***).

Murein The type of **peptidoglycan** that forms bacterial cell walls.

Mutagen A chemical or physical agent that causes **mutations**.

Mutation A change in the sequence of an organism's DNA that is inherited by its progeny.

Mutualism Intimate interaction between two organisms of different species in which both benefit.

Mycelium The vegetative component of a fungus composed of many **hyphae**.

Mycolic acid A **wax** found in the membrane of tubercle bacilli that contains long, branched fatty acids.

Mycoplasma Bacteria that lack a cell wall and, in many instances, incorporate sterols in their membrane for strength and rigidity.

Mycorrhiza (pl. mycorrhizae) A fungus that lives in symbiotic association with plant roots, enhancing the uptake of water and some minerals.

Myxospores Dormant **spores** produced by the myxobacteria.

NADP transhydrogenase An enzyme that catalyzes the reduction of NAD by NADPH and also the reduction of NADP by NADH.

Natural transformation Naturally occurring **transformation**, in contrast to the introduction of DNA by human-directed techniques.

Negative sense Refers to RNA whose complementary strand can be translated into protein, or to DNA with the same sequence. *See also **positive sense***.

Negative supercoiling Extra twists of a double-stranded DNA molecule in the direction opposite to the twist of the DNA double helix.

Nitrification The microbial oxidation of ammonia to nitrate carried out in two steps (ammonia to nitrite, then nitrite to nitrate) by two different organisms.

Neutrophil The most abundant class of **phagocytes**, which are the first responders in an innate immune response.

Nitrogen fixation The conversion of atmospheric dinitrogen gas (N_2) into ammonia or nitrate.

Nitrogenase A key enzyme in the **nitrogen fixation** pathway that is inhibited by oxygen.

Nodulation factor During establishment of **root nodules**, a chemical produced by the nitrogen-fixing bacteria in response to signals from the plant and that, in turn, induces nodule formation by the plant.

NOD-like receptor The nucleotide-binding oligomerization domain receptors in the cytoplasm that recognize **MAMPs** that escape the **phagosomes** as well as other cellular damage and trigger **innate immune** responses.

Noncomposite transposon A **transposon** that encodes a **transposase** flanked by **inverted repeats** and one or more genes not required for transposition; is not flanked by **insertion sequences.**

Nonsense codon A triplet codon that does not correspond to the anticodon of any tRNA and thus can serve as a termination or stop codon.

Noncyclic photosynthesis Photosynthesis in which a variety of electron donors are used to replace the electron donated by the excited chlorophyll to the **electron transport chain**.

Nuclear factor κB (NF-κB) A eukaryotic transcription factor activated by Toll-like receptors and that, in turn, induces expression of **cytokine** genes.

Nuclear occlusion A model in which the **nucleoid** spatially inhibits the formation of the **Z-ring** in the midcell to time chromosome replication with cell division.

Nucleocapsid A viral **capsid** with the enclosed nucleic acid.

Nucleoid In the cytoplasm of prokaryotes, the region that contains the highly condensed chromosome and excludes ribosomes.

Nucleosome A repeating structure of eukaryotic chromatin formed by coiling a portion of the DNA around a histone core, and thereby affecting the frequency of transcription of that DNA region. *See also* **heterochromatin** *and* **euchromatin**.

Nutrient Organic or inorganic chemical required for metabolism or growth.

Nutritional immunity A tactic to fight microbial infection by sequestration of essential metals, such as iron.

O-antigen Long carbohydrate chains that extend from the **LPS core** to swathe the external surface of Gram-negative bacteria. It can be highly immunogenic and is quite diverse, varying between species or even between strains.

Obligate intracellular parasite An organism or a virus that is capable of replication only when inside a suitable host cell.

Okazaki fragments During DNA replication, the short segments produced during **lagging strand** synthesis.

Oncogenes Genes that have the potential to cause **cell transformation** and cancer by disrupting normal cellular control mechanisms.

Oncogenesis The formation of a tumor.

Open reading frame (ORF) A DNA or RNA sequence presumed to encode one protein.

Operational taxonomic unit (OTU) A microbial taxonomic group, replacing the species, defined based on genetic sequence using an arbitrary threshold, typically >97% identity of their 16S rRNA genes. *See also* **phylotype**.

Operator A regulatory sequence in DNA upstream of a protein coding region, overlapping the **promoter,** that when bound by a specific protein repressor regulates transcription of the downstream gene(s).

Operon A contiguous group of protein-coding genes that are transcribed together under the control of a single regulatory site (**promoter**).

Opportunistic infection A microbial infection in which a pathogen that normally cannot mount an infection in that host takes advantage of the host's compromised immune system, disrupted microbiota, or other conditions that weaken host defenses.

Opportunistic pathogen A microbe that normally cannot mount an infection but that takes advantage of the opportunity presented by weakened host defenses to do so.

Opsonin A molecule that binds to an antigen and thereby designates the cell for destruction by **phagocytes**.

Optical density *See* **absorbance**.

ORF *See* **open reading frame.**

Origin of replication In prokaryotes, a specific chromosomal site where DNA replication is initiated.

Origin sequestration The temporary shielding of the origin of replication by SeqA following initiation of a cycle of replication, possibly to prevent premature initiation of another cycle of replication.

Outer membrane The outer lipid bilayer of Gram-negative bacteria, consisting of an outer **lipopolysaccharide** leaflet and an inner **phospholipid** leaflet with embedded proteins.

Oxidative phosphorylation In **chemotrophs**, the phosphorylation of ADP to generate ATP using the oxidation of chemical substrates as the energy source.

Oxygenic photosynthesis A process in which captured light energy is used to generate ATP and **reducing power**, with water being oxidized in the process to produce O_2.

P site One of the three tRNA binding sites on ribosomes, specifically the one bound by the tRNA attached via its amino acid to the growing peptide chain.

Pandemic A global **epidemic**.

Panton-Valentine leukocidin (PVL) A **cytotoxin** produced by some strains of *Staphylococcus aureus* that kills **neutrophils**.

Parasitism Intimate interaction between two organisms of different species in which one (the parasite) lives at the expense of the other (the host).

Parthenogenesis In animals, the development of a new individual from an unfertilized ovum.

Passive transport Movement of solutes across a membrane driven by a concentration gradient and requiring no other energy input.

Pasteurization Moderate heat treatment intended to prevent spoilage of foods, typically a liquid, by killing the microbes responsible but without undesirable alteration of the food itself.

Pathogen An agent, such as a microbe or a virus, that causes disease under some conditions.

Pathogenicity The ability of an organism (i.e., a pathogen) to cause disease.

Pattern recognition receptor (PRR) A group of proteins produced by the **innate immune** system that sense tissue damage caused by infection or other agents.

Pausing site A region in a DNA template that when transcribed forms a **hairpin loop** in the RNA transcript that in turn **attenuates** transcription.

PCR *See* **polymerase chain reaction.**

Pelagic microbe A microbe that occupies any region in the water column of the oceans. *See also* **benthic microbe**.

Pentose phosphate pathway A fueling pathway that converts glucose into 5-carbon and 4-carbon **precursor metabolites** needed to synthesize nucleotides, while producing **reducing power** (NADPH).

Peptide bond The bond joining two amino acids in a polypeptide chain that is formed between the amino group of one amino acid and the carboxyl group of the adjacent amino acid.

Peptidoglycan A complex polymer composed of sugar chains cross-linked via amino acid side chains that is found in many prokaryotic **cell walls**.

Periplasm The cellular compartment of Gram-negative bacteria bounded by the cell membrane and the **outer membrane**.

Peritrichous (adj.) Having cellular appendages, such as **flagella**, over the entire surface.

Phage A common name for **bacteriophage**, i.e., for viruses that infect *Bacteria*.

Phagocyte A type of white blood cell that contributes to **innate immunity** by engulfing and digesting bacteria and other foreign particles.

Phagocytosis The engulfment of foreign particles, including bacteria and viruses, by amoeboid cells such as the **phagocytes** of the **innate immune** system.

Phagolysosomal vesicle A vesicle in a **phagocyte** formed by the fusion of a **phagosome** with a **lysosome**.

Phagosome A membrane-bounded vesicle in the cytoplasm of a **phagocyte** that forms around foreign particles as they are engulfed.

Pheromone A chemical released by a microbe or multicellular organism that triggers a social response in other members of the same species.

Phosphatase An enzyme that removes a phosphate group from a compound, such as from a phosphorylated protein in a **signal transduction** pathway.

Phosphate An inorganic phosphorus-containing molecule with phosphorus in the +5 oxidation state.

Phospholipid A class of molecules found in membranes and composed of glycerol carrying a phosphate group and linked to two hydrophobic hydrocarbon chains (fatty acids in *Bacteria* and **isoprenoids** in *Archaea*).

Phosphotransferase An enzyme that transfers a phosphorus-containing group from one compound to another, such as the kinases that transfer phosphate groups to proteins.

Photoautotroph An organism that gains energy for growth from light (electromagnetic radiation) and uses an inorganic carbon molecule such as carbon dioxide as sole carbon source.

Photoheterotroph An organism that gains energy for growth from light (electromagnetic radiation) and uses an organic carbon molecule as a carbon source.

Photophosphorylation In photosynthetic microbes, the phosphorylation of ADP to generate ATP using light as the energy source.

Photosystem In **phototrophs**, all of the components of the light-harvesting apparatus.

Phototroph An organism that obtains its energy from light (electromagnetic radiation).

Phylogenetic analysis The comparison-based deciphering of the phylogeny of groups of genes or organisms.

Phylogeny A hypothetical history of the evolution of a gene or of a species or other taxonomic group.

Phylotype A species identified solely based on DNA sequence similarity. *See also* **operational taxonomic unit**.

Phytoplankton Floating **photoautotrophic** and **photoheterotrophic** microbes in aquatic environments.

Piezophile An organism that grows optimally at pressures exceeding 1 atm (*see* **barophile**).

Pilin The peptide subunit that self-assembles to form a **pilus**.

Pilus (plural, pili) Thin, filamentous appendages of some prokaryotic cells that carry out one of a variety of functions such as attachment to host cells or surfaces, transfer of DNA between cells, and motility.

Planktonic (adj.) Free-living, individual cells typically in an aquatic environment.

Plaque When culturing a **phage**, a clear zone in a "lawn" of host bacteria on an agar plate formed where an infectious phage particle has completed repeated **lytic infection cycles**, lysing the bacteria in that region.

Plasmid Independently replicating, extrachromosomal DNA genetic element found in many prokaryotes that may carry genes that benefit the cell under specific circumstances; also may be a nonintegrated **prophage**.

Plasmodium A genus of parasitic protozoa within the phylum *Apicomplexa* with complex life cycles that include two host species.

Point mutation A **mutation** that alters a single base pair in the DNA.

Pol I *See* **DNA polymerase I**.

Pol III *See* **DNA polymerase III**.

Polar (adj.) Located at or near either or both ends of a rod-shaped cell, e.g., polar flagella.

Polar septum During **endospore** differentiation, the septum formed close to one pole during the unequal cell division that gives rise to the **mother cell** and **forespore**.

Polycistronic (adj.) Prokaryotic mRNAs that encode more than one peptide chain.

Polymerase chain reaction (PCR) A technique that uses a heat-tolerant DNA polymerase to amplify (make many copies of) the DNA located between two known "bookend" sequences using primers complementary to the bookends.

Polyribosome Multiple ribosomes attached to and simultaneously translating the same mRNA.

Pore-forming toxin A **toxin** that punctures the cell membrane of target cells, often a lethal event.

Porins Large proteins found in the **outer membrane** of Gram-negative bacteria. They associate to form transmembrane channels that allow the passive diffusion of specific metabolites such as sugars and amino acids.

Positive sense Refers to RNA that can be translated into a protein, or DNA with the same sequence. *See also* **negative sense**.

Positive supercoiling Extra twists of a double-stranded DNA molecule in the same direction as the twist of the DNA double helix, found in thermophiles.

ppGpp *See* **guanosine tetraphosphate**.

Prebiotic phase A postulated early phase in the evolution of life on Earth during which necessary **building blocks** were synthesized by abiotic processes.

Precursor metabolites The 13 compounds from which all **building blocks** can be synthesized.

Predation Ecological interaction between two organisms of different species in which one uses the other as food.

Primary TB The first stage of a tuberculosis infection, typically noninfectious and asymptomatic in healthy people.

Primase An RNA polymerase that is a component of the **replisome** and that creates the short segment of RNA required to prime initiation of a new DNA strand.

Prion An infectious agent that lacks any nucleic acid and is composed solely of misfolded protein.

Programmed cell death *See **apoptosis** and **pyroptosis.***

Promoter A region in the DNA upstream of each gene or **operon**, and the site of initiation of transcription.

Prophage The DNA of a **temperate phage** in a **lysogen**, typically inserted in the chromosome of the host cell. Some of its genes may be transcribed but genes for lytic replication are repressed.

Prophage induction The process in which a change in the state of a **lysogen**, such as DNA damage caused by UV irradiation, prompts the **prophage** to excise from the host chromosome and enter the **lytic cycle**.

Prostheca A cellular projection containing a filamentous stalk and a terminal organ called a holdfast that anchors the cell to a solid surface.

Protective immunity Immunity resulting from a previous infection by a pathogen and conferred by **antibodies** against specific surface structures of that pathogen.

Protein domain A structural and/or functional unit within a protein that can evolve as an independent unit and either function independently or be mixed-and-matched with other domains to produce proteins with more complex functions.

Protein superfamily The largest grouping of related proteins that can be traced to a common ancestor, i.e., that form a clade.

Protein-synthesizing system (PSS) All the cellular components required for the synthesis of proteins.

Proteobacteria The largest bacterial phylum, its members comprising six classes of Gram-negative bacteria.

Proteome All the different proteins that a cell can synthesize or is currently synthesizing.

Proteorhodopsin A membrane protein in some bacteria that uses energy from sunlight to pump protons out of the cell, thus creating a **proton motive force** to drive the production of ATP.

Protist The informal name for an exceedingly diverse group of mostly unicellular eukaryotes.

Protocell A hypothetical primitive cellular entity, composed of RNA enclosed within a membrane, postulated to have arisen at an early stage in the evolution of life and being capable of self-replication and evolving further.

Proton motive force (PMF) The force generated by the electrochemical gradient across a cell membrane that drives protons into the cell and is used to power the generation of ATP.

Protoplasts Spherical cells produced experimentally by treating Gram-positive bacteria with **lysozyme**.

Prototroph An organism that can make all the metabolic **building blocks** from **precursor metabolites**.

Protoxin An inactive precursor of a **toxin** that becomes toxic when further modified, sometimes within the intended target organism.

Protozoa The informal name for a group of **protists** that exhibit animal-like proper-ties such as motility and predation.

Provirus The DNA of a virus that has inserted into the chromosome of a eukaryotic host cell.

Pseudomurein The type of **peptidoglycan** that forms archaeal cell walls.

PSS See **protein-synthesizing system**.

Psychrophile An organism that thrives in cold environments and that typically grows optimally at ~15°C and may grow below 0°C.

Punctuated equilibrium A theory of evolution positing that, when offered new or unoccupied ecological niches, evolution makes sudden jumps that yield increased complexity or diversity.

Purple bacteria A group of anoxygenic phototrophs that contain carotenoid pigments.

Purple membrane A region in the cell membrane of some **haloarchaea** where the **bacteriorhodopsins** are located.

Pus The accumulation of live and dead white blood cells, debris, and tissue fluids at a site of **inflammation**.

Pyroptosis A form of programmed cell death (**apoptosis**) mediated by **inflammasomes** and associated with **inflammation**.

Quorum sensing Regulation of gene expression in response to the sensed cell population density.

R plasmid (R factor) A **plasmid** that carries functional genes not required for its own transmission, such as antibiotic resistance genes.

Reaction center In **phototrophs**, an organized association of light-harvesting pigments, proteins, and cofactors that facilitates the transfer of electrons to the **electron transport chain**.

Reaction coupling A tactic for driving essential but unfavorable chemical reactions by coupling them to energetically favorable reactions via energy transfer in the form of high-energy bonds.

Reactivation The process in which a **latent** virus shifts to active replication.

Reading frame The grouping of bases in mRNA into three-letter "words" that converts the continuous base sequence in mRNA into a series of **codons**.

RecA A bacterial protein that carries out homologous **recombination** and is involved in the regulation of DNA repair.

Receptor During **adsorption** of a virus to a potential host cell, the specific molecule or structure on the surface of the cell that the virus recognizes and attaches to.

Recombinant A cell or organism that has acquired DNA by the process of **recombination**.

Recombinase One of a group of **enzymes** that catalyze **recombination** by cutting and splicing the DNA.

Recombination In prokaryotes, the generation of new combinations of genes by a reciprocal exchange between two, usually homologous, DNA molecules from different sources.

Redox potential The affinity of a molecule for electrons. The more electropositive the potential, the greater the tendency to acquire electrons and thereby be reduced.

Redox reaction An oxidation-reduction reaction in which one molecule is oxidized (loses electrons) and another is reduced (gains electrons).

Reducing power A source of electrons (e.g., reduced coenzymes such as NADH and NADPH) used, for example, to make the more reduced **building blocks** from **precursor metabolites**.

Regulatory cascade A regulatory network in which regulators govern regulators, which in turn govern other regulators.

Regulon A group of unlinked genes that are controlled by the same regulatory element and that are thus coordinately activated or repressed.

Replica plating A technique to transfer representatives of all the microbial colonies present on a master agar plate to a number of replica plates, preserving the spatial distribution of the colonies on all plates.

Replicase *See **RNA-dependent RNA polymerase**.*

Replicative transposition A mode of transposition in which the **transposon** is duplicated and the copy is inserted in a new location.

Replisome The large protein complex that carries out the replication of DNA starting at the origin of replication.

Repressor A regulatory protein that turns off expression of one or more specific genes. In a **lysogen**, a prophage-encoded protein that turns off the **prophage** genes required for **lytic phage replication**.

Reservoir The normal habitat of an infectious agent where its population is maintained and from which the pathogen can be transmitted to other hosts.

Respiration A process by which electrons, either alone or as hydrogen atoms, move from an electron donor to a terminal electron acceptor while simultaneously pumping protons across the membrane to generate the **proton motive force**.

Response regulator protein The element in a two-component regulatory system that receives the signal from the environmental sensor and triggers the response.

Restriction endonuclease A group of **enzymes** that protect *Bacteria* from foreign DNA by checking specific DNA sequences for the correct, strain-specific pattern of base methylation. If the site is not correctly methylated, the DNA is cleaved.

Retrovirus A class of RNA viruses that replicates via a DNA intermediate. The viral **reverse transcriptase** makes a DNA copy that is then inserted into the chromosome of the host cell.

Reverse gyrase ATP-dependent topoisomerase type I of hyperthermophilic archaea that introduces **positive supercoiling** into DNA.

Reverse transcriptase An enzyme that synthesizes complementary DNA (cDNA) using RNA as a template.

Ribonucleoside Precursor of a **ribonucleotide** that contains a ribose bound to a nitrogenous base.

Ribonucleotide A phosphorylated **ribonucleoside** that serves as a precursor of nucleic acids.

Ribosome-associated trigger factor A **chaperone** that interacts with a nascent polypeptide chain while still associated with the ribosome and oversees the correct rotation of proline peptide bonds.

Riboswitch Sequence encoded within an mRNA molecule that regulates its transcription or translation.

Ribozyme An RNA with catalytic activity, thus an <u>ribo</u>nucleic acid en<u>zyme</u>.

Ribulose bisphosphate carboxylase *See **RuBisCO**.*

RNA chaperone A class of proteins that bind **small RNAs**, escort them to their regulatory targets, and stabilize their specific interaction with mRNA.

RNA polymerase *See **DNA-dependent RNA polymerase**.*

RNA silencing A regulatory mechanism in which expression of specific genes is inhibited by small, noncoding RNAs. *See also **small RNA**.*

RNAP *See **DNA-dependent RNA polymerase**.*

RNA-dependent RNA polymerase (replicase) An enzyme found in RNA viruses that catalyzes the synthesis of a complementary strand of RNA from an RNA template.

Rolling-circle replication A mode of unidirectional replication of circular DNA or RNA molecules that is used by some viruses and plasmids.

Root nodule A growth on plant roots caused by infection with *Rhizobium* species that is the site of bacterial nitrogen fixation, the nitrogen being then shared with the plant.

Rough LPS A type of **lipopolysaccharide** that lacks the **O-antigen** component of the polysaccharide chain and thus is relatively hydrophobic.

RuBisCO (ribulose bisphosphate carboxylase) The key enzyme in carbon dioxide fixation and a constituent of bacterial **carboxysomes**.

Rumen An anaerobic zone within the gut of ruminant animals that contains a diverse population of microbes that **ferment** ingested food.

Run During motility, the productive vectorial motion of a bacterial cell that is generated by coordinated counterclockwise rotation of all the cell's **flagella**.

Sacculus A bag-like molecule formed by the cross-linked **peptidoglycan** of the **cell wall** that surrounds the cell membrane and provides shape and rigidity to the cell.

Saprophyte An organism (plant or microbe) that uses dead or decaying organic matter as its carbon and/or energy source.

Sec system A group of proteins that together secrete selected proteins across the cell membrane, thus to the outside of Gram-positive bacteria and to the periplasm or outer membrane of Gram-negative bacteria.

Secondary metabolites Organic compounds made by organisms which are not directly involved in the growth or reproduction of that organism.

Secondary response The rapid response mediated by **memory cells** of the **adaptive immune** system to repeated exposure to an **antigen**.

Secondary TB The second stage of tuberculosis infection, typically infectious and accompanied by classic symptoms such as a persistent cough, fever, and emaciation.

Self-assembly The process whereby structural subunits associate noncovalently to form a more complex structure, e.g., the formation of a viral **capsid** from **capsomers**.

Sepsis The presence of bacteria and/or toxins in the blood.

Serogroup A group of bacteria that have one **antigen** in common and that may otherwise be quite diverse.

Serotype A group within a microbial species or a virus classified by their dominant surface **antigen**.

Sessile (adj.) When cells are attached to a surface.

Sex pilus A **pilus** specialized to serve as the bridge between two **conjugating** bacteria through which DNA is passed from the donor to the recipient cell.

Shine-Dalgarno sequence The ribosome binding site on an mRNA molecule, typically ~8 bases upstream of the start **codon**.

Siderophore A small molecule secreted by microbes that has exceptionally high affinity for iron and that assists with import of iron into the cell.

Sigma factor (σ) A subunit of the RNAP enzyme that binds to specific **promoter** regions to initiate transcription of the downstream gene or **operon**.

Signal peptidase An enzyme that cleaves proteins at specific locations so as to clip off the N-terminal signal sequence.

Signal peptide *See* **signal sequence.**

Signal recognition particle (SRP) A protein-RNA complex that recognizes a specific signal sequence of a newly synthesized protein and escorts the protein to dedicated secretory proteins in the cell membrane.

Signal sequence A 15- to 20-amino-acid sequence at the N-terminus of a protein that specifies the intracellular or extracellular destination of the protein, sometimes by being recognized by a **signal recognition particle**.

Signal transduction The process by which a sensed environmental condition is converted into a cellular response.

Simple diffusion The basic mechanism of **passive transport** of solutes across a membrane without any increase in speed or specificity due to membrane components.

Single amplified genome (SAG) A genome sequence obtained from the amplified DNA of a single cell.

Single-cell genomics The analysis of genome sequences each derived from an individual cell.

Single-hit kinetics When a single random "hit" causes an effect such as cell death or infection of a host.

Single-stranded binding protein (SSB) Proteins that bind single-stranded DNA, such as that produced by the local melting of DNA during replication, to prevent the two strands from reannealing.

Site-directed mutagenesis A laboratory procedure that changes the nucleotide sequence in a specific target segment of the DNA.

Site-specific DNA recombinase An enzyme that catalyzes **site-specific recombination**.

Site-specific DNA recombination The insertion or excision of a genetic element at a specific chromosomal site, thus facilitating the horizontal transfer of that element between cells.

S-layer *See **crystalline surface layer**.*

Slime layer A diffuse layer of secreted **exopolysaccharide** that is loosely attached to the cell but serves to attach the cell to a surface or provide the matrix for a **biofilm**.

Small RNA (sRNA) A class of small, noncoding RNA molecules in prokaryotes that regulate gene expression during transcription or translation.

Small ribosomal subunit The subunit of the ribosome that recognizes and binds mRNA, then initiates translation and synthesis of the encoded protein.

Smooth LPS A type of **lipopolysaccharide** with the **O-antigen** component added to the polysaccharide chain that is therefore more hydrophilic.

Social motility In myxobacteria, the coordinated movement of individual cells within a collective raft.

SOS response In prokaryotes, a coordinated response to sensed DNA damage.

Specialized transduction **Transduction** that transfers a host cell DNA fragment that had been adjacent to a **prophage** and thus was packaged in the **transducing particle** due to imprecise prophage excision.

Spheroplasts Spherical cells produced experimentally by treating Gram-negative bacteria, e.g., with **lysozyme**.

Spo0A phosphorelay system The phosphorylation cascade that activates Spo0A, a necessary step in triggering **sporulation**.

Spongiform encephalopathy A group of neurodegenerative diseases caused by **prions** that produce sponge-like lesions in the brain.

Sporangioles Globular, **spore**-containing structures at the tips of the fruiting bodies formed by myxobacteria.

Spore A metabolically inert structure formed by some bacteria and fungi in response to adverse conditions and capable of long-term survival under those conditions. *See also* **exospore** *and* **endospore**.

Spore germination The conversion of a dormant spore into a metabolically active vegetative cell.

Sporulation The differentiation of a cell into a dormant spore.

sRNA *See* **small RNA**.

Stable RNA A generic term for RNA species, such as rRNAs and tRNAs, that have a relatively long lifetime in the cell.

Stationary phase A growth phase in which the rate of cell growth matches the rate of cell death and therefore there is no change in cell number.

Stem-loop *See* **hairpin loop**.

Sterilize Remove or kill all microorganisms.

Stress response A strategy used to cope with fluctuating environmental stresses.

Strict aerobes Organisms that require oxygen as terminal electron acceptor for **respiration**.

Strict anaerobes Organisms that cannot use oxygen as electron acceptor for respiration and thus use alternative electron acceptors to carry out **anaerobic respiration**.

Stringent control system A multi**operon** gene network regulated by **guanosine tetraphosphate** that coordinates the cellular response to amino acid or energy starvation or other stresses.

Substrate-level phosphorylation A process that generates ATP by transferring a high-energy phosphoryl bond from an oxidized metabolic substrate to ADP.

Superantigen A class of **toxins** that stimulates a nonspecific, inflammatory, immune response by bridging MHC class II molecules on antigen-presenting cells and T cell receptors.

Supercoiling The over- or underwinding of a double-stranded DNA double helix.

Superfamily *See* **protein superfamily**.

Super-resolution microscopy Any of a number of innovative techniques that increase the resolution obtainable with light microscopy beyond the theoretical limits.

Swarming A type of **flagellar** motility exhibited by "rafts" of cells of some types of bacteria on solid surfaces.

Symbiosis Intimate interaction between two organisms of different species, usually with benefits to both partners.

Syncytins A class of proteins that function during placental morphogenesis in humans and some other mammals, and that were evolutionarily derived from a **retroviral** envelope protein.

Synthetic biology The branch of biology that strives to construct artificial biological systems of use to humans.

Syntrophy An obligate, **mutualistic** metabolic relationship between different organisms in which the metabolic products of one are utilized as food by the other, with the net result of harvesting energy from individually energetically unfavorable reactions.

Systemic listeriosis A serious systemic infection that results when invading *Listeria monocytogenes* organisms escape from the gut tract and enter the blood, the central nervous system, etc.

Systems biology The computational modeling of very complex biological systems, such as cells or ecosystems, to understand the interactions between all the components.

T helper cell A class of lymphocytes that make **cytokines** that regulate the activity of both antibody-producing **B lymphocytes** and **cytotoxic T lymphocytes**.

T lymphocyte *See* **cytotoxic T lymphocyte.**

TCA cycle A fueling pathway that converts acetyl-coenzyme A (derived from the pyruvate produced by **glycolysis**) to carbon dioxide, and yields both **reducing power** and four of the **precursor metabolites** used in the synthesis of fatty acids and amino acids.

Teichoic acids Chains of glycerol phosphate or ribitol phosphate linked by phospho-diester bonds and bound covalently to the **murein** cell wall of Gram-positive bacteria.

Temperate phage A **phage** that can postpone lytic infection and instead **lysogenize** its host.

Terminal inverted repeats **Inverted repeats** found at the ends of genetic elements, including in transposons, where they function as recognition sites for the **transposase**.

Terminator A sequence in DNA that indicates the end of a gene or **operon**, i.e., the point where transcription of each RNA ends.

Terminus of replication In prokaryotes, specific chromosomal sites where DNA replication is terminated.

Thermoacidophile A **thermophile** that is also an **acidophile**.

Thermophile An organism that typically grows optimally between 60°C and 85°C.

Thioredoxin A class of small proteins that are maintained in a reduced state and then function as cellular antioxidants by reducing other compounds.

Thylakoid Membranous internal compartments in cyanobacteria and chloroplasts whose membrane topology provides increased surface area for the photosynthetic machinery without compromising the exchange of molecules with the cytoplasm.

Toll-like receptor (TLR) A class of **pattern recognition receptors** (PRRs), each specific for a particular **microbe-associated molecular pattern** (MAMP), the binding of which activates the **nuclear factor κB**.

Topoisomerase An enzyme that changes the topology of double-stranded DNA, specifically its **supercoiling**.

Topoisomerase I A **topoisomerase** that makes single-stranded nicks that relax DNA supercoiling and is often ATP-independent. For an exception, *see* **reverse gyrase**.

Toxigenic strains Bacterial strains capable of producing a **toxin** or toxic effect.

Toxin A poison produced by living organisms, often a protein, capable of causing disease in low concentrations.

Toxin-coregulated pilus (TCP) A filamentous appendage of vibrios that facilitates *V. cholerae* pathogenesis by enabling their aggregation and colonization of the human intestine; the receptor for **CTX phage**.

Toxin-mediated disease A disease whose symptoms can be mimicked by the purified toxin.

Transamination The transfer of an amino group from one molecule to another.

Transcript An RNA molecule synthesized from a DNA template and whose base sequence is complementary to that template.

Transducing particle A phage **virion** that contains some host DNA and that can potentially transfer that DNA to a recipient cell.

Transduction The transfer of DNA between bacteria mediated by a **phage**.

Transformation A mechanism of acquisition of genetic material used by some bacteria in which they take up naked DNA from their surroundings and then incorporate it into their chromosome by **recombination**.

Transition A **mutation** in which one pyrimidine is substituted for another, or one purine is substituted for another.

Translational repression The regulation of gene expression by blocking the initiation of translation from specific mRNAs.

Translocation A **mutation** that moves a fragment of DNA from one location to another.

Translocon A secretory complex composed of multiple proteins that spans all layers of the **cell envelope** of prokaryotes and transports hydrophilic molecules across the cell membrane and/or the outer membrane, in some cases polymerizing the molecules en route.

Transmembrane ion gradient The electrical potential across a membrane due to a difference in chemical concentration. In cells, typically protons are exported across the membrane and the resultant difference in their concentration generates the **proton motive force** used to drive ATP production.

Transposase An enzyme encoded by a **transposon** that catalyzes its transposition to a different location.

Transposon Common mobile elements that move from one location to another, either within a chromosome or to another chromosome, by utilizing any of a variety of transposition mechanisms.

Tricarboxylic acid cycle *See **TCA cycle***.

Trophosome The structure inside tube worms living near deep-sea hydrothermal vents that houses the sulfide-oxidizing bacteria that supply the worm with nutrients.

Tumble The undirected motion in place of a bacterial cell that is produced by the coordinated clockwise rotation of all the cell's **flagella**.

Turbidity The cloudiness of a liquid due to the presence of small, invisible particles such as cells. Measurement of culture turbidity as **absorbance** or optical density instantly assays the bacterial biomass present (weight/culture volume).

Twitching motility A type of microbial motility across a surface in which a **pilus** is extended, its tip adheres to the surface ahead, and then the pilus is retracted to draw the cell forward.

Type IV secretion system Large protein complexes found in many bacteria that cross the **cell envelope** and serve as channels for the passage of proteins or protein-DNA complexes. A subset of these are used for **conjugation**.

Universal primer Primers used for the **polymerase chain reaction** (PCR) that, due to sequence conservation of the target gene, are applicable to all members of a large group of organisms.

Universal second messenger *See **cyclic di-GMP***.

Unstable RNA A generic term for mRNAs that have a relatively short lifetime.

Vesicle A membrane-bounded structure in the cytoplasm that may be filled with a variety of materials for different purposes, e.g., with gas for buoyancy, with nitrate for a reserve supply.

Viable count The number of cells in a sample that are capable of reproducing and forming a visible colony in culture, in contrast to the total number of cells present.

Viral latency The ability of a virus to infect a host cell and then remain present in a dormant state for an extended period of time until a change in cellular conditions prompts its **reactivation**.

Virion A viral particle composed of the viral chromosome(s) enclosed in a protective shell or **capsid** that, after assembly in a host, is released and can transport the viral chromosome(s) to the next host.

Viroid Infectious agents that infect plants and that consist of only a circular, single-stranded molecule of noncoding RNA without a **capsid**.

Virome All the viruses that inhabit a particular environment, such as the human body, or all of the genetic information that they contain.

Virulence The ability of a pathogen to cause disease.

Virulence factor A molecule produced by a pathogen that contributes to its **pathogenicity**.

Wax A class of hydrophobic organic compounds that melt above 45°C, including **mycolic acid**.

Wobble When more than one codon specifies the same amino acid but the codons differ only in the third position, the ambiguous reading of that base that allows one tRNA to service more than one codon.

YAC *See **yeast artificial chromosome***.

Yeast artificial chromosome (YAC) Large constructed plasmids, up to 800 kb, that can be introduced into yeasts and replicated by the cells.

Zetaproteobacteria A very small class of *Proteobacteria* currently with only one species, an iron oxidizer.

Zoonosis A disease of other vertebrates that can be transmitted to and infect humans.

Zygote During sexual reproduction, a diploid cell formed by the fusion of two **haploid gametes**.

Z-ring The constricting ring formed by **FtsZ** at the site of the future septum during the division of most prokaryotic cells.

Index